DICTIONARY
OF
SCIENTIFIC
AND TECHNICAL
TERMINOLOGY

A. S. MARKOV, V. A. ROMANOV, V. I. RYDNIK,
E. G. RYS, V. M. BORSCH, S. N. KORCHEMKIN,
N. S. SKOROKHOD, YU. V. FEDIRKO, N. B. SHAYKHON

SPRINGER-SCIENCE+BUSINESS MEDIA, B.V.
1984

DICTIONARY OF SCIENTIFIC AND TECHNICAL TERMINOLOGY

ENGLISH
GERMAN
FRENCH
DUTCH
RUSSIAN

1984
SPRINGER-SCIENCE+BUSINESS MEDIA, B.V.

ISBN 978-90-481-8248-0   ISBN 978-94-017-1255-2 (eBook)
DOI 10.1007/978-94-017-1255-2

Originally published by Kluwer Academic Publishers in 1984
Softcover reprint of the hardcover 1st edition 1984

PREFACE

Scientific and technical contacts between nations have necessitated the publication of various language textbooks, manuals and reference books. Particularly important among them are multilingual scientific and technical dictionaries. This English—German—French—Dutch—Russian Dictionary of Scientific and Technical Terms contains some 9000 entries. The main feature of the Dictionary is that it includes first and foremost general scientific terms needed by an engineer working in any branch of science and technology. Besides, the Dictionary includes the basic terms used in physics, mathematics, the fundamentals of electrical engineering and chemistry, and also the most essential terms pertaining to manufacturing processes, machine design, testing methods, etc.

The Compilers were confronted with a difficult task, as nowadays science and technology are developing rapidly and the minimum scientific and technical vocabulary required by a specialist is increasing accordingly.

The Compilers have taken special pains to include the entire basic modern technical vocabulary, omitting superfluous words and phrases. They have tried to solve this problem by selecting mainly those scientific and technical terms which constitute the basic of a specialised vocabulary. Therefore, the Dictionary includes the vocabulary pertaining to general study courses in mathematics, physics and chemistry, and also in electrical engineering, electronics and machine design, given in technical colleges irrespective of their specification. This lends the Dictionary an «all-purpose» character, making it equally useful to scientists and engineers of different countries, who have graduated from colleges with different curricula.

In the compilation of the English vocabulary of the Dictionary, use has been made of various English textbooks, manuals and reference books widely recognised by English-speaking users. The Dictionary is therefore both comprehensive and up-to-date.

The Appendix contains Indexes of German, French, Dutch and Russian terms which makes it possible to use the Dictionary when translating texts from or into any of the languages it deals with.

The Compilers and the Publishers will be grateful for comments and suggestions. They should be forwarded to Kluwer Technische Boeken BV, Postbox 23 7400 GA Deventer, Holland or to 103012 Moskva, Staropanski per., 1/5, Izdatelstvo «Russki yazyk».

USING THE DICTIONARY

The English entries are arranged in the Dictionary in alphabetic order, compound and composite terms being regarded as though they were spelt as one word, e. g.:

arm	**armature winding**
armature core	**arm of bridge**

Each Dictionary article consists of the English, German, French, Dutch and Russian terms separated from one another by two parallel strokes (||). To refer the user to a synonymous English term, the word *see* is used. All German, French, Dutch and Russian terms are marked *m*, *f*, or *n* for the plural number (where necessary). All the English terms are preceded by numbers and letters corresponding to those following their German, French, Dutch and Russian equivalents in the Indexes given at the end of the Dictionary. The English entries are printed in bold-face type.

Thus a Dictionary article appears as follows:

O4 **object language** || Objektsprache *f* || langage *m* objet || objecttaal *f* || объектный язык *m*.

Translations which render different meanings are separated by numbers, translations which convey similar meanings are separated by a semicolon, and translations which are synonymous are separated by a comma.

To save space, the interchangeable parts of synonymous translations are given in square brackets, e. g.: линейная [плоская] поляризация, which is equivalent to: линейная поляризация, плоская поляризация.

For the same reason, the optional parts of a term are given in parentheses, e. g.: скорость (распространения) звука, which is equivalent to: скорость распространения звука, скорость звука.

Grammatical homonyms are given as separate entries, each homonym being preceded by a Roman number, e. g.:

I **polar** || Polare *f* || polair *f* || polair *f* || поляра *f*
II **polar** || polar, Polar... || polair || polair || полярный

Translators of German, French, Dutch or Russian texts should use the Indexes given at the end of the Dictionary, in which all the terms are followed by figures and letters corresponding to those preceding their English equivalents in the body of the Dictionary.

If a word, e.g. the adjective **абсолютный** or the substantive **подпрограмма**, is followed by several consecutive numbers, the user should look up all the terms given under these numbers. He will find phrases incorporating the word concerned, e. g. **абсолютный максимум, абсолютная температура** or **библиотека подпрограмм**. If he fails to find the phrase he needs, he should be able to form it on the analogy to the phrases given in the Dictionary.

SPECIAL ABBREVIATIONS

phys., *Phys.*, *физ.* — physics
math., *Math.*, *мат.* — mathematics
chem., *Chem.*, *chim.*, *хим.* — chemistry
el., *El.*, *él.*, *эл.* — electrical engineering

VORWORT

Der Ausbau multilateraler Beziehungen in Wissenschaft und Technik macht die Entwicklung von neuen Sprachhilfen erforderlich. Dabei wird den mehrsprachigen wissenschaftlich-technischen Wörterbüchern ein besonderer Platz eingeräumt. Das vorliegende «Englisch—deutsch—französisch—niederländisch—russische Wörterbuch der wissenschaftlichen und technischen Lexik» umfaßt etwa 9000 Termini. Insbesondere wurde der allgemeinwissenschaftliche Wortschatz berücksichtigt, den Ingenieure beliebiger Fachrichtungen beherrschen sollten. Darüber hinaus sind Grundtermini zur Physik, Mathematik, zu den Grundlagen der Elektrotechnik und Chemie sowie die wesentlichsten technischen Termini zur Technologie des Maschinenbaus, den allgemeinen Maschinenelementen, Prüfmethoden u. a. aufgenommen.

Die Verfasser standen vor einer schwierigen Aufgabe. Die Wissenschaft und die Technik durchmachen heutzutage eine stürmische Entwicklung, und ebenso rasch wächst das Wortschatzminimum, über das ein Fachmann verfügen muß.

Die Autoren waren bemüht, eine «Momentaufnahme» des wichtigsten technischen Wortschatzes der Gegenwart vorzunehmen, nur das Wesentlichste ins Wörterbuch aufzunehmen und auf alles Überflüssige zu verzichten. Bei der Erarbeitung des Wörterbuchs wurde in erster Linie ein solcher wissenschaftlich-technischer Wortschatz ausgewählt, der die Grundlage aller technischen Disziplinen bildet. Aus diesem Grunde wurde der Wortschatz zu den Grundlagenfächern Mathematik, Physik, Chemie, Elektrotechnik, Elektronik, Maschinen und Maschinenteile aufgenommen, die an technischen Hochschulen beliebiger Fachrichtungen unterrichtet werden. Das macht das Wörterbuch universell, so gut wie unabhängig von unterschiedlicher Ausbildung der Wissenschaftler und Ingenieure in verschiedenen Ländern und an verschiedenen Hochschulen.

Bei der Erarbeitung des englischen Wörterverzeichnisses stützte man sich auf zahlreiche, schon bewährte und anerkannte Lehrmaterialien und Nachschlagewerke. Das bietet Garantie für entsprechende Vollständigkeit und Aktualität des aufgenommenen Wortschatzes.

Im Anhang sind Register der Termini in Deutsch, Französisch, Niederländisch und Russisch gegeben, was die Übersetzung der Fachliteratur in diesen Sprachen ermöglicht.

Die Verfasser und Herausgeber sind für Vorschläge und Hinweise jeglicher Art dankbar und bitten sie an folgende Adressen zu senden: Kluwer Technische Boeken BV, Postfach 23 7400 GA Deventer, Holland oder 103012 Moskva, Staropanski per., 1/5, Izdatelstvo «Russki jasyk»

HINWEISE FÜR DIE BENUTZUNG

Die englischen Stichwörter sind alphabetisch geordnet, wobei ein aus mehreren Wörtern bestehender Terminus wie ein zusammengeschriebenes Wort aufgefaßt wird, z. B.:

arm	**armature winding**
armature core	**arm of bridge**

Zum Wortartikel gehören Termini in der englischen, deutschen, französischen, niederländischen und russischen Sprache, die voneinander durch senkrechten Doppelstrich (||) getrennt sind. Als Hinweis auf einen synonymischen englischen Terminus wird der Vermerk *see* (siehe) verwendet. Alle deutschen, französischen, niederländischen und russischen Termini sind mit einer Geschlechtsangabe (*m*, *f*, *n*) und, wenn nötig, einer Pluralformangabe (*pl*) versehen. Alle englischen Termini sind numeriert, um eine Benutzung der deutschen, französischen, niederländischen und russischen Register zu ermöglichen. Die Numerierung erfolgt für jeden Buchstaben getrennt. Die englischen Stichwörter sind fettgedruckt. Demzufolge sieht ein Wortartikel folgendermaßen aus:

O4 **object language** || Objektsprache *f* || langage *m* objet || objecttaal *f* || объектный язык *m*

In den Übersetzungen werden verschiedene Bedeutungen durch Ziffern, sinnverwandte Varianten durch ein Semikolon, synonymische Varianten durch ein Komma abgetrennt.

Austauschbare Teile synonymischer Varianten werden bei der Übersetzung in eckige Klammern geschlossen, z. B.: линейная [плоская] поляризация, was gelesen wird: линейная поляризация, плоская поляризация.

Der fakultative Teil des Terminus steht in runden Klammern, z. B. скорость (распространения) звука, was gelesen wird: скорость распространения звука oder скорость звука.

Grammatische Homonyme bilden selbständige Wortartikel und sind mit römischen Ziffern gekennzeichnet, z. B.:

I **polar** || Polare *f* || polaire *f* || polair *f* || поляра *f*
II **polar** || polar, Polar... || polaire || polair || полярный.

Für die Übersetzung aus dem Deutschen, Französischen, Niederländischen bzw. Russischen sind die Register im Anhang des Wörterbuches zu verwenden. Hier sind ebenfalls alle Termini dem englischen Stichwort entsprechend numeriert.

Wenn z. B. beim Adjektiv **абсолютный** oder dem Substantiv **подпрограмма** einige aufeinander folgende Ziffern angegeben sind, so muß bei der Wahl eines Terminus mit diesem Wort bei allen angegebenen Nummern nachgeschlagen werden. Dort findet man Wortverbindungen, die mit diesem Wort gebildet wurden, z. B. **абсолютный максимум, абсолютная температура** oder **библиотека подпрограмм**. Wenn der gesuchte Terminus nicht vorhanden ist, kann er in Analogie zu den im Wörterbuch enthaltenen gebildet werden.

VERWENDETE ABKÜRZUNGEN

phys., *Phys.*, *физ.* — Physik
math., *Math.*, *мат.* — Mathematik
chem., *Chem.*, *chim.*, *хим.* — Chemie
el., *El.*, *él.*, *эл.* — Elektrotechnik

PRÉFACE

La multiplication des contacts internationaux dans le domaine de la science et de la technique nécessite la création de divers ouvrages linguistiques dont les dictionnaires polyglottes des termes techniques et scientifiques occupent une place de choix. Le présent dictionnaire anglais-allemand-français-néerlandais-russe des termes techniques et scientifiques comporte environ 9000 mots. Le trait particulier de cet ouvrage consiste en ce qu'il renferme en premier lieu le vocabulaire scientifique de base indispensable aux ingénieurs de n'importe quelle discipline. En plus, il contient des termes de base relevant de la physique, des mathématiques, de la chimie, des principes de l'électrotechnique ainsi que de la technologie des constructions mécaniques, des éléments de machines et des méthodes des essais.

Les auteurs se trouvaient devant une tâche difficile. Vu le développement impétueux de la science et de la technique de l'époque contemporaine le vocabulaire technique de base s'élargit rapidement.

Les auteurs se sont posés pour but de faire de leur ouvrage une sorte de photographie instantanée de la terminologie principale du jour en y introduisant tout ce qui est essentiel et en éliminant le superflu. Afin d'atteindre cet objectif ils ont pris la décision d'inclure dans le dictionnaire les termes techniques et scientifiques constituant les fondements de la terminologie la plus spécialisée. Aussi le dictionnaire comporte-t-il les termes utilisés dans les cours que l'on fait dans tout établissement d'enseignement supérieur technique, c'est à dire les cours fondamentaux de mathématiques, de physique, de chimie, d'électrotechnique, d'électronique, de pièces mécaniques.

Cela confère au dictionnaire un caractère universel permettant de l'utiliser dans tous les systèmes de formation des scientifiques et des techniciens indépendemment du pays ou de l'établissement d'enseignement.

Les auteurs de la partie anglaise du présent dictionnaire ont utilisé au cours de leur travail les dictionnaires les plus variés ainsi que des ouvrages pédagogiques en anglais déja approuvés par lecteurs. Une telle méthode de sélection des termes rend le contenu du dictionnaire exhaustif et conforme aux exigeances de l'époque contemporaine.

L'annexe renferme les index des termes en allemand, en français, en néerlandais et en russe ce qui permet de traduire la littérature en différentes langues.

Notre Maison d'édition ainsi que les auteurs seront reconnaissants pour toutes suggestions qui puissent être envoyées à l'adresse:
Kluwer Technische Boeken BV, Boîte postale 23 7400 GA Deventer, Pays Bas ou 103012 Moskva, Staropanski per., Izdatelstvo «Russki yazyk»

INDICATIONS POUR L'UTILISATION DE CE DICTIONNAIRE

Les termes vedettes anglais sont disposés de sorte que les termes composés soient considérés comme un seul mot, par exemple:

arm	**armature winding**
armature core	**arm of bridge**

Les articles du dictionnaire comportent les termes anglais, allemands, français, néerlandais et russes. Chaque langue est séparée de l'autre par deux traits parallèles (||) Afin de se référer à un autre synonyme anglais on utilise le signe *see* (voir). Les termes allemands, français, néerlandais et russes sont accompagnés de l'indication du genre (m, f, n) et en cas de besoin, de celle du nombre (pl). Tous les termes anglais sont numérotés à l'intérieur de chaque rubrique littérale en vue de faciliter la recherche du terme allemand, français, néerlandais ou russe correspondants. Les termes vedettes anglais sont composés en caractères demi-gras. La présentation des articles est donc la suivante:

> O4 **object language** || Objektsprache *f* || langage *m* objet || objecttaal *f* || объектный язык *m*.

Dans les articles les sens différents sont présentés sous des numéros différents, les variantes proches sont séparées par un point virgule et les synonymes par une virgule.

Dans la traduction les équivalents synonymiques peuvent être mis en crochets, par exemple: линейная [плоская] поляризация ce qui est identique à: линейная поляризация, плоская поляризация.

La partie facultative du terme peut être également mise entre parenthèses, par exemple: скорость (распространения) звука, ce qui est égal à: скорость распространения звука, скорость звука.

Les homonymes grammaticaux forment un article indépendant portant un chiffre romain, par exemple:

> I **polar** || Polare *f* || polaire *f* || polair *f* || поляра *f*
> II **polar** || polar, Polar... || polaire || polair || полярный

Pour traduire à partir de l'allemand, du français, du néerlandais ou du russe il faut consulter les index à la fin du dictionnaire où tous les termes ont des indications numériques ou littérales et correspondent aux termes anglais vedettes.

Si un mot comme, par exemple, l'adjectif **абсолютный** ou le substantif **подпрограмма** porte plusieurs numéros consécutifs l'usager du dictionnaire doit parcourir toutes les indications afin de trouver le terme avec un de ces mots. Sous chaque numéro il trouvera un groupe de mots comportant le vocable en question, par exemple: **абсолютный максимум, абсолютная температура** ou **библиотека подпрограмм**. S'il n'arrive pourtant pas à trouver le groupe de mots recherchés il peut le former par analogie avec ceux cités dans le dictionnaire.

LISTE DES ABREVIATIONS

phys., Phys., физ. — physique

math., Math., мат. — mathématiques

chem., Chem., chim., хим. — chimie

el., El., él., эл. — électrotechnique

VOORWOORD

De toename van de multilaterale contacten op het gebied van wetenschap en techniek doet de behoefte opkomen aan diverse taalkundige hulpmiddelen, waaronder meertalige technisch-wetenschappelijke woordenboeken een bijzondere plaats innemen.

Het onderhavige Engels-Duits-Frans-Nederlands-Russisch woordenboek van de wetenschappelijke en technische terminologie bevat ca 9 000 termen. Het voornaamste kenmerk van dit woordenboek is, dat het vooral wetenschappelijke termen bevat van meer algemene aard en als zodanig van toepassing voor alle beta-disciplines. Bovendien bevat het de terminologie van de grondslagen van wis- en natuurkunde, scheikunde en elektrotechniek, alsmede de belangrijkste terminologie op het gebied van de machinebouw en de beproevingstechniek.

De samenstellers hebben er naar gestreefd, een momentopname te maken van de voornaamste hedendaagse technische woordenschat, en daarvan slechts de essentie in het woordenboek op te nemen. Dientengevolge hebben zij zich beperkt tot die termen, welke geacht mogen worden te behoren tot de grondslag van de gespecialiseerde terminologie. Zij hebben gemeend dit probleem het beste te kunnen benaderen door in het woordenboek lexica op te nemen uit algemene leergangen van de wiskunde, de fysica en de chemie. Op gelijke wijze werd aandacht besteed aan de elektronica, de elektrotechniek en de machinebouw. Aangezien het hier leergangen betreft, die toegepast worden bij het technisch onderwijs op verschillend niveau, verleent deze methode het woordenboek een universeel karakter. Het is hiermee onafhankelijk geworden van de onderscheiden opleidingen voor wetenschappelijk en technisch personeel, zoals deze worden toegepast in verschillende landen en aan de verschillende opleidingsinstituten.

Voor het samenstellen van de Engelstalige woordenlijst werd geput uit talrijke leerboeken en naslagwerken van naam hetgeen een waarborg zal zijn voor de toereikendheid en de actualiteit van de opgenomen terminologie.

Het woordenboek is voorzien van een bijlage met alfabetische registers voor het Duits, Frans, Nederlands en Russisch, zodat het woordenboek ook gebruikt kan worden bij het vertalen vanuit een van deze talen.

Zowel de samenstellers als beide uitgeverijen zullen dankbaar zijn voor suggesties ter verbetering, welke gestuurd kunnen worden aan een van beide onderstaande adressen:
Kluwer Technische Boeken BV, Postbus 23 7400 GA Deventer, Holland
of 103012 Moskva, Staropanski per., 1/5, Izdatelstvo «Roesski jazyk».

AANWIJZINGEN VOOR HET GEBRUIK

De termen zijn gerangschikt in een strikt alfabetische volgorde, dus bijvoorbeeld als volgt:

arm	armature winding
armature core	arm of bridge

Naast de Engelse term, bevat ieder artikel vertalingen in het Duits, Frans, Nederlands en Russisch, van elkaar gescheiden door middel van ·||. Van de zelfstandige naamwoorden in deze vier talen wordt tevens het genus vermeld, waarbij de afkortingen *m*, *f*, *n*, en *pl* worden gebruikt ter aanduiding van masculinum, femininum, neutrum, respectievelijk pluralis. Vanuit de alfabetisch gerangschikte indices wordt naar de onderscheiden artikels verwezen met behulp van een cijfersysteem.

Slechts wanneer er sprake is van grammaticale homonymie, zijn de Engelse trefwoorden twee of meer malen opgenomen, in welk geval zij van elkaar zijn onderscheiden door middel van romeinse cijfers, bijvoorbeeld

I polar (als substantivum)
II polar (als adiectivum)

Als gevolg hiervan is het mogelijk, dat een Engels trefwoord meer dan een betekenis kan hebben, en dientengevolge door meer dan één vertaling kan worden gevolgd. Deze, in betekenis verschillende vertalingen worden van elkaar onderscheiden door middel van arabische cijfers, bijvoorbeeld
A58 accelerator — 1. versneller *m* 2. versnellingselektrode *f* 3. gaspedaal *n*
Vertalingen met geringe verschillen in betekenis onderling, zijn van elkaar gescheiden door middel van een puntkomma; synonieme vertalingen zijn gescheiden door middel van een komma.
In samengestelde termen kan een van de componenten tussen rechte haken zijn geplaatst, dit om aan te geven, dat deze component de daaraan voorafgaande component kan vervangen, bijvoorbeeld
— werkelijke maat [afmeting]
Componenten tussen ronde haken kunnen aan de term worden toegevoegd dan wel daaruit worden weggelaten, zoals in
— trapezium(schroef)draad
Cursief gezette toevoegingen tussen haakjes dienen te worden geïnterpreteerd als nadere toelichting op de eraan voorafgaande vertaling. In dit verband is er gebruik gemaakt van de volgende afkortingen:

chem. (chemie)
el. (elektrotechniek en elektronica)
math. (wiskunde)
phys. (fysica)

Voor het vertalen uit het Duits, Frans, Nederlands of Russisch dient gebruik gemaakt te worden van het achter in het woordenboek geplaatste register, waar alle termen van een letter- en cijfersindex voorzien zijn die met de desbetreffende voerende Engelstalige term overeenkomt.

Als bij een woord, bijvoorbeeld, bij een bijvoeglijk naamwoord **абсолютный** of bij een zelfstandig naamwoord **подпрограмма** zijn enkele opeenvolgende cijfers gegeven, dann moet de lezer, die zoekt naar een term met deze woord, alle deze nummers controleren. Daar treft hij de semenstellingen met deze woord, bijvoorbeeld: **абсолютный максимум, абсолютная температура** of **библиотека подпрограмм.** Als er tussen geen nodige samenstelling is, dann kan de lezer het vormen in overeenkomst met de samenstallingen die aangegeven zijn in het woordenboek.

ПРЕДИСЛОВИЕ

Многосторонние международные связи в области науки и техники вызывают потребность в создании различных языковых пособий, среди которых особое место занимают многоязычные научно-технические словари. Предлагаемый читателям Англо-немецко-французско-нидерландско-русский словарь научной и технической лексики содержит около 9 000 терминов. Основной особенностью словаря является то, что он включает прежде всего общенаучную лексику, которой должен овладеть инженер любой специальности. Кроме того, он охватывает основную терминологию по физике, математике, основам электротехники и химии, а также наиболее важную техническую терминологию по технологии машиностроения, общим деталям машин, методам испытаний и др.

Перед авторами стояла трудная задача. В наше время наука и техника стремительно развиваются, и столь же быстро возрастает тот минимум лексики, которым должен владеть специалист.

Авторы стремились дать «моментальный снимок» основной технической лексики сегодняшнего дня, включить в словарь все основное и не включать ничего лишнего. Эту задачу авторы пытались решить, отбирая для словаря главным образом ту научно-техническую лексику, которая является фундаментом для специальной терминологии. Поэтому в словарь включена лексика общих курсов математики, физики, химии, а также курсов электротехники, электроники, машин и их деталей, которые читаются в технических вузах любого профиля. Это делает словарь универсальным, практически почти не зависящим от разнообразных систем подготовки ученых и инженеров, принятых в различных странах и высших учебных заведениях.

Для составления английского словника были привлечены многие известные и получившие признание читателей англоязычные учебные и справочные пособия. Это дает гарантию достаточной полноты и современности лексики, включенной в словарь.

В приложении к словарю даны указатели терминов на немецком, французском, нидерландском и русском языках, что обеспечивает возможность перевода литературы в различных языковых направлениях.

Авторы и издательство будут благодарны за отзывы и пожелания, которые можно направлять по адресу: Kluwer Technische Boeken BV, Postbox 23 7400 GA Deventer, Holland или 103012 Москва, Старопанский пер., д. 1/5, издательство «Русский язык».

О ПОЛЬЗОВАНИИ СЛОВАРЕМ

Ведущие английские термины расположены в словаре в алфавитном порядке, причем составные термины рассматриваются как слитно написанные слова, например:

arm	**armature winding**
armature core	**arm of bridge**

Словарная статья состоит из терминов английского, немецкого, французского, нидерландского и русского языков, причем один язык отделен от другого параллельками (||). Для ссылки с одного синонимичного английского термина на другой используется помета *see* (смотри). Все немецкие, французские, нидерландские и русские термины имеют указание рода (*m*, *f*, *n*) и, при необходимости, множественного числа (*pl*). Все английские термины пронумерованы по буквам для возможности использования указателей немецких, французских, нидерландских и русских терминов. Ведущие английские термины даны полужирным шрифтом. Таким образом, словарная статья имеет следующий вид:

O4 **object language** || Objektsprache *f* || langage *m* objet || objecttaal *f* || объектный язык *m*.

В переводах разные значения разделяются цифрами, близкие по смыслу варианты — точкой с запятой, синонимичные варианты — запятой.

Для экономии места взаимозаменяемые части синонимичных вариантов перевода могут быть заключены в квадратные скобки, например: линейная [плоская] поляризация, что равнозначно записи: линейная поляризация, плоская поляризация.

Факультативная часть термина может быть в тех же целях заключена в круглые скобки, например: скорость (распространения) звука, что равнозначно записи: скорость распространения звука, скорость звука.

Грамматические омонимы выделены в самостоятельные статьи и обозначены римскими цифрами, например:

I **polar** || Polare *f* || polaire *f* || polair *f* || поляра *f*
II **polar** || polar, Polar... || polaire || polair || полярный

Для перевода с немецкого, французского, нидерландского или русского языков следует пользоваться помещенными в конце словаря указателями, в которых все термины имеют буквенно-цифровое обозначение, соответствующее ведущему английскому термину.

Если при слове, например, прилагательном **абсолютный** или существительном **подпрограмма**, указаны несколько цифр, идущих подряд, то читатель при поиске термина с этим словом должен просмотреть все эти номера. Там он найдет словосочетания, образованные с этим словом, например: **абсолютный максимум, абсолютная температура** или **библиотека подпрограмм**. Если среди них не будет нужного ему словосочетания, он сможет образовать его по аналогии с имеющимися в словаре.

ПОМЕТЫ В СЛОВАРЕ

phys., *Phys.*, *физ.* — физика
math., *Math.*, *мат.* — математика
chem., *Chem.*, *chim.*, *хим.* — химия
el., *El.*, *él.*, *эл.* — электротехника

A

A1 **Abelian group** || abelsche Gruppe *f* || groupe *m* abélien || abelse groep *f* (*m*) || абелева группа *f*

A2 **Abel's test** || Abelsches Konvergenzkriterium *n* || critère *m* d'Abel || Abel-test *m* || признак *m* (сходимости) Абеля

A3 **aberration** || Aberration *f* || aberration *f* || aberratie *f* || аберрация *f*

A4 **ablation** || Ablation *f* || ablation *f* || ablatie *f* || абляция *f*

A5 **abnormal dispersion** || anomale Dispersion *f* || dispersion *f* anormale || anomale dispersie *f* || аномальная дисперсия *f*

A6 **abnormality** || Anomalie *f* || anomalie *f* || anomalie *f* || аномалия *f*

A7 **above-critical** || überkritisch || su(pe)rcritique || overkritisch || сверхкритический, надкритический

A8 **abrasion** || **1.** Verschleiß *m*; Abrieb *m* **2.** Schleifen *n*; Polieren *n* || abrasion *f* || **1.** afslijting *f*; afwrijving *f* **2.** slijpen *n*; polijsten *n* || **1.** истирание *n*; абразивный износ *m* **2.** шлифовка *f*; полировка *f*

A9 **abrasion resistance** || Abriebwiderstand *m*, Abriebfestigkeit *f*, Abnutzungswiderstand *m* || résistance *f* à l'abrasion || afslijtingsbestendigheid *f*, slijtvastheid *f* || сопротивление *n* истиранию, износостойкость *f*

A10 **abrasion-resisting** || abriebfest || résistant à l'abrasion || slijtvast || износостойкий

A11 **abrasion test** || Abschleifversuch *m* || essai *m* d'abrasion || afslijtingstoets *m* || испытание *n* на износ

A12 I **abrasive** || Schleifmittel *n*, Schleifstoff *m* || abrasif *m*, produit *m* abrasif || slijpmiddel *n* || абразив *m*, абразивный материал *m*

A13 II **abrasive** || Schleif... || abrasif || slijp ... || абразивный

A14 **abrasive hardness** *see* **scratch hardness**

A15 **abrasive paper** || Sandpapier *n*, Schmirgelpapier *n* || papier *m* abrasif || schuurpapier *n* || абразивная [наждачная] бумага *f*

A16 **abrasive powder** || Schleifpulver *n* || poudre *f* abrasive || slijppoeder *n*, *m* || абразивный порошок *m*

A17 **abrasive surface** || Abriebfläche *f* || surface *f* d'abrasion || afwrijvingsoppervlak *n* || поверхность *f* истирания

A18 **abrasive tool** || Schleifkörper *m*, Schleifwerkzeug *n* || outil *m* d'abrasion || slijpwerktuig *n* || абразивный инструмент *m*

A19 **abrasive tumbling** *see* **tumbling**

A20 **abrasive wear** || Verschleiß *m*; Abrieb *m* || usure *f* par abrasion || afwrijving *f* || абразивный износ *m*

A21 **abrasive wheel** *see* **grinding wheel**

A22 **abscissa** || Abszisse *f* || abscisse *f* || abscis *f* (*m*) || абсцисса *f*

A23 **absolute** || absolut, Absolut... || absolu || absoluut || абсолютный

A24 **absolute acceleration** || Absolutbeschleunigung *f* || accélération *f* absolue || absolute versnelling *f* || абсолютное ускорение *n*

A25 **absolute address** || absolute Adresse *f* || adresse *f* absolue || absoluut adres *n* || абсолютный [истинный] адрес *m*

A26 **absolute completeness** || absolute Vollständigkeit *f* || complétude *f* absolue || absolute volledigheid *f* || абсолютная полнота *f*

A27 **absolute consistency** || absolute Widerspruchsfreiheit *f* || non-contradiction *f* absolue || absolute niet-tegenstrijdigheid *f* || абсолютная непротиворечивость *f*

A28 **absolute convergence** || absolute Konvergenz *f* || convergence *f* absolue || absolute convergentie *f* || абсолютная сходимость *f*

A29 **absolute degree** || absoluter Grad *m* || degré *m* absolu || absolute graad *m* || кельвин *m*, градус *n* (по шкале) Кельвина *(единица СИ)*

A30 **absolute error** || absoluter Fehler *m* || erreur *f* absolue || absolute fout *f* (*m*) || абсолютная погрешность *f*

A31 **absolute motion** || absolute Bewegung *f*, Absolutbewegung *f* || mouvement *m* absolu || absolute beweging *f* || абсолютное движение *n*

A32 **absolute pressure** || absoluter Druck *m*, Absolutdruck *m* || pression *f* absolue || absolute druk *m* || абсолютное давление *n*

A33 **absolute probability** || absolute [unbedingte] Wahrscheinlichkeit *f* || probabilité *f* absolue || absolute waarschijnlijkheid *f* || безусловная вероятность *f*

A34 **absolute system of units** || absolutes Einheitensystem *n*, CGS-System *n* || système *m* d'unités absolu || absoluut eenhedenstelsel *n* || абсолютная система *f* единиц

A35 **absolute temperature** || absolute Temperatur *f* || température *f* absolue || absolute temperatuur *f* || абсолютная [термодинамическая] температура *f*

A36 **absolute time** || absolute Zeit *f* || temps *m* absolu || absolute tijd *m* || абсолютное время *n*

A37 **absolute value** || absoluter Wert *m*, Absolutwert *m*; Betrag *m* || valeur *f* absolue || absolute waarde *f* || абсолютное значение *n*, модуль *m*

A38 **absolute velocity** || Absolutgeschwindigkeit *f* || vitesse *f* absolue || absolute snelheid *f* || абсолютная скорость *f*

A39 **absolute zero (temperature)** || absoluter Nullpunkt *m* || zéro *m* absolu || absoluut nulpunt *n* || абсолютный нуль *m* (температуры)

A40 **absorbance** || spektrales Absorptionsmaß *n* || pouvoir *m* absorbant spectral || spectraal absorptievermogen *n* || спектральная поглощательная способность *f*

A41 **absorptance** || Absorptionsvermögen *n* || absorptivité *f*, pouvoir *m* absorbant || absorptievermogen *n* || лучепоглощательная способность *f*

A42 **absorption** || Absorption *f* || absorption *f* || absorptie *f* || поглощение *n*, абсорбция *f*

A43 **absorption band** || Absorptions bande *f* || bande *f* d'absorption || absorptieband *m* || полоса *f* поглощения

A44 **absorption dynamometer** || Bremsdynamometer *n* ||, dynamomètre *m* d'absorption || remdynamometer *m* || тормозной динамометр *m*

A45 **absorption line** || Absorptionslinie *f* || raie *f* d'absorption || absorptielijn *f* (*m*) || линия *f* (спектра) поглощения, абсорбционная линия *f*

A46 **absorption spectrum** || Absorptionsspektrum *n* || spectre *m* d'absorption || absorptiespectrum *n* || спектр *m* поглощения, абсорбционный спектр *m*

A47 **absorptivity** || integrales Absorptionsvermögen *n* || absorptivité *f*, pouvoir *m* absorbant intégral || integraal absorptievermogen *n* || интегральная лучепоглощательная способность *f*

A48 **abstraction** || Abstraktion *f* || abstraction *f* || abstractie *f* || абстракция *f*; абстрактность *f*

A49 **abstract mathematics** || abstrakte [reine] Mathematik *f* || mathématiques *f pl* pures || abstracte wiskunde *f* || чистая математика *f*

A50 **abstractness** || Abstraktheit *f*, abstrakter Charakter *m* || abstraction *f* || abstractie *f* || абстрактность *f*

A51 **accelerated motion** || beschleunigte Bewegung *f* || mouvement *m* accéléré || versnelde beweging *f* || ускоренное движение *n*

A52 **accelerated test** || Schnellanalyse *f*; Schnellversuch *m*, Kurzzeitversuch *m* || essai *m* accéléré || snelanalyse *f*; acceleratieproef *f* (*m*) || ускоренное испытание *n*

A53 **accelerating field** || Beschleunigungsfeld *n*, beschleunigendes Feld *n* || champ *m* accélérateur || versnellingsveld *n* || ускоряющее поле *n*

A54 **accelerating voltage** || Beschleunigungsspannung *f* || tension *f* d'accélération || versnellingsspanning *f* || ускоряющее напряжение *n*

A55 **acceleration** || Beschleunigung *f* || accélération *f* || versnelling *f* || ускорение *n*

A56 **acceleration due to gravity** || Erdbeschleunigung *f*, Fallbeschleunigung *f* || accélération *f* de la pesanteur || zwaartekrachtversnelling *f* || ускорение *n* силы тяжести, ускорение *n* свободного падения

A57 **acceleration of convergence** || Konvergenzverbesserung *f* || accélération *f* de la convergence || convergentieverbetering *f* || улучшение *n* сходимости

A58 **accelerator** || 1. Beschleuniger *m*, Teilchenbeschleuniger *m* 2. Beschleunigungselektrode *f* 3. Beschleuniger *m*; Drosselklappe *f* 4. *chem.* Katalysator *m* || 1. accélérateur *m* 2. électrode *f* d'accélération 3. accélerateur *m*; papillon *m* (d'étranglement) 4. *chim.* catalysateur *m* || 1. versneller *m* 2. versnellings-elektrode *f* 3. gaspedaal *n* 4. *chem.* katalysator *m* || 1. ускоритель *m* частиц 2. ускоряющий электрод *m* 3. акселератор *m*; дроссельная заслонка *f* 4. *хим.* катализатор *m*

A59 **accelerogram** || Akzelerogramm *n* || accélérogramme *m* || accelerogram *n* || акселерограмма *f*, запись *f* ускорений

A60 **accelerograph** || Beschleunigungsschreiber *m* || accélérographe *m*, enregistreur *m* d'accélération || accelerograaf *m* || акселерограф *m*, самописец *m* ускорений

A61 **accelerometer** || Beschleunigungsmesser *m* || accéléromètre *m* || accelerometer *m*, versnellingsmeter *m* || акселерометр *m*; датчик *m* ускорений

A62 **acceptance test** || Abnahmeprüfung *f* || essai *m* de réception || afnametoets *m* || приёмочное испытание *n*

A63 **accepted tolerance** || zulässige Toleranz *f* || écart *m* toléré; tolérance *f* adoptée || toelaatbaare tolerantie *f* || допускаемое отклонение *n*

A64 **acceptor** || Akzeptor *m* || accepteur *m* || acceptor *m* || акцептор *m* *(электронов)*

A65 **acceptor atom** || Akzeptoratom *n* || atome *m* accepteur || acceptoratoom *n* || атом-акцептор *m*, акцепторный атом *m*

A66 **acceptor level** || Akzeptorniveau *n* || niveau *m* accepteur || acceptorniveau *n* || акцепторный (энергетический) уровень *m*

A67 **access** || Zugriff *m* || accès *m* || toegang *m* || доступ *m*

A68 **accessibility** || Zugang *m*, Zugänglichkeit *f* *(z. B. zwecks Wartung)* || accessibilité *f* *(p. ex. pour l'entretien)* || toegankelijkheid *f* *(bijv. voor onderhoud)* || доступность *(напр. для технического обслуживания)*

A69 **accessible state** || erreichbarer Zustand *m* || état *m* accessible || bereikbare toestand *m* || достижимое состояние *n*

A70 **accessory** || Hilfsvorrichtung *f*; Zusatzgerät *n* || dispositif *m* auxiliaire || toestel *n*, toebehoren *n*, hulpinrichting *f* || приспособление *n*, принадлежность *f*, вспомогательное устройство *n*

A71 **accessory condition** || zusätzliche Bedingung *f* || condition *f* complémentaire || aanvullingsvoorwaarde *f* || дополнительное условие *n*

A72 **access time** || Zugriffszeit *f* || temps *m* d'accès || toegangstijd *m* || время *n* доступа

A73 **accomodation** || Akkomodation *f* || accomodation *f* || accomodatie *f* || аккомодация *f*

A74 **accumulation** || Anhäufung *f*, Speicherung *f*, Akkumulation *f* || accumulation *f*, emmagasinage *m*, stockage *m*, collection *f* || accumulatie *f* *m* накопление *n*, аккумулирование *n*

A75 **accumulator** || Akkumulator *m*, Sammler *m*; Speicher *m* || accumulateur *m* || accumulator *m* || аккумулятор *m*; аккумуляторная батарея *f*

A76 **accuracy** || 1. Genauigkeit *f* 2. Fehler *m* || 1. précision *f*, exactitude *f* 2. erreur *f* || 1. nauwkeurigheid *f*; betrouwbaarheid *f* 2. fout *f* (*m*) || 1. точность *f*, правильность *f*, достоверность *f*; надёжность *f* *(данных)* 2. погрешность *f* *(измерения)*

A77 **accuracy of measurement** || Meßgenauigkeit *f* || précision *f* de mesure || meetnauwkeurigheid *f* || точность *f* измерения

A78 **acetic acid,** CH_3COOH || Essigsäure *f*, Azetsäure *f*, Äthansäure *f* || acide *m* acétique, acide *m* éthanoïque || azijnzuur *n* || уксусная [этановая] кислота *f*

A79 **acetic anhydride,** $(CH_3CO)_2O$ || Essigsäureanhydrid *n*, Äthansäureanhydrid *n* || anhydride *m* acétique ||

azijnzuuranhydride *n* || ангидрид *m* уксусной кислоты

A80 **acetylene,** CH:CH || Azetylen *n*, Äthin *n* || acétylène *m*, éthyne *m* || acetyleen *n* || ацетилен *m*, этин *m*

A81 **achromatic** || achromatisch || achromatique || achromatisch || ахроматический *(о линзе)*; нейтральный *(о светофильтре)*

A82 **achromatism** || Achromatismus *m* || achromatisme *m* || achromatisme *n* || ахроматизм *m*, ахроматичность *f*

A83 I **acid** || Säure *f* || acide *m* || zuur *n* || кислота *f*

A84 II **acid** || sauer || acide || zuur || кислый

A85 **acid-resistant** || säurefest, säurebeständig || résistant aux acides, antiacide || zuurvast || кислотостойкий, кислотоупорный

A86 **acid-resisting alloy** || säurebeständige Legierung *f* || alliage *m* résistant aux acides || zuurvaste legering *f* || кислотостойкий сплав *m*

A87 **acme thread** || Trapezgewinde *n* || filetage *m* Acmé, filet *m* trapézoïdal || trapezium(schroef)draad *m* || резьба *f* Акме, (стандартная американская) трапецеидальная резьба *f* с углом профиля 29°

A88 **acoustic** || akustisch || acoustique, phonique || akoestisch || акустический, звуковой

A89 **acoustic mass** || akustische Masse *f* || masse *f* acoustique || akoestische massa *f* *(m)* || акустическая масса *f*

A90 **acoustics** || Akustik *f* || acoustique *f* || akoestiek *f* || акустика *f*

A91 **acoustic transmission** || Schalldurchlässigkeit *f* || transmission *f* acoustique || geluiddoorlating *f* || звукопроницаемость *f*

A92 **acoustic wave** || Schallwelle *f* || onde *f* sonore || geluidgolf *f* *(m)* || звуковая волна *f*

A93 **acousto-electric** || elektroakustisch || électro-acoustique || elektroakoestisch || электроакустический

A94 **action at a distance** || Fernwirkung *f* || action *f* à distance || afstandswerking *f* || дальнодействие *n*, действие *n* на расстоянии

A95 **action integral** || Wirkungsintegral *n* || intégrale *f* d'action || werkingsintegraal *f* *(m)* || интеграл *m* действия, фазовый интеграл *m*

A96 **activation** || Aktivierung *f* || activation *f* || activering *f* || активация *f*; возбуждение *n*

A97 **activation analysis** || Aktivierungsanalyse *f* || analyse *f* par activation || activeringsanalyse *f* || (радио)активационный анализ *m*

A98 **activation energy** || Aktivierungsenergie *f* || énergie *f* d'activation || activeringsenergie *f* || энергия *f* активации

A99 **activator** || Aktivator *m* || activateur *m* || activator *m* || активатор *m*

A100 **active component** || Wirkstromkomponente *f*, ohmsche Komponente *f* || composante *f* active [wattée] || actieve component *m* || активная [омическая] составляющая *f*

A101 **active power** || Wirkleistung *f* || puissance *f* active || actief vermogen *n* || активная мощность *f*

A102 **active zone** || Spaltzone *f*, aktive Zone *f* || zone *f* active *(d'un réacteur)* || reactorkern *n* || активная зона *f* *(реактора)*

A103 **activity** || Aktivität *f* || activité *f* || activiteit *f* || **1.** активность **2.** радиоактивность

A104 **activity coefficient** || Aktivitätskoeffizient *m* || coefficient *m* d'activité || activiteitscoëfficiënt *m* || коэффициент *m* активности

A105 **actual efficiency** || realer Wirkungsgrad *m* || rendement *m* réel || werkelijk rendement *n*, reële nutsfactor *m* || действительный кпд *m*

A106 **actual load** || Wirklast *f* || charge *f* actuelle || werkelijke belasting *f* || действующая нагрузка *f*

A107 **actual size** || Istmaß *n* || cote *f* effective || werkelijke maat *f* *(m)* [afmeting *f*] || действительный размер *m*

A108 **actual stress** || wirksame Spannung *f* || contrainte *f* effective || werkelijke spanning *f* || действующее [истинное] напряжение *n*

A109 **actuator** || Antrieb *m*; Stellvorrichtung *f*; Stellantrieb *m* || commande *f*; entraînement *m*; servomécanisme *m* || aandrijfinrichting *f*; aandrijver *m*; servomechanisme *n* || привод *m*; исполнительный механизм *m*

A110 **acyclic** || azyklisch || acyclique || acyclisch || ациклический

A111 **adaptation** || Adaptation *f*, Anpassung *f* || adaptation *f* || adaptatie *f* || адаптация *n*, приспособление *n*

A112 **addend** || Addend *m* || nombre *m* d'addition || op te tellen getal *n* || слагаемое *n*

A113 **addendum** || Zahnkopfhöhe *f* || hauteur *f* de tête de la dent || tandkophoogte *f* || высота *f* головки зуба *(колеса)*

A114 **addendum circle** || Kopfkreis *m*, Außenkreis *m* || cercle *m* extérieur *(engrenage)* || kopcirkel *m* *(tandwiel)* || окружность *f* головок *или* выступов зубьев *(колеса)*

A115 **adder** || Addierer *m* || additionneur *m* || additietrap *m* || сумматор *m*

A116 **addition formulas** || Additionsformeln *f pl* || formules *f pl* de sommation || additieformules *f (m) pl* || формулы *f pl* сложения

A117 **addition modulo N** || Addition *f* modulo N || addition *f* modulo N || additie *f* modulo N || сложение *n* по модулю N

A118 **addition theorem** || Additionstheorem *n* || théorème *m* d'addition || additietheorema *n* || теорема *f* сложения

A119 **additive** || Zusatz *m*; Beimengung *f* || additif *m*; admixture *f* || toeslag *m*; bijmengsel *n* || присадка *f*; добавка *f*; примесь *f*

A120 **additive group** || Additionsgruppe *f* || groupe *m* additif || additiegroep *f (m)* || аддитивная группа *f*

A121 **additivity** || Additivität *f* || additivité *f* || additie *f* || аддитивность *f*

A122 **address** || Adresse *f* || adresse *f* || adres *n* || адрес *m*

A123 **address arithmetic** || Adressenarithmetik *f* || arithmétique *f* d'adresse || adreswiskunde *f* || адресная арифметика *f*

A124 **address constant, adcon** || Adressenkonstante *f* || constante *f* d'adresse || adresconstante *f (m)* || адресная константа *f*

A125 **addressing** || Adressierung *f* || adressage *m* || adresseren *n* || адресация *f*

A126 **address register** || Adressenregister *n* || registre *m* d'adresse || adresregister *n* || регистр *m* адреса, адресный регистр *m*

A127 **address space** || Adressenraum *m* || espace *m* adressable || adresruimte *f* || адресное пространство *n*

A128 **adhesion** || Adhäsion *f* || adhésion *f* || adhesie *f* || адгезия *f*; сцепление *n*

A129 **adhesive force** || Adhäsionskraft *f*; Haftvermögen *n* || force *f* d'adhésion || adhesiekracht *f (m)* || сила *f* адгезии

A130 **adiabat** || Adiabate *f* || adiabatique *f*, courbe *f* adiabatique || adiabaat *f* || адиабата *f*

A131 **adiabatic** || adiabatisch || adiabatique || adiabatisch || адиабатический, адиабатный

A132 **adiabatic compression** || adiabatische Kompression *f* || compression *f* adiabatique || adiabatische compressie *f* || адиабатическое сжатие *n*

A133 **adiabatic expansion** || adiabatische Ausdehnung *f* || expansion *f* adiabatique || adiabatische ontspanning *f* || адиабатическое [изэнтропическое] расширение *n*

A134 **adiabatic invariant** || adiabatische Invariante *f* || invariant *m* adiabatique || adiabatische invariant *m* || адиабатический инвариант *m*

A135 **adiabatic process** || adiabatischer Vorgang *m* || procédé *m* adiabatique || adiabatisch proces *n* || адиабатический процесс *m*

A136 **adjacency matrix** || Adjazenzmatrix *f* || matrice *f* associée || aanliggende matrix *f (m)* || матрица *f* смежностей

A137 **adjoint kernel** || adjungierter Kern *m* || noyau *m* adjoint || geadjugeerde kern *f* || сопряжённое ядро *n*

A138 **adjoint matrix** || adjungierte Matrix *f* || matrice *f* adjointe [conjuguée] || geadjugeerde matrix *f* || сопряжённая матрица *f*

A139 **adjoint operator** || adjungierter Operator *m* || opérateur *m* adjoint [conjugué] || geadjugeerde operator *m* || сопряжённый оператор *m*

A140 **adjoint variables** || adjungierte Variablen *f pl* || variables *f pl* adjointes || geadjugeerde variabelen *f* (*m*) *pl* || сопряжённые переменные *f pl*

A141 **adjugate matrix** *see* **adjoint matrix**

A142 **adjustable** || einstellbar, nachstellbar, regelbar || ajustable, réglable || regelbaar || регулируемый

A143 **adjustable reamer** || nachstellbare Reibahle *f* || alésoir *m* expansible || regelbare opruimer *m* || разжимная развёртка *f*

A144 **adjustable speed** || einstellbare Drehzahl *f* || vitesse *f* réglable || afstelbare snelheid *f* || регулируемая скорость *f*

A145 **adjustable valve** || einstellbares Ventil *n* || soupape *f* ajustable || afstelbare klep *f* (*m*) || регулируемый клапан *m*

A146 **adjustable wrench** || verstellbarer Schraubenschlüssel *m* || clef *f* [clé *f*] (à overture) réglable || verstelbare [Engelse] sleutel *m* || разводной гаечный ключ *m*

A147 **adjusting screw** || Einstellschraube *f*; Justierschraube *f* || vis *f* de réglage; vis *f* d'ajustage || justeerschroef *f* (*m*) || установочный винт *m*; регулировочный винт *m*; юстировочный винт *m*

A148 **adjusting shim** || verstellbares Abstandsstück *n* || rondelle *f* de calage || regelbaar vulplaatje *n* || регулировочная прокладка *f*

A149 **adjustment** || Einstellung *f*; Justierung *f*; Anspassung *f* || ajustage *m*; réglage *m* || adjusteren *n* || **1.** регулировка *f*, настройка *f*, юстировка *f* **2.** подгонка *f* (*кривой под данные*)

A150 **admissible transformations** || zulässige Transformationen *f pl* || transformations *f pl* admissibles || toelaatbare transformaties *f pl* || допустимые преобразования *n pl*

A151 **admission** || Zuführung *f*, Zufuhr *f* (*z. B. Luft, Wasser*) || admission *f* (*p. ex. air, eau*) || toevoer *m*, toelaat *m* (*bijv. lucht, water*) || впуск *m*, подвод *m* (*напр. воздуха, воды*)

A152 **admission port** || Einlaßöffnung *f*; Zulauf *m*; Einlaufkanal *m* || lumière *f* [orifice *m*] d'admission || toelaatopening *f*; toevoerkanaal *n* || впускное отверстие *n*; впускной канал *m*

A153 **admittance** || Admittanz *f*, Scheinleitwert *m* || admittance *f* || admittantie *f* || полная проводимость *f* (*электрической цепи*)

A154 **adsorbate** || Adsorbat *n*, Adsorptiv *n* || adsorbat *m* || adsorbaat *n* || адсорбат *m*

A155 **adsorbent** || Adsorbens *n*, Adsorptionsmittel *n*, Adsorber *m* || adsorbant *m* || adsorbens *n*, adsorptiemiddel *n* || адсорбент *m*

A156 **adsorption** || Adsorption *f*, Adsorbierung *f* || adsorption *f* || adsorptie *f* || адсорбция *f*

A157 **adsorptivity** || Adsorptionsvermögen *n*, Adsorptionsfähigkeit *f* || pouvoir *m* d'adso rption || adsorptievermogen *n* || адсорбционная способность *f*

A158 **advance of spark** || Frühzündung *f* || avance *f* à l'allumage || voorontsteking *f* || опережение *n* зажигания

A159 **advance of waves** || Wellenausbreitung *f*, Wellenfortpflanzung *f* || propagation *f* d'ondes || golfuitbreiding *f* || распространение *n* волн

A160 **aerial** || Antenne *f* || antenne *f*, aérien *m* || antenne *f* (*m*) || антенна *f*

A161 **aerodynamic drag** || aerodynamischer Widerstand *m* || résistance *f* aérodynamique; traînée *f* || aërodynamische weerstand *m* || аэродинамическое сопротивление *n*

A162 **aerodynamic heating** || aerodynamische Erwärmung *f* || réchauffement *m* aérodynamique || aërodynamische verwarming *f* || аэродинамический нагрев *m*

A163 **aerodynamic lift** || aerodynamischer Auftrieb *m* || portance *f* aérodynamique || aërodynamische stijgkracht *f* (*m*) || аэродинамическая подъёмная сила *f*

A164 **aerodynamics** || Aerodynamik *f* || aérodynamique *f* || aërodynamica *f* || аэродинамика *f*

A165 **aerosol** || Aerosol *n* || aérosol *m* || aërosol *n* || аэрозоль *m*

A166 **aerostatics** || Aerostatik *f* || aérostatique *f* || aërostatica *f* || аэростатика *f*

A167 **affine transformations** || affine Transformationen *f pl* || transformations *f pl* affines || affine transformaties *f pl* || аффинные преобразования *n pl*

A168 **affinity** || Affinität *f* || affinité *f* || affiniteit *f* || 1. аффинность *f* 2. *хим.* сродство

A169 **afocal** || afokal || afocal || afocaal || афокальный

A170 **afterburner** || Nachbrennkammer *f*, Nachbrenner *m* || chambre *f* [tuyère *f*] de postcombustion || nabrander *m* || камера *f* дожигания; форсажная камера *f*

A171 **aftercurrent** || Reststrom *m* || courant *m* résiduel || reststroom *f* || остаточный ток *m*

A172 **afterglow** || Nachleuchten *n*, Nachglimmen *n* || postluminescence *f* || nalichting *f* || послесвечение *n*

A173 **age-hardening** || Ausscheidungshärtung *f* || trempe *f* par saturation || verouderingsharden *n* || дисперсионное твердение *n*

A174 **ag(e)ting** || Alterung *f* || vieillissement *m* || veroudering *f* || старение *n*

A175 **air brake** || Druckluftbremse *f* || frein *m* pneumatique [à air comprimé] || luchtdrukrem *m* || тормоз *m* с пневматическим приводом

A176 **air compressor** || Luftverdichter *m* || compresseur *m* (d'air) || luchtperspomp *f* (*m*) || воздушный компрессор *m*

A177 **air-cooled engine** || luftgekühlter Motor *m* || moteur *m* à refroidissement par air || luchtgekoelde motor *m* || двигатель *m* с воздушным охлаждением

A178 **air cooler** || Luftkühler *m* || refroidisseur *m* d'air || luchtkoeler *m* || воздухоохладитель *m*

A179 **air cooling** || Luftkühlung *f* || refroidissement *m* par air [à air] || luchtkoeling *f* || воздушное охлаждение *n*

A180 **air-core magnet** || kernloser Magnet *m* || aimant *m* sans noyau || magneet *m* zonder kern || магнит *m* без сердечника

A181 **air-driven** || druckluftangetrieben, Druckluft..., Preßluft... || commandé par air comprimé [à air comprimé] || persluchtaangedreven, perslucht... || с пневматическим приводом, пневматический

A182 **air duct** || Luftleitung *f* || conduit *m* d'air || luchtleiding *f*, luchtkanaal *n* || воздухопровод *m*, воздушный канал *m*

A183 **air engine** || Druckluftmotor *m* || moteur *m* à air comprimé || persluchtmotor *m* || пневматический двигатель *m*

A184 **air filter** || Luftfilter *n* || filtre *m* àir || luchtfilter *n*, *m* || воздушный фильтр *m*, воздухоочиститель *m*

A185 **air gap** || Luftspalt *m* || entrefer *m*, espace *m* d'air || luchtspleet *f* (*m*) || воздушный зазор *m*

A186 **air gauge** || Luftdruckmesser *m* || manomètre *m* à air || luchtdrukmanometer *m* || воздушный манометр *m*

A187 **air hammer** || Preßlufthammer *m* || marteau *m* pneumatique || persluchthamer *m* || 1. пневматический молот *m* 2. пневматический молоток *m*

A188 **air-hardening steel** || lufthärtender Stahl *m* || acier *m* trempant à l'air || luchthardend staal *n* || самозакаливающаяся сталь *f*

A189 **air humidity** || Luftfeuchtigkeit *f*, Luftfeuchte *f* || humidité *f* d'air || luchtvochtigheid *f* || влажность *f* воздуха

A190 **air jet** || Luftstrahl *m* || jet *m* d'air || luchtstraal *m*, *f* || воздушная струя *f*

A191 **air pipe** || Entlüftungsrohr *n*; Luftleitung *f* || tube *m* d'air || luchtkoker *m*; luchtleiding *f* || вентиляционная труба *f*; воздушная труба *f*; воздухопровод *m*

A192 **air pressure** || Luftdruck *m*, atmosphärischer Druck *m* || pression *f* d'air, pression *f* atmosphérique || luchtdruk *m* || давление *n* воздуха, атмосферное давление *n*

A193 **air-pressure gauge** *see* **air gauge**

A194 **air pump** || Luftpumpe *f* || pompe *f* à air (comprimé), pompe *f* pneumatique || luchtpomp *f* (*m*) || воздушный насос *m*

A195 **air spring** || Luftfeder *f* || amortisseur *m* à air [pneumatique] || luchtveer *f* (*m*) || пневматический амортизатор *m*

A196 **air stream** || Luftstrom *m*, Luftströmung *f* || courant *m* d'air || luchtstroom *m* || воздушный поток *m*

A197 **air turbine** || Druckluftturbine *f* || turbine *f* à air || luchtturbine *f* || воздушная турбина *f*

A198 **air valve** || Luftventil *n* || soupape *f* à air || luchtklep *f* (*m*) || воздушный клапан *m*

A199 **albedo** || Albedo *f*, Rückstrahlvermögen *n* || albédo *m* || albedo *f* (*m*) || альбедо

A200 **aldehyde** || Aldehyd *m* || aldéhyde *m* || aldehyde *f* || альдегид *m*

A201 **algebraic complement** || algebraisches Komplement *n* || complément *m* algébrique || algebraïsch complement *n* || алгебраическое дополнение *n*

A202 **algebraic equation** || algebraische Gleichung *f* || équation *f* algébrique || algebraïsche vergelijking *f* || алгебраическое уравнение *n*

A203 **algebraic expression** || algebraischer Ausdruck *m* || expression *f* algébrique || algebraïsche uitdrukking *f* || алгебраическое выражение *n*

A204 **algebraic function** || algebraische Funktion *f* || fonction *f* algébrique || algebraïsche functie *f* || алгебраическая функция *f*

A205 **algebraic multiplicity** || algebraische Multiplizität *f* || multiplicité *f* algébrique || algebraïsche multipliciteit *f* || алгебраическая кратность *f*

A206 **algebraic numbers** || algebraische Zahlen *f pl* || nombres *m pl* algébriques || algebraïsche getallen *n pl* || алгебраические числа *n pl*

A207 **algebra of sets** || Mengenalgebra *f* || algèbre *f* des ensembles || algebra *f* van menigten || алгебра *f* множеств

A208 **algorithm** || Algorithmus *m* || algorithme *m* || algorithme *n* || алгоритм *m*

A209 **algorithmic language** || algorithmische Sprache *f* || langage *m* algorithmique || algoritmische taal *f* || алгоритмический язык *m*

A210 **algorithmic presentation** || algorithmische Darstellung *f* || algorithmisation *f* || algoritmisatie *f* || алгоритмизация *f*

A211 **alicyclic** || alizyklisch || alicyclique, cyclanique || alicyclisch || алициклический

A212 **alignment** || Ausrichtung *f*; Fluchtung *f*; Flucht *f* || alignement *m*; centrage *m* || het richten; het in een lijn brengen; centrering *f* || выверка *f*; центрирование *n*; соосность *f*

A213 **aliphatic** || aliphatisch || aliphatique, cyclanique || alifatisch || алифатический

A214 **alkali** || Alkali *n*, Alkalilauge *f* || alcali *m* || alkali *n* || щёлочь *f*

A215 **alkane** || Alkan *n*, Paraffin *n* || alcane *m*, hydrocarbure *m* saturé || alkaan *n* || алкан *m*, насыщенный алифатический углеводород *m*

A216 **alkene** || Alken *n*, Olefin *n* || alcène *m*, oléfine *f* || alkeen *n*, olefine *f* (*m*) || алкен *m*, олефин *m*

A217 **alkylation** || Alkylieren *n*, Alkylierung *f*, Alkylation *f* || alcoylation *f*, alkylation *f* || alkylering *f* || алкилирование *n*

A218 **alligator shears** || Hebelschere *f* || cisaille *f* à levier || hefboomschaar *f* (*m*) || рычажные [аллигаторные] ножницы *pl*

A219 **allotropic transformation** || allotrope Umwandlung *f* || transformation *f* allotropique || allotrope transformatie *f* || аллотропическое превращение *n*

A220 **allotropy** || Allotropie *f* || allotropie *f* || allotropie *f* || аллотропия *f*

A221 **allowable load** || zulässige Belastung *f* || charge *f* admissible || toelaatbare belasting *f* || допускаемая нагрузка *f*

A222 **allowable stress** || zulässige Spannung *f*, zulässige Beanspruchung *f* || contrainte *f* admissible || toelaatbare spanning *f* || допускаемое напряжение *n*

A223 **allowance** || Zugabe *f*; Toleranz *f* || surépaisseur *f*; tolérance *f* || toegave *f*; toelating *f*, tolerantie *f* || припуск *m* (на обработку); допуск *m*

A224 **allowed energy band** || erlaubtes Energieband *m* || bande *f* d'énergie permise || gepermitteerde energieband *m* || разрешённая энергетическая зона *f*

A225 **allowed transition** || erlaubter Übergang *m* || transition *f* permise || gepermitteerde transitie *f* || разрешённый [дозволенный] переход *m*

A226 **alloy** || Legierung *f* || alliage *m* || legering *f* || сплав *m*

A227 **alloyed steel** || legierter Stahl *m* || acier *m* allié || gelegeerd staal *n* || легированная сталь *f*

A228 **alloyed transistor** || legierter Transistor *m*, Legierungstransistor *m* || transistor *m* allié || gelegeerde transistor *m* || сплавной транзистор *m*

A229 **alloying** || Legieren *n* || alliage *m* || legeren *n* || легирование *n*

A230 **alloying element** || Legierungselement *n*, Legierungszusatz *m* || élément *m* [additif *m*] d'alliage || legeringselement *n*, legeringstoevoeging *f* || легирующий элемент *m*, легирующая добавка *f*

A231 **all-pass** || alle Frequenzen durchlassend, Allpaß... || passant toutes les fréquences || alle frequenties doorlatend || пропускающий все частоты

A232 **all-purpose** || universal, Universal... || universel || universeel || универсальный *(о приборе)*

A233 **alpha-active** || alpha-aktiv || alpha-actif || alfa-actief || альфа-(радио)активный

A234 **alpha decay** || Alpha-Zerfall *m* || désintégration *f* alpha || alfa-desintegratie *f*, alfa-verval *n* || альфа-распад *m*

A235 **alpha emitter** || Alpha-Strahler *m*, Alpha-Strahlungsquelle *f* || émetteur *m* alpha || alfa-emissor *m* || альфа-излучатель *m*, альфа-источник *m*

A236 **alphameric, alphanumeric** || alpha-numerisch || alphanumérique || alfanumeriek || алфавитно-цифровой

A237 **alphanumeric code** || alphanumerscher Kode *m* || code *m* alphanumérique || alfanumerieke code *m* || алфавитно-цифровой код *m*

A238 **aplha particle** || Alpha-Teilchen *n* || particule *f* alpha || alfadeeltje *n* || альфа-частица *f*

A239 **alternate load** || wechselnde Belastung *f*, Wechsellast *f* || charge *f* alternative || wisselende belasting *f* || знакопеременная нагрузка *f*

A240 **alternate stress** || Dauerschwingbeanspruchung *f* || contrainte *f* alternative || wisselende spanning *f* || знакопеременное напряжение *n*

A241 **alternating current** || Wechselstrom *m* || courant *m* alternatif || wisselstroom *m* || переменный ток *m*

A242 **alternating-current circuit** || Wechselstromkreis *m* || circuit *m* de courant alternatif || wisselstroomcircuit *n* || цепь *f* переменного тока

A243 **alternating-current generator** || Wechselstromgenerator *m* || alternateur *m* || wisselstroomdynamo *m* || (электромашинный) генератор *m* переменного тока

A244 **alternating-current motor** || Wechselstrommotor *m* || moteur *m* à courant alternatif || wisselstroommotor *m* || электродвигатель *m* переменного тока

A245 **alternating group** || alternierende Gruppe *f* || groupe *m* alterné || wisselgroep *f* (*m*), afwisselende groep *f* (*m*) || знакопеременная группа *f*

A246 **alternating matrix** || alternierende Matrix *f* || matrice *f* alternée || afwisselende matrix *f* || альтернирующая матрица *f*

A247 **alternating voltage** || Wechselspannung *f* || tension *f* alternative || wisselspanning *f* || переменное напряжение *n*

A248 **altitude** || Höhe *f* || altitude *f* || hoogte *f* || высота *f*

A249 **aluminium,** Al || Aluminium *n* || aluminium *m* || aluminium *n* || алюминий *m*

A250 **aluminium(-base) alloy** || Aluminiumlegierung *f*, Al-Legierung *f* || alliage *m* d'aluminium || aluminiumlegering *f* || алюминиевый сплав *m*

A251 **aluminium foil** || Aluminiumfolie *f* || feuille *f* mince d'aluminium || aluminiumfolie *f* (*m*) || алюминиевая фольга *f*

A252 **ambient conditions** || äußere Bedingungen *f pl* || conditions *f pl* ambiantes || buitenvoorwaarden *f pl* || окружающие [внешние] условия *n pl*

A253 **ambiguity** || Mehrdeutigkeit *f* || ambiguïté *f* || meerduidigheid *f* || неоднозначность *f*

A254 **ambiguous language** || mehrdeutige Sprache *f* || langage *m* ambigu || voor meer dan één uitleg vatbare taal *f* || неоднозначный язык *m*

A255 **ambipolar** || ambipolar || ambipolaire || ambipolair || амбиполярный

A256 **amide** || Amid *n* || amide *m* || amide *f* || амид *m*

A257 **amine** || Amin *n* || amine *f* || amine *f* (*m*) || амин *m*

A258 **amino-acid** || Aminosäure *f* || amino-acide *m* || aminozuur *n* || аминокислота *f*

A259 **ammeter** || Amperemeter *m*, Strommesser *m* || ampèremètre *m* || ampèremeter *m* || амперметр *m*

A260 **ammonia,** NH_3 || Ammoniak *n* || ammoniac *m* || ammoniak *m* || аммиак *m*

A261 **amorphous** || amorph || amorphe || amorf || аморфный

A262 **amorphous state** || amorpher Zustand *m*, Gestaltlosigkeit *f* || état *m* amorphe || amorfe toestand *m* || аморфное состояние *n*

A263 **amperage** || Stromstärke *f* in Amperes || ampérage *m* || ampèrage *f* || сила *f* тока (в амперах)

A264 **ampere** || Ampere *n* || ampère *m* || ampère *m* || ампер *m*

A265 **ampere-balance** || Stromwaage *f*, Kelvin-Stromwaage *f* || balance *f* électrodynamique || stroombalans *f* (*m*) van Kelvin || токовые весы *pl* Кельвина

A266 **ampere-hour** || Amperestunde *f* || ampère-heure *f* || ampère-uur *n* || ампер-час *m* *(3600 кл)*

A267 **ampere-lux characteristic** || Fotostromkennlinie *f*, Lux-Ampere-Kennlinie *f* || caractéristique *f* lux-ampère || lux-ampèrekarakteristiek *f* || люкс-амперная характеристика *f*

A268 **ampere-second** || Amperesekunde *f* || ampère-seconde *m* || ampère-seconde *f* (*m*) || ампер-секунда *f*, кулон *m*

A269 **ampere-turn** || Amperewindung *f* || ampère-tour *m* || ampère-winding *f* || ампер-виток *m*

A270 **ampere-voltmeter** || Volt- und Amperemeter *n*, Spannungs- und Strommesser *m* || ampèrevoltmètre *m* || voltampèremeter *m* || ампер-вольтметр *m*, вольтамперметр *m*, авометр *m*

A271 **amplidyne** || Amplidyne *f*, Amplidyneverstärker *m* || amplidyne *f* || amplidyne *m* || амплидин *m*

A272 **amplification** || Verstärkung *f* || amplification *f* || versterking *f* || усиление *n* *(сигналов)*

A273 **amplification factor** || Verstärkungskoeffizient *m*, Verstärkungsfaktor *m* || coefficient *m* d'amplification || versterkingsfactor *m* || коэффициент *m* усиления *(сигнала)*

A274 **amplifier** || Verstärker *m* || amplificateur *m* || versterker *m* || усилитель *m*

A275 **amplitude** || Amplitude *f*, Schwingungsweite *f* || amplitude *f* || amplitude *f* || амплитуда *f*

A276 **amplitude-modulated** || amplitudenmoduliert || modulé en amplitude || amplitude-gemoduleerd || амплитудно-модулированный

A277 **amplitude modulation** || Amplitudenmodulation *f* || modulation *f* en amplitude || amplitudemodulatie *f* || амплитудная модуляция *f*

A278 **amplitude of a complex number** || Argument *n* einer komplexen Zahl || amplitude *f* [argument *m*] d'un nombre complexe || amplitude *f* van een complex getal || аргумент *m* комплексного числа

A279 **analog computer** || Analogrechner *m* || calculateur *m* analogique || analoge rekenmachine *f* || аналоговая [моделирующая] вычислительная машина *f*

A280 **analog-digital converter** || Analog-Digital-Umsetzer *m*, A/D-Umsetzer *m* || convertisseur *m* analogique-numérique || analoog-digitaalomzetter *m* || аналого-цифровой преобразователь *m*

A281 **analogous** *see* **analogue** II

A282 **analog representation** || analoge Darstellung *f* || représentation *f* analogique || analoge voorstelling *f* || аналоговое представление *n*

A283 I **analogue** || Analogon *n* || analogue *m*; modèle *m* || analoog *m* || аналог *m*, модель *f*

A284 II **analogue** || analog, Analog..., Analogie... || analogique || analoog... || аналоговый

A285 **analogue-digital** || analog-digital || analogique-digital || analoog-digitaal... || аналого-цифровой

A286 **analogy** || Analogie *f*, Ähnlichkeit *f* || analogie *f* || analogie *f* || аналогия *f*

A287 **analogy parameter** || **1.** Analogieparameter *m* **2.** Ähnlichkeitskriterium *n* || **1.** paramètre *m* d'analogie **2.** argument *m* de similitude || **1.** analogieparameter *m* **2.** gelijkvormigheidscriterium *n* || **1.** параметр *m* аналогии **2.** критерий *m* подобия

A288 **analyser** || Analysator *m* || analyseur *m* || analysator *m* || анализатор *m*

A289 **analysis** || Analyse *f* || analyse *f* || analyse *f* || анализ *m*, расчёт *m*

A290 **analysis of variance** || Varianzanalyse *f* || analyse *f* de variance || дисперсионный анализ *m*

A291 **analytical expression** || analytischer Ausdruck *m* || expression *f* analytique || analytische uitdrukking *f* || аналитическое выражение *n*

A292 **analytical geometry** || analytische Geometrie *f* || géométrie *f* analytique || analytische geometrie [meetkunde *f*] || аналитическая геометрия *f*

A293 **analytic, analytical** || analytisch || analytique || analytisch || аналитический

A294 **analytic continuation** || analytische Fortsetzung *f* || continuation *f* analytique || analytische voortzetting *f* || аналитическое продолжение *n*

A295 **analytic function** || analytische Funktion *f* || fonction *f* analytique || analytische functie *f* || аналитическая функция *f*

A296 **analyzer** || Analysator *m*, Analysiergerät *n* || analyseur *m* || analysator *m* || анализатор *m*

A297 **anastigmat** || Anastigmat *m* || anastigmat *m* || anastigmaat *m* || анастигмат *m*

A298 **anchor** || Anker *m* || ancre *m* || anker *m*; (scheeps)anker *n* || анкер *m*; якорь *m*

A299 **anchorage** || Verankerung *f* || ancrage *m* || verankering *f* || анкеровка *f*, анкерное крепление *n*, анкер *m*

A300 **anchor bolt** || Ankerbolzen *m*, Fundamentschraube *f* || boulon *m* d'ancrage || verankeringsbout *m*, ankerbout *m* || анкерный [фундаментальный] болт *m*

A301 **AND** || UND *n* || ET, AND || EN || И *(логическая функция или операция)*

A302 **AND-circuit, AND-element** || UND-Schaltung *f* || élément ET, circuit ET || EN-poort *f* (*m*), EN-poortschakeling *f* || схема *f* И

A303 **anelasticity** || Inelastizität *f* || anélasticité *f* || anelasticiteit *f* || неупругость *f*

A304 **angle** || Winkel *m*; Ecke *f* || angle *m* || hoek *m* || угол *m*

A305 **angle bar** *see* **angle iron**

A306 **angle gauge** || Winkelmesser *m* || goniomètre *m* || hoekmeter *m* || угломер *m*

A307 **angle iron** || Winkelstahl *m* || acier *m* en L; cornière *f* || hoekstaal *n* || угловая сталь *f*

A308 **angle lever** || Winkelhebel *m* || levier *m* coudé || kniehefboom *m* || коленчатый рычаг *m*

A309 **angle of attack** || Anstellwinkel *m*; Anströmwinkel *m* || angle *m* d'attaque || invalshoek *m* || угол *m* атаки

A310 **angle of beam** || Bündelöffnungswinkel *m* || angle *m* d'ouverture de faisceau || bundelopeningshoek *m* || угол *m* раствора [раскрытия] пучка

A311 **angle of contact** || Kontaktwinkel *m*, Randwinkel *m* || angle *m* de contact || contacthoek *m* || краевой угол *m*

A312 **angle of declination** || magnetische Deklination *f* || déclinaison *f* magnétique || magnetische declinatie *f* || магнитное склонение *n*

A313 **angle of friction** || Reibungswinkel *m* || angle *m* de frottement || wrijvingshoek *m* || угол *m* трения

A314 **angle of incidence** || **1.** Einfallswinkel *m*; Auftreffwinkel *m* **2.** Anstellwinkel *m* || **1.** angle *m* d'incidence **2.** angle *m* d'attaque || invalshoek *m* || **1.** угол *m* падения *(луча)* **2.** угол *m* атаки

A315 **angle of prism** || Ablenkwinkel *m* eines Prismas || angle *m* de déviation d'un prisme || afwijkingshoek *m* van een prisma || отклоняющий угол *m* призмы

A316 **angle of rake** *see* **rake angle**

A317 **angle of repose** || natürlicher Böschungswinkel *m* || angle *m* de repos [de talus naturel] || natuurlijke taludhoek *m* || угол *m* естественного откоса

A318 **angle of rest** *see* **angle of repose**

A319 **angle of rotation** || Drehwinkel *m*; Schwenkwinkel *m* || angle *m* de rotation || draaiingshoek *m* || угол *m* поворота

A320 **angle of shear** || Schubwinkel *m*; Verschiebungswinkel *m* || angle *m* de cisaillement || verschuivingshoek *m* || угол *m* сдвига

A321 **angle of torsion** || Verdreh(ungs)winkel *m* || angle *m* de torsion || wringingshoek *m* || угол *m* кручения; угол *m* закручивания

A322 **angle of twist** || Torsionswinkel *m*, Verdrehungswinkel *m* || angle *m* de torsion || torsiehoek *m* || угол *m* закручивания

A323 **angstrom** || Angström *n* || unité *f* Angström || Ångström *m*, Angström-eenheid *f* || ангстрем *m* *(10⁻¹⁰ м)*

A324 **angular acceleration** || Winkelbeschleunigung *f* || accélération *f* angulaire || hoekversnelling *f* || угловое ускорение *n*

A325 **angular ball bearing** || Schrägkugellager *n* || palier *m* radial et de butée à billes || radiaalkogellager *n* || радиально-упорный шариковый подшипник *m*

A326 **angular dispersion** || Winkeldispersion *f* || dispersion *f* angulaire || hoekdispersie *f* || угловая дисперсия *f*

A327 **angular displacement** || Winkelverschiebung *f* || déplacement *m* angulaire || draaiingshoek *m* || угол *m* поворота; угловое смещение *n*

A328 **angular distance** || Winkelabstand *m* || distance *f* angulaire || hoekafstand *m* || угловое расстояние *n*

A329 **angular frequency** || Winkelfrequenz *f* || fréquence *f* angulaire || hoekfrequentie *f* || угловая частота *f*

A330 **angular momentum** *see* **moment of momentum**

A331 **angular resolution** || Winkelauflösung *f* || résolution *f* angulaire || hoekoplossing *f* || угловое разрешение *n*

A332 **angular thread** || Spitzgewinde *n* || filet *m* triangulaire || driehoekige schroefdraad *m* || резьба *f* треугольного профиля, треугольная резьба *f*

A333 **angular velocity** || Winkelgeschwindigkeit *f* || vitesse *f* angulaire || hoeksnelheid *f* || угловая скорость *f*

A334 **anharmonicity** || Anharmonizität *f* || anharmonicité *f* || anharmoniciteit || ангармоничность *f*, нелинейность *f* колебаний

A335 **anharmonic oscillations** || anharmonische Schwingungen *f pl* || oscillations *f pl* anharmoniques || anharmonische trillingen *f pl* || ангармонические колебания *n pl*

A336 **anharmonic ratio** || Doppelverhältnis *n* || rapport *m* anharmonique || biquotiënt *n*, dubbelverhouding *f* || ангармоническое [двойное инвариантное] отношение *n*

A337 **anhydride** || Anhydrid *n* || anhydride *m* || anhydride *n* || ангидрид *m*

A338 **anhysteretic** || hysteresefrei, hysteresislos || sans hystérésis || hysteresisvrij || безгистерезисный

A339 **anion** || Anion *n* || anion *m* || anion *n* || анион *m*

A340 **anisotropic** || anisotrop || anisotrope || anisotroop || анизотропный

A341 **anisotropy** || Anisotropie *f* || anisotropie *f* || anisotropie *f* || анизотропия *f*, анизотропность *f*

A342 **annihilation** || Annihilation *f* || annihilation *f* || annihilatie *f* || аннигиляция *f*

A343 **annulus** || Innenrad *n*, innenverzahntes Rad *n* || engrenage *m* intérieur || tandwie *n* met inwendige vertanding || зубчатое колесо *n* с внутренним зацеплением

A344 **anode** || Anode *f* || anode *f* || anode *f* || анод *m*

A345 **anomaly** || Anomalie *f* || anomalie *f* || anomalie *f* || аномалия *f*

A345 a **antecedent** || 1. Antezedent *n* 2. erstes Proportionsglied *n* || 1. antécédent *m* 2. premier terme *m* d'une proportion || 1. antecedent *n* 2. tweede proportieglid *n* || 1. антецедент *m* 2. второй член *m* пропорции

A346 **antenna** *see* **aerial**

A347 **anticathode** || Antikatode *f* || anticathode *f* || antikathode *f* || антикатод *m* *(рентгеновской трубки)*

A348 **anticoincidence** || Antikoinzidenz *f* || anticoïncidence *f* || anticoincidentie *f* || антисовпадение *n*

A349 **anticommutativity** || Antikommutativität *f* || anticommutativité *f* || anticommutativiteit *f* || антикоммутативность *f*

A350 **antiferromagnet** || Antiferromagnetikum *n*, antiferromagnetischer Stoff *m* || antiferromagnétique *m* || antiferromagneet *m* || антиферромагнетик *m*

A351 **antiferromagnetism** || Antiferromagnetismus *m* || antiferromagnétisme *m* || antiferromagnetisme *n* || антиферромагнетизм *m*

A352 **antifrictional property** || Gleiteigenschaft *f* || propriété *f* antifriction || antifrictie-eigenschap *f* || антифрикционное свойство *n*

A353 **antifriction bearing** *see* **rolling**

A354 **antifriction metal** || Lagermetall *n*, Antifriktionsmetall *n* || métal *m* antifriction || antifrictiemetaal *n* || антифрикционный сплав *m*

A355 **antilog** *see* **antilogarithm**

A356 **antilogarithm** || Antilogarithmus *m* || antilogarithme *m* || antilogaritme *f* *(m)* || антилогарифм *m*

A357 **antimatter** || Antimaterie *f* || antimatière *f* || antistof *f* *(m)* || антивещество *n*

A358 **antimonic acid** || Antimonsäure *f*, Antimon(V)-Säure *f* || acide *m* antimonique || antimoonzuur *n* || сурьмяная кислота *f*

A359 **antimonious acid** || antimonige Säure *f*, Antimon(III)-Säure *f* || acide *m* antimonieux || antimonigzuur *n* || сурьмянистая кислота *f*

A360 **antimony, Sb** || Antimon *n* || antimoïne *m* || antimoon *n* || сурьма *f*

A361 **antinode** || Bauch *m* || antinœud *m* [ventre *m*] d'onde || buik *m* || пучность *f* *(волны)*

A362 **antinomy** || Antinomie *f* || antinomie *f* || antinomie *f* || антиномия *f*

A363 **antiparallel** || antiparallel || antiparallèle || antiparallel || антипараллельный

A364 **antiparticle** || Antiteilchen *n* || antiparticule *f*, anticorpuscule *m* || antideeltje *n* || античастица *f*

A365 **antiperiodic function** || antiperiodische Funktion *f* || fonction *f* antipériodique || antiperiodische functie *f* || антипериодическая функция *f*

A366 **antipole** || Antipol *m* || antipôle *m* || antipool *f* *(m)* || противоположный полюс *m* *(магнита)*

A367 **antiproton** || Antiproton *n* || antiproton *m* || antiproton *n* || антипротон *m*

A368 **antiresonance** || Antiresonanz *f* || antirésonance *f* || antirésonantie *f* || антирезонанс *m*

A369 **antisymmetric matrix** || antisymmetrische Matrix *f* || matrice *f* antisymétrique || antisymmetrische matrix *f* (*m*) || антисимметрическая матрица *f*

A370 **antisymmetric relation** || antisymmetrisches Verhältnis *n* || relation *f* antisymétrique || antisymmetrische verhouding *f* || антисимметричное отношение *n*

A371 **antisymmetry** || Antisymmetrie *f* || antisymétrie *f* || antisymmetrie *f* || антисимметрия *f*

A372 **aperiodic** || aperiodisch || apériodique || aperiodisch || апериодический

A373 **aperiodic motion** || aperiodische Bewegung *f* || mouvement *m* apériodique || aperiodische beweging *f* || апериодическое [непериодическое] движение *n*

A374 **aperture** || Apertur *f*, Öffnung *f* || ouverture *f* || apertuur *f* || апертура *f*, диафрагма *f*

A375 **aperture angle** || Winkelapertur *f*, Öffnungswinkel *m* || ouverture *f* angulaire || hoekapertuur *f* || угловая апертура *f*

A376 **aperture diaphragm** || Aperturblende *f* || diaphragme *f* d'ouverture || apertuurdiafragma *f* || апертурная диафрагма *f*

A377 **aperture ratio** || 1. Öffnungsverhältnis *n*, relative Öffnung *f* 2. Lichtstärke *f* || 1. ouverture *f* relative 2. clarté *f* || 1. openingsverhouding *f* 2. lichtsterkte *f* || 1. относительное отверстие *n* 2. светосила *f*

A378 **apochromat** || Apochromat *m* || apochromat *m* || apochromaat *m* || апохромат *m*

A379 **a posteriori probability** || Wahrscheinlichkeit *f* a posteriori || probabilité *f* a posteriori || waarschijnlijkheid *f* a posteriori || апостериорная вероятность *f*

A380 **apothem** || Apothema *f* || apothème *m* || apothema *n* || апофема *f*

A381 **apparatus** || Apparat *m*; Gerät *n* || appareil *m*; dispositif *m* || toestel, apparaat *n* || аппарат *m*; прибор *m*

A382 **apparent resistance** || Scheinwiderstand *m*, Impedanz *f* || résistance *f* totale, impédance *f* || schijnweerstand *m* || полное сопротивление *n*, импеданс *m*

A383 **apparent variable** || scheinbare Variable *f* || variable *f* apparente || schijnbare variabele *f* || фиктивная переменная *f*

A384 **application of force** || Kraftangriff *m*, Angreifen *n* der Kraft || application *f* de la force || krachtaanwending *f* || приложение *n* сил

A385 **application program** || Anwendungsprogramm *n* || programme *n* d'application || aanwendingsprogramma *n* || прикладная программа *f*

A386 **applied field** || eingeprägtes Feld *n* || champ *m* appliqué || aangelegd veld *n* || приложенное [внешнее] поле *n*

A387 **applied force** || aufgeprägte Kraft *f* || force *f* appliquée || aangelegd kracht *f* (*m*) || внешняя [приложенная] сила *f*

A388 **applied load** || angreifende Belastung *f* || charge *f* appliquée || aangelegde belasting *f* || приложенная нагрузка *f*

A389 **applied mathematics** || angewandte Mathematik *f* || mathématiques *f pl* appliquées || toegepaste wiskunde *f* || прикладная математика *f*

A390 **applied mechanics** || angewandte Mechanik *f* || mécanique *f* appliquée || toegepaste mechanica *f* || прикладная механика *f*

A391 **applied physics** || angewandte Physik *f* || physique *f* appliquée || toegepaste fysica *f* || прикладная [техническая] физика *f*

A392 **applied voltage** || angelegte [zugeführte] Spannung *f* || tension *f* appliquée || aangelegde spanning *f* || приложенное напряжение *n*

A393 **approximate** || angenähert, Näherungs... || approximatif, approché || benaderend || приближённый

A394 **approximation** || Approximation *f*, Approximierung *f*, Näherung *f* || approximation *f* || benadering *f* || аппроксимация *f*, приближение *n*

A395 **approximation of functions** || Approximierung *f* von Funktionen || approximation *f* de fonctions || benadering *f* van de functies || приближение *n* функций

A396 **a priori distribution** || Apriori-Verteilung *f* || distribution *f* a priori || a-priori-verdeling *f* || априорное распределение *n*

A397 **a priori probability** || Apriori-Wahrscheinlichkeit *f* || probabilité *f* a priori || a-priori-waarschijnlijkheid *f* || априорная вероятность *f*

A398 **arbitrary value** || willkürlicher Wert *m* || valeur *f* arbitraire || willekeurige waarde *f* || произвольное значение *n*

A399 **arbor** || **1.** Welle *f*, Achse *f*; Spindel *f* **2.** Dorn *m* || **1.** arbre *m*; broche *f* **2.** mandrin *m* || **1.** as *f* (*m*) **2.** doorn *m* || **1.** ось *f*; вал *m*; шпиндель *m* **2.** оправка *f*

A400 **arc** || **1.** Bogenentladung *f* **2.** *Math.* Bogen *m* || **1.** décharge *f* d'arc **2.** *math.* arc *m* || **1.** boogontlading *f* **2.** *math.* boog *m* || **1.** дуговой разряд, дуга **2.** *мат.* дуга *f*

A401 **arc discharge** || Bogenentladung *f* || décharge *f* d'arc || boogontlading *f* || дуговой разряд *m*, дуга *f*

A402 **arc-discharge lamp** || Bogenlampe *f* || lampe *f* à arc || booglamp *f* (*m*) || дуговая лампа *f*

A403 **arc function** || Arkusfunktion *f* || fonction *f* de l'arc || arcusfunctie *f* || аркфункция *f*

A404 **arc gap** || Lichtbogenstrecke *f* || distance *f* d'éclatement [de décharge] || lichtboogruimte *f* || дуговой промежуток *m*

A405 **Archimedian assumption, Archimedian axiom** || Archimedisches Axiom *n* || axiome *m* d'Archimède || Archimedisch axioma *n* || аксиома *f* Архимеда

A406 **Archimedes' spiral** || Archimedische Spirale *f* || spirale *f* d'Archimède || Archimedische spiraal *f* (*m*) || спираль *f* Архимеда

A407 **arc length** || Bogenlänge *f* || longueur *f* de l'arc || booglengte *f* || длина *f* дуги

A408 **arc-over** || Bogendurchbruch *m*, Bogenüberschlag *m* || recouvrement *m* par arc || boogoverslag *m* || перекрытие *n* дугой

A409 **arc welding** *see* **electric-arc welding**

A410 **area** || Fläche *f*; Oberfläche *f* || aire *f*; superficie *f*; surface *f* || vlak *n*; oppervlak *n*, oppervlakte *f* || площадь *f*; поверхность *f*

A411 **areal velocity** || Flächengeschwindigkeit *f* || vitesse *f* aréolaire || sectorsnelheid *f* || секторная [секториальная] скорость *f*

A412 **area of contact** || Berührungsfläche *f*, Kontaktfläche *f* || surface *f* de contact || contactvlac *n* || площадь *f* контакта [соприкосновения]

A413 **area of triangle** || Dreiecksfläche *f* || aire *f* d'un triangle || driehoekoppervlakte *f* || площадь *f* треугольника

A414 **areometer** || Aräometer *n*, Senkwaage *f*, Senkspindel *f* || densimètre *m*, aréomètre *m* || areometer *m* || ареометр *m*

A415 **argentous oxide,** Ag_2O || Silberoxid *n* || oxyde *m* d'argent || zilveroxyde *n* || окись *f* серебра

A416 **argon,** Ar || Argon *n* || argon *m* || argon *n* || аргон *m*

A417 **argument** || Argument *n* || argument *m* || argument *n* || аргумент *m*, независимая переменная *f*

A418 **argument of a complex number** || Argument *n* einer komplexen Zahl || argument *m*, d'un nombre complexe || argument *n* van een complex getal || аргумент *m* комплексного числа

A419 **argument of function** || Argument *n* einer Funktion || argument *m* d'une fonction || argument *n* van een functie || аргумент *m* функции

A420 **arithmetical average** || arithmetisches Mittel *n* || moyenne *f* arithmétique || aritmetisch middel *n* || среднее арифметическое *n*

A421 **arithmetical mean** *see* **arithmetical average**

A422 **arithmetic progression** || arithmetische Reihe *f* || progression *f* arithmétique || rekenkundige reeks *f* (*m*) || арифметическая прогрессия *f*

A423 **arithmetization** || Arithmetisierung *f* || arithmétisation *f* || aritmetisatie *f* || арифметизация *f*

A424 **arm** || **1.** Arm *m*; Hebel *m* **2.** Zeiger *m* *(eines Meßinstrumentes)* **3.** Ausleger *m*, Arm *m* || **1.** bras *m*;

levier *m* 2. aiguille *f* 3. bras *m* || 1. arm *m* 2. wijzer *m* 3. arm *m* || 1. плечо *n* *(рычага)*: рычаг *m* 2. стрелка *f* *(прибора)* 3. кронштейн *m*

A425 **armature core** || Ankerkern *m* || noyau *m* d'induit || ankerkern *m* || сердечник *m* [тело *n*] якоря

A426 **armature winding** || Ankerwicklung *f* || enroulement *m* d'induit || ankerwikkeling *f* || обмотка *f* якоря

A427 **arm of bridge** || Brückenzweig *m* || bras *m* du pont || brugarm *m* || плечо *n* (измерительного) моста

A428 **arm of couple** || Kraftarm *m*, Arm *m* des Kräftepaares || bras *m* de levier d'un couple || krachtarm *m* || плечо *n* пары сил

A429 **arm of force** || Kraftarm *m* || bras *m* de force || krachtarm *m* || плечо *n* силы

A430 **armored glass** || Drahtglas *n* || verre *m* armé || gewapend glas *n* || армированное стекло *n*

A431 **array** || Feld *n*, Datenfeld *n* || tableau *m* || rij *f* (*m*) || массив *m* *(данных)*

A432 **arrester** || 1. Arretiervorrichtung *f* 2. Entlader *m*, Entladungsstrecke *f* || 1. dispositif *m* [mécanisme *m*] de blocage 2. déchargeur *m* || 1. arrêteerhefboom *m*, arrêteerinrichting *f* 2. ontlader *m* || 1. арретир *m* 2. разрядник *m*

A433 **arsenic,** As || Arsen *n* || arsenic *m* || arsenicum *n* || мышьяк *m*

A434 **arsenic acid** || Arsensäure *f* || acide *m* arsénique || arseenzuur *n* || мышьяковая кислота *f*

A435 **artificial age hardening** || künstliche Alterung *f* || vieillissement *m* artificiel || kunstmatige veroudering *f* || искусственное старение *n*

A436 **artificial ag(e)ing** *see* **artificial age hardening**

A437 **artificial language** || künstliche Sprache *f* || langage *m* artificiel || kunstmatige taal *f* (*m*) || искусственный язык *m*

A438 **artificial radioactivity** || künstliche Radioaktivität *f* || radioactivité *f* artificielle || geïnduceerde [kunstmatige] radioactiviteit *f* || искусственная радиоактивность *f*

A439 **artificial variable** || künstliche Variable *f* || variable *f* artificielle || kunstmatige variabele *f* || искусственная переменная *f*

A440 **aspherical** || asphärisch || asphérique || asferisch || асферический

A441 **assembler** || Assembler *m*, Assemblierer *m*, Assembly-Programm *n* || assembleur *m* || assembler *m* || ассемблер *m*

A442 **assembly** || 1. Montage *f* 2. Bandgruppe *f*; Aggregat *n* 3. Ensemble *n*, Gesamtheit *f* || 1. montage *m*, assemblage *m* 2. ensemble *m* 3. groupement *m* || 1. samenvoeging *f*, samenstelling *f*, montage *f* 2. knooppunt *n*; aggregaat *n* 3. groepering *f* || 1. сборка *f*, монтаж *m* 2. узел *m*; агрегат *m* 3. ансамбль *m* *(напр. частиц)*

A443 **assertion** || Behauptung *f* || assertion *f* || bevestiging *f* || утверждение *n*

A444 **associate operator** || adjungierter Operator *m* || opérateur *m* associé || geadjugeerde operator *m* || сопряжённый оператор *m*

A445 **associative** || assoziativ || associatif || associatief || ассоциативный

A446 **associative law** || Assoziativgesetz *n* || loi *f* associative || associatieve wet *f* (*m*) || ассоциативный закон *m*

A447 **associative matrix** *see* **adjoint matrix**

A448 **associativeness, associativity** || Assoziativität *f* || associativité *f* || associativiteit *f* || ассоциативность *f*

A449 **astatic** || astatisch || astatique || astatisch || астатический, астазированный

A450 **astigmatism** || Astigmatismus *m* || astigmatisme *m* || astigmatisme *n* || астигматизм *m*

A451 **astroid** || Astroide *f* || astroïde *f* || astroïde *f* || астроида *f*

A452 **astronomy** || Astronomie *f* || astronomie *f* || astronomie *f*, sterrenkunde *f* || астрономия *f*

A453 **astrophysics** || Astrophysik *f* || astrophysique *f* || astrofysica *f* || астрофизика *f*

A454 **asymmetric(al)** || asymmetrisch || asymétrique || asymmetrisch || асимметричный, асимметрический

A455 **asymmetric relation** || asymmetrisches Verhältnis *n* || relation *f* asymétrique || asymmetrische verhouding *f* || асимметричное отношение *n*

A456 **asymmetry** || Asymmetrie *f* || asymétrie *f* || asymmetrie *f* || асимметрия *f*, несимметричность *f*

A457 **asymptote** || Asymptote *f* || asymptote *f* || asymptoot *m* || асимптота *f*

A458 **asymptotic(al)** || asymptotisch || asymptotique || asymptotisch || асимптотический

A459 **asymptotic line** *see* **asymptote**

A460 **asymptotic series** || asymptotische Reihe *f* || série *f* asymptotique || asymptotische reeks *f (m)* || асимптотический ряд *m*

A461 **asymptotic stability** || asymptotische Stabilität *f* || stabilité *f* asymptotique || asymptotische stabiliteit *f* || асимптотическая устойчивость *f*

A462 **asymptotic value** || asymptotischer Wert *m* || valeur *f* asymptotique || asymptotische waarde *f* || асимптотическое значение *n*

A463 **asynchronous machine** || Asynchronmaschine *f* || machine *f* asynchrone || asynchrone machine *f* || асинхронная машина *f*

A464 **asynchronous motor** || Asynchronmotor *m* || moteur *m* asynchrone || asynchrone motor *m* || асинхронный электродвигатель *m*

A465 **atmosphere** || Atmosphäre *f*; gasförmiges Milieu *n* || atmosphère *f* || atmosfeer *f (m)* || атмосфера *f*; газовая среда *f*

A466 **atmospheric electricity** || Luftelektrizität *f* || électricité *f* atmosphérique || luchtelektriciteit *f* || атмосферное электричество *n*

A467 **atmospheric optics** || atmosphärische Optik *f*, Optik *f* der Atmosphäre || optique *f* atmosphérique || atmosferische optiek *f* || оптика *f* атмосферы, атмосферная оптика *f*

A468 **atmospheric pressure** || atmosphärischer Druck *m*, Luftdruck *m* || pression *f* atmosphérique || atmosferische druk *m*, luchtdruk *m* || атмосферное давление *n*

A469 **atmospheric refraction** || atmosphärische Refraktion *f*, atmosphärische Strahlenbrechung *f* || refraction *f* atmosphérique || atmosferische refractie *f* || атмосферная рефракция *f*

A470 **atmospherics** || atmosphärische Störungen *f pl* || perturbations *f pl* [parasites *m pl*] atmosphériques || atmosferische storingen *f pl* || атмосферики *m pl*, атмосферные помехи *f pl*

A471 **atom** || Atom *n* || atome *m* || atoom *n* || атом *m*

A472 **atom core** || Atomrumpf *m* || noyau *m* d'atome || atoomkern *f (m)* || атомный остов *m*; атом *m* без валентных электронов

A473 **atomic** || Atom..., atomar || atomique, élémentaire || atoom.., atomair || 1. атомный 2. атомарный

A474 **atomic beam** || Atomstrahl *m* || faisceau *m* atomique || atoombundel *m* || атомный пучок *m*

A475 **atomic binding** || Atombindung *f*, kovalente Bindung *f* || liaison *f* atomique || atoombinding *f* || ковалентная [атомная] связь *f*

A476 **atomic clock** || Atomuhr *f* || chronomètre *m* atomique || atoomklok *f (m)* || атомные часы *pl*; атомный эталон *m* частоты

A477 **atomic crystal** || Atomkristall *m*, Valenzkristall *m* || cristal *m* covalent || atoom-kristal *n* || ковалентный кристалл *m*

A478 **atomic energy** || Atomenergie *f*, Kernenergie *f* || énergie *f* atomique || atoomenergie *f* || атомная [ядерная] энергия *f*

A479 **atomic ion** || Atomion *n* || ion *m* atomique || atoomion *n* || атомарный [иониз(ир)ованный] атом *m*

A480 **atomic mass** || Atommasse *f* || masse *f* atomique || atoommassa *f (m)* || 1. масса *f* атома 2. массовое число *n (ядра)*

A481 **atomic mass unit** || atomare Masseneinheit *f*, Atommasseneinheit *f* || unité *f* de masse atomique || ame *m*, eenheid *f* van atoommassa || атомная единица *f* массы

A482 **atomic nucleus** || Atomkern *m* || noyau *m* atomique || atoomkern *f* || атомное ядро *n*

A483 **atomic number** || Atomnummer *f*, Kernladungszahl *f* || numéro *m* [nombre *m*] atomique || atoomnummer *n* || атомный [порядковый] номер *m*

A484 **atomic orbit** || Atomorbital *n* || orbite *f* atomique || atoombaan *f (m)* || атомная *(электронная)* орбита *f*

A485 **atomic physics** || Atomphysik *f* || physique *f* atomique || atoomfysica *f* || атомная физика *f*, физика *f* атома

A486 **atomic spectroscopy** || Atomspektroskopie *f* || spectrografie *f* atomique || atoomspectrografie *f* || атомная спектроскопия *f*

A487 **atomic spectrum** || Atomspektrum *n* || spectre *m* atomique || atoomspectrum *n* || атомный спектр *m*

A488 **atomic structure** || Atombau *m*, Atomstruktur *f* || structure *f* atomique || atoomstructuur *f* || 1. атомистическое строение *n (вещества)* 2. структура *f* атома

A489 **atomic weight** || relative Atommasse *f* || masse *f* atomique relative || relatieve atoommassa *f (m)* || (относительная) атомная масса *f*

A490 **atomism** || Atomistik *f* || atomistique *f* || atomistiek *f* || атомистика *f*, атомистическая теория *f*

A491 **atomizer** || Zerstäuber *m* || atomiseur *m*; buse *f* || verstuiver *m* || форсунка *f*, распылитель *m*

A492 **attachment** || 1. Zusatzeinrichtung *f*; Anbaugerät *n* 2. Befestigung *f*; Anbringung *f* || attachement *m* || 1. inrichting *f* 2. bevestiging *f* || 1. приспособление *n*; приставка *f* 2. крепление *n*; присоединение *n*

A493 **attendance** || Wartung *f* || entretien *m*, maintenance *f* || onderhoud *m* || (техническое) обслуживание *n*, уход *m*

A494 **attenuation** || Abschwächung *f*, Schwächung *f*, Dämpfung *f* || atténuation *f* || verzwakking *f* || демпфирование *n*, ослабление *n*, затухание *n*

A495 **attenuation factor** || Schwächungsfaktor *m* || coefficient *m* d'atténuation; facteur *m* d'affaiblissement || verzwakkingscoëfficiënt *m* || коэффициент *m* ослабления

A496 **attenuator** || Abschwächer *m*, Dämpfungsglied *n* || atténuateur *m* || verzwakker *m* || ослабитель *m*, аттенюатор *m*

A497 **atto...** || Atto... || atto... || atto... || атто... *(приставка в наименовании дольной единицы, соответствующая* 10^{-18}*)*

A498 **attraction** || Anziehung *f*, Attraktion *f* || attraction *f* || aantrekking *f* || притяжение *n*

A499 **attractive force** || Anziehungskraft *f* || force *f* d'attraction || aantrekkingskracht *f (m)* || сила *f* притяжения

A500 **attractive interaction** || gegenseitige Anziehung *f* || attraction *f* mutuelle || wederzijdse aantrekking *f* || (взаимное) притяжение *n*

A501 **attractive potential** || Anziehungspotential *n* || potentiel *m* d'attraction || aantrekkingspotentiaal *m* || потенциал *m* притяжения, притягивающий потенциал *m*

A502 **audibility threshold** || Hörschwelle *f*, Hörbarkeitsschwelle *f* || seuil *m* d'audibilité || hoorbaarheidsdrempel *m* || порог *m* слышимости, порог *m* слухового ощущения

A503 **audibility zone** || Hörbarkeitsgebiet *n*, Hörbarkeitszone *f* || zone *f* d'audibilité || hoorbaarheidszone *f* || зона *f* слышимости

A504 **audibleness** || Hörbarkeit *f* || audibilité *f* || hoorbaarheid *f* || слышимость *f*

A505 **audible range** || Hörbereich *m* || zone *f* d'audibilité || hoorbereik *n* || (частотные) пределы *m pl* слышимости

A506 **audio frequency** || Tonfrequenz *f*, Hörfrequenz *f* || audiofréquence *f*, fréquence *f* sonore || hoorfrequentie *f* || звуковая [низкая] частота *f*

A507 **audio generator** || Ton(frequenz)generator *m* || générateur *m* à audiofréquence, générateur *m* de fréquence sonore || hoorfrequentiegenerator *m* || генератор *m* звуковой частоты

A508 **audio tape** || Magnettonband *n* || bande *f* magnétique || magneetband *m* || магнитная лента *f* звукозаписи

A509 **augend** || Augend *m*, erster Summand *m* || premier opérande *m* d'une addition || eerste op te tellen getal *n* || первое слагаемое *n*

A510 **aural perception** || Hörempfindung *f* || perception *f* acoustique || hoorwaarneming *f* || слуховое восприятие *n*

A511 **aurora** || Polarlicht *n*, polare Aurora *f* || aurore *f* polaire || poollicht *n* || полярное сияние *n*

A512 **autocode** || Autokode *m* || autocode *m* || autocode *m* || автокод *m*

A513 **autocollimator** || Autokollimator *m* || autocollimateur *m* || autocollimator *m* || автоколлиматор *m*

A514 **autoconnected transformer** *see* **autotransformer**

A515 **autocorrelation** || Autokorrelation *f* || autocorrélation *f* || autocorrelatie *f* || автокорреляция *f*

A516 **autocorrelation function** || Autokorrelationsfunktion *f* || fonction *f* d'autocorrélation || autocorrelatiefunctie *f* || автокорреляционная функция *f*

A517 **autocovariance** || Autokovarianz *f* || autocovariance *f* || autocovariantie *f* || автоковариация *f*

A518 **autodyne** || Autodyne *f* || amplidyne *f* || autodyne *f* || автодин *m*, электромашинный усилитель *m*, амплидин *m*

A519 **autoionization** || Autoionisation *f*, Autoionisierung *f* || autoionisation *f* || auto-ionisatie *f* || автоионизация *f*

A520 **automata theory** || Automatentheorie *f* || théorie *f* des automates || automatentheorie *f* || теория *f* автоматов

A521 **automatic control** || automatische Steuerung *f*; automatische Regelung *f* || commande *f* automatique; réglage *m* automatique || automatische sturing *f*; automatische regeling *f* || автоматическое управление *n*; автоматическое регулирование *n*

A522 **automatic copying lathe** || Nachformdrehautomat *m* || tour *m* à copier automatique || kopiëerdraaiautomaat *m* || копировально-токарный автомат *m*

A523 **automatic cycle milling machine** || Fräsautomat *m* || fraiseuse *f* automatique || freesautomaat *m* || фрезерный автомат *m*

A524 **automatic data processing** || automatische Datenverarbeitung *f* || traitement *m* automatique de données || automatische gegevensverwerking *f* || автоматическая обработка *f* данных

A525 **automatic gearbox** || Getriebeautomat *m* || boîte *f* de vitesses automatique || automatische tandwielkast *f* (*m*) || автоматическая коробка *f* передач

A526 **automatic lathe** || Drehautomat *m* || tour *m* automatique || draaiautomaat *m* || токарный автомат *m*

A527 **automatic machine** || Automat *m* || machine-outil *f* automatique || automaat *m* || станок-автомат *m*

A528 **automatic milling machine** *see* **automatic cycle milling machine**

A529 **automatic oiling** || selbsttätige Schmierung *f* || autograissage *m*, graissage *m* automatique || automatische smering *f* || автоматическая смазка *f*

A530 **automatics** || Automatisierungstechnik *f* || automatique *f* || automatie *f* || автоматика *f*

A531 **automatic screw machine** || Schraubendrehautomat *m* || tour *m* à fileter automatique || schroefdraadautomaat *m* || токарно-винторезный автомат *m*

A532 **automatic steel** || Automatenstahl *m* || acier *m* de décolletage || automatenstaal *n* || автоматная сталь *f*

A533 **automatic tool changer** || automatische Werkzeugwechseleinrichtung *f* || dispositif *m* automatique de changement d'outil || automatische werktuigverwisselingsinrichting *f* || автомат-избиратель *m* инструмента *(по заданной программе)*

A534 **automatic transfer machine** || Transfermaschine *f* || machine-transfert *f* automatique, machine *f* à transfert automatique || automatische transfermachine *f* || многопозиционный станок-автомат *m*

A535 **automatic turret lathe** || Revolverdrehautomat *m* || tour *m* (à) revolver automatique || revolver(draai)automaat *m* || токарно-револьверный автомат *m*

A536 **automatic valve** || selbsttätiges Ventil *n* || soupape *f* automatique || zelfinstellende klep *f* (*m*) || автоматический клапан *m*

A537 **automatic welding** || Automatenschweißung *f*, automatische Schweißung *f* || soudage *m* automatique || automatisch lassen *n* || автоматическая сварка *f*

A538 **automation, automatization** || Automatisierung *f* || automatisation *f* || automatisering *f* || автоматизация *f*

A539 **automaton** || Automat *m* || automate *m* || automaat *m* || автомат *m*

A540 **automorphism** || Automorphismus *m* || automorphisme *m* || automorfisme *n* || автоморфизм *m*

A541 **automorphism group** || Automorphiegruppe *f* || groupe *m* des automorphismes || automorfismegroep *f* (*m*) || группа *f* автоморфизмов

A542 **autonomous** || autonom, selbständig || autonome || autonoom || автономный

A543 **autonomous system** || autonomes System *n* || système *m* autonome || autonoom systeem *n* || автономная система *f*

A544 **autoradiography** || Autoradiographie *f* || autoradiographie *f* || autoradiografie *f* || авторадиография *f*, радиоавтография *f*

A545 **autotransformer** || Autotransformator *m*, Spartransformator *m* || autotransformateur *m* || autotransformator *m* || автотрансформатор *m*

A546 **autoxidation** || Autoxydation *f* || auto-oxydation *f* || auto-oxydatie *f* || самоокисление *n*

A547 **auxiliary optical axis** || optische Nebenachse *f* || axe *m* optique auxiliaire || optische bijas *f* (*m*) || побочная оптическая ось *f*

A548 **auxiliary scale** || Hilfsskale *f*; Nonius *m* || échelle *f* auxiliaire; nonius *m* || hulpschaal *f* (*m*) || вспомогательная шкала *f*; нониус *m*

A549 **available energy** || **1.** freie Energie *f* **2.** verfügbare Energie *f* || **1.** énergie *f* libre **2.** énergie *f* abordable || **1.** vrije energie *f* **2.** toegankelijke energie *f* || **1.** свободная энергия *f* **2.** доступная энергия *f*

A550 **avalanche** || Lawine *f* || avalanche *f* || lawine *f* || лавина *f*

A551 **avalanche-type** || Lawinen... || à avalanche || lawine... || лавинный *(о разряде, приборе)*

A552 **average** || Mittel *n* || moyenne *f* || gemiddelde *n*, gemiddelde waarde *f* || среднее *n*

A553 **average load** || durchschnittliche Belastung *f* || charge *f* moyenne || gemiddelde belasting *f* || средняя нагрузка *f*

A554 **average speed** || mittlere Geschwindigkeit *f* || vitesse *f* moyenne || gemiddelde snelheid *f* || средняя скорость *f*

A555 **avometer** || Avometer *n*, Ampere-Volt-Ohm-Meter *n* || ampère-volt-ohm-mètre *m* || avometer *m* || авометр *m*, ампервольтметр *m*, вольтамперметр *m*

A556 **axial** || axial, Axial..., Achsen... || axial || assen..., axiaal || осевой, аксиальный

A557 **axial angle** || axialer Winkel *m* || angle *m* axial || optische assenhoek *m* || угол *m* с оптической осью

A558 **axial clearance** || axialer Spalt *m* || jeu *m* axial || axiale spleet *f* (*m*) || осевой зазор *m*

A559 **axial compression** || zentrischer [axialer] Druck *m*, axiale Kompression *f* || compression *f* axiale || assencompressie *f* || осевое сжатие *n*; одноосное сжатие *n*

A560 **axial compressor** || Axialgebläse *n*, Axialverdichter *m* || compresseur *m* axial || axiale compressor *m* || осевой компрессор *m*

A561 **axial-flow turbine** || Axialturbine *f* || turbine *f* axiale || axiale turbine *f* || осевая турбина *f*

A562 **axial force** || Axialkraft *f* || force *f* axiale || axiale kracht *f* || (*m*) осевая сила *f*; продольная сила *f*

A563 **axial load** || Axiallast *f* || charge *f* axiale || axiale belasting *f* || осевая нагрузка *f*; продольная нагрузка *f*

A564 **axially symmetric(al)** || axialsymmetrisch, achsensymmetrisch || axialement symétrique || assensymmetrisch || аксиально-симметричный, осесимметричный

A565 **axial moment of inertia** || axiales Trägheitsmoment *n* || moment *m* d'inertie axial || assentraagheidsmoment *n*, axiaal traagheidsmoment *n* || осевой момент *m* инерции

A566 **axial pitch** || Steigung *f* (*Gewinde*) || pas *m* axial (*de filetage*) || axiale spoed *m* (*v. e. schroef*) || осевой шаг *m* (*резьбы*)

A567 **axial plane** || Axialschnitt *m* || plan *m* (de section) axial || axiale doorsnede *f* (*m*) || плоскость *f* осевого сечения, осевая плоскость *f*

A568 **axial symmetry** || Axialsymmetrie *f*, Achsensymmetrie *f* || symétrie *f* axiale || axiale symmetrie *f* || осевая [аксиальная, вращательная] симметрия *f*

A569 **axial thrust** || Axialdruck *m* || pression *f* axiale || axiale druk *m* || осевое давление *n*

A570 **axial vector** || axialer Vektor *m* || vecteur *m* axial || axiale vector *m* || аксиальный вектор *m*

A571 **axiom** || Axiom *m* || axiome *m* || axioma *n* || аксиома *f*

A572 **axiomatic method** || axiomatische Methode *f* || méthode *f* axiomatique || axiomatische methode *f* || аксиоматический метод *m*

A573 **axiomatics** || Axiomatik *f* || axiomatique *f* || axiomatiek *f* || аксиоматика *f*

A574 **axiomatization** || Axiomatisierung *f* || axiomatisation *f* || axiomatisatie *f* || аксиоматизация *f*

A575 **axiom of Archimedes** || Archimedisches Axiom *n* || axiome *m* d'Archimède || Archimedes-axioma *n* || аксиома *f* Архимеда

A576 **axiom of choice** || Auswahlaxiom *n* || axiome *m* de choix || keuraxioma *n* || аксиома *f* выбора

A577 **axiom of complete induction** || Axiom *n* der mathematischen Induktion || axiome *m* d'induction complète [d'induction mathématique] || axioma *n* van wiskundige inductie || аксиома *f* математической индукции

A578 **axiom of completeness** || Vollständigkeitsaxiom *n* || axiome *m* de complétude || volledigheidsaxioma *n* || аксиома *f* полноты

A579 **axiom of infinity** || Unendlichkeitsaxiom *n* || axiome *m* de l'infini || infiniteitsaxioma *n* || аксиома *f* бесконечности

A580 **axiom of perfect induction** *see* **axiom of complete induction**

A581 **axiom of power** || Mächtigkeitsaxiom *n* || axiome *m* de puissance || vermogensaxioma *n* || аксиома *f* мощности

A582 **axiom of separability** || Separabilitätsaxiom *n* || axiome *m* de séparation || afscheidbaarheidsaxioma *n* || аксиома *f* отделимости

A583 **axis** || Achse *f* || axe *m* || as *f* (*m*) || ось *f*

A584 **axis of gravity** || Schwer(punkt)achse *f* || axe *m* de gravité || zwaartepuntas *f* (*m*) || центральная ось *f*

A585 **axis of inertia** || Trägheitsachse *f* || axe *m* d'inertie || traagheidsas *f* (*m*) || ось *f* инерции

A586 **axis of lens** || Linsenachse *f* || axe *m* optique d'une lentille || optische as *f* (*m*) van een lens || (главная) оптическая ось *f* линзы

A587 **axis of rotation** || Rotationsachse *f*, Drehachse *f* || axe *m* de rotation || draaiingsas *f* (*m*) || ось *f* вращения

A588 **axis of symmetry** || Symmetrieachse *f* || axe *m* de symétrie ||

symmetrieas *f* (*m*), draaiingsas *f* (*m*) || ось *f* симметрии

A589 **axissymmetrical** || axialsymmetrisch, achsensymmetrisch || axialement symétrique || assensymmetrisch || осесимметричный

A590 **axle** || Achse *f* || arbre *m*; essieu *m* || as *f* (*m*) || ось *f*

A591 **axoid** || Axoid *n* || axoïde *f* || axoïde *n* || аксоид *m*

A592 **axonometry** || Axonometrie *f* || axonométrie *f* || axonometrie *f* || аксонометрия *f*

A593 **azeotropy** || Azeotropie *f* || azéotropie *f* || azeotropie *f* || азеотропия *f*

A594 **azide** || Azid *n* || azoture *m*, azide *m* || azide *n* || азид *m*, соль *f* азотистоводородной кислоты

A595 **azimuthal quantum number** || azimutale Quantenzahl *f* || nombre *m* quantique azimutal || azimutaal quantumgetal *n* || азимутальное [орбитальное] квантовое число *n*

A596 **azo-compound** || Azoverbindung *f* || composé *m* azoïque || azoverbinding *f* || азосоединение *n*

A597 **azo-dye** || Azofarbstoff *m* || colorant *m* azoïque || azokleurstof *f* (*m*) || азокраситель *m*

B

B1 **babbitt metal** || Weißmetall *n* Lagerweißmetall *n* || babbit(t) *m*, métal *m* blanc, métal *m* antifriction || babbitt(metaal) *n*, antifrictiemetaal *n* || баббит *m*, белый антифрикционный сплав *m*

B2 **babbit-metal bearing** || Weißmetallager *n* || palier *m* à coussinet antifrictionné || kogellager *n* met babbitkussen, || подшипник *m* с баббитовым вкладышем

B3 **back-and-forth motion** || hin- und hergehende Bewegung *f*, Vor- und Rückwärtsbewegung *f* || mouvement *m* de va-et-vient [alternatif] || heen-en -weergaande beweging *f* || возвратно-поступательное движение *n*

B4 **back coupling** || Rückkopplung *f*: Ruckführung *f* || rétroaction *f*; réaction *f* || terugkoppeling *f* || обратная связь *f*

B5 **background** || Hintergrund *m* || fond *m* || achtergrond *m* || фон *m*

B6 **background radiation** || Hintergrundstrahlung *f*, Strahlungshintergrund *m* || radiation *f* de fond || achtergrondstraling *f* || фоновое излучение *n*, фон *m*

B7 **backing-off lathe** || Hinterdrehmaschine *f* || tour *m* à détalonner || achterdraaibank *f* (*m*) || токарно-затыловочный станок *m*

B8 **backlash** || 1. toter Gang *m*, 2. Spiel *n*, Flankenspiel *n* *(Zahnrad)* 1. course *f* morte 2. jeu *m*; jeu *m* entre les flancs *(de dents)* || 1. dode gang *m*; blinde gang *m* 2. flankspeling *f* || 1. мёртвый ход *m* 2. зазор *m*; боковой зазор *m* *(между зубьями колёс)*

B9 **back pressure** || Gegendruck *m* || contre-pression *f*, sous-pression *f* || tegendruk *m* || противодавление *n*

B10 **back-pressure valve** || Rückschlagventil *n* || soupape *f* de retenue || tegendrukklep *f* (*m*) || обратный клапан *m*

B11 **backscattering** || Rück(wärts)-streuung *f* || diffusion *f* vers l'arrière [en retour], dispersion *f* réfléchie || terugstrooiing *f* || обратное рассеяние *n*

B12 **backward difference** || Rückwärtsdifferenz *f* || différence *f* inverse || omgekeerde differentie *f* || обратная разность *f*

B13 **backward flow** || Rückstrom *m*, Rückströmung *f* || contre-courant *m* || terugstroom *m*, terugstroming *f* || противоток *m*, обратное течение *n*

B14 **backward motion** || rückläufige Bewegung *f*, Rücklauf *m*; Rückwärtsbewegung *f* || mouvement *m* de retour; mouvement *m* inverse || terugloop *m*, terugkeer *m* || обратное движение *n*

B15 **backward scattering** *see* **backscattering**

B16 **balance** || 1. Waage *f* 2. Gleichgewicht *n*; Balance *f* || 1. balance *f* 2. équilibre *m* || 1. waag *f* (*m*) 2. evenwicht *n*; balans *f* (*m*) || 1. весы *pl* 2. равновесие *n*; баланс *m*

B17 **balanced force** || ausgeglichene Kraft *f* || force *f* équilibrée || evenwichtige kracht *f* *(m)* || уравновешенная сила *f*

B18 **balanced load** || symmetrische Belastung *f* || charge *f* équilibrée || symmetrische belasting *f* || уравновешенная нагрузка *f*

B19 **balance-weight lever** || Ausgleichhebel *m* || levier *m* d'équilibrage || tegengewichthefboom *m* || уравнительный рычаг *m*

B20 **balancing** || 1. Ausgleich *m*, Ausgleichen *n*; Kompensierung *f* 2. Balancieren *n* 3. Auswuchten *n*; Auswuchtung *f* || 1. équilibrage *m*, compensation *f* 2. égalisation *f* 3. equilibrage *m* || balancering *f* || 1. уравновешивание *n*, компенсация *f* 2. выравнивание *n*, симметрирование *n* 3. уравновешивание *n*, балансировка *f*

B21 **balancing force** || Ausgleichskraft *f*, ausgleichende Kraft *f* || force *f* équilibrante || evenwichtskracht *f* *(m)*, balanceerkracht *f* *(m)* || уравновешивающая сила *f*

B22 **balancing load** || Ausgleichslast *f*, ausgleichende Belastung *f* || charge *f* équilibrante || evenwichtsbelasting *f* || уравновешивающая нагрузка *f*

B23 **balancing machine** || Auswuchtmaschine *f* || machine *f* à équilibrer, équilibreuse *f* || balanseermachine *f* || балансировочный станок *m*

B24 **balancing torque** || Ausgleichsdrehmoment *n* || couple *m* de torsion équilibrant || gebalanseerd draaimoment *n* || уравновешивающий крутящий момент *m*

B25 **balancing transformer** || Symmetriertransformator *m* || transformateur *m* dissymétrique-symétrique, transformateur *m* de symétrisation || balanceertransformator *m* || симметрирующий трансформатор *m*

B26 **ball** || 1. Kugel *f* 2. Kugelspurzapfen *m* || 1. *math.* sphère *f* 2. bille *f*, boule *f* 3. pivot *m* sphérique [à rotule] || 1. kogel *m* 2. kogelvoet *m* || 1. шар *m*; шарик *m* 2. шаровая пята *f*

B27 **ball-and-socket joint** || Kugelgelenk *n* || articulation *f* [joint *m*] à rotule, rotule *f* || kogelgewricht *n*, kogelscharnier *n* || шаровой шарнир *m*

B28 **ball bearing** || Kugellager *n* || palier *m* [roulement *m*] à billes || kogellager *n* || шариковый подшипник *m*

B29 **ball cutter** || Kugelfräser *m* || fraise *f* sphérique || kogelfrees *f* *(m)* || шаровая фреза *f*

B30 **ballistic flight** || ballistischer Flug *m* || vol *m* balistique || ballistische vlucht *f* *(m)* || полёт *m* по баллистической траектории

B31 **ballistic galvanometer** || ballistisches Galvanometer *n* || galvanomètre *m* balistique || ballistische galvanometer *m* || баллистический гальванометр *m*

B32 **ballistic pendulum** || ballistisches Pendel *n*, Stoßpendel *n* || pendule *f* balistique || ballistisch pendel *n* || баллистический маятник *m*

B33 **ballistics** || Ballistik *f* || balistique *f* || ballistiek *f* || баллистика *f*

B34 **ballistic trajectory** || ballistische Flugbahn *f* || trajectoire *f* balistique || ballistische vliegbaan *f* *(m)* || баллистическая траектория *f*, траектория *f* свободного полёта

B35 **ball joint** *see* **ball-and-socket joint**

B36 **ball vavle** || Kugelventil *n* || soupape *f* à bille || kogelklep *f* *(m)* || шаровой клапан *m*

B37 **Banach space** || Banachraum *m* || espace *m* de Banach || Banachruimte *f* || банахово пространство *n*

B38 **band** || Band *n* || bande *f* || band *m* || 1. полоса *f*, диапазон *m* (частот) 2. зона *f* (уровней энергии)

B39 **band brake** || Bandbremse *f* || frein *m* à bande [à ruban] || bandrem *f* *(m)* || ленточный тормоз *m*

B40 **banded spectrum** *see* **band spectrum**

B41 **band-pass filter** || Bandfilter *n* || filtre *m* à bande || bandfilter *m*, *n* || полосовой [полосный] фильтр *m*

B42 **band saw** || Bandsäge *f* || scie *f* à ruban || bandzaag *f* *(m)* || ленточная пила *f*

B43 **band-sawing machine** || Bandsägemaschine *f* || machine *f* à scier ||

bandzaagmachine *f* || ленточнопильный станок *m*

B44 **band spectrum** || Bandspektrum *n* || spectre *m* cannelé [de bandes] || gestreept spectrum *n* || полосатый спектр *m*

B45 **band theory** || Bändertheorie *f*, Bandtheorie *f* || théorie *f* des bandes || bandentheorie *f* || зонная теория *f* *(твёрдого тела)*

B46 **bandwidth** || Frequenzband *n* || bande *f* de fréquences || frequentieband *m* || полоса *f* частот, ширина *f* полосы частот, полоса *f* пропускания

B47 **bar** || **1.** Bar *n* **2.** Stange *f*; Stab *m* **3.** Barren *m*; Block *m* || **1.** bar *m* **2.**, **3.** barre *f* || **1.** bar *m* **2.** staaf *f* (*m*) **3.** baar *f* (*m*), blok *n* || **1.** бар *(10^5 Па)* **2.** пруток *m*; стержень *m* **3.** болванка *f*, заготовка *f*

B48 **bare conductor** || blanker Draht *m*, blanke Leitung *f* || conducteur *m* nu || naakte leiding *f* || голый [неизолированный] провод *m*

B49 **bariron** || Stabstahl *m* || barres *f pl* (d'acier), ronds *m pl* (d'acier) || staafstaal *n* || прутковая сталь *f*

B50 **barium,** Ba || Barium *n* || baryum *m* || barium *n* || барий *m*

B51 **bar magnet** || Stabmagnet *m* || barreau *m* aimanté, aimant *m* droit || staafmagneet *m* || стержневой магнит *m*

B52 **barn** || Barn *n* || barn *m* || barn *m* || барн *(10^{-28} м²)*

B53 **barograph** || Barograf *m*, Luftdruckschreiber *m* || barographe *m*, baromètre *m* enregistreur || barograaf *m* || барограф *m*, самопишущий барометр *m*

B54 **barometer** || Barometer *n*, Luftdruckmesser *m* || baromètre *m* || barometer *m* || барометр *m*

B55 **barometric pressure** || barometrischer Druck *m*, Luftdruck *m* || pression *f* barométrique || luchtdruk *m* || барометрическое давление *n*

B56 **barotropic** || barotrop(isch) || barotropique || barotroop || баротропный, с баротропным распределением массы

B57 **barotropy** || Barotropie *f* || barotropie *f* || barotropie *f* || баротропность *f*

B58 **barrel** || Trommel *f*; Zylinder *m* || tambour *m*, trommel *m*; barillet *m*, cylindre *m* || trommel *f* (*m*); cilinder *m* || барабан *m*; цилиндр *m*

B59 **barrel-shaped distortion** || tonnenförmige Verzeichnung *f*, Tonnenverzeichnung *f* || distorsion *f* en barillet || tonvormige distorsie *f* || бочкообразная дисторсия *f*

B60 **barrier layer** || Sperrschicht *f* || couche *f* d'arrêt || barrièrelaag *f* (*m*) sperlaag *f* (*m*) || запирающий слой *m*

B61 **barycenter** || Baryzentrum *n* || barycentre *m* || barycentrum *n* || барицентр *m*; центр *m* массы

B62 **baryon** || Baryon *n* || baryon *m* || baryon *n* || барион *m*

B63 **base** || **1.** Grundlage *f* **2.** Basis *f* **3.** *Math.* Basis *f*, Grundzahl *f* **4.** *Math.* Grundfläche *f* **5.** *Chem.* Base *f* || base *f* || **1.** grond *m* **2.**, **3.**, **4.** basis *f* **5.** *chem.* base *f* || **1.** основание *n* **2.** база *f*; базис *m* **3.**, **4.** *мат.* основание *n* *(степени, геометрической фигуры)* **5.** *хим.* основание *n*

B64 **base address** || Basisadresse *f*, Grundadresse *f* || adresse *f* de base || grondadres *n* || базовый адрес *m*

B65 **base circle** || Grundkreis *m* *(Zahnrad)* || cercle *m* de base *(engrenage)* || grondcirkel *m* *(tandwiel)* || основная окружность *f* *(зубчатого колеса)*

B66 **base of logarithm** || Basis *f* eines Logarithmus || base *f* du logarithme || basis *f* van een logarithme || основание *n* логарифма

B67 **base of power** || Basis *f* einer Potenz || base *f* de la puissance || basis *f* van de macht || основание *n* степени

B68 **base pitch** || Eingriffsteilung *f*, Grundkreisteilung *f* || pas *m* du cercle de base || grondsteek *m* (*tandwiel*) || шаг *m* зацепления по основной окружности (*зубчатого колеса*)

B69 **base plane** || **1.** Bezugsebene *f* *(z. B. Werkzeugbezugsebene)* 2. Stützfläche *f* || **1.** plan *m* de référence *(p. ex. d'un outil)* **2.** surface *f* d'appui || **1.** grondvlakte *f* *(bijv. van de snede)* **2.** steunoppervlak *n* || **1.** основная плоскость *f* *(напр, резца)* **2.** опорная поверхность *f*

B70 **base plate** || Grundplatte *f*, Fundamentplatte *f* || plaque *f* d'appui [de base], plaque-support *f* || steunplaat *f* || опорная [фундаментная] плита *f*

B71 **base vector** || Basisvektor *m* || vecteur *m* de base || basisvector *m* || базисный вектор *m*

B72 **basic circuit** || Stromlaufplan *m*, Schaltplan *m* || schéma *m* de principe (électrique) || schakelschema *n* || принципиальная (электрическая) схема *f*

B73 **basic size** || Nennmaß *n* *(im Toleranzsystem)* || cote *f* nominale || nominale maat *f* (*m*) *(in een tolerantiesysteem)* || номинальный размер *m* *(в системе допусков)*

B74 **basic tolerance** || Grundtoleranz *f* || tolérance *f* principale || basistolerantie *f* || основной допуск *m*

B75 **basic variable** || Basisvariable *f* || variable *f* de base || basisvariabele *f* || базисная переменная *f*

B76 **basis** || Basis *f* || base *f* || basis *f* || базис *m*

B77 **basis of a vector space** || Basis *f* eines Vektorraums || base *f* de l'espace vectoriel || basis *f* van een vectorruimte || базис *m* векторного пространства

B78 **basis vector** *see* **base vector**

B79 **batch processing** || Batch-Verarbeitung *f* || traitement *m* par lots || batch-verwerking *f* || пакетная обработка *f* *(данных)*

B80 **battery** || Batterie *f* || batterie *f* || batterij *f* || (аккумуляторная) батарея *f*

B81 **baud** || Baud *n* || baud *m* || baud *m* || бод *m*

B82 **Baudot code** || Baudot-Kode *m* || code *m* Baudot || Baudot-code *m* || код *m* Бодо

B83 **Bayes estimation** || Bayessche Schätzung *f* || estimation *f* de Bayes || Bayes-schatting *f* || оценка *f* по Бейесу

B84 **Bayes test** || Bayesscher Test *m* || test *m* de Bayes || Bayes-test *m* || критерий *m* Бейеса

B85 **beam** || **1.** Strahlenbündel *n*, Bündel *n*, Strahl *m* **2.** Waagebalken *m* **3.** Träger *m*, Balken *m* || **1.** rayon *m*; faisceau *m* **2.** fléau *m* **3.** poutre *f* || **1.** straal *m*, *f*, bundel *m* **2.** evenaar *m* *(van balans)* **3.** balk *m* || **1.** луч *m*; пучок *m* лучей; пучок *m* частиц **2.** коромысло *n* *(весов)* **3.** балка *f*; брус *m*

B86 **beam current** || Strahlstrom *m* || courant *m* du faisceau || bundelstroom *m* || ток *m* (первичного) пучка *(электронов)*

B87 **beam splitting** || Strahlspaltung *f* || division *f* du faisceau || bundelverdeling *f* || разделение *n* [расщепление *n*] пучка

B88 **bearing** || **1.** Lager *n* **2.** Lagerung *f* || **1.** palier *m*; roulement *m* **2.** appui *m*, support *m* || **1.** lager *n* **2.** steun *m*, lager *n* || **1.** подшипник *m* **2.** опора *f*

B89 **bearing area** *see* **bearing surface**

B90 **bearing bracket** || Lagerbock *m* || console *f* du palier || lagersteun *n* || кронштейн *m* подшипника

B91 **bearing clearance** || Lagerspiel *n* || jeu *m* du palier || lagerspeling *f* || зазор *m* подшипника

B92 **bearing housing** || Lagergehäuse *n* || corps *m* de palier || lagerhuis *n* || корпус *m* подшипника

B93 **bearing liner** || Lagerschale *f* || coussinet *m* [garniture *f*] de palier || lagerschaal *f* (*m*) || вкладыш *m* подшипника

B94 **bearing load** || Lagerbelastung *f* || charge *f* sur appui [à l'appui] || lagerbelasting *f* || нагрузка *f* на опору

B95 **bearing seat** || Lagersitz *m* || logement *m* du palier || kogellagerzitting *f* || гнездо *n* подшипника

B96 **bearing shell** *see* **bearing housing**

B97 **bearing surface** || Auflagefläche *f* || surface *f* d'appui || draagvlak *n* || опорная поверхность *f*

B98 **beats** || Schwebungen *pl* || battements *pl* || zweving *f* || биения *pl*

B99 **bed** || **1.** Bett *n*, Maschinenbett *n*; Grundplatte *f*; Untergestell *n* **2.** Tragkörper *m*; Fundament *n* || **1.** bâti *m* **2.** lit *m*; couche *f*, assiette *f* || **1.** bed *n*; machinegestel *n* **2.** grond *m*; fundamentraam *n* || **1.** станина *f*; рама *f* **2.** основание *n*; фундамент *m*

B100 **bed plate** || Grundplatte *f* || plaque *f* d'assise || steunplaat *f* (*m*) || опорная плита *f*; станина *f*
B101 **Beilstein's test** || Beilstein-Probe *f*, Beilsteinsche Probe *f* || test *m* de Beilstein || Beilsteintest *m* || проба *f* Бейльштейна
B102 **bel** || Bel *n* || bel *m* || bel *f* (*m*) || бел *m*
B103 **belt** || Riemen *m*; Band *n*, Gurt *m* || courroie *f*; bande *f* || riem *m*; band *m* || ремень *m*; лента *f*
B104 **belt fork** || Riemenschalter *m*, Riemenausrücker *m* || fourche *f* à courroie || riemvork *f* (*m*), riemhefboom *m* || ремённая отводка *f*
B105 **belt pulley** || Riemenscheibe *f*; Gurtscheibe *f* || poulie *f* à courroie || riemschijf *f* (*m*) || ремённый шкив *m*
B106 **belt slip** || Riemenschlupf *m* || glissement *m* de courroie || slip *m* van de drijfriem || проскальзывание *n* ремня
B107 **bench** || Werktisch *m*; Werkbank *f* || banc *m*, établi *m* || werkbank *f* (*m*); bank *f* (*m*) || верстак *m*; станок *m*
B108 **bench vice** || Bankschraubstock *m* || étau *m* d'établi || bankschroef *f* (*m*) || верстачные тиски *pl*; слесарные тиски *pl*
B109 **bending** || Biegung *f* || flexion *f* || buiging *f* || изгиб *m*, изгибание *n*
B110 **bending invariant** || Biegungsinvariante *f* || invariant *m* de flexion || buigingsinvariante *f* || инвариант *m* изгибания
B111 **bending machine** || Biegemaschine *f* || cintreuse *f*, machine à cintrer || buigmachine *f* || гибочная машина *f*; гибочный станок *m*
B112 **bending moment** || Biegemoment *n* || moment *m* de flexion [fléchissant] || buigmoment *n* || изгибающий момент *m*
B113 **bending press** || Biegepresse *f* || presse *f* à cintrer [cintreuse] || buigpers *f* (*m*) || гибочный пресс *m*
B114 **bending strain** || Biegeverformung *f*, Biegeformänderung *f* || déformation *f* de flexion || buigingsdeformatie *f* || деформация *f* изгиба
B115 **bending strength** || Biegefestigkeit *f* || résistance *f* à la flexion || buigvastheid *f* || прочность *f* на изгиб
B116 **bending stress** || Biegebeanspruchung *f* || contrainte *f* de flexion || buigspanning *f* || напряжение *n* изгиба
B117 **bending test** || Biegeprüfung *f*, Biegeversuch *m* || essai *m* de flexion [à la flexion] || buigproef *f* (*m*) || испытание *n* на изгиб
B118 **bent tool** || gebogener Meißel *m* || outil *m* coudé || gebogen beitel *m* || изогнутый резец *m*
B119 **benzene,** C_6H_6 || Benzol *n* || benzène *m*, benzol *m* || benzeen *n* || бензол *m*
B120 **Bernoulli numbers** || Bernoullische Zahlen *f pl* || nombres *m pl* bernoulliens || Bernoulli-getallen *n pl* || числа *n pl* Бернулли
B121 **beryllium,** Be || Beryllium *n* || béryllium *m*, glucinium *m* || beryllium *n* || бериллий *m*
B122 **Bessel functions** || Besselsche Funktionen *f pl* || fonctions *f pl* de Bessel || Besselfuncties *f pl* || функции *f pl* Бесселя
B123 **Bessemer process** || Bessemer-Verfahren *n* || procédé *m* Bessemer || bessemerproces *n* || бессемеровский процесс *m*
B124 **beta-active** || beta-aktiv || bêta-actif || beta-actief || бета-(радио)-активный
B125 **beta decay** || Beta-Zerfall *m* || désintégration *f* bêta || beta-desintegratie *f*, beta-verval *n* || бета-распад *m*
B126 **beta distribution** || Beta-Verteilung *f* || distribution *f* bêta || beta-verdeling *f* || бета-распределение *n*
B127 **beta emitter** || Beta-Strahler *m* || émetteur *m* bêta || beta-emissor *m* || бета-излучатель *m*, бета-источник *m*
B128 **beta function** || Beta-Funktion *f* || fonction *f* bêta || beta-functie *f* || бета-функция *f*
B129 **betatron** || Betatron *n*, Elektronenschleuder *f* || bêtatron *m* || betatron *n* || бетатрон *m*
B130 **bevel angle** || Abschrägungswinkel *m* || angle *m* du chanfrein ||

afschuiningshoek *m* || угол *m* скоса; угол *m* скашивания

B131 **bevel gear** || Kegel(zahn)rad *n* || roue *f* conique [d'angle], pignon *m* conique || kegelwiel *n* || коническое зубчатое колесо *n*, коническая шестерня *f*

B132 **bias** || 1. Vorspannung *f* 2. Verschiebung *f* || 1. polarisation *f* 2. biais *m* || 1. polarisatie *f* 2. verschuiving *f* || 1. смещение *n*, смещающее напряжение *n* 2. смещение *n*

B133 **bias current** || Verschiebungsstrom *m* || courant *m* de polarisation || polarisatiestroom *m* || ток *m* смещения

B134 **bias voltage** || Vorspannung *f* || tension *f* de polarisation || polarisatiespanning *f* || напряжение *n* смещения

B135 **biatomic** || zweiatomig || biatomique || tweeatomig || двухатомный

B136 **biaxial** || zweiachsig || biaxial || tweeassig || дву(х)осный

B137 **bicompact, bicompactum** || Bikompaktum *n* || bicompact *m* || bicompact *n* || бикомпакт *m*

B138 **biconcave** || bikonkav || biconcave || biconcaaf || двояковогнутый

B139 **biequivalence** || Biäquivalenz *f* || biéquivalence *f* || bi-equivalentie *f* || биэквивалентность *f*

B140 I **bifilar** || 1. bifilare Leitung *f* 2. Doppelfaden *m* || bifilaire *m* || bifilaire leiding *f* || бифиляр *m*

B141 II **bifilar** || bifilar || bifilaire || bifilair || бифилярный

B142 **bifocal** || bifokal, Doppelfokus... || bifocal || bifocaal || двухфокусный, бифокальный

B143 **bifurcation** || Bifurkation *f*, Gabelung *f*, Verzweigung *f* || bifurcation *f* || bifurcatie *f* || бифуркация *f*, разветвление *n*

B144 **bijective function** || bijektive Funktion *f* || fonction *f* bijective || bi-jective functie *f* || биективная функция *f*

B145 **bilateral tolerance** || Plus- und Minusabweichung *f* || tolérance *f* bilatérale || bilaterale tolerantie *f* || симметричный наружный допуск *m*

B146 **bilens** || Bilinse *f*, Doppellinse *f* || bilentille *f* || bilens *f* (*m*) || билинза *f*

B147 **bilinear form** || bilineare Form *f* || forme *f* bilinéaire || bilineaire vorm *f* || билинейная форма *f*

B148 **bilinear transformation** || bilineare Transformation *f* || transformation *f* bilinéaire || bilineaire transformatie *f* || билинейное преобразование *n*

B149 **bimetal** || Bimetall *n* || bimétal *m* || bimetaal *n* || биметалл *m*

B150 **bimolecular reaction** || bimolekulare Reaktion *f* || réaction *f* bimoléculaire || bimoleculaire reactie *f* || бимолекулярная реакция *f*

B151 **binary** || binär, Binär... || binaire || binair || бинарный; двоичный

B152 **binary alphabet** || binäres Alphabet *n* || alphabet *m* binaire || binair alfabet *n* || двоичный алфавит *m*

B153 **binary function** || binäre Funktion *f* || fonction *f* binaire || binaire functie *f* || бинарная [двоичная] функция *f*

B154 **binary number system** || Dualsystem *n* || système *m* de numération binaire || tweetallig stelsel *n* || двоичная система *f* счисления

B155 **binary operator** || Binäroperator *m* || opérateur *m* binaire || binaire operator *m* || бинарный оператор *m*

B156 **binder** || 1. Bindemittel *n* 2. Verbindungsteil *n*, Kopplungselement *n* || 1. liant *m* 2. pièce *f* de raccordement || 1. bindmiddel *n* 2. verbindingsstuk *n* || 1. вяжущее (вещество) *n* 2. соединительное устройство *n*; связывающая деталь *f*

B157 **binding** || Bindung *f* || liaison *f* || binding *f* || связь *f* (*между частицами*)

B158 **binding energy** || Bindungsenergie *f* || énergie *f* de liaison || bindingsenergie *f* || энергия *f* связи

B159 **binding screw** *see* **clamping screw**

B160 **binocular microscope** || Doppelmikroskop *n*, Stereomikroskop *n*, binokulares Mikroskop *n* || microscope *m* binoculaire || binoculaire microscoop *m* || бинокулярный микроскоп *m*

B161 I **binomial** || Binom *n* || binôme *m* || binomium *n* || бином *m*
B162 II **binomial** || binomial || binomial, binôme || binominaal || биномиальный
B163 **binomial coefficient** || Binomialkoeffizient *m* || coefficient *m* binomial || binominale coëfficiënt *m* || биномиальный коэффициент *m*
B164 **binomial differential** || Differentialbinom *n* || binôme *m* différentiel || differentiaalbinomium *n* || дифференциальный бином *m*
B165 **binomial distribution** || Binomialverteilung *f* || distribution *f* binomiale || binominale verdeling *f* || биномиальное распределение *n*
B166 **binomial series** || Binomialreihe *f* || série *f* binomiale || binominale reeks *f* (*m*) || биномиальный ряд *m*
B167 **binomial theorem** || binomischer Lehrsatz *m* || binôme *m* de Newton || Newton-binomium *n* || бином *m* Ньютона
B168 **binormal** || Binormale *f* || binormale *f* || binormaal *f* (*m*) || бинормаль *f*
B169 **biocybernetics** || Biokybernetik *f* || biocybernétique *f* || biocybernetica *f* || биокибернетика *f*
B170 **bionics** || Bionik *f* || bionique *f* || bionica *f* || бионика *f*
B171 **biophysics** || Biophysik *f* || biophysique *f* || biofysica *f* || биофизика *f*
B172 **biphase** || zweiphasig, Zweiphasen... || biphasé || tweefasig, tweefasen... || двухфазный
B173 **bipolar coordinates** || Bipolarkoordinaten *f pl* || coordonnées *f pl* bipolaires || bipolaire coördinaten *m pl* || биполярные координаты *f pl*
B174 **biprism** || Biprisma *n*, Doppelprisma *n* || biprisme *m* || biprisma *n* || бипризма *f*
B175 **biquadratic** || biquadratisch || biquadratique || bikwadratisch || биквадратный
B176 **birefringence** || Doppelbrechung *f* || biréfringence *f* || dubbelbreking *f* || двойное лучепреломление *n*, дву(луче)преломление *n*
B177 **birefringent** || doppelbrechend || de biréfringence || dubbelbrekend || двоякопреломляющий
B178 **bisector, bisectrix** || Bisektrix *f*, Halbierende *f* || bissectrice *f* || bissectrice *f* || биссектриса *f*
B179 **bistable** || bistabil || bistable || bistabiel || бистабильный, с двумя устойчивыми состояниями
B180 **bit** || **1.** Bit *n* **2.** Schneide *f*; Schneidkante *f* **3.** Bohrer *m* || **1.** bit *m* **2.** lame *f* **3.** foret *m*, mèche *f* || **1.** bit *n* **2.** snede *f* (*m*), snijkant *m* **3.** boor *f* (*m*) || **1.** бит *m*, двоичная единица *f* (количества) информации **2.** лезвие *n* **3.** сверло *n*; бурав *m*
B181 **biunique correspondence** || eineindeutige Zuordnung *f* || correspondance *f* biunivoque || wederkerige overeenstemming *f* met een cijfer || взаимно-однозначное соответствие *n*
B182 **bivector** || Bivektor *m* || bivecteur *m* || bivector *m* || бивектор *m*
B183 **black body** || (absolut) schwarzer Körper *m*, Schwarzkörper *m*, Schwarzstrahler *m* || corps *m* noir || zwart lichaam *n* || абсолютно чёрное тело *n*
B184 **blackbody radiation** || schwarze Strahlung *f*, Schwarzkörperstrahlung *f* || rayonnement *m* de corps noir || straling *f* van een zwart lichaam || излучение *n* абсолютно чёрного тела
B185 **blackbody spectrum** || Schwarzkörper-Emissionsspektrum *n* || spectre *m* de corps noir || spectrum *n* van een zwart lichaam || спектр *m* излучения абсолютно чёрного тела
B186 **black hole** || schwarzes Loch *n* || trou *m* noir || zwart gat *n* || чёрная дыра *f*, коллапсар *m*
B187 **blade** || **1.** Schneide *f* **2.** Schaufel *f* *(z. B. einer Turbine)* || **1.** lame *f* **2.** aile(tte) *f* || **1.** snijkant *m*, snede *f* (*m*) **2.** blad *n*, schoep *f* (*m*) *(turbine)* || **1.** лезвие *n*; режущая часть *f* резца **2.** лопатка *f* *(напр. турбины)*
B188 **blade wheel** || Schaufelrad *n*; Laufrad *n* || roue *f* à palettes || schoepenrad *n* || лопастное колесо *n*; крыльчатка *f*
B189 **blank** || **1.** Rohteil *n*, Rohling *m* **2.** Zwischenraum *m* || **1.** ébauche *f*

2. espace *m*, blanc *m* || 1. werkstuk *n* 2. leegte *f* || 1. заготовка *f* 2. пробел *m*

B190 **blanking** || Ausschneiden *n*, Stanzen *n* || 1. découpage *m* 2. estampage *m* || 1. uitsnijding *f* 2. persing *f* || 1. вырубка *f* 2. штамповка *f*

B191 **blast** || Explosion *f* || explosion *f* || explosie *f* || взрыв *m*

B192 **blaster** || Sandstrahlgebläse *n*; Kugelstrahler *m* || sableuse *f* || zandslinger *m*; kogelstraler *m* || пескоструйный аппарат *m*; дробемётный аппарат *m*

B193 **blast wave** || Expansionswelle *f*, Druckwelle *f* || onde *f* explosive || explosiegolf *f* (*m*) || взрывная волна *f*

B194 **blind** || Blende *f* || écran *m* pare-lumière; parasoleil *m* || blende *f* || бленда *f*

B195 **Bloch wall** || Bloch-Wand *f*, Blochsche Wand *f* || paroi *f* de Bloch || Bloch-wand *m* || блоховская [доменная] стенка *f*

B196 **block** || Block *m*; Einheit *f* || bloc *m* || blok *n* || блок *m*

B197 **block-and-tackle** || Flaschenzug *m*, Rollenzug *m* || palan *m* || talie *f* || полиспаст *m*

B198 **block diagram** || Blockdiagramm *n*, Blockschaltbild *n* || bloc-diagramme *m*, schéma *m* fonctionnel || blokdiagram *n* || блок-схема *f*

B199 **block gauge** || Parallelendmaß *n* || cale-étalon *f*, cale *f* Johanson || paralleleindmaat *f* (*m*) || плоскопараллельная концевая мера *f* длины, плитка *f* Иогансона

B200 **block pulley** *see* **block-and-tackle**

B201 **blooming mill** || Blockwalzwerk *n* || blooming *m*, laminoir *m* à lingots [à blooms] || walsbaan *f*, blooming *m* || блюминг *m*

B202 **blower** || Gebläse *n* || soufflante *f*; ventilateur *m* || ventilator *m* || вентилятор *m*; воздуходувка *f*

B203 **blowoff valve** || Abblaseventil *n* || soupape *f* de purge || uitblaasklep *f* (*m*) || продувочный клапан *m*

B204 **blue magnetism** || Südmagnetismus *m* || magnétisme *m* du sud || zuidmagnetisme *n* || южный магнетизм *m*

B205 **blue shift** || Blauverschiebung *f*, Violettverschiebung *f* || déplacement *m* violet || blauwverschuiving *f* || фиолетовое смещение *n* (*спектра*)

B206 **bluff (body)** || Phantom *n*, Körper *m* mit hohem Strömungswiderstand || corps *m* à haute résistance hydraulique || lichaam *n* met hoge stromingsweerstand || неудобообтекаемое [затупленное] тело *n*, тело *n* необтекаемой формы

B207 **body** || 1. *Phys.* Körper *m* 2. Körper *m* 3. Ständer *m*, Gestell *n* 4. Aufbau *m*, Karosserie *f* || 1. *phys.* corps *m* 2. corps *m* 3. bâti *m* 4. carrosserie *f* || 1. *phys.* lichaam *n* 2. lichaam *n* 3. gestel *n*; frame *n* 4. carrosserie *f* || 1. *физ.* тело *n* 2. корпус *m*; тело *n* 3. станина *f* 4. кузов *m*

B208 **body axes system** || körperfestes Koordinatensystem *n* || système *m* d'axes de coordonnées lié au corps || coördinatenstelsel *n* verbonden met een voortbewegend lichaam || система *f* координат, связанная с движущимся телом

B209 **body seat** *see* **valve seat**

B210 **bogie** || Fahrgestell *n* || chariot *m*; bogie *f*; train *m* de roulement || bogie *f*; onderstel *n*; draaistel *n* || тележка *f*; ходовая часть *f*

B211 **Bohr magneton** || Bohrsches Magneton *n* || magnéton *m* de Bohr || Bohr-magneton *n* || магнетон *m* Бора

B212 **Bohr orbit** || Bohrsche Bahn *f*, Bohrscher Kreis *m* || orbite *f* de Bohr || Bohr-baan *f* (*m*) || боровская орбита *f*

B213 **Bohr radius** || Bohrscher Radius *m* || rayon *m* de Bohr || Bohr-radius *m* || боровский радиус *m*

B214 **boiler** || Kessel *m*; Dampfkessel *m* || chaudière *f* (à vapeur) || stoomketel *m* || котёл *m*; паровой котёл *m*; бойлер *m*

B215 **boiling** || Sieden *n* || bouillonnement *m*, ébullition *f* || koken *n* || кипение *n*

B216 **boiling point** || Siedepunkt *m* || point *m* d'ébullition || kookpunt *n* || точка *f* кипения

B217 **boiling temperature** || Siedetemperatur *f* || température *f* d'ébullition || kooktemperatuur *f* || температура *f* [точка *f*] кипения

B218 **bolometer** || Bolometer *n* || bolomètre *m* || bolometer *m* || болометр *m*

B219 **bolt** || Bolzen *m* || boulon *m* || bout *m* || болт *m*; шпилька *f*

B220 **bolt head** || Bolzenkopf *m*, Schraubenkopf *m* || tête *f* de boulon || boutkop *m* || головка *f* болта

B221 **bolting** || Verbolzung *f* || boulonnage *m* || boutverbinding *f* || 1. болтовое соединение *n* 2. сболчивание *n*

B222 **bolt threading machine** || Bolzengewindeschneidmaschine *f*, Schraubenschneidmaschine *f* || machine *f* à fileter les boulons || boutsnijmachine *f* || болторезный станок *m*

B223 **Boltzmann constant** || Boltzmann-Konstante *f* || constante *f* de Boltzmann || Boltzmann-constante *f* (*m*) || постоянная *f* Больцмана

B224 **Boltzmann statistics** || Boltzmann-Statistik *f* || statistique *f* de Boltzmann || Boltzmann-statistiek *f* || статистика *f* Больцмана, классическая статистика *f*

B225 **bombardment** || Beschuß *m*, Beschießung *f* || bombardement *m* || beschieting *f*, bombardement *n* || бомбардировка *f* (*частицами*)

B226 **bomb calorimeter** || kalorimetrische Bombe *f*, Bombenkalorimeter *n* || bombe *f* calorimétrique || calorimetrische bom *f* (*m*) || калориметрическая бомба *f*

B227 **bond** || (chemische) Bindung *f* || liaison *f* (chimique) || chemische binding *f* || (химическая) связь *f*

B228 **bonding agent** || Bindemittel *n* || liant *m* || bindmiddel *n* || связующее (вещество) *n*

B229 **bonding force** || Bindungskraft *f* || force *f* de liaison || bindingskracht *f* (*m*) || сила *f* связи

B230 **Boolean algebra** || Boolesche Algebra *f*, Algebra *f* der Logik || algèbre *f* booléenne || Boole-algebra *f* (*m*) || алгебра *f* логики, булева алгебра *f*

B231 **Boolean function** || Boolesche [logische] Funktion *f* || fonction *f* booléenne || Boole-functie *f* || булева функция *f*

B232 **boost pressure** || Ladedruck *m* || pression *f* de suralimentation; surpression *f* || aanjaagdruk *m* || давление *n* наддува; добавочное давление *n*

B233 **borane** || Boran *n*, Borwasserstoff *m*, Borhydrid *n* || borane *m* || boraan *n* || боран *m*, бороводород *m*, гидрид *m* бора

B234 **bore** || Bohrung *f* || trou *m* (*alésé*) || boring *f* || отверстие *n* (*расточенное или рассверленное*)

B235 **Borel set** || Borelsche Menge *f* || ensemble *m* borélien || Borel-menigte *f* || борелево множество *n*

B236 **borer** || 1. Bohrer *m* 2. Ausbohrmaschine *f* || 1. foret *m* 2. machine *f* à aléser, aléseuse *f* || 1. boor *f* (*m*) 2. uitboormachine *f* || 1. сверло *n*; бурав *m* 2. расточной станок *m*

B237 **boric acid**, H_3BO_3 || Borsäure *f* || acide *m* borique || boorzuur *n* || борная кислота *f*

B238 **boring** || Aufbohren *n*, Bohren *n* || alésage *m* || uitboring *f* || расточка *f*, рассверливание *n*, высверливание *n*, сверление *n*

B239 **boring bar** || Bohrstange *f* || barre *f* d'alésage, porte-outil *m* à aléser || boorstang *f* (*m*) || расточная оправка *f*, борштанга *f*

B240 **boring head** || Bohrkopf *m*; Bohreinheit *f* || tête *f* d'alésage; bloc *m* d'alésage || boorkop *m* || расточная головка *f*; расточной блок *m*

B241 **boring lathe** || Aufbohrdrehmaschine *f* || tour-alésoir *m* || boordraaibank *f* || токарно-расточной станок *m*

B242 **boring machine** || Aufbohrmaschine *f*; Bohrmaschine *f* || aléseuse *f*; perceuse *f* || uitboormachine *f*; boormachine *f* || расточной станок *m*; сверлильный станок *m*

B243 **boring mill** || Senkrechtbohrwerk *n*, Ständerbohrwerk *n* || aléseuse *f* verticale || verticale boormachine *f* met draaibare tafel || вертикально-расточной станок *m*

B244 **boring tool** || Bohrmeißel *m* || outil *m* à aléser || boorbeitel *m* || расточной резец *m*

B245 **boron**, B || Bor *n* || bore *m* || boor *n* || бор *m*

B246 **Bose-Einstein statistics** || Bose—Einstein-Statistik *f* || statistique *f* de Bose et Einstein || Bose-Einstein-statistiek *f* || (квантовая) статистика *f* Бозе(—Эйнштейна)

B247 **Bose gas** || Bose-Gas *n* || gaz *m* de Bose || Bose-gas *n* || бозе-газ *m*. газ *m* бозе-частиц

B248 **Bose particle, boson** || Bose-Teilchen *n*, Boson *n* || particule *f* de Bose, boson *m* || Bose-deeltje *n*, boson *n* || бозе-частица *f*, бозон *m*

B249 **bottom of band** || Unterkante *f* des Energiebandes || limite *f* inférieure de la zone énergétique || onderkant *m* van energieband || нижняя граница *f* энергетической зоны

B250 **bound, boundary** || Grenze *f*; Rand *m* || limite *f* || grens *f* (*m*) || граница *f*; предел *m*

B251 **boundary conditions** || Grenzbedingungen *f pl* || conditions *f pl* aux limites, conditions *f pl* frontières || grensvoorwaarden *f pl* || граничные условия *n pl*

B252 **boundary friction** || Feststoffreibung *f* || frottement *m* de limite || grenswrijving *f* || граничное трение *n*

B253 **boundary layer** || Grenzschicht *f* || couche *f* frontière [limite] || grenslaag *f* (*m*) пограничный слой *m*

B254 **boundary lubrication** || Grenz(schicht)schmierung *f* || lubrification *f* pelliculaire || grenslaagsmering *f* || тонкоплёночная [граничная] смазка *f*

B255 **boundary maximum** || Grenzmaximum *n* || maximum *m* frontière || grensmaximum *n* || граничный максимум *m*

B256 **boundary minimum** || Grenzminimum *n* || minimum *m* frontière || grensminimum *n* || граничный минимум *m*

B257 **boundary of a set** || Mengengrenze *f*, Grenze *f* einer Menge || frontière *f* d'un ensemble || grens *f* (*m*) van een menigte || граница *f* множества

B258 **boundary point** || Grenzpunkt *m* || point *m* frontière || grenspunt *f* (*m*) || граничная точка *f*

B259 **boundary surface** || Grenzfläche *f*; Trennfläche *f* || surface *f* limite || grensvlak *n* || поверхность *f* раздела, граничная поверхность *f*

B260 **boundary-value problem** || Randwertaufgabe *f*, Randwertproblem *n* || problème *m* aux limites || grenswaardeopgave *f* || краевая задача *f*

B261 **boundedly compact** || lokal kompakt || localement compact || lokaal compact || локально компактный

B262 **bounded operator** || beschränkter Operator *m* || opérateur *m* borné || begrensde operator *m* || ограниченный оператор *m*

B263 **bounded quantity** || beschränkte Größe *f* || quantité *f* bornée || begrensde grootheid *f* || ограниченная величина *f*

B264 **bounded region** || beschränktes Gebiet *n* || région *f* bornée || begrensd gebied *n* || ограниченная область *f*

B265 **bounded set** || beschränkte Menge *f* || ensemble *m* borné || begrensde menigte *f* || ограниченное множество *n*

B266 **bounded variation** || beschränkte Variation *f* || variation *f* bornée || begrensde variatie *f* || ограниченная вариация *f*

B267 **bound energy** || gebundene Energie *f* || énergie *f* liée || gebonden energie *f* || связанная энергия *f*

B268 **bound state** || gebundener Zustand *m* || état *m* lié || gebonden toestand *m* || связанное состояние *n*

B269 **bound vortex** || gebundener Wirbel *m* || tourbillon *m* attaché [adjoint] || gebonden wervel *m* || присоединённый вихрь *m*

B270 **Bourdon gauge** || Röhrenfedermanometer *n* || manomètre *m* de Bourdon || buisveermanometer *m*, Bourdonmanometer *m* || манометр *m* Бурдона

B271 **bow wave** || Bugwelle *f* || onde *f* de choc frontale || boeggolf *f* (*m*) || головная ударная волна *f*

B272 **box** || **1.** Gehäuse *n*; Magazin *n* **2.** Gehäuse *n* **3.** Kasten *m* || boîte *f*, boîtier *m* || **1.** magazijn *n* **2.** huis *n* **3.** kist *f* (*m*) || **1.** магазин *m* (*напр. ёмкостей*) **2.** коробка *f*; кожух *m* **3.** ящик *m*

B273 **box frame** || kastenförmiger Rahmen *m* || châssis-caisson *m* || kistvormig raam *n* || коробчатая рама *f*

B274 **brace** || **1.** Bohrer *m*; Bohrwinde *f* **2.** Binder *m*; Verspannung *f*; Strebe *f* || **1.** perceuse *f*; vilebrequin *m* **2.** tirant *m*; liaison *f* || **1.** drilboor *f* (*m*); boorratel *m* **2.** binding *f* || **1.** дрель *f*; коловорот *m* **2.** связь *f*; обвязка *f*; скрепление *n*

B275 **brachistochrone** || Brachistochrone *f*, Brachystochrone *f* || brachystochrone *f* || brachystochron *f* || брахистохрона *f*

B276 **bracket** || **1.** Kragarm *m*; Konsole *f* **2.** Armhängelager *n* || **1.** support *m*; bras *m*, console *f* **2.** palier-console *m* || **1.** console *f* (*m*), kraagsteen *m* **2.** consolelager *n*, kraaglager *n* || **1.** кронштейн *m*; консоль *f* **2.** консольный подшипник *m*

B277 **bracket drill** || Auslegerbohrmaschine *f* || perceuse *f* à console || kraagsteenboormachine *f* || консольный сверлильный станок *m*

B278 **brackets** || Klammern *f pl* || paranthèses *f pl* || haakjes *n pl* || скобки *f pl*

B279 **Bragg angle** || Braggscher Reflexionswinkel *m* || angle *m* de Bragg || Bragg-hoek *m*, reflectiehoek *m* van Bragg || угол *m* брэгговского отражения, брэгговский угол *m*

B280 **Bragg reflection** || Braggsche Reflexion *f*, Bragg-Reflexion *f* || réflexion *f* de Bragg || Bragg-reflectie *f*, reflectie *f* van Bragg || брэгговское отражение *n*

B281 **brake** || Bremse *f* || frein *m* || rem *f* (*m*) || тормоз *m*

B282 **brake band** || Bremsband *n* || bande *f* de frein || remband *m* || тормозная лента *f*

B283 **brake horsepower** || Bremsleistung *f* || puissance *f* au frein || remvermogen *n* || тормозная [полезная] мощность *f* *(двигателя)*

B284 **brake lining** || Bremsbelag *m* || garniture *f* de frein || remvoering *f* || фрикционная тормозная накладка *f*

B285 **brake shoe** || Bremsbacke *f*; Bremsklotz *m*, Bremsschuh *m* || mâchoire *f* de frein; sabot *m* de frein || remblok *n*, remschoen *m* || тормозная колодка *f*; тормозной башмак *m*

B286 **brake torque** *see* **braking torque**

B287 **brake wheel** || Bremstrommel *f* || poulie *f* de frein || remschijf *f*; remtrommel *f* (*m*) || тормозной шкив *m*; тормозной барабан *m*

B288 **braking** || Bremsung *f*, Bremsen *n* || freinage *m* || remming *f* || торможение *n*

B289 **braking distance** || Bremsstrecke *f*, Bremsweg *m* || distance *f* d'arrêt, chemin *m* de freinage || remweg *m* || тормозной путь *m*

B290 **braking force** || Bremskraft *f* || force *f* de freinage || remkracht *f* (*m*) || тормозное усилие *n*

B291 **braking length** *see* **braking distance**

B292 **braking radiation** || Bremsstrahlung *f* || rayonnement *m* de freinage || remstraling *f* || тормозное излучение *n*

B293 **braking torque** || Bremsmoment *n* || moment *m* [couple *m*] de freinage || remmoment *n* || тормозной момент *m*

B294 **branch** || **1.** Zweig *m*; Ast *m* **2.** Kante *f* *(eines Graphen)* || **1.** branche *f* **2.** arête *f* *(du graphe)* || **1.** tak *m*, branche *f* **2.** kant *m* *(van een graaf)* || **1.** ветвь *f* **2.** ребро *n* *(графа)*

B295 **branched circuit** || verzweigter Stromkreis *m* || circuit *m* à branches || vertakt circuit *n* || разветвлённая цепь *f*

B296 **branching** || Verzweigung *f* || branchement *m* || vertakking *f* || ветвление *n*; условный переход *m*

B297 **branch of a curve** || Kurvenast *m* || branche *f* d'une courbe || tak *m* van een kromme || ветвь *f* кривой

B298 **branch of a tree** || Baumast *m* || branche *f* d'arbre || boomtak *m* || ветвь *f* дерева

B299 **branch point** || Verzweigungspunkt *m* || point *m* de branchement || vertakkingspunt *f* (*m*) || точка *f* (раз)ветвления; узловая точка *f*

B300 **brass** || **1.** Messing *n* **2.** Schale *f* *(Lagerschale)* || **1.** laiton *m* **2.** coussinet *m* || **1.** messing *n* **2.** kussenblokbus *f* (*m*), lagerbus *f* (*m*) || 1. латунь *f* 2. вкладыш *m* *(подшипника)*

B301 **brazed-tip tool** || Plättchenmeißel *m*, hartmetallbestücktes Werkzeug *n* || outil *m* à pastille rapportée || gesoldeerde beitel *m* || резец *m* с твердосплавной режущей пластинкой

B302 **braze welding** || Schweißlöten *n* || soudo-brasure *f* || las-solderen *n* || пайка-сварка *f*

B303 **brazing** || Hartlötung *f* || brasure *f* forte || hard solderen *n* || пайка *f* тугоплавким [твёрдым] припоем

B304 **breakaway torque** || Höchstlastmoment *n* || couple *m* de torsion limite || grensdraaimoment *m* || предельный крутящий момент *m*

B305 **breakdown** || 1. Havarie *f*, Ausfall *m*; Störung *f* 2. Durchschlag *m* || 1. défaut *m* 2. claquage *m* (électrique) || 1. averij *f*, gebrek *n* || 1. авария *f*; сбой *m* 2. (электрический) пробой

B306 **breakdown torque** *see* **breakaway torque**

B307 **breakdown voltage** || Durchschlagsspannung *f*, Durchbruchsspannung *f* || tension *f* disruptive [de claquage] || doorslagspanning *f* || пробивное напряжение *n*, напряжение *n* пробоя

B308 **breaker** || Unterbrecher *m* || rupteur *m* || onderbreker *m* || прерыватель *m* *(тока)*

B309 **breaking** || 1. Trennung *f*, Trennen *n*, Öffnen *n* 2. Bruch *m*, Bindungsbruch *m* || 1. coupure *f* 2. rupture *f* || 1. onderbreking *f* 2. afbreking *f*; breking *f* || 1. размыкание *n* *(цепи)* 2. разрушение *n*; обрыв *m*

B310 **breaking arc** || Öffnungsbogen *m*; Unterbrechungsbogen *m* || arc *m* de rupture || onderbrekingslichtboog *m* || (электрическая) дуга *f* при размыкании

B311 **breaking elongation** *see* **elongation at fracture**

B312 **breaking load** || Bruchbelastung *f*, Bruchlast *f* || charge *f* de rupture || breukbelasting *f* || разрушающая нагрузка *f*

B313 **breaking machine** || Zerreißmaschine *f* || machine *f* d'essais à la rupture || breukmachine *f* || машина *f* для испытания на разрыв

B314 **breaking strain** || Bruchverformung *f* || déformation *f* à la limite de rupture || rek *m* bij breukspanning || разрушающая (относительная) деформация *f*

B315 **breaking strength** || Zerreißfestigkeit *f*, Reißfestigkeit *f*, Reißfestigkeitsgrenze *f* || limite *f* de rupture || breukvastheid *f* || предел *m* прочности при разрыве, временное сопротивление *n* разрыву

B316 **breaking stress** || Bruchbeanspruchung *f*, Bruchspannung *f* || contrainte *f* de rupture || breukspanning *f* || разрушающее [разрывающее] напряжение *n*; предел *m* прочности при разрыве

B317 **breeder (reactor)** || Brutreaktor *m*, Brüter *m* || réacteur *m* (auto)-régénérateur || aanmaakreactor *m*, kweekreactor *m* || бридерный реактор *m*, реактор-размножитель *m*, бридер *m*

B318 **bremsstrahlung (radiation)** *see* **braking radiation**

B319 **bridge** || elektrische Brücke *f* || pont *m* || brug *f* (*m*) || (электрический) мост(ик) *m*

B320 **bridge joint** || Brückenschaltung *f* || couplage *m* en pont || brugschakeling *f* || соединение *n* по мостовой схеме

B321 **brightness** || Helligkeit *f*; Leuchtdichte *f* || brillance *f*; luminance *f* || helderheid *f* || яркость *f*

B322 **brightness contrast** || Leuchtdichtekontrast *m*, Helligkeitskontrast *m* || contraste *m* de brillances [de luminance] || helderheidscontrast *n* || яркостный контраст *m*

B323 **brightness temperature** || Luminanztemperatur *f* || température *f* de brillance || helderheidstemperatuur *f* || яркостная температура *f*

B324 **Brinell hardness** || Brinellhärte *f* || dureté *f* Brinell || Brinellhardheid *f* || твёрдость *f* по Бринеллю

B325 **briquette press** || Brikettpress *f* || presse *f* à briqueter || briketpers *f* || брикетировочный пресс *m*

B326 **brittle** || spröde; brüchig || fragile; friable || breekbaar, bros || хрупкий

B327 **brittleness** || Sprödigkeit *f*; Brüchigkeit *f* || fragilité *f*; friabilité *f* || breekbaarheid *f*, brosheid *f* || хрупкость *f*, ломкость *f*

B328 **broach** || Räumwerkzeug *n* || broche *f*, outil *m* à brocher || trekfrees *f* (*m*) || протяжка *f* (*инструмент*)

B329 **broaching** || 1. Räumen *n* 2. Lochen *n* || brochage *m* || trekfrezing *f* || 1. протягивание *n* 2. прошивка *f*

B330 **broaching machine** || Räummaschine *f* || machine *f* à brocher || trekfreesbank *f* (*m*) || протяжной станок *m*

B331 **broaching tool** *see* **broach**

B332 **broad-band** || breitbandig, Breitband... || à bande large || breedbandig, breedband... || 1. широкополосный (*об усилителе*) 2. широкозонный (*о полупроводнике*)

B333 **broadening** || Verbreiterung *f*, Erweiterung *f* || élargissement *m* || uitbreiding *f* || уширение *n* (*спектральных линий*)

B334 **broken circuit** || offener [geöffneter] Stromkreis *m* || circuit *m* ouvert || open circuit *n* || разомкнутая [незамкнутая] цепь *f*

B335 **bromic acid,** $HBrO_3$ || Bromsäure *f* || acide *m* bromique || broomzuur *n* || бромноватая кислота *f*

B336 **bromide** || Bromid *n* || bromure *m* || bromide *n* || бромид *m*

B337 **bromine,** Br || Brom *n* || brome *m* || bromium *n*, broom *n* || бром *m*

B338 **bronze** || Bronze *f* || bronze *m* || brons *n* || бронза *f*

B339 **Brownian motion** || Brownsche Bewegung *f* || mouvement *m* brownien || Brownse beweging *f* || броуновское движение *n*

B340 **brush** || Bürste *f* || balai *m*; frotteur *m* || borstel *m* || щётка *f*

B341 **brushing** || Büschelentladung *f* || aigrette *f* (électrique) || bundelontlading *f* || кистевой разряд *m*

B342 **bubble** || Blase *f* || bulle *f* || blaas *f*, bel *m* || пузырь *m*, пузырёк *m*

B343 **bubble chamber** || Blasenkammer *f* || chambre *f* à bulles || bellenkamer *f* (*m*) || пузырьковая камера *f*

B344 **bubble domain** || zylinderförmiger Weißscher Bezirk *m* || domaine *m* cylindrique magnétique || cilindervormig magneetdomein *n* || цилиндрический магнитный домен *m*, ЦМД

B345 **buckling** || Knickung *f* || flambage *m*, flambement *m* || knik *m* || продольный изгиб *m*

B346 **buckling load** || kritische Last *f*, Knicklast *f* || charge *f* de flambage || kniklast *m* || критическая [вызывающая потерю устойчивости] нагрузка *f*

B347 **buckling stress** || Knickspannung *f* || contrainte *f* de flambage [de flambement] || knikspanning *f* || напряжение *n* при продольном изгибе

B348 **buckling test** || Knickversuch *m* || essai *m* de flambage [de flambement] || knikbeproeving *f* || испытание *n* на продольный изгиб

B349 **buckling vector** || Krümmungsvektor *m* || vecteur *m* de courbure || krommingsvector *m* || вектор *m* кривизны

B350 **buffer** || Puffer *m*; Dämpfer *m* || buffer *m*; tampon *m*; amortisseur *m* de choc, pare-chocs *m* || buffer *m*; schokdemper *m*, schokbreker *m* || буфер *m*; демпфер *m*

B351 **buffer memory** *see* **buffer storage**

B352 **buffer solution** || Pufferlösung *f* || solution *f* tampon || bufferoplossing *f* || буферный раствор *m*

B353 **buffer storage** || Pufferspeicher *m* || mémoire *f* tampon || buffergeheugen *n* || буферная память *f*, буферное запоминающее устройство *n*

B354 **built-in beam** || (beidseitig) eingespannter Balken *m* [Träger *m*] || poutre *f* encastrée aux deux extrémités || (aan beide zijden) ingeklemde balk *m* || защемлённая балка *f*

B355 **bulb** || 1. Kolben *m* 2. Kugel *f*, Gefäß *n* (eines *Thermometers*) || 1. bouteille *f* 2. bille *f* || 1. kolf *f* 2. kogeltje *n* || 1. баллон *m*, колба *f* 2. шарик *m* (*термометра*)

B356 **bulk concentration** || Volumenkonzentration *f* || concentration *f* volumique || volumeconcentratie *f* || объёмная концентрация *f*

B357 **bulk density** || Schüttdichte *f* || densité *f* en vrac || stortdichtheid *f* || насыпная плотность *f*

B358 **bulk modulus** || Kompressionsmodul *m*, Triaxialdruckmodul *m* || module *m* de compression triaxiale [volumétrique] || volume-elasticiteitsmodulus *m*, compressiemodulus *m* || модуль *m* объёмного сжатия

B359 **bulk viscosity** || Volumenviskosität *f*, zweite Viskosität *f* || viscosité *f* volumétrique || volumeviscositeit *f* || объёмная [вторая] вязкость *f*

B360 **bunching** || Paketbildung *f*, Paketierung *f* || groupement *m* || groepering *f* || группирование *n* *(частиц)*, банчирование *n*; модуляция *f* пучка (частиц) по плотности

B361 **bundle of electrons** || Elektronenpaket *n* || groupe *m* d'électrons || elektronengroep *f* *(m)* || сгусток *m* электронов

B362 **bundle of rays** || Strahlenbündel *n* || faisceau *m* de rayons || stralenbundel *m* || пучок *m* лучей

B363 **buoyancy** || Auftrieb *m* || flottabilité *f*; force *f* d'Archimède || verdringing *f* || плавучесть *f*; выталкивающая *(гидро- или аэростатическая)* сила *f*

B364 **buoyant force** || Auftriebskraft *f*, Auftrieb *m* || force *f* d'Archimède || verdringingskracht *f* *(m)* || выталкивающая [архимедова] сила *f*

B365 **buried wiring** || verdeckte Leitung *f*, Unterputzleitung *f* || canalisation *f* (posée) sous crépi || verborgen leiding *f* || скрытая (электро)проводка *f*

B366 **burner** || Brenner *m*; Düse *f* || brûleur *m* || brander *m*; sproeier *m* || горелка *f*; форсунка *f*

B367 **burning temperature** || Brenntemperatur *f* || température *f* de combustion || verbrandingstemperatuur *f* || температура *f* горения

B368 **burning velocity** || Brenngeschwindigkeit *f* || vitesse *f* de combustion || brandsnelheid *f* || скорость *f* горения

B369 **burnishing** || **1.** Polieren *n* **2.** Polierdrücken *n*, Polierrollen *n* || **1.** polissage *m* **2.** brunissage *m* || **1.** polijsting *f* **2.** bruinering *f* || **1.** полирование *n*, полировка *f* **2.** воронение *n*

B370 **burnishing wheel** || Polierscheibe *f* || disque *m* de polissage || polijstschijf *f* *(m)* || полировальный круг *m*

B371 **bus** || Bus *m*, Datenbus *m* || bus *m* || (data) bus *f* *(m)* || шина *f* *(линия связи, допускающая подсоединение одного или нескольких абонентов для приёма и передачи данных)*

B372 **busbar** || Schiene *f*, Sammelschiene *f* || barre *f* || collector *m* || (электрическая) шина *f*

B373 **bush(ing)** || Buchse *f*; Schale *f*, Lagerschale *f* || douille *f*; manchon *m*; coussinet *m* *(de palier)* || (lager)bus *f* *(m)*; (lager)schaal *f* *(m)* || втулка *f*, вкладыш *m* *(подшипника)*

B374 **business machine** || Büromaschine *f* || calculateur *m* de bureau || bureaumachine *f* || бухгалтерская машина *f*

B375 **butterfly screw** || Flügelschraube *f* || vis *f* [boulon *m*] à papillons || beugelschroef *f* *(m)* || винт-барашек *m*

B376 **butterfly valve** || **1.** Drosselklappe *f* *(Vergaser)* **2.** Flügelventil *n* || **1.** papillon *m* *(de carburateur)* **2.** soupape *f* à papillon || **1.** vleugelventiel *n* **2.** vleugelklep *f* *(m)* || **1.** дроссельная заслонка *f* *(карбюратора)* **2.** мотыльковый клапан *m*

B377 **button head** || Halbrundkopf *m* *(z. B. am Niet)*, Rundkopf *m* || tête *f* ronde *(p. ex. d'un rivet)* || halfronde kop *m* *(bijv. van de klinknagel)* || полукруглая головка *f* *(напр. заклёпки)*

B378 **buttress screw thread** || Sägengewinde *n* || filet *m* d'appui || trapeziumdraad *m* || упорная резьба *f*

B379 **butt weld** || Stumpfnaht *f* || soudure *f* en bout || stompnaad *m*; stompe lasverbinding *f* || стыковой шов *m*; стыковое сварное соединение *n*

B380 **butt welding** || Stumpfschweißung *f* || soudage *m* en bout || stomp-lassen *n* || стыковая сварка *f*

B381 **by-effect** || Nebeneffekt *m* || effet *m* secondaire || bijeffect *n* || побочный эффект *m*, побочное явление *n*

B382 **bypass** || 1. Beipaß *m*, Nebenleitung *f*, Umgehungsleitung *f* 2. Umleitkanal *m* 3. Umgehungsventil *n* || 1. dérivation *f*, by-pass *m* 2. canal *m* de dérivation 3. soupape *f* de by-pass || 1. loop *m* 2. omloopkanaal *n*, omloopleiding *f* 3. omloopventiel *n*, omloopklep *f* (*m*) || 1. байпас *m* 2. перепускной канал *m*; боковой канал *m* 3. перепускной клапан *m*

B383 **bypass valve** || Umgehungsventil *n* || soupape *f* de by-pass || omloopventiel *n*, omloopklep *f* (*m*) || перепускной клапан *m*

B384 **byte** || Byte *n* || byte *m* || byte *n* || байт *m*

C

C1 **cable** || 1. Kabel *n* 2. Seil *n* || câble *m* || 1. kabel *m* 2. touw *n*, kabel *m* || 1. кабель *m*, многожильный провод *m* 2. трос *m*

C2 **cadmium**, Cd || Kadmium *n* || cadmium *m* || cadmium *n* || кадмий *m*

C3 **cadmium (normal) cell** || Kadmium-Normalelement *n* || élément *m* normal de cadmium || cadmiumnormaalelement *n* || нормальный элемент *m* Вестона, эталонный кадмиевый элемент *m*

C4 **calcium**, Ca || Kalzium *n* || calcium *m* || calcium *n* || кальций *m*

C5 **calcium carbonate**, $CaCO_3$ || Kalziumkarbonat *n* || carbonate *m* de calcium || calciumcarbonaat *n* || карбонат *m* кальция, углекислый кальций *m*

C6 **calcium hydroxide**, $Ca(OH)_2$ || Kalziumhydroxid *n* || hydroxyde *m* de calcium || calciumhydroxyde *n* || гидроокись *f* кальция

C7 **calcium sulphate**, $CaSO_4$ || Kalziumsulfat *n* || sulfate *m* de calcium || calciumsulfaat *n* || сульфат *m* кальция, сернокислый кальций *m*

C8 **calculable** || berechenbar || calculable || berekenbaar || вычислимый

C9 **calculable set** || berechenbare Menge *f* || ensemble *m* calculable || berekenbare menigte *f* || вычислимое множество *n*

C10 **calculator** || Rechner *m*, Rechenmaschine *f* || calculateur *m*, calculatrice *f* || rekenmachine *f* || калькулятор *m*, счётная машина *f*

C11 **calculus of variations** || Variationsrechnung *f* || calcul *m* variationnel || variatierekening *f* || вариационное исчисление *n*

C12 **calibration** || Kalibrierung *f* || calibrage *m* || kalibrering *f* || калибрование *n*, калибровка *f*

C13 **calipers** || Taster *m*, Greifzirkel *m* || compas *m* à balustre; compas *m* à verge; compas *m* d'intérieur || krompasser *m*, voetjespasser *m* || кронциркуль *m*; штангенциркуль *m*; нутромер *m*

C14 **call** || Aufruf *m* || appel *m* || oproep *m* || вызов *m* (*напр. подпрограммы*)

C15 **calorie** || Kalorie *f* || calorie *f* || calorie *f* || калория *f*

C16 **calorific value** || Heizwert *m* || valeur *f* [pouvoir *m*] calorifique || verbrandingswarmte *f*, calorieënwaarde *f* || теплота *f* сгорания

C17 **calorimeter** || Kalorimeter *n* || calorimètre *m* || calorimeter *m* || калориметр *m*

C18 **calorimetric bomb** || kalorimetrische Bombe *f*, Bombenkalorimeter *n* || bombe *f* calorimétrique || calorimetrische bom *f* (*m*) || калориметрическая бомба *f*

C19 **calorimetry** || Kalorimetrie *f*, Wärmemessung *f* || calorimétrie *f* || calorimetrie *f* || калориметрия *f*

C20 **cam** || 1. Nocken *m*, Kurve *f*; Exzenter *m* 2. Kopiernocken *m* || came *f*; excentrique *m* 2. gabarit *m* || 1. nok *f* (*m*) 2. sjabloon *f* (*m*) || 1. кулачок *m*; эксцентрик *m* 2. копирный кулачок *m*

C21 **camera** || Kamera *f* || caméra *f* || camera *f* (*m*) || камера *f* (*напр. фото- или кинокамера*)

C22 **camera objective** || Fotoobjektiv *n* || objectif *m* photographique ||

foto-objectief *n* || фотографический объектив *m*

C23 **camshaft** || Nockenwelle *f*, Steuerwelle *f* || arbre *m* à cames || nokkenas *f* (*m*) || распределительный вал *m*

C24 **canalization** *see* **channel(l)ing**

C25 **cancellation** || 1. *Math.* Aufhebung *f* 2. Löschen *n* || 1. *math.* réduction *f* (*des termes semblables*) 2. effacement *m* (*p. ex. d'enregistrement*) || 1. opheffing *f*; reductie *f* 2. uitwissen *n* || 1. *мат.* сокращение *n*; взаимное уничтожение *n* 2. гашение *n*; стирание *n* (*напр. записи*)

C26 **cancellation law** || Aufhebungsregel *f* || loi *f* de réduction (*des termes semblables*) || reductieregel *m* || правило *n* взаимного уничтожения (*подобных членов*)

C27 **candela,** cd || Candela *f* || candela *f* || candela *f* || кандела *f* (*единица силы света в СИ*)

C28 **candle** || Kerze *f* || candela *f* || kaars *f* || свеча *f* (*единица силы света*)

C29 **candle power** || Kerzenstärke *f* || intensité *f* lumineuse || kaarssterkte *f* (*m*) || сила *f* света (в свечах)

C30 **canonical equations** || kanonische Gleichungen *f pl* || équations *f pl* canoniques || canonieke vergelijkingen *f pl* || канонические уравнения *n pl*

C31 **canonical equations of motion** || kanonische Bewegungsgleichungen *f pl* || équations *f pl* canoniques de la mécanique || canonieke bewegingsvergelijkingen *f pl* || канонические уравнения *n pl* механики, уравнения *n pl* Гамильтона

C32 **canonical form** || kanonische Form *f* || forme *f* canonique || canonieke vorm *m* || канонический вид *m*, каноническая форма *f*

C33 **canonical transformation** || kanonische Transformation *f* || transformation *f* canonique || canonieke transformatie *f* || каноническое преобразование *n*

C34 **cantilever beam** || Freiträger *m*, Kragträger *m* || poutre-console *f*, poutre *f* en porte-à-faux || overhangende ligger *m* || консольная балка *f*

C35 **cap** || Kappe *f*; Kopf *m* || 1. couvercle *m* 2. chapeau *m*, tête *f* || 1. cap *f* (*m*); deksel *n* 2. gek *m*; kop *m* || 1. крышка *f* 2. шляпка *f*, головка *f*

C36 **capacitance** || Kapazität *f* || capacité *f* || capaciteit *f* || (электрическая) ёмкость *f*

C37 **capacitive** || kapazitiv || capacitif || capacitief || ёмкостный

C38 **capacitive component** || kapazitive Komponente *f* || composante *f* capacitive || capacitieve component *m* || ёмкостная составляющая *f*

C39 **capacitive coupling** || kapazitive Kopplung *f* || couplage *m* capacitif || capacitieve koppeling *f* || ёмкостная [конденсаторная] связь *f*

C40 **capacitive resistance** || kapazitiver Widerstand *m* || résistance *f* capacitive || capacitieve weerstand *m* || ёмкостное сопротивление *n*

C41 **capacitivity** || dielektrische Konstante *f*, Dielektrizitätskonstante *f* || constante *f* diélectrique, permittivité *f* || diëlektrische constante *f* (*m*) || диэлектрическая проницаемость *f*

C42 **capacitor** || Kondensator *m* || condensateur *m* || condensator *m* || (электрический) конденсатор *m*

C43 **capacitor bank** || Kondensatorenbatterie *f* || batterie *f* [banc *m*] de condensateurs || condensatorbank *f* (*m*) || батарея *f* конденсаторов

C44 **capacity** || 1. Kapazität *f*, Fassungsvermögen *n* 2. Arbeitsvolumen *n*; Arbeitsbereich *m* 3. Leistungsfähigkeit *f*; Leistung *f* 4. Tragfähigkeit *f*; Ladefähigkeit *f* || capacité *f*; pouvoir *m* || 1. volume *n*, inhoud *m* 2. bevattingsvermogen *m* 3. capaciteit *f* 4. draagvermogen *n*; laadvermogen *n* || 1. вместимость *f*, ёмкость *f* 2. рабочий объём *m* 3. производительность *f*; мощность *f* 4. грузоподъёмность *f*

C45 **capacity bridge** || Kapazitätsbrücke *f* || pont *m* (de mesure) de capacités || capaciteitsbrug *f* (*m*) || ёмкостный мост *m*

C46 I **capillary** || Kapillare *f*, Haarröhrchen *n* || tube *m* capillaire || capillair *n*, haarvat *n* || капилляр *m*

C47 II **capillary** || Kapillar... || capillaire || capillair || капиллярный

C48 **capillary ascent** || kapillare Steighöhe *f*, kapillare Hebung *f* || montée *f* [ascension *f*] capillaire || capillaire stijghoogte *f* || капиллярное поднятие *n* *(жидкости)*

C49 **capillary correction** || Kapillaritätskorrektion *f*, Kapillarkorrektion *f* || correction *f* capillaire || capillariteitscorrectie *f* || капиллярная поправка *f*, поправка *f* на капиллярность

C50 **capillary depression** || Kapillardepression *f*, kapillare Senkung *f* || dépression *f* capillaire || capillaire depressie *f* || капиллярная депрессия *f* *(уровня жидкости)*

C51 **capillary pressure** || Kapillardruck *m* || pression *f* capillaire || capillaire druk *m* || капиллярное давление *n*

C52 **capture** || Einfang *m* || capture *f* || invangen *n* || захват *(частицы)*

C53 **carbide** || Karbid *n* || carbure *m* || carbid(e) *n* || карбид *m*

C54 **carbide tipped tool** || Hartmetallwerkzeug *n* || outil *m* en carbure métallique || hardmetaalgereedschap *n* || твердосплавный (режущий) инструмент *m*, режущий инструмент *m* с твердосплавной пластинкой

C55 **carbide tipping** || Hartmetallauflage *f* || revêtement *m* en carbure métallique || hardmetaalbedekking *f* || покрытие *n* *(напр. режущей кромки)* твёрдым сплавом

C56 **carbinol,** CH_3OH || Methanol *n*, Karbinol *n*, Methylalkohol *m* || carbinol *m*, méthanol *m* || methanol *n*, methylalcohol *m* || карбинол *m*, метиловый спирт *m*, метанол *m*

C57 **carbon,** C || Kohlenstoff *m* || carbone *m* || koolstof *f* *(m)* || углерод *m*

C58 **carbonate** || Karbonat *n* || carbonate *m* || carbonaat *n* || карбонат *m*

C59 **carbon dioxide,** CO_2 || Kohlen(stoff)dioxid *n* || bioxyde *m* de carbone, anhydride *m* carbonique || koolstofdioxyde *n* || двуокись *f* углерода

C60 **carbon disulfide,** CS_2 || Schwefelkohlenstoff *m* || sulfure *m* de carbone || zwavelkoolstof *f* *(m)* || сероуглерод *m*

C61 **carbonic acid,** H_2CO_3 || Kohlensäure *f* || acide *m* carbonique || koolzuur *n* || угольная кислота *f*

C62 **carbon monoxide,** CO || Kohlen(mon)oxid *n*, Kohlenstoffmonoxid *n* || oxyde *m* de carbone || koolstofoxyde *n* || окись *f* углерода

C63 **carbon steel** || Kohlenstoffstahl *m* || acier *m* au carbone || koolstofstaal *n* || углеродистая сталь *f*

C64 **carbonyl group,** $= C = O$ || Karbonylgruppe *f* || groupe *m* carbonyle || carbonylgroep *f* *(m)* || карбонильная группа *f*

C65 **carboxyl group,** $=O - COOH$ || Karboxylgruppe *f* || groupe *m* carboxylique || carboxylgroep *m* || карбоксильная группа *f*

C66 **carburet(t)or** || Vergaser *m* || carburateur *m* || carburateur *m* || карбюратор *m*

C67 **carburization** *see* **carburizing**

C68 **carburized steel** || Einsatzstahl *m* || acier *m* cémenté || gecarboneerd staal *n* || науглероженная сталь *f*

C69 **carburizing** || Aufkohlung *f*; Einsatzhärtung *f* || carburation *f*; cémentation *f* || carboneren *n* || науглероживание *n*; цементация *f*

C70 **cardan drive** || Kardanantrieb *m* || transmission *f* à cardan || cardanaandrijving *f* || карданная передача *f*

C71 **cardan joint** || Kardangelenk *n* || cardan *m*, joint *m* universel (de Cardan) || cardan *m*; cardankoppeling *f* || карданный [универсальный] шарнир *m*, карданный механизм *m*

C72 **cardan shaft** || Kardanwelle *f*, Gelenkwelle *f* || arbre *m* à cardan || cardanas *f* *(m)* || карданный вал *m*

C73 **cardinal, cardinal number** || Kardinalzahl *f* || cardinal *m*, nombre *m* cardinal || kardinaal getal *n* || кардинальное число *n*, мощность *f* *(математического множества)*

C74 **cardinal plane** || Kardinalebene *f*, Grundebene *f* || plan *m* principal || grondvlak *n* || кардинальная плоскость *f*

C75 **cardioid** || Kardioide *f* || cardioïde *f* || cardioïde *f* || кардиоида *f*

C76 **Carnot cycle** || Carnotscher Kreisprozeß *m*, Carnot-Prozeß *m* ||

cycle *m* de Carnot || Carnot-proces *n* || цикл *m* Карно, идеальный цикл *m*

C77 **Carnot's theorem** || Carnotscher Satz *m* || théorème *m* de Carnot || Carnotische wet *f* (*m*), wet van Carno || теорема *f* Карно

C78 **carriage** || 1. Support *m*; Schlitten *m*; Wagen *m* 2. Fahrgestell *n* || chariot *m* || 1. support *n* 2. wagen *m* || 1. суппорт *m*; каретка *f* 2. тележка *f*

C79 **carrier** || 1. Träger *m* 2. Tragelement *n* (*einer Konstruktion*); Konsole *f* 3. Laufwerk *n*; Gleitstück *n*; Schlitten *m* || porteur *m*; élément *m* porteur || 1., 2. drager *m* 3. slede *f* (*m*) || 1. носитель *m* 2. несущий элемент *m* конструкции; кронштейн *m* 3. ходовой механизм *m*; ползун *m*; салазки *pl*

C80 **carrier frequency** || Trägerfrequenz *f* || fréquence *f* porteuse || draaggolffrequentie *f* || несущая (частота) *f*

C81 **carrier mobility** || Trägerbeweglichkeit *f* || mobilité *f* de porteurs || dragerbeweeglijkheid *f* || подвижность *f* носителей (*тока*)

C82 **cartesian coordinates** || Cartesische [rechtwinklige] Koordinaten *f pl* || coordonnées *f pl* cartésiennes || cartesiaanse [rechthoekige] coördinaten *m pl* || декартовы [прямоугольные] координаты *f pl*

C83 **cartesian coordinate system** || Cartesisches Koordinatensystem *n* || système *m* de coordonnées cartésiennes || cartesiaans coördinatenstelsel *n* || декартова система *f* координат

C84 **cartesian product** || kartesisches Produkt *n* || produit *m* cartésien || cartesiaans produkt *n* || декартово [прямое] произведение *n*

C85 **cascade** || Kaskade *f*; Stufe *f* || cascade *f*, étage *m* || cascade *f* || каскад *m*

C86 **cascading** || Kaskadenschaltung *f* || connexion *f* en étages || cascadeschakeling *f* || каскадное [ступенчатое] включение *n*

C87 **case** || Gehäuse *n*; Mantel *m* || boîte *f* || 1. omhulsel *n*; mantel *m* 2. carter *n*; gestel *n* || 1. кожух *m*, чехол *m*; коробка *f*; оболочка *f*; 2. картер *m*; корпус *m*

C88 **case-hardened steel** *see* **cemented steel**

C89 **cast** || 1. Schmelze *f* 2. Guß *m* 3. Gußstück *n* || fusion *f*; coulée *f* || 1. smelting *f* 2. gietstuk *n* 3. gieten *n* || 1. плавка *f* 2. литьё *n* 3. отливка *f*

C90 **casting** || 1. Gießen *n* 2. Gußstück *n* || coulée *f* || gieten *n*, gietstuk *n* || 1. литьё *n* 2. отливка *f*

C91 **casting alloy** || Gußlegierung *f* || alliage *m* de fonderie || gietlegering *f* || литейный сплав *m*

C92 **casting properties** || Gießeigenschaften *f pl* || propriétés *f pl* de fonderie || gieteigenschappen *f pl* || литейные свойства *n pl*

C93 **cast iron** || Gußeisen *n* || fonte *f* || gietijzer *n* || чугун *m*

C94 **cast steel** || Gußstahl *m* || acier *m* fondu [coulé] || gietstaal *n* || литая сталь *f*; стальное литьё *n*

C95 **catalysis** || Katalyse *f* || catalyse *f* || katalyse *f* || катализ *m*

C96 **catalyst** || Katalysator *m* || catalyseur *m* || katalysator *m* || катализатор *m*

C97 **catenary** || Kettenlinie *f* || caténaire *f* || kettinglijn *f* (*m*) || цепная линия *f*

C98 **caterpillar** || 1. Raupe *f*, Raupenkette *f* 2. Raupe *f*, Raupenschlepper *m* || 1. chenille *f* 2. tracteur *m* à chenilles || 1. rupsband *m*; rupsketting *f* 2. rupstrekker *m* || 1. гусеница *f*; гусеничная цепь *f* 2. гусеничный трактор *m*

C99 **cathode** || Katode *f* || cathode *f* || kathode *f* || катод *m*

C100 **cathode-ray oscillograph** || Elektronenstrahloszillograf *m*, Katodenstrahloszillograf *m* || oscillographe *m* cathodique || kathodestraaloscillograaf *m* || электронно-лучевой осциллограф *m*

C101 **cathode-ray tube** || Elektronenstrahlröhre *f*, Katodenstrahlröhre *f* || tube *m* cathodique || kathodestraalbuis *f* (*m*) || электроннолучевая трубка *f*

C102 **cathodoluminescence** || Katodolumineszenz *f* || luminescence *f* cathodique || kathodeluminescentie *f* || катодолюминесценция *f*

C103 **cation** || Kation *n* || cation *m* || kation *n* || катион *m*

C104 **Cauchy integral test** || Integralkriterium *n (Reihe)* || critère *m* de Cauchy-Maclaurin || integrale test *m* van Cauchy || интегральный признак *m* сходимости Коши

C105 **Cauchy problem** || Anfangswertproblem *n*, Cauchysches Problem *n* || problème *m* aux valeurs initiales, problème *m* de Cauchy || Cauchy-opgave *f* || задача *f* с начальными значениями, задача *f* Коши

C106 **Cauchy radical test, Cauchy root test** || zweites Wurzelkriterium *n*, (Cauchysches) Konvergenzkriterium *n* für Zahlenfolgen || critère *m* de Cauchy || Cauchy-worteltest *m* || признак *m* Коши *(сходимости числовых рядов)*

C107 **Cauchy's test for convergence** || Cauchysches Konvergenzkriterium *n* || critère *m* de convergence de Cauchy || Cauchy-convergentietest *m* || признак *m* Коши сходимости последовательностей *или* рядов

C108 **caustic** || **1.** *s.* caustic soda **2.** *Math.* Kaustik *f*, Kaustikfläche *f* || **1.** *v.* caustic soda **2.** *math.* caustique *f*, surface *f* caustique || **1.** *zie* caustic soda **2.** *math.* caustische opprevlakte *f* || **1.** *см.* **caustic soda** **2.** *мат.* каустика *f*, каустическая поверхность *f*

C109 **caustic soda,** NaOH || kaustische Soda *f*, Ätznatron *n*, Natriumhydroxid *n* || caustique *m*, soude *f* (caustique), hydroxyde *m* de sodium || caustische soda *m, f* || каустическая сода *f*, каустик *m*, гидроокись *f* натрия, едкий натрий *m*

C110 **caustic surface** *see* **caustic 2.**

C111 **cavitation** || Kavitation *f*, Hohlraumbildung *f* || cavitation *f* || cavitatie *f* || кавитация *f*

C112 **cavity** || **1.** Hohlraum *m* **2.** Topfresonator *m*, Topfkreis *m* || **1.** cavité *f* **2.** cavité *f* (résonnante), résonateur *m* à volume || **1.** ruimte *f* **2.** potresonator *m* || **1.** (замкнутая) полость *f* **2.** (объёмный) резонатор *m*

C113 **celestial body** || Himmelskörper *m* corps *m* céleste || hemellichaam *n* || небесное тело *n*

C114 **celestial mechanics** || Himmelsmechanik *f* || mécanique *f* céleste || hemelmechanica *f* || небесная механика *f*

C115 **cell** || **1.** galvanisches Element *n*, galvanische Zelle *f*, Zelle *f* **2.** Elementarzelle *f* || **1.** élément *m* [pile *f*] galvanique **2.** cellule *f* élémentaire || **1.** galvanisch element *n* **2.** elementaire cel *f (m)* || **1.** гальванический элемент *m* **2.** (элементарная) ячейка *f*

C116 **cell polarization** || galvanische Polarisation *f* || polarisation *f* galvanique || galvanische polarisatie *f* || гальваническая поляризация *f*, поляризация *f* гальванического элемента

C117 **cellulose** || Zellulose *f*, Zellstoff *m* || cellulose *f*, pâte *f* chimique || cellulose *f (m)* || целлюлоза *f*

C118 **cement** || Zement *m*; Bindemittel *n*, Kitt *m* || ciment *m* || cement *n, m*; bindmiddel *n*, lijm *m* || цемент *m*; вяжущее вещество *n*

C119 **cemented steel** || Einsatzstahl *m* || acier *m* cémenté || gecarboneerd staal *n* || цементированная сталь *f*

C120 **center** || Zentrum *n*; Mittelpunkt *m* || centre *m* || centrum *n* || центр *m*; середина *f*

C121 **center drill** || Zentrierbohrer *m* || foret *m* à centrer || centreerboor *f (m)* || центровочное сверло *n*

C122 **centering** || Zentrieren *n* || centrage *m* || centrering *f* || **1.** центрирование *n* **2.** центрование *n*, центровка *f*

C123 **centering machine** || Zentriermaschine *f* || machine *f* à centrer || centreermachine *f* || центровальный станок *m*

C124 **center lathe** || Spitzendrehmaschine *f* || tour *m* parallèle || centerdraaibank *f* || центровый токарный станок *m*

C125 **centerless grinding machine** || spitzenlose Schleifmaschine *f* || machine *f* à rectifier sans centres || centerloze slijpmachine *f* || бесцентрово-шлифовальный станок *m*

C126 **center of gravity** || Schwerpunkt *m* || centre *m* de gravité || gravitatiecentrum *n* || центр *m* тяжести

C127 **center rest** || Mittelsupport *m* *(Werkzeugmaschine)* || lunette *f* fixe || vaste bril *m* *(draaibank)* || неподвижный люнет *m* *(токарного станка)*

C128 **centi-** || Zenti... || centi... || centi... || санти... *(приставка в наименовании дольной единицы, соответствующая 10^{-2})*

C129 **centigrade degree** || Grad *m* Celsius, Celsius-Grad *m*, Zentigrad *m* || degré *m* Celsius || graad *m* Celsius || градус *m* (по шкале) Цельсия

C130 **centigrade scale** || Celsius-Skale *f* || échelle *f* Celsius || Celsius-schaal *f* (*m*) || (температурная) шкала *f* Цельсия

C131 **centigrade temperature** || Celsius-Temperatur *f* || température *f* Celsius || Celsius-temperatuur *f* || температура *f* по шкале Цельсия, температура *f* по Цельсию

C132 **centimeter** || Zentimeter *n* || centimètre *m* || centimeter *m* || сантиметр *m* *(10^{-2} м)*

C133 **central** || zentral, Zentral... || central || centraal || центральный

C134 **central collision** || Zentralstoß *m* || choc *m* central || centrale schok *m* || центральный удар *m*

C135 **central force** || Zentralkraft *f* || force *f* centrale || centrale kracht *f* (*m*) || центральная сила *f*

C136 **central force field** || Zentralkraftfeld *n*, Zentralkräftefeld *n* || champ *m* de forces centrales || centraal krachtveld *n* || поле *n* центральных сил

C137 **central moment of inertia** || zentrales Trägheitsmoment *n* || moment *m* d'inertie central || centraal traagheidsmoment *n* || центральный момент *m* инерции

C138 **central oiling** || Zentralschmierung *f* || graissage *m* centralisé, lubrification *f* centralisée || centrale smering *f* || централизованная смазка *f*

C139 **central processor** || Zentraleinheit *f* || processeur *m* central || centrale processor *m* || центральный процессор *m*

C140 **centre** || Zentrum *n*; Mittelpunkt *m* || centre *m* || centrum *n*, middelpunt *n* || центр *m*

C141 **centre of curvature** || Krümmungsmittelpunkt *m* || centre *m* de courbure || krommingsmiddelpunt *n* || центр *m* кривизны

C142 **centre of force** || Kraftangriffspunkt *m* || point *m* d'application d'une force || krachtcentrum *n* || точка *f* приложения силы

C143 **centre of gravity** || Schwerpunkt *m* || centre *m* de gravité || zwaartepunt *n* || центр *m* тяжести

C144 **centre of gyration** || Drehpunkt *m*, Rotationspunkt *m* || centre *m* de rotation || draaipunt *n* || центр *m* (свободного) вращения

C145 **centre of inertia** || Trägheitsmittelpunkt *m*, Massenmittelpunkt *m* || centre *m* d'inertie [des masses] || traagheidsmiddelpunt *n*, massamiddelpunt *n* || центр *m* инерции, центр *m* масс

C146 **centre of lens** || optischer Mittelpunkt *m* einer Linse || centre *m* optique d'une lentille || optisch middelpunt *n* van lens || оптический центр *m* линзы

C147 **centre of mass** *see* **centre of inertia**

C148 **centre-of-mass system** || Massenmittelpunktsystem *n* || système *m* du centre de masses || massamiddelpuntsysteem *n* || система *f* отсчёта, связанная с центром масс, система *f* центра масс

C149 **centre of oscillation** || Schwingungsmittelpunkt *m* || centre *m* d'oscillations || slingerpunt *n* || центр *m* качаний *(физического маятника)*

C150 **centre of percussion** || Stoßzentrum *n* || centre *m* de choc || stootcentrum *n* || центр *m* удара

C151 **centre of symmetry** || Symmetriezentrum *n* || centre *m* de symétrie || symmetriecentrum *n* || центр *m* симметрии

C152 **centrifugal** || zentrifugal, Zentrifugal..., Fliehkraft... || centrifuge || centrifugaal || центробежный

C153 **centrifugal casting** || Schleuderguß *m* || coulée *f* centrifuge || centrifugaal gieten *n* || центробежное литьё *n*

C154 **centrifugal clutch** || Fliehkraftkupplung *f* || embrayage *m* centrifuge || centrifugaalkoppeling *f* || центробежное сцепление *n*

C155 **centrifugal compressor** || Kreiselverdichter *m* || compresseur *m* centrifuge || centrifugaal compressor *m* || центробежный компрессор *m*

C156 **centrifugal force** || Fliehkraft *f*, Zentrifugalkraft *f* || force *f* centrifuge || centrifugaalkracht *f* (*m*) || центробежная сила *f*

C157 **centrifugal lubrication** || Schleuderschmierung *f* || graissage *m* [lubrification *f*] centrifuge || centrifugale smering *f* || центробежная смазка *f*

C158 **centrifugal pump** || Kreiselpumpe *f* || pompe *f* centrifuge || centrifugaalpomp *f* (*m*) || центробежный насос *m*

C159 **centrifuge** || Zentrifuge *f*, Trennschleuder *f* || centrifugeuse *f* || centrifuge *f* (*m*) || центрифуга *f*

C160 **centripetal** || zentripetal, Zentripetal... || centripète || centripetaal || центростремительный

C161 **centripetal force** || Zentripetalkraft *f* || force *f* centripète || centripetale kracht *f* (*m*) || центростремительная сила *f*

C162 **centroid (line)** || Zentroide *f* || centroïde *f* || centroïde *f* || центроида *f*, полодия *f*

C163 **ceramic cutting tool** || Hartmetallwerkzeug *n*, Sinterkarbidwerkzeug *n* || outil *m* céramique || hardmetaalgereedschap *n* || режущий инструмент *m* из керметов

C164 **ceramic material** *see* **cermet**

C165 **ceramic tool** *see* **ceramic cutting tool**

C166 **cermet** || Kermet *m*, Cermet *m* || cermet *m*, céramique *f* à métal || cermet *n* || кермет *m* (*высокопрочный материал, полученный спеканием металлических и керамических порошков*)

C167 **certain event** || sicheres Ereignis *n* || événement *m* certain || zekere gebeurtenis *f* || достоверное событие *n*

C168 **cesium,** Cs || Zäsium *n* || césium *m* || cesium *m* || цезий *m*

C169 **chain** || Kette *f* || chaîne *f* || keten *f* (*m*) || цепь *f*

C170 **chain conveyor** || Kettenförderer *m* || transporteur *m* [convoyeur *m*] à chaîne || kettingtransporteur *m* || цепной конвейер *m*

C171 **chain gear(ing)** || Kettenübersetzung *f* || transmission *f* par chaîne || kettingdrijfwerk *n* || цепная передача *f*

C172 **chain pump** || Kettenpumpe *f* || pompe *f* à chaîne || kettingpomp *f* (*m*) || цепной насос *m*; водоотливная нория *f*

C173 **chain reaction** || Kettenreaktion *f* || réaction *f* en chaîne || kettingreactie *f* || цепная реакция *f*

C174 **chain wheel** || Kettenrad *n* || roue *f* à chaîne || kettingrad *n* || цепное колесо *n*; (цепная) звёздочка *f*

C175 **chamber** || Kammer *f* || chambre *f* || kamer *f* (*m*) || камера *f*

C176 **change gearbox** *see* **gear box**

C177 **change-gear lever** || Schalthebel *m* || levier *m* de changement de vitesse || handgreep *f* (*m*) van de snelheidsregelaar || рычаг *m* переключения скоростей

C178 **change of state** || Zustandsänderung *f*; Phasenübergang *m* || transformation *f* d'état [de phase] || faseovergang *m* || фазовый переход *m*, фазовое превращение *n*

C179 **change wheel** || Wechselrad *n* || pignon *m* démontable [de rechange] || wisselwiel *n* || сменное зубчатое колесо *n*, сменная шестерня *f*

C180 **channel** || Kanal *m* || canal *m* || kanaal *n* || канал *m*

C181 **channel iron** || U-Stahl *m* || profilé *m* [fer *m*] en U || U-staal *n*, U-balk *m* || швеллер *m*

C182 **channel(l)ing** || Kanalisieren *n*, Kanalieren *n* || canalisation *f* || kanaliseren *n* || каналирование *n* (*энергии, частиц*)

C183 **chaotic motion** || ungeordnete [chaotische] Bewegung *f* || mouvement *m* chaotique [en désordre, an hasard] || chaotische [wanordelijke] beweging *f* || беспорядочное [хаотическое] движение *n*

C184 I **characteristic** || Charakteristik *f*, Kennlinie *f* || caractéristique *f* || karakteristiek *f* || характеристика *f*

C185 II **characteristic** || charakteristisch || caractéristique || karakteristiek || характеристический

C186 **characteristic curve** || charakteristische Kurve *f*, Kennlinie *f* || courbe *f* caractéristique || karakteristieke kromme *f* (*m*) || характеристическая кривая *f*, характеристика *f*

C187 **characteristic direction** || charakteristische Richtung *f* || direction *f* caractéristique || karakteristieke richting *f* || характеристическое направление *n*

C188 **characteristic equation** || charakteristische Gleichung *f* || équation *f* caractéristique || karakteristieke vergelijking *f* || характеристическое уравнение *n*

C189 **characteristic function** || charakteristische Funktion *f* || fonction *f* caractéristique || karakteristieke functie *f* || характеристическая функция *f*

C190 **charakteristic radiation** || charakteristische Strahlung *f* || rayonnement *m* caractéristique || karakteristieke straling *f* || характеристическое излучение *n*

C191 **characteristic spectrum** || charakteristisches Spektrum *n* || spectre *m* caractéristique || karakteristiek spectrum *n* || характеристический спектр *m*

C192 **characteristic temperature** || charakteristische Temperatur *f* || température *f* caractéristique || karakteristieke temperatuur *f* || характеристическая *(дебаевская)* температура *f*

C193 **characteristic time** || charakteristische Zeit *f* || temps *m* caractéristique || karakteristieke tijd *m* || характерное [характеристическое] время *n*

C194 **characteristic vector** || Eigenvektor *m* || vecteur *m* caractéristique || eigenvector *m* || собственный вектор *m*; характеристический вектор *m*

C195 **characteristic X-rays** || charakteristische Röntgenstrahlung *f* || rayonnement *m* X caractéristique || karakteristieke röntgenstraling *f* || характеристическое рентгеновское излучение *n*

C196 **charge** || **1.** Ladung *f* **2.** Beanspruchung *f* || charge *f* || **1.** lading *f* **2.** last *f* (*m*) || **1.** заряд *m* **2.** нагрузка *f*

C197 **charge carrier** || Ladungsträger *m* || porteur *m* de la charge || ladingsdrager *m* || носитель *m* заряда

C198 **charged particle** || geladenes Teilchen *n* || particule *f* chargée [électrisée] || geladen deeltje *n* || заряженная частица *f*

C199 **charged particle accelerator** || Teilchenbeschleuniger *m* || accélérateur *m* des particules chargées || deeltjesversneller *m* || ускоритель *m* заряженных частиц

C200 **charge leak** || Ladungsverlust *m* || fuite *f* de charge || ladingsverlies *n* || утечка *f* заряда

C201 **charge-to-mass ratio** || spezifische Ladung *f*, Ladungs-Masse-Verhältnis *n* || charge *f* spécifique || verhouding *f* van lading tot massa || удельный заряд *m*, отношение *n* заряда к массе

C202 **charge transfer** || Ladungstransport *m* || transfert *m* de la charge || ladingsoverdracht *f* (*m*) || перенос *m* заряда

C203 **charging** || Laden *n*, Aufladen *n* || charge *f* || lading *f* || зарядка *f* *(напр. конденсатора)*

C204 **charging current** || Ladestrom *m*, Aufladestrom *m* || courant *m* de charge || laadstroom *m* || зарядный ток *m*

C205 **charging voltage** || Ladespannung *f*, Aufladespannung *f* || tension *f* de charge || laadspanning *f* || зарядное напряжение *n*

C206 **charm** || Charme *m*, Charmezahl *f* || charme *f* || charme *f* || чарм *m* *(характеристика кварка)*

C207 **chart** || Diagramm *n*; Kurvenbild *n* || diagramme *m*, graphique *m* || diagram *n*, grafiek *f* || график *m*

C208 **chaser** || Gewindestrehler *m*, Strehler *m* || peigne *m* à fileter || schroefbeitel *m*, draadbeitel *m* || резьбовая [резьбонарезная] гребёнка *f*

C209 **chasing lathe** *see* **screw cutting lathe**

C210 **chasing tool** *see* **chaser**

C211 **chassis** || Chassis *n*; Fahrgestell *n* || châssis *m* || chassis *n*, frame *n*; onderstel *n* || шасси *n*; ходовая часть *f*

C212 **Chebyshev polynomials** || Tschebyscheffsche Polynome *n pl* || polynômes *m pl* de Chebyshev || Tschebyscheff-polynomen *n pl* || многочлены *m pl* Чебышева

C213 **Chebyshev's unequality** || Tschebyscheffsche Ungleichung *f* || inégalité *f* de Chebyshev || Tschebyscheff-ongelijkheid *f* || неравенство *n* Чебышева

C214 **check** || Kontrolle *f*, Prüfung *f* || vérification *f*, contrôle *m* || beproeving *f*, controle *f* (*m*) || проверка *f*, контроль *m*

C215 **check gauge** || Prüflehre *f*, Prüfmaß *n* || jauge-étalon *m* || proefkaliber *n*, proefmaat *f* (*m*) || контрольный калибр *m*

C216 **check screw** || Stellschraube *f* || vis *f* de réglage [d'ajustage] || stelschroef *f* (*m*) || регулировочный винт *m*

C217 **check test** || Kontrollprüfung *f*, Test *m* || essai *m* [épreuve *f*] de contrôle || controle-beproeving *f*, toets *m* || контрольное испытание *n*

C218 **check valve** || **1.** Rückschlagventil *n*; Absperrventil *n* **2.** Kontrollventil *n* || **1.** soupape *f* de retenue; soupape *f* d'arrêt [de fermeture] **2.** soupape *f* de contrôle || **1.** afsluiter *m*; terugslagklep *f* (*m*) **2.** controleklep *f* || **1.** обратный клапан *m*; запорный [стопорный] клапан *m* **2.** контрольный клапан *m*

C219 **chelate compound** || Chelat(verbindung *f*) *n*, Chelatkomplex *m* || composé *m* chélaté, chélate *m* || chelaatverbinding *f* || хелатное [клешневидное, внутрикомплексное] соединение *n*, хелат *m*

C220 **chemical affinity** || chemische Affinität *f* || affinité *f* chimique || chemische affiniteit *f* || химическое сродство *n*

C221 **chemical binding, chemical bond** || chemische Bindung *f* || liaison *f* chimique || chemische binding *f* || химическая связь *f*

C222 **chemical compound** || chemische Verbindung *f* || composé *m* [combinaison *f*] chimique || chemische verbinding *f* || химическое соединение *n*

C223 **chemical energy** || chemische Energie *f* || énergie *f* chimique || chemische energie *f* || химическая энергия *f*

C224 **chemical equation** || chemische Gleichung *f* || équation *f* de réaction || reactievergelijking *f*, chemische vergelijking *f* || химическое уравнение *n*

C225 **chemical equilibrium** || chemisches Gleichgewicht *n* || équilibre *m* chimique || chemisch evenwicht *n* || химическое равновесие *n*

C226 **chemical formula** || chemische Formel *f* || formule *f* chimique || chemische formule *f* (*m*) || химическая формула *f*

C227 **chemical kinetics** || chemische Kinetik *f* || cinétique *f* chimique || chemische kinetica *f* || химическая кинетика *f*

C228 **chemical milling** || Formätzen *n* || attaque *f* en bas-relief || vormetsing *f* || рельефное травление *n*

C229 **chemical potential** || chemisches Potential *n* || potentiel *m* chimique || chemische potentiaal *m* || химический потенциал *m*

C230 **chemical properties** || chemische Eigenschaften *f pl* || propriétés *f pl* chimiques || scheikundige eigenschappen *f pl* || химические свойства *n pl*

C231 **chemical reaction** || chemische Reaktion *f* || réaction *f* chimique || chemische reactie *f* || химическая реакция *f*

C232 **chemiluminescence** || Chemilumineszenz *f*, Chemolumineszenz *f* || chimiluminescence *f* || chemoluminescentie *f* || хемилюминесценция *f*

C233 **chemisorption** || Chemisorption *f*, Chemosorption *f* || chimisorption *f* || chemosorptie *f* || хемосорбция *f*

C234 **chemistry** || Chemie *f* || chimie *f* || scheikunde *f*, chemie *f* || химия *f*

C235 **chevron gear** || Pfeilrad *n* || roue *f* à denture à chevrons || schuinvertand tandwiel *n* || шевронное зубчатое колесо *n*

C236 **chill casting** || Kokillenguß *m* || coulée *f* en coquille || coquillegieten *n* || литьё *n* в кокиль, кокильное литьё *n*

C237 **chilled cast iron** || Hartguß *m* || fonte *f* trempée || gehard gietijzer *n* || отбелённый чугун *m*

C238 **chill mo(u)ld** || Kokille *f* || coquille *f* à mouler || coquille *f* || кокиль *n*, металлическая (литейная) форма *f*

C239 **chip** || Span *m* || copeau *m* || spaan *f* (*m*) || стружка *f*

C240 **chip breaker** || Spanbrecher *m*, Spanbrechernut *f* || brise-copeaux *m* || spaanbreker *m* || стружколом *m*

C241 **chip compression ratio** || Spanstauchung *f* || retrait *m* du copeau || spaanzetting *f* || усадка *f* стружки

C242 **chisel** || Meißel *m*; Beitel *m*; Stemmeisen *n* || ciseau *m*; bédane *m*, burin *m* || beitel *m* || долото *n*; стамеска *f*; зубило *n*

C243 **chi-square distribution** || χ^2-Verteilung *f* || distribution *f* chicarré || χ^2-verdeling *f* || распределение *n* хи-квадрат, χ^2-распределение *n*

C244 **chi-square test** || χ^2**-Test** *m*, || test *m* chi-carré || χ^2-test *m* || критерий *m* хи-квадрат

C245 **chloric acid**, $HClO_3$ || Chlorsäure *f* || acide *m* chlorique || chloorzuur *n* || хлорноватая кислота *f*

C246 **chloride** || Chlorid *n* || chlorure *m* || chloride *n* || хлорид *m*

C247 **chlorine**, Cl || Chlor *n* || chlore *m* || chloor *m*, *n* || хлор *m*

C248 **chlorous acid**, $HClO_2$ || chlorige Säure *f* || acide *m* chloreux || chlorigzuur *n* || хлористая кислота *f*

C249 **choke** || Drossel *f*, Drosselklappe *f* || étrangleur *m* || gasklep *f* (*m*) || заслонка *f*; заглушка *f*

C250 **choke (coil)** || Drosselspule *f* || inductance *f* [bobine *f*] d'arrêt || smoorspoel *f* (*m*) || дроссельная катушка *f*, дроссель *m*

C251 **choking** || Drosseln *n*, Drosselung *f* || étranglement *m*; laminage *m* *(de vapeur)* || smoren *n* || дросселирование *n*

C252 **chopper** || Unterbrecher *m* || rupteur *m* || onderbreker *m* || прерыватель *m*

C253 **chord** || Sehne *f* || corde *f* || koorde *f* (*m*) || хорда *f*

C254 **chromatic aberration** || chromatische Aberration *f*, Farbabweichung *f* || aberration *f* chromatique || chromatische aberratie *f* || хроматическая аберрация *f*

C255 **chromatography** || Chromatografie *f* || chromatographie *f* || chromatografie *f* || хроматография *f*

C256 **chrome-manganese steel** || Chrom-Mangan-Stahl *m* || acier *m* au chrome-manganèse || chroommangaanstaal *n* || хромомарганцевая сталь *f*

C257 **chrome-molybdenum steel** || Chrom-Molybdän-Stahl *m* || acier *m* au chrome-molybdène || chroommolybdeenstaal *n* || хромомолибденовая сталь *f*

C258 **chrome-nickel steel** || Chrom-Nickel-Stahl *m* || acier *m* au chrome-nickel || chroomnikkelstaal *n* || хромоникелевая сталь *f*

C259 **chrome-plated** || galvanisch verchromt || chromé || verchroomd || хромированный

C260 **chrome steel** || Chromstahl *m* || acier *m* au chrome || chroomstaal *n* || хромистая сталь *f*

C261 **chromium**, Cr || Chrom *n* || chrome *m* || chromium *n* || хром *m*

C262 **chuck** || Futter *n*, Spannfutter *n* || mandrin *m* (de serrage) || klauwplaat *f* (*m*) || (зажимной) патрон *m*

C263 **circle** || Kreis *m* || cercle *m* || cirkel *m* || круг *m*; окружность *f*

C264 **circle of curvature** || Krümmungskreis *m* || cercle *m* de courbure || krommingscirkel *m* || круг *m* кривизны

C265 **circle shears** || Kreisschere *f* || cisaille *f* circulaire || cirkelschaar *f* (*m*) || дисковые ножницы *pl*

C266 **circuit** || 1. Kreis *m*, Stromkreis *m* 2. Schaltung *f* 3. (geschlossener) Kantenzug *m (Graph)* || circuit *m* 1. stroomkring *m*, stroomketen *f (m)* 2. schakeling *f* 3. cyclus *m (graf)* || 1. цепь *f*; контур *m*; схема *f* 2. *мат.* цикл *m (графа)*

C267 **circuit analysis** || Netzwerkanalyse *f*, Stromkreisanalyse *f* || analyse *f* de circuit || netwerkanalyse *f* || схемный анализ *m*; теория *f* цепей

C268 **circuitry** || elektrische Schaltung *f* || circuit *m* électrique || elektrische schakeling *f* || электрическая схема *f*

C269 **circuit synthesis** || Netzwerksynthese *f* || synthèse *f* de circuit || netwerksynthese *f* || синтез *m* цепей; схемный синтез *m*

C270 **circular accelerator** || Kreisbeschleuniger *m*, Ringbeschleuniger *m* || accélérateur *m* circulaire || kringversneller *m* || кольцевой ускоритель *m*

C271 **circular loop** || kreisförmige Windung *f* || spire *f* circulaire || kringvormige winding *f* || круглый виток *m* (с током)

C272 **circularly polarized wave** || zirkular polarisierte Welle *f* || onde *f* à polarisation circulaire || circulair gepolariseerde golf *f (m)* || циркулярно-поляризованная волна *f*, волна *f* с круговой поляризацией

C273 **circular motion** || Kreisbewegung *f*, kreisförmige Bewegung *f* || mouvement *m* circulaire || kringbeweging *f* || круговое движение *n*, движение *n* по окружности

C274 **circular orbit** || Kreisbahn *f*, kreisförmige Umlaufbahn *f* || orbite *f* circulaire || cirkelbaan *f (m)* || круговая орбита *f*

C275 **circular pitch** || Wälzkreisteilung *f* || pas *m* circonférentiel d'engrenage || tandsteek *m* || окружной шаг *m* зацепления

C276 **circular polarization** || zirkulare Polarisation *f*, Zirkularpolarisation *f* || polarisation *f* circulaire [rotatoire] || circulaire polarisatie *f* || круговая [циркулярная] поляризация *f*

C277 **circular saw** || Kreissäge *f* || scie *f* circulaire || circelzaag *f (m)* || дисковая пила *f*

C278 **circular sawing machine** || Kreissägemaschine *f*, Kreissäge *f* || machine *f* à scier circulaire || cirkelzaagmachine *f* || круглопильный станок *m*

C279 **circular velocity** || Kreisbahngeschwindigkeit *f* || vitesse *f* circonférentielle [périphérique], vitesse *f* circulaire || kringbewegingssnelheid *f* || окружная [круговая] скорость *f*

C280 **circular vortex** || zirkularer Wirbel *m*, Kreiswirbel *m* || tourbillon *m* circulaire || circulaire wervelwind *m* || кольцевой вихрь *m*, вихревое кольцо *n*

C281 **circular wave** || Kreiswelle *f* || onde *f* circulaire || cirkelvormige golf *f (m)* || кольцевая [круговая] волна *f*

C282 **circulating current** || Kreisstrom *m*, Ringstrom *m* || courant *m* circulaire || cirkelstroom *m* || циркулирующий ток *m*

C283 **circulation** || Zirkulation *f*, Kreislauf *m*, Umlauf *m* || circulation *f* || circulatie *f* || циркуляция *f*

C284 **circulation flow** || Zirkulationsströmung *f* || écoulement *m* circulatoire || circulatiestroming *f* || циркуляционное течение *n*, обтекание *n* с циркуляцией

C285 **circumferential speed, circumferential velocity** || Umfangsgeschwindigkeit *f* || vitesse *f* circonférentielle [périphérique] || omtreksnelheid *f* || окружная скорость *f*

C286 **circumscribed circle** || umbeschriebener Kreis *m* || cercle *m* circonscrit || omgeschreven circel *m* || описанная окружность *f*

C287 **circumscribed triangle** || umbeschriebenes Dreieck *n* || triangle *m* circonscrit || omgeschreven driehoek *m* || описанный треугольник *m*

C288 **cissoid** || Zissoide *f* || cissoïde *f* || cissoïde *f* || циссоида *f*

C289 **clack valve** *see* **flap valve**

C290 **clamp** || Klammer *f*; Bügel *m* || crampon *m*, clameau *m*; collier *m*; étrier *m* || klamp *m*; klem *f (m)* || зажим *m*; скоба *f*; хомут *m*

C291 **clamping device** || Spannvorrichtung *f* || dispositif *m* de serrage || spaninrichting *f* || зажимное приспособление *n*

C292 **clamping screw** || Klemmschraube *f* || vis *f* de serrage || klembout *m* || зажимной винт *m*

C293 **clapper box** *see* **flapper box**

C294 **classical mechanics** || klassische Mechanik *f* || mécanique *f* classique [rationnelle] || klassieke mechanica *f* || теоретическая [классическая] механика *f*

C295 **classical physics** || klassische Physik *f* || physique *f* classique || klassieke fysica *f* || классическая физика *f*

C296 **classical statistics** || klassische Statistik *f*, Boltzmann-Statistik *f* || statistique *f* classique || klassieke statistiek *f* || классическая статистика *f*, статистика *f* Больцмана

C297 **classical theory** || klassische Theorie *f* || théorie *f* classique || klassieke theorie *f* || классическая теория *f*

C298 **clathrate compound** || Klathrat(verbindung *f*) *n*, Käfigeinschlußverbindung *f* || clathrate *m* || clathraat *n* || клатрат *m*, соединение *n* включения

C299 **clause** || Klausel *f* || clause *f* || clausule *f* (*m*) || условие *n*; оговорка *f*

C300 **clearance** || **1.** Spiel *n*; Spalt *m*; Abstand *m* **2.** Löschen *n* *(Zähler)* || **1.** jeu *m*; espace *m* libre **2.** effacement *m* || **1.** tussenruimte *f*, speling *f* **2.** uitwissing *f* || **1.** зазор *m*; промежуток *m* **2.** гашение *n* *(счётчика)*

C301 **clearance angle** || Freiwinkel *m* || (angle *m* de) dépouille *f* *(outil coupant)* || stelhoek *m* *(beitel)* || задний угол *m* *(режущего инструмента)*

C302 **clearance fit** || Laufsitz *m*, Spielsitz *m*, Spielpassung *f* || ajustement *m* avec jeu [tournant] || losse passing *f* || подвижная посадка *f*

C303 **cleavage strength** || Spaltfestigkeit *f* || résistance *f* au fendage || splitsingsweerstand *m* || сопротивление *n* раскалыванию

C304 **clock** || Zeitmesser *m*; Uhr *f*; Chronometer *n* || horloge *f*, montre *f*; chronomètre *m* || uurwerk *n*, horloge *n*, klok *f*; chronometer *m* || часы *pl;* хронометр *m*

C305 **clock(-)paradox** || Uhrenparadoxon *n* || paradoxe *m* des horloges || klokparadox *m* || парадокс *m* часов, парадокс *m* близнецов *(в теории относительности)*

C306 **clock-pulse generator** || Taktimpulsgeber *m* || générateur *m* de rythme || tactimpulsgever *m*, tactimpulsgenerator *m* || генератор *m* тактовых импульсов

C307 **clockwise direction** || Uhrzeigersinn *m* || sens *m* [direction *f*] horaire || uurwijzerrichting *f* || направление *n* по часовой стрелке

C308 **clockwise-polarized wave** || rechtspolarisierte Welle *f* || onde *f* polarisée à droite || rechtsgepolariseerde golf *f* (*m*) || правополяризованная волна *f*

C309 **closed circuit** || geschlossener Kreis *m* || circuit *m* fermé [bouclé] || gesloten kring *m* || замкнутая цепь *f*

C310 **closed interval** || abgeschlossenes Intervall *n* || intervalle *m* fermé || gesloten interval *n* || замкнутый интервал *n*

C311 **closed loop** || **1.** geschlossener Stromkreis *m* **2.** geschlossene Schleife *f* || **1.** circuit *m* fermé [bouclé] **2.** boucle *f* fermée || **1.** gesloten stroomkring *m* **2.** gesloten lus *f*(*m*) || **1.** замкнутый контур *m* (с током), виток *m* тока

C312 **closed set** || abgeschlossene Menge *f* || ensemble *m* fermé || gesloten menigte *f* || замкнутое множество *n*

C313 **closed shell** || abgeschlossene [vollbesetzte] Schale *f* || couche *f* complète [saturée] || bezette schil *f* (*m*) || заполненная (электронная) оболочка *f*

C314 **closed-shell atom** || Atom *n* mit Elektronenschalen || atome *m* à couches saturées || atoom *n* met bezette schillen || атом *m* с заполненными электронными оболочками

C315 **closed system** || abgeschlossenes System *n* || circuit *m* fermé || gesloten systeem *n* || замкнутая система *f*

C316 **close-packed structure** || dichtgepackte Struktur *f*, dichte Packungsstruktur *f* || structure *f* compacte [serrée] || dichte pakkingstructuur *f* || плотноупакованная структура *f* *(кристалла)*

C317 **close tolerance** || enge Toleranz *f* || tolérance *f* serrée || vaste tolerantie *f* || малый [жёсткий] допуск *m*

C318 **closing switch** || Hebelschalter *m*, Messerschalter *m* || interrupteur *m* à couteaux || messchakelaar *m* || рубильник *m*

C319 **closure** || *Math.* **1.** Abschluß *m*, Abschließen *n* **2.** Abgeschlossenheit *f* || *math.* fermeture *f* || *math.* **1.** sluiting *f* **2.** afgeslotenheid *f* || *мат.* **1.** замыкание *n* **2.** замкнутость *f*

C320 **cloud chamber** || Nebelkammer *f* || chambre *f* de Wilson || Wilsoncamera *f* *(m)* || камера *f* Вильсона

C321 **cluster** || Cluster *m*, Schwarm *m*, Haufen *m* || essaim *m* de particules || bundel *m* || сгусток *m*, кластер *m* *(напр. частиц)*

C322 **cluster mill** || Vielwalzen-Walzwerk *n* || laminoir *m* multicylindre || walserij *f* met vele rollen || многовалковый прокатный стан *m*

C323 **clutch** || Kupplung *f* || embrayage *m*, manchon *m* à embrayage || mof *f* *(m)*; koppeling *f* || муфта *f*; сцепление *n*

C324 **clutch coupling** || Kupplungsverbindung *f*, Kupplung *f* || embrayage *m* || mofverbinding *f* || муфтовое соединение *n*

C325 **clutch plate** || Kupplungsscheibe *f* || disque *m* d'embrayage || koppelingsplaat *f* *(m)* || диск *m* сцепления

C326 **clutch release sleeve** || Kupplungsausrückmuffe *f* || manchon *m* de débrayage || ontkoppelingsmof *f* *(m)* || муфта *f* выключения сцепления

C327 **CNC [computer numerical control] machine** || CNC-Werkzeugmaschine *f* || machine-outil *f* commandée par un ordinateur numérique || CNC-werktuigmachine *f* || станок *m* с числовым программным управлением от ЦВМ [с микропроцессорным ЧПУ]

C328 **coacervation** || Koazervation *f*, Koazervierung *f* || coacervation *f* || coacervatie *f* || коацервация *f*

C329 **coagulation** || Koagulieren *n*, Koagulation *f* || coagulation *f*, floculation *f* || coagulatie *f* || коагуляция *f*

C330 **coaxial** || koaxial, Koaxial... || coaxial || coaxiaal || коаксиальный, соосный

C331 **coaxial cable** || Koaxialkabel *n* || câble *m* coaxial || coaxkabel *m* || коаксиальный кабель *m*

C332 **cobalt,** Co || Kobalt *n* || cobalt *m* || cobalt *n*, kobalt *n* || кобальт *m*

C333 **cock** || Hahn *m* || robinet *m* || kraan *f* *(m)* || кран *m* *(трубопроводный)*

C334 **code** || Kode *m*, Code *m* || code *m* || code *m* || код *m*

C335 **coefficient** || Koeffizient *m*, Faktor *m*; Beiwert *m* || coefficient *m*, facteur *m*; rapport *m* || coëfficiënt *m* || коэффициент *m*

C336 **coefficient of cubical expansion** || kubischer Ausdehnungskoeffizient *m* || coefficient *m* de dilatation cubique || kubische uitzettingscoëfficiënt *m* || температурный коэффициент *m* объёмного расширения

C337 **coefficient of efficiency** || Wirkungsgrad *m* || (coefficient *m* de) rendement *m* || rendement *n* || коэффициент *m* полезного действия

C338 **coefficient of elasticity** *see* **modulus of elasticity**

C339 **coefficient of expansion** || **1.** *s.* coefficient of cubical expansion **2.** *Math.* Entwicklungskoeffizient *m* || **1.** *v.* coefficient of cubical expansion **2.** *math.* coefficient *m* de développement || **1.** *zie* coefficient of cubical expansion **2.** *math.* ontwikkelingscoëfficiënt *m* || **1.** *см.* coefficient of cubical expansion **2.** *мат.* коэффициент *m* разложения

C340 **coefficient of friction** || Reibungskoeffizient *m*, Reibungszahl *f*, Reibwert *m* || coefficient *m* de frottement [de friction] || wrijvingscoëfficiënt *m* || коэффициент *m* трения

C341 **coefficient of linear expansion** || linearer Ausdehnungskoeffizient *m* || coefficient *m* de dilatation linéare || lineaire uitzettingscoëfficiënt *m* || температурный коэффициент *m* линейного расширения

C342 **coefficient of performance** *see* **coefficient of efficiency**

C343 **coefficient of viscosity** || Viskositätskoeffizient *m*, Viskositätskonstante *f* || coefficient *m* de viscosité || viscositeitscoëfficiënt *m* || коэффициент *m* вязкости

C344 **coercive force** || Koerzitivkraft *f* || force *f* coercitive || coërcitiefkracht *f* (*m*) || коэрцитивная сила *f*

C345 **cofactor** || **1.** Factor *m* **2.** Adjunkte *f*, algebraisches Komplement *n* || cofacteur *m* || cofactor *m* || **1.** сомножитель *m* **2.** адъюнкта *f*

C346 **cogging mill** *see* **blooming mill**

C347 **coherence** || Kohärenz *f* || cohérence *f* || coherentie *f* || когерентность *f*

C348 **coherent** || kohärent, Kohärenz... || cohérent || coherent || когерентный

C349 **coherent radiation** || kohärente Strahlung *f* || radiation *f* cohérente || coherente straling *f* || когерентное излучение *n*

C350 **coherent waves** || kohärente Wellen *f pl* || ondes *f pl* cohérentes || coherente golven *f* (*m*) *pl* || когерентные волны *f pl*

C351 **cohesion** || Kohäsion *f* || kohésion *f* || cohesie *f* || когезия *f*, сцепление *n*

C352 **cohesive force** || Kohäsionskraft *f* || force *f* de cohésion || cohesiekracht *f* (*m*) || сила *f* сцепления

C353 **coil** || Spule *f* || bobine *f* || spoel *f* (*m*) || катушка *f* (индуктивности)

C354 **coil (friction) clutch** || Reibungskupplung *f* mit Schraubenfeder || accouplement *m* à friction avec un ressort hélicoïdal || wrijvingsmof *m* met een spiraalveer *f* (*m*) || фрикционная муфта *f* с винтовой пружиной

C355 **cold-cathode** || Kaltkatoden... || à cathode froide || koude-kathode... || ненакальный, с холодным катодом

C356 **cold-drawn steel** || kaltgezogener Stahl *m* || acier étiré *m* [tréfilé] à froid || koudgetrokken staal *n* || холоднотянутая сталь *f*

C357 **cold forging** || Kaltschmieden *n* || forgeage *m* à froid || koud smeden *n* || холодная ковка *f*

C358 **cold forming** || Kaltumformen *n* || formage *m* [travail *m*] à froid || koudvormen *n* || холодная обработка *f* давлением, холодная обработка *f* с формоизменением

C359 **cold light** || kaltes Leuchten *n*, Lumineszenz *f* || luminescence *f*, émission *f* luminescente || luminescentie *f* || люминесценция *f*, холодное свечение *n*

C360 **cold-rolled** || kaltgewalzt || laminé à froid || koud gewalst || холоднокатаный

C361 **cold-rolling mill** || Kaltwalzwerk *n* || laminoir *m* à froid || koudwalserij *f* || стан *m* холодной прокатки

C362 **cold welding** || Kaltschweißen *n* || soudage *m* (par pression) à froid || koudlassen *n* || холодная сварка *f* (давлением)

C363 **cold working** || Kaltumformen *n*, Kaltbearbeitung *f* || usinage *m* [travail *m*] à froid || mechanische bewerking *f* || холодная [механическая] обработка *f*

C364 **collapsar** || schwarzes Loch *n* || trou *m* noir || zwart gat *n* || коллапсар *m*, чёрная дыра *f*

C365 **collapsing tap** || selbstauslösender Gewindeschneidkopf *m* || taraud *m* ajustable [réglable] || uittrekbare schroefdraadtap *m* || раздвижной метчик *m*, метчик *m* с убирающимися резьбовыми гребёнками

C366 **collar** || Ring *m*; Bund *m* || anneau *m*; bague *f*; collet *m*; collier *m*; bride *m* || ring *m*; kraag *m*; askraag *m* || кольцо *n*; буртик *m*; заплечик *m*

C367 **collar bearing** || Halslager *n* || pallier *m* à collets || halskussenblok *n* || подшипник *m* с упорным кольцом

C368 **collective interaction** || kollektive Wechselwirkung *f* || interaction *f* collective || collectieve wisselwerking *f* || коллективное взаимодействие *n*

C369 **collector** || Kollektor *m*; Kommutator *m*, Stromwender *m* || collecteur *m* || collector *m* || коллектор *m*; токосниматель *m*

C370 **collet chuck** || Zangenspannfutter *n* || mandrin *m* à pince || tangvormige klauwplaat *f* (*m*) || цанговый патрон *m*

C371 **colliding beams** || kollidierende Bündel *n pl* || faisceaux *m pl* incidents || botsende bundels *m pl* || встречные пучки *m pl* (*частиц в ускорителе*)

C372 **collimated beam** || kollimiertes Bündel *n* || faisceau *m* collimé || gecollimeerde bundel *m* || коллимированный пучок *m*

C373 **collimator** || Kollimator *m* || collimateur *m* || collimator *m* || коллиматор *m*

C374 **collinear** || kollinear || colinéaire || collineair || коллинеарный

C375 **collision** || Zusammenstoß *m*, Kollision *f* || collision *f* || botsing *f* || соударение *n*, столкновение *n*

C376 **collisional ionization** || Stoßionisation *f* || ionisation *f* par choc || stootionisatie *f* || ударная ионизация *f*

C377 **collocation** || Kollokation *f* || collocation *f* || collocatie *f* || коллокация *f*

C378 **colloid** || Kolloid *n* || colloïde *m* || colloïde *f* (*m*) || коллоид *m*

C379 **colorimetry** || Kolorimetrie *f*, Farbmessung *f* || colorimétrie *f* || colorimetrie *f* || колориметрия *f*

C380 **colour balance** || Farbgleichgewicht *n* || équilibre *m* des couleurs || kleurenevenwicht *n* || баланс *m* цветов, цветовое равновесие *n*

C381 **colour centre** || Farbzentrum *n*, F-Zentrum *n* || centre *m* de couleurs || kleurcentrum *n* || центр *m* окраски, F-центр *m*

C382 **colour contrast** || Farbkontrast *m* || contraste *m* des couleurs || kleurcontrast *n* || цветовой контраст *m*

C383 **colour filter** || Farbfilter *n* || filtre *m* coloré || kleur(en)filter *m*, *n* || цветофильтр *m*, (цветной) светофильтр *m*

C384 **colour mixture** || Farbmischung *f* || mélange *m* des couleurs || kleurenmengeling *f*, kleurmenging *f* || смесь *f* цветов

C385 **column** || **1.** Säule *f*; Ständer *m* **2.** Spalte *f* (*einer Matrix*) || **1.** colonne *f*; poteau *m* **2.** colonne *f* || **1.** zuil *f* (*m*); kolom *f* (*m*); pilaar *m* **2.** kolom *f* (*m*) || **1.** колонна *f*; стойка *f* **2.** столбец (*матрицы*)

C386 **column matrix** || Spaltenmatrix *f* || matrice-colonne *f* || matrijskolom *f* (*m*) || матрица-столбец *m*

C387 **coma** || Koma *n* || coma *m* || coma *f* || кома *f*

C388 **combination** *math.* || Kombination *f* || combinaison *f* || combinatie *f* || сочетание *n*

C389 **combination tool block** *see* **multiple tool block**

C390 **combination turbine** || Gleichdruck-Überdruckturbine *f* || turbine *f* mixte (*à action et réaction*) || gecombineerde turbine *f* || комбинированная (*активно-реактивная*) турбина *f*

C391 **combinatorial analysis** || kombinatorische Analyse *f* || analyse *f* combinatoire || combinatorische analyse *f* || комбинаторный анализ *m*

C392 **combinatory logic** || kombinatorische Logik *f* || logique *f* combinatoire || combinatorische logica *f* || комбинаторная логика *f*

C393 **combined oscillations** || Kombinationsschwingungen *f pl* || oscillations *f pl* combinées || combinatietrillingen *f pl* || комбинационные колебания *n pl*

C394 **combined parity** || kombinierte Parität *f* || parité *f* combinée || gecombineerde pariteit *f* || комбинированная чётность *f*

C395 **combustible gas** || Brenngas *n*; Heizgas *n* || gaz *m* combustible || brandbaar gas *n* || горючий газ *m*

C396 **combustible mixture** || brennbares Gemisch *n*, Brenngemisch *n*; Gas-Luft-Gemisch *n* || mélange *m* combustible || brandbaar mengsel *n* || горючая смесь *f*; рабочая смесь *f*

C397 **combustion** || Verbrennung *f* || combustion *f* || branden *n*, verbranding *f* || горение *n*; сгорание *n*

C398 **combustion chamber** || Brennkammer *f*; Verbrennungsraum *m* || chambre *f* de combustion || verbrandingskamer *f* (*m*) || камера *f* сгорания

C399 **combustion engine** || Verbrennungsmotor *m* || moteur *m* à combustion interne [à explosion] || verbrandingsmotor *m* || двигатель *m* внутреннего сгорания

C400 **combustion heat** || Verbrennungswärme *f* || chaleur *f* de combustion || verbrandingswarmte *f* || теплота *f* сгорания

C401 **combustion stroke** || Arbeitstakt *m* || course *f* active [de détente] || expansieslag *m* || рабочий ход *m*, рабочий такт *m*; ход *m* [такт *m*] расширения

C402 **combustion temperature** || Verbrennungstemperatur *f* || température *f* de combustion || verbrandingstemperatuur *f* || температура *f* сгорания

C403 **command** || Befehl *m*, Instruktion *f* || commande *f* || commando *n* || команда *f*

C404 **commensurable** || kommensurabel; vergleichbar || commensurable || commensurabel || соизмеримый

C405 **common area** || gemeinsames Gebiet *n* || aire *f* [zone *f*] commune || gemeenschappelijk gebied *n* || общая область *f*

C406 **common divisor** || gemeinsamer Teiler *m* || diviseur *m* commun || gemene deler *m* || общий делитель *m*

C407 **commutation** || **1.** Kommutierung *f*, Kommutation *f*, Stromwendung *f* **2.** Umschalten *n*, Schalten *n* || commutation *f* || commutatie *f* || коммутация *f*; переключение *n*

C408 **commutative algebra** || kommutative Algebra *f* || algèbre *f* commutative || commutatieve algebra *f* (*m*) || коммутативная алгебра *f*

C409 **commutative field** || kommutatives Feld *n* || corps *m* commutatif || commutatief veld *n* || коммутативное поле *n*

C410 **commutative group** || kommutative Gruppe *f* || groupe *m* commutatif || commutatieve groep *f* (*m*) || коммутативная группа *f*

C411 **commutative law** || Kommutativgesetz *n*, Kommutationsgesetz *n* || loi *f* commutative || commutative wet *f* (*m*) || коммутативный закон *m*

C412 **commutativity** || Kommutativität *f* || commutativité *f* || commutativiteit *f* || коммутативность *f*

C413 **commutator** || **1.** Kommutator *m*, Stromwender *m* **2.** Umschalter *m*, Schalter *m* || commutateur *m* || commutator *m* || коммутатор *m*

C414 **commuting operators** || vertauschbare Operatoren *m pl* || opérateurs *m pl* commutatifs || commutatieve operatoren *m pl* || перестановочные операторы *m pl*

C415 **compact group** || kompakte Gruppe *f* || groupe *m* compact || compacte groep *f* (*m*) || компактная группа *f*

C416 **compactness** *math.* || Kompaktheit *f*; Bikompaktheit *f* || compacité *f*; bicompacité *f* || compactheid *f*; bicompactheid *f* || компактность *f*; бикомпактность *f*

C417 **compact set** || kompakte Menge *f* || ensemble *m* compact || compacte menigte *f* || компактное множество *n*

C418 **compact space** || kompakter Raum *m* || espace *m* compact || compacte ruimte *f* || компактное пространство *n*

C419 **comparator** || Komparator *m*, Vergleicher *m* || comparateur *m* || comparator *m* || компаратор *m*

C420 **comparison bridge** || Wheatstonesche Brücke *f*, Wheatstone-Brücke *f* || pont *m* de Wheatstone, pont *m* de mesure || brug *f* (*m*) van Wheatstone || мост *m* Уитстона, измерительный мост *m*

C421 **compatibility** *math.* || Kompatibilität *f* || compatibilité *f* || compatibiliteit *f* || совместность *f*, совместимость *f*

C422 **compatibility conditions** || Kompatibilitätsbedingungen *f pl* || conditions *f pl* de compatibilité || compatibiliteitsvoorwaarden *f pl* || условия *n pl* совместности

C423 **compensator** || Kompensator *m*, Ausgleicher *m* || compensateur *m* || compensator *m* || компенсатор *m*

C424 **competitive processes** || konkurrierende Prozesse *m pl*, Konkurrenzprozesse *m pl* || processus *m pl* compétitifs || concurrentieprocessen *n pl* || конкурирующие процессы *m pl*

C425 **compilation** || Kompilieren *n*, Übersetzen *n* || compilation *f* || compilatie *f* || компиляция *f*, трансляция *f*

C426 **compiler, compiling program, compiling routine** || Kompilierer *m*, Compiler *m* || compilateur *m*, programme *m* de compilation || compiler *m* || компилятор *m*, транслятор *m*, компилирующая программа *f*

C427 **complement** || Komplement *n* || complément *m* || complement *n* || дополнение *n*

C428 **complementarity principle** || Komplementaritätsprinzip *n* || principe *m* de complémentarité || complementariteitsprincipe *n* || принцип *m* дополнительности (Бора)

C429 **complementary colour** || Komplementärfarbe *f*, Ergänzungsfarbe *f* || couleur *f* complémentaire || complementaire kleur *f* (*m*) || дополнительный цвет *m*

C430 **complementation** || Komplementieren *n* || complémentation *f* || complementeren *n* || дополнение *n*

C431 **complement of an event** || Komplement *n* eines Ereignisses || complément *m* d'un événement || complement *n* van een gebeurtenis || дополнение *n* события

C432 **complement of a set** || Komplement *n* einer Menge || complément *m* d'un ensemble || complement *n* van een menigte || дополнение *n* множества

C433 **complete additivity** || vollständige Additivität *f* || additivité *f* complète || volledige additiviteit *f* || полная аддитивность *f*

C434 **complete induction** || vollständige [mathematische] Induktion *f* || induction *f* complète || volledige inductie *f* || полная [математическая] индукция *f*

C435 **complete integral** || vollständiges Integral *n* || intégrale *f* complète || volledige integraal *f* (*m*) || полный интеграл *m* *(общее решение дифференциального уравнения)*

C436 **completely defined** || überall definiert || défini complètement, défini partout || overal gedefiniëerd || всюду определённый

C437 **completeness** || Vollständigkeit *f* || complétude *f* || volledigheid *f* || полнота *f*

C438 **completeness axiom** || Vollständigkeitsaxiom *n* || axiome *m* de complétude || volledigheidsaxioma *n* || аксиома *f* полноты

C439 **complete radiator** || Schwarzkörper *m*, Schwarzstrahler *m* || corps *m* noir, radiateur *m* intégral || zwart lichaam *n* || абсолютно чёрное тело *n*

C440 **complete revolution** || Gesamtdrehung *f* || révolution *f* || totale omwenteling *f* || полный оборот *m*

C441 **complete solution** || vollständige Lösung *f* || solution *f* complète || volledige oplossing *f* || полное решение *n*

C442 **complete space** || vollständiger Raum *m* || espace *m* complet || volledige ruimte *f* || полное пространство *n*

C443 **complete system** || vollständiges System *n* || système *m* complet || volledig systeem *n* || полная система *f*

C444 **complete thread** || Vollprofilgewinde *n* || filet *m* plein || schroefdraad *m* met volledig profiel || резьба *f* полного профиля

C445 **complex** || **komplex** || complexe || complex || комплексный

C446 **complex amplitude** || komplexe Amplitude *f*, Komplexamplitude *f* || amplitude *f* complexe || complexe amplitude *f* || комплексная амплитуда *f*

C447 **complex-conjugate matrix** || konjugiert komplexe Matrix *f* || matrice *f* complexe-conjuguée || complex-geconjugeerde matrix *f* (*m*) || комплексно-сопряжённая матрица *f*

C448 **complex impedance** || komplexer Scheinwiderstand *m* || impédance *f* complexe || complexe impedantie *f* || комплексный импеданс *m*

C449 **complex number** || komplexe Zahl *f* || nombre *m* complexe || complex getal *n* || комплексное число *n*

C450 **complex quantity** || komplexe Größe *f* || quantité *f* [valeur *f*] complexe || complexe grootheid *f* || комплексная величина *f*

C451 **compliance** || reziproke Steifigkeit *f*, Nachgiebigkeit *f* || déformabilité *f*, compressibilité *f*, flexibilité *f* || elastische vervormbaarheid *f* || податливость *f*

C452 **component** || 1. Komponente *f*; Bestandteil *m* 2. Komponente *f* || 1. composant *m* 2. composante *f* || 1. component *m*, bestanddeel *n* 2. component *m* || 1. компонент *m*; элемент *m* 2. составляющая *f*; компонента *f*

C453 **composite lens** || zusammengesetzte Linse *f*, Verbundlinse *f* || lentille *f* complexe || samengestelde lens *f* (*m*) || сложная [составная] линза *f*

C454 **composition** || Struktur *f*, Aufbau *m*; Zusammensetzung *f* || composition *f*, structure *f* || structuur *f*; compositie *f* || состав *m*; структура *f*, строение *n*; композиция *f*

C455 **compound nucleus** || Compoundkern *m*, Verbundkern *m* || noyau *m* composé [intermédiaire] || compoundkern *f* || составное ядро *n*, компаунд-ядро *n*

C456 **compound slide rest** || Kreuzsupport *m* || support *m* à chariot croisé || kruissupport *n* || крестовый суппорт *m*

C457 **compound state** || Zwischenzustand *m*, Compoundzustand *m*, Verbundzustand *m* || état *m* intermédiaire || tussentoestand *m* || промежуточное состояние *n* ядра, компаунд-состояние *n*

C458 **compound winding** || Compoundwicklung *f*, Doppelschlußwicklung *f* || enroulement *m* compound || compoundwikkeling *f* || компаундная обмотка *f* (возбуждения)

C459 **compressed gas** || Druckgas *n*, Preßgas *n* || gaz *m* comprimé || persgas *n*, drukgas *n* || сжатый газ *m*

C460 **compressibility** || Kompressibilität *f* || compressibilité *f* || compressibiliteit *f* || сжимаемость *f*

C461 **compression** || Kompression *f*, Verdichtung *f*, Zusammendrückung *f* || compression *f* || compressie *f*, samendrukking *f* || сжатие *n*

C462 **compression-dilatation wave** || Verdichtungs-Verdünnungs-Welle *f* || onde *f* de compression-dépression || verdichtings-verdunningsgolf *f* (*m*) || волна *f* сжатия — разрежения

C463 **compression ignition engine** || Motor *m* mit Selbstzündung, Dieselmotor *m* || moteur *m* à allumage par compression, moteur *m* diesel || dieselmotor *m* || двигатель *m* с воспламенением от сжатия, дизель *m*

C464 **compression ratio** || Kompressionsgrad *m* || taux *m* de compression || compressieverhouding *f*, verdichtingsgraad *m* || степень *f* сжатия

C465 **compression spring** || Druckfeder *f* || ressort *m* à compression || drukveer *f* (*m*) || пружина *f* сжатия

C466 **compression stroke** || Verdichtungstakt *m* || course *f* de compression || compressieslag *m* || ход *m* [такт *m*] сжатия

C467 **compression test** || Druckversuch *m*, Druckprobe *f* || essai *m* de compression || drukproef *f* (*m*) || испытание *n* на сжатие

C468 **compression wave** || Verdichtungswelle *f*, Kompressionswelle *f* || onde *f* de compression || verdichtingsgolf *f* (*m*) || волна *f* сжатия

C469 **compressive force** || Druckkraft *f*; Kompressionskraft *f* || effort *m* de compression || drukkracht *f* (*m*) || сжимающее усилие *n*

C470 **compressive strain** || Druckverformung *f*, Druckdeformation *f* || déformation *f* de compression || drukvervorming *f* || относительная деформация *f* сжатия

C471 **compressive strength** || Druckfestigkeit *f* || résistance *f* à la compression || samendrukkingsgrens *f* (*m*) || временное сопротивление *n* сжатию; предел *m* прочности при сжатии

C472 **compressive stress** || Druckbeanspruchung *f*, Druckspannung *f* || contrainte *f* de compression || drukspanning *f* || сжимающее напряжение *n*; напряжение *n* сжатия

C473 **compressive yield point** || Quetschgrenze *f*, Fließgrenze *f* bei Druck || limite *f* élastique à la compression || vloeigrens *f* (*m*) bij druk || предел *m* текучести при сжатии

C474 **compressor** || Verdichter *m*, Kompressor *m* || compresseur *m* || compressor *m* || компрессор *m*

C475 **Compton scattering** || Compton-Streuung *f* || diffusion *f* Compton || Compton-effect *n* || комптоновское рассеяние *n*, эффект *m* Комптона

C476 **Compton wavelength** || Compton-Wellenlänge *f* || longueur *f* d'onde de Compton || Compton-golflengte *f* || комптоновская длина *f* волны

C477 **computational algorithm** || Rechenalgorithmus *m* || algorithme *m* de calcul || rekenalgoritme *n* || вычислительный алгоритм *m*

C478 **computations** || Rechnung *f*, Rechnen *n* || calculs *m pl* || berekening *f*, rekenen *n* || выкладки *f pl*; вычисления *n pl*

C479 **computer** || Rechner *m*, Rechenmaschine *f*, Computer *m* || ordinateur *m*, calculateur *m* || computer *m* || (электронная) вычислительная машина *f*, (Э)ВМ

C480 **computer language** || Maschinensprache *f* || langage *m* (de) machine || machinetaal *f* (*m*) || машинный язык *m*

C481 **computer-oriented language** || rechnerorientierte [maschinenorientierte] Sprache *f* || langage *m* orienté vers l'ordinateur || computer-georiënteerde taal *f* (*m*) || машино-ориентированный язык *m*

C482 **computer program** || Rechnerprogramm *n* || programme *m* pour un ordinateur || computerprogramma *n* || программа *f* для вычислительной машины

C483 **computing system** || Rechnersystem *n* || système *m* calculatrice || computersysteem *n* || вычислительная система *f*

C484 **concatenation** || Verkettung *f* || concaténation *f* || concatenatie *f* || конкатенация *f*, сцепление *n*

C485 **concave lens** || 1. Konkavlinse *f*, Hohllinse *f* 2. Zerstreuungslinse *f* || 1. lentille *f* concave 2. lentille divergente || 1. concave lens *f* (*m*) 2. divergerende lens *f* (*m*) || 1. вогнутая линза *f* 2. рассеивающая линза *f*

C486 **concave mirror** || Konkavspiegel *m*, Hohlspiegel *m* || miroir *m* concave || holle spiegel *m* || вогнутое зеркало *n*

C487 **concavity of a curve** || Konkavität *f* einer Kurve || concavité *f* d'une courbe || concaviteit *f* van een kromme || вогнутость *f* кривой

C488 **concavo-convex lens** || konkavkonvexe Linse *f* || lentille *f* convexe-concave || concaaf-convexe lens *f* (*m*) || вогнуто-выпуклая линза *f*

C489 **concentrated force** || konzentrierte Kraft *f*, Einzelkraft *f* || force *f* ponctuelle [concentrée] || geconcentreerde kracht *f* (*m*) || сосредоточенное усилие *n*

C490 **concentrated load** || konzentrierte Last *f*, Punktlast *f*, Einzellast *f* || charge *f* concentrée || geconcentreerde belasting *f* || сосредоточенная нагрузка *f*

C491 **concentrated solution** || konzentrierte Lösung *f* || solution *f* concentrée *f* || geconcentreerde oplossing *f* || концентрированный раствор *m*

C492 **concentration** || 1. Konzentration *f* 2. Konzentrieren *n*, Anreichern *n* || concentration *f* || concentratie *f* || 1. концентрация *f* (*величина*) 2. концентрация *f*, концентрирование *n*

C493 **concentric (jaw) chuck** || selbstzentrierendes Spannfutter *n* || mandrin *m* autocentreur || zelfcentrerende klauwplaat *f* (*m*) || самоцентрирующий патрон *m*

C494 **conchoid** || Konchoide *f* || conchoïde *f* || conchoïde *f* || конхоида *f*

C495 **conclusion** || Rückschluß *m* || conclusion *f* || conclusie *f* || заключение *n*, вывод *m*

C496 **condensate** || Kondensat *n*, Niederschlag *m* || condensat *m* || condensaat *n* || конденсат *m*, продукт *m* конденсации

C497 **condensation** || Kondensation *f*, Kondensierung *f* || condensation *f* || condensatie *f*, || конденсация *f*

C498 **condensation centre** || Kondensationszentrum *n*, Kondensationskern *m* || centre *m* de condensation || condensatiekern *f* (*m*) || центр *m* конденсации

C499 **condensed phase** || kondensierte Phase *f* || phase *f* condensée || gecondenseerde fase *f* || конденсированная фаза *f*

C500 **condenser** || 1. Kondensator *m*, Verflüssiger *m* 2. Kondensator *m*, Kapazität *f* 3. Kondensor *m* || 1. condenseur *m* 2. cuve *f* collectrice 3. condenseur *m* || 1. condensor *m* 2. condensator *m* 3. condensor *m* || 1. конденсатор *m*, холодильник *m* 2. (электрический) конденсатор *m*, ёмкость *f* 3. конденсор *m* *(оптическая система)*

C501 **condensing temperature** || Verflüssigungstemperatur *f*, Kondensationstemperatur *f* || température *f* de condensation || condensatietemperatuur *f* || температура *f* ожижения [конденсации]

C502 **conditional branch** *see* **conditional jump**

C503 **conditional entropy** || bedingte Entropie *f* || entropie *f* conditionnelle || voorwaardelijke entropie *f* || условная энтропия *f*

C504 **conditional jump** || bedingter Sprung *m* || saut *m* conditionnel || voorwaardelijke sprong *m* || условный переход *m*

C505 **conditional probability** || bedingte Wahrscheinlichkeit *f* || probabilité *f* conditionnelle || voorwaardelijke waarschijnlijkheid *f* || условная вероятность *f*

C506 **conditional risk** || bedingtes Risiko *n* || risque *m* conditionnel || voorwaardelijk risico *n*, *m* || условный риск *m*

C507 **conductance** || Wirkleitwert *m*, Konduktanz *f* || conductance *f* || conductantie *f* || (активная) проводимость *f*

C508 **conductiometric titration** || konduktometrische Titration *f*, Leitfähigkeitstitration *f* || titrage *m* conductométrique || conductometrische titratie *f* || кондуктометрическое титрование *n*

C509 **conduction** || Leitfähigkeit *f*, Leitvermögen *n*, Leitwert *m* || conductibilité *f*, conduction *f* || conductie *f*, geleiding *f* || проводимость *f*

C510 **conduction band** || Leitungsband *n*, Leitfähigkeitsband *n* || bande *f* de conduction || geleidingsband *m* || зона *f* проводимости

C511 **conduction current** || Leitungsstrom *m* || courant *m* de conduction || geleidingsstroom *m* || ток *m* проводимости

C512 **conduction electron** || Leitungselektron *n*, Leitfähigkeitselektron *n* || électron *m* de conduction || geleidingselektron *n* || электрон *m* проводимости

C513 **conductivity** || spezifische Leitfähigkeit *f*; elektrische Leitfähigkeit *f* || conductibilité *f* spécifique, conductivité *f* || specifieke geleiding *f* || удельная проводимость *f*; электропроводность *f*

C514 **conductor** || Leiter *m* || conducteur *m* || geleider *m* || проводник *m*; провод *m*

C515 **cone** || Kegel *m*, Konus *m* || cône *m* || kegel *m* || конус *m*

C516 **cone clutch** || Kegelkupplung *f* || embrayage *m* à cône || kegelkoppeling *f* || 1. конусная муфта *f* 2. конусное сцепление *n*

C517 **cone-shaped** || konisch, kegelig, Kegel... || conique, à cône || kegelvormig || конический

C518 **confidence interval** || Vertrauensintervall *n*, Konfidenzintervall *n* || intervalle *m* de confiance || vertrouwelijk interval *n* || доверительный интервал *m*

C519 **confidence level** || Irrtumswahrscheinlichkeit *f*, Sicherheitswahrscheinlichkeit *f* || niveau *m* de confiance || zekerheidsniveau *m* || доверительный уровень *m*, уровень *m*, значимости

C520 **confined plasma** || eingeschlossenes [isoliertes] Plasma *n* || plasma *m* confiné || geisoleerd plasma *n* || удерживаемая [изолированная] плазма *f*

C521 **confinement** || Einschließung *f*, Halterung *f* || confinement *m* || insluiting *f* || удерживание *n* *(от разлёта)*, конфайнмент *m*

C522 **confluent** || konfluent || confluent || confluent || конфлюентный, вырожденный
C523 **conformal mapping** || konforme Abbildung *f* || application *f* conforme || conforme afbeelding *f* || конформное отображение *n*
C524 **congruent** || kongruent || congruent || congruent || конгруэнтный
C525 **conic** || **1.** Konik *f*, Kurve *f* **2.** Ordnung *f* || conique *f* || conica *f* || коника *f*, кривая *f* второго порядка
C526 **conical** *see* **cone-shaped**
C527 **conical gear** || Kegelrad *n* || roue *f* [engrenage *m*] conique || kegelwiel *n* || коническое зубчатое колесо *n*, коническая шестерня *f*
C528 **conical value** || Kegelventil *n* || soupape *f* à cône || kegelklep *f* (*m*) || клапан *m* с коническим седлом
C529 **conic section** || Kegelschnitt *m* || section *f* conique || kegelsnede *f* (*m*) || коническое сечение *n*
C530 **conjugate** || konjugiert; adjungiert || conjugué || geconjugeerd || сопряжённый
C531 **conjugate matrix** || adjungierte Matrix *f* || matrice *f* conjuguée || geconjugeerde matrix *f* (*m*) || сопряжённая матрица *f*
C532 **conjugate operator** || adjungierter Operator *m* || opérateur *m* geconjugué || geconjugeerde operator *m* || сопряжённый оператор *m*
C533 **conjunction** || Konjunktion *f* || conjonction *f* || conjunctie *f* || конъюнкция *f*
C534 **connected graph** || zusammenhängender Graph *m* || graphe *m* connexe || samenhangende graaf *f* (*m*) || связный граф *m*
C535 **connected set** || zusammenhängende Menge *f* || ensemble *m* connexe || samenhangende menigte *f* || связное множество *n*
C536 **connecting rod** || Pleuelstange *f*; Schubstange *f*; Verbindungsstange *f* || bielle *f*; barre *f* d'accouplement || drijfstang *f* (*m*) || шатун *m*; соединительная тяга *f*
C537 **connection** || Verbinden *n*; Anschalten *n*; Schalten *n* || connexion *f*; jonction *f* || aansluiting *f* || соединение *n*, присоединение *n*; включение *n*
C538 **connection hose** || Verbindungsschlauch *m* || manche *f* de raccordement [de liaison] || aansluitslang *f* (*m*), verbindingsslang *f* (*m*) || соединительный шланг *m*
C539 **consequence** || Folge *f* || conséquence *f* || gevolg *n* следствие *n*
C540 **conservation** || Erhaltung *f*, Beibehaltung *f*, Aufrechterhaltung *f* || conservation *f* || bewaren *n* || сохранение *n*
C541 **conservation equation** || Erhaltungssatzgleichung *f* || équation *f* de conservation || conservatievergelijking *f* || уравнение *n* (закона) сохранения
C542 **conservation of angular momentum** || Erhaltung *f* des Drehimpulses, Drehimpulserhaltung *f* || conservation *f* du moment angulaire || conservatie *f* van het kinetisch moment || сохранение *n* момента количества движения
C543 **conservation of charge** || Erhaltung *f* der Ladung, Ladungserhaltung *f* || conservation *f* de la charge || bewaren *n* van lading || сохранение *n* заряда
C544 **conservation of energy** || Energieerhaltung *f*, Erhaltung *f* der Energie || conservation *f* de l'énergie || energiebewaring *f* || сохранение *n* энергии
C545 **conservation of linear momentum** || Erhaltung *f* des linearen Impulses, Erhaltung *f* der Bewegungsgröße *f* || conservation *f* de la quantité de mouvement || conservatie *f* van bewegingshoeveelheid || сохранение *n* количества движения
C546 **conservation of mass** || Erhaltung *f* der Masse, Massenerhaltung *f* || conservation *f* de la masse || behoud *n* van massa || сохранение *n* массы
C547 **conservation of matter** || Erhaltung *f* der Materie || conservation *f* de la matière || conservatie *f* van stof || сохранение *n* материи
C548 **conservatism** || Konservatismus *m* || conservatisme *m* || conserva-

tisme *n* || консервативность *f* *(сил, систем)*

C549 **conservative** || konservativ || conservatif || conservatief || консервативный *(о силе, системе)*

C550 **conservative force** || konservative Kraft *f* || force *f* conservative || conservatieve kracht *f* (*m*) || консервативная сила *f*

C551 **conservative system** || konservatives System *n* || système *m* conservatif || conservatief systeem *n* || консервативная система *f*

C552 **consistency** || 1. Widerspruchsfreiheit *f* 2. Konsistenz, *f*, Kompatibilität *f* *(von Gleichungen)* || 1. non-contradiction *f* 2. compatibilité *f* *(des équations)* || consistentie *f* || 1. непротиворечивость *f* 2. совместность *f* *(уравнений)*

C553 **consistent** || 1. widerspruchsfrei 2. konsistent || 1. non-contradictoire 2. compatible || consistent || 1. непротиворечивый 2. совместный

C554 **consistent equations** || konsistente Gleichungen *f pl* || équations *f pl* compatibles || consistente vergelijkingen *f pl* || совместные уравнения *n pl*

C555 **consistent estimate** || konsistente Schätzung *f* || estimation *f* valable || consistente schatting *f* || состоятельная оценка *f*

C556 **consistent fat** || Schmierfett *n* || graisse *f* consistante || consistentvet *n* || консистентная смазка *f*

C557 **console** || 1. Steuerpult *n* 2. Konsole *f* || console *f* || 1. bedieningspaneel *n* 2. console *f* (*m*) || 1. пульт *m* (управления) 2. консоль *f*

C558 **constant** || Konstante *f* || constante *f* || constante *f* (*m*) || константа *f*, постоянная *f*

C559 **constant component** || Gleichkomponente *f*, konstante Komponente *f* || (composante *f*) constante *f* || constante component *m* || постоянная составляющая *f*

C560 **constant force** || konstante [unveränderliche] Kraft *f* || force *f* constante || constante kracht *f* (*m*) || постоянная сила *f*

C561 **constant of integration** || Integrationskonstante *f* || constante *f* d'intègration || integratieconstante *f* (*m*) || постоянная *f* интегрирования

C562 **constant of the motion** || Bewegungskonstante *f* || constante *f* du mouvement || bewegingsconstante *f* (*m*) || постоянная *f* движения

C563 **constant pressure** || konstanter Druck *m* || pression *f* constante || constante druk *m* || постоянное давление *n*

C564 **constant-speed motor** || Motor *m* mit konstanter Drehzahl || moteur *m* à vitesse constante || motor *m* met constant toerental *n* || двигатель *m* с постоянным числом оборотов

C565 **constraint** || 1. Bindung *f* 2. Beschränkung *f* || 1. liaison *f* 2. contrainte *f* || 1. binding *f* 2. beperking *f* || 1. (механическая) связь *f* 2. ограничение *n*

C566 **constraint force** || Bindungsreaktion *f* || réaction *f* [force *f*] de liaison || bindingsreactie *f* || реакция *f* связи

C567 **construction** || 1. Konstruktion *f*; Bauweise *f*; Bau *m* 2. Bau *m* 3. *Math.* Konstruktion *f* *(einer Figur, Kurve)* || bâtiment *m*; construction *f* || constructie *f* || 1. конструкция *f*; сооружение *n* 2. строительство *n* 3. *мат.* построение *n* *(геометрической фигуры, графика)*

C568 **constructive** || konstruktiv || constructif || constructief || конструктивный

C569 **constructive definition** || konstruktive Definition *f* || définition *f* constructive || constructieve definitie *f* || конструктивное определение *n*

C570 **constructive interference** || konstruktive Interferenz *f* || interférence *f* constructive || constructieve interferentie *f* || конструктивная интерференция *f*, интерференция *f* с взаимным усилением

C571 **contact** || 1. Berührung *f*, Kontakt *m* 2. Kontakt *m*; Kontaktstück *n* || contact *m* || contact *n* || 1. соприкосновение *n*, контакт *m* 2. (электрический) контакт *m*

C572 **contact area** *see* **area of contact**

C573 **contact electromotive force, contact potential difference** || Kontaktpotentialdifferenz *f*, Kontaktspannung *f* || différence *f* de potentiel de contact || contactpotentiaal *n* || контактная разность *f* потенциалов

C574 **contact pressure** || Kontaktdruck *m* || pression *f* de contact || contactdruk *m* || контактное давление *n*

C575 **contact surface** || Berührungsfläche *f* || surface *f* de contact || aanrakingsoppervlak *n* || поверхность *f* соприкосновения [контакта]

C576 **container** || Behälter *m*; Container *m* || container *m*; récipient *m* || transportkist *f* (*m*), container *m* || контейнер *m*

C577 **container transportation** || Containerverkehr *m* || transport *m* par container || containerverkeer *n* || контейнерные перевозки *f pl*

C578 **context-free grammar** || kontextfreie Grammatik *f* || grammaire *f* «context-free» || contextvrije grammatica *f* || бесконтекстная [контекстно-свободная] грамматика *f*

C579 **context-free language** || kontextfreie Sprache *f* || langage *m* «context-free» || contextvrije taal *f* || бесконтекстный [контекстно-свободный] язык *m*

C580 **context-insensitive grammar** *see* **context-free grammar**

C581 **context-sensitive grammar** || kontextempfindliche Grammatik *f* || grammaire *f* «context-sensitive» || contextafhankelijke grammatica *f* || грамматика *f* непосредственных составляющих, контекстно-зависимая грамматика *f*

C582 **context-sensitive language** || kontextempfindliche Sprache *f* || langage *m* «context-sensitive» || contextafhankelijke taal *f* || контекстно-зависимый язык *m*

C583 **contiguity** || Benachbartsein *n*; Benachbarkeit *f* || contiguïté *f* || nabijheid *f* || смежность *f*

C584 **contingency** || Kontingenz *f* || contingence *f* || contingentie *f* || сопряжённость *f* признаков, контингенция *f*

C585 **continued fractions** || Kettenbrüche *m pl* || fraction *f* continue || repeterende breuken *f pl* || непрерывные [цепные] дроби *f pl*

C586 **continuity** || Kontinuität *f* || continuité *f* || continuïteit *f* || непрерывность *f*

C587 **continuity axiom** || Kontinuitätsaxiom *n* || axiome *m* de continuité || continuïteitsaxioma *n* || аксиома *f* непрерывности

C588 **continuity equation** || Kontinuitätsgleichung *f* || équation *f* de continuité || continuïteitsvergelijking *f* || уравнение *n* непрерывности [неразрывности]

C589 **continuity theorem** || Kontinuitätssatz *m* || théorème *m* de continuité || continuïteitstheorema *n* || теорема *f* непрерывности

C590 **continuous beam** || Durchlaufträger *m*, Mehrfeldträger *m* || poutre *f* continue || op enige steunpunten opgelegde balk *m* || неразрезная балка *f*

C591 **continuous chip** || Fließspan *m* || copeaux *m pl* continus || vloeispaan *f* (*m*) || сливная стружка *f*

C592 **continuous flow** || Kontinuumströmung *f* || écoulement *m* continu || continustroming *f* || неразрывное течение *n*

C593 **continuous function** || kontinuierliche Funktion *f* || fonction *f* continue || continufunctie *f* || непрерывная функция *f*

C594 **continuous lubrication** || beständige Schmierung *f* || lubrification *f* continue || continue smering *f* || непрерывная смазка *f*

C595 **continuous medium** || kontinuierliches Medium *n* || continuum *m*, milieu *m* continu || continu medium *n* || сплошная среда *f*

C596 **continuous mill** || kontinuierliches Walzwerk *n* || laminoir *m* continu || continu-walserij *f* || непрерывный прокатный стан *m*

C597 **continuous oscillations** || ungedämpfte Schwingungen *f pl*, Dauerschwingungen *f pl* || oscillations *f pl* entretenues || ongedempte trillingen *f pl* || незатухающие колебания *n pl*

C598 **continuous radiation** || kontinuierliche Strahlung *f* || émission *f* [radiation *f*] continue || continue straling *f* || сплошное [непрерывное] излучение *n*

C599 **continuous spectrum** || kontinuierliches Spektrum *n*, Kontinuum *n* || spectre *m* continu || continu spectrum *n* || сплошной [непрерывный] спектр *m*

C600 **continuous wave** || ungedämpfte [kontinuierliche] Welle *f* || onde *f* entretenue || ongedempte [doorlopende] golf *f* (*m*) || незатухающая [непрерывная] волна *f*

C601 continuous-wave laser || Dauerstrichlaser *m* || laser *m* à fonctionnement continu [à fonctionnement permanent] || continue laser *m* || лазер непрерывного режима работы

C602 **continuous X-rays** || kontinuierliche Röntgenstrahlung *f*, Bremsröntgenstrahlung *f* || rayons *m pl* X continus || continue röntgenstralen *m. f pl* || непрерывное [сплошное, тормозное] рентгеновское излучение *n*

C603 **continuum** || **1.** kontinuierliches Medium *n*, Kontinuum *n* **2.** kontinuierliches Spektrum *n*, Kontinuum *n* || **1.** continuum *m*, milieu *m* continu **2.** spectre *m* continu || **1.** continu medium *n* **2.** continu spectrum *n* || **1.** сплошная среда *f* **2.** непрерывный спектр *m*

C604 **contour** || Kontur *f* || contour *m* || contour *m* || контур *m*

C605 **contour grinding** *see* **profile grinding**

C606 **contour integral** || Umlaufintegral *n* || intégrale *f* de contour || contourintegraal *f* (*m*) || контурный интеграл *m*

C607 **contour milling** *see* **profile milling**

C608 **contraction** || **1.** Kontraktion *f* **2.** *Math.* Kontraktion *f*; Faltung *f* || contraction *f* || **1.** contractie *f* **2.** *math.* inkrimping *f* || **1.** сжатие *n* **2.** *мат.* сжатие *n*; свёртывание *n*

C609 **contraction mapping** || kontrahierende Abbildung *f* || application *f* contractante || contractieafbeelding *f* || сжатое отображение *n*

C610 **contradiction** || Widerspruch *m* || contradiction *f* || tegenspraak *f* || противоречие *n*

C611 **contragredient** || kontragredient || contragrédient || contragrediënt || контрагредиентный

C612 **contrast** || Kontrast *m* || contraste *m* || contrast *n* || контраст *m*, контрастность *f*

C613 **contravariant vector** || kontravarianter Vektor *m* || vecteur *m* contravariant || contravariante vector *m* || контравариантный вектор *m*

C614 **control** || Steuerung *f*; Regelung *f*; Überwachung *f* || commande *f*; réglage *m*; contrôle *m* || sturing *f*; regeling *f*; controle *f* (*m*) || управление *n*; регулирование *n*; контроль *m*

C615 **control accuracy** || Regelgenauigkeit *f* || précision *f* de réglage || regelnauwkeurigheid *f* || точность *f* регулирования

C616 **control algorithm** || Steueralgorithmus *m* || algorithme *m* de contrôle || controle-algoritme *n* || алгоритм *m* управления

C617 **control computer** || Steuerrechner *m* || calculateur *m* de contrôle || controlecomputer *m* || управляющая вычислительная машина *f*

C618 **control gauge** || **1.** Prüflehre *f* **2.** Normal *n*, Vergleichsnormal *n* || **1.** jauge-étalon *m* **2.** étalon *m* || **1.** controlekaliber *n* **2.** etalon *m* || **1.** контрольный калибр *m* **2.** эталон *m*

C619 **controlled quantity** *see* **controlled variable**

C620 **controlled reaction** || gesteuerte Reaktion *f* || réaction *f* contrôlable [ménagée] || beheerste reactie *f* || управляемая реакция *f*

C621 **controlled variable** || Steuergröße *f*, gesteuerte Größe *f* || variable *f* contrôlée || gecontroleerde variabele *f* || управляемая переменная *f*, управляемая величина *f*

C622 **control lever** || Schalthebel *m*, Steuerhebel *m* || levier *m* de commande || bedieningshefboom *m* || рычаг *m* управления

C623 **controlling quantity** *see* **controlling variable**

C624 **controlling variable** || Stellgröße *f* || variable *f* de contrôle || controlevariabele *f* || управляющая величина *f*, управляющее воздействие *n*

C625 **control panel** || Schalttafel *f*, Bedienungspult *n* || pupitre *m* de commande; tableau *m* d'instruments || schakelbord *n* || пульт *m* управления; приборная доска *f*

C626 **control valve** || Steuerventil *n* || valve *f* pilote; soupape *f* de commande || regelafsluiter *m* || клапан *m* управления; распределительный клапан *m*; регулировочный клапан *m*

C627 **convection** || Konvektion *f* || convection *f* || convectie *f* || конвекция *f*

C628 **convection current** || (elektrischer) Konvektionsstrom *m* || courant *m* de convection || convectiestroom *m* || конвекционный (электрический) ток *m*

C629 **conventional milling** || Gegenlauffräsen *n* || fraisage *m* en opposition || tegenlopend frezen *n* || фрезерование *n* против (направления) подачи *(заготовки)*

C630 **convergence** || Konvergenz *f* || convergence *f* || convergentie *f* || сходимость *f*

C631 **convergence acceleration** || Konvergenzverbesserung *f* || accélération *f* de convergence || convergentieverbetering *f* || улучшение *n* сходимости

C632 **convergence criterion** || Konvergenzkriterium *n* || critère *m* de convergence || convergentiecriterium *n* || признак *m* сходимости

C633 **convergent beam** || konvergentes [konvergierendes] Bündel *n* || faisceau *m* convergent || convergerende bundel *m* || сходящийся пучок *m*

C634 **convergent sequence** || konvergierende [konvergente] Folge *f* || séquence *f* convergente || convergente volgorde *f* *(m)* || сходящаяся последовательность *f*

C635 **converging lens** || Sammellinse *f*, Positivlinse *f* || lentille *f* convergente || convergerende lens *f* *(m)* || собирающая [положительная] линза *f*

C636 **converging waves** || konvergierende Wellen *f pl* || ondes *f pl* convergentes || convergerende golven *f* *(m) pl* || сходящиеся волны *f pl*

C637 **conversational language** || Dialogsprache *f* || langage *m* conversationnel || dialoogtaal *f* || язык *m* диалога

C638 **conversion** || **1.** Umwandlung *f*, Umformung *f*, Konversion *f* **2.** Umrechnung *f* *(von Maßeinheiten)* || conversion *f*, transformation *f* || **1.** conversie *f* **2.** omrekening *f* || **1.** преобразование *n*, конверсия *f* **2.** перевод *m* *(единиц измерения)*

C639 **conversion factor** || Umrechnungsfaktor *m* || facteur *m* de conversion || omrekeningsfactor *m* || переводной множитель *m*

C640 **conversion ratio** || **1.** Koeffizient *m* der inneren Konversion **2.** Konversionsgrad *m* || **1.** coefficient *m* de conversion interne **2.** taux *m* de conversion || **1.** coëfficiënt *m* van inwendige conversie **2.** conversiegraad *m* || **1.** коэффициент *m* внутренней конверсии **2.** коэффициент *m* воспроизводства *(ядерного топлива)*

C641 **conversion table** || Umrechnungstafel *f* || table *f* de conversion des unités || omrekeningstabel *f* *(m)* || таблица *f* перевода единиц (измерения)

C642 **converter** || **1.** Wandler *m*, Umformer *m*, Umsetzer *m* **2.** Konvertor *m* || convertisseur *m* || **1.** omzetter *m* **2.** convertor *m* || **1.** преобразователь *m* **2.** конвертер *m*

C643 **converter steel** || Konverterstahl *m* || acier *m* de convertissage || convertorstaal *n* || конвертерная сталь *f*

C644 **convex** || konvex || convexe || convex || выпуклый

C645 **convex lens** || **1.** Konvexlinse *f* **2.** Sammellinse *f* || lentille *f* convexe || convexe lens *f* *(m)* || **1.** выпуклая линза *f* **2.** собирающая линза *f*

C646 **convex mirror** || Konvexspiegel *m*, Wölbspiegel *m* || miroir *m* convexe || bolle spiegel *m* || выпуклое зеркало *n*

C647 **convexo-concave** || konvexkonkav || convexo-concave || convexconcaaf || выпукло-вогнутый

C648 **convexo-concave lens** || konvex-konkave Linse *f* || lentille *f* concave-convexe || convex-concave lens *f* (*m*) || выпукло-вогнутая линза *f*

C649 **convex programming** || konvexe Programmierung *f* || programmation *f* convexe || convexe programmering *f* || выпуклое программирование *n*

C650 **convex set** || konvexe Menge *f* || ensemble *m* convexe || convexe menigte *f* || выпуклое множество *n*

C651 **conveyor** || Förderer *m* || convoyeur *m*, transporteur *m*, transporteuse *f* || transporteur *m*, transportband *m* || конвейер *m*, транспортёр *m*

C652 **conveyor belt** || Förderband *n* || courroie *f* transporteuse [de transport] || transportband *m* || конвейерная [транспортёрная] лента *f*

C653 **convolution** || Faltung *f* || convolution *f* || convolutie *f* || свёртка *f*

C654 **convolution integral** || Faltungsintegral *n* || intégrale *f* de convolution || convolutie-integraal *f* (*m*) || интеграл *m* свёртки, свёртка *f*

C655 **convolution theorem** || Faltungssatz *m* || théorème *m* de convolution || convolutietheorema *n* || теорема *f* о свёртке

C656 **coolant** || Kühlmittel *n*, Kühlflüssigkeit *f* || agent *m* refroidisseur [réfrigérant], liquide *f* de refroidissement || koelmiddel *n* || охладитель *m*; хладагент *m*; холодоноситель *m*

C657 **cooler** || Kühler *m* || réfrigérateur *m* || koeler *m* || холодильник *m*

C658 **cooling** || **1.** Kühlung *f*, Abkühlung *f* **2.** Abklingen *n* (*der Aktivität*) || **1.** refroidissement *m* **2.** décroissance *f* (*de la radio-activité*) || koeling *f* || **1.** охлаждение *n* **2.** спадание *n* (*радиоактивности*)

C659 **cooling medium** || Kühlmittel *n* || milieu *m* refroidisseur [réfrigérant], || koelmiddel *n* || охлаждающая среда *f*

C660 **cooling surface** || Abkühlungsfläche *f*, Kühlfläche *f* || surface *f* de refroidissement || koelend oppervlak *n* || поверхность *f* охлаждения

C661 **cooling system** || Kühlsystem *n* || système *m* de réfrigération [de refroidissement] || koelsysteem *n* || система *f* охлаждения

C662 **Cooper pair** || Cooper-Paar *n* || couple *m* de Cooper || Cooper-koppel *n* || куперовская пара *f* (*электронов в сверхпроводнике*)

C663 **coordinate axes** || Koordinatenachsen *f pl* || axes *m pl* de coordonnées || coördinaatassen *f* (*m*) *pl* || оси *f pl* координат, координатные оси *f pl*

C664 **co-ordinate bond** || koordinative Bindung *f*, Koordinationsbindung *f* || liaison *f* coordinée || gecoördineerde binding *f* || координационная связь *f*

C665 **coordinate boring machine** || Koordinatenbohrmaschine *f* || aléseuse-pointeuse *f* || coördinatenboormachine *f* || координатно-расточный станок *m*

C666 **coordinate grinding head** || Koordinatenschleifkopf *m* || tête *f* de rectification entre pointes || coördinatenslijpkop *m* || шлифовальная головка *f* с измерительным устройством (для особо точной обработки)

C667 **coordinate plane** || Koordinatenebene *f* || plan *m* de coordonnées || coördinaatvlak *n* || координатная плоскость *f*

C668 **coordinate surface** || Koordinatenfläche *f* || surface *f* coordonnée || coördinaatoppervlakte *f* || координатная поверхность *f*

C669 **coordinate system** || Koordinatensystem *n* || système *m* de coordonnées || coördinatenstelsel *n* || система *f* координат

C670 **coordinate transformation** || Koordinatenumwandlung *f* || transformation *f* des coordonnées || coördinatentransformatie *f* || преобразование *n* координат

C671 **coplanar vectors** || ko(m)planare Vektoren *m pl* || vecteurs *m pl* coplanaires || coplanaire vectoren *m pl* || компланарные векторы *m pl*

C672 **copper,** Cu || Kupfer *n* || cuivre *m* || koper *n* || медь *f*

C673 **copper alloy** || Kupferlegierung *f* || alliage *m* de cuivre || koperlegering *f* || медный сплав *m*

C674 **copper nickel alloy** || Kupfer-Nickel-Legierung *f* || alliage *m* de cuivre-nickel || kopernikkellegering *f* || медно-никелевый сплав *m*

C675 **copying** || Kopieren *n*, Kopierarbeit *f* || copiage *m* || kopiëeren *n*; kopiëerbewerking *f* || обработка *f* по копиру

C676 **copying attachment** || Kopiereinrichtung *f* || copieur *m*, dispositif *m* à copier || kopieërinrichting *f* || копировальное устройство *n*

C677 **copying grinding machine** || Kopierschleifmaschine *f* || rectifieuse *f* à copier || kopiëerslijpmachine *f* || копировально-шлифовальный станок *m*

C678 **copying lathe** || Kopierdrehmaschine *f* || tour *m* à copier || kopiëerdraaibank *f* (*m*) || копировально-токарный станок *m*

C679 **copying milling machine** || Kopierfräsmaschine *f* || fraiseuse *f* à copier || kopiëerfreesmachine *f* || копировально-фрезерный станок *m*

C680 **core** || **1.**, **2.** Kern *m* **3.** Atomrumpf *m* **4.** Spaltzone *f*, aktive Zone *f* || **1.** âme *f*, cœur *m* **2.** noyau *m* **3.** carcasse *f* d'atome **4.** noyau *m* central du réacteur || **1.** hart *n* **2.** kern *f* (*m*) **3.** atoomgeraamte *n* **4.** reactorhart *n* || **1.** ядро *n*, сердцевина *f* **2.** сердечник *m* **3.** атомный остов *m* **4.** активная зона *f* *(ядерного реактора)*

C681 **cored winding** || Wicklung *f* mit Kern || enroulement *m* à noyau || wikkeling *f* met kern || обмотка *f* с сердечником

C682 **coreless** || kernlos || sans noyau || zonder kern || без сердечника

C683 **core loss** || Verluste *m pl* im Eisenkern, Kernverluste *m pl* || pertes *f pl* dans le noyau || kernverliezen *n pl* || потери *f pl* в сердечнике

C684 **core region** *see* **core 4.**

C685 **core storage** || Ferritkernspeicher *m* || mémoire *f* à noyau en ferrite || ferrietkerngeheugen *n* || память *f* на ферритовых сердечниках

C686 **Coriolis acceleration** || Coriolis-Beschleunigung *f* || accélération *f* de Coriolis || Coriolis-versnelling *f* || кориолисово [поворотное] ускорение *n*

C687 **Coriolis force** || Coriolis-Kraft *f* || force *f* de Coriolis || Corioliskracht *f* (*m*) || сила *f* Кориолиса, кориолисова сила *f*

C688 **corkscrew rule** || Korkenzieherregel *f*, Rechtsschraubenregel *f*, Rechte-Hand-Regel *f* || règle *f* de tire-bouchon [de la main droite] || rechterhandregel *m* || «правило *n* буравчика», правило *n* правой руки

C689 **corner weld** || Kehlnaht *f* || soudure *f* d'angle || hoeklas *f* (*m*) || угловой сварной шов *m*; угловое сварное соединение *n*

C690 **corona (discharge)** || Koronaentladung *f*, Korona *f* || corona *f*; effluve *m* (électrique), décharge *f* en couronne || corona-ontlading *f*, corona *f* || коронный разряд *m*, корона *f*, коронирование *n*

C691 **corona voltage** || Anfangsspannung *f* der Korona, Koronaanfangsspannung *f* || tension *f* du début de décharge en couronne || corona-doofspanning *f* || напряжение *n* начала коронирования

C692 **corpuscle** || Korpuskel *n*, Teilchen *n* || corpuscule *m*, particule *f* || deeltje *n* || корпускула *f*, частица *f*

C693 **corpuscular** || korpuskular || corpusculaire || corpusculair || корпускулярный

C694 **corpuscular radiation** || Korpuskularstrahlung *f*, Teilchenstrahlung *f* || rayonnement *m* corpusculaire || corpusculaire straling *f* || корпускулярное излучение *n*

C695 **corpuscular theory** || Korpuskulartheorie *f* *(des Lichtes)* || théorie *f* corpusculaire || corpusculaire theorie *f* || корпускулярная теория *f* *(света)*

C696 **correction** || Korrektion *f*, Berichtigung *f*; Korrektur *f* || correction *f* || correctie *f* || **1.** коррекция *f*, исправление *n* **2.** поправка *f*

C697 **correlation** || Korrelation *f* || corrélation *f* || correlatie *f* || корреляция *f*

C698 **correlation analysis** || Korrelationsanalyse *f* || analyse *f* corrélative || correlatieanalyse *f* || корреляционный анализ *m*

C699 **correlation coefficient** || Korrelationskoeffizient *m* || coefficient *m* de corrélation || correlatiecoëfficiënt *m* || коэффициент *m* корреляции

C700 **correlation function** || Korrelationsfunktion *f* || fonction *f* corrélative || correlatiefunctie *f* || корреляционная функция *f*

C701 **correlation matrix** || Korrelationsmatrix *f* || matrice *f* de corrélation || correlatiematrix *f* || корреляционная матрица *f*

C702 **correspondence principle** || Korrespondenzprinzip *n* (von Bohr) || principe *m* de correspondance || overeenkomstbeginsel *n* || принцип *m* соответствия (Бора)

C703 **corresponding states** || übereinstimmende [korrespondierende] Zustände *m pl* || états *m pl* correspondants || overeenkomstige toestanden *m pl* || соответственные состояния *n pl*

C704 **corrosion** || Korrosion *f*; Angriff *m* || corrosion *f* || corrosie *f* || коррозия *f*; разъедание *n*

C705 **corrosion fatigue** || Korrosionsermüdung *f* || fatigue *f* due à la corrosion || corrosievermoeiïng *f* || коррозионная усталость *f*

C706 **corrosion rate** || Korrosionsgeschwindigkeit *f*; Korrosionsgrad *m* || vitesse *f* d'attaque corrosive || corrosiesnelheid *f*; corrosiegraad *m* || скорость *f* коррозии; степень *f* коррозии

C707 **corrosion resistance** || Korrosionsfestigkeit *f*, Korrosionsbeständigkeit *f* || résistance *f* à la corrosion || corrosie-vastheid *f*, corrosiebestendigheid *f* || устойчивость *f* против коррозии

C708 **corrosion-resistant** || korrosionsfest, korrosionsbeständig || résistant à la corrosion || corrosiebestendig || коррозиеустойчивый, антикоррозийный

C709 **corrosive wear** || Korrosionsverschleiß *m* || usure *f* corrosive || corrosieslijtage *f* || коррозионный износ *m*

C710 **cosec, cosecant** || Kosekans *m* || cosécante *f* || cosecans *m* || косеканс *m*

C711 **coset** || Nebenklasse *f* *(Mengentheorie)* || classe *f* suivante un sousgroupe || nevenklasse *f* || смежный класс *m* *(в теории множеств)*

C712 **cosine** || Kosinus *m* || cosinus *m* || cosinus *m* || косинус *m*

C713 **cosine integral** || Integralkosinus *m* || cosinus *m* intégral || integraalcosinus *m* || интегральный косинус *m*

C714 **cosmic** || kosmisch, Kosmos..., Weltraum... || cosmique, spatial || kosmisch || космический

C715 **cosmic rays** || kosmische Strahlen *pl*, kosmische Strahlung *f* || rayons *m pl* cosmiques || kosmische stralen *m, f, pl* || космические лучи *pl*

C716 **cosmos** || Kosmos *m*, Weltall *n*, Weltraum *m* || cosmos *m*, espace *m* || kosmos *m* || космос *m*; космическое пространство *n*

C717 **cotangent** || Kotangens *m* || cotangente *f* || cotangens *f* || котангенс *m*

C718 **cotter** || Keil *m*, Querkeil *m*; Splint *m* || goupille *f*; clavette *f* || splitpen *f* *(m)* || шплинт *m*; чека *f*

C719 **coulomb**, C || Coulomb *n* || coulomb *m* || coulomb *m* || кулон *m*

C720 **Coulomb field** || Coulomb-Feld *n*, Coulombsches Feld *n* || champ *m* coulombien || Coulomb-veld *n* || кулоновское поле *n*

C721 **Coulomb interaction** || Coulombsche Wechselwirkung *f* || interaction *f* coulombienne || Coulombwisselwerking *f* || кулоновское [электростатическое] взаимодействие *n*

C722 **Coulomb repulsion** || Coulomb-Abstoßung *f*, Coulombsche Abstoßung *f* || répulsion *f* coulombienne || Coulomb-afstoting *f* || кулоновское [электростатическое] отталкивание *n*

C723 **coulometer** || Coulometer *n*, Voltameter *n* || voltamètre *m* || coulometer *m* || вольтаметр *m*

C724 **count** || Einzelimpuls *m*, Zählstoß *m* || impulsion *f* unique [isolée, monocycle] || enkelimpuls *m* || одиночный импульс *m* (счётчика)

C725 **countable set** || abzählbare Menge *f* || ensemble *m* dénombrable || telbare menigte *f* || счётное множество *n*

C726 **counter** || Zählrohr *n*, Zähler *m* || compteur *m* || teller *m* || счётчик *m* *(частиц)*

C727 **counteracting force** || Gegenkraft *f* || contre-force *f* || tegenkracht *f* (*m*) || противодействующая сила *f*

C728 **counteraction** || Gegenwirkung *f* || réaction *f* || tegenwerking *f* || противодействие *n*

C729 **counterbalance** *see* **counterweight**

C730 **counterbore** || Senker *m* || outil *m* à lamer || verzinkboor *m* || зенкер *m*

C731 **counterclockwise direction** || Gegenuhrzeigersinn *m* || sens *m* antihoraire || tegenuurwijzerrichting *f* || направление *n* против часовой стрелки

C732 **counterclockwise-polarized wave** || linkspolarisierte Welle *f* || onde *f* polarisée à gauche || linksgepolariseerde golf *f* (*m*) || левополяризованная волна *f*

C733 **countercurrent** || Gegenstrom *m* || contre-courant *m* || tegenstroom *m* || противоток *m*, ток *m* обратного направления; экстраток *m* замыкания

C734 **countercurrent capacitor** || Sperrkondensator *m* || condensateur *m* d'arrêt || sluitcondensator *m* || заградительный конденсатор *m*

C735 **counterflow** || Gegenströmung *f* || contre-courant *m* || tegenstroom *m* || обратное течение *n*, противоток *m*

C736 **counterforce** *see* **counteracting force**

C737 **counterpressure** || Gegendruck *m* || contre-pression *f* || tegendruk *m* || противодавление *n*

C738 **counterschaft** || Vorgelegewelle *f* || contre-arbre *m* || overbrengas *f* (*m*); tussenas *f* (*m*) || контрпривод *m*; промежуточный вал *m*

C739 **countersunk head** || Senkkopf *m* || tête *f* noyée *(p. ex. d'une vis)* || verzonken kop *m* || потайная головка *f* *(напр. винта)*

C740 **counterweight** || Gegengewicht *n* || contre-poids *m* || tegengewicht *n* || противовес *m*

C741 **counting** || Zählung *f* || calcul *m*, comptage *m* || telling *f* || счёт *m*; подсчёт *m* *(частиц, импульсов)*

C742 **couple** || Kräftepaar *n* || couple *m* (des forces) || krachtmoment *n* || пара *f* сил; момент *m* пары сил

C743 **coupled circuits** || gekoppelte Kreise *m pl* || circuits *m pl* couplés || gekoppelde schakelingen *f pl* || связанные контуры *m pl*

C744 **coupled oscillations** || gekoppelte Schwingungen *f pl* || oscillations *f pl* couplées || gekoppelde trillingen *f pl* || связанные колебания *n pl*

C745 **couple of forces** || Kräftepaar *n* || couple *f* (des forces) || krachtenpaar *n* || пара *f* сил

C746 **coupler** *see* **coupling coil**

C747 **coupling** || 1. Kopplung *f*; Bindung *f* 2. Kupplung *f*; Kopplung *f* 3. Kopplung *f*, Verbindung *f* || 1. couplage *m* 2. accouplement *m*; raccord *m*; manchon *m* 3. attelage *m* || 1. koppeling *f*; koppelen *n*; koppeling *f* 2. verbindingsmof *f* (*m*) 3. verbinding *f* || 1. связь *f* 2. (соединительная) муфта *f* 3. сцепление *n*; сцепное устройство *n*

C748 **coupling bolt** || Kupplungsbolzen *m* || boulon *m* d'attelage || koppelbout *m* || соединительный болт *m*; стяжной болт *m*

C749 **coupling coil** || Kopplungsspule *f* || bobine *f* de couplage || koppelspoel *f* (*m*) || катушка *f* связи

C750 **coupling flange** || Kupplungsflansch *m* || bride *f* d'accouplement || koppelflens *m* || соединительный фланец *m*

C751 **coupling transformer** || Kopplungstransformator *m* || transformateur *m* de couplage || koppeltransformator *m* || трансформатор *m* связи

C752 **covalent binding, covalent bond** || kovalente Bindung *f*, Kovalenz *f*, Atombindung *f* || liaison *f* covalente [atomique] || covalente binding *f*, atoombinding *f* || ковалентная [атомная] связь *f*

C753 **covariance** || Kovarianz *f* || covariance *f* || covariantie *f* || 1. ковариация *f* 2. ковариантность *f*

C754 **covariant** || kovariant || covariant || covariant || ковариантный

C755 **covariant vector** || kovarianter Vektor *m* || vecteur *m* covariant || covariante vector *m* || ковариантный вектор *m*

C756 **crack** || Riß *m* || fente *f*; fissure *f*; crique *f* || scheur *f* (*m*); spleet *f* (*m*) || трещина *f*

C757 **crack front** || Ermüdungsbruchfront *f* || front *m* de la fissure || vermoeiingsbreukfront *n* || фронт *m* (усталостной) трещины

C758 **crane** || Kran *m*; Hubeinrichtung *f*, Hebezeug *n* || grue *f* || kraan *f* (*m*); hijskraan *f* (*m*) || (грузоподъёмный) кран *m*

C759 **crank** || Kurbel *f* || manivelle *f* || kruk *f* (*m*) || кривошип *m*, кривошипный рычаг *m*; коленчатое соединение *n*

C760 **crankcase** || Getriebegehäuse *n*, Kurbelgehäuse *n* || carter *m* (du moteur) || carter *n*; krukkast *f* (*m*) || картер *m* (двигателя)

C761 **cranked tool** *see* **bent tool**

C762 **crankpin** || Kurbelzapfen *m*; Hubzapfen *m* *(Kurbelwelle)* || maneton *m* || krukashals *m*; krukastap *m* || палец *m* кривошипа; шатунная шейка *f* *(коленчатого вала)*

C763 **crank press** || Kurbelpresse *f* || presse *f* à manivelle || krukpers *f* (*m*) || кривошипный пресс *m*

C764 **crankshaft** || Kurbelwelle *f* || arbre *m* manivelle, vilebrequin *m* || krukas *f* (*m*) || коленчатый вал *m*

C765 **creep** || **1.** Kriechen *n* **2.** Auflauf *m* *(z. B. Riemen)*; Schlupf *m* *(z. B. Riemen)* || **1.** fluage *m* **2.** heurt *m* *(de la courroie)*; glissement *m* *(p. ex. de la courroie)* || **1.** kruip *m* **2.** inloop *m*; afglijding *f* || **1.** ползучесть *f*, крип *m* **2.** набегание *n* *(напр. ремня)*; скольжение *n* *(напр. ремня)*

C766 **creepage** || Gleitentladung *f* || décharge *f* glissante [rampante] || glijdende ontlading *f* || скользящий разряд *m*

C767 **creep limit** || Kriechgrenze *f* || limite *m* de fluage || kruipgrens *f* (*m*) || предел *m* ползучести

C768 **creep resistance** *see* **creep strength**

C769 **creep strain** || Kriechdehnung *f* || déformation *f* due au fluage || kruipdeformatie *f* || деформация *f* ползучести

C770 **creep strength** || Kriechfestigkeit *f* || résistance *f* au fluage || kruipbestendigheid *f* || сопротивление *n* ползучести

C771 **creep test** || Kriechprüfung *f* || essai *m* de fluage || kruipbeproeving *f* || испытание *n* на ползучесть

C772 **crest** || **1.** Kamm *m*; Krone *f* **2.** Gewindespitze *f*; Zahnkrone *f* || **1.** crête *f* **2.** sommet *m* *(du filet, de la dent)* || **1.** kam *m*; kruin *f* (*m*) **2.** schroefdraadkruin *f* (*m*); tandtop *m* || **1.** гребень *m*, пик *m*, максимум *m* **2.** гребень *m* *(резьбы)*; вершина *f* *(головки зуба)*

C773 **crest of thread** || Gewindespitze *f* || sommet *m* du filet || schroefdraadkruin *f* (*m*) || гребень *m* резьбы

C774 **crest voltage** || Spitzenspannung *f*, Scheitelspannung *f* || tension *f* de crête || topspanning *f* || пиковое [амплитудное] напряжение *n*

C775 **crest voltmeter** || Spitzenspannungsmesser *m*, Spitzenvoltmeter *n* || voltmètre *m* de crête || topspanningsmeter *m* || пиковый [амплитудный] вольтметр *m*

C776 **criterion** || Kriterium *n* || critère *m* || kenmerk *n*, criterium *n* || критерий *m*

C777 **critical** || kritisch || critique || kritisch || критический

C778 **critical angle** || Grenzwinkel *m*; kritischer Winkel *m* || angle *m* limite; angle *m* critique || grenshoek *m*; kritische hoek *m* || предельный угол *m*; критический угол *m*

C779 **critical fluctuations** || kritische Fluktuationen *f* || fluctuations *f pl* critiques || kritische fluctuaties *f pl* || флуктуации *f pl* в критической точке

C780 **criticality** || Kritizität *f* || criticalité *f* || kritikaliteit *f* || критичность *f*

C781 **critical mass** || kritische Masse *f* || masse *f* critique || kritische massa *f* (*m*) || критическая масса *f* *(делящегося изотопа)*, критмасса *f*

C782 **critical point** || kritischer Punkt *m* || point *m* critique || kritisch punt *n* || критическая точка *f*

C783 **critical pressure** || kritischer Druck *m* || pression *f* critique || kritische druk *m* || критическое давление *n*

C784 **critical region** || kritischer Bereich *m* || région *f* critique || kritisch gebied *n* || критическое множество *n*, критическая область *f*; область *f* непринятия *(гипотезы)*

C785 **critical state** || kritischer Zustand *m* || état *m* critique || kritische toestand *m* || критическое состояние *n*

C786 **critical temperature** || kritische Temperatur *f* || température *f* critique || kritische temperatuur *f* || критическая температура *f*

C787 **critical value** || kritischer Wert *m* || valeur *f* critique || kritische waarde *f* || критическое значение *n* *(величины)*

C788 **critical velocity** || kritische Geschwindigkeit *f* || vitesse *f* critique || kritische snelheid *f* || критическая скорость *f*

C789 **cross** || Fadenkreuz *n* || croix *f* || kruis *n* || перекрестие *n*, крест *m* (нитей)

C790 **crossbar, crossbeam** || Querträger *m*, Traverse *f*; Strebe *f* || traverse *f*; poutre *f* transversale; contrefiche *f* || dwarshout *n*; dwarsligger *m*, dwarsbalk *m*; dwarsstaaf *f* (*m*) || поперечина *f*; поперечная балка *f*; раскос *m*

C791 **cross coupling** || Kreuzkopplung *f* || couplage *m* en croix || kruiskoppeling *f* || перекрёстная связь *f*

C792 **cross drive** || Antrieb *m* in Querrichtung || transmission *f* transversale || aandrijving *f* in dwarsrichting || поперечная передача *f*

C793 **crossed fields** || gekreuzte Felder *n pl* || champs *m pl* croisés || gekruiste velden *n pl* || скрещённые поля *n pl*

C794 **cross feed** || Quervorschub *m*, Planvorschub *m* || avance *f* transversale || dwarsbeweging *f*, dwarsvoeding *f* || поперечная подача *f*

C795 **crosshead** || Kreuzkopf *m* || crosse *f*, crossette *f* || kruishoofd *n*, kruiskop *m* || ползун *m*, крейцкопф *m*

C796 **cross joint** || Kreuzschaltung *f* || connexion *f* en croix || kruisverbinding *f* || перекрёстное соединение *n*

C797 **cross product** || Vektorprodukt *n* || produit *m* vectoriel || vectorprodukt *n* || векторное произведение *n*

C798 **cross ratio** || Doppelverhältnis *n* || rapport *m* anharmonique || invariante dubbelverhouding *f* || инвариантное двойное отношение *n*, ангармоническое [сложное] отношение *n*

C799 **cross section** || **1.** Querschnitt *m* **2.** Wirkungsquerschnitt *m* || **1.** section *f* transversale **2.** section *f* efficace || **1.** dwarssnede *f* (*m*) **2.** effectieve doorsne(d)e *f* (*m*) || **1.** поперечное сечение *n* **2.** эффективное сечение *n* *(процесса)*

C800 **cross-sectional area** || Querschnittsfläche *f* || aire *f* de la section transversale || doorsneevlak *n* || площадь *f* поперечного сечения

C801 **crown gear** || Kron(en)rad *n* || roue-couronne *f*, roue *f* à couronne || kroonwiel *n* || коронное (зубчатое) колесо *n*, коронная шестерня *f*

C802 **crucible steel** || Tiegelstahl *m* || acier *m* au creuset || kroezenstaal *n* || тигельная сталь *f*

C803 **crude structure** || Grobstruktur *f* || structure *f* grossière || grofstructuur *f* || грубая [общая] структура *f*

C804 **crushing load** *see* **breaking load**

C805 **crushing test** || Druckversuch *m* || essai *m* d'écrasement || brekingsbeproeving *f* || испытание *n* на раздавливание

C806 **cryogenic** || kryogen, Kryogen... || cryogénique || cryogeen || криогенный, низкотемпературный

C807 **cryogenics** || Kryogenik *f* || cryogénique *f* || cryogeniek *f* || криогеника *f*, физика *f* и техника *f* низких температур

C808 **cryostat** || Kryostat *m*, Kühlanlage *f* || cryostat *m* || cryostaat *m* || криостат *m*

C809 **crystal** || Kristall *m* || cristal *m* || kristal *n* || кристалл *m*
C810 **crystal axis** || Kristallachse *f* || axe *m* de cristal || kristalas *f* (*m*) || ось *f* кристалла
C811 **crystal detector** || Kristalldetektor *m*, Kristallgleichrichter *m* || détecteur *m* à cristal [à galène] || kristaldetector *m* || кристаллический детектор *m*
C812 **crystal grain** || Kristallkorn *n* || cristallite *f* || kristalkorrel *m*, kristalliet *m* || кристаллит *m*, кристаллическое зерно *n*
C813 **crystal lattice** || Kristallgitter *n*, Gitter *n* || réseau *m* cristallin || kristalrooster *m*, *n* || кристаллическая решётка *f*
C814 **crystal lattice parameter** || Kristallgitterkonstante *f*, Gitterkonstante *f* || constante *f* de réseau cristallin || kristalroosterconstante *f* (*m*) || параметр *m* [постоянная *f*] кристаллической решётки
C815 **crystalline** || kristallin, Kristall... || cristallin || kristallijn || кристаллический
C816 **crystalline state** || kristalliner Zustand *m* || état *m* cristallin || kristallijne toestand *m* || кристаллическое состояние *n*
C817 **crystalline structure** || kristalline Struktur *f* || structure *f* cristalline || kristallijne structuur *f* || кристаллическая структура *f*
C818 **crystallinity** || Kristallinität *f* || cristallinité *f* || kristalliniteit *f* || кристаллическое состояние *n*, кристалличность *f*
C819 **crystallite** || Kristallit *m*, Kristallkorn *n* || cristallite *f* || kristalliet *m* || кристаллит *m*
C820 **crystallization** || Kristallisation *f*, Kristallisieren *n* || cristallisation *f* || kristallisatie *f* || кристаллизация *f*
C821 **crystallization centre** || Kristallisationszentrum *n*, Kristallkeim *m*, Kristallkern *m* || centre *m* [noyau *m*] de cristallisation || kristallisatiekern *f* (*m*) || центр *m* [ядро *n*] кристаллизации
C822 **crystallographic** || kristallografisch || cristallographique || kristallografisch || кристаллографический
C823 **crystallography** || Kristallografie *f*, Kristallkunde *f* || cristallographie *f* || kristallografie *f* || кристаллография *f*
C824 **crystal optics** || Kristalloptik *f* || optique *f* cristalline || kristaloptica *f* || кристаллооптика *f*, оптика *f* кристаллов
C825 **crystal plane** || Kristallebene *f* || plan *m* cristallographique || kristalvlak *n* || кристаллографическая плоскость *f*
C826 **crystal physics** || Kristallphysik *f* || cristallophysique *f* || kristalfysica *f* || кристаллофизика *f*, физика *f* кристаллов
C827 **crystal rectifier** || Kristallgleichrichter *m*, Kristalldetektor *m* || redresseur *m* à cristal || kristalgelijkrichter *m* || кристаллический выпрямитель *m*, кристаллический детектор *m*
C828 **crystal spectrometer** || Kristallspektrometer *n*, Braggsches Kristallspektrometer *n* || spectromètre *m* à cristal || kristalspectrometer *m* || кристалл-спектрометр *m*
C829 **crystal system** || Kristallsystem *n*, Syngonie *f* || système *m* cristallin || kristalstelsel *n* || кристаллографическая система *f*, сингония *f*
C830 **cube** || Würfel *m* || cube *m* || kubus *m* || куб *m*
C831 **cubical parabola** || kubische Parabel *f* || parabole *f* cubique || kubische parabool *f* (*m*) || кубическая парабола *f*
C832 **cubic equation** || kubische Gleichung *f* || équation *f* cubique || kubische vergelijking *f* || кубическое уравнение *n*, уравнение *n* третьей степени
C833 **cumulant** || Semiinvariante *f* || cumulant *m* || semiinvariante *f* (*m*) || семиинвариант *m*
C834 **cumulative** || kumulativ || cumulatif || cumulatief || кумулятивный
C835 **cumulative frequency** || kumulative Frequenz *f* || fréquence *f* cumulative || cumulatieve frequentie *f* || накопленная частота *f*

C836 **cup head bolt** || Halbrundkopfbolzen *m* || boulon *m* à tête demi-ronde || halfrondkopbout *m* || болт *m* с полукруглой головкой

C837 **cup wheel** || Topfschleifscheibe *f* || meule *f* boisseau || komvormige slijpschijf *f* (*m*) || чашечный шлифовальный круг *m*

C838 **curie** || Curie *n* || curie *m* || curie *f*, *m* || кюри *f*, *m* *(3,7 10^{10} распадов/с)*

C839 **Curie point, Curie temperature** || Curie-Temperatur *f*, Curie-Punkt *m* || température *f* [point *m*] de Curie || Curie-temperatuur *f*, Curie-punt *f* (*m*) || точка *f* [температура *f*] Кюри

C840 **curl** || 1. Wirbelring *m* 2. *Math.* Rotation *f* *(Vektorfeld)* || 1. tourbillon *m* 2. *math.* rotationnel *m* || 1. wervel *m* 2. *math.* rotor *m* || 1. вихрь *m*, вихревое кольцо *n* 2. *мат.* ротор *m* *(векторной функции)*

C841 **curl field** || Wirbelfeld *n*, Drehfeld *n* || champ *m* rotationnel [tourbillonnaire] || wervelveld *n* || вихревое поле *n*

C842 **current** || 1. Strom *m*, elektrischer Strom *m*; Stromstärke *f* 2. Strömung *f*; Fluß *m* || courant *m* || 1. stroomsterkte *f* 2. stroom *m*, stroming *f* || 1. (электрический) ток *m*, сила *f* тока 2. поток *m*, течение *n*

C843 **current amplification** || Stromverstärkung *f* || amplification *f* de courant || stroomversterking *f* || усиление *n* по току

C844 **current balance** || Stromwaage *f* || balance *f* électrodynamique || stroombalans *f* (*m*) van Kelvin || токовые весы *pl*, весы *pl* Кельвина

C845 **current-carrying** || stromführend, stromdurchflossen || conducteur || stroomvoerend || токонесущий

C846 **current circuit** || Stromkreis *m* || circuit *m* de courant || stroomketen *f* (*m*) || цепь *f* тока

C847 **current collector** || Stromabnehmer *m* || collecteur *m* de courant || stroomafnemer *m* || токосниматель *m*; коллектор *m*

C848 **current-conducting** || stromleitend, stromführend || conducteur || stroomgeleidend || токопроводящий

C849 **current density** || Stromdichte *f* || densité *f* de courant || stroomdichtheid *f* || плотность *f* тока

C850 **current gain** || Stromverstärkung *f*; Stromgewinn *m* || amplification *f* de courant || stroomversterking *f* || усиление *n* по току

C851 **current generator** || Stromgenerator *m* || générateur *m* de courant || stroomgenerator *m* || генератор *m* тока

C852 **current intensity** || Stromstärke *f* || intensité *f* de courant || stroomsterkte *f* || сила *f* тока

C853 **current loop** || geschlossener Stromkreis *m* || circuit *m* fermé [bouclé] || geslotene stroomketen *f* (*m*) || виток *m* тока, замкнутый контур *m* (с током)

C854 **current noise** || Stromrauschen *n* || bruits *m pl* de courant || stroomruis *m* || токовые шумы *m pl*, шумы *m pl* тока

C855 **current output** || Ausgangsstrom *m* || courant *m* de sortie || uitgangsstroom *m* || выходной ток *m*, ток *m* на выходе

C856 **current overload** || Stromüberlastung *f* || surcharge *f* de courant || stroomoverbelasting *f* || перегрузка *f* по току

C857 **current pulse** || Stromimpuls *m*, Stromstoß *m* || impulsion *f* de courant || stroomimpuls *m* || импульс *m* тока

C858 **current ratio** || Stromübersetzung *f* || rapport *m* de transformation de courant || stroomoverzetting *f* || коэффициент *m* трансформации по току

C859 **current resonance** || Stromresonanz *f*, Parallelresonanz *f* || résonance *f* de courant || stroomresonantie *f* || резонанс *m* токов, параллельный резонанс *m*

C860 **current sense** || Stromrichtung *f* || sens *m* de courant || stroomrichting *f* || направление *n* тока

C861 **current source** || Stromquelle *f* || source *f* de courant || stroombron *f* (*m*) || источник *m* тока

C862 **current supply** || Stromversorgung *f* || alimentation *f* en courant || stroomvoorziening *f* || питание *n* током; подвод *m* тока

C863 **current transformer** || Stromtransformator *m*, Stromwandler *m* || transformateur *m* de courant || stroomtransformator *m* || трансформатор *m* тока

C864 **current-voltage characteristic** || Stromspannungskennlinie *f*, Strom-Spannungs-Charakteristik *f* || caractéristique *f* volt-ampère || stroom-spanningskarakteristiek *f* || вольтамперная характеристика *f*

C865 **current winding** || Stromwicklung *f* || enroulement *m* de courant || stroomwikkeling *f* || токовая обмотка *f*

C866 **curvature** || Krümmung *f* || courbure *f* || kromming *f* || кривизна *f*

C867 **curvature of field** || Bildfeldwölbung *f* || courbure *f* de champ || beeldveldkromming *f* || кривизна *f* поля (изображения)

C868 **curvature tensor** || Krümmungstensor *m* || tenseur *m* de courbure || krommingstensor *m* || тензор *m* кривизны

C869 **curvature vector** || Krümmungsvektor *m* || vecteur *m* de courbure || krommingsvector *m* || вектор *m* кривизны

C870 **curve** || Kurve *f* || courbe *f* || kromme *f* (*m*) || кривая *f*

C871 **curve gauge** || **1.** Schablone *f* **2.** Profillehre *f* || **1.** pistolet *m* à dessin, règle *f* courbe **2.** gabarit *m* de contrôle || **1.** bochtmal *m* **2.** profielkaliber *n* || **1.** лекало *n* **2.** профильный калибр *m*

C872 **curve plotter** || Kurvenzeichner *m*, Kurvenschreiber *m*, Plotter *m* || traceur *m* de courbes || krommentekenaar *m* || графопостроитель *m*

C873 **curvilinear coordinates** || krummlinige Koordinaten *f pl* || coordonnées *f pl* curvilignes || kromlijnige coördinaten *m pl* || криволинейные координаты *f pl*

C874 **curvilinear motion** || krummlinige Bewegung *f* || mouvement *m* curviligne || kromlijnige beweging *f* || криволинейное движение *n*

C875 **cushion** || (elastische) Auflage *f*, Polster *n*, Puffer *m*; Dämpfer *m* || plaque *f* [garniture *f*] élastique || kussen *n*, buffer *m*; schokdemper *m* || упругая прокладка *f*; амортизатор *m*

C876 **cushion-shaped distortion** || kissenförmige Verzeichnung *f*, Kissenverzeichnung *f* || distorsion *f* en coussinet || kussendistorsie *f*, kussenvervorming *f* || подушкообразная дисторсия *f*

C877 **cusp** || Rückkehrpunkt *m* || point *m* de rebroussement || terugkeerpunt *f* (*m*) || точка *f* возврата

C878 **cut** || **1.** Spanen *n*, Zerspanen *n* **2.** Hieb *m* *(Feile)* **3.**, **4.** Schnitt *m* || **1.** coupe *f*, coupage *m* **2.** taille *f* **3.** section *f* **4.** coupe *f* || **1.** snede *f* (*m*), snijding *f* (*m*) **2.** vijlen *n* **3.**, **4.** snede *f* (*m*) || **1.** резание *n*, обработка *f* резанием **2.** насечка *f* *(напильника)* **3.** разрез *m*, распил *m* **4.** сечение *n*

C879 **cutoff** || **1.** Abschneidung *f*, Trennung *f* **2.** Abschaltung *f* || coupure *f* || afsluiting *f* || **1.** отсечка *f* **2.** отключение *n*

C880 **cutter** || Schneidwerkzeug *m*; Fräser *m* || outil *m* coupant [tranchant, de coupe] || 1. snijgereedschap *n*; frees *f* (*m*) || режущий инструмент *m*; фреза *f*

C881 **cutter head** || Fräskopf *m*; Messerkopf *m* || tête *f* à tailler || freeskop *m*; snijkop *m* || фрезерная головка *f*; резцовая головка *f*

C882 **cutting angle** || Schnittwinkel *m* || angle *m* de coupe || snijhoek *m* || угол *m* резания

C883 **cutting depth** || Schnittiefe *f*; Spantiefe *f* || profondeur *f* de coupe || snijdiepte *f* || глубина *f* резания

C884 **cutting edge** || Schneide *f*, Schneidkante *f* || tranchant *m*, arête *f* coupante || snijkant *f* (*m*) || режущая кромка *f*

C885 **cutting edge side rate** || Spanwinkel *m* (der Hauptschneide) || angle *m* de pente de l'outil || spaanhoek *m* || передний угол *m* *(резца)*

C886 **cutting face** *see* **major cutting edge**

C887 **cutting fluid** || Schneidflüssigkeit *f* || huile *f* de coupe, liquide *m* d'arrosage || smeerkoelvloeistof *m* || смазочно-охлаждающая жидкость *f*, эмульсия *f*

C888 **cutting force** || Schnittkraft *f* || effort *m* [puissance *f*] de coupe || snijkracht *f* (*m*) || усилие *n* резания

C889 **cutting lip** *see* **cutting edge**

C890 **cutting metal** || Schneidmetall *n* || métal *m* dur pour outil coupant || hardmetaal *n* || твёрдый сплав *m* для обработки резанием

C891 **cutting-off machine** || Abstechmaschine *f* || tronçonneuse *f*; machine *f* [tour *m*] à tronçonner || afkortmachine *f* || отрезной станок *m*

C892 **cutting-off tool** || Stechmeißel *m*; Abstechwerkzeug *n* || outil *m* à tronçonner || snijwerktuig *n*, snijgereedschap *n* || отрезной резец *m*

C893 **cutting speed** || Schnittgeschwindigkeit *f* || vitesse *f* de coupe || snijsnelheid *f* || скорость *f* резания

C894 **cutting tool** || Schneidwerkzeug *n*; Meißel *m* || outil *m* coupant || snijgereedschap *n*, snijwerktuig *n* || режущий инструмент *m*

C895 **cycle** || **1.** Zyklus *m*; Periode *f* **2.** Kreisprozeß *m*; Kreislauf *m*; Lastspiel *n* || cycle *m* || cyclus *m*, kringloop *m*; kringproces *n* || **1.** цикл *m*, период *m* **2.** цикл *m*, круговой [замкнутый] процесс *m*; такт *m*

C896 **cycle group** || zyklische Gruppe *f* || groupe *m* cyclique || cyclische groep *f* (*m*) || циклическая группа *f*

C897 **cycle per second, Hz** || Hertz *n* || hertz *m* || hertz *m* || герц *m*, гц

C898 **cyclic** || zyklisch, kreisläufig || cyclique || cyclisch || циклический, круговой

C899 **cyclic frequency** || Kreisfrequenz *f* || fréquence *f* circulaire [cyclique] || cyclische frequentie *f* || круговая [циклическая] частота *f*

C900 **cyclic graph** || zyklischer Graph *m* || graphe *m* cyclique || cyclische graaf *f* (*m*) || циклический граф *m*

C901 **cyclic loading** || Dauerschwingbeanspruchung *f*, zyklische Beanspruchung *f* || chargement *m* cyclique || cyclische belasting *f* || периодическое нагружение *n*

C902 **cyclic process** || Kreisprozeß *m* || processus *m* [procédé *m*] cyclique || cyclisch proces *n* || круговой [циклический] процесс *m*, цикл *m*

C903 **cyclic stresses** || Schwingspannungen *f pl* || contraintes *f pl* cycliques || cyclusspanningen *f pl* || циклические напряжения *n pl*, напряжения *n pl* цикла

C904 **cycloid** || Zykloide *f* || cycloïde *f* || cycloïde *f* || циклоида *f*

C905 **cyclotron** || Zyklotron *n* || cyclotron *m* || cyclotron *n* || циклотрон *m*

C906 **cyclotron radiation** || Zyklotronstrahlung *f* || rayonnement *m* cyclotron [synchrotron] || cyclotronstraling *f* || циклотронное [магнитотормозное] излучение *n*

C907 **cyclotron resonance** || Zyklotronresonanz *f*, gyromagnetische Resonanz *f* || résonance *f* cyclotronique || cyclotronresonantie *f* || циклотронный резонанс *m*

C908 **cylinder** || Zylinder *m* || cylindre *m* || cilinder *m* || цилиндр *m*

C909 **cylinder coordinates** || zylindrische Koordinaten *f pl* || coordonnées *f pl* cylindriques || cilindrische coördinaten *m pl* || цилиндрические координаты *f pl*

C910 **cylinder function** || zylindrische Funktion *f* || fonction *f* cylindrique || cilindrische functie *f* || цилиндрическая функция *f*

C911 **cylinder head** || Zylinderkopf *m* || culasse *f* || cilinderkop *m*, cilinderdeksel *n* || головка *f* цилиндра; головка *f* блока цилиндров

C912 **cylinder liner, cylinder sleeve** || Zylinderlaufbuchse *f* || fourrure *f* de cylindre || cilinderbus *f* (*m*) || гильза *f* цилиндра

C913 **cylindrical gear** || Stirnrad *n* || roue *f* cylindrique || cilindervormig tandwiel *n*; || цилиндрическое зубчатое колесо *n*, цилиндрическая шестерня *f*

C914 **cylindrical grinding** || Rundschleifen *n* || rectification *f* cylindrique || rondslijpen *n* || круглое шлифование *n*

C915 **cylindrical grinding machine** || Rundschleifmaschine *f* || rectifieuse *f* cylindrique || rondslijpmachine *f* || круглошлифовальный станок *m*

C916 **cylindrical surface** || zylindrische Oberfläche *f* || surface *f* cylindrique || rolvormig oppervlak *n* || цилиндрическая поверхность *f*

D

D1 **damped** || gedämpft || amorti || gedempt || демпфированный; амортизированный

D2 **damped oscillation** || gedämpfte Schwingungen *f pl* || oscillations *f pl* décroissantes [amorties] || gedempte trillingen *f pl* || затухающие колебания *n pl*

D3 **damped wave** || gedämpfte Welle *f* || onde *f* atténuée [amortie] || gedempte golf *f* (*m*) || затухающая волна *f*

D4 **damper** || Dämpfer *m* || damper *m*, amortisseur *m* || demper *m* || демпфер *m*; амортизатор *m*; успокоитель *m*

D5 **damping** || Dämpfung *f*, Dämpfen *n* || amortissement *m* || demping *f* || демпфирование *n*; амортизация *f*

D6 **damping coefficient** *see* **damping factor**

D7 **damping constant** || Dämpfungskonstante *f* || constante *f* d'amortissement || dempingsconstante *f* (*m*) || постоянная *f* затухания

D8 **damping decrement** || Dämpfungsdekrement *n*, Dämpfungsabnahme *f* || décrément *m* d'amortissement || dempingsdecrement *n* || декремент *m* затухания

D9, D10 **damping factor, damping ratio** || Dämpfungsfaktor *m*, Dämpfungsverhältnis *n* || coefficient *m* d'amortissement || dempingscoëfficiënt *m* || коэффициент *m* затухания

D11 **dark adaptation** || Dunkeladaptation *f*, Dunkelanpassung *f* || adaptation *f* à l'obscurité || donkeradaptatie *f* || темновая адаптация *f*

D12 **dark current** || Dunkelstrom *m* || courant *m* d'obscurité || donkerstroom *m* || темновой ток *m*

D13 **darkening** || Schwärzung *f* || noircissement *m* (photographique) || zwarting *f* || (фотографическое) почернение *n*

D14 **dark field** || Dunkelfeld *n* || champ *m* noir || donkerveld *n* || тёмное поле *n* (*микроскопа*)

D15 **data** || Daten *pl* || données *pl* || data *pl*, gegevens *pl* || данные *pl*

D16 **data bank** || Datenbank *f* || banque *f* de données || databank *f* (*m*) || банк *m* данных

D17 **data cell** || Magnetblatt *n* || cellule *f* de données || datacel *f* (*m*) || ячейка *f* хранения данных, магнокард *m*

D18 **data flowchart** || Datenflußplan *m* || diagramme *m* de flot des données || datastroomschema *n* || диаграмма *f* (потоков) данных

D19 **data input** || Dateneingabe *f* || entrée *f* de données || datainvoer *m* || ввод *m* данных

D20 **data output** || Datenausgabe *f* || sortie *f* de données || datauitvoer *m* || вывод *m* данных

D21 **data processing** || Datenverarbeitung *f* || traitement *m* de données || dataverwerking *f* || обработка *f* данных

D22 **data set** || Datensatz *m*; Datei *f* || ensemble *m* de données || gegevensbestand *n* || набор *m* данных

D23 **datum plane** || Bezugsfläche *f*, Bezugsebene *f* || plan *m* [surface *f*] de référence || referentie-oppervlakte *f* || базовая поверхность *f*, базовая плоскость *f*

D24 **daughter isotope** || Tochterisotop *n* || isotope *m* descendant || dochterisotoop *m* || дочерний изотоп *m*

D25 **daughter nucleus** || Folgekern, *m*, Tochterkern *m* || noyau *m* suivant || dochterkern *f* (*m*) || дочернее ядро *n*

D26 **d-chart** *see* **flowchart**

D27 **dead contact** || geöffneter Kontakt *m* || contact *m* ouvert ||

open contact *n* || разомкнутый контакт *m*

Д28 **dead load** || statische Belastung *f*, ruhende Last *f* || charge *f* permanente || statische belasting *f* || постоянная нагрузка *f*

D29 **dead space** || toter Raum *m* || espace *m* mort || dode ruimte *f* || мёртвое пространство *n*

D30 **dead time** || Totzeit *f*; Unempfindlichkeitszeit *f* || temps *m* mort || doodtijd *m* || мёртвое время *n*; время *n* нечувствительности

D31 **dead turns** || Leerwindungen *f pl*, tote Windungen *f pl* || spires *f pl* mortes || dode windingen *f pl* || холостые витки *m pl* *(обмотки)*

D32 **dead wire** || stromlose Leitung *f* || conducteur *m* mis hors tension || stroomloze leiding *f* || обесточенный провод *m*

D33 **dead zone** || tote Zone *f*, Totzone *f*; Unempfindlichkeitsbereich *m* || zone *f* morte || dode zone *f* || мёртвая область *f*, область *f* нечувствительности *(детектора)*

D34 **de Broglie wavelength** || De-Broglie-Wellenlänge *f*, de Brogliesche Wellenlänge *f* || longueur *f* d'onde de Broglie || De-Broglie-golflengte *f* || дебройлевская длина *f* волны

D35 **debug, debugging** || Fehlerbeseitigung *f*, Ausprüfen *n* || mise *f* au point || beproeving *f* || отладка *f* *(программы)*

D36 **deburring** || Entgraten *n* || ébavurage *m* || afbramen *n* || снятие *n* [удаление *n*] заусенцев

D37 **deburring machine** || Entgratmaschine *f* || machine *f* à ébavurer || afbraammachine *f* || станок *m* для удаления заусенцев

D38 **debye** || Debye *n* || Debye *m* || Debije *m* || дебай *m* *(3,33 10^{-30} Кл·м)*

D39 **debye temperature** || Debye-Temperatur *f*, Debyesche charakteristische Temperatur *f* || température *f* de Debye || Debijetemperatuur *f* || дебаевская *(характеристическая)* температура *f*

D40 **decagon** || Dekagon *n*, Zehneck *n* || décagone *m* || tienhoek *m* || десятиугольник *m*

D41 **decay** || **1.** Abklingen *n*, Abfall *m* **2.** Zerfall *m*, Kernzerfall *m*, radioaktiver Zerfall *m* || **1.** décroissance *f* de luminescense **2.** désintégration *f* || verval *n*; desintegratie *f* || **1.** спадание *n* свечения **2.** (радиоактивный) распад *m*

D42 **decay sequence** || (radioaktive) Zerfallsreihe *f*, Zerfallskette *f* || chaîne *f* de désintégration || desintegratiereeks *f*, vervalreeks *f* *(m)* || цепочка *f* радиоактивных распадов, радиоактивная цепочка *f*

D43 **decay time** || 1. Abklingzeit *f*, Abklingdauer *f* 2. Zerfallzeit *f* || temps *m* de désintégration || vervaltijd *m* || 1. время *n* затухания 2. время *n* распада

D44 **decelerated motion** || verzögerte Bewegung *f* || mouvement *m* retardé || vertraagde beweging *f* || замедленное движение *n*

D45 **decelerating field** || Bremsfeld *n* || champ *m* retardateur [de freinage] || remveld *n* || тормозящее поле *n*

D46 **deceleration** || Verzögerung *f*, negative Beschleunigung *f* || décélération *f* || vertraging *f*, negatieve versnelling *f* || замедление *n*, отрицательное ускорение *n*

D47 **deci-** || Dezi... || déci... || deci... || деци... *(приставка к наименованию дольной единицы, соответствующая 10^{-1})*

D48 **decibel,** dB || Dezibel *n* || décibel *m* || decibel *f* *(m)* || децибел *m*, дБ

D49 **decidable problem** || lösbares Problem *n* || problème *m* décidable || oplosbaar problem *n* || разрешимая проблема *f*

D50 **decile** || Dezil *n* || décile *m* || decile *n* || дециль *m*

D51 **decimal notation, decimal number system** || dezimales Zahlensystem *n* || système *m* décimal, système de numération de base 10 || tientallig stelsel *n*, decimaal stelsel *n* || десятичная система *f* счисления

D52 **decision procedure** || Entscheidungsprozedur *f* || procédure *f* de la décision [de la résolution] ||

beslissingswerkwijze *f* (*m*) || разрешающая процедура *f*

D53 **decision table** || Entscheidungstabelle *f* || table *f* de décision || beslissingstabel *f* (*m*) || таблица *f* (принятия) решений, решающая таблица *f*

D54 **declinometer** || Deklinationsmesser *m*, Deklinationsnadel *f* || déclinomètre *m* || declinometer *m* || деклинометр *m*, деклинатор *m*, укломер *m*

D55 **decomposition** || Zerlegung *f*; Zersetzung *f* || décomposition *f*; désintégration *f* || ontbinding *f*; desintegratie *f* || разложение *n*; распад *m*

D56 **decomposition heat** || Zersetzungswärme *f*, Zerfallswärme *f* || chaleur *f* de désagrégation || ontbindingwarmte *f* || теплота *f* разложения

D57 **decompression** || Dekompression *f* || décompression *f* || decompressie *f* || декомпрессия *f*

D58 **decoupling** || Entkopplung *f* || découplage *m* || ontkoppeling *f* || развязывание *n*, развязка *f* (*контуров*)

D59 **decrease** || Verringerung *f*, Verminderung *f*, Abnahme *f* || décroissance *f* || vermindering *f* || уменьшение *n*, убывание *n*

D60 **Dedekind section** || Dedekindscher Schnitt *m* || section *f* dédékindienne || Dedekind-snede *f* (*m*) || дедекиндово сечение *n*

D61 **dedendum** || Zahnfußhöhe *f* || hauteur *f* du pied de dent || tandvoethoogte *f* || высота *f* ножки зуба (*колеса*)

D62 **dedendum circle** || Fußkreis *m* || cercle *m* de pied, cercle *m* intérieur (*engrenage*) || voetcirkel *m*, binnencirkel *n* (*tandwiel n*) || окружность *f* ножек *или* впадин зубьев (*колеса*)

D63 **deduction** || **1.** Ableiten *n*, Herleiten *n* **2.** Deduktion *f* || déduction *f* || **1.** afleiden *n* **2.** deductie *f* || **1.** вывод *m* **2.** дедукция *f*

D64 **dee** || Duant *m*, Dee *n* || dé *m* || d-elektrode *f* || дуант *m* (*циклотрона*)

D65 **deep freezing** || Tiefkühlung *f* || refroidissement *m* poussé || diepvriezen *n* || глубокое охлаждение *n*

D66 **definite integral** || bestimmtes Integral *n* || intégrale *f* définie || bepaalde integraal *f* (*m*) || определённый интеграл *m*

D67 **definition** || **1.** Definition *f*, Schärfe *f* (*eines Bildes*) **2.** Definition *f* (*einer Größe*) || **1.** netteté *f* **2.** définition *f* || definitie *f* || **1.** резкость *f* (*изображения*) **2.** определение *n* (*величины, понятия*)

D68 **deflecting force** || Ablenkungskraft *f* || force *f* de déviation || afwijkingskracht *f* (*m*) || отклоняющая сила *f*

D69 **deflection** || **1.** Ablenkung *f*, Auslenkung *f*; Ausschlag *m* **2.** Durchbiegung *f*; Durchhang *m* || déflexion *f* || **1.** afwijking *f* **2.** doorbuiging *f* || **1.** отклонение *n* **2.** прогиб *m*; провес *m*

D70 **deflection coil** || Ablenkspule *f*, Auslenkspule *f* || bobine *f* de déflexion [de déviation] || afwijkingsspoel *f* (*m*) || отклоняющая катушка *f*

D71 **deflection plate** || Ablenkplatte *f* || électrode *f* de déviation || afwijkingselektrode *f* || отклоняющий электрод *m*, отклоняющая пластина *f*

D72 **deflection system** || Ablenksystem *n* || système *m* de déflexion || afwijkingssysteem *n* || отклоняющая система *f*

D73 **deflexion** *see* **deflection**

D74 **defocusing** || Defokussierung *f*, Entbündelung *f* || déconcentration *f*, défocalisation *f* || defocusseren *n* || дефокусировка *f*, расфокусировка *f*

D75 **deformability** || Verformbarkeit *f*, Formänderungsfähigkeit *f* || déformabilité *f* || vervormbaarheid *f* || деформируемость *f*, (упругая) податливость *f*

D76 **deformation** || Deformation *f*, Verformung *f*, Formänderung *f* || déformation *f* || deformatie *f*, vervorming *f* || деформация *f*, деформирование *n*

D77 **defroster** || Enteiser *m*, Entfroster *m* || dégivreur *m* || ontdooier *m* || дефростер *m*; антиобледенитель *m*

D78 **degenerate** || entartet, ausgeartet || dégénéré || gedegenereerd || вырожденный

D79 **degenerate level** || entartetes Niveau *n* || niveau *m* dégénéré || gedegenereerd niveau *n* || вырожденный (энергетический) уровень *m*

D80 **degenerate quadric** || ausgeartete Quadrik *f* || quadrique *f* dégénérée || degenereerde kwadriek *f* || вырожденная поверхность *f* второго порядка

D81 **degenerate state** || entarteter Zustand *m* || état *m* dégénéré || gedegenereerde toestand *m* || вырожденное состояние *n*

D82 **degeneration** || **1.** Entartung *f*, Ausartung *f*, Degeneration *f* **2.** negative Rückkopplung *f*, Gegenkopplung *f* || **1.** dégénération *f* **2.** réaction *f* négative || **1.** degeneratie *f* **2.** negative terugkoppeling *f* || **1.** вырождение *n* *(системы)* **2.** отрицательная обратная связь *f*

D83 **degradation of energy** || Abwertung *f* der Energie *f*, Energieentwertung *f*, Energiedegradation *f* || dégradation *f* d'énergie || energiedegradatie *f* || деградация *f* [обесценение *n*] энергии

D84 **degree** || Grad *m* || degré *m* || graad *m* || **1.** градус *m* **2.** степень *f*

D85 **degree of a polynomial** || Grad *m* eines Polynoms || degré *m* d'un polynôme || graad *m* van een polynoom || степень *f* многочлена

D86 **degree of a representation** || Darstellungsrang *m* || degré *m* de représentation || voorstellingsgraad *m* || ранг *m* представления

D87 **degree of compression** || Kompressionsgrad *m* || degré *m* de compression || compressiegraad *m* || степень *f* сжатия

D88 **degree of expansion** || Expansionsgrad *m*; Ausdehnungsgrad *m* || degré *m* d'expansion || uitzettingsgraad *m* || степень *f* расширения

D89 **degree of freedom** || Freiheitsgrad *m* || degré *m* de liberté || vrijheidsgraad *m* || степень *f* свободы

D90 **degree of vacuum** || Güte *f* des Vakuums || degré *m* de vide || vacuumgraad *m* || степень *f* разрежения

D91 **deka-** || Deka... || déca... || deca... || дека... *(приставка в наименовании кратной единицы, соответствующая 10^1)*

D92 **dekalitre** || Dekaliter *n* || décalitre *m* || decaliter *m* || декалитр *m*

D93 **delay** || (zeitliche) Verzögerung *f*, Verzug *m*, Zeitverzug *m* || retard *m* || vertraging *f* || запаздывание *n*, задержка *f*

D94 **delay line** || Verzögerungsleitung *f* || ligne *f* à retard [de délai] || vertragingslijn *f* (*m*) || линия *f* задержки

D95 **delay time** || Verzögerungszeit *f*, Verzugszeit *f* || temps *m* de retard || vertragingstijd *m* || время *n* задержки, время *n* запаздывания *(напр. сигнала)*

D96 **delayed neutron** || verzögertes [verspätetes] Neutron *n* || neutron *m* retardé || vertraagd neutron *n* || запаздывающий нейтрон *m* (деления)

D97 **delta circuit** || Dreieckschaltung *f* || couplage *m* en triangle || driehocksschakeling *f* || схема *f* соединения треугольником

D98 **delta connection** || Dreieckschaltung *f* || couplage *m* en triangle || driehoeksschakeling *f* || соединение *n* треугольником

D99 **delta function** || Delta-Funktion *f* || fonction-delta *f* || deltafunctie *f* || дельта-функция *f*

D100 **delta voltage** || Dreieckspannung *f*, Deltaspannung *f* || tension *f* (entre fils) de la ligne || driehoekspanning *f* || линейное напряжение *n* *(в трёхфазной системе)*

D101 **demagnetization** || Entmagnetisierung *f*, Demagnetisierung *f* || démagnétisation *f*, désaimantation *f* || demagnetisatie *f* || размагничивание *n*

D102 **demagnetization factor** || Entmagnetisierungsfaktor *m* || facteur *m* de démagnétisation || demagnetisatiefactor *m* || коэффициент *m* размагничивания

D103 **denominate quantity** || dimensionsbehaftete Größe *f* || grandeur *f* nommée || dimensionale [dimensionele] grootheid *f* || именованная [размерная] величина *f*

D104 **dense** || dicht || dense || dicht || плотный; густой

D105 **dense packing** || dichte Packung *f*, Dichtpackung *f* || assemblage *m* compact || dichte pakking *f* || плотная упаковка *f (атомов в кристалле)*

D106 **dense set** || überall dichte Menge *f* || ensemble *m* partout dense || overal dichte menigte *f* || всюду плотное множество *n*

D107 **densimeter** || Dichtemesser *m*, Dichtemeßgerät *n*; Aräometer *n* || densimètre *m* || dichtheidsmeter *m* || плотномер *m*; ареометр *m*; пикнометр *m*

D108 **densitometer** || Densitometer *n*, Schwärzungsmesser *m* || densi(to)mètre *m* || **1.** densimeter *m* **2.** dichtheidsmeter *m* || **1.** денситометр *m* **2.** плотномер *m*

D109 **density** || Dichte *f* || densité *f* || dichtheid *f*, densiteit *f* || плотность *f*

D110 **density gauge** || Dichtemesser *m* || densimètre *m* || dichtheidsmeter *m*, densimeter *m* || плотномер *m*

D111 **density impulse** || spezifischer Impuls *m* || impulsion *f* spécifique || specifieke impuls *m* || удельный импульс *m (на единицу объёма топлива)*

D112 **denumerability axiom** || Abzählbarkeitsaxiom *n* || axiome *m* de dénombrabilité || telbaarheidsaxioma *n* || аксиома *f* счётности

D113 **denumerable set** || abzählbare Menge *f* || ensemble *m* dénombrable || telbare menigte *f* || счётное множество *n*

D114 **dependent variable** || abhängige Variable *f* || variable *f* dépendente || afhankelijke variabele *f (m)* || зависимая переменная *f*

D115 **dephlegmation** || Dephlegmation *f* || déflegmation *f* || deflegmatie *f* || дефлегмация *f*

D116 **dephlegmator** || Dephlegmator *m* || déflegmateur *m* || deflegmator *m* || дефлегматор *m*

D117 **depolarization** || Depolarisation *f* || dépolarisation *f* || depolarisatie *f* || деполяризация *f*

D118 **depolarizer** || Depolarisator *m* || dépolarisateur *m* || depolarisator *m* || деполяризатор *m*

D119 **depression** || **1.** Senkung *f*, Depression *f* **2.** Unterdruck *m*, Vakuum *n* || dépression *f* || depressie *f* || **1.** понижение *n (уровня)* **2.** разрежение *n*, вакуум *m*

D120 **depression of wet bulb** || psychrometrische Differenz *f*, Psychrometerdifferenz *f* || différence *f* psychrométrique || psychrometrische differentie *f* || психрометрическая разность *f*

D121 **depressurization** || Druckentlastung *f* || dépressurisation *f* || drukontlasting *f* || сброс *m* [резкое понижение *n*] давления

D122 **depth finder** || Echolot *n* || écho-sondeur *m*, échomètre *m* || echolood *n* || эхолот *m*

D123 **depth of focus** || Schärfentiefe *f*, Tiefenschärfe *f* || profondeur *f* de foyer || scherptediepte *f* || глубина *f* резкости

D124 **depth of modulation** || Modulationstiefe *f* || profondeur *f* [taux *m*] de modulation || modulatiediepte *f* || глубина *f* модуляции

D125 **depth of thread** || Gewindetiefe *f* || profondeur *f* de filetage || draaddiepte *f* || глубина *f* нарезки

D126 **derivation** || **1.** Herleiten *n (Formel, Gleichung)* **2.** Derivation *f* || **1.** déduction *f (d'une formule, d'une équation)* **2.** dérivation *f* || **1.** afleiding *f (van een formule, vergelijking)* **2.** derivatie *f* || **1.** вывод *m (формулы, уравнения)* **2.** деривация *f*

D127 **derivative** || Ableitung *f* || dérivée *f* || afgeleide *f* || производная *f*

D128 **derived set** || abgeleitete Menge *f* || ensemble *m* dérivé || afgeleide menigte *f* || производное множество *n*

D129 **derived unit** || abgeleitete Einheit *f* || unité *f* dérivée || afgeleide eenheid *f* || производная единица *f*

D130 **derrick** || Derrick(kran) *m* || grue *f* derrick, derrick *m* || mastkraan *f* (*m*) || мачтовый [мачтово-стреловой] кран *m*

D131 **descriptive geometry** || darstellende Geometrie *f* || géométrie *f* descriptive || beschrijvende meetkunde *f* || начертательная геометрия *f*

D132 **design** || **1.** Entwurf *m*; Plan *m* **2.** Design *n* || **1.** projet *m*, plan *m*, calcul *m* 2. design *m* || ontwerp *n*; tekening *f* || **1.** проект *m*, план *m*, расчёт *m* **2.** дизайн *m*

D133 **designation** || Bezeichnung *f*; Bezeichnungsweise *f* || désignation *f* || aanwijzing *f* || обозначение *n* (*величины*)

D134 **design data** || Entwurfsdaten *pl*, Auslegungsdaten *pl* || données *f pl* de calcul || rekengegevens *n pl*, ontwerpgegevens *m pl* || расчётные данные *pl*, данные *pl* (для) проектирования

D135 **design load** || rechnerische Belastung *f*, Lastannahme *f* || charge *f* de projet || berekende belasting *f* || расчётная нагрузка *f*

D136 **design power** || Solleistung *f* || puissance *f* calculée || berekend vermogen *n* || расчётная [номинальная] мощность *f*

D137 **design stress** || rechnerische Spannung *f*, rechnerische Beanspruchung *f* || contrainte *f* de calcule; contrainte *f* calculée || berekende spanning *f* || расчётное напряжение *n*

D138 **design value** || rechnerischer Wert *m*, Auslegungswert *m* || valeur *f* calculée [nominale] || berekende grootheid *f* || расчётная величина *f*, расчётное [номинальное] значение *n*

D139 **desk (computing) machine** || Tischrechner *m* || machine *f* à additionner, additionneuse *f* || rekenmachine *f* || настольная вычислительная машина *f*, арифмометр *m*

D140 **desorption** || Desorption *f* || désorption *f* || desorptie *f* || десорбция *f*

D141 **destination** || Adresse *f*, Bestimmung *f*, Ziel *n* || destination *f* || destinatie *f*, adres *n*, bestemming *f* || адрес *m* записи

D142 **destructive interference** || destruktive Interferenz *f* || interférence *f* destructive || destructieve interferentie *f* || деструктивная интерференция *f*, интерференция *f* со взаимным ослаблением

D143 **detection** || **1.** Nachweis *m*, Detektion *f*; Feststellung *f* 2. Gleichrichtung *f*, Demodulation *f* || détection *f* || **1.** detectie *f* 2. gelijkrichten *n* || **1.** обнаружение *n* 2. детектирование *n*, выпрямление *n* (*тока*)

D144 **detection of errors** || Fehlerermittlung *f* || détection *f* des erreurs || foutdetectie *f* || обнаружение *n* [выявление *n*] ошибок

D145 **detector** || **1.** Detektor *m*, Demodulator *m*, Gleichrichter *m* 2. Nachweisgerät *n* || détecteur *m* || **1.** detector *m* 2. indicator *m* || **1.** детектор *m*; датчик *m* 2. указатель *m*, индикатор *m*

D146 **detergent** || Detergens *n*, Detergent *n*, Reinigungsmittel *n* || détergent *m*, détersif *m* || detergens *n*, wasmiddel *n* || моющее средство *n*

D147 **determinant** || Determinante *f* || déterminant *m* || determinant *m* || детерминант *m*, определитель *m*

D148 **determinant of a matrix** || Determinante *f* einer Matrix || déterminant *m* d'une matrice || determinant *m* van een matrix || определитель *m* матрицы

D149 **determinism** || Determinismus *m* || déterminisme *m* || determinisme *n* || детерминизм *m*

D150 **deterministic** || determiniert || déterminé || deterministisch || детерминированный

D151 **detonation** || Detonation *f* || détonation *f* || detonatie *f* || детонация *f*

D152 **detuning** || Verstimmung *f* || désaccord *m* || onenigheid *f* || нарушение *n* резонанса; расстройка *f*

D153 **deuterium,** D || Deuterium *n* || deutérium *m*, hydrogène *m* lourd || deuterium *n* || дейтерий *m*

D154 **deuteron** || Deut(e)ron *n* || deutéron *m* || deuteron *n* || дейтрон *m*, дейтон *m*

D155 **developer** || Entwickler *m* || révélateur *m* || ontwikkelaar *m* || проявитель *m* *(изображения)*

D156 **development** || **1.** *Math.* Entwicklung *f* **2.** Herleitung *f*, Ableitung *f* *(Formel)* **3.** Entwicklung *f*, Fotoentwicklung *f* **4.** Ausarbeitung *f*, Entwicklung *f* || **1.** *math.* développement *m* **2.** déduction *f* **3.** développement *m* **4.** élaboration *f* || **1.** *math.* ontwikkeling *f* **2.** afleiding *v* **3.** ontwikkeling *f* **4.** uitwerking *f* || **1.** *мат.* разложение *n* **2.** вывод *m* *(формулы)* **3.** (фотографическое) проявление *n* **4.** разработка *f*, развитие *n* *(напр. теории)*

D157 **deviation** || Ablenkung *f*, Deviation *f* || déviation *f* || deviatie *f* || отклонение *n*; девиация *f*

D158 **device** || Vorrichtung *f*, Gerät *n* || dispositif *m*, appareil *m* || inrichting *f*, toestel *n* || приспособление *n*; устройство *n*; прибор *m*

D159 **Dewar flask** || Dewar-Gefäß *n*, Dewarsches Gefäß *n* || vase *m* Dewar || Dewar-vat *n* || сосуд *m* Дьюара, дьюар *m*

D160 **dew point** || Taupunkt *m* || point *m* de rosée || dauwpunt *n* || точка *f* росы

D161 **dextrorotation** || Rechtsdrehung *f* || rotation *f* à droite || rechtsdraaiing *f* || правое вращение *n* *(плоскости поляризации)*

D162 **diad** || Diade *f* || diade *f* || diade *f* || диада *f*

D163 **diagonalization of matrices** || Diagonalisierung *f* von Matrizen || diagonalisation *f* des matrices || diagonalisering *f* van de matrizen || диагонализация *f* матриц

D164 **diagonal matrix** || Diagonalmatrix *f* || matrice *f* diagonale || diagonaalmatrix *f* || диагональная матрица *f*

D165 **diagram** || Diagramm *n*, Schaubild *n* || diagramme *m* || diagram *n* || диаграмма *f*, схема *f*

D166 **diagram of gears** || Getriebeplan *m* || schéma *m* cinématique || diagram *n* van de tandwieloverbrenging *f* || кинематическая схема *f*

D167 **dial** || **1.** Rundskala *f* **2.** Skalenscheibe *f* || cadran *m*; limbe *m* || ronde schaal *f* *(m)* (met schaalverdeling); schaalplaat *f* *(m)*; wijzerplaat *f* *(m)* || **1.** циферблат *m*, круговая шкала *f* **2.** лимб *m*

D168 **dial gauge** || Meßuhr *f* || indicateur *m* à cadran || meetuurwerk *n*, meetklokje *n* || индикатор *m* часового типа

D169 **dialysis** || Dialyse *f* || dialyse *f* || dialyse *f* || диализ *m*

D170 **diamagnet** || Diamagnetikum *n* || diamagnétique *m* || diamagnetiek *m* || диамагнетик *m*

D171 **diamagnetic** || diamagnetisch || diamagnétique || diamagnetisch || диамагнитный

D172 **diamagnetism** || Diamagnetismus *m* || diamagnétisme *m* || diamagnetisme *n* || диамагнетизм *m*

D173 **diameter** || Durchmesser *m* || diamètre *m* || diameter *m* || диаметр *m*

D174 **diameter at bottom of thread** || Kerndurchmesser *m* *(Gewinde)* || diamètre *m* intérieur *(du filet)* || inwendige schroefdraaddiameter *m* || внутренний диаметр *m* резьбы

D175 **diameter of screw** || Außendurchmesser *m* *(Gewinde)* || diamètre *m* extérieur du filet || uitwendige schroefdraaddiameter *m* || наружный диаметр *m* резьбы

D176 **diameter plane** || Diametralebene *f* || plan *m* diamétral || diametraal vlak *n* || диаметральная плоскость *f*

D177 **diametral pitch** || Modulkehrwert *m* || pas *m* diamétral, module *m* de la roue dentée || tandwielmodulus *m* || диаметральный питч *m*, модуль *m* (зубчатого колеса)

D178 **diamond** || Diamant *m* || diamant *m* || diamant *n* || алмаз *m*

D179 **diamond pyramid hardness** || Vickershärte *f* || dureté *f* Vickers ||

Vickers-hardheid *f* || твёрдость *f* по Виккерсу

D180 **diamond tool** || Diamantwerkzeug *n* || outil *m* (à) diamant || diamantbeitel *m* || алмазный резец *m*; алмазный инструмент *m*

D181 **diamond wheel** || Diamantschleifscheibe *f* || meule *f* diamantée; meule *f* diamant || diamantslijpschijf *f* (*m*) || алмазный шлифовальный круг *m*

D182 **diaphragm** || **1.** Blende *f*, Diaphragma *n* **2.** Membran *f* || membrane *f*, diaphragme *m* || diafragma *n* || диафрагма *f*; перегородка *f*; мембрана *f*

D183 **diaphragm pump** || Membranpumpe *f* || pompe *f* à diaphragme || membraanpomp *f* (*m*) || диафрагменный [мембранный] насос *m*

D184 **diaphragm valve** || Membranventil *n* || soupape *f* à membrane || membraanklep *f* (*m*) || диафрагменный клапан *m*

D185 **dichroism** || Dichroismus *m*, Zweifarbigkeit *f* || dichroïsme *m* || dichroïsme *n* || дихроизм *m*

D186 **die** || **1.** Matrize *f*; Gesenkoberteil *n*; Gesenk *n* **2.** Gewindeschneidkluppe *f*; Gewindeschneidkopf *m* || matrice *f*; étampe *f*; filière *f* || **1.** matrijs *f* (*m*); stempel *m* **2.** schroefsnijkop *m* || **1.** матрица *f*; пуансон *m*; штамп *m* **2.** клупп *m*; винто(на)резная головка *f*

D187 **die-casting machine** || Spritzgußmaschine *f* || machine *f* à couler sous pression || persgietmachine *f* || машина *f* для литья под давлением

D188 **die forging** || Gesenkschmieden *n* || matriçage *m*, étampage *m* à chaud || matrijzensmeden *n* || ковка *f* в штампах

D189 **die hammer** || Gesenkschmiedehammer *m* || pilon *m* estampeur, marteau *m* estampeur || gegroefde hamer *m* || штамповочный молот *m*

D190 **die head** || Gewindeschneidkopf *m* || tête *f* à fileter || schroefsnijkop *m* || винто(на)резная головка *f*; резьбонарезная головка *f*

D191 I **dielectric** || Dielektrikum *n*, Nichtleiter *m* || diélectrique *m* || diëlectricum *n* || диэлектрик *m*

D192 II **dielectric** || dielektrisch, nichtleitend || diélectrique || diëlektrisch || диэлектрический

D193 **dielectric constant** || Dielektrizitätskonstante *f* || constante *f* diélectrique, permittivité *f* || diëlektrische constante *f* (*m*) || диэлектрическая проницаемость *f*, диэлектрическая постоянная *f*

D194 **dielectric loss** || dielektrische Verluste *m pl* || pertes *f pl* diélectriques || diëlektrische verliezen *n pl* || диэлектрические потери *f pl*, потери *f pl* в диэлектрике

D195 **dielectric polarization** || dielektrische Polarisation *f* || polarisation *f* diélectrique || diëlektrische polarisatie *f* || диэлектрическая поляризация *f*. поляризация *f* диэлектрика

D196 **dielectric strength** || dielektrische Festigkeit *f*, Durchschlagsfestigkeit *f* || résistance *f* diélectrique || doorslagsterkte *f* || электрическая прочность *f*; пробивная напряжённость *f* диэлектрика

D197 **dielectric susceptibility** || dielektrische Suszeptibilität *f* || susceptibilité *f* diélectrique || diëlektrische susceptibiliteit *f* || электрическая восприимчивость *f*, коэффициент *m* электризации

D198 **diene** || Dien *n* || diène *m* || diëen *n* || диен *m*

D199 **diesel** *see* **diesel engine**

D200 **Diesel cycle** || Diesel-Prozeß *m*, Kreisprozeß *m* nach Diesel, Kreisprozeß *m* mit Selbstzündung || cycle *m* de Diesel || diesel-cyclus *m* || цикл *m* Дизеля, цикл *m* с воспламенением от сжатия

D201 **diesel engine** || Dieselmotor *m* || moteur *m* diesel || dieselmotor *m* || дизель *m*, дизельный двигатель *m*

D202 **diesel fuel** || Dieselkraftstoff *m* || huile *f* diesel || dieselbrandstof *f* (*m*) || дизельное топливо *n*

D203 **die-sinking machine** || Gesenk(kopier)fräsmaschine *f* || machine *f* à fraiser les matrices || matrijzenfreesmachine *f* || станок *m* для фрезерования ручьёв штампов

D204 **difference equation** || Differenz(en)gleichung *f* || équation *f* aux différences finies || differentievergelijking *f* || уравнение *n* в конечных разностях, разностное уравнение *n*

D205 **difference operator** || Differenzenoperator *m* || opérateur *m* de différences || differentieoperator *m* || разностный оператор *m*

D206 **differentiable function** || differenzierbare Funktion *f* || fonction *f* différentiable || differentiabele functie *f* || дифференцируемая функция *f*

D207 **differential** || 1. Ausgleichgetriebe *n*, Differential *n* 2. *Math.* Differential *n* || 1. différentiel *m* 2. *math.* dafférentielle *f* || differentiëel *n* || дифференциал *m*

D208 **differential calculus** || Differentialrechnung *f* || calcul *m* différentiel || differentiaalrekening *f* || дифференциальное исчисление *n*

D209 **differential gear** || Ausgleichgetriebe *n*, Differential *n* || différentiel *m*, engrenage *m* différentiel || differentiëel *n* || дифференциал *m*, дифференциальный механизм *m*

D210 **differential geometry** || Differentialgeometrie *f* || géométrie *f* différentielle || differentiële geometrie *f* [meetkunde *f*] || дифференциальная геометрия *f*

D211 **differential invariant** || Differentialinvariante *f* || invariant *m* différentiel || differentiële invariante *f* (*m*) || дифференциальный инвариант *m*

D212 **differential lever** || Ausgleichhebel *m* || levier *m* différentiel || differentiële hefboom *m* || дифференциальный рычаг *m*

D213 **differential motion** || Differentialbewegung *f* || mouvement *m* différentiel || differentiële beweging *f* || результирующее движение *n* в дифференциальном механизме

D214 **differential operator** || Differentialoperator *m* || opérateur *m* différentiel || differentiële operator *m* || дифференциальный оператор *m*

D215 **differential pressure** || Druckgefälle *n* || pression *f* différentielle || drukval *m* || перепад *m* [разность *f*] давлений

D216 **differential pressure gauge** || Differentialmanometer *n*, Differenzdruckmesser *m* || manomètre *m* différentiel || differentiaaldrukmeter *m* || дифференциальный манометр *m*, дифманометр *m*

D217 **differential resistance** || Differentialwiderstand *m* || résistance *f* différentielle || differentiaalweerstand *m* || дифференциальное сопротивление *n*

D218 **differential thermometer** || Differentialthermometer *n* || thermomètre *m* différentiel || differentiaalthermometer *m* || дифференциальный термометр *m*

D219 **differentiation** || Differentiation *f*, Differenzierung *f* || différentiation *f* || differentiatie *f* || дифференцирование *n*

D220 **diffracted beam** || (ab)gebeugtes Bündel *n* || faisceau *m* diffracté || gediffracteerde bundel *m* || дифрагированный пучок *m* [луч *m*]

D221 **diffracted wave** || (ab)gebeugte Welle *f* || onde *f* diffractée || gediffracteerde golf *f* (*m*) || дифрагированная [огибающая] волна *f*

D222 **diffraction** || Diffraktion *f*, Beugung *f* || diffraction *f* || diffractie *f* || дифракция *f*

D223 **diffraction analysis** || Beugungsanalyse *f* || analyse *f* radiocristallographique, radiocristallographie *f* || diffractieanalyse *f* || структурный анализ *m*

D224 **diffraction by aperture** || Beugung *f* an der Öffnung || diffraction *f* par une fente [par une ouverture] || diffractie *f* bij een opening || дифракция *f* на отверстии [на краях отверстия]

D225 **diffraction by disk** || Beugung *f* am runden Schirm || diffraction *f* par un écran rond || diffractie *f* bij een rond scherm || дифракция *f* на круглом экране

D226 **diffraction fringe** || Beugungsstreifen *m* || frange *f* de diffraction || diffractiestreep *f* (*m*) || дифракционная полоса *f*

D227 **diffraction grating** || Diffraktionsgitter *n*, Beugungsgitter *n* || réseau *m* de diffraction || diffractierooster *m*, *n* || дифракционная решётка *f*

D228 **diffraction image** || Beugungsbild *n* || image *f* de diffraction || diffractiebeeld *n* || дифракционное изображение *n*

D229 **diffraction pattern** || Beugungsbild *n* || figure *f* de diffraction || diffractiepatroon *n* || дифракционная картина *f*

D230 **diffuse image** || verwaschenes Bild *n* || image *f* floue || wazig beeld *n* || размытое [расплывчатое] изображение *n*

D231 **diffuse light** || Streulicht *n*, gestreutes [diffuses] Licht *n* || lumière *f* diffuse || diffuus licht *n* || рассеянный свет *m*

D232 **diffuseness** || Verwaschenheit *f*, Unschärfe *f* || flou *m* || wazigheid *f* || размытость *f*, расплывчатость *f*

D233 **diffuser** || Diffusor *m* || diffuseur *m* || diffusor *m* || диффузор *m*

D234 **diffuse reflection** || diffuse [gestreute] Reflexion *f*, Streureflexion *f* || réflexion *f* diffuse || diffuse reflectie *f* || диффузное отражение *n*

D235 **diffusion** || 1. Diffusion *f*; Diffundieren *f* 2. Verwaschung *f* || 1. diffusion *f* 2. affouillement *m* || 1. diffusie *f* 2. uitspoelen *n* || 1. диффузия *f* 2. размывание *n*, расплывание *n*

D236 **diffusion cloud chamber** || Diffusionsnebelkammer *f* || cellule *f* de diffusion, chambre *f* à diffusion || diffusiekamer *f* (*m*) || диффузионная камера *f*

D237 **diffusivity** || 1. Diffusionskoeffizient *m* 2. Temperaturleitfähigkeit *f*, Temperaturleitzahl *f* || 1. coefficient *m* de diffusion 2. conductibilité *f* thermique || 1. diffusiecoëfficiënt *m* 2. temperatuurgeleidingsvermogen *n* || 1. коэффициент *m* диффузии 2. температуропроводность *f*; коэффициент *m* температуропроводности

D238 **digit** || Ziffer *f* || digit *m*, chiffre *m* || cijfer *n* || цифра *f*

D239 **digital** || digital, Digital..., Ziffern... || digital || digitaal || цифровой; дискретный

D240 **digital computer** || Digitalrechner *m* || calculatrice *f* digitale || digitale rekenmachine *f* || цифровая вычислительная машина *f*, ЦВМ

D241 **digital device** || Digitalgerät *n* || appareil *m* digital || digitaal toestel *n* || цифровой [дискретный] прибор *m*

D242 **digital voltmeter** || Digitalvoltmeter *n*, Digitalspannungsmesser *m* || voltmètre *m* digital || digitaalvoltmeter *m* || цифровой вольтметр *m*

D243 **dilat(at)ion** *math.* || Homothetie *f* || dilatation *f* || dilatatie *f* || гомотетия *f*

D244 **dilatometer** || Dilatometer *n*, Ausdehnungsmesser *m* || dilatomètre *m* || dilatometer *m* || дилатометр *m*

D245 **dilemma** || Dilemma *n* || dilemme *m* || dilemma *n* || дилемма *f*

D246 **dilute solution** || verdünnte Lösung *f* || solution *f* diluée [faible] || verdunde oplossing *f* || разбавленный раствор *m*

D247 **dimension** || Dimension *f* || dimension *f* || dimensie *f* || 1. размерность *f* *(величины)* 2. размер *m*; габарит *m*

D248 **dimensional** || dimensional || dimensionnel || dimensionaal || размерный *(о величине)*

D249 **dimensional analysis** || Dimensionsanalyse *f* || analyse *f* dimensionnelle || dimensieanalyse *f* || анализ *m* размерностей

D250 **dimensional tolerance** || Maßtoleranz *f* || tolérance *f* dimensionnelle || maattolerantie *f* || допуск *m* на размер

D251 **dimensionless** || dimensionslos || sans-dimension, non dimensionnel || dimensieloos || безразмерный *(о величине)*

D252 **dimensionless quantity** || dimensionslose Größe *f*, Dimensionslose *f* || valeur *f* non dimensionnelle || dimensieloze grootheid *f* || безразмерная величина *f*

D253 **diminished image** || verkleinertes Bild *n*, verkleinerte Abbildung *f* || image *f* réduite || verkleind beeld *n* || уменьшенное изображение *n*

D254 **dimorphism** || Dimorphie *f*, Dimorphismus *m* || dimorphisme *m* || dimorfisme *n* || диморфизм *m*

D255 **diode** || Diode *f* || diode *f* || diode *f* (*m*) || диод *m*

D256 **diopter** || **1.** Diopter *n* **2.** Dioptrie *f* || **1.** viseur *m*, pinnule *f* **2.** dioptrie *f* || **1.** diopter *m* **2.** dioptrie *f* || **1.** диоптр *m* **2.** диоптрия *f*

D257 **dipole** || Dipol *m* || dipôle *m* || dipool *f* (*m*) || **1.** (электрический) диполь *m* **2.** диполь *m*, вибратор *m*

D258 **dipole moment** || Dipolmoment *n* || moment *m* dipolaire || dipoolmoment *n* || дипольный момент *m*

D259 **direct current** || Gleichstrom *m* || courant *m* continu || gelijkstroom *m* || постоянный ток *m*

D260 **direct-current circuit** || Gleichstromkreis *m* || circuit *m* de courant continu || gelijkstroomschakeling *f* || цепь *f* постоянного тока

D261 **direct-current generator** || Gleichstromgenerator *m* || générateur *m* à courant continu || gelijkstroomgenerator *m* || (электромашинный) генератор *m* постоянного тока

D262 **direct-current motor** || Gleichstrommotor *m* || moteur *m* à courant continu || gelijkstroommotor *m* || электродвигатель *m* постоянного тока

D263 **direct drive** || unmittelbarer Antrieb *m*, Direktantrieb *m* || commande *f* directe, entraînement *m* direct || direct drijfwerk *n*; onmiddellijke [directe] overbrenging *f* || прямой [непосредственный] привод *m*; прямая передача *f*

D264 **direct-excitation** || direkt [unmittelbar] erregt || à excitation directe || direct opgewekt || с непосредственным [прямым] возбуждением

D265 **direct-flow** || Gleichstrom... || à écoulement direct || gelijkstroom... || прямоточный

D266 **direction** || Richtung *f*, Sinn *m* || direction *f*, sens *m* || richting *f* || направление *n*

D267 **directional derivative** || Richtungsableitung *f* || dérivée *f* dans une direction || afleiding *f* van richting || производная *f* по направлению

D268 **directional diagram** || Richtdiagramm *n*, Richtungsdiagramm *n* || diagramme *m* directionnel || richtingsdiagram *n* || диаграмма *f* направленности *(излучения)*

D269 **directional radiation** || Richtstrahlung *f*, gerichtete Strahlung *f* || émission *f* directionnelle, radiation *f* dirigée || gerichte straling *f* || направленное излучение *n*

D270 **direction cosine** || Richtungskosinus *m* || cosinus *m* directeur || richtingscosinus *m* || направляющий косинус *m*

D271 **direction of force** || Kraftwirkungsrichtung *f* || direction *f* de la force || krachtrichting *f* || направление *n* (действия) силы

D272 **direct measurement** || direkte [unmittelbare] Messung *f*, Direktmessung *f* || mesure *f* directe [immédiate] || direct meten *n* || прямое [непосредственное] измерение *n*

D273 **direct method** || direkte Methode *f* || méthode *f* directe || directe methode *f* || прямой метод *m*

D274 **direct operation** || Direktsteuerung *f* || commande *f* directe || directe sturing *f* || непосредственное управление *n*

D275 **direct product** || direktes [cartesisches] Produkt *n* || produit *m* direct || direct produkt *m* || прямое [декартово] произведение *n*

D276 **directrix** || Direktrix *f* || directrice *f* || directrice *f* || директриса *f*

D277 **direct sum** || direkte Summe *f* || somme *f* directe || directe som *f* || прямая сумма *f*

D278 **direct voltage** || Gleichspannung *f* || tension *f* continue || gelijkspanning *f* || постоянное напряжение *n*

D279 **Dirichlet integral** || Dirichletsches Integral *n* || intégrale *f* de Dirichlet || Dirichlet-integraal *f* (*m*) || интеграл *m* Дирихле

D280 **disbalance** || 1. Unausgeglichenheit *f*; Unbalance *f* 2. Gleichgewichtsstörung *f* || déséquilibre *m* || onevenwicht *n* || 1. неуравновешенность *f* 2. нарушение *n* равновесия

D281 **disc** *see* **disk**

D282 **discharge** || 1. (elektrische) Entladung *f* 2. Abflußmenge *f* || 1. décharge *f* 2. débit *m* || 1. ontlading *f* 2. afvloeiing *f*; debiet *n* || 1. (электрический) разряд *m* 2. расход *m*; сток *m*

D283 **discharge current** || Entladungsstrom *m*, Entladestrom *m* || courant *m* de décharge || ontladingsstroom *m* || разрядный ток *m*

D284 **discharge gap** || Entladungsstrecke *f*, Entladestrecke *f* || distance *f* d'éclatement [de décharge] || ontladingsinterval *n* || разрядный промежуток *m*

D285 **discharge lamp** || Entladungslampe *f*, Entladungsröhre *f* || lampe *f* à décharge [à décharge gazeuse, à décharge luminescente] || ontladingsbuis *f* (*m*) || газоразрядная лампа *f*

D286 **discharger** || Funken(entladungs)strecke *f* || éclateur *m* || vonkbrug *f* (*m*) || разрядник *m*; искровой промежуток *m*

D287 **discharge valve** || Entlastungsventil *n* || soupape *f* de refoulement || ontladingsklep *f* (*m*) || разгрузочный клапан *m*

D288 **discharge voltage** || Entladespannung *f*, Entladungsspannung *f* || tension *f* de décharge || ontladingsspanning *f* || разрядное напряжение *n*

D289 **discharging** || Entladen *n*, Entladung *f* || décharge *f* || ontlading *f* || разрядка *f* (*напр. конденсатора*)

D290 **discharging jet** || Ausflußstrahl *m* || jet *m* d'écoulement || uitvloeistraal *m* (*f*) || вытекающая струя *f*

D291 **disconnecting switch** || Trennschalter *m*, Trenner *m* || sectionneur *m* || uitschakelaar *m* || разъединитель *m*

D292 **discontinuity** || Unstetigkeit *f* || discontinuité *f* || onderbreking *f* || разрыв *m*; нарушение *n* непрерывности; скачок *m*

D293 **discontinuity of the first kind** || Unstetigkeitspunkt *m* erster Art || discontinuité *f* de première espèce || onderbrekingspunt *n* van de eerste aard || точка *f* разрыва первого рода

D294 **discontinuous chip** || Bruchspan *m*; Reißspan *m* || copeaux *m pl* à déchirement, copeaux *m pl* discontinus || afgehakte spaan *f* (*m*); discontinue spaan *f* (*m*) || стружка *f* надлома; стружка *f* скалывания

D295 **discontinuous flow** || Nichtkontinuumströmung *f* || écoulement *m* non continu || discontinue stroming *f* || разрывное течение *n*

D296 **discrete (energy) level** || diskretes Energieniveau *n* || niveau *m* (énergétique) discret || discreet energieniveau *n* || дискретный (энергетический) уровень *m*

D297 **discrete quantity** || diskrete Größe *f* || quantité *f* discrete || discrete grootheid *f* || дискретная величина *f*

D298 **discrete random process** || diskreter Zufallsprozeß *m* || processus *m* aléatoire discret || discreet toevalsproces *n* || дискретный случайный процесс *m*

D299 **discrete set** || diskrete Menge *f* || ensemble *m* discret || discrete menigte *f* || дискретное множество *n*

D300 **discriminant** || Diskriminante *f* || discriminant *m* || discriminante *f* || дискриминант *m*

D301 **discriminant of an algebraic equation** || Diskriminante *f* einer algebraischen Gleichung || discriminant *m* d'une équation algébrique || discriminante *f* van een algebraïsche vergelijking || дискриминант *m* алгебраического уравнения

D302 **discriminator** || Diskriminator *m* || discriminateur *m* || discriminator *m* || дискриминатор *m*

D303 **disintegrating nucleus** || spaltbarer Kern *m* || noyau *m* fissile || splijtbare kern *f* || расщепляющееся ядро *n*

D304 **disintegration** || Spaltung *f* *(Atomkern)* || désintégration *f* || verval *n* || расщепление *n*, дезинтеграция *f* *(атомного ядра)*

D305 **disjoint elements** || disjunkte Elemente *n pl* || éléments *m pl* disjonctifs || disjonctieve elementen *n pl* || дизъюнктивные элементы *m pl*

D306 **disjoint events** || unvereinbare Ereignisse *n pl* || événements *m pl* disjoints || onverenigbare gebeurtenissen *f pl* || несовместные события *n pl*

D307 **disjunction** || Disjunktion *f* || disjonction *f* || disjunctie *f* || дизъюнкция *f*, логическое сложение *n*

D308 **disk** || Scheibe *f*; Diskus *m* || disque *m* || discus *m*; schijf *f* (*m*) || диск *m*

D309 **disk-and-drum turbine** *see* **combination turbine**

D310 **disk brake** || Scheibenbremse *f* || frein *m* à disque || schijfrem *f* (*m*) || дисковый тормоз *m*

D311 **disk clutch** || Scheibenkupplung *f* || embrayage *m* à disque || schijfkoppeling *f* || дисковое сцепление *n*

D312 **disk cutter** || Scheibenfräser *m* || fraise *f* plate [à disque] || schijffrees *f* (*m*) || дисковая фреза *f*

D313 **disk of least confusion** || kleinster Zerstreuungskreis *m* || cercle *m* de confusion minimale || kleinste verstrooiingscirkel *m* || кружок *m* наименьшего рассеяния

D314 **disk valve** || Tellerventil *n* || soupape *f* à plateau [à disque] || schijfventiel *n* || тарельчатый клапан *m*

D315 **dislocation** || Versetzung *f* || dislocation *f* || dislocatie *f* || дислокация *f* *(в кристалле)*

D316 **disorder** || Fehlordnung *f*, Ungeordnetsein *n* || désordre *m* || wanorde *f* || беспорядок *m*, неупорядоченность *f*

D317 **disordered structure** || ungeordnete Struktur *f* || structure *f* désordonnée || ongeordende [ongeregelde] structuur *f* || неупорядоченная [разупорядоченная] структура *f*

D318 **disperse system** || disperses System *n* || système *m* dispersé || dispers systeem *n* || дисперсная система *f*

D319 **dispersion** || **1.** Dispersion *f* **2.** Dispersität *f* || dispersion *f* || dispersie *f* || **1.** дисперсия *f* **2.** дисперсность *f* *(среды)*

D320 **dispersion of light** || Lichtzerlegung *f* || dispersion *f* de lumière || lichtdispersie *f* || дисперсия *f* света

D321 **dispersive power** || relative Dispersion *f* || pouvoir *m* dispersif || dispersievermogen *n* || относительная дисперсия *f*

D322 **displacement** || **1.** Verschiebung *f*; Verlagerung *f* **2.** Wasserverdrängung *f* **3.** Hubraum *m* *(Zylinder)* || **1.** déplacement *m* **2.** déplacement *m*, tonnage *m* de déplacement **3.** cylindrée *f* || **1.** verplaatsing *f*, verschuiving *f* **2.** waterverplaatsing *f* **3.** (cilinder)inhoud *m* || **1.** смещение *n*, перемещение *n* **2.** водоизмещение *n* **3.** литраж *m*; рабочий объём *m* *(цилиндра двигателя)*

D323 **displacement compressor** || Verdränger *m* || compresseur *m* volumétrique || zuigercompressor *m* || поршневой компрессор *m*

D324 **displacement current** || Verschiebungsstrom *m* || courant *m* de déplacement || verschuivingsstroom *m* || ток *m* смещения

D325 **displacement operator** || Verschiebungsoperator *m* || opérateur *m* de déplacement || verschuivingsoperator *m* || оператор *m* смещения

D326 **disproportionation** || Disproportionierung *f* || disproportionation *f* || disproportionatie *f* || диспропорционирование *n*

D327 **dissipated power** || Zerstreuungsleistung *f* zerstreute Leistung *f* || puissance *f* dissipée || dissipatievermogen *n* || рассеиваемая мощность *f*

D328 **dissipation** || Dissipation *f*, Zerstreuung *f* || dissipation *f* || verstrooiing *f*, dissipatie *f* || диссипация *f*, рассеяние *n*
D329 **dissipative** || dissipativ || dissipatif || dissipatief || диссипативный, неконсервативный
D330 **dissipative force** || dissipative Kraft *f* || force *f* dissipative || dissipatieve kracht *f* (*m*) || диссипативная сила *f*
D331 **dissipative system** || dissipatives System *n* || système *m* dissipatif || dissipatief systeem *n* || диссипативная система *f*
D332 **dissociation** || Dissoziation *f* || dissociation *f* || dissociatie *f* || диссоциация *f*
D333 **distance** || Abstand *m*, Entfernung *f*, Distanz *f* || distance *f* || afstand *m* || расстояние *n*
D334 **distance of closed approach** || geringster Abstand *m* bei der Annäherung, dichteste Annäherung *f* || distance *f* d'approche immédiate || kleinste toenaderingsafstand *m* || расстояние *n* наибольшего сближения
D335 **distance of normal vision** || Normsehweite *f*, deutlichste Sehweite *f* || distance *f* de vision normale || duidelijkste gezichtsafstand *m* || расстояние *n* наилучшего зрения
D336 **distillation** || Destillation *f*, Destillieren *n* || distillation *f* || destillatie *f* || дистилляция *f*, перегонка *f*
D337 **distortion** || 1. Verzerrung *f* 2. Verzeichnung *f*, Verzeichnungsfehler *m*, Distorsion *f* || distorsion *f* || 1. verwringing *f* 2. distorsie *f* || 1. искажение *n* 2. дисторсия *f*
D338 **distortion correction** || Entzerrung *f*, Verzerrungskompensation *f* || correction *f* des distorsions || verwringingscorrectie *f* || коррекция *f* искажений
D339 **distributed force** || verteilte Kraft *f* || force *f* distribuée || verdeelde kracht *f* (*m*) || распределённое усилие *n*
D340 **distributed load** || verteilte Last *f*, (stetig) verteilte Belastung *f* || charge *f* distribuée || verdeelde belasting *f* || распределённая нагрузка *f*
D341 **distributed mass** || verteilte Masse *f* || masse *f* distribuée || verdeelde massa *f* (*m*) || распределённая масса *f*
D342 **distribution function** || Verteilungsfunktion *f* || fonction *f* de distribution || verdelingsfunctie *f* || функция *f* распределения
D343 **distribution valve** || Steuerventil *n* || soupape *f* de distribution || verdelingsklep *f* (*m*) || распределительный клапан *m*
D344 **distributive law** || Distributivgesetz *n* || loi *f* distributive || distributieve wet *f* (*m*) || дистрибутивный закон *m*
D345 **distributivity** || Distributivität *f* || distributivité *f* || distributiviteit *f* || дистрибутивность *f*
D346 **distributor** || 1. Verteiler *m* 2. Leitapparat *m* (*Turbine*) || distributeur *m* || 1. verdeler *m* 2. turbineleiwiel *n* || 1. распределитель *m* 2. направляющий аппарат *m* (*турбины*)
D347 **disturbed motion** || gestörte Bewegung *f* || mouvement *m* perturbé || gestoorde beweging *f* || возмущённое движение *n*
D348 **disturbed orbit** || gestörte Bahn *f* || orbite *f* perturbée || verstoorde baan *f* (*m*) || возмущённая орбита *f*
D349 **divector** || Bivektor *m* || bivecteur *m* || bivector *m* || бивектор *m*
D350 **divergence** || Divergenz *f* || divergence *f* || divergentie *f* || дивергенция *f*
D351 **divergent** || divergent || divergent || divergerend, uiteenlopend || расходящийся
D352 **divergent beam** || divergierendes [divergentes] Bündel *n* || faisceau *m* divergent || divergerende bundel *m* || расходящийся пучок *m*
D353 **divergent series** || divergente Reihe *f* || série *f* divergente || uiteenlopende reeks *f* (*m*) || расходящийся ряд *m*
D354 **diverging lens** || Zerstreuungslinse *f*, Negativlinse *f* || lentille

f divergente || divergerende lens *f* (*m*) || рассеивающая [отрицательная] линза *f*

D355 **diverging waves** || ablaufende [fortlaufende] Wellen *f pl* || ondes *f pl* divergentes || divergerende golven *f* (*m*) *pl* || расходящиеся волны *f pl*

D356 **divided bearing** *see* **split bearing**

D357 **divided differences** || dividierte Differenzen *f pl* || différences *f pl* divisées || verdeelde differenties *f pl* || разделённые разности *f pl*

D358 **divider** || Spannungsteiler *m* || diviseur *m* de tension || spanningsdeler *m* || делитель *m* напряжения

D359 **dividing device** || Teiler *m*, Teilvorrichtung *f* || dispositif de division || deler *m*; deelinrichting *f* || делительное устройство *n*

D360 **dividing head** || Teilkopf *m* || diviseur *m*, poupée *f* diviseuse || verdeelkop *m* || делительная головка *f* (*станка*)

D361 **dividing machine** || Teilmaschine *f* || machine *f* à diviser || verdeelmachine *f* || делительная машина *f*

D362 **division** || **1.** *Math.* Teilung *f*, Division *f* **2.** Skalenteil *m*, Teilstrich *m*, Skalenstrich *m* || division *f* || **1.** *math.* verdeling *f* **2.** schaaldeel *n* || **1.** *мат.* деление *n* **2.** деление *n* (шкалы)

D363 **division algebra** || Divisionsalgebra *f* || algèbre *f* à division || verdelingsalgebra *f* (*m*) || алгебра *f* с делением

D364 **division algorithm** || Divisionsalgorithmus *m* || algorithme *m* de division || verdelingsalgoritme *n* || алгоритм *m* деления

D365 **divisior** || Teiler *m*, Divisor *m* || diviseur *m* || deler *m* || делитель *m*

D366 **dodecahedron** || Dodekaeder *n*, Zwölfflach *n* || dodécaèdre *m* || dodecaëder *m* || двенадцатигранник *m*

D367 **dog clutch** *see* **jaw clutch**

D368 **domain** || Domäne *f*, Bezirk *m*, Bereich *m* || domaine *m* || domein *n* || домен *m*

D369 **domain of definition** || Definitionsbereich *m (einer Funktion)* || domaine *m* de définition || definitiegebied *n (van een functie)* || область *f* определения *(функции)*

D370 **domain wall** || Domänengrenzfläche *f*, Domänengrenze *f* || limite *f* de domaine || domeingrens *f* (*m* || граница *f* [стенка *f*] домена)

D371 **donor** || Don(at)or *m*, Spender *m* || donateur *m*, donneur *m* || donor *m* || донор *m (электронов)*

D372 **donor atom** || Don(at)oratom *n* || atome *m* donneur || donoratoom *n* || атом-донор *m*, донорный атом *m*

D373 **donor level** || Don(at)orniveau *n*, Don(at)orstufe *f* || niveau *m* donneur || donnorniveau *n* || донорный уровень *m* (энергии)

D374 **doped semiconductor** || dotierter [legierter] Halbleiter *m* || semi-conducteur *m* dopé [à impurtés] || extrinsieke halfgeleider *m* || легированный полупроводник *m*

D375 **doping** || Dotierung *f*, Dopung *f*, Legierung *f* || dopage *m* || doping *f* || легирование *n (полупроводника)*

D376 **Doppler shift** || Doppler-Verschiebung *f* || déplacement *m* Doppler-Fizeau || Doppler-verschuiving *f* || доплеровское смещение *n*

D377 **dose** || Dosis *f* || dose *f* || dosis *f* || доза *f*

D378 **dose rate** || Dosisleistung *f* || taux *m* de dose || dosis-vermogen *n*, dosiscapaciteit *f* || мощность *f* дозы (излучения)

D379 **dosimeter** || Dosimeter *n*, Dosismesser *m* || dosimètre *m* || dosismeter *m* || дозиметр *m*

D380 **dosimetry** || Dosimetrie *f*, Dosismessung *f* || dosimétrie *f* || dosimetrie *f* || дозиметрия *f*

D381 **dot product** || Skalarprodukt *n* || produit *m* scalaire || scalair produkt *n* || скалярное произведение *n*

D382 **double-acting engine** || doppeltwirkender Motor *m* || moteur *m* à double effet || dubbelwerkende machine *f* || двигатель *m* двойного действия

D383 **double amplitude** || Doppelamplitude *f* || amplitude *f* totale [crête à crête] || doubbelamplitude *f* || двойная амплитуда *f*, размах *m*

D384 **double-arm lever** || zweiarmiger Hebel *m* || levier *m* à deux bras || tweearmige hefboom *m* || двуплечий рычаг *m*

D385 **double-beat valve** || Doppelsitzventil *n* || soupape *f* à double siège || ventiel *n* met dubbelzitting || двухопорный [двухседельный] клапан *m*

D386 **double bond** || Doppelbindung *f*, doppelte Bindung *f* || double liaison *f* || dubbelbinding *f* || двойная связь *f*

D387 **double bridge** || Thomson-Brücke *f*, Thomsonsche Doppelbrücke *f* || pont *m* (double) de Thomson || Thomson-brug *f* (*m*) || мост *m* Томсона

D388 **double-column planing machine** || Zweiständerhobelmaschine *f* || raboteuse *f* à deux montants || tweekoloms(slee)schaafbank *f* (*m*) || двухстоечный (продольно-)строгальный станок *m*

D389 **double-concave lens** || Bikonkavlinse *f*, bikonkave Linse *f* || lentille *f* biconcave || biconcave lens *f* (*m*) || двояковогнутая линза *f*

D390 **double-convex lens** || Bikonvexlinse *f*, bikonvexe Linse *f* || lentille *f* biconvexe || biconvexe lens *f* (*m*) || двояковыпуклая линза *f*

D391 **double electrical layer** || elektrischer Doppelschicht *f* || couche *f* bipolaire || elektrische dubbellaag *f* (*m*) || двойной электрический слой *m*

D392 **double-end bolt** || Stiftschraube *f* || goujon *m*, prisonnier *m* || stifttap *m* || шпилька *f*

D393 **double-end wrench** || doppelmäuliger Schraubenschlüssel *m* || clé *f* [clef *f*] double || dubbele schroefsleutel *m*, gepaarde schroefsleutel *m*, dubbele moersleutel *m* || двусторонний гаечный ключ *m*

D394 **double helical gear** *see* **herringbone gear**

D395 **double helical spur gear** || Stirnrad *n* mit Pfeilverzahnung || engrenage *m* cylindrique à chevrons || rechttandwiel *n* met pijltanden || шевронное цилиндрическое зубчатое колесо *n*

D396 **double line** || Zweidrahtleitung *f*, Doppelleitung *f* || ligne *f* bifilaire || dubbeldraadsleiding *f*, tweedraadsverbinding *f*, tweedraadscircuit *n* || двухпроводная линия *f*

D397 **double-magic nucleus** || doppeltmagischer Kern *m* || noyau *m* double-magique || dubbelmagische kern *f* (*m*) || дважды магическое ядро *n*

D398 **double point** || Doppelpunkt *m* || point *m* double || dubbelpunt *f* (*m*) || двойная точка *f*

D399 **doubler** || Verdoppler *m* || duplicateur *m*, doubleur *m* || duplicator *m* || удвоитель *m* *(частоты, напряжения)*

D400 **double-range bearing, double-row bearing** || zweireihiges Lager *n* || palier *m* double || dubbelrijig lager *n* || двухрядный подшипник *m*

D401 **double series** || Doppelreihe *f* || série *f* double || dubbelrij *f* (*m*) || двойной ряд *m*

D402 **doublet** || 1. Dublett *n*, Doppellinie *f* 2. Duplet *n*, Zweilinsensystem *n* 3. Dipol *m* || 1. doublet *m* 2. dipôle *m* || 1. doublet *m* 2. dipool *f* (*m*) || 1. (спектральный) дублет *m* 2. линзовый дублет *m* 3. диполь *m*, вибратор *m*

D403 **double tee** || Doppel-T-Träger *m* || poutre *f* en double T [en H] || dubbel-T-balk *m* || двутавровая балка *f*

D404 **double thread** || zweigängiges [doppelgängiges] Gewinde *n* || filet *m* à deux pas || dubbele schroefdraad *m* || двухзаходная резьба *f*

D405 **double-wire** || Zweileiter..., Zweidraht... || bifilaire || tweedraads... || двухпроводный

D406 **doubling** || Verdopplung *f* || duplication *f* || dubbelen *n* || удвоение *n* *(напряжения, частоты)*

D407 **doubly periodic function** || doppeltperiodische Funktion *f* || fonction *f* doublement périodique ||

dubbelperiodische functie *f* || двоякопериодическая функция *f*

D408 **dovetail** || Schwalbenschwanz *m* || queue *f* d'aronde || zwaluwstaart *m* || ласточкин хвост *m* *(вид соединения)*

D409 **dowel** || Dübel *m* || cheville *f*, goujon *m* || deuvel *m*; paspen *f* (*m*); dook *f* (*m*) || штырь *m*, круглая шпонка *f*, нагель *m*

D410 **down time** || Stillstandszeit *f*; Wartezeit *f* || temps *m* d'inactivité, temps *m* mort || uitvaltijd *m* || время *n* простоя, простой *m*

D411 **downward transition** || Übergang *m* nach unten || transition *f* à un niveau plus bas || overgang *m* op een lager (energie)niveau || переход *m* на более низкий (энергетический) уровень

D412 **drag** || Widerstand *m* *(eines Mediums)* || résistance *f* || weerstand *m* || сопротивление *n* *(среды)*

D413 **drag-bar** || Kupplungsstange *f* || bielle *f* d'accouplement; barre *f* d'attelage || koppelstang *f* (*m*) || тяга *f*, сцепной брус *m*

D414 **draw-cut shaper** || Keilnutenziehmaschine *f* || machine *f* à brocher les rainures || gleuventrekmachine *f* || шпоночно-протяжной станок *m*

D415 **drawing** || **1.** Ziehen *n* **2.** Anlassen *n* *(Stahl)* **3.** Zeichnung *f* **4.** Zeichnen *n* || **1.** tréfilage *m* *(p. ex. du fil)* **2.** revenu *m* *(de l'acier)* **3.** dessin *m* **4.** dessin *m* || **1.** trekken *n* *(draden, buizen)* **2.** ontlating *(staal)*, uitgloeiing *f* **3.** tekening *f* **4.** tekenen *n* || **1.** волочение *n*, протягивание *n* *(напр. проволоки)* **2.** отпуск *m* *(стали)* **3.** чертёж *m* **4.** черчение *n*

D416 **drawing press** || Ziehpresse *f* || presse *f* à étirer || trekpers *m* || вытяжной пресс *m*

D417 **drift** || Drift *f*, Wanderung *f* || dérive *f* || drift *f* (*m*) || дрейф *m*

D418 **drift mobility** || Driftbeweglichkeit *f* || mobilité *f* de dérive || driftmobiliteit *f* || дрейфовая подвижность *f* *(носителей тока)*

D419 **drift velocity** || Driftgeschwindigkeit *f* || vitesse *f* de dérive || driftsnelheid *f* || скорость *f* дрейфа

D420 **drill** || **1.** Bohrer *m*; Bohrmaschine *f* **2.** Bohrmaschine *f* *(Werkzeugmaschine)* || **1.** foret *m* **2.** perceuse *f* **3.** machine *f* à percer, perceuse *f* || **1.**, **2.** boor *f* (*m*) **3.** boormachine *f* || **1.** сверло *n* **2.** дрель *f* **3.** сверлильный станок *m*

D421 **drill chuck** || Bohrfutter *n* || manchon *m* porte-foret || boorhouder *m* || сверлильный патрон *m*

D422 **drill head** || Bohrkopf *m* || tête *f* de perçage || boorkop *m* || сверлильная головка *f*

D423 **drilling** || Bohren *n* || forage *m*; perçage *m* || boren *n*; boring *f* || **1.** сверление *n* **2.** бурение *n*

D424 **drilling machine** || Bohrmaschine *f* || machine *f* à percer, perceuse *f* || boormachine *f* || сверлильный станок *m*

D425 **drilling table** || Bohrtisch *m* || table *f* de perceuse || draaibanktafel *m* || стол *m* сверлильного станка

D426 **drilling unit** *see* **drill head**

D427 **drill press** || Säulenbohrmaschine *f* || perceuse *f* verticale || kolomboormachine *f* || вертикально-сверлильный станок *m*

D428 **drill spindle** || Bohrspindel *f* || arbre *m* porte-foret || boorspil *f* (*m*), boorspindel *f* (*m*) || **1.** сверлильный шпиндель *m* **2.** шпиндель *m* бурового станка

D429 **drive** || Antrieb *m* || commande *f*, entraînement *m* || aandrijving *f*; overbrenging *f* || привод *m*; передача *f*

D430 **drive gear** || Antriebsrad *n* || roue *f* de commande || aandrijftandrad *n*, aandrijftandwiel *n* || ведущее (зубчатое) колесо *n*; ведущая шестерня *f*

D431 **driven gear** || getriebenes Rad *n* || roue *f* menée [entraînée, commandée] || gedreven tandrad *n*, gedreven tandwiel *n* || ведомое (зубчатое) колесо *n*; ведомая шестерня *f*

D432 **driving axle** || Treibachse *f* || essieu *m* moteur || aandrijvendet as *f* (*m*) || ведущая ось *f*

D433 **driving belt** || Treibriemen *m* || courroie de commande [de transmission, d'entraînement] || drijfriem *m* || приводной ремень *m*

D434 **driving force** || Triebkraft *f*, Antriebskraft *f* || force *f* motrice; force *f* mouvante; force *f* propulsive || drijfkracht *f* || движущая сила *f*, сила *f* тяги

D435 **driving machine** || Antriebswerk *n*, Antrieb *m* || mécanisme *m* de commande || aandrijfmechanisme *n* || приводной механизм *m*, привод *m*

D436 **driving pulley** || Antriebsscheibe *f* || poulie *f* menante [de commande] || (aan)drijfschijf *f* (*m*) || ведущий [приводной] шкив *m*

D437 **driving shaft** || Antriebswelle *f*, Hauptwelle *f* || arbre *m* de commande [primaire, moteur, d'entraînement] || aandrijfas *f* (*m*) || ведущий [первичный] вал *m*

D438 **drop** || **1.** Abfall *m*, Fallen *n* **2.** Tropfen *m* || **1.** chute *f*, abaissement *m* **2.** goutte *f* || **1.** daling *f* **2.** druppel *m* || **1.** падение *n*, понижение *n* **2.** капля *f*

D439 **drop forging** *see* **die forging**

D440 **drop hammer** || Fallhammer *m* || mouton *m*, marteau *m* frontal, marteau *m* battant || valhamer *m*, spoorhamer *m* || падающий молот *m*; копёр *m*

D441 **drum** || Trommel *f* || tambour *m*; cylindre *m* || trommel *f* (*m*); cilinder *m*, *f* || барабан *m*; цилиндр *m*

D442 **dry-bulb thermometer** || Trokkenthermometer *n* || thermomètre *m* sec || droogthermometer *m* || сухой термометр *m*

D443 **dry cell** || Trockenelement *n* || pile *f* [cellule *f*] sèche || droogelement *n* || сухой (гальванический) элемент *m*

D444 **dry friction** || Trockenreibung *f* || frottement *m* sec || droge wrijving *f* || сухое трение *n*

D445 **duality** || Dualität *f* || dualité *f* || dualiteit *f* || двойственность *f*, дуализм *m*

D446 **duct** || Kanal *m*; Rohrleitung *f* || conduit *m*; conduite *f* || kanaal *n*, doorgang *m*; leiding *f* || канал *m*, проход *m*; трубопровод *m*

D447 **ductile metal** || schmiedbares Metall *n* || métal *m* ductile || smeedbaar metaal *n* || ковкий металл *m*

D448 **dummy** || Schein..., Pseudo... || fictif, pseudo- || pseudo... fictief... || фиктивный; пустой

D449 **dummy address** || Scheinadresse *f*, Pseudoadresse *f* || adresse *f* fictive || fictief adres *n* || фиктивный адрес *m*

D450 I **duplex** || Duplexschaltung *f*; Duplexbetrieb *m* || duplex *m* || duplex *n* || дуплекс *m*

D451 II **duplex** || Duplex... || duplex || duplex... || дуплексный

D452 **duplicating attachment** || Kopiereinrichtung *f*, Nachformeinrichtung *f* || dispositif *m* de copiage || kopiëerinrichting *f* || копировальное приспособление *n*

D453 **duplicating milling machine** || Nachformfräsmaschine *f* || fraiseuse *f* à copier || kopiëerfreesmachine *f* || копировально-фрезерный станок *m*

D454 **durability** || Haltbarkeit *f*; Dauerhaftigkeit *f*; Standzeit *f* || durabilité *f* || duurzaamheid *f* || долговечность *f*

D455 **duralumin(ium)** || Duraluminium *n*, Dural *n* || duralumin *m* || duralumin(ium) *n* || дуралюмин *m*, дюраль *m*, дюралюмин(ий) *m*

D456 **duration** || Dauer *f* || durée *f* || duur *m* || длительность *f*, продолжительность *f*

D457 **duration of vision** || Gesichtswahrnehmungsdauer *f* || durée *f* de vision || gezichtsduur *m* || длительность *f* зрительного восприятия

D458 **dust collector** || Staubabscheider *m* || dépoussiéreur *m*, collecteur *m* [capteur *m*] de poussières || stofafscheider *m* || пылеуловитель *m*; пылесборник *m*

D459 **dust-proof** || staubdicht, staubgeschützt || étanche aux poussières, protégé contre les poussières || stofdicht || пылезащитный, пыленепроницаемый

D460 **dyad** || Dyade *f* || dyade *f* || dyade *f* || диада *f*

D461 **dyadic** || dyadisch || dyadique || dyadisch || диадный

D462 **dying-out** || Abklingen *n*, Schwund *m*, Schwunderscheinung *f* || évanouissement *m* (du signal) || sluiering *f* (signaal) || замирание *n* (*сигнала*)

D463 **dynamic balance** || dynamisches Gleichgewicht *n* || équilibre *m* dynamique || dynamisch evenwicht *n* || динамическое равновесие *n*

D464 **dynamic balancing** || (dynamische) Auswuchtung *f* || balancement *m* [équilibrage *m*] dynamique || dynamische vereffening *f*, dynamische compensatie *f* || динамическая балансировка *f*

D465 **dynamic braking** || dynamische Bremsung *f* || freinage *m* dynamique || dynamische afremming *f* || динамическое торможение *n*

D466 **dynamic characteristic** || dynamische Charakteristik *f* || caractéristique *f* dynamique || dynamische karakteristiek *f* || динамическая характеристика *f*

D467 **dynamic conditions** || dynamischer Betrieb *m* || régime *m* dynamique || dynamisch regime *n* || динамический режим *m*

D468 **dynamic equation** || Formel *f* der Dynamik || équation *f* de dynamique || dynamicavergelijking *f* || уравнение *n* динамики

D469 **dynamic equilibrium** || dynamisches Gleichgewicht *n* || équilibre *m* dynamique || dynamisch evenwicht *n* || динамическое [подвижное] равновесие *n*

D470 **dynamic friction** || Bewegungsreibung *f* || frottement *m* de mouvement || kinetische [dynamische] wrijving *f* || кинетическое трение *n*

D471 **dynamic hardness** || Shore-Harte *f*; Herbert-Härte *f* || dureté *f* Shore ou Herbert || Shore-hardheid *f*; Herbert-hardheid *f* || твёрдость *f* по Шору; твёрдость *f* по Герберту

D472 **dynamic hardness number** || Shore-Hartezahl *f*; Herbert-Härtezahl *f* || nombre *m* (de dureté) Shore ou Herbert || Shore-hardheidsgraad *m*; Herbert-hardheidsgraad *m* || число *n* твёрдости по Шору; число *n* твёрдости по Герберту

D473 **dynamic lift** || dynamischer Auftrieb *m* || portance *f* dynamique || dynamische stijgkracht *f* (*m*) || динамическая подъёмная сила *f*

D474 **dynamic load** || dynamische Belastung *f* || charge *f* dynamique || dynamische belasting *f* || динамическая нагрузка *f*

D475 **dynamic pressure** || Staudruck *m*, Geschwindigkeitsdruck *m* || pression *f* dynamique || dynamische druk *m* || динамическое давление *n*, скоростной напор *m*

D476 **dynamic programming** || dynamische Programmierung *f* || programmation *f* dynamique || dynamische programmering *f* || динамическое программирование *n*

D477 **dynamics** || Dynamik *f* || dynamique *f* || dynamica *f* || динамика *f*

D478 **dynamic similarity** || dynamische Ähnlichkeit *f* || similitude *f* dynamique || dynamische gelijkvormigheid *f* || динамическое подобие *n*

D479 **dynamic strength** || dynamische Festigkeit *f* || résistance *f* dynamique || dynamische vastheid *f* || динамическая прочность *f*

D480 **dynamic test** || Schlagversuch *m* || essai *m* dynamique || dynamische proef *f* (*m*) || динамическое [ударное] испытание *n*

D481 **dynamic viscosity** || dynamische Viskosität *f* || viscosité *f* dynamique || dynamische viscositeit *f* || (динамическая) вязкость *f*

D482 **dynamo** || selbsterregter Generator *m* || génératrice *f* à auto-excitation *f* || dynamo *m* || генератор *m* с самовозбуждением

D483 **dynamometer** || Dynamometer *n* || dynamomètre *m* || dynamometer *m* || динамометр *m*

D484 **dynamometer ammeter** || elektrodynamisches Amperemeter *n* || ampèremètre *m* électrodynamique || elektrodynamische ampère-meter *m* || электродинамический амперметр *m*

D485 **dynamometer test** || Dynamometertest *m* || essai *m* dynamométrique || dynamometrische proef *f* (*m*) || динамометрическое испытание *n*

D486 **dynamotor** || Motorgenerator *m* || moteur-générateur *m* || motorgenerator *m* || двигатель-генератор *m*

D487 **dyne** || Dyn *n* || dyne *f* || dyne *m* || дина *f* (*10^{-5} H*)

E

E1 **earth circuit** || Erd(ungs)leitung *f* || circuit *m* de mise à la terre || aardingscircuit *n* || цепь *f* заземления

E2 **earthing** || Erdung *f* || mise *f* à la terre || aarding *f* || заземление *n*

E3 **earth magnetism** || Erdmagnetismus *m* || magnétisme *m* terrestre || aardmagnetisme *n* || земной магнетизм *m*, геомагнетизм *m*

E4 **Earth's attraction** || Erdanziehung *f* || attraction *f* terrestre || aardaantrekking *f* || земное притяжение *n*, притяжение *n* Земли

E5 **E-capture** || E-Einfang *m*, Elektroneneinfang *m* || piégeage *m* d'électron || elektronenvangst *f* || электронный захват *m*, E-захват *m*

E6 I **eccentric** || Exzenter *m* || excentrique *m* || excentriek *n* || эксцентрик *m*

E7 II **eccentric** || exzentrisch, außermittig || excentrique || excentrisch || эксцентричный, внецентренный

E8 **eccentricity** || Exzentrizität *f* || excentricité *f* || excentriciteit *f* || эксцентриситет *m*

E9 **eccentric load** || außermittige Belastung *f* || charge *f* excentrée || excentrische belasting *f* || внецентренная нагрузка *f*

E10 **echo** || Echo *n*, Echosignal *n* || (signal *m* d')écho *m* || echo *m* || отражённый сигнал *m*, эхо *n*

E11 **echoing** || Echoerscheinung *f* || apparition *f* d'écho || echoverschijnsel *n* || явление *n* эха, отражение *n* сигналов

E12 **echo signal** || Echosignal *n* || signal *m* d'écho || echo *m* || эхосигнал *m*, отражённый импульс *m*

E13 **eddy** || Wirbel *m* || tourbillon *m* || wervel *m* || вихрь *m*, завихрение *n*

E14 **eddy-current heating** || Induktionserwärmung *f*, HF-Erwärmung *f* || chauffage *m* par induction haute fréquence || inductieve verwarming *f*, HF-verwarming *f* || индукционный [высокочастотный] нагрев *m*

E15 **eddy-current loss(es)** || Wirbelstromverlust *m* || pertes *f pl* tourbillonnaires || wervelstroomverliezen *n pl* || вихревые потери *f pl*, потери *f pl* на вихревые токи

E16 **eddy currents** || Wirbelströme *pl*, Foucault-Ströme *pl* || courants *m pl* tourbillonnaires [de Foucault] || wervelstromen *pl* || вихревые токи *pl*, токи *pl* Фуко

E17 **eddy motion** || turbulente Strömung *f*; Wirbelbewegung *f* || mouvement *m* tourbillonnaire || wervelbeweging *f* || турбулентное течение *n*; турбулентное движение *n*; вихревое движение *n*

E18 **edge** || **1.** Kante *f* **2.** Schneidkante *f*; Schneide *f* || **1.** arête *f* **2.** taillant *m*, tranchant *m* || **1.** kant *m* **2.** snede *f*, snee *f* || **1.** ребро *n* **2.** режущая кромка *f*; лезвие *n*

E19 **edm, edm-machine** *see* **spark eroding machine**

E20 **effect** || **1.** Effekt *m*, Erscheinung *f* **2.** Wirkung *f*, Einfluß *m* || effet *m* || effect *n* || **1.** эффект *m*, явление *n* **2.** влияние *n*, действие *n*

E21 **effective** || effektiv, wirksam || effectif || effectief || эффективный; действующий

E22 **effective address** || effektive Adresse *f* || adresse *f* effective || effectief adres *n* || исполнительный адрес *m*

E23 **effective current** || Effektivstromwert *m* || courant *m* effectif || effectieve stroom *m* || эффективное [действующее] значение *n* тока

E24 **effective diameter** || Flankendurchmesser *m* *(Gewinde)* || diamètre *m* moyen *(du filet)* || flankdiameter *m* (schroefdraad) || средний диаметр *m* *(резьбы)*

E25 **effective force** || Effektivkraft *f* || force *f* effective || effective kracht *f* *(m)* || среднее значение *n* силы

E26 **effective power** || effektive Leistung *f*, Effektivleistung *f* || puissance *f* effective || effectief [nuttig] vermogen *n* || эффективная мощность *f*

E27 **effectiveness** || Effektivität *f* || effectivité *f* || **effectiviteit** *f* || эффективность *f*

E28 **effective pressure** || Effektivdruck *m*, Wirkdruck *m* || pression *f* effective || nuttige druk *n* || эффективное [полезное] давление *n*

E29 **effective value** || Effektivwert *m* || valeur *f* effective || effectieve waarde *f* || эффективное [действующее] значение *n*

E30 **effective voltage** || Effektivspannung *f* || tension *f* effective || effectieve spanning *f* || эффективное [действующее] напряжение *n*

E31 **efficiency** || **1.** Nutzeffekt *m*, Effektivität *f* **2.** Wirkungsgrad *m* || || **1.** efficacité *f* **2.** rendement *m*, coefficient *m* d'efficacité [de rendement] || efficiency *f*, rendement *n* **1.** эффективность *f* **2.** коэффициент *m* полезного действия, кпд

E32 **efficient estimate** || effektive Schätzung *f* || estimateur *m* efficace || effectieve schatting *f* || эффективная оценка *f*

E33 **effusion** || Effusion *f* || effusion *f* || effusie *f* || эффузия *f*

E34 **eigenfunction** || Eigenfunktion *f* || fonction *f* propre || eigenfunctie *f* || собственная функция *f*

E35 **eigenvalue** || Eigenwert *m* || valeur *f* propre || eigenwaarde *f* || собственное значение *n*

E36 **eigenvalue problem** || Eigenwertproblem *n* || problème *m* de valeurs propres || eigenwaardeprobleem *n* || задача *f* на собственные значения

E37 **eigenvector** || Eigenvektor *m* || vecteur *m* propre || eigenvector *m* || собственный вектор *m*

E38 **eighty-column card** || 80-spaltige Lochkarte *f* || carte *f* à quatre-vingts colonnes || (pons)kaart *f* (*m*) met 80 kolommen || 80-колонная перфокарта *f*

E39 **ejection stroke** || Auslaßtakt *m* || course *f* d'échappement || uitlaatslag *m* || ход *m* [такт *m*] выпуска

E40 **ejector** || Ejektor *m*; Auswerfer *m*; Strahlpumpe *f* || éjecteur *m*; pompe *f* à jet || ejecteur *m*; straalpomp *f* (*m*) || эжектор *m*; выбрасыватель *m*; струйный насос *m*

E41 **elastic** || elastisch || élastique || elastisch || упругий

E42 **elastic clutch** || elastische Kupplung *f* || accouplement *m* élastique || elastische koppeling || упругая муфта *f*

E43 **elastic collision** *f* || elastischer Stoß *m* || collision *f* élastique || elastische collisie *f* || упругое столкновение *n*, упругий удар *m*

E44 **elastic deformation** || elastische Deformation *f*, elastische Verformung *f* || déformation *f* élastique || elastische deformatie *f* || упругая деформация *f*

E45 **elastic equilibrium** || elastisches Gleichgewicht *n* || équilibre *m* élastique || elastisch evenwicht *n* || упругое равновесие *n*

E46 **elastic force** || elastische Kraft *f* || force *f* élastique || elastische kracht *f* (*m*) || упругая сила *f*

E47 **elastic impact** || elastischer Stoß *m* || choc *m* élastique || elastische stoot *m* || упругий удар *m*

E48 **elasticity** || Elastizität *f* || élasticité *f* || elasticiteit *f* || упругость *f*; эластичность *f*

E49 **elasticity of form** || Formelastizität *f* || élasticité *f* de forme vormelasticiteit *f* || упругость *f* формы

E50 **elastic limit** *f* Elastizitätsgrenze *f* || limite *f* d'élasticité || elasticiteitsgrens *f* (*m*) || предел *m* упругости

E51 **elastic modulus** *see* **modulus of elasticity**

E52 **elastic recovery** || elastische Erholung *f*, elastische Rückbildung *f*; Rückfederung *f* || restitution *f* élastique *f* || elastische restitutie *f* || упругое восстановление *n*

E53 **elastic relaxation** || elastische Relaxation *f* || relaxation *f* élastique || elastische relaxatie *f* || релаксация *f* упругих напряжений

E54 **elastic scattering** || elastische Streuung *f* || diffusion *f* élastique || elastische diffusie *f* || упругое рассеяние *n*

E55 **elastic strain** || elastische Dehnung *f* || déformation *f* (relative) élastique || elastische deformatie *f* || (относительная) упругая деформация *f*

E56 **elastic strength** *see* **proportional limit**
E58 **elastic vibrations** || elastische Schwingungen *f pl* || oscillations *f pl* élastiques || elastische trillingen *f pl* || упругие колебания *n pl*
E59 **elastic washer** || Federring *m* || rondelle *f* élastique; rondelle-ressort *f* || veerring *m* || пружинная шайба *f*
E60 **elastic wave** || elastische Welle *f* || onde *f* élastique || elastische golf *f* (*m*) || упругая волна *f*, волна *f* упругой деформации
E61 **elastomer** *f* Elastomer(es) *n*, Elast *m* || élastomère *m* || elastomeer *f* (*m*) || эластомер *m*
E62 **electret** || Elektret *n* || électret *m* || elektreet *n* || электрет *m*
E63 **electrical** || elektrisch || électrique || elektrisch || электрический
E64 **electrical conduction** || elektrische Leitfähigkeit *f* || conductibilité *f* électrique || elektrisch geleidingsvermogen *n* || электрическая проводимость *f*, электропроводность *f*
E65 **electrical contact** || elektrischer Kontakt *m* || contact *m* électrique || elektrisch contact *n* || электрический контакт *m*
E66 **electrical power** || 1. elektrische Leistung *f* 2. elektrische Energie *f*, Elektroenergie *f* || 1. puissance *f* électrique 2. énergie *f* électrique || 1. electrisch vermogen *n* 2. elektrische energie *f* || 1. электрическая мощность *f* 2. электрическая энергия *f*, электроэнергия *f*
E67 **electrical power factor** || Leistungsfaktor *m* || coefficient *m* [facteur *m*] de puissance électrique || vermogensfactor *m* || коэффициент *m* (электрической) мощности
E68 **electrical energy** || elektrische Energie *f*, Elektroenergie *f* || énergie *f* électrique || elektrische energie *f* || электрическая энергия *f*, электроэнергия *f*
E69 **electrical engineering** || Electrotechnik *f* || électrotechnique *f* || elektrotechniek *f* || электротехника *f*
E70 **electrical insulation** || elektrische Isolation *f* || isolation *f* électrique || elektrische isolatie *f* || электрическая изоляция *f*
E71 **electrical machine** || elektrische Maschine *f* || machine *f* électrique || elektrische machine *f* || электрическая машина *f*
E72 **electrical neutrality** || elektrische Neutralität *f* || neutralité *f* électrique || elektrische neutraliteit *f* || (электро)нейтральность *f*
E73 **electrical resistance** || elektrischer Widerstand *m* || résistance *f* électrique || elektrische weerstand *m* || электрическое сопротивление *n*
E74 **electric arc** || Bogenentladung *f*, (elektrischer) Lichtbogen *m* || décharge *f* d'arc || boog *m*, boogontlading *f* || дуговой разряд *m*, электрическая дуга *f*
E75 **electric-arc welding** || Lichtbogenschweißen *n* || soudage *m* à arc électrique || vlambooglassen *n* || (электро)дуговая сварка *f*
E76 **electric balance** || 1. Stromwaage *f* 2. Wheatstone-Brücke *f* || 1. balance *f* électrodynamique 2. pont *m* de Wheatstone || 1. stroombalans *f* (*m*) 2. brug *f* (*m*) van Wheatstone || 1. токовые весы *pl* 2. мост(ик) *m* Уитстона
E77 **electric cell voltage** || EMK *f* eines galvanischen Elementes || force *f* électromotrice d'une pile électrique || electromotorische kracht *f* (*m*) van een galvanisch element || электродвижущая сила *f* гальванического элемента
E78 **electric charge** || elektrische Ladung *f* || charge *f* électrique || elektrische lading *f* || электрический заряд *m*
E79 **electric current** || elektrischer Strom *m* || courant *m* électrique || elektrische stroom *m* || электрический ток *m*
E80 **electric discharge machine** *see* **spark eroding machine**
E81 **electric-discharge machining** *see* **spark machining**
E82 **electric displacement** || (di)-electrische Verschiebung *f* || déplacement *m* électrique || elektrische verschuiving *f* || электрическое смещение *n*, электрическая индукция *f*

E83 **electric drive** || elektrischer Antrieb *m* || entraînement *m* électrique || elektrische aandrijving *f* || электропривод *m*

E84 **electric field** || elektrisches Feld *n* || champ *m* électrique || elektrisch veld *n* || электрическое поле *n*

E85 **electric furnace** || Elektroofen *m* || électrofour *m* || elektrische oven *m* || электрическая печь *f*

E86 **electric generator** || Elektrogenerator *m*, Stromerzeuger *m* || générateur *m* électrique || elektrische generator *m* || электрогенератор *m*

E87 **electricity** || Elektrizität *f* || électricité *f* || elektriciteit *f* || электричество *n*

E88 **electric moment** || elektrisches Moment *n*, Dipolmoment *n* || moment *m* de dipôle || dipoolmoment *n* || электрический (дипольный) момент *m*

E89 **electric motor** || Elektromotor *m* || électromoteur *m* || elektromotor *m* || электрический двигатель *m*, электродвигатель *m*

E90 **electric network** || (elektrisches) Netzwerk *n*; elektrische Schaltung *f* || circuit *m* électrique || elektrisch netwerk *n*, elektrische schakeling *f* || электрическая (разветвлённая) цепь *f*; электрическая схема *f*

E91 **electric spark** || Funkenentladung *f*, (elektrischer) Funken *m* || décharge *f* par étincelles || vonkontlading *f* || искровой разряд *m*, электрическая искра *f*

E92 **electric strength** || (di)elektrische Festigkeit *f* || résistance *f* diélectrique || diëlektrische stevigheid *f* || (ди)электрическая прочность *f*

E93 **electric vector** || elektrischer Vektor *m*, elektrischer Feldstärkevektor *m* || vecteur *m* du champ électrique || elektrische vector *m* || вектор *m* напряжённости электрического поля, электрический вектор *m*

E94 **electric wave** || elektrische Welle *f*, Welle *f* des elektrischen Feldstärkevektors || onde *f* électrique || elektrische golf *f* (*m*) || волна *f* электрического вектора (напряжённости)

E95 **electrification** || 1. Elektrisierung *f* 2. Elektrifizierung *f* || 1. électrisation *f* 2. électrification *f* || 1. elektrisering *f* 2. elektrificatie *f* || 1. электризация *f* 2. электрификация *f*

E96 **electroacoustics** || Elektroakustik *f* || électro-acoustique *f* || elektroakoestiek *f* || электроакустика *f*

E97 **elektrochemical polarization** || elektrochemische Polarisation *f* || polarisation *f* électrochimique || elektrochemische polarisatie *f* || электрохимическая [гальваническая] поляризация *f*

E98 **electrochemical series** || elektrochemische Spannungsreihe *f* || série *f* de tensions || spanningsreeks *f* (*m*) || ряд *m* напряжений

E99 **electrochemistry** || Elektrochemie *f* || électrochimie *f* || elektrochemie *f* || электрохимия *f*

E100 **electrode** || Elektrode *f* || électrode *f* || elektrode *f* || электрод *m*

E101 **electrodeposition** || 1. elektrolytische Fällung *f*, elektrolytische Abscheidung *f* 2. Elektroabscheidung *f* (*Staub u. dgl.*) || 1. précipitation *f* par électrolyse 2. dépôt *m* électrolitique || 1. elektrolytisch neerslaan *n* 2. elektrisch neerslaan *n* || 1. электролитическое осаждение *n*, гальваностегия *f* 2. электроосаждение *n* (*напр. пыли*)

E102 **electrode voltage** || Elektrodenpotential *n* || potentiel *m* d'électrode || elektrode-potentiaal *m* || потенциал *m* электрода

E103 **electrodialysis** || Elektrodialyse *f* || électrodialyse *f* || elektrodialyse *f* || электродиализ *m*

E104 **electrodynamic** || elektrodynamisch || électrodynamique || elektrodynamisch || электродинамический

E105 **electrodynamics** || Elektrodynamik *f* || électrodynamique *f* || elektrodynamica *f* || электродинамика *f*

E106 **electroluminescence** || Elektrolumineszenz *f* || électroluminescence *f* || elektroluminescentie *f* || электролюминесценция *f*

E107 **electrolysis** || Elektrolyse *f* || électrolyse *f* || elektrolyse *f* || электролиз *m*

E108 **electrolyte** || Elektrolyt *m* || électrolyte *m* || elektrolyt *m* || электролит *m*

E109 **electrolytic** || elektrolytisch || électrolytique || elektrolytisch || электролитический

E110 **electrolytic capacitor** || elektrolytischer Kondensator *m*, Elko *m* || condensateur *m* électrolytique || elektrolytische condensator *m* || электролитический конденсатор *m*

E111 **electrolytic dissociation** || electrolytische Dissoziation *f* || dissociation *f* électrolytique || elektrolytische dissociatie *f* || электролитическая диссоциация *f*

E112 **electrolytic tank** || elektrolytisches Bad *n* || électrolyseur *m*, cuve *f* électrolytique, bain *m* d'électrolyse || elektrolytisch bad *n* || электролитическая ванна *f*

E113 **electromagnet** || Elektromagnet *m* || électro-aimant *m* || elektromagneet *m* || электромагнит *m*

E114 **electromagnetic** || elektromagnetisch || électromagnétique || elektromagnetisch || электромагнитный

E115 **electromagnetic energy** || elektromagnetische Energie *f* || énergie *f* électromagnétique || elektromagnetische energie *f* || электромагнитная энергия *f*, энергия *f* электромагнитного поля

E116 **electromagnetic field** || elektromagnetisches Feld *n* || champ *m* électromagnétique || elektromagnetisch veld *n* || электромагнитное поле *n*

E117 **electromagnetic induction** || elektromagnetische Induktion *f* || induction *f* électromagnétique || elektromagnetische inductie *f* || электромагнитная индукция *f*

E118 **electromagnetic interaction** || elektromagnetische Wechselwirkung *f* || interaction *f* électromagnétique || elektromagnetische wisselwerking *f* || электромагнитное взаимодействие *n*

E119 **electromagnetic momentum** || elektromagnetischer Impuls *m* || impulsion *f* électromagnétique || elektromagnetische impuls *m* || импульс *m* электромагнитных волн, электромагнитный импульс *m*

E120 **electromagnetic oscillations** || elektromagnetische Schwingungen *f pl* || oscillations *f pl* électromagnétiques || elektromagnetische trillingen *f pl* || электромагнитные колебания *n pl*

E121 **electromagnetic radiation** || elektromagnetische Strahlung *f* || rayonnement *m* électromagnétique || elektromagnetische straling *f* || электромагнитное излучение *n*

E122 **electromagnetic separation** || elektromagnetische Trennung *f* || séparation *f* électromagnétique || elektromagnetische scheiding *f* || электромагнитное разделение *n* (*напр. изотопов*)

E123 **electromagnetic theory** || elektromagnetische Theorie *f* || théorie *f* électromagnétique || elektromagnetische theorie *f* || электромагнитная теория *f*

E124 **electromagnetic wave** || elektromagnetische Welle *f* || onde *f* électromagnétique || elektromagnetische golf *f* (*m*) || электромагнитная волна *f*

E125 **electromagnetism** || Elektromagnetismus *m* || électromagnétisme *m* || elektromagnetisme *n* || электромагнетизм *m*

E126 **electromechanical** || elektromechanisch || électromécanique || elektromechanisch || электромеханический

E127 **electromechanical analogy** || elektromechanische Analogie *f* || analogie *f* électromécanique || elektromechanische analogie *f* || электромеханическая аналогия *f*

E128 **electrometer** || Elektrometer *n* || électromètre *m* || elektrometer *m* || электрометр *m*

E129 **electrometry** || Elektrizitätsmessung *f* || électrométrie *f* || elektrometrie *f* || электрометрия *f*

E130 **electromotive force** || elektromotorische Kraft *f*, EMK *f* || force *f* électromotrice || elektromotorische kracht *f* (*m*) || электродвижущая сила *f*, эдс

E131 **electron** || Elektron *n* || électron *m* || elektron *n* || электрон *m*

E132 **electron accelerator** || Elektronenbeschleuniger *m* || accélérateur *m* d'électrons || elektronenversneller *m* || ускоритель *m* электронов

E133 **electron affinity** || Elektronenaffinität *f* || affinité *f* électronique || elektronenaffiniteit *f* || сродство *n* к электрону, электронное сродство *n*

E134 **electron avalanche** || Elektronenlawine *f* || avalanche *f* électronique || elektronenlawine *f* || электронная лавина *f*

E135 **electron beam** || Elektronenbündel *n*, Elektronenstrahl *m* || faisceau *m* électronique || elektronenbundel *m* || пучок *m* электронов, электронный луч *m*

E136 **electron-beam machining** || Elektronenstrahlbearbeitung *f* || usinage *m* par faisceau électronique || elektronenstraalbewerking *f* || электроннолучевая обработка *f*

E137 **electron-beam welding** || Elektronenstrahlschweißen *n* || soudage *m* par faisceau d'électrons || elektronenstraallassen *n* || электроннолучевая сварка *f*

E138 **electron cloud** || Elektronenwolke *f* || nuage *m* électronique || elektronenwolk *f* (*m*) || электронное облако *n*

E139 **electron conduction** || Elektronenleitfähigkeit *f*; Überschußleitung *f*, n-Leitung *f* || conductibilité *f* électronique || elektronengeleiding *f* || электронная проводимость *f*

E140 **electron configuration** || Elektronenkonfiguration *f* (*Atom*) || configuration *f* élektronique || elektronenconfiguratie *f* || электронная конфигурация *f* (*атома*)

E141 **electron diffraction analysis** || Elektronenbeugungsuntersuchung *f*, Elektronographie *f* || radiocristallographie *f* [diffraction *f*] électronique || elektronendiffractieanalyse *f* || электронография *f*

E142 **electronegative** || elektronegativ || électronégatif || elektronegatief || электроотрицательный

E143 **electronegativity** || Elektronegativität *f* || électronégativité *f* || elektronegativiteit *f* || электроотрицательность *f*

E144 **electron emission** || Elektronenemission *f* || émission *f* électronique *f* || elektronenemissie *f* || эмиссия *f* электронов, электронная эмиссия *f*

E145 **electron escape** || Austritt *m* eines Elektrons || fuite *f* de l'électron || elektronenvlucht *f* (*m*) || вылет *m* электрона

E146 **electron gas** || Elektronengas *n* || gaz *m* électronique || elektronengas *n* || электронный газ *m*

E147 **electron gun** || Elektronenkanone *f* || canon *m* à électrons || elektronenkanon *n* || электронная пушка *f*

E148 **electron-hole junction** || n-p-Übergang *m* || jonction *f* n-p, jonction *f* électron-trou || n-p-overgang *m* || электронно-дырочный переход *m*, *n-p*-переход *m*

E149 **electron-hole pair** || Exciton *n*, Elektron-Defektelektronpaar *n* || paire *f* électron-trou || elektron-defectelektronpaar *n* || электронно-дырочная пара *f* (*в полупроводнике*), экситон *m*

E150 **electronic conductivity** || elektronische Leitfähigkeit *f* || conductibilité *f* électronique || elektronengeleiding *f* || электронная проводимость *f*

E151 **electronic image transformer** || elektron(enopt)ischer Bildwandler *m* || convertisseur *m* d'images électronique || elektronenbeeldomzetter *m* || электроннооптический преобразователь *m*; электронный преобразователь *m* изображения

E152 **electronic ionization** || Elektronenionisation *f* || ionisation *f* électronique || elektronenionisatie *f* || ионизация *f* электронами

E153 **electronics** || Elektronik *f* || électronique *f* || elektronica *f* || электроника *f*

E154 **electronic transition** || Elektronenübergang *m*, Elektronensprung *m* || transition *f* électronique || elektronensprong *m* || электронный переход *m*, переход *m* электрона

E155 **electron microscope** || Elektronenmikroskop *n* || microscope *m* électronique || elektronenmicroscoop *m* || электронный микроскоп *m*

E156 **electron multiplier** || Sekundärelektronenvervielfacher *m*, SEV *m* || multiplicateur *m* électronique || elektronenvermenigvuldiger *m* || электронный умножитель *m*

E157 **electron noise** || elektronisches Rauschen *n* || bruits *m pl* électroniques || elektronenlawaai *n* || электрические шумы *m pl* [помехи *f pl*]

E158 **electron-optical** || elektronenoptisch || électrono-optique || elektronenoptisch || электроннооптический

E159 **electron optics** || Elektronenoptik *f* || optique *f* électronique, électrono-optique *f* || elektronenoptiek *f* || электронная оптика *f*

E160 **electron shell** || Elektronenschale *f* || couche *f* électronique || elektronenshil || электронная оболочка *f* *(атома)*

E161 **electron spin resonance** || Elektronenspinresonanz *f* || résonance *f* paramagnétique électronique || elektronenspinresonantie *f* || электронный парамагнитный резонанс *m*, ЭПР

E162 **electron temperature** || Elektronentemperatur *f* || température *f* électronique || elektronentemperatuur *f* || электронная температура *f*

E163 **electron term** || Elektronenterm *m* || terme *m* électronique || elektronenterm *m* || электронный терм *m*

E164 **electron volt** || Elektronenvolt *n*, eV || électron-volt *m* || elektronvolt *m* || электронвольт *m* *(1,6 10^{-19}, Дж)*

E165 **electro-optic** || elektrooptisch || électro-optique || elektro-optisch || электрооптический

E166 **electro-optics** || Elektrooptik *f* || électro optique *f* || elektro-optiek *f* || электрооптика *f*

E167 **electrophilic reagent** || elektrophiles Reagens *n*, Elektrophil *n* || réactif *m* électrophile || elektrofiel reagens *n* || электрофильный реагент *m*, электрофил *m*

E168 **electrophoresis** || Elektrophorese *f* || électrophorèse *f* || elektroforese *f* || электрофорез *m*

E169 **electrophotography** || Elektrographie *f*, Xerographie *f* || électrographie *f*, xérographie *f* || elektrografie *f*, xerografie *f* || электро(фото)графия *f*, ксерография *f*

E170 **electroplating** || Galvanostegie *f*, Elektroplattierung *f* || précipitation *f* par électrolyse, dépôt *m* électrolytique || galvanostegie *f* || электролитическое осаждение *n*, гальваностегия *f*

E171 **electropositive** || elektropositiv || électropositif || elektropositief || электроположительный

E172 **electroscope** || Elektroskop *m* || électroscope *m* || elektroscoop *m* || электроскоп *m*

E173 **electrostatic** || elektrostatisch || électrostatique || elektrostatisch || электростатический

E174 **electrostatic deflection** || elektrostatische Ablenkung *f* || déflexion *f* électrostatique || elektrostatische deflectie *f* || электростатическое отклонение *n* *(пучка)*

E175 **electrostatic energy** || elektrostatische Energie *f* || énergie *f* électrostatique || elektrostatische energie *f* || электростатическая энергия *f*, энергия *f* электростатического взаимодействия

E176 **electrostatic field** || elektrostatisches Feld *n* || champ *m* électrostatique || elektrostatisch veld *n* || электростатическое поле *n*

E177 **electrostatic generator** || 1. elektrostatische Maschine *f* 2. Van de Graafscher Generator *m* || 1. générateur *m* électrostatique 2. générateur *m* [accélérateur *m*] Van de Graaf || 1. electrostatische machine *f* 2. vandegraafgenerator *m* || 1. электростатическая машина *f* 2. генератор *m* Ван-де-Граафа

E178 **electrostatic induction** || elektrostatische Induktion *f*, Influenz *f* || induction *f* électrostatique || elektrostatische inductie *f* || электростатическая индукция *f*, электрическое смещение *n*

E179 **electrostatic lens** || elektrostatische Linse *f* || lentille *f* électrostatique *f* || elektrostatische lens *f* (*m*) || электростатическая линза *f*

E180 **electrostatics** || Elektrostatik *f* || électrostatique *f* || elektrostatica *f* || электростатика *f*

E181 **electrostatic voltmeter** || elektrostatisches Voltmeter *n* || voltmètre *m* électrostatique || elektrostatische voltmeter *m* || электростатический вольтметр *m*

E182 **electrostriction** || Elektrostriktion *f* || électrostriction *f* || elektrostrictie *f* || электрострикция *f*

E183 **elektrotransmission** || Stromübertragung *f* || transport *m* d'électricité || stroomoverdracht *f* (*m*) || передача *f* электрического тока, электропередача *f*

E184 **element** || Element *n* || élément *m* || element *n* || элемент *m*

E185 **elementary** || elementar, Elementar... || élémentaire || elementair || элементарный

E186 **elementary algebra** || elementare Algebra *f* || algèbre *f* élémentaire || elementaire algebra *f* (*m*) || элементарная алгебра *f*

E187 **elementary analysis** || Elementaranalyse *f* || analyse *f* élémentaire || elementairanalyse *f* || элементарный анализ *f*

E188 **elementary charge** || Elementarladung *f* || charge *f* élémentaire || elementaire lading *f* || элементарный заряд *m*, заряд *m* электрона

E189 **elementary event** || Elementarereignis *n* || événement *m* élémentaire || elementaire gebeurtenis *f* || элементарное событие *n*

E190 **elementary mathematics** || Elementarmathematik *f* || mathématiques *f pl* élémentaires || elementaire mathematica *f* || элементарная математика *f*

E191 **elementary operation** || Elementaroperation *f* || opération *f* élémentaire || elementaire operatie *f* || элементарная операция *f*

E192 **elementary particle** || Elementarteilchen *n* || particule *f* élémentaire || elementair deeltje *n* || элементарная [фундаментальная] частица *f*

E193 **elevated temperature** || erhöhte Temperatur *f* || température *f* élevée || verhoogde temperatuur *f* || повышенная температура *f*

E194 **elevating force** || Auftrieb *m*, Auftriebskraft *f* || portance *f*, force *f* sustentatrice || stijgkracht *f* (*m*) || подъёмная сила *f*

E195 **elevation** || **1.** Höhe *f* **2.** Anstieg *m*, Zunahme *f* **3.** Aufriß *m* || **1.** altitude *f*, hauteur *f* **2.** ascension *f* **3.** élévation *f* || **1.** hoogte *f* **2.** verhevenheid *f*, opstijging *f* **3.** verticale projektie *f* || **1.** высота *f* **2.** подъём *m*, повышение *n* **3.** вертикальная проекция *f*

E196 **elevation above sea level** || Höhe *f* über dem Meeresspiegel [über Normalnull] || altitude *f* au-dessus du niveau de la mer || hoogte *f* boven zeeniveau || высота *f* над уровнем моря

E197 **elevation of boiling point** || Siedepunktserhöhung *f* || élévation *f* du point *m* d'ébullition || kookpuntsverhoging *f* || повышение *n* точки кипения

E198 **elevator** || Aufzug *m*; Elevator *m* || élévateur *m* || elevator *m* || лифт *m*; подъёмник *m*; элеватор *m*

E199 **elimination of unknowns** || Elimination *f* von Unbekannten || élimination *f* des inconnues || eliminatie *f* van onbekende grootheden || исключение *n* неизвестных

E200 **ellipse** || Ellipse *f* || ellipse *f* || ellips *f* (*m*) || эллипс *m*

E201 **ellipsoid** || Ellipsoid *n* || ellipsoïde *m* || ellipsoïde *f* || эллипсоид *m*

E202 **ellipsoid of elasticity** || Elastizitätsellipsoid *n*, Spannungsellipsoid *n* || ellipsoïde *m* des contraintes || spanningsellipsoïde *f* || эллипсоид *m* (тензора) напряжений

E203 **ellipsoid of inertia** || Trägheitsellipsoid *n* || ellipsoïde *m* d'inertie || traagheidsellipsoïde *f* || эллипсоид *m* инерции

E204 **ellipsoid of polarization** || Polarisationsellipsoid *n* || ellipsoïde *m* de polarisation || polarisatie-ellipsoïde *f* || эллипсоид *m* поляризации

E205 **ellipsoid of revolution** || Rotationsellipsoid *n*, Sphäroid *n* ||

ellipsoïde *m* de révolution, sphéroïde *m* || rotatie-ellipsoïde *f* || эллипсоид *m* вращения, сфероид *m*

E206 **elliptically polarized wave** || elliptisch polarisierte Welle *f* || onde *f* polarisée elliptiquement || elliptisch gepolariseerde golf *f* (*m*) || эллиптически поляризованная волна *f*, волна *f* с эллиптической поляризацией

E207 **elliptic function** || elliptische Funktion *f* || fonction *f* elliptique || elliptische functie *f* || эллиптическая функция *f*

E208 **elliptic integral** || elliptisches Integral *n* || intégrale *f* elliptique || elliptische integraal *f* (*m*) || эллиптический интеграл *m*

E209 **elliptic orbit** || elliptische Bahn *f* || orbite *f* elliptique || elliptische baan *f* (*m*) || эллиптическая орбита *f*

E210 **elliptic point** || elliptischer Punkt *m* || point *m* elliptique || elliptisch punt *n* || эллиптическая точка *f*

E211 **elliptic polarization** || elliptische Polarisation *f* || polarisation *f* elliptique || elliptische polarisatie *f* || эллиптическая поляризация *f*

E212 **elongation** || Dehnung *f*, Längenänderung *f* || allongement *m* || (uit)rekking *f*, verlenging *f* || удлинение *n*; относительное удлинение *n*

E213 **elongation at fracture, elongation at rupture** || Bruchdehnung *f* || allongement *m* de rupture || (relatieve) breuk(uit)rekking *f* || (относительное) удлинение *n* при разрыве

E214 **elongation (per unit length)** || relative Dehnung *f* || allongement *m* relatif || relatieve verlenging *f* || относительное удлинение *n*

E215 **emergency** || Notfall *m*; Havarie *f*; Störung *f* || défaut *m*; panne *f* || averij *f*, storing *f* || авария *f*; сбой *m*

E216 **emergency stop** || Havarieabschaltung *f*, Notabschaltung *f* || arrêt *m* d'urgence [de sécurité] || noodstop *m* || авост *m*, аварийный останов *m*

E217 **emery wheel** || Schmirgelscheibe *f* || disque *m* abrasif || amarilschijf *f* (*m*) || наждачный круг *m*

E218 **emission** || Emission *f*, Ausstrahlung *f* || émission *f* || emissie *f* || эмиссия *f*, испускание *n*

E219 **emission line** || Emissionslinie *f* || ligne *f* d'émission || emissielijn *f* (*m*) || линия *f* (спектра) испускания, эмиссионная линия *f*

E220 **emission spectrum** || Emissionsspektrum *n* || spectre *m* d'émission || emissiespectrum *n* || спектр *m* испускания, эмиссионный спектр *m*

E221 **emissive power** || Emissions vermögen *n*, Ausstrahlungsvermögen *n* || puissance *f* d'émission || emissievermogen *n* || **1.** лучеиспускательная способность *f* **2.** мощность *f* излучения

E222 **emissivity** || Emissionsvermögen *n* || émissivité *f* || emissiviteit *f* || излучательная [эмиссионная] способность *f*

E223 **emittance** || **1.** Emittanz *f* **2.** spezifische Ausstrahlung *f* || **1.** émittance *f*, émissivité *f* **2.** luminosité *f* || **1.** emittentie *f* **2.** specifieke uitstraling *f* **1.** эмиттанс, *m* лучеиспускательная способность *f* **2.** светимость *f*

E224 **emitted wave** || ausgestrahlte Welle *f* || onde *f* émise || uitgestraalde golf *f* (*m*) || излучаемая [испускаемая] волна *f*

E225 **emitter** || Emitter *m* || émetteur *m* || emitter *m* || эмиттер *m*

E226 **emitting surface** || Emissionsfläche *f*, Strahlungsfläche *f* || surface *f* rayonnante || stralingsoppervlakte *f* || излучающая поверхность *f*

E227 **empirical distribution** || empirische Verteilung *f* || distribution *f* empirique || empirische verdeling *f* || эмпирическое распределение *n*

E228 **empirical formula** || empirische Formel *f* || formule *f* empirique || empirische formule *f* (*m*) || эмпирическая формула *f*

E229 **empty band** || unbesetztes Energieband *n* || bande *f* vide || lege band *m* || пустая [незаполненная] зона *f*

E230 **empty class** || leere Klasse *f* || classe *f* vide || lege klasse *f* || пустой класс *m*

E231 **empty level** || unbesetztes Energieniveau *n* || niveau *m* vide || leeg niveau *n* || незаполненный [незанятый] уровень *m* энергии

E232 **empty set** || leere Menge *f* || ensemble *m* vide || lege menigte *f* || пустое множество *n*

E233 **emulsification** || Emulgieren *n*, Emulgierung *f*, Emulgation *f* || émulsification *f*, émulsionnement *m* || emulgering *f* || эмульгирование *n*

E234 **emulsion** || Emulsion *f* || émulsion *f* || emulsie *f* || эмульсия *f*

E235 **encoder** || Verschlüßler *m* || codeur *m* || codeerinrichting *f* || кодирующее устройство *n*

E237 **endless screw** || endlose Schnecke *f* || vis *f* sans fin || schroef *f* (*m*) zonder einde, worm *m* || червяк *m*, червячный винт *m*

E238 **end mill** || Schaftfräser *m* || fraise *f* en bout, fraise *f* à queue || vingerfrees *f* (*m*) || концевая фреза *f*

E239 **endoergic process** *see* **endothermic process**

E240 **endomorphism** || Endomorphie *f*, Endomorphismus *m* || endomorphisme *m* || endomorfisme *n* || эндоморфизм *m*

E241 **endothermic process** || endothermer [endoenergetischer] Prozeß *m* || processus *m* endothermique [endoénergétique] || endothermisch [endoenergetisch] proces *n* || эндотермический [эндоэнергетический] процесс *m*

E242 **endurance** *see* **fatigue**

E243 **endurance limit** *see* **fatigue limit**

E244 **endurance strength** *see* **fatigue strength**

E245 **energetic** || 1. Hochenergie... 2. energetisch, Energie... || 1. energique 2. énergétique || 1. hoogenergie... 2. energie... || 1. энергичный, с большой энергией 2. энергетический

E246 **energetically favourable** || energetisch günstig || énergétiquement favorable || energetisch voordelig [gunstig] || энергетически выгодный

E247 **energetically unfavourable** || energetisch ungünstig || énergétiquement non favorable || energetisch ongunstig || энергетически невыгодный

E248 **energetics** || Energetik *f* || énergétique *f* || energetica *f* || энергетика *f*

E249 **energization** || Energiezufuhr *f*, Erregung *f*, Anfachung *f* || alimentation *f* en énergie, excitation *f* || energietoevoer *m*, excitatie *f* || питание *n* *(энергией)*, возбуждение *n*

E250 **energy** || 1. Energie *f* 2. Energetik *f* || 1. énergie *f* 2. énergétique *f* || 1. energie *f* 2. energetica *f* || 1. энергия *f* 2. энергетика *f*

E251 **energy amplification** || Leistungsverstärkung *f* || gain *m* en puissance || vermogensversterking *f* || усиление *n* мощности *(напр. сигнала)*

E252 **energy-analyzing magnet** || Magnetanalysator *m* || analyseur *m* à aimant || magneetanalysator *m* || магнитный анализатор *m* *(частиц по энергиям)*

E253 **energy band** || Energieband *n* || bande *f* d'énergie || energieband *m* || энергетическая зона *f*

E254 **energy barrier** || Energieschwelle *f* || barrière *f* d'énergie || energiebarrière *f* || энергетический [потенциальный] барьер *m*

E255 **energy conservation** || 1. Energieerhaltung *f* 2. Energieeinsparung *f* || 1. conservation *f* d'énergie 2. économie *f* d'énergie || 1. energiebehoud *n* 2. energiebesparing *f* || 1. сохранение *n* энергии 2. экономия *f* энергии

E256 **energy consumption** || Energieverbrauch *m* || consommation *f* d'énergie || energieverbruik *n* || потребляемая энергия *f*; потребление *n* энергии

E257 **energy content** || Energieinhalt *m* || pouvoir *m* énergétique || energie-inhoud *m* || энергосодержание *n*

E258 **energy conversion** || Energieumwandlung *f*, Energieumformung *f* || conversion *f* d'énergie; transformation *f* d'énergie || energieomzetting *f* || превращение *n* энергии

E259 **energy density** || Energiedichte *f* || densité *f* d'énergie || energiedichtheid *f* || плотность *f* энергии
E260 **energy dissipation** || Energiestreuung *f*, Energiedissipation *f* || dissipation *f* d'énergie || energiedissipatie *f* || диссипация *f* [рассеяние *n*] энергии
E261 **energy distribution** || Energieverteilung *f*; Energiespektrum *n* || distribution *f* d'énergie; spectre *m* energétique || energiedistributie *f*; energiespectrum *n* || энергетическое распределение *n*; энергетический спектр *m*
E262 **energy exposure** || Bestrahlung *f* || exposition *f* d'énergie || bestraling *f* || лучистая экспозиция *f*
E263 **energy flux** || Energiefluß *m*, Energieströmung *f* || flux *m* d'énergie || energiestroom *m* || поток *m* энергии
E264 **energy gap** || Energielücke *f* || bande *f* interdite || energiespleet *f* (*m*) || энергетическая щель *f*, запрещённая (энергетическая) полоса *f*
E265 **energy input** || zugeführte Energie *f* || énergie *f* amenée || toegevoerde energie *f* || подводимая энергия *f*
E266 **energy integral** || Energieintegral *n* || intégrale *f* d'énergie || energie-integraal *f* (*m*) || интеграл *m* энергии
E267 **energy level** || Energieniveau *n* || niveau *m* énergétique || energieniveau *n* || уровень *m* энергии, энергетический уровень *m*
E268 **energy level diagram** || Energie(niveau)schema *n*, Termschema *n* || diagramme *m* des niveaux énergétiques || energieniveaudiagram *n*, energieniveauschema *n* || диаграмма *f* уровней энергии
E269 **energy-level splitting** || Energieniveauaufspaltung *f* || fission *f* des niveaux énergétiques || energieniveaudesintegratie *f*, energiekloof *f* (*m*) || расщепление *n* уровней энергии
E270 **energy liberation** || Energiefreisetzung *f* || dégagement *m* d'énergie || energievrijmaking *f* || высвобождение *n* энергии, энерговыделение *n*
E271 **energy limit** || Grenzenergie *f* || limite *f* d'énergie || grensenergie *f* || граничная энергия *f*
E272 **energy loss** || Energieverlust *m* || perte *f* d'énergie || energieverlies *n* || потеря *f* энергии; энергетические потери *f pl*
E273 **energy of state** || Zustandsenergie *f* || énergie *f* d'état || toestandsenergie *f* || энергия *f* состояния
E274 **energy quantum** || Energiequant(um) *n* || quantum *m* d'énergie || energiequantum *n* || квант *m* энергии
E275 **energy range** || Energiebereich *m* || domaine *m* [gamme *f*] d'énergie || energiebereik *n* || область *f* [диапазон *m*, интервал *m*] энергий
E276 **energy release** || Energieerzeugung *f* || dégagement *m* d'énergie || energieontgrendeling *f* || энерговыделение *n*, выделение *n* энергии
E277 **energy scheme** *see* **energy level diagram**
E278 **energy source** || Energiequelle *f* || source *f* d'énergie || energiebron *f* (*m*) || источник *m* энергии
E279 **energy spectrum** || Energiespektrum *n* || spectre *m* énergétique || energiespectrum *n* || спектр *m* энергий; энергетический спектр *m*
E280 **energy state** || Energiezustand *m* || état *m* énergétique || energietoestand *n* || энергетическое состояние *n*
E281 **energy storage** || Energiespeicherung *f* || accumulation *f* d'énergie || energieopslag *m* || аккумулирование *n* [накопление *n*] энергии
E282 **energy threshold** || Energieschwelle *f*, Schwellenenergie *f* || seuil *m* énergétique || drempel-energie *f* || энергетический порог *m*, пороговая энергия *f*
E283 **energy transfer** || Energieübertragung *f*, Energietransport *m* || transfert *m* [migration *f*] d'énergie || energieoverbrenging *f* || передача *f* [перенос *m*] энергии
E284 **energy transformation** || Energieumwandlung *f* || transformation *f* d'énergie || energieomzet *m* || превращение *n* энергии
E285 **engine** || Motor *m* || moteur *m* || motor *m* || двигатель *m*

E286 **engineering** || 1. Technik *f* 2. Entwurfs- und Projektierungsarbeiten *f pl*; Konstruieren *n* 3. Maschinenbau *m* || 1. technique *f* 2. études *f pl* 3. constructions *f pl* || 1. techniek *n* 2. construeren *n* 3. machinenbouw *m* || 1. техника *f* 2. проектно-конструкторские работы *f pl*; конструирование *n* 3. машиностроение *n*

E287 **engine speed** || Motordrehzahl *f* || vitesse *f* du moteur || motortoerental *n* || частота *f* вращения двигателя

E288 **enlarged image** || vergrößertes Bild *n*, vergrößerte Abbildung *f* || image *f* grossie || vergroot beeld *n* || увеличенное изображение *n*

E289 **enlargement** || Vergrößerung *f* || grossissement *m* || vergroting *f* || увеличение *n*

E290 **ensemble** || Gesamtheit *f*, Ensemble *n* || ensemble *m* || ensemble *n* || ансамбль *m*

E291 **enthalpy** || Enthalpie *f* || enthalpie *f* || enthalpie *f* || энтальпия *f*

E292 **entropy** || Entropie *f* || entropie *f* || entropie *f* || энтропия *f*

E293 **entropy production** || Entropieerzeugung *f* || production *f* de l'entropie || entropieproduktie *f* || производство *n* энтропии

E294 **entry** || Eingang *m* || entrée *f* || ingang *m* || вход *m* *(в программу, подпрограмму)*

E295 **enumerable set** || abzählbare Menge *f* || ensemble *m* dénombrable || aftelbare menigte *f* || счётное множество *n*

E296 **envelope** *math.* || Einhüllende *f*, Hüllkurve *f* || enveloppe *f* || envelop(pe) *f* *(m)* || огибающая *f*

E297 **environment** || Umwelt *f*; Umgebung *f* || environnement *m* || milieu *n*; omgeving *f* || окружающее пространство *n*; окружающая среда *f*

E298 **environmental conditions** || äußere Bedingungen *pl*; Umweltbedingungen *pl* || conditions *f pl* ambiantes || milieuomstandigheden *pl*, omgevingsvoorwaarden *pl* || окружающие условия *pl*; внешние условия *pl*

E299 **epicyclic gear** || Planetenrad *n* || roue *f* planétaire, planétaire *m* || planeettandwiel *n* || планетарное (зубчатое) колесо *n*

E300 **epicyclic gearing** || Planetengetriebe *n*, Umlaufgetriebe *n* || engrenage *m* planétaire || planeetdrijfwerk *n* || планетарная передача *f*

E301 **epicycloid** || Epizykloide *f* || épicycloïde *f* || epicycloïde *f* || эпициклоида *f*

E302 **epitaxy** || Epitaxie *f* || épitaxie *f* || epitaxie *f* || эпитаксия *f*, эпитаксиальный рост *m* *(кристалла)*

E303 **E-plane** || E-Ebene *f* || E-plan *m* || E-vlak *n* || плоскость *f* колебаний электрического вектора

E304 **equal-arm balance** || gleicharmige Waage *f* || balance *f* à bras égaux || gelijkarmige balans *f* *(m)* || равноплечие весы *pl*

E305 **equality** || Gleichung *f*; Gleichheit *f* || égalité *f* || gelijkheid *f* || равенство *n*

E306 **equalization** || Ausgleich *m*, Abgleich *m* || égalisation *f* || gelijkmaking *f*; correctie *f* || уравнивание *n*, выравнивание *n*; коррекция *f*

E307 **equatorial moment of inertia** || äquatoriales Trägheitsmoment *n* || moment *m* d'inertie équatorial || equatoriaal traagheidsmoment *n* || экваториальный момент *m* инерции

E308 **equation of continuity** || Kontinuitätsgleichung *f* || équation *f* de continuité || continuïteitsvergelijking *f* || уравнение *n* непрерывности [неразрывности]

E309 **equation of motion** || Bewegungsgleichung *f* || équation *f* de mouvement || bewegingsvergelijking *f* || уравнение *n* движения

E310 **equation of state** || Zustandsgleichung *f*, Phasengleichung *f* || équation *f* d'état || toestandsvergelijking *f* || уравнение *n* состояния

E311 **equiareal mapping** || flächentreue Abbildung *f* || application *f* équivalente || vlakbehoudende afbeelding *f* || сохраняющее площади отображение *n*

E312 **equilateral triangle** || gleichseitiges Dreieck *n* || triangle *m* équilatéral || gelijkzijdige driehoek *m* || равносторонний треугольник *m*

E313 **equilibrium** || Gleichgewicht *n* || èquilibre *m* || evenwicht *n* || равновесие *n*

E314 **equilibrium concentration** || Gleichgewichtskonzentration *f* || concentration *f* d'èquilibre || evenwichtsconcentratie *f* || равновесная концентрация *f*

E315 **equilibrium condition** || **1.** Gleichgewichtsbedingung *f* **2.** Gleichgewichtszustand *m* || **1.** condition *f* d'èquilibre **2.** état *m* d'èquilibre || **1.** evenwichtsvoorwaarde *f* **2.** evenwichtstoestand *m* || **1.** условие *n* равновесия **2.** состояние *n* равновесия

E316 **equilibrium of forces** || Kräftegleichgewicht *n* || èquilibre *m* des forces || krachtenevenwicht *n* || равновесие *n* сил

E317 **equilibrium orbit** || Sollbahn *f*, Gleichgewichtsbahn *f* || orbite *f* d'èquilibre || evenwichtsbaan *f* (*m*) || равновесная [устойчивая] орбита *f*

E318 **equilibrium point** || Gleichgewichtspunkt *m* || point *m* d'èquilibre || evenwichtspunt *n* || точка *f* равновесия

E319 **equilibrium polygon** || Gleichgewichtspolygon *n* || polygone *m* de forces fermé || gesloten krachtenveelhoek *m* || замкнутый многоугольник *m* сил

E320 **equilibrium position** || Gleichgewichtslage *f* || position *f* d'èquilibre || evenwichtspositie *f* || равновесное положение *n*, положение *n* равновесия

E321 **equilibrium process** || Gleichgewichtsprozeß *m* || procédé *m* d'èquilibre || evenwichtsproces *n* || равновесный процесс *m*

E322 **equilibrium separation** || Gleichgewichtsabstand *m* || distance *f* d'èquilibre || evenwichtsafstand *m* || равновесное расстояние *n* (*между частицами*)

E323 **equilibrium state** || Gleichgewichtszustand *m* || état *m* d'èquilibre || evenwichtstoestand *m* || состояние *n* равновесия, равновесное состояние *n*

E324 **equilibrium temperature** || Gleichgewichtstemperatur *f* || température *f* d'èquilibre || evenwichtstemperatuur *f* || равновесная температура *f*

E325 **equilibrium time** || Gleichgewichtszeit *f*, Relaxationszeit *f* || temps *m* d'èquilibre || evenwichtstijd *m* || время *n* установления равновесия

E326 **equilibrium value** || Gleichgewichtswert *m* || valeur *f* d'èquilibre || evenwichtswaarde *f* || равновесное значение *n*

E327 **equipartition of energy** || Energiegleichverteilung *f* || èquirépartition *f* de l'ènergie || gelijke verdeling *f* van energie || равнораспределение *n* энергии (по степеням свободы)

E328 **equipotential** || äquipotential, Äquipotential... || èquipotentiel || equipotentiaal || эквипотенциальный

E329 **equipotential line** || Äquipotentiallinie *f* || ligne *f* èquipotentielle || equipotentiaallijn *f* (*m*) || эквипотенциальная линия *f*

E330 **equipotential plane** || Äquipotentialebene *f* || plan *m* èquipotentiel || equipotentiaalvlak *n* || эквипотенциальная плоскость *f*

E331 **equipotential surface** || Äquipotentialfläche *f* || plan *m* èquipotentiel || equipotentiaalvlak *n* || эквипотенциальная поверхность *f*

E332 **equivalence** || Äquivalenz *f* || èquivalence *f* || equivalentie *f* || эквивалентность *f*

E333 **equivalence relation** || Äquivalenzrelation *f* || relation *f* d'èquivalence || equivalentierelatie *f* || отношение *n* эквивалентности

E334 **equivalent network** || äquivalente Schaltung *f*, Ersatzschaltung *f* || circuit *m* èquivalent || equivalente schakeling *f* || эквивалентная схема *f*

E335 **equivalent representations** || äquivalente Darstellungen *f pl* || représentations *f pl* èquivalentes || equivalente voorstellingen *f pl* || эквивалентные представления *n pl*

E336 **erect image** || aufrechtes Bild *n* || image *f* droite || recht beeld *n* || прямое изображение *n*

E337 **erecting prism** || Umkehrprisma *n* || prisme *m* d'inversion || omkeerprisma *n* || оборачивающая призма *f*

E338 **erection** || Montage *f* || montage *m* || montage *f*, opstelling *f*, oprichting *f* || монтаж *m*, возведение *n*

E339 **erection bolt** || Heftbolzen *m*, Montagebolzen *m* || boulon *m* de montage [d'assemblage] || montagebout *m* || монтажный болт *m*

E340 **erg** || Erg *n* || erg *m* || erg *n* || эрг *m* *(10⁻⁷ Дж)*

E341 **ergodic process** || ergodischer Prozeß *m* || processus *m* ergodique || ergodisch proces *n* || эргодический процесс *m*

E342 **ergodic theorem** || Ergodensatz *m*, Ergodentheorem *n* || théorème *m* ergodique || ergodisch theorema *n* || эргодическая теорема *f*

E343 **error** || Fehler *m* || erreur *f* || fout *f* *(m)* || ошибка *f*; погрешность *f*

E344 **error-correcting code** || Fehlerkorrekturkode *m* || code *m* autocorrecteur || foutencorrectiecode *m* || код *m* с исправлением ошибок

E345 **error correction** || Fehlerberichtigung *f*, Fehlerkorrektur *f* || correction *f* des erreurs || foutencorrectie *f* || исправление *n* ошибок

E346 **error estimate** || Fehlerabschätzung *f* || estimation *f* d'erreur || foutschatting *f* || оценка *f* погрешности

E347 **error function** || Fehlerfunktion *f* || fonction *f* des erreurs; intégrale *f* d'erreurs || foutfunctie *f* || функция *f* ошибок; интеграл *m* вероятности ошибок

E348 **error voltage** || Verstimmungsspannung *f* || tension *f* d'écart || ontstemmingsspanning *f* || напряжение *n* рассогласования

E349 **escape valve** || Sicherheitsventil *n*; Auslaßventil *n* || soupape *f* d'échappement [de décharge] || uitlaatklep *f* *(m)* || выпускной клапан *m*

E350 **escape velocity** || Fluchtgeschwindigkeit *f*, zweite kosmische Geschwindigkeit *f* || deuxième vitesse *f* cosmique, vitesse de libération || tweede kosmische snelheid *f* || скорость *f* убегания, вторая космическая скорость *f*

E351 **essential singularity** || wesentliche Singularität *f* || singularité *f* essentielle || wezenlijke singulariteit *f* || существенно особая точка *f*

E352 **ester** || Ester *m* || ester *m*, éther-sel *m*, éther *m* composé || ester *m* || сложный эфир *m*

E353 **estimate** || Schätzwert *m* || estimation *f* || schatting *f* || оценка *f*

E354 **estimation** || Abschätzung *f* || estimation *f* || schatting *f* || оценка *f*

E355 **ether** || Äther *m* || éther *m* || ether *m* || (физический) эфир *m*

E356 **ether, R-O-R** || Äther *m* || éther *m*, éther-oxyde *m* || ether *m* || простой эфир *m*

E357 **Euclidean algorithm** || Euklidischer Algorithmus *m* || algorithme *m* d'Euclide || euclidisch algoritme *n* || алгоритм *m* Евклида

E358 **Euclidean geometry** || euklidische Geometrie *f* || géometrie *f* euclidienne || euclidische meetkunde [geometrie] *f* || евклидова геометрия *f*

E359 **Eulerian axes** || Eulersche Achsen *f pl* || axes *m pl* d'Euler || Euler-assen *f* *(m)* *pl* || эйлеровы оси *f pl*

E360 **evaporation** || Verdampfung *f* || évaporation *f*, vaporisation *f* || verdamping *f* || испарение *n*

E361 **even** || gerade, geradzahlig || pair || even || чётный

E362 **even-even nucleus** || gg-Kern *m*, Gerade-gerade-Kern *m* || noyau *m* pair-pair || even-even-kern *f* || чётно-чётное ядро *n*

E363 **even function** || gerade Funktion *f* || fonction *f* paire || even functie *f* || чётная функция *f*

E364 **evenness** || **1.** Geradzahligkeit *f* **2.** Parität *f* || parité *f* || pariteit *f* || чётность *f*

E365 **even-odd nucleus** || gu-Kern *m*, Gerade-ungerade-Kern *m* || noyau *m* pair-impair || even-oneven-kern *f* || чётно-нечётное ядро *n*

E366 **event** || Ereignis *n* || événement *m* || gebeurtenis *f* || событие *n*

E367 **event algebra** || Ereignisalgebra *f* || algèbre *f* des événements ||

gebeurtenisalgebra *f* (*m*) || алгебра *f* событий

E368 **everywhere dense** || überall dicht *(Menge)* || partout dense || overal dicht || всюду плотный *(о множестве)*

E369 **evolute** || Evolute *f* || développée *f* || evolute *f* || эволюта *f*

E370 **E-wave** || E-Welle *f*, TM-Welle *f* || E-onde *f* || E-golf *f* (*m*) || волна *f* электрического вектора (напряжённости)

E371 **exa...** || Exa... || exa... || exa... || экса... *(приставка в наименовании кратной единицы, соответствующая 10^{18})*

E372 **excess** || 1. Exzeß *m*; Überschuß *m* 2. *Math.* Exzeß *m* || excès *m* || 1. overmaat *f* (*m*) 2. *math.* exces *n* || 1. избыток *m* 2. *мат.* эксцесс *m*

E373 **excess concentration** || Überschußkonzentration *f* || concentration *f* en excès || overmaatconcentratie *f* || избыточная концентрация *f*

E374 **excess electron** || Überschußelektron *n* || électron *m* en excès || overmaatelektron *n* || избыточный электрон *m*

E375 **excess energy** || Überschußenergie *f*, überschüssige Energie *f* || énergie *f* en excès || overmaatenergie *f* || избыточная энергия *f*

E376 **excess entropy** || Überschußentropie *f* || accroissement *m* d'entropie || overmaatentropie *f* || приращение *n* энтропии

E377 **excess load** || Überlastung *f*, Überlast *f* || charge *f* en excès || overbelasting *f* || перегрузка *f*

E378 **excess pressure** || Überdruck *m* || surpression *f* || overdruk *m* || избыточное давление *n*

E379 **excess voltage** || Überspannung *f* || surtension *f* || overspanning *f* || избыточное напряжение *n*; перенапряжение *n*

E380 **exchange** || Austausch *m*; Wechsel *m* || échange *m* || uitwisseling *f* || обмен *m*

E381 **exchange energy** || Austauschenergie *f* || énergie *f* d'échange || uitwisselingsenergi *f* || энергия *f* обменного взаимодействия

E382 **exchange forces** || Austauschkräfte *f pl* || forces *f pl* d'échange || uitwisselingskrachten *f* (*m*) *pl* || обменные силы *f pl*

E383 **exchange interaction** || Austauschwechselwirkung *f* || interaction *f* d'échange *f* || uitwisselwerking *f* || обменное взаимодействие *n*

E384 **exchanger** || Wärme(aus)-tauscher *m* || échangeur *m* (thermique) || warmte(uit)wisselaar *m* || (тепло)-обменник *m*

E385 **excitation** || Erregung *f*, Anregung *f* || excitation *f* || opwekking *f*, excitatie *f* || возбуждение *n*

E386 **excitation energy** || Anregungsenergie *f* || énergie *f* d'excitation || excitatie-energie *f* || энергия *f* возбуждения

E387 **excitation voltage** || Erregungsspannung *f*; Anregungsspannung *f* || tension *f* d'excitation || opwekkingsspanning *f* || напряжение *n* возбуждения

E388 **excited atom** || erregtes Atom *n* || atome *m* excité || geëxciteerd atoom *n* || возбуждённый атом *m*

E389 **excited level** || angeregtes Niveau *n*, Anregungsniveau *n* || niveau *m* d'excitation || excitatieniveau *n* || возбуждённый (энергетический) уровень *m*

E390 **excited state** || angeregter Zustand *m*, Anregungszustand *m* || état *m* d'excitation || excitatietoestand *m* || возбуждённое состояние *n*

E391 **exciter** || Erreger *m* || excitateur *m* || opwekker *m* || возбудитель *m*

E392 **exciting circuit** || Erreger-(strom)kreis *m* || circuit *m* d'excitation || excitatiecircuit *n* || цепь *f* возбуждения

E393 **exciting coil** || Erregerspule *f* || bobine *f* d'excitation || excitatiespoel *f* (*m*) || катушка *f* возбуждения

E394 **exciting winding** || Erregerwicklung *f* || enroulement *m* d'excitation || excitatiewikkeling *f* || обмотка *f* возбуждения

E395 **exclusion principle** || (Paulisches) Ausschließungsprinzip *n*, Pauli-Prinzip *n* || principe *m* d'exclusion (de Pauli) || Pauli-principe *n* || принцип *m* запрета (Паули)

E396 **exclusive OR** || exklusives Oder *n*, Nichtäquivalenz *f* || OU *m* exclusif || exclusief OF *n* || неэквивалентность *f*; исключающее ИЛИ

E397 **execution address** || Exekutivadresse *f* || adresse *f* d'exécution || definitief adres *n* || исполнительный адрес *m*

E398 **exhaust** || Auspuff *m* || échappement *m* || uitlaat *m* || выхлоп *m*, выпуск *m*

E399 **exhaust gas** || Abgas *n* || gaz *m* brûlé [d'échappement] || uitlaatgas *n* || отработавший [выхлопной] газ *m*

E400 **exhaust manifold** || Abgassammelleitung *f* || collecteur *m* d'échappement || uitlaatspruitstuk *n* || выпускной коллектор *m*

E401 **exhaust silencer** || Auspuffschalldämpfer *m* || silencieux *m* || knaldemper *m* || глушитель *m*

E402 **exhaust stroke** *see* **ejection stroke**

E403 **exhaust valve** || Auslaßventil *n*; Auspuffventil *n* || soupape *f* d'échappement [de décharge] || uitlaatklep *f* (*m*) || выпускной [выхлопной] клапан *m*

E404 **exhaust velocity** || Ausströmgeschwindigkeit *f* || vitesse *f* d'écoulement || uitstromingssnelheid *f* || скорость *f* истечения

E405 **existence theorem** || Existenzsatz *m* || théorème *m* d'existence || existentietheorema *n* || теорема *f* существования

E406 **existential quantifier** || Existenzquantor *m* || quantificateur *m* d'existence || existentiequantor *m* || квантор *m* существования

E407 **exit** || Ausgang *m* || sortie *f* || uitgang *m* || выход *m* *(из программы, подпрограммы)*

E408 **exoergic** || exergonisch, exoenergetisch; exotherm, wärme(ab)gebend || exoénergétique, exothermique || exo-energetisch, exothermisch || экзоэнергетический, экзотермический

E409 **exoergic process** *see* **exothermic process**

E410 **exothermal** || exotherm, wärme(ab)gebend || exothermique || exothermisch || экзотермический

E411 **exothermic process** || exothermer [exoenergetischer] Prozeß *m* || processus *m* exothermique [exoénergétique] || exothermisch [exo-energetisch] proces *n* || экзотермический [экзоэнергетический] процесс *m*

E412 **expanding clutch** || Spreizkupplung *f* || accouplement *m* extensible || expansiekoppeling *f*, uitzetbare koppeling *f* || разжимная муфта *f*; разжимное сцепление *n*

E413 **expanding reamer** *see* **adjustable reamer**

E414 **expansion** || Ausdehnung *f*, Expansion *f* || expansion *f* || uitzetting *f*, expansie *f* || расширение *n*; растяжение *n*

E415 **expansion ratio** || Expansionsgrad *m*, Ausdehnungsgrad *m* || degré *m* d'expansion || expansiegraad *m*, uitzettingsgraad *m* || степень *f* расширения

E416 **expansion stroke** *see* **combustion stroke**

E417 **expansion work** || Dehnungsarbeit *f*, Expansionsarbeit *f* || travail *m* d'expansion || expansiearbeid *m* || работа *f* расширения *(газа)*

E418 **expected risk** || erwartetes Risiko *n* || risque *m* espéré || verwacht risico *n*, *m* || ожидаемый риск *m*

E419 **expected value** || Erwartungswert *m* || espérance *f* mathématique || verwachte waarde *f* || математическое ожидание *n*

E420 **experiment** || Experiment *n*, Versuch *m* || essai *m*, expérience *f* || experiment *n*, proef *f* (*m*) || опыт *m*, эксперимент *m*

E421 **experimental data** || experimentelle Werte *pl*, Versuchswerte *pl*, Versuchsdaten *pl* || données *pl* expérimentales || experimentele gegevens *pl* || экспериментальные данные *pl*

E422 **experimental evidence** || experimenteller Beweis *m* || témoignage *m* expérimental || experimenteel bewijs *n* || экспериментальное доказательство *n*, экспериментальные данные *pl*

E423 **experimental physics** || Experimentalphysik *f* || physique *f* expérimentale || experimentele fysica *f* || экспериментальная физика *f*

E424 **experimental value** || experimentell ermittelter [experimenteller] Wert *m* || valeur *f* expérimentale || experimentele waarde *f* || экспериментальное [наблюдаемое] значение *n*

E425 **explicit definition** || explizite Definition *f* || définition *f* explicite || expliciete definitie *f* || явное определение *n*

E426 **explicit equation** || explizite Gleichung *f* || équation *f* explicite || expliciete vergelijking *f* || уравнение *n* в явном виде

E427 **explosion** || Explosion *f* || explosion *f* || explosie *f* || взрыв *m*

E428 **explosive forming** || Explosionsumformen *n* || formage *m* par explosion || explosieve vervorming *f* || взрывная штамповка *f*

E429 **explosive wave** || Explosionswelle *f* || onde *f* explosive || explosiegolf *f* (*m*) || взрывная волна *f*

E430 **exponent** || Exponent *m* || exposant *m* || exponent *m* || экспонента *f*; порядок *m* (*числа*); показатель *m* (*степени*)

E431 **exponential function** || Exponentialfunktion *f* || fonction *f* exponentielle || exponentiële functie *f* || показательная функция *f*

E432 **exponential integral** || Exponentialintegral *n* || intégrale *f* exponentielle || exponentiële integraal *f* (*m*) || 1. интегральная показательная функция, Ei, *f* 2. интеграл *m* от показательной функции

E433 **exposed wiring** || offen verlegte Leitung *f* || canalisation *f* (posée) sur crépi || open gelegde leiding *f* || открытая (электро)проводка *f*

E434 **exposure** || Belichtung(szeit) *f* || exposition *f* || expositie *f*, belichting *f* || экспозиция *f*, количество *n* освещения

E435 **exposure dose** || Exposure *f*, Exposition *f* || dose *f* d'exposition || expositiedosis *f* || экспозиционная доза *f* (*облучения*)

E436 **exposure time** || Belichtungszeit *f*, Belichtungsdauer *f* || durée *f* d'exposition || expositietijd *m*, belichtingstijd *m* || экспозиция *f*, длительность *f* экспозиции, выдержка *f*

E437 **expression** || Ausdruck *m*; Term *m* || expression *f* || uitdrukking *f* || выражение *n*

E438 **extension** || **1.** Dehnung *f* **2.** *Math.* Erweiterung *f* || extension *f* || **1.** dilatatie *f* **2.** *math.* uitbreiding *f* || **1.** удлинение *n*; растяжение *n* **2.** *мат.* расширение *n*

E439 **extension spring** || Zugfeder *f* || ressort *m* de traction || trekveer *f* (*m*) || пружина *f*, работающая на растяжение

E441 **exterior Lebesgue measure** || äußeres Lebesguesches Maß *n* || mesure *f* extérieure lebesguienne || uitwendige Lebesgue-maat *f* (*m*) || внешняя мера *f* Лебега

E442 **external** || Außen...; extern || externe || buiten..., uitwendig || внешний

E443 **external broaching machine** || Außenräummaschine *f* || machine *f* à brocher extérieur || buitentrekfreesmachine *f* || станок *m* для наружного протягивания

E444 **external circuit** || äußerer Stromkreis *m*, Außen(strom)kreis *m* || circuit *m* extérieur || buitencircuit *n* || внешняя цепь *f* (*относительно источника тока*)

E445 **external current** || Außenstrom *m* || courant *m* extérieur || buitenstroom *m* || ток *m* во внешней цепи

E446 **external cylindrical grinding machine** || Außenrundschleifmaschine *f* || rectifieuse *f* cylindrique extérieure || uitwendige rondslijpmachine *f* || круглошлифовальный станок *m*

E447 **external force** || äußere Kraft *f* || force *f* extérieure || buitenkracht *f* (*m*), uitwendige kracht *f* (*m*) || внешняя сила *f*

E448 **external friction** || äußere Reibung *f* || frottement *m* extérieur || buitenwrijving *f* || внешнее [контактное] трение *n*

E449 **external pressure** || äußerer Druck *m*, Außendruck *m* || pression *f* extérieure || buitendruck *m* || внешнее [приложенное] давление *n*

E450 **external resistance** || äußerer Widerstand *m*, Außenwiderstand *m* || résistance *f* externe || buitenweerstand *m*, uitwendige weerstand *m* || сопротивление *n* внешней цепи, внешнее сопротивление *n*

E451 **external splines** || Keilwellenprofile *n pl* || cannelures *f pl* extérieures || uitwendige glijribbels *f(m) pl* || наружные шлицы *m pl*

E452 **external thread** *see* **male thread**

E453 **external voltage** || Außenspannung *f* || tension *f* extérieure || buitenspanning *f* || напряжение *n* во внешней цепи

E454 **extinction** || Extinktion *f*, Abschwächung *f* der Strahlungsleistung || extinction *f*; atténuation *f* || extinctie *f* || экстинкция *f*, ослабление *n* *(излучения)*

E455 **extract** || Extrakt *m* (*n*) Auszug *m* || extrait *m* || extract *n* || экстракт *m*

E456 **extraction** || Extraktion *f*, Extrahieren *n* || extraction *f* || extractie *f* || экстракция *f*

E457 **extra current** || Induktionsstrom *m*, Selbstinduktionsstrom *m* || extra-courant *m* || extrastroom *m* || экстраток *m*

E458 **extraordinary ray** || außerordentlicher [extraordinärer] Strahl *m* || rayon *m* extraordinaire || buitengewone straal *m*, *f* || необыкновенный луч *m*

E459 **extraordinary wave** || außerordentliche Welle *f* || onde *f* extraordinaire || buitengewone golf *f* (*m*) || необыкновенная волна *f*

E460 **extrapolation** || Extrapolation *f* || extrapolation *f* || extrapolatie *f* || экстраполяция *f*

E461 **extremal** || Extremale *f* || extrémale *f* || extremale *f* || экстремаль *f*

E462 **extreme** || extrem, Extrem...; Rand... || extrême || extreem || экстремальный; крайний

E463 **extreme value** || Extremwert *m* || valeur *f* extrême || extreme waarde *f* || экстремум *m*

E464 **extrem** || Extremum *n*, Extremwert *m* || extremum *m* || extremum *n* || экстремум *m*

E465 **extrinsic photoeffect** || äußerer Photoeffekt *m*, äußerer lichtelektrischer Effekt *m*; photoelektrische Emission *f* || effet *m* photo-émissif [photo-électrique] externe || foto-emissie *f*, buitenfoto-effect *n* || внешний фотоэффект *m*, фотоэмиссия *f*

E466 **extrusion** || **1.** Fließpressen *n* **2.** Extrusion *f*, Strangpressen *n* || extrusion *f* || **1.** uitpersing *f* **2.** extrusie *f* || **1.** выдавливание *n* **2.** экструзия *f*

E467 **eye bolt** || Ringbolzen *m*, Ösenbolzen *m* || piton *m*, boulon *m* œil(let) || ringbout *m* || рым-болт *m*, болт *m* с проушиной

E468 **eyepiece** || Okular *n* || oculaire *m* || oculair *n* || окуляр *m*

F

F1 **face** || **1.** Vorderfläche *f*; Stirnfläche *f* **2.** Spanfläche *f* *(Meißel)* || **1.** face *f* **2.** (sur)face *f* d'attaque; plan *m* de coupe || **1.** voorvlak *n*; spaanvlak *n* || **1.** лицевая [передняя] сторона *f* **2.** передняя поверхность *f*, передняя грань *f* *(резца)*

F2 **face cam** || Plankurve *f* || came *f* plate || platte nok *f* (*m*) || плоский кулачок *m*

F3 **face chuck** || Plankurvenfutter *n* || mandrin *m* à serrage en bout || klauwplaat *f* (*m*) || зажимной торцевой патрон *m*

F4 **face cutter** || Stirnfräser *m* || fraise *f* en bout || vlakfrees *m* || торцовая фреза *f*

F5 **face plate** || **1.** Planscheibe *f* *(Drehmaschine)* **2.** Frontplatte *f* || **1.** plateau *m*; table *f* tournante **2.** panneau *m* avant || **1.** stelplaat *f* (*m*) **2.** frontplaat *f* (*m*) || **1.** планшайба *f*; круглый стол *m* *(карусельного станка)* **2.** передняя панель *f*

F6 **face width of tooth** || Zahnbreite *f*; Kranzbreite *f* || largeur *m* de denture || tanddiepte *f* || длина *f* зуба;

ширина *f* зубчатого колеса; ширина *f* зубчатого обода

F7 **facing** || 1. Plandrehen *n* 2. Auskleidung *f* || tournage *m* [surfaçage *m*] de face 2. revêtement *m* || 1. vlakdraaien *n* 2. beschoeiing *f*; bekleding *f* || 1. торцевое [лобовое] точение *n*; поперечное точение *n*, подрезка *f* 2. облицовка *f*, обшивка *f*

F8 **facing lathe** || Plandrehmaschine *f* || tour *m* en l'air [à plateau vertical] || planeer(draai)bank *f* (*m*) || лобото-карный станок *m*

F9 **facing tool** || Plandrehmeißel *m* || outil *m* à saigner || planeerijzer *m* || резец *m* для торцевого точения; подрезной резец *m*

F10 **factor** || 1. Faktor *m* 2. Koeffizient *m*, Faktor *m* || facteur *m* || factor *m* || множитель *m*; коэффициент *m*

F11 **factor group** || Faktorgruppe *f* || groupe-facteur *m* || factorgroep *f* (*m*) || фактор-группа *f*

F12 **factorial** || Fakultät *f* || factorielle *f* || faculteit *f* || факториал *m*

F13 **factorial polynomial** || faktorielles Polynom *n* || polynôme *m* factoriel || factoriëel polynoom *n* || факториальный многочлен *m*

F14 **factor of safety** || Sicherheitsfaktor *m* || facteur *m* [coefficient *m*] de sécurité || veiligheidscoëfficiënt *m*, veiligheidsfactor *m* || коэффициент *m* запаса прочности, запас *m* прочности, коэффициент *m* безопасности

F15 **factor theorem** || Faktorisierungssatz *m* || théorème *m* de factorisation || factorisatietheorema *n* || теорема *f* о разложении многочлена на множители

F16 **factory test** || Werkprüfung *f* || essai *m* en usine || bedrijfsproef *f* (*m*) || заводское испытание *n*

F17 **fading** || Fading *n*, Schwund *m* || fading *m*, évanouissement *m* || fading *n*, *f* || фединг *m*, замирание *n* сигнала

F18 **Fahrenheit scale** || Fahrenheit-Skala *f* || échelle *f* Fahrenheit || Fahrenheit-schaal *f* (*m*) || (температурная) шкала *f* Фаренгейта

F19 **failure** || 1. Bruch *m*, Zerstörung *f*; Beschädigung *f* 2. Ausfall *m* || 1. destruction *f*; rupture *f*; défaut *m* 2. défaut *m*; défaillance *f*; panne *f* || 1. breuk *f* (*m*); vernieling *f*, beschadiging *f* 2. uitval *m* || 1. разрушение *n*; повреждение *n* 2. отказ *m*

F20 **failure prediction** || Vorhersage *f* von Ausfällen || prévision *f* des défauts || uitvalvoorspelling *f* || прогнозирование *n* отказов

F21 **failure rate** || Ausfallrate *f* || taux *m* de défaillance || intensiteit *f* van de storingen || интенсивность *f* отказов

F22 **failure stress** *see* **fracture stress**

F23 **faithful representation** || treue Darstellung *f* || représentation *f* véritable || werkelijke voorstelling *f* || точное [истинное] представление *n*

F24 **fall** || 1. Fallen *n*, Fall *m* 2. Abfall *m*, Abnahme *f*, Absinken *n* || 1. chute *f* 2. décroissance *f* || 1. vallen *n*, val *m* 2. daling *f*, afneming *f* || 1. падение *n* 2. спад *m*

F25 **false(hood)** || Falschheit *f* || faux *m* || valsheid *f* || ложь *f*; ложность *f*

F26 **fan** || Lüfter *m*, Ventilator *m*, Flügel *m* || ventilateur *m*; ailette *f* (*d'un ventilateur*) || waaier *m*, ventilator *m*; vleugel *m* (*van de ventilateur*) || вентилятор *m*; лопасть *f* (*вентилятора*)

F27 **farad**, F || Farad *n* || farad *m* || farad *m* || фарад *m*, Ф (*1 Кл/В*)

F28 **faraday** || Faraday *n*, Faraday-Konstante *f*, Faraday-Zahl *f* || faraday *m* || faraday *m* || фарадей *m*, число *n* Фарадея (*96250 Кл*)

F29 **far infrared** || fernes Infrarot *n*, FIR || infrarouge *m* lointain || verre-infraroodgebied *n* || дальняя инфракрасная область *f* (*спектра*)

F30 **far ultraviolet** || fernes Ultraviolett *n* || ultraviolet *m* lointain || ver ultraviolet *n* || дальняя ультрафиолетовая область *f* (*спектра*), далёкий ультрафиолет *m*

F31 **fast Fourier transformation** || schnelle Fouriertransformation *f* || transformation *f* rapide de Fourier || snelle Fouriertransformatie *f* || быстрое преобразование *n* Фурье, БПФ

F32 **fast neutron** || schnelles Neutron *n* || neutron *m* rapide || snel neutron *n* || быстрый нейтрон *m*

F33 **fast neutron reactor** || schneller Reaktor *m*, Schnellreaktor *m* || réacteur *m* à neutrons rapides || snelle reactor *m* || реактор *m* на быстрых нейтронах, быстрый реактор *m*

F34 **fatigue** || Ermüdung *f* || fatigue *f* || vermoeidheid *f*, vermoeiing *f* || усталость *f*

F35 **fatigue bend(ing) test** || Dauerbiegeversuch *m* || essai *m* de fatigue en flexion || vermoeiingsproef *f* (*m*) bij veelvoudige buigingen || испытание *n* на усталость (при многократном изгибе)

F36 **fatigue crack** || Ermüdungsriß *m*, Dauerriß *m* || fissure *f* de fatigue || vermoeidheidsscheur *f* (*m*) || усталостная трещина *f*

F37 **fatigue damage** || Schaden *m* durch Dauerschwingbeanspruchung || détérioration *f* de fatigue || vermoeidheidsstoring *f* || усталостное повреждение *n*

F38 **fatigue failure, fatigue fracture** || Ermüdungsbruch *m*, Dauerbruch *m* || rupture *f* par fatigue [d'endurance] || vermoeidheidsbreuk *f* (*m*) || усталостное разрушение *n*

F39 **fatigue life** || Lebensdauer *f* bei Dauerschwingbeanspruchung || nombre *m* de cycles qui provoquent la rupture || vermoeidheidsduurzaamheid *f* || циклическая долговечность *f*

F40 **fatigue limit** || Dauerschwingfestigkeit *f* || limite *f* de fatigue || vermoeidheidsgrens *f* (*m*) || предел *m* выносливости *(при бесконечно большом числе циклов напряжений)*

F41 **fatigue rupture** *see* **fatigue failure**

F42 **fatigue strength** || Zeitschwingfestigkeit *f* || limite *f* de fatigue conventionnelle || taaiheidsgrens *f* (*m*), vermoeiingsvastheid *(bij een gegeven getal van spanningscyclussen)* || предел *m* выносливости *(при заданном числе циклов напряжений)*

F43 **fatigue test** || Dauerschwingversuch *m* || essai *m* de fatigue || vermoeidheidsproef *f* (*m*) || испытание *n* на усталость

F44 **fatigue wear** || (berührungsbedingter) Ermüdungsverschleiß *m* || usure *f* superficielle due à la fatigue || contact-vermoeidheidsslijtage *f* || контактно-усталостный износ *m*

F45 **fatty acid** || Fettsäure *f* || acide *m* aliphatique || vetzuur *n* || жирная кислота *f*, кислота *f* жирного ряда

F46 **fault** || Ausfall *m* || défaut *m*; panne *f* || defect *n*, uitval *m* || отказ *m*, неисправность *f*

F47 **feather** || **1.** Paßfeder *f* **2.** Feder *f (z. B. einer Keilwellenverbindung)* || clavette *f*; languette *f* || **1.** veer *f* (*m*) 2. spie *f* (*m*) *(bijv. van een glijribbelverbinding)* || **1.** призматическая шпонка *f* **2.** выступ *m (напр. шлица)*

F48 **feather key** || Paßfeder *f* || clavette *f* prismatique *f* (*m*) || veer *f* (*m*) || призматическая шпонка *f*

F49 **feed** || Vorschub *m*; Zuführung *f*, Zufuhr *f*; Speisung *f* || avance *f*; amenée *f*, amenage *m*; alimentation *f* || aanvoer *m*, toevoer *m*, voeding *f* || подача *f*; питание *n*

F50 **feedback** || Rückkopplung *f*; Rückführung *f* || rétroaction *f*; réaction *f* || terugkoppeling *f* || обратная связь *f*

F51 **feedback loop** || Rückkopplungskreis *m*, Rückführ(ungs)kreis *m* || boucle *f* de ré(tro)action || terugkoppellus *f* (*m*) || контур *m* обратной связи

F52 **feed change gearbox** || Vorschubräderkasten *m (Werkzeugmaschine)* || boîte *f* des avances *(d'une machine-outil)* || toevoertandwielkast *f* (*m*) || коробка *f* подач *(станка)*

F53 **feed-change lever** || Vorschubwechselhebel *m* || levier *m* de commande d'avance || stuurhefboom *m* van de toevoertandwielkast || рычаг *m* управления коробкой подач *(станка)*

F54 **feeder** || **1.** Zuteiler *m* **2.** Speiseleitung *f*, Feeder *m* || alimenteur *m*, feeder *m*, conducteur *m* [ligne *f*] d'alimentation || feeder *m* || **1.** питатель *m* 2. фидер *m*, питающий провод *m*

F55 **feed force** || Vorschubkraft *f* || force *f* d'avance || toevoerkracht *m* || усилие *n* подачи

F56 **feed line** *see* **feeder 2.**

F57 **feed mechanism** || Vorschubgetriebe *n* || mécanisme *m* d'avance [d'alimentation] || toevoermechanisme *n* || механизм *m* подачи

F58 **feed screw** || **1.** Leitspindel *f* **2.** Schnecke *f* || vis *f* de mouvement *(d'une machine-outil)* **2.** hélice *f* transporteuse || **1.** leispindel *m* **2.** toevoerschroef *f* (*m*) || **1.** ходовой винт *m (станка)* **2.** шнек *m*

F59 **feed valve** || Speiseventil *n* || soupape *f* d'alimentation || voedingsklep *f* (*m*) || питательный клапан *m*, клапан *m* подачи

F60 **Féjer's integral** || Féjersches Integral *n* || intégrale *f* de Féjer || Fejer-integraal *f* (*m*) || интеграл *m* Фейера

F61 **felt seal** || Filzdichtung *f* || garniture *f* en feutre || vilt-afdichting *f* || фетровое уплотнение *n*

F62 **female thread** || Innengewinde *n* || filet *m* intérieur, filet *m* femelle || moer(schroef)draad *m*, binnendraad *m* || внутренняя резьба *f*

F63 **femto...** || Femto... || femto...|| femto... || фемто *(приставка в наименовании дольной единицы, соответствующая 10^{-15})*

F64 **fermi,** F || Fermi *n* || fermi *m* || fermi *n* || ферми, Ф *(10^{-15} м; внесистемная единица длины, заменённая фемтометром: 1Ф = 1 фм)*

F65 **Fermi-Dirac statistics** || Fermi-(Dirac-)Statistik *f* || statistique *f* de Fermi et Dirac || Fermi-Diracstatistiek *f* || статистика *f* Ферми(-Дирака)

F66 **Fermi energy** || Fermi-Energie *f*, Fermische Energie *f* || énergie *f* de Fermi || Fermi-energie *f* || энергия *f* [граница *f*] Ферми

F67 **Fermi gas** || Fermi-(Dirac-)Gas *n* || gaz *m* de Fermi (-Dirac) || Fermi-Dirac-gas *n* || ферми-газ *m*, газ *m* ферми-частиц

F68 **Fermi level** || Fermi-Grenze *f*, Fermi-Niveau *n* || niveau *m* de Fermi || Fermi-niveau *n* || уровень *m* [граница *f*] Ферми, фермиевская энергия *f*

F69 **fermion** || Fermion *n*, Fermi-Teilchen *n* || Fermion *m* || Fermi-ion *m* || фермион *m*, ферми-частица *f*

F70 **Fermi particle** || Fermi-Teilchen *n*, Fermion *n* || particule *m* de Fermi || Fermi-deeltje *n* || ферми-частица *f*, фермион *m*

F71 **Fermi surface** || Fermi-Fläche *f*, Fermi-Oberfläche *f* || surface *f* de Fermi || Fermi-vlak *n* || поверхность *f* Ферми, ферми-поверхность *f*

F72 **ferrimagnetic** || ferrimagnetisch || ferromagnétique || ferrimagnetisch || ферримагнитный

F73 **ferrimagnetism** || Ferrimagnetismus *m* || ferromagnétisme *m* || ferrimagnetisme *n* || ферримагнетизм *m*

F74 **ferrite** || Ferrit *n*; Ferritwerkstoff *m* || ferrite *m* || ferriet *n* || феррит *m*, магнитодиэлектрик *m*

F75 **ferritic steel** || Ferritstahl *m* || acier *m* ferritique || ferritisch staal *n* || ферритная сталь *f*

F76 **ferroelectric** || Ferroelektrikum *n*, Seignette(di)elektrikum *n* || matériau *m* ferro-électrique, ferroélectrique *m* || ferro-elektricum *n* || сегнетоэлектрик *m*

F77 **ferroelectricity** || Ferroelektrizität *f*, Seignetteelektrizität *f* || ferro-électricité *f* || ferro-elektriciteit *f* || сегнетоэлектричество *n*

F78 **ferromagnet** || Ferromagnetikum *n*, ferromagnetischer Stoff *m* || ferromagnétique *m* || ferromagneticum *n* || ферромагнетик *m*

F79 **ferromagnetism** || Ferromagnetismus *m* || ferromagnétisme *m* || ferromagnetisme *f* || ферромагнетизм *m*

F80 **fiberglass** || Glasfaser *f* || fibre *f* de verre || fiberglas *n*, glasvezel *f* (*m*) || стекловолокно *n*

F81 **Fibonacci numbers** || Fibonaccische Zahlen *f pl* || nombres *m pl* de Fibonacci || Fibonacci-getallen *n pl* || числа *n pl* Фибоначчи

F82 **fibre optics** || Glasfaseroptik *f*, Faseroptik *f*, Fiberoptik *f* || optique *f* des fibres || glasvezeloptiek *f* || волоконная оптика *f*

F83 **fictitious force** || Scheinkraft *f*, fiktive Kraft *f* || force *f* fictive || fictieve kracht *f* (*m*) || фиктивная сила *f*, сила *f* инерции

F84 **fiducial limits** || Vertrauensintervall *n* || limites *f* fiduciaires || vertrouwensinterval *n* || доверительный интервал *m*

F85 **field** || Feld *n* || champ *m* || veld *n* || поле *n*

F86 **field coil** || Feldspule *f* || bobine *f* d'excitation || veldspoel *f* || катушка *f* возбуждения

F87 **field emission** || Feldemission *f*, Kaltemission *f*, Autoemission *f* || émission *f* froide [électrostatique], auto-émission *f* || veldemissie *f* || холодная [полевая, автоэлектронная] эмиссия *f*, автоэмиссия *f*

F88 **field equation** || Feldgleichung *f* || équation *f* de champ || veldvergelijking *f* || уравнение *n* поля, полевое уравнение *n*

F89 **field force** || Feldstärke *f* || force *f* de champ || veldkracht *f* (*m*) || сила *f* (со стороны) поля

F90 **field gradient** || Potentialgradient *m*, Spannungsgefälle *n* || gradient *m* de potentiel, intensité *f* de champ || potentiaalgradiënt *m* || градиент *m* потенциала, напряженность *f* [градиент *m*] поля

F91 **field intensity** || Feldstärke *f*, Feldintensität *f* || intensité *f* de champ || veldsterkte *f*, veldgrootheid *f* || напряженность *f* поля

F92 **field line** || Feldlinie *f*, Kraftlinie *f* || ligne *f* de force || veldlijn *f* (*m*) || силовая линия *f* *(поля)*

F93 **field magnet** || Feldmagnet *m* || aimant *m* de champ || veldmagneet *m* || возбуждающий магнит *m*, индуктор *m*

F94 **field of matrices** || Matrizenfeld *n* || corps *m* des matrices || matrixveld *n* || поле *n* матриц

F95 **field of vision** || Sichtfeld *n*, Gesichtsfeld *n* || champ *m* de vision || gezichtsveld *n* || поле *n* зрения *(прибора)*

F96 **field pattern** || Feld(linien)-bild *n*, Kraftlinienbild *n* || configuration *f* du champ || veldpatroon *n* || конфигурация *f* поля, картина *f* силовых линий *(поля)*

F97 **field quantum** || Feldquant *n* || quantum *m* de champ || veldquantum *n* || квант *m* поля

F98 **field strength** || Feldstärke *f*, Feldintensität *f* || intensité *f* de champ || veldsterkte *f* || напряжённость *f* поля

F99 **field theory** || Feldtheorie *f* || théorie *f* de champ || veldtheorie *f* || теория *f* поля

F100 **field vector** || Feldvektor *m* || vecteur *m* de champ || veldvector *m* || вектор *m* (напряжённости) поля

F101 **field winding** || Erregerwicklung *f*, Feldwicklung *f* || enroulement *m* d'excitation || veldwikkeling *f*, excitatiewikkeling *f* || обмотка *f* возбуждения

F102 **figure** || Figur *f* || figure *f* || figuur *f* (*m*) || фигура *f*; рисунок *m*

F103 **figure of merit** || Güte(zahl) *f* || facteur *m* de qualité [de surtension] || hoedanigheid *f* || добротность *f* *(резонатора)*

F104 **figure of noise** || Rauschzahl *f*, Rauschfaktor *m*, Geräuschfaktor *m* || coefficient *m* de bruit || lawaaicoëfficiënt *m* || коэффициент *m* шумов, шум-фактор *m*

F105 **filament** || **1.** Faden *m* **2.** Glühfaden *m*, Glühdraht *m*; Heizfaden *m* || filament *m* || filament *n* || **1.** нить *f* **2.** нить *f* накала

F106 **filament voltage** || Fadenspannung *f*, Heizspannung *f* || tension *f* de filament || filamentspanning *f* || напряжение *n* накала

F107 **file** *(data processing)* || Datei *f* || fichier *m* || file *m* || файл *m*

F108 **file test** || Feilhärteprüfung *f*; Anfeilhärteprüfung *f* || dureté *f* à la lime || vijlhardheidsproef *f* (*m*) || испытание *n* твёрдости по напильнику; испытание *n* на опиливаемость

F109 **filing machine** || Feilmaschine *f* || machine *f* à limer, limeuse *f* || vijlmachine *f* || опиловочный станок *m*

F110 **filled band** || (voll)besetztes Energieband *n*, (voll)beetztes Bandn || bande f remplie || (vol)bezette band m || (полностью) заполненная зона f

F111 **filled level** || (voll)besetztes Niveau n, (voll)besetztes Energieniveau n || niveau m rempli || (vol)bezet niveau n || заполненный [занятый] уровень m энергии

F112 **film** || Film m, Schicht f || film m || film m || плёнка f

F113 **film of lubricant** || Schmierfilm m || film m lubrifiant || oliefilm m || масляная плёнка f

F114 **filter** || Filter n || filtre m || filter n, m || фильтр m

F115 **filtration** || Filtration f, Filtrierung f || filtrage m, filtration f || filtratie f || фильтрация f, фильтрование n

F116 **final energy level** || Endniveau n || niveau m final || eindniveau n || конечный уровень m (перехода)

F117 **final product** || Enderzeugnis n || produit m final || eindprodukt n || конечный продукт m

F118 **final velocity** || (stabile) Endgeschwindigkeit f || vitesse f stabilisée || eindsnelheid f || установившаяся скорость f *(падения)*

F119 **fine boring machine** || Feinbohrmaschine f || aléseuse f de précision || fijnboormachine f || прецизионный расточный станок m; алмазно-расточный станок m

F120 **fine feed** || Feinvorschub m || avance f fine [sensitive] || fijnaanzetting f || точная [микрометрическая] подача f

F121 **fine filter** || Feinreinigungsfilter n || filtre m finisseur || fijnreinigingsfilter n || фильтр m тонкой очистки

F122 **fine-focused** || scharffokussiert || focalisé de manière aiguë || scherp gefocusseerd || остро сфокусированный

F123 **fine-pitch cutter** || feingezahnter Fräser m || fraise f à denture fine || fijntandige frees m || фреза f с мелкими зубьями

F124 **fine-pitch screw** || Feingewindeschraube f || vis f à filet fin || schroef f (m) met fijne schroefdraad || винт m с резьбой мелкого шага

F125 **fine-pitch thread** || Feingewinde n || filet m fin || fijne schroefdraad m || резьба f с мелким шагом, мелкая резьба f

F126 **fine structure** || Feinstruktur f || structure f fine || fijnstructuur f || 1. микроструктура f 2. тонкая структура f *(спектра)*

F127 **fine tool** || Präzisionswerkzeug n || outil m de précision, outil m de finition || precisie-instrument n || прецизионный инструмент m

F128 **fine thread** *see* **fine-pitch thread**

F129 **finishing** || Fertigbearbeitung f; Schlichten n || finissage m, finition f || eindbewerking f, fijnbewerking f || (окончательная) чистовая обработка f; доводка f

F130 **finishing cutter** || Fertigfräser m, Schlichtfräser m || fraise f de finition || eindfrees m || чистовая фреза f

F131 **finishing lathe** || Feinbearbeitungsdrehmaschine f || tour m de finition || nadraaibank f (m) || токарный станок m для чистовой обработки

F132 **finishing tool** || Schlichtmeißel m || outil m à dresser || gladijzer n || чистовой резец m

F133 **finite automaton** || endlicher Automat m || automate m fini || eindige automaat m || конечный автомат m

F134 **finite-difference equation** || Differenzengleichung f || équation f aux différences finies || differentiesvergelijking f || уравнение n в конечных разностях, конечно-разностное [разностное] уравнение n

F135 **finite-difference method** || Differenzverfahren n || méthode f des différences finies || differentiesmethode f || метод m конечных разностей, разностный метод m

F136 **finite graph** || endlicher Graph m || graphe m fini || eindige graaf f (m) || конечный граф m

F137 **finite group** || endliche Gruppe f || groupe m fini || eindige groep f (m) || конечная группа f

F138 **finite interval** || endliches Intervall *n* || intervalle *m* fini || eindig interval *n* || конечный интервал *m*
F139 **finite matrix** || endliche Matrix *f* || matrice *f* finie || eindige matrix *f* || конечная [конечномерная] матрица *f*
F140 **finite-memory automaton** || Automat *m* mit endlichem Speicher || automate *m* à mémoire finie || automaat *m* met eindig geheugen || автомат *m* с конечной памятью
F141 **finite motion** || endliche Bewegung *f* || mouvement *m* fini || eindige beweging *f* || финитное движение *n*
F142 **finite quantity** *see* **finite value**
F143 **finite region** || beschränktes Gebiet *n* || région *f* finie || beperkt gebied *n* || ограниченная область *f*
F144 **finite set** || endlose Menge *f* || ensemble *m* fini || eindige menigte *f* || конечное множество *n*
F145 **finite state machine** || endlicher Automat *m* || machine *f* aux états finis || eindige automaat *m* || конечный автомат *m*
F146 **finite transducer** || endlicher Wandler *m* || transducteur *m* fini || eindomvormer *m* || конечный преобразователь *m*
F147 **finite value** || Endwert *m* || valeur *m* finie || eindwaarde *f* || конечное значение *n*
F148 **fireproof** || feuerbeständig, feuerfest; feuersicher || résistant au feu || vuurvast || огнестойкий
F149 **firing** || Zünden *n*, Anzünden *n* || allumage *m*; amorçage *m*; ignition *f* || ontsteking *f* || зажигание *n*; воспламенение *n*
F150 **firing voltage** || Zündpotential *n*, Zündspannung *f* || potentiel *m* d'amorçage || ontstekingspotentiaal *m* || потенциал *m* зажигания
F151 **firmware** || Firmware *f* || firmware *m* || firmware *f* || микропрограммное обеспечение *n*
F152 **first curvature vector** || Vektor *m* der mittleren Krümmung *f* || vecteur *m* de la première courbure || vector *m* van de eerste kromming || вектор *m* первой кривизны
F153 **first flank** || Spanfläche *f* || face *f* d'attaque *(d'un outil coupant)* || voorvlak *n* *(snijgereedschap)* || передняя поверхность *f*, передняя грань *f* *(режущего инструмента)*
F154 **first harmonic** || Grundharmonische *f*, erste Harmonische *f* || premier harmonique *m* || grondharmonische *f* || основная [первая] гармоника *f*
F155 **first integral** || erstes Integral *n*, Vorintegral *n* || intégrale *f* première || eerste integraal *f* (*m*) || первый интеграл *m*
F156 **fissile** || spaltbar, spaltfähig || fissile, fissible || splijtbaar || делящийся *(напр. о ядре)*
F157 **fission** || Spaltung *f*, Kernspaltung *f* || fission *f* || splijting *f* || деление *n* *(атомного ядра)*
F158 **fissionable nucleus** || Spaltkern *m* || noyau *m* fissile || splijtkern *f* || делящееся ядро *n*
F159 **fission energy** || Spalt(ungs)-energie *f* || énergie *f* de fission || splijtingsenergie *f* || энергия *f* деления *(ядра)*
F160 **fission fragment** || Spaltbruchstück *n*, Spaltfragment *n* || fragment *m* de fission || splijtbrokstuk *n* || осколок *m* деления *(ядра)*
F161 **fission neutron** || Spaltneutron *n* || neutron *m* de fission || splijtingsneutron *n* || нейтрон *m* деления *(ядра)*
F162 **fission product** || Spaltprodukt *n* || produit *m* de fission || splijtingsprodukt *n* || продукт *m* деления *(ядра)*
F163 **fission-reaction** || Spaltreaktion *f* || réaction *f* de fission || splijtreactie *f* || реакция *f* деления *(ядра)*
F164 **fission yield** || Spalt(produkt)-ausbeute *f* || rendement *m* de fission || splijtingsopbrengst *f* || выход *m* продуктов деления *(ядра)*
F165 **fit** || Passung *f* || ajustement *m*; ajustage *m*; || passing *f* || посадка *f*; пригонка *f*
F166 **fit tolerance** || Passungstoleranz *f* || tolérance *f* d'ajustement || pas-

singtolerantie *f* || допуск *m* посадки

F167 **fitting** || 1. Formstück *n*, Rohrverbindungsstück *n* 2. Paßarbeit *f*; Anpassen *n*; Einpassen *n* 3. Zusammenbau *m*, Montage *f* || 1. raccord *m*, pièce *f* de raccordement 2. ajustage *m* 3. assemblage *m* || 1. fitting *m*; verbindingsstuk *f* *(leiding)* 2. passing *f* 3. montage *f* || 1. фитинг *m*, соединительная деталь *f* (трубопровода) 2. пригонка *f* 3. сборка *f*, монтаж *m*

F168 **fixed address** || konstante Adresse *f* || adresse *f* fixe || constant adres *n* || постоянный адрес *m*

F169 **fixed axis** || feste [ruhende] Achse *f* || axe *m* fixe || vaste as *f* *(m)* || неподвижная ось *f*

F170 **fixed axoid** || Spurkegel *m*, ruhende Achsenfläche *f* || axoïde *f* fixe, surface *f* fixe des axes instantanés || vast assenvlak *n* || неподвижный аксоид *m*

F171 **fixed capacitor** || Festkondensator *m* || condensateur *m* fixe || vaste condensator *m* || постоянный конденсатор *m*, конденсатор *m* постоянной ёмкости

F172 **fixed coordinate system** || ortsfestes Koordinatensystem *n*, festes System *n* || système *m* de coordonnés immobile || vast coördinatenstelsel *n* || неподвижная система *f* координат

F173 **fixed end** || eingespanntes Trägerende *n* || extrémité *f* encastrée *(d'une poutre)* || ingeklemd balkeinde *n* || защемлённый конец *m* *(балки)*

F174 **fixed load** || konstante [ständige] Belastung *f*, Dauerbelastung *f*, Dauerlast *f* || charge *f* constante || constante belasting *f* || постоянная нагрузка *f*

F175 **fixed point** || 1. *Math.* Fixpunkt *m* 2. Festkomma *n*, Festpunkt *m* *(Rechenmathematik)* || 1. point *m* fixe 2. virgule *f* fixe || 1. onbeweeglijk punt *n*, *f* *(m)* 2. vaste komma *f* *(m)*, *n* || 1. *мат.* неподвижная точка *f* 2. фиксированная точка *f*, фиксированная запятая *f* *(в вычислительной математике)*

F176 **fixed-point arithmetics** || Festkommarechnung *f*, Festpunktrechnung *f* || arithmétique *f* en virgule fixe || rekenkunde *f* met vaste komma || арифметика *f* с фиксированной запятой

F177 **fixed rest** *see* **center rest**

F178 **fixing** || Befestigung *f*, Anbringung *f* || fixation *f* || bevestiging *f*, fixeren *n* || крепление *n*; фиксация *f*; установка *f*

F179 **fixing screw** || Befestigungsschraube *f* || vis *f* [boulon *m*] de fixation || bevestigingsschroef *f* *(m)* || крепёжный винт *m*, крепёжный болт *m*

F180 **flame cutting** || Gasbrennschneiden *n* || oxycoupage *m* || autogeen snijbranden *n* || газопламенная [кислородная] резка *f*

F181 **flammability** || Entzündbarkeit *f*, Entflammbarkeit *f* || inflammabilité *f* || ontvlambaarheid *f* || воспламеняемость *f*

F182 **flame** || Flamme *f* || flamme *f* || vlam *f* *(m)* || пламя *n*

F183 **flange** || Flansch *m* || 1. bride *m* 2. aile *f*, membrure *f* || flens *m* || 1. фланец *m* 2. полка *f* *(балки)*

F184 **flange bearing** || Flanschlager *n* || palier *m* à collerettes || flenslager *n* || фланцевый подшипник *m*

F185 **flange coupling** || Flanschkupplung *f* || accouplement *m* [raccord *m*] à brides || flenskoppeling *f* || фланцевая муфта *f*

F186 **flanging press** || Bördelpresse *f* || presse *f* à border || flensmachine *f* || отбортовочный пресс *m*

F187 **flank** || 1. Fußflanke *f*; Zahnflanke *f*; Flanke *f* *(Gewinde)* 2. Freifläche *f* *(Meißel)* || 1. flanc *m* *(de la dent, du filet)* 2. face *f* d'appui [dorsale] *(du tranchant)* || 1. tandflank *f* *(m)* 2. hulpvrijloopvlak *n* *(beitel)* || 1. профиль *m* ножки *(зуба)*; профиль *m* *(резьбы)* 2. задняя поверхность *f* *(напр. резца)*

F189 **flap valve** || Klapp(en)ventil *n* || soupape *f* [clapet *m*] à opercule || scharnierklep *f* *(m)* || шарнирный (обратный) клапан *m*; откидной клапан *m*

F190 **flash lamp** || Impulslampe *f*, Blitzlampe *f* || lampe *f* pulsée, lampe-éclair *f* || flitslamp *f* || импульсная лампа *f*, лампа-вспышка *f*

F191 **flash point** || Flammpunkt *m* || température *f* d'inflammabilité || ontbrandingspunt *f* (*m*) || температура *f* вспышки

F192 **flash test** || Entflammungsprobe *f* || essai *m* d'inflammabilité || ontvlammingspuntsproef *f* (*m*) || определение *n* температуры воспламенения *(напр. рабочей смеси)*

F193 **flat countersunk head** *see* **countersunk head**

F194 **flat grinding machine** *see* **surface grinding machine**

F195 **flat key** || Flachkeil *m* || clavette *f* plate || vlakke spie *f* (*m*) || шпонка *f* на лыске

F196 **flat space** || ebener [nichtgekrümmter] Raum *m* || espace *m* plat || platte ruimte *f* || плоское пространство *n*

F197 **flat steel** || Flachstahl *m* || fer *m* plat || vlak staal *n* || полосовая сталь *f*

F198 **flat surface grinding machine** *see* **surface grinding machine**

F199 **flattening** || Glätten *n*, Glättung *f* || aplanissement *m* || effening *f* || сглаживание *n*, выравнивание *n*

F200 **flat trajectory** || Flachbahn *f*, gestreckte Flugbahn *f* || trajectoire *f* tendue [rasante] || gestrekte baan *f* (*m*) || настильная траектория *f*

F201 **flexible coupling** || elastische Kupplung *f*; Ausgleichskupplung *f* || accouplement *m* élastique || elastische koppeling *f* || упругая муфта *f*

F202 **flexible drive** || Gelenkwellenantrieb *m* || transmission *f* flexible [élastique] || flexibele aandrijving *f* || гибкий привод *m*

F203 **flexural strain** || Biegedeformation *f* || déformation *f* de flexion || buigingsvervorming *f* || деформация *f* изгиба

F204 **flexural stress** || Biegespannung *f* || contrainte *f* de flexion || buigspanning *f* || напряжение *n* изгиба

F205 **flexure** || Biegung *f*; Durchbiegung *f* || flexion *f* || buiging *f*, kromming *f* || изгиб *m*

F206 **flight** || Flug *m* || vol *m* || vlucht *f* (*m*) || полёт *m*

F207 **flip-flop** || Flip-flop *n*, Trigger *m* || bascule *f* || flip-flop *m*, trigger *m* || триггер *m*

F208 **float(-controlled) valve** || Schwimmerventil *n* || soupape *f* flotteur || vlotterafsluiter *m*, vlotterklep *f* (*m*) || поплавковый клапан *m*

F209 **floating point** || Gleitkomma *n*, Gleitpunkt *m* || virgule *f* flottante || drijvende komma *f* || плавающая точка *f*, плавающая запятая *f*

F210 **floating-point arithmetics** || Gleitkommarechnung *f*, Gleitpunktrechnung *f* || arithmétique *f* en virgule flottante [en point flottant] || rekenkunde *f* met drijvende komma || арифметика *f* с плавающей точкой [с плавающей запятой]

F211 **floating reamer** || Pendelreibahle *f* || alésoir *m* flottant || opruimer *m* met zwemmende opname || плавающая развёртка *f*

F212 **flotation** || Flotation *f*, Flotieren *n* || flottation *f*, flottage *m* || flotatie *f* || флотация *f*

F213 **flow** || Fließen *n*, Fluß *m*; Strömen *n*, Strömung *f* || flux *m* || stroming *f*, vloed *m* || течение *n*, поток *m*

F214 **flowchart** || Flußdiagramm *n* || organigramme *m*, schéma *m* synoptique || blokdiagram *n* || блок-схема *f*

F215 **flow chip** *see* **continuous chip**

F216 **flow indicator** || Durchfluß-(mengen)messer *m* || débitmètre *m* || doorstromingsmeter *m* || расходомер *m*

F217 **flow rate** || Durchfluß *m*, Durchflußmenge *f*, Durchsatz *m* || débit *m* || doorstromingshoeveelheid *f*, doorstroomcapaciteit *f* || расход *m* *(количество вещества, проходящее через сечение в единицу времени)*

F218 **flow velocity** || Strömungsgeschwindigkeit *f*, Fließgeschwindigkeit *f*, Durchflußgeschwindigkeit *f* || vitesse *f* d'écoulement || stroomsnelheid *f* || скорость *f* течения [потока]

F221 **fluctuation** || Fluktuation *f*; Schwankung *f* || fluctuation *f* || fluctuatie *f* || флуктуация *f*; колебание *n*, изменение *n*

F222 **fluid** || strömungsfähiges Medium *n*, Fluid(um) *n* || fluide *m* || fluïdum *n* || текучая среда *f*; жидкость *f*

F223 **fluid cooling** || Flüssigkeitskühlung *f* || refroidissement *m* par liquide || vloeistofkoeling *f* || жидкостное охлаждение *n*

F224 **fluid clutch flywheel** || Strömungskupplung *f* || embrayage *m* [coupleur *m*] hydraulique || hydraulische koppeling *f* || гидромуфта *f*

F225 **fluid drive** *see* **hydraulic drive**

F226 **fluid dynamics** || Strömungsdynamik *f*, Aerohydrodynamik *f* || hydroaérodynamique *f* || hydroaërodynamica *f* || динамика *f* жидкостей и газов, гидроаэродинамика *f*

F227 **fluid filament** || Flüssigkeitsfaden *m* || ligne *f* d'écoulement || stromingslijn *f* (*m*) || линия *f* тока (*жидкости или газа*)

F228 **fluid film** || Flüssigkeitsfilm *m* || film *m* de luquide || vloeistoffilm *m* || плёнка *f* жидкости

F229 **fluid flow** || Strömung *f*, Strömen *n* || flux *m* de fluide || stroming *f* || поток *m* текучей среды

F229a **fluid flywheel** *see* **fluid clutch flywheel**

F230 **fluid friction** || Flüssigkeitsreibung *f* || frottement *m* fluide || vloeistofwrijving *f* || гидродинамическое трение *n*

F231 **fluid head** || Flüssigkeitsdruck *m* || charge *f* hydraulique || vloeistofdruk *m* || напор *m* жидкости

F232 **fluidity** || Fluidität *f*, Fließfähigkeit *f* || fluidité *f* || vloeibaarheid *f* || текучесть *f*, обратная величина *f* вязкости

F233 **fluid jet** || Flüssigkeitsstrahl *m* || jet *m* liquide || vloeistofstraal *m* || струя *f* жидкости

F234 **fluid lubrication** || hydrodynamische Schmierung *f* || lubrification *f* hydrodynamique || hydrodynamische smering *f* || гидродинамическая [жидкостная] смазка *f*

F235 **fluid mechanics** || Strömungsmechanik *f*, Mechanik *f* der Flüssigkeiten und Gase || mécanique *f* des fluides || hydroaëromechanica *f* || механика *f* жидкостей и газов, гидроаэродинамика *f*, гидроаэромеханика *f*

F236 **fluid medium** || **1.** strömungsfähiges Medium *n*, Fluid(um) *n* **2.** Druckflüssigkeit *f*, Druckmittel *n* || **1.** milieu *m* fluide **2.** liquide *f* hydraulique || **1.** vloeibaar medium *n* **2.** hydraulische vloeistof *f* (*m*) || **1.** текучая среда *f* **2.** жидкость *f*

F237 **fluid statics** || Statik *f* der Flüssigkeiten und Gase || hydroaérostatique *f* || hydroaërostatica *f* || статика *f* жидкостей и газов, гидроаэростатика *f*

F238 **fluid viscosity** || Viskosität *f* des Fluidums || viscosité *f* des fluides || fluïdumsviscositeit *f* || вязкость *f* текучей среды [жидкости]

F239 **fluorescence** || Fluoreszenz *f* || fluorescence *f* || fluorescentie *f* || флуоресценция *f*

F240 **fluorescence yield** || Fluoreszenzausbeute *f* || rendement *m* de fluorescence || fluorescentieopbrengst *f* || выход *m* флуоресценции

F241 **fluorine**, F || Fluor *n* || fluor *m* || fluor *n* || фтор *m*

F242 **flute** || **1.** Riefe *f* **2.** Rille *f*; Hohlkehle *f* || cannelure *f*; gorge *f* || **1.** groef *f* (*m*) **2.** ribbel *f* (*m*) || **1.** рифление *n* **2.** канавка *f*; галтель *f*

F243 **flywheel** || Schwungrad *n* || volant *m* (d'inertie) || vliegwiel *n* || маховое колесо *n*, маховик *m*

F244 **flywheel moment** || Schwungmoment *n* || moment *m* d'inertie de masse (du volant) || zwaaimoment *n* || маховой момент *m*

F245 **flux** || Fluß *m* || flux *m* || vloed *m* || поток *m* (*величина*)

F246 **flux density** || Flußdichte *f*, Stromdichte *f* || densité *f* du flux || stroomdichtheid *f* || плотность *f* потока

F247 **flux linkage** || Flußverkettung *f* || adhérence *f* de courant || stroomcohesie *f* || потокосцепление *n*

F248 **foam rubber** || Schaumgummi *m* || caoutchouc *m* spongieux || schuimrubber *m*, *n* || губчатая резина *f*
F249 **focal distance, focal length** || Brennpunktabstand *m*, Fokaldistanz *f* || distance *f* focale || brandpuntsafstand *m* || фокусное расстояние *n*
F250 **focal plane** || Brennebene *f*, Fokalebene *f* || plan *m* focal || brandvlak *n* || фокальная плоскость *f*
F251 **focal point, focus** || **1.** Brennpunkt *m* **2.** *Math.* Strudelpunkt *m* *(Differentialgleichung)* || point *m* focal || brandpunt *f* (*m*), focus *m* || фокус *m*
F252 **focusing** || Fokussierung *f*, Bündelung *f*; Scharfeinstellung *f* || focalisation *f* || focussering *f* || фокусировка *f*; наведение *n* на фокус
F253 **focusing coil** || Fokussier(ungs)spule *f*, Sammelspule *f* || bobine *f* de focalisation || focusseringsspoel *f* (*m*) || фокусирующая катушка *f*
F254 **focusing magnet** || Fokussiermagnet *m* || aimant *m* de focalisation || focusseringsmagneet *m* || фокусирующий магнит *m*
F255 **follow system** || Folgesystem *n* || système *m* asservi [d'asserissement] || volgsysteem *m*, *n* || следящая система
F256 **follower** || **1.** Abtriebsglied *n*; (an)getriebenes Rad *n*; angetriebener Mechanismus *m* **2.** Folgeregler *m* || **1.** élément *m* entraîné (d'une transmission) **2.** servorégulateur *m* || **1.** gedreven element *n*; aangedreven mechanisme *n* **2.** volgregelaar *m* || **1.** ведомый элемент *m* передачи; ведомый механизм *m* **2.** сервορегулятор *m*
F257 **foot brake** || Fußbremse *f* || frein *m* à pédale [à pied] || voetrem *f* (*m*) || ножной [педальный] тормоз *m*
F258 **foot pedal** || Pedal *n*, Fußhebel *m* || pédale *f* || pedaal *n*, *m* || педаль *f*
F259 **forbidden energy band** || verbotenes Energieband *n*, verbotenes Band *n* || bande *f* d'énergie interdite || verboden energieband *m* || запрещённая энергетическая зона *f*
F260 **forbiddenness** || Verbot *n* || interdiction *f* || verbod *n* || запрет *m*
F261 **forbidden transition** || verbotener Übergang *m* || transition *f* interdite || verboden overgang *m* || запрещённый переход *m*
F262 **force** || Kraft *f* || force *f* || kracht *f* (*m*) || сила *f*; усилие *n*
F263 **forced circulation** || Zwangsumlauf *m* || circulation *f* forcée || gedwongen circulatie *f* || принудительная циркуляция *f*
F264 **force constant** || Kraftkonstante *f* || constante *f* des forces || krachtconstante *f* (*m*) || силовая постоянная *f*
F265 **forced convection** || erzwungene Konvektion *f* || convection *f* forcée || gedwongen convectie *f* || вынужденная конвекция *f*
F266 **forced cooling** || Zwangskühlung *f*, erzwungene Kühlung *f* || refroidissement *m* forcé || gedwongen [geforceerde] koeling *f* || принудительное охлаждение *n*
F267 **forced-lubricated bearing** || druckgeschmiertes Lager *n* || palier *m* à graissage sous pression || lager *n* met druksmering || подшипник *m* с принудительной смазкой (под давлением)
F268 **forced motion** || zwang(s)läufige [erzwungene] Bewegung *f*, Zwangsbewegung *f* || mouvement *m* lié || gedwongen beweging *f* || несвободное движение *n*
F269 **forced oscillation** || erzwungene Schwingung *f* || oscillation *f* forcée || gedwongen trilling *f* || вы нужденные колебания *n pl*
F270 **forced ventilation** || Zwangslüftung *f* || ventilation *f* forcée || geforceerde ventilatie *f* || принудительная вентиляция *f*
F271 **forced vibration** *see* **forced oscillation**
F271a **force field** || Kraftfeld *n*, Kräftefeld *n* || champ *m* de forces || krachtenveld *n* || силовое поле *n*, поле *n* сил(ы)
F272 **force fit** || Preßpassung *f* || ajustement *m* sans jeu || perspassing *f* || неподвижная посадка *f*; тугая посадка *f*

F273 **force function** || Kräftefunktion *f* || fonction *f* des forces || krachtenfunctie *f* || силовая функция *f*

F274 **force of compression** *see* **compressive force**

F275 **force of constraint** || Zwangskraft *f*, Reaktionskraft *f*, Führungskraft *f* || réaction *f* [force *f*] de liaison || dwangkracht *f* (*m*), reactiekracht *f* (*m*) || реакция *f* связи

F276 **force of friction** *see* **frictional force**

F277 **force of gravity** || Schwerkraft *f* || force *f* gravitaire, gravité *f*, pesanteur *f* || zwaartekracht *f* (*m*) || сила *f* тяжести

F278 **force pump** || Druckpumpe *f* || pompe *f* refoulante [de refoulement] || perspomp *f* (*m*) || нагнетательный насос *m*

F279 **force ratio** || Kraftgewinn *m* || gain *m* de force || krachtwinst *f* || выигрыш *m* в силе

F280 **force resolution** || Zerlegung *f* der Kraft, Kraftzerlegung *f* || décomposition *f* de la force || krachtontbinding *f* || разложение *n* силы

F281 **force system** || Kräftesystem *n* || système *m* des forces || krachtenstelsel *n* || система *f* сил

F282 **forge hammer** || Schmiedehammer *m* || marteau *m* de forge [à forger] || smeedhamer *m*, voorhamer *m* || ковочный молот *m*; кузнечный молот *m*

F283 **forging** || Schmieden *n* || forgeage *m* || smeding *f*, smeedwerk *n* || ковка *f*

F284 **forging machine** || Schmiedemaschine *f*; Schmiedepresse *f* || machine *f* à forger || smeedmachine *f*; smeedpers *f* (*m*) || ковочная машина *f*; ковочный пресс *m*

F285 **forging press** || Schmiedepresse *f* || presse *f* à forger || smeedpers *f* (*m*) || ковочный пресс *m*

F286 **forging steel** || schmiedbarer Stahl *m* || acier *m* forgeable [ductile] || smeedstaal *n* || ковкая сталь *f*

F287 **forevacuum** || Vorvakuum *n*, Anfangsvakuum *n* || prévide *m* || voorvacuum *n* || форвакуум *m*

F288 **fork** || Stimmgabel *f* || diapason *m* || stemvork *f* || камертон *m*

F289 **fork-lift truck** || Gabelstapler *m* || chargeur *m* à fourches || vorkheftruck *m* || автопогрузчик *m* с вилочным захватом, вилочный погрузчик *m*

F290 **form** || Form *f* || forme *f* || vorm *m* || форма *f*

F291 **formal grammar** || formale Grammatik *f* || grammaire *f* formelle || formele grammatica *f* || формальная грамматика *f*

F292 **formal logic** || formale Logik *f* || logique *f* formelle || formele logica *f* || формальная логика *f*

F293 **formation heat** || Bildungswärme *f*, Entstehungswärme *f* || chaleur *f* de combinaison [de formation] || vormingswarmte *f* || теплота *f* образования

F294 **form cutter** || Formfräser *m* || fraise *f* profilée, outil *m* à profiler || profielfrees *f* (*m*) || фасонная фреза *f*

F295 **form diameter** || Grundkreisdurchmesser *m* *(Zahnrad)* || diamètre *m* du cercle de base *(engrenage)* || grondmiddellijn *f* (*m*) *(tandwiel)* || основной диаметр *m* *(зубчатого колеса)*

F296 **form factor** || Formfaktor *m* || coefficient *m* de la forme || **1.** vormcoëfficiënt *m* **2.** vormfactor *m* || **1.** коэффициент *m* формы **2.** форм-фактор *m*

F297 **form grinding** *see* **profile grinding**

F298 **forming** || **1.** Formgebung *f*; Umformen *n* **2.** Blechbiegen *n* || **1.** façonnage *m*; formage *m* **2.** cintrage *m* de la tôle || **1.** vorming *f* **2.** buiging *f* || **1.** формообразование *n* **2.** гибка *f* листового металла *(в листогибочной машине)*

F299 **form tolerance** || Formtoleranz *f* || tolérance *f* de forme [sur les formes] || vormtolerantie *f* || допуск *m* на форму [профиль *m*] изделия

F300 **form tool** || Formmeißel *m* || lame *f* profilée; outil *m* à profiler || profielbeitel *m* || фасонный резец *m*

F301 **formula** || Formel *f* || formule *f* || formule *f* (*m*) || формула *f*

F302 **forward flow** || Gleichstrom *m* || équicourant *m* || gelijkstroom *m* || прямоток *m* *(поток одного направления)*

F303 **forward motion** || Vorwärtsbewegung *f*, fortschreitende Bewegung *f*, Fortbewegung *f* || mouvement *m* de translation; mouvement *m* en avant || voortgaande beweging *f* || поступательное движение *n*, движение *n* вперёд

F304 **Foucalt pendulum** || Foucaltsches Pendel *n*, Foucault-Pendel *n* || pendule *m* de Foucault || Foucaultpendel *n* || маятник *m* Фуко

F305 **foundation bolt** *see* **anchor bolt**

F306 **foundry** || Gießerei *f* || fonderie *f* || (metaal)gieterij *f* || **1.** литейный цех *m* **2.** литейное производство *n*

F307 **foundry pig (iron)** || Gießereiroheisen *n* || fonte *f* de moulage || ruw gietijzer *n* || литейный чугун *m*

F308 **four-dimensional** || vierdimensional || quadridimensionnel, tétramétrique || vierdimensionaal || четырёхмерный

F309 **four-high mill** || Doppelduowalzwerk *n* || train *m* [laminoir *m*] quarto || walserij *f* met vier rollen || стан кварто *m*, четырёхвалковый прокатный стан *m*

F310 **Fourier analysis** || Fourier-Analyse *f*, harmonische Analyse *f* || analyse *f* de Fourier, analyse *f* harmonique || Fourier-analyse *f*, harmonische analyse *f* || гармонический анализ *m*

F311 **Fourier series** || Fourier-Reihe *f* || série *f* de Fourier || Fourierreeks *f* *(m)* || ряд *m* Фурье

F312 **four-pole** || Vierpol *m* || quadripôle *m* || vierpool *f* *(m)* || четырёхполюсник *m*

F313 **four-stroke cycle** || Viertakt-Arbeitsspiel *n* *(Verbrennungsmotor)* || cycle *m* à quatre temps || viertaktcyclus *m* || четырёхтактный цикл *m*

F314 **four-terminal network** || vierpoliges Netzwerk *n*, Vierpol *m* || circuit quadripôle || vierpool *f* *(m)*, vierpoolnetwerk *m* || четырёхполюсник *m*

F315 **fourth power** || Biquadrat *n* || bicarré *m* || bikwadraat *n* || биквадрат *m*, четвёртая степень *f*

F316 **fractile** || Quantil *n* || fractile *m* || fractile *m* || квантиль *m*

F317 **fractional distillation** || fraktionierte Destillation *f* || distillation *f* fractionnée || gefractioneerde distillatie *f* || фракционная перегонка *f*

F318 **fractional unit** || Anteileinheit *f* || unité *f* fractionnaire || aandeeleenheid *f* || дольная единица *f*

F319 **fracture** || Bruch *m* || fracture *f*; rupture *f* || breuk *f* *(m)* || разрыв *m*

F320 **fracture stress** || Bruchspannung *f* || contrainte *f* de rupture || breukspanning *f*; breukvastheid *f* || разрушающее напряжение *n*; сопротивление *n* разрушению

F322 **frame** || Rahmen *m*; Gerüst *n*; Rumpf *m*; Ständer *m* *(Presse)* || châssis *m*; bâti *m*; cadre *m* *(presse)* || gestel *n*; raam *n*; onderstel *n* || рама *f*; каркас *m*; корпус *m*; остов *m*; станина *f* *(пресса)*

F323 **frame construction** || Skelettbauweise *f* || ossature *f*, charpente *f* || skeletbouw *m* || рамная каркасная конструкция *f*

F324 **frame (of reference)** || Bezugssystem *n* || système *m* de référence || referentiesysteem *n* || система *f* отсчёта; система *f* координат

F325 **Fraunhofer diffraction** || Fraunhofersche Beugung *f* || diffraction *f* de Fraunhofer || Fraunhofer-diffractie *f* || дифракция *f* Фраунгофера

F326 **free** || frei; ungebunden || libre || vrij || свободный

F327 **free convection** || freie [natürliche] Konvektion *f* || convection *f* libre [naturelle] || vrije convectie *f* || свободная [естественная] конвекция *f*

F328 **free electron** || freies Elektron *n* || électron *m* libre || vrij elektron *n* || свободный электрон *m*

F329 **free energy** || freie Energie *f* || énergie *f* libre || vrije energie *f* || свободная энергия *f*

F330 **free fall** || freier Fall *m* || chute *f* libre || vrije val *m* || свободное падение *n*

F331 **free-fall acceleration** || Fallbeschleunigung *f*, Erdbeschleunigung *f* || accélération *f* de la pesanteur || valversnelling *f* || ускорение *n* свободного падения, ускорение *n* силы тяжести

F332 **free gyroscope** || kräftefreier Kreisel *m* || gyroscope *m* libre || vrije gyroscoop *m* || свободный [уравновешенный] гироскоп *m*

F333 **free-jet turbine** || Freistrahlturbine *f* || turbine *f* à veine libre || straalturbine *f* || свободноструйная [активная] турбина *f*

F334 **free motion** || freie Bewegung *f* || mouvement *m* libre || vrije beweging *f* || свободное движение *n*, движение *n* свободного тела

F335 **free oscillation** || freie Schwingung *f* || oscillation *f* libre || vrije trilling *f* || свободные [собственные] колебания *n pl*

F336 **free radical** || (freies) Radikal *n* || radical *m* libre || vrij radicaal *n* || свободный радикал *m*

F337 **free rotation** || freie Rotation *f*, (freie) Drehbarkeit *f* || rotation *f* libre || vrife rotatie *f* || свободное вращение *n*

F338 **free running** || Freilauf *m* || marche *f* libre; rotation *f* libre; marche *f* à vide; mouvement *m* de roulement || vrijloop *m* || свободное движение *n*; холостой ход *m*

F339 **free state** || ungebundener Zustand *m* || état *m* libre || vrije toestand *m* || свободное [несвязанное] состояние *n*

F340 **free variable** || freie Variable *f* || variable *f* libre || vrije variabele *f* || свободная переменная *f*

F341 **free vibration** *see* **free oscillation**

F342 **free-wheel(ling) clutch** || Freilaufkupplung *f* || roue *f* libre, mécanisme *m* à roue libre || vrijloopkoppeling *f* || муфта *f* свободного хода, обгонная муфта *f*

F343 **freezing** || Einfrieren *n*, Gefrieren *n* || congélation *f* || bevriezing *f* || замерзание *n*; замораживание *n*

F344 **freezing machine** || Kältemaschine *f* || machine *f* frigorifique; congélateur || koudemachine *f* || холодильная машина *f*

F345 **freezing point** || Gefriertemperatur *f*, Gefrierpunkt *m* || température *f* [point *m*] de congélation || vriespunt *n* || температура *f* [точка *f*] замерзания

F346 **freon** || Freon *n* || fréon *m* || freon *n* || фреон *m*

F347 **frequency** || Frequenz *f* || fréquence *f* || frequentie *f* || частота *f*

F348 **frequency analysis** *see* **Fourier analysis**

F349 **frequency band** || Frequenzband *n*, Frequenzbereich *m* || bande *f* de fréquences || frequentieband *m* || полоса *f* частот; диапазон *m* частот

F350 **frequency changer** || Frequenzwandler *m*, Frequenzumformer *m* || convertisseur [changeur *m*] de fréquence || frequentieomvormer *m*, frequentieomzetter *m* || преобразователь *m* частоты, смеситель *m*

F351 **freguency channel** || Frequenzkanal *m* || voie *f* de fréquence || frequentiekanaal *n* || частотный канал *m*

F352 **frequency conversion** *see* **frequency transformation**

F353 **frequency distribution** || Häufigkeitsverteilung *f* || distribution *f* de fréquences || frequentiedistributie *f* || частотное распределение *n*

F354 **frequency function** || Verteilungsdichte *f*, Wahrscheinlichkeit *f* || fonction *f* de fréquence || verdelingsdichtheid *f* || плотность *f* распределения (вероятности), функция *f* плотности

F355 **frequency indicator** || Frequenzmesser *m*, Frequenzanzeiger *m* || fréquencemètre *m* || frequentiemeter *m* || частотомер *m*

F356 **frequency-modulated** || frequenzmoduliert || modulé en fréquence || frequentie-gemoduleerd || частотно-модулированный

F357 **frequency modulation** || Frequenzmodulation *f* || modulation *f* en fréquence || frequentiemodulatie *f* || частотная модуляция *f*

F358 **frequency range** || Frequenzbereich *m*, Frequenzgebiet *n* || domaine *m* de fréquences || frequentiegebied *n* || диапазон *m* частот

F359 **frequency response** || Frequenzcharakteristik *f*, Frequenzkennlinie *f*, Frequenzgang *m* || caractéristique *f* fréquentielle || frequentiekarakteristiek *f* || частотная характеристика *f*

F360 **frequency stabilization** || Frequenzstabilisierung *f* || stabilisation *f* de fréquence || frequentiestabilisatie *f* || стабилизация *f* частоты

F361 **frequency standard** || Frequenznormal *n* || étalon *m* de fréquence || frequentiestandaard *m* || эталон *m* [стандарт *m*] частоты

F362 **frequency transformation** || Frequenztransformation *f*, Frequenzwandlung *f*, Frequenzumsetzung *f* || conversion *f* [transformation *f*] de frequence || frequentieconversie *f* || преобразование *n* частоты

F363 **frequency transformer** || Frequenzwandler *m*, Frequenzumformer *m* || convertisseur *m* [transformateur *m*] de frequence || frequentieomvormer *m* || преобразователь *m* частоты

F364 **fresnel** || Fresnel *n* || fresnel *m* || fresnel *m* || френель *m* *(10^{12} Гц)*

F365 **Fresnel diffraction** || Fresnelsche Beugung *f* || diffraction *f* de Fresnel || Fresnel-diffractie *f* || дифракция *f* Френеля

F366 **friction** || Reibung *f* || frottement *m*; friction *f* || wrijving *f* || трение *n*

F367 **frictional electricity** *see* **triboelectricity**

F368 **frictional force** || Reibungskraft *f* || force *f* de frottement || wrijvingskracht *f* *(m)* || сила *f* трения

F369 **frictional losses** || Reibungsverlust *m* || pertes *f pl* par frottement || wrijvingsverliezen *n pl* || потери *f pl* на трение

F370 **frictional wear** || reibungsbedingter Verschleiß *m* || usure *f* par frottement || wrijvingsslijtage *f* || фрикционный износ *m*, износ *m* поверхностного слоя при трении

F371 **friction brake** || Reibungsbremse *f* || frein *m* à friction [à frottement] || wrijvingsrem *f* *(m)* || фрикционный тормоз *m*

F372 **friction clutch** || Reib(ungs)-kupplung *f* || embrayage *m* à friction || wrijvingskoppeling *f* || фрикционное сцепление *n*; фрикционная муфта *f*

F373 **friction coefficient** *see* **coefficient of friction**

F374 **friction cone** || Reibungskegel *m* || cône *m* de friction [de frottement] || wrijvingskegel *m* || конус *m* трения

F375 **friction coupling** || Reib(ungs)kupplung *f* || accouplement *m* à friction || wrijvingskoppeling *f* || фрикционная муфта *f*

F376 **friction drive** || Reib(scheiben)antrieb *m* || entraînement *m* par friction || wrijvingsdrijfwerk *n* || фрикционная передача *f*

F377 **friction of motion** || Bewegungsreibung *f* || frottement *m* au mouvement || kinetische wrijving *f* || кинетическое трение *n*

F378 **friction of rest** || Ruhereibung *f* || frottement *m* au repos || rustwrijving *f* || трение *n* покоя

F379 **friction resistance** || Reibungswiderstand *m* || résistance *f* au frottement || wrijvingsweerstand *m* || сопротивление *n* трения

F380 **friction surface** || Reibungsfläche *f* || surface *f* de frottement || wrijvingsoppervlak *n*, wrijvingsvlak *n* || поверхность *f* трения

F381 **friction welding** || Reibungsschweißen *n* || soudage *m* par frottement || wrijvingslassen *n* || сварка *f* трением

F382 **fringe** || Interferenzstreifen *m* || frange *f* d'interférence || interferentiestreep *f* *(m)* || интерференционная полоса *f*

F383 **front axle** || Vorderachse *f* || essieu *m* avant || vooras *f* *(m)* || передняя ось *f*

F384 **front-end computer** || Front-End-Rechner *m*, Satellitenrechner *m* || ordinateur *m* satellite || front-end-computer *m* || периферийная [сателлитная] ЭВМ *f*

F385 **front focus** || Dingbrennpunkt *m*, vorderer Brennpunkt *m* || foyer *m* objet, foyer *m* antérieur || voorbrandpunt *n* || передний фокус *m*

F386 **fuel** || Brennstoff *m*; Kraftstoff *m*, Treibstoff *m* || combustible *m*; carburant *m* || brandstof *f* (*m*) || топливо *n*, горючее *n*

F387 **fuel calorimeter** || Verbrennungskalorimeter *n* || bombe *f* calorimétrique || calorimetrische bom *f* (*m*) || калориметрическая бомба *f*

F388 **fuel cell** || Brennstoffelement *n* || pile *f* à combustible || brandstofcel *f* (*m*) || топливный элемент *m*

F389 **fuel element** || Brenn(stoff)-element *n* || élément *m* combustible || splijtstofelement *n* || тепловыделяющий элемент *m*, твэл *m*

F390 **fuel gas** || Brenngas *n*; Treibgas *n* || combustible *m* gazeux, gaz *m* combustible || gasvormige brandstof *f* (*m*) || горючий газ *m*; газовое топливо *n*

F391 **fuel nozzle** || Brennstoffdüse *f*; Kraftstoffdüse *f* || injecteur *m* de combustible, gicleur *m* d'essence [de carburant] || brandstofsproeier *m* || топливная форсунка *f*; топливный жиклёр *m*

F392 **fuel pump** || Kraftstoffeinspritzpumpe *f* || pompe *f* à carburant [à combustible]; pompe *f* d'injection *(diesel)* || brandstofpomp *f* (*m*) || топливный насос *m*

F393 **fuel rod** || Brennstoffstab *m* || barre *f* combustible || splijtstofstaaf *f* || топливный стержень *m*, тепловыделяющий элемент *m*, твэл *m*

F394 **fuel system** || Kraftstoffanlage *f*, Kraftstoffördersystem *n* || système *m* d'alimentation [de carburant]; système *m* d'injection de combustible || brandstofsysteem *n* || топливная система *f*, система *f* подачи топлива

F395 **fugacity** || Fugazität *f*, Flüchtigkeit *f* || fugacité *f*, volatilité *f* || vluchtigheid *f* || фугитивность *f*, летучесть *f*

F396 **full-automatic turret lathe** *see* **automatic turret lathe**

F397 **full linear group** || volle lineare Gruppe *f* || groupe *m* linéaire complet || volle lineaire groep *f* (*m*) || полная линейная группа *f*

F398 **full load** || Vollast *f*, Vollbelastung *f* || pleine charge *f* || volle belasting *f* || полная нагрузка *f*

F399 **full power** || Volleistung *f* || pleine puissance *f*, puissance *f* totale || vol vermogen *n* || полная мощность *f*

F400 **full wave** || volle Welle *f*, Vollwelle *f* || pleine onde *f* || volle golf *f* (*m*) || полное колебание *n*, полная волна *f*

F401 **function** || Funktion *f* || fonction *f* || functie *f* || функция *f*

F402 **function space** || Funktionalraum *m* || espace *m* de fonction || functieruimte *f* || функциональное пространство *n*, пространство *n* функций

F403 I **functional** || Funktional *n* || fonctionnelle *f* || functionaal *m* || функционал *m*

F404 II **functional** || funktional || fonctionnel || functioneel || функциональный

F405 **functional analysis** || Funktionalanalyse *f* || analyse *f* fonctionnelle || functionaalanalyse *f* || функциональный анализ *m*

F406 **functional calculus** || Funktionenkalkül *m*, Prädikatenkalkül *m* || calcul *m* fonctionnel || functionaalrekening *f* || функциональное исчисление *n*

F407 **functional dependence** || funktionale Abhängigkeit *f*, funktionale Beziehung *f* || dépendance *f* fonctionnelle || functionaalafhankelijkheid *f* || функциональная зависимость *f*

F408 **functional equation** || Funktionalgleichung *f* || équation *f* fonctionnelle || functionaalvergelijking *f* || функциональное уравнение *n*

F409 **functional transformation** || Funktionaltransformation *f* || transformation *f* fonctionnelle || functionaaltransformatie *f* || функциональное преобразование *n*

F410 **fundamental form** || Grundform *f*, Fundamentalform *f*; Urform || forme *f* fondamentale || grondvorm *m* || основная форма *f*

F411 **fundamental frequency** || Grundfrequenz *f*, Fundamentalfrequenz *f* || fréquence *f* fondamentale || grondfrequentie *f* || основная [фундаментальная] частота *f*

F412 **fundamental mode** || Grundschwingung *f* || vibration *f* fondamentale || grondtrilling *f* || основная мода *f* колебаний; основное [нормальное] колебание *n*

F413 **fundamental oscillation** || Grundschwingung *f* || oscillation *f* fondamentale || grondtrilling *f* || основное колебание *n*

F414 **fundamental particle** || Elementarteilchen *n* || particule *f* fondamentale [élémentaire] || elementair deeltje *n* || фундаментальная [элементарная] частица *f*

F415 **fundamental quantity** || Grundgröße *f*, Basisgröße *f* || valeur *f* principale || grondeenheid *f* || основная величина *f*

F416 **fundamental solution** || Grundlösung *f*, Fundamentallösung *f* || solution *f* fondamentale || grondoplossing *f* || основное [фундаментальное] решение *n*

F417 **fundamental theorem of algebra** || Fundamentalsatz *m* der Algebra || théorème *m* fondamental de l'algèbre || grondtheorema *n* van algebra || основная теорема *f* алгебры

F418 **fundamental tone** || Grundton *m* || ton *m* fondamental || grondtoon *m* || основной тон *m* *(колебаний)*

F419 **fundamental unit** || Grundeinheit *f*, Basiseinheit *f* || unité *f* fondamentale || grondeenheid *f* || основная единица *f*

F420 **funicular polygon** || Seilpolygon *n*, Seileck *n* || polygone *m* funiculaire || kabelpolygoon *m*, kabelveelhoek *m* || верёвочный многоугольник *m*

F421 **furan** || Fur(fur)an *n* || furan(n)e *m*, furfuran(n)e *m* || furaan *n* || фуран *m*

F422 **fuse** || Schmelzsicherung *f* || fusible *m* || smeltzekering *f* || плавкий предохранитель *m*

F423 **fusion** || 1. Schmelzen *n*, Schmelzung *f* 2. Fusion *f*, Verschmelzung *f* *(von Kernen)* || fusion *f* || smelting *f* || 1. плавление *n* 2. слияние *n* (ядер); (ядерный) синтез *m*

F424 **fusion point** || Schmelzpunkt *m* || point *m* de fusion || smeltpunt *n* || точка *f* [температура *f*] плавления

F425 **fusion reaction** || Fusionsreaktion *f*, Kernverschmelzungsreaktion *f* || réaction *f* de fusion || versmeltingsreactie *f* || реакция *f* ядерного синтеза

F426 **fusion reactor** || Fusionsreaktor *m*, Kernfusionsreaktor *m* || réacteur *m* de fusion || kernfusiereactor *m* || (термо)ядерный реактор *m*

F427 **fusion temperature** || 1. Schmelztemperatur *f*, Schmelzpunkt *m* 2. Fusionstemperatur *f*, Kernverschmelzungstemperatur *f* || température *f* [point *m*] de fusion || smeltpunt *n* || 1. температура *f* [точка *f*] плавления 2. температура *f* (ядерного) синтеза

G

G1 **gain** || 1. Verstärkung *f*; Verstärkungsfaktor *m* 2. Gewinn *m*; Zunahme *f* || 1. amplification *f* 2. gain *m* || 1. versterking *f* 2. winst *f*; toename *f* || 1. усиление *n*; коэффициент *m* усиления 2. выигрыш *m*; приращение *n*

G2 **gain factor** || Verstärkungsfaktor *m*, Verstärkungskoeffizient *m* || coefficient *m* d'amplification || versterkingsfactor *m* || коэффициент *m* усиления *(сигнала)*

G3 **gain in energy** || Energiegewinn *m* || gain *m* d'énergie || energiewinst *f* || выигрыш *m* в энергии

G4 **gallium**, Ga || Gallium *n* || gallium *m* || gallium *n* || галлий *m*

G5 **galvanic cell** || galvanisches Element *n*, galvanische Zelle *f* || élément *m* [pile *f*] galvanique || galvanisch element *n* || гальванический элемент *m*

G6 **galvanometer** || Galvanometer *n* || galvanomètre *m* || galvanometer *m* || гальванометр *m*

G7 **game theory** || Spieltheorie *f* || théorie *f* des jeux || speltheorie *f* || теория *f* игр

G8 **gamma** || 1. Gamma-Wert *m*, Kontrastfaktor *m* 2. Gamma *n* || 1. gamma *m* *(facteur de contraste)* 2. gamma *m* || 1. contrastverhouding *f* 2. gamma *n* || 1. коэффициент *m* контрастности 2. гамма *f* *(единица напряжённости магнитного поля, 7,96 10⁻⁴ А/м)*

G9 **gamma-active** || gamma-aktiv || gamma-actif || gamma-actief || (радио)активный

G10 **gamma decay** || Gamma-Zerfall *m* || désintégration *f* gamma || gamma-verval *n* || гамма-распад *m*

G11 **gamma function** || Gamma-Funktion *f* || fonction *f* gamma || gamma-functie *f* || гамма-функция *f*

G12 **gamma-induced** || gamma-induziert || gamma-induit || gamma-geïnduceerd || вызванный гамма-излучением

G13 **gamma quantum** || Gamma-Quant *n* || quantum *m* [rayon *m*, photon *m*] gamma || gammaquantum *n*, gamma-straal *m*, gammafoton *n* || гамма-квант *m*

G14 **gamma radiation** || Gamma-strahlung *f* || rayonnement *m* gamma || gamma-straling *f* || гамма-излучение *n*

G15 **gamma-ray emitter** || Gamma-Strahler *m* || émetteur *m* de rayons gamma || gamma-straler *m* || гамма-излучатель *m*, гамма-источник *m*

G16 **gamma rays** *see* **gamma radiation**

G17 **gamma resonance** || Gamma-Resonanz *f* || résonance *f* gamma || gamma-resonantie *f* || гамма-резонанс *m*, эффект *m* Мёссбауэра

G18 **gang saw** || Gattersäge *f* || scie *f* à cadre à plusieurs lames || raamzaagmachine *f* || лесопильная рама *f*

G19 **gap** || 1. Spalt *m*; Lücke *f*; Abstand *m* 2. Funkenstrecke *f* || 1. jeu *m*; intervalle *m*; distance *f* 2. distance *f* d'éclatement, éclateur || 1. spleet *f* (*m*); speelruimte *f*; speling *f* 2. vonkbrug *f* (*m*) || 1. зазор *m*; интервал *m*; промежуток *m* 2. искровой промежуток *m*

G20 **gas** || Gas *n* || gaz *m* || gas *n* || газ *m*

G21 **gas amplification** || Gasverstärkung *f* || amplification *f* gazeuse || gasversterking *f* || газовое [ионное] усиление *n*

G22 **gas constant** || (universelle) Gaskonstante *f* || constante *f* (universelle) de gaz || (universele) gasconstante *f* (*m*) || (универсальная) газовая постоянная *f*

G23 **gas cutting** || Brennschneiden *n* || oxycoupage *m* || snijbranden *m* || газовая [кислородная, автогенная] резка *f*

G24 **gas-cutting machine** || Brennschneider *m* || machine *f* d'oxycoupage || snijbrandmachine *f* || газорезательная машина *f*

G25 **gas discharge** || Gasentladung *f* || décharge *f* dans le gaz || gasontlading *f* || газовый разряд *m*

G26 **gas-discharge** || Gasentladungs... || à décharge gazeuse || gasontladings... || газоразрядный

G27 **gas dynamics** || Gasdynamik *f* || mécanique *f* de gaz || gasdynamica *f* || газовая динамика *f*, газодинамика *f*

G28 **gas engine** || Gasmotor *m* || moteur *m* à gaz || gasmotor *m* || газовый двигатель *m*

G29 **gaseous** || gasförmig || gazeux || gasvormig || газообразный

G30 **gaseous ion** || Gasion *n* || ion *m* gazeux || gasion *n* || газовый ион *m*

G31 **gaseous medium** || gasförmiges Medium *n*, Gasmedium *n* || milieu *m* gazeux || gasvormig medium *n* || газовая [газообразная] среда *f*

G32 **gaseous mixture** || Gasgemisch *n* || mélange *m* gazeux || gasmengsel *n* || смесь *f* газов, газовая смесь *f*

G33 **gaseous phase** || gasförmige Phase *f*, Gasphase *f* || phase *f* gazeuse || gasface *f* || газовая [газообразная] фаза *f*

G34 **gaseous state** || gasförmiger Zustand *m* || état *m* gazeux || gasvormige toestand *m* || газообразное состояние *n*

G35 **gas-filled (filament) lamp** || Gasfüllungslampe *f*, gasgefüllte Glühlampe *f* || lampe *f* à atmosphère gazeuse || gasvullingslamp *f* (*m*) || газополная лампа *f*

G36 **gas fuel** || Gasbenzin *n* || essence *f* (de gaz) naturel; combustible *m* gazeux || gasvormige brandstof *f* (*m*) || газовый бензин *m*; газовое (моторное) топливо *n*

G37 **gas generator** || Gasgenerator *m*, Gaserzeuger *m* || gazogène *m*, générateur *m* de gaz || gasgenerator *m* || газогенератор *m*

G38 **gas hole** || Gasblase *f* *(im Guß)* || soufflure *f* de gaz *(défaut de moulage)* || gasholte *f*; gasbel *m* || газовый пузырь *m*, раковина *f* *(в отливках)*

G39 **gasket** || Dichtung *f*; Packung *f* || joint *m*, garniture *f* d'étanchéité || dichting *f*; pakking *f* || прокладка *f*, набивка *f*, уплотнение *n*

G40 **gas-kinetic theory** || gaskinetische Theorie *f* || théorie *f* cinétique des gaz || gaskinetische theorie *f* || кинетическая теория *f* газов

G41 **gas lift** || aerostatische Auftriebskraft *f* || portance *f* aérostatique || opstuwende kracht *f* (*m*) || аэростатическая выталкивающая сила *f*

G42 **gasoline engine** || Benzinmotor *m* || moteur *m* à essence || benzinemotor *m* || бензиновый двигатель *m*

G43 **gasproof** || gasdicht || imperméable [étanche] aux gaz || gasdicht || газонепроницаемый

G44 **gas thermometer** || Gasthermometer *n* || thermomètre *m* à gaz || gasthermometer *m* || газовый термометр *m*

G45 **gas thread** || Gas(rohr)gewinde *n* || filetage *m* (au pas du) gaz || gasdraad *m* || трубная [газовая] резьба *f*

G46 **gas turbine** || Gasturbine *f* || turbine *f* à gaz || gasturbine *f* || газовая турбина *f*

G47 **gas-turbine engine** || Gasturbinentriebwerk *n* || turbomoteur *m*, turbomachine *f* || gasturbinemotor *m* || газотурбинный двигатель *m*

G48 **gas welding** || Gasschweißung *f* || soudage *m* aux gaz || gaslassen *n* || газовая сварка *f*

G49 **gate** || Tor *n*, Torschaltung *f* || porte *f* || ventiel *n* || вентиль *m* *(схема пропускания)*

G50 **gate valve** || Schieber *m*, Absperrschieber *m* || robinet-vanne *m*, tiroir *m*, vanne *f* d'arrêt || schuifafsluiter *m* || задвижка *f*, заслонка *f*

G51 **gauge** || **1.** Meßgerät *n*, Meßinstrument *n* **2.** Lehre *f*; Normalmaß *n*; Schablone *f* || **1.** appareil *m* de mesure **2.** calibre *m*; jauge *f*; étalon *m* || **1.** meetapparaat *n* **2.** kaliber *n*; maat *f* (*m*); sjabloon *f* (*m*) || **1.** контрольно-измерительный прибор *m* **2.** калибр *m*; шаблон *m*

G52 **gauge barometer** || Manometer *n*, Druckmesser *m* || manomètre *m* || manometer *m* || манометр *m*

G53 **gauge pressure** || Überdruck *m* || surpression *f* || overdruk *m* || манометрическое [избыточное] давление *n*

G54 **gauss,** G || Gauß *n* || gauss *m* || gauss *m* || гаусс *m*, Гс *(10^{-4} Тл)*

G55 **Gaussian curvature** || Gaußsche Krümmung *f* || courbure *f* gaussienne [totale] || Gauss-kromming *f* || полная [гауссова] кривизна *f*

G56 **Gaussian distribution** || Normalverteilung *f*, Gaußverteilung *f* || distribution *f* gaussienne || Gaussverdeling *f* || нормальное [гауссово] распределение *n*

G57 **gear** || Zahnrad *n* || roue *f* dentée; engrenage *m* || tandwiel *n*, tandrad *n* || зубчатое колесо *n*, шестерня *f*

G58 **gear box** || Getriebe *n*; Schaltgetriebe *n* || boîte *f* de vitesses || versnellingsbak *m*; tandwielkast *f* (*m*); || коробка *f* скоростей; коробка *f* передач

G59 **gear clutch** || Zahnkupplung *f* || accouplement *m* denté; embrayage *m* à dents || tandkoppeling *f* || зубчатая муфта *f*

G60 **gear cutter** || Zahnradfräser *m* || fraise *f* à tailler les engrenages || tandwielfrees *f* (*m*) || зуборезная фреза *f*

G61 **gear cutting** || Zahnradschneiden *n* || taillage *m* d'engrenages || tandwielsnijden *n* || нарезание *n* зубчатых колёс

G62 **gear cutting machine** || Verzahn(ungs)maschine *f* || machine *f* à tailler les engrenages || vertandingsmachine *f* || зуборезный станок *m*

G63 **gear drive** || Zahnradgetriebe *n*, Zahnradtrieb *m* || engrenage *m*; entraînement *m* [commande *f*] par engrenages || tandwielaandrijving *f* || зубчатая передача *f*

G64 **gear grinding machine** || Zahnflankenschleifmaschine *f*, Zahnradschleifmaschine *f* || machine *f* à rectifier [à roder] les engrenages || tandwielslijpmachine *f* || зубошлифовальный станок *m*

G65 **gearing** || Zahngetriebe *n*, Zahnradübersetzung *f*; Eingriff *m* || engrenage *m* || raderwerk *n*; vertanding *f* || зубчатая передача *f*; зубчатое зацепление *n*

G66 **gear lever** || Schalthebel *m* || levier *m* de changement de vitesses || versnellingshandel *n*, *m* || рычаг *m* переключения передач

G67 **gear milling machine** || Zahnradfräsmaschine *f* || machine *f* à fraiser les engrenages || tandfreesmachine *f* || зубофрезерный станок *m*

G68 **gear planing machine** || Zahnradhobelmaschine *f* || machine *f* à raboter les dents || tandwielschaafmachine *f* || зубострогальный станок *m*

G69 **gear pump** || Zahnradpumpe *f* || pompe *f* à engrenages || tandradpomp *f* (*m*) || шестерёнчатый насос *m*

G70 **gear rack** || Zahnstange *f* || crémaillère *f* || tandstang *f* (*m*) || зубчатая рейка *f*

G71 **gear ratio** || Übersetzungsverhältnis *n* || rapport *m* d'engrenage [de transmission] || overbrengingsverhouding *f* || передаточное число *n*; передаточное отношение *n*

G72 **gear shaping machine** || Wälzstoßmaschine *f*; Zahnradhobelmaschine *f* || machine *f* à mortaiser les engrenages; machine *f* à raboter les engrenages || tandwielsteekmachine *f*; tandwielschaafmachine *f* || зубострогальный станок *m*; зубодолбёжный станок *m*

G73 **gear shaving cutter** || Zahnradschabwerkzeug *n* || couteau *m* de shaving [à shaver] || tandwielschaaf *f* (*m*) || шевер *m*

G74 **gear shaving machine** || Zahnradschabmaschine *f* || machine *f* à shaver || tandwielschraapmachine *f* || шевинговальный станок *m*

G75 **gear slotting machine** || Zahnradstoßmaschine *f* || machine *f* à mortaiser les engrenages || tandwielsteekmachine *f* || зубодолбёжный станок *m*

G76 **Geiger counter** || Geiger-Zähler *m* || compteur *m* de Geiger || geigerteller *m* || счётчик *m* Гейгера

G77 **gel** || Gel *n* || gel *m* || gel *n*, *m* || гель *m*

G78 **gelatine** || Gelatine *f* || gélatine *f* 2. gélatine *f* détonante || gelatine *f* (*m*) || 1. желатин *m* 2. нитроглицерин *m*

G79 **general chemistry** || allgemeine Chemie *f* || chimie *f* générale || algemene chemie *f* || общая химия *f*

G80 **general integral** || allgemeine Lösung *f* *(einer Gleichung)*, allgemeines Integral *n* || intégrale *f* générale || algemene oplossing *f*, algemene integraal *f* (*m*) || общий интеграл *m*, общее решение *n* (уравнения)

G81 **generality quantifier** || Allquantor *m* || quantificateur *m* universel || gemeenschappelijke quantor *m* || квантор *m* общности

G82 **generalized** || verallgemeinert, generalisiert || généralisé || gegeneraliseerd || обобщённый *(напр. о координате)*

G83 **generalized coordinate** || verallgemeinerte [generalisierte] Koordinate *f* || coordonnée *f* généralisée || gegeneraliseerde coördinaat *m* || обобщённая координата *f*

G84 **generalized force** || verallgemeinerte [generalisierte] Kraft *f* || force *f* généralisée || gegeneraliseerde kracht *f* (*m*) || обобщённая сила *f*

G85 **generalized momentum** || verallgemeinerter [generalisierter] Impuls *m* || impulsion *f* généralisée || gegeneraliseerde impuls *m* || обобщённый импульс *m*

G86 **generalized variance** || verallgemeinerte Streuung *f* || variance *f* généralisée || gegeneraliseerde variantie *f* || обобщённая дисперсия *f*

G87 **generalized velocity** || verallgemeinerte [generalisierte] Geschwindigkeit *f* || vitesse *f* généralisée || gegeneraliseerde snelheid *f* || обобщённая скорость *f*

G88 **general relativity** || allgemeine Relativitätstheorie *f* || relativité *f* générale || algemene relativiteitstheorie *n* || общая теория *f* относительности

G89 **general solution** || allgemeine Lösung *f* || solution *f* générale || algemene oplossing || || общее решение *n*

G90 **general theory of communication** || Nachrichtentheorie *f* || théorie *f* générale des communications || algemene communicatietheorie *f* || теория *f* связи

G91 **generated heat** || entwickelte Wärme *f* || chaleur *f* dégagée || ontwikkelde warmte *f* || выделяемая теплота *f*

G92 **generating function** || erzeugende Funktion *f* || fonction *f* génératrice || voortbrengende functie *f* || производящая функция *f*

G93 **generation** || **1.** Generation *f* **2.** Bildung *f*, Erzeugung *f* || **1.** génération *f* **2.** formation *f* || **1.** generatie *f* **2.** formatie *f* || **1.** генерация *f*, генерирование *n* **2.** рождение *n*, образование *n*

G94 **generation of random numbers** || Generierung *f* von Zufallszahlen || génération *f* des nombres aléatoires || generatie *f* van toevalsgetallen || генерирование *n* случайных чисел

G95 **generator** || Generator *m*, Erzeuger *m* || générateur *m*, génératrice *f* || generator *m* || генератор *m* (*энергии, частоты*)

G96 **generatrix** || Erzeugende *f* || génératrice *f* || voortbrengende *f* || образующая *f*

G97 I **geodesic** || Geodätische *f*, geodätische Linie *f* || géodésique *f* || geodetische lijn *f* (*m*) || геодезическая (линия) *f*

G98 II **geodesic** || geodätisch || géodésique || geodetisch || геодезический

G99 **geodesic curvature** || geodätische Krümmung *f* || courbure *f* géodésique || geodetische kromming *f* || геодезическая кривизна *f*

G100 **geodesic triangle** || geodätisches Dreieck *n* || triangle *m* géodésique || geodetische driehoek *m* || геодезический треугольник *m*

G101 **geoid** || Geoid *n* || géoïde *m* || geoïde *n* || геоид *m*

G102 **geomagnetic field** || Magnetfeld *n* der Erde, erdmagnetisches Feld *n* || champ *m* géomagnétique || aardmagnetisch veld *n* || магнитное поле *n* Земли, геомагнитное поле *n*

G103 **geometrical mean** || geometrisches Mittel *n* || moyenne *f* géométrique || geometrisch gemiddelde *n* || среднее геометрическое *n*

G104 **geometrical optics** || geometrische Optik *f*, Strahlenoptik *f* || optique *f* géométrique || geometrische optiek *f* || геометрическая [лучевая] оптика *f*

G105 **geometric progression** || geometrische Reihe *f* || progression *f* géométrique || geometrische reeks *f* (*m*) || геометрическая прогрессия *f*

G106 **geometry** || Geometrie *f* || géométrie *f* || meetkunde *f*, geometrie *f* || геометрия *f*

G107 **geophysics** || Geophysik *f* || géophysique *f* || geofysica *f* || геофизика *f*

G108 **geostationary orbit** || geostationäre Bahn *f* || orbite *f* géostationnaire || geostationaire baan *f* (*m*) || геостационарная орбита *f*

G109 **germanium,** Ge || Germanium *n* || germanium *m* || germanium *n* || германий *m*

G110 **getter** || Getter *m*, Getterstoff *m*, Fangstoff *m* || getter *m*, fixateur *m* de gaz || getter *m* || геттер *m*, газопоглотитель *m*

G111 **ghost (image)** || störende Abbildung *f*, Geisterabbildung *f* || image *f* fantôme, fantôme *m* || fantoombeeld *n* || паразитное [ложное, побочное] изображение *n*

G112 **gib-headed key** || Nasenkeil *m* || clavette *f* à talon || kopspie *f* (*m*) || шпонка *f* с головкой

G113 **giga...** || Giga... || giga... || giga... || гига... (*приставка в наименовании кратной единицы, соответствующая* 10^9)

G114 **gigaelectronvolt** || Gigaelektronenvolt *n* || giga-electron-volt *m* || gigaelektronvolt *m* || гигаэлектронвольт *m* (10^9 *эВ*)

G115 **glancing collision** || schiefer Stoß *m* || collision *f* glissante || glijdende collisie *f* || скользящее столкновение *n*; косой удар *m*

G116 **gland** || Stopfbuchse *f* || boîte *f* à étoupes, presse-garniture *f*, presse-étoupe *m* || pakkingbus *f* (*m*), drukstuk *n* || сальник *m*

G117 **glasscloth** || Glasgewebe *n* || tissu *m* de verre || glasvezeldoek *n*, *m* || стеклоткань *f*

G118 **glass electrode** || Glaselektrode *f* || électrode *f* de verre || glaselektrode *f* || стеклянный электрод *m*

G119 **glassy state** || Glaszustand *m*, glasartiger Zustand *m* || état *m* amorphe || glasachtige toestand *m* || аморфное состояние *n*

G120 **globe photometer** || Kugelphotometer *n* || photomètre *m* sphérique || kogelfotometer *m* || шаровой фотометр *m*

G121 **glow** || Glühen *n*, Glimmen *n* || chauffage *m*, lueur *f* || gloeiing *f*, lichten *n* || накал *m*, свечение *n*

G122 **glow discharge** || Glimmentladung *f* || décharge *f* luminescente [en lueur] || gloeiontlading *f* || тлеющий разряд *m*

G123 **gluon** || Gluon *n* || gluon *m* || gluon *n* || глюон *m*

G124 **go gauge** || Gutlehre *f* || calibre *m* entre || lopende mal *m*, doorlopend kaliber *n* || проходной калибр *m*

G125 **gold,** Au || Gold *n* || or *m* || goud *n* || золото *n*

G126 **gold leaf electroscope** || Goldblattelektroskop *n* || électroscope *m* à plaques en or || goudbladelektroscoop *m* || электроскоп *m* с золотыми листочками

G127 **goniometer** || Goniometer *n*, Winkelmesser *m* || goniomètre *m* || goniometer *m* || гониометр *m*

G128 **gooseneck tool** || gekröpfter Meißel *m* || outil *m* coudé [de côté] || gebogen beitel *m* || изогнутый резец *m*

G129 **governor** || Regler *m* || régulateur *m* || regelaar *m* || регулятор *m*

G130 **grad(ient)** || Gefälle *n*, Gradient *m* || gradient *m* || gradient *m* || градиент *m*

G131 **gradient line** || Gradientenlinie *f* || ligne *f* de gradient || gradiëntenlijn *f* (*m*) || линия *f* градиента

G132 **gradient method** || Gradientenverfahren *n* || méthode *f* de gradient || gradiëntenmethode *f* || градиентный метод *m*

G133 **gradient operator** || Nablaoperator *m*, ∇-Operator *m* || opérateur *m* gradient, opérateur *m* nabla || gradiëntoperator *m*, nablaoperator *m* || набла-оператор *m*, ∇-оператор *m*

G134 **graduation** || Graduieren *n*, Graduierung *f*, Gradeinteilung *f* || graduation *f* || graduatie *f* || градуировка *f*

G135 **gram** || Gramm *n* || gramme *m* || gram *n*, *m* || грамм *m* (10^{-3} *кг*)

G136 **gram atom** || Grammatom *n* || gramme-atome *m* || gramatoom *n* || грамм-атом *m*

G137 **gram eguivalent** || Grammäquivalent *n* || gramme-équivalent *m* || gramequivalent *n* || граммэквивалент *m*

G138 **gram molecule** || Grammol(ekül) || *n* gramme-molécule *m*, molécule-gramme *f* || grammolecule *f* (*m*) || грамм-молекула *f*

G139 **gram-weight** || Pond *m* || gramme-force *m*, gramme-poids *m* || pond *n* || грамм-сила *f*, гс *(9,80665·$\cdot 10^{-3}$ H)*

G140 **granular medium** || körniges Medium *n*; rolliges Material *n* || milieu *m* pulvérulant || korrelig [gekorreld] medium *n* || сыпучая среда *f*

G141 **granulation** || Granulation *f*, Granulierung *f* || granulation *f* || granulatie *f* || грануляция *f*; дробление *n*

G142 **graph** || **1.** Diagramm *n* **2.** Graph *m* || graphe *m* || **1.** diagram *n* **2.** graf af *f* (*m*) || **1.** график *m* **2.** граф *m*

G143 **graphics** || Graphik *f* || arts *m plg* raphiques || grafische kunst *f* || графика *f*

G144 **graphite** || Graphit *m* || graphite *m*, plombagine *f* || grafiet *n* || графит *m*

G145 **graphostatics** || graphische Statik *f*, Graphostatik *f* || statique *f* graphique, graphostatique *f* || grafostatica *f* || графическая статика *f*, графостатика *f*

G146 **graph theory** || Graphentheorie *f* || théorie *f* des graphes || theorie *f* der graafen || теория *f* графов

G147 **grating** || Beugungsgitter *n*, Gitter *n* || réseau *m* de diffraction || diffractierooster *m*, *n* || дифракционная решётка *f*

G148 **grating spectrograph** || Gitterspektrograf *m*, Beuguugsspektrograf *m* || spectrographe *m* à réseaux de diffraction || diffractiespectrograaf *m* || спектрограф *m* с дифракционной решёткой, дифракционный спектрограф *m*

G149 **gravitation** || Gravitation *f*, Anziehungskraft *f* || gravitation *f*, attraction *f* gravitationnelle || gravitatie *f* || гравитация *f*, тяготение *n*

G150 **gravitational** || Gravitations..., Schwerkraft..., Schwere... || gravitationnel, de gravitation || gravitatie... || гравитационный

G151 **gravitational attraction** || (allgemeine) Gravitation *f*, Newtonsche Gravitationskraft *f* || attraction *f* gravitationnelle [universelle] || gravitatie *f* || гравитационное притяжение *n*, гравитация *f*, (всемирное) тяготение *n*

G152 **gravitational collapse** || Gravitationskollaps *m* || collapsus *m* de gravitation || gravitatiecollaps *m* || гравитационный коллапс *m*

G153 **gravitational field** || Gravitationsfeld *n*, Schwerefeld *n* || champ *m* gravitationnel || gravitatieveld *n* || гравитационное поле *n*, поле *n* тяготения

G154 **gravitational force** || Massenanziehungskraft *f*, Gravitationskraft *f* || force *f* gravitationnelle [de gravitation] || gravitatiekracht *f* (*m*) || сила *f* тяготения; гравитационная сила *f*

G155 **gravitational interaction** || Gravitationswechselwirkung *f*, gravitative Wechselwirkung *f* || interaction *f* gravitationnelle || gravitatiewisselwerking *f* || гравитационное взаимодействие *n*

G156 **gravitational mass** || schwere Masse *f* || masse *f* pesante [gravitante] || zware massa *f* (*m*), gravitatiemassa *f* (*m*) || тяжёлая [гравитационная] масса *f*

G157 **gravitational waves** || Gravitationswellen *f pl*, Schwerewellen *f pl* || ondes *f pl* gravifiques [gravitationnelles] || gravitatiegolven *f* (*m*) *pl*, zwaartekrachtgolven *f* (*m*) *pl* || гравитационные волны *f pl*

G158 **graviton** || Graviton *n*, Gravitationsquant *n* || graviton *m* || graviton *n* || гравитон *m*, квант *m* гравитационного поля

G159 **gravity** || Schwerkraft *f*; Schwere *f* || gravité *f* || zwaartekracht *f* (*m*) || тяжесть *f*, сила *f* тяжести; гравитация *f*

G160 **gravity feed** || Schwerkraftförderung *f*; Freifallbeschickung *f* || alimentation *f* par gravité || zwaartekrachttoevoer *m* || гравитационная [самотёчная] подача *f*

G161 **gravity force** || Schwerkraft *f* || force *f* de gravité || zwaartekracht *f* (*m*) || сила *f* тяжести

G162 **gravity waves** *see* **gravitational waves**

G163 **gray** || grau; neutralgrau || gris || grijs; grauw || серый, нейтральный *(в оптике)*

G164 **gray body** || Graustrahler *m*, grauer Körper *m* || corps *m* gris || grauw lichaam *n* || серое тело *n*

G165 **Gray code** || Gray-Kode *m*, zyklisch permutierter Kode *m* || code *m* de Gray || Gray-code *m* || код *m* Грея

G166 **gray iron** *see* **grey iron**

G167 **gray radiation** || graue Strahlung *f* || rayonnement *m* gris || grauwe straling *f* || серое излучение *n*

G168 **gray wedge** || Graukeil *m*, Neutralkeil *m* || coin *m* gris || grauwe wig *f* (*m*) || серый клин *m*

G169 **grazing incidence** || streifender Einfall *m* || incidence *f* rasante || glijdende inval *m* || скользящее падение *(луча) n*, падение *(луча) n* под малым углом

G170 **grease** || Schmierfett *n* || graisse *f* || smeervet *n* || консистентная смазка *f*

G171 **grease gun** || Fettspritze *f* || graisseur *m* à refoulement || vetspuit *f* (*m*) || смазочный шприц *m*

G172 **greatest common divisor** || größter gemeinsamer Teiler *m* || le plus grand diviseur *m* commun || grootste gemene deler *m* || наибольший общий делитель *m*

G173 **greatest lower boun** || Infimum *n* || borne *f* inférieure (la plus grande) || grootste benedenkant *m* || точная нижняя грань *f*

G174 **grey iron** || **1.** graues Roheisen *n* **2.** Graueisen *n* || fonte *f* grise [graphitique] || **1.** grijs ruw ijzer *n* **2.** grijs gietijzer *n* || серый чугун *m*

G175 **grid** || Gitter *n* || grille *f* || rooster *m*, *n* || сетка *f (электронной лампы)*

G176 **grid leak** || Gitterableitung *f*; Gitterableitwiderstand *m* || fuite *f* de grille || roosterlek *n* || сеточная утечка *f*; смещающее сеточное сопротивление *n*

G177 **grinder** *see* **grinding machine**

G178 **grinding** || Schleifen *n* || rectification *f*, meulage *m* || slijpen *n*; polijsten *n* || шлифование *n*

G179 **grinding head** || Schleifmaschinenspindelstock *m*; Schleifkopf *m* || poupée *f* [tête *f*] porte-meule || slijpkop *m* || бабка *f* шлифовального станка, шлифовальная бабка *f*; шлифовальная головка *f*

G180 **grinding machine** || Schleifmaschine *f* || machine *f* à rectifier, rectifieuse *f* || slijpmachine *f* || шлифовальный станок *m*

G181 **grinding paste** || Schleifpaste *f* || pâte *f* abrasive || slijppasta *m*, *n* || шлифовальная паста *f*

G182 **grinding powder** || Schleifpulver *n* || poundre *f* abrasive || slijppoeder *n*, *m* || шлифовальный [абразивный] порошок *m*

G183 **grinding slide** || Schleifschlitten *m* || chariot *m* porte-meule [à rectifier] || slijpkopslede *f* (*m*) || каретка *f* шлифовальной бабки

G184 **grinding wheel** || Schleifscheibe *f* || meule *f* || slijpschijf *f* (*m*) || шлифовальный круг *m*

G185 **grindstone** || Schleifstein *m* || pierre *f* à polir [à affûter], affiloir *m* || slijpsteen *m* || точило *n*, точильный камень *m*

G186 **grip vice** || Schraubstock *m* || étau *m* || (klem)bankschroef *f* (*m*) || тиски *pl*

G187 **groove** || Nut *f*; Aussparung *f*; Schlitz *m* || encoche *f*; rainure *f*; cannelure *f* || groef *f* (*m*); sponning *f* || канавка *f*, паз *m*; выточка *f*; прорезь *f*

G188 **gross load** || Gesamtmasse *f*; Bruttomasse *f*; Gesamtlast *f* || poids *m* total, masse *f* totale; charge *f* totale || bruto-last *f* (*m*), volle last *f* (*m*) || полный вес *m*, полная масса *f*; полная нагрузка *f*

G189 **ground** *see* **earthing**

G190 **grounded circuit** || geerdete Leitung *f* || circuit *m* mis à la terre || geaard circuit *n* || заземлённая цепь *f*

G191 **grounding** *see* **earthing**

G192 **ground level** || Grundniveau *n* || niveau *m* énergétique fondamental [normal] || grondniveau *n* || основной [нормальный] уровень *m* энергии

G193 **ground state** || Grundzustand *m* || état *m* fondamental || grondtoestand *m* || основное состояние *n*
G194 **group** || Gruppe *f* || groupe *m* || groep *f* (*m*) || группа *f*
G195 **group of transformations** || Transformationsgruppe *f* || groupe *m* de transformations || transformatiegroep *f* (*m*) || группа *f* преобразований
G196 **group representation** || Gruppendarstellung *f* || représentation *f* du groupe || groepenvoorstelling *f* || представление *n* группы
G197 **group theory** || Gruppentheorie *f* || théorie *f* des groupes || groepentheorie *f* || теория *f* групп
G198 **group velocity** || Gruppengeschwindigkeit *f* || vitesse *f* de groupe || groepsnelheid *f* || групповая скорость *f*
G199 **grub screw** || versenkte Schraube *f* || vis *f* noyée; vis *f* sans tête || verzonken schroef *f* (*m*) || потайной винт *m*; штифт *m*
G200 **guard** || Sicherungseinrichtung *f*, Schutzeinrichtung *f* || dispositif *m* de protection [de sécurité, de sûreté] || zekeringsinrichting *f* || предохранительное [защитное] устройство *n*
G201 **guard ring** || Schutzring *m* || anneau *m* de garde || beschermingsring *m* || предохранительное кольцо *n*
G202 **gudgeon pin** *see* **piston pin**
G203 **guide** || Führung *f* || guide *m*; guidage *m* || geleiding *f*, geleider *m* || направляющая *f*
G204 **guide pulley** || Leitscheibe *f*; Leitrolle *f* || poulie *f* de guidage, poulie *f* guide || leirol *f* (*m*), leischijf *f* (*m*) || направляющий шкив *m*; направляющий ролик *m*
G205 **guide sleeve** || Führungsbuchse *f* || manchon *m* guide; douille *f* de gûidage || leibus *f* (*m*); || направляющая втулка *f*
G206 **guide wheel** || Leitrad *n* (*Turbine*) || roue *f* (de turbine) directrice || leiwiel *n* || направляющее колесо *n* (*турбины*)
G207 **guillotine shears** || Blechtafelschere *f*, Tafelschere *f* || cisaille *f* à guillotine || guillotineschaar *f* (*m*) || гильотинные ножницы *pl*
G208 **gun** || Spritze *f* || pompe *f* à graisse || spuit *f* (*m*); oliepomp *f* (*m*) || шприц *m*; раздаточное приспособление *n* (*для топлив и масел*)
G209 **gyration** || Kreiselbewegung *f*, Gyration(sbewegung) *f* || giration *f* || rotatie *f* || движение *n* по окружности, вращение *n*
G210 **gyromagnetic ratio** || gyromagnetisches Verhältnis *n* || rapport *m* gyromagnétique || gyromagnetische verhouding *f* || гиромагнитное отношение *n*
G211 **gyroscope** || Kreisel *m*, Gyroskop *n* || gyroskope *m* || gyroscoop *m* || гироскоп *m*
G212 **gyroscopic precession** || Kreiselpräzession *f* || précession *f* du gyroscope || gyroscoopprecessie *f* || прецессия *f* гироскопа
G213 **gyroscopic stabilization** || Kreiselstabilisierung *f*, gyrostatische Stabilisierung *f* || stabilisation *f* gyroscopique || gyroscopische stabilisatie *f* || гироскопическая стабилизация *f*
G214 **gyroskopic torque** || Kreiselmoment *n* || moment *m* gyroscopique || gyroscopisch moment *n* || гироскопический момент *m*
G215 **gyrostabilized** || kreiselstabilisiert, gyrostatisch stabilisiert || gyrostabilisé || gyro-gestabiliseerd || гиростабилизированный, гироскопически стабилизированный

H

H1 **hadron** || Hadron *n* || hadron *m* || hadron *n* || адрон *m* (*сильновзаимодействующая частица*)
H2 **half-breadth** || Halbwertsbreite *f* || demi-largeur *f* || halfbreedte *f* || полуширина *f* (*напр. распределения*)
H3 **half-countersunk head** || Linsensenkkopf *m* || tête *f* mi-fraisée || halfverzonken kop *m* || полупотайная головка *f* (*напр. винта*)
H4 **half-integer spin** || halbzahliger Spin *m* || spin *m* demi-entier || halfhele spin *m* || полуцелый спин *m*

H5 **half-life** || Halbwert(s)zeit *f* || période *f* de demi-vie || halveringstijd *m* || период *m* [время *n*] полураспада

H6 **half-period** || Halbperiode *f* || demi-période *f* || halfperiode *f* || полупериод *m*

H7 **half-plane** || Halbebene *f* || demi-plan *m* || halfvlak *n* || полуплоскость *f*

H8 **half-round** || Halbrund..., halbrund || demi-sphérique || halfrond || полукруглый

H9 **half-shadow** || Halbschatten *m* || pénombre *f* || halfschaduw *f* (*m*) || полутень *f*

H10 **half-value layer** || Halbwertsschicht *f*, Halbwertsdicke *f* || couche *f* de demi-absorption [de valeur moyenne] || halveringsdikte *f* || слой *m* половинного ослабления

H11 **half-value thickness** || Halbwertsschichtdicke *f* || épaisseur *f* de la couche de demi-absorption || halveringsdikte *f* || толщина *f* слоя половинного поглощения *(излучения)*

H12 **half-wave detection** || Einweggleichrichtung *f*, Halbweggleichrichtung *f* || redressement *m* en demi-onde [à une seule alternance] || halvegolfgelijkrichting *f* || однополупериодное выпрямление *n*

H13 **half wave (length)** || halbe Wellenlänge *f*, Halbwelle *f* || demi-onde *f* || halve golflengte *f* || половина *f* длины волны; полуволна *f*

H14 **half-wave plate** || Halbwellenplättchen *n* || lame *f* demi-onde || halfgolfplaat *f* (*m*) || пластинка *f* в полволны

H15 **half-width** || Halb(werts)breite *f* || demi-largeur *f* || halfbreedte *f* || полуширина *f*; ширина *f* на полувысоте (максимума)

H16 **halo** || Halo *m*, Hof *m* || auréole *f*, halo *m* || halo *m* || ореол *m*; гало *n*

H17 **halogen** || Halogen *n* || halogène *m*, haloïde *m* || halogeen *n* || галоген *m*

H18 **Hamiltonian (function)** || Hamilton-Funktion *f*, Hamiltonsche Funktion *f* || fonction *f* de Hamilton, hamiltonien *m* || Hamilton-functie *f* || функция *f* Гамильтона, гамильтониан *m*

H19 **Hamiltonian operator** || Hamilton-Operator *m*, Nabla-Operator *m* || hamiltonien *m*, opérateur *m* nabla || Hamilton-operator *m*, nabla-operator *m* || набла-оператор *m*

H20 **hammer** || Hammer *m* || 1. marteau *m* 2. linguet *m* || 1. hamer *m* 2. onderbrekerhamer *m* || 1. молот *m*; молоток *m* 2. молоточек *m* *(прерывателя)*

H21 **hammering** || Schmieden *n* || martelage *m* || hameren *n* || ковка *f* молотом

H22 **Hamming code** || Hamming-Kode *m* || code *m* de Hamming || Hamming-code *m* || код *m* Хэмминга

H23 **hand brace, hand drill** || Handbohrmaschine *f* || perceuse *f* à main || handboor *f* (*m*) || (ручная) дрель *f*

H24 **handle** || Griff *m* || poignée *f*; manche *f*; levier *m* || handgreep *m*, handvat *n*, greep *m* || ручка *f*; рукоятка *f*

H25 **hand tool** || Handwerkzeug *n* || outil *m* à main || handgereedschap *n* || ручной инструмент *m*; слесарный инструмент *m*

H26 **hand vice** || Handfeilkloben *m*; Handschraubstock *m* || étau *m* à main || handschroef *f* (*m*) || ручные тиски *pl*

H27 **hard** || hart || dur || hard || 1. твёрдый, жёсткий 2. жёсткий, проникающий *(об излучении)*

H28 **hardenability** || Härtbarkeit *f* || aptitude *f* à la trempe, trempabilité *f* || hardbaarheid *f* || 1. закаливаемость *f* 2. отверждаемость *f*

H29 **hardened steel** || gehärteter Stahl *m* || acier *m* trempé || gehard staal *n* || закалённая сталь *f*

H30 **hardening** || Härten *n* || trempe *f*; durcissement *m* || harding *f* || 1. закалка *f* 2. упрочнение *n* 3. твердение *n* *(напр. бетона)*; отверждение *n* *(пластмасс)*

H31 **hard facing** || Auftragschweißen *n* *(einer Aufschweißlegierung)*; Verstählung *f* || rechargement *m* dur || oplassen *n* van hardmetaal || наплавка *f* твёрдым сплавом

H32 **hard-magnetic** || hartmagnetisch, magnetisch hart || magnétique dur || hardmagnetisch || магнитнотвёрдый

H33 **hardness** || Härte *f* || dureté *f* || hardheid *f* || твёрдость *f*

H34 **hardness number** || Härtezahl *f* || nombre *m* de dureté || hardheidsgraad *m* || число *n* твёрдости

H35 **hardness test** || Härteprüfung *f* || essai *m* de dureté || hardheidsproef *f* (*m*) || определение *n* твёрдости

H36 **hard radiation** || harte [durchdringende] Strahlung *f* || rayonnement *m* dur || harde straling *f* || жёсткое [проникающее] излучение *n*

H37 **hard surfacing** *see* **hard facing**

H38 **hardware** || Hardware *f* || matériel *m* || hardware *f* || аппаратные средства *n pl*, аппаратура *f*

H39 **hard X-rays** || harte Röntgenstrahlung *f*, harte Strahlen *pl* || rayons X *m pl* durs || harde X-stralen *m*, *f pl* || жёсткое рентгеновское излучение *n*

H40 I **harmonic** || Harmonische *f* || onde *f* harmonique, harmonique *f* || harmonische *f* || гармоника *f* ||

H41 II **harmonic** || harmonisch || harmonique || harmonisch || гармонический

H42 **harmonic analysis** || harmonische Analyse *f*, Fourier-Analyse *f* || analyse *f* harmonique || harmonische analyse *f* || гармонический анализ *m*

H43 **harmonic frequency** || Oberschwingungsfrequenz *f* || fréquence *f* d'harmonique || frequentie *f* van harmonischen || частота *f* гармоники

H44 **harmonic function** || harmonische Funktion *f* || fonction *f* harmonique || harmonische functie *f* || гармоническая функция *f*

H45 **harmonic oscillations** || harmonische Schwingungen *f pl* || oscillations *f pl* harmoniques || harmonische trillingen *f pl* || гармонические колебания *n pl*

H46 **harmonic oscillator** || 1. harmonischer Oszillator *m* 2. harmonischer Generator *m* || 1. oscillateur *m* harmonique 2. générateur *m* d'harmoniques || 1. harmonische oscillator *m* 2. harmonische generator *m* || 1. гармонический осциллятор *m* 2. генератор *m* гармоник

H47 **head** || 1. Druckhöhe *f* 2. Kopf *m*, Meßkopf *m* || 1. chute *f* totale 2. tête *f* || 1. drukhoogte *f*, toeloop *m* 2. kop *m* || 1. напор *m* 2. головка *f*

H48 **heading machine** || Kopfstauchgesenk *n* || presse *f* pour formage des têtes || pers *f* (*m*) voor koudsmeden van boutskoppen || пресс *m* для высадки головок *(напр. болтов)*

H49 **headless bolt** *see* **stud bolt**

H50 **head-on collision** || zentraler Stoß *m*, Zentralstoß *m* || collision *f* frontale || frontale collisie *f* || лобовой удар *m*, лобовое столкновение *n*

H51 **headstock** || Spindelstock *m* *(Werkzeugmaschine)* || poupée *f* avant [fixe] *(d'une machine-outil)* || spilkop *m (gereedsch.-mach.)* || передняя бабка *f (станка)*

H52 **heat** || 1. Wärme *f*, Wärmewirkung *f* 2. spezifische Wärme *f* || 1. chaleur *f* 2. chaleur *f* spécifique || 1. warmte *f* 2. soortelijke warmte *f* || 1. теплота *f*, тепловой эффект *m* 2. удельная теплота *f*

H53 **heat-absorbing** || wärmeabsorbierend, wärmeaufnehmend || à absorption thermique || warmteopnemend || теплопоглощающий; эндотермический

H54 **heat balance** || Wärmebilanz *f*, Wärmehaushalt *m* || bilan *m* thermique || warmtebalans *f* (*m*) || тепловой баланс *m*

H55 **heat capacity** || Wärmekapazität *f* || capacité *f* calorifique [thermique] || warmtecapaciteit *f* || теплоёмкость *f*

H56 **heat consumption** || Wärmeverbrauch *m* || consommation *f* de la chaleur || warmteafname *f* (*m*) || 1. потребляемая теплота *f* 2. потребление *n* тепла

H57 **heat content** *see* **enthalpy**

H58 **heat density** || Wärmedichte *f* || densité *f* thermique || warmtedichtheid *f* || теплонапряжённость *f*

H59 **heat efficiency** || thermischer Wirkungsgrad *m*, Wärmewirkungsgrad *m* || rendement *m* thermique || thermisch rendement *n* || термический [тепловой] коэффициент *m* полезного действия

H60 **heat energy** || Wärmeenergie *f* || énergie *f* calorifique [thermique] ||

warmteënergie *f* || тепловая энергия *f*

H61 **heat engine** || Wärmekraftmaschine *f* || machine *f* thermique || warmtemachine *f*; warmtemotor *m* || тепловая машина *f*; тепловой двигатель *m*

H62 **heat engineering** || Wärmetechnik *f* || technique *f* de la chaleur || warmtetechniek *f* || теплотехника *f*

H63 **heater** || Heizgerät *n*; Heizkörper *m* || réchaud *m*, réchauffeur *m* || verwarmingsapparaat *n* || нагреватель *m*; нагревательный прибор *m*

H64 **heat exchange** || Wärmeaustausch *m* || échange *f* calorifique || warmtewisseling *f* || теплообмен *m*

H65 **heat exchanger** || Wärme(aus)-tauscher *m* || échangeur *m* de chaleur || warmte(uit)wisselaar *m* || теплообменник *m*

H66 **heat flux** || Wärmestrom *m*, Wärmefluß *m* || flux *m* de chaleur || warmtestroom *m* || тепловой поток *m*

H67 **heating** || **1.** Erwärmung *f*; Erhitzung *f* **2.** Heizung *f*, Heizen *n* || chauffage *m* || verwarming *f* || нагрев *m*, нагревание *n*

H68 **heating surface** || Heizfläche *f* || surface *f* de chauffe || verwarmingsvlak *n* || поверхность *f* нагрева

H69 **heat-insulated** || wärmeisoliert || à isolement calorifique || warmtegeïsoleerd || теплоизолированный

H70 **heat insulation** || Wärmeisolierung *f*, Wärmeisolation *f*, Wärmedämmung *f* || calorifugeage *m*; isolation *f* calorifique || warmteïsolatie *f* || тепловая изоляция *f*, теплоизоляция *f*

H71 **heat load** || Wärmebelastung *f*, Wärmebeanspruchung *f* || charge *f* thermique || warmtebelasting *f* || тепловая нагрузка *f*

H72 **heat loss** || Wärmeverluste *m pl* || pertes *f pl* de chaleur || warmteverliezen *n pl* || тепловые потери *f pl*, потери *f pl* тепла

H73 **heat motion** || Wärmebewegung *f*, thermische Bewegung *f* || mouvement *m* thermique || thermische beweging *f* || тепловое движение *n*

H74 **heat of condensation** || Kondensationswärme *f* || chaleur *f* de condensation || condensatiewarmte *f* || (удельная) теплота *f* конденсации

H75 **heat of fusion, heat of melting** || (latente) Schmelzwärme *f* || chaleur *f* de fusion || (latente) smeltwarmte *f* || удельная [скрытая] теплота *f* плавления

H76 **heat of process** || Prozeßwärme *f* || effet *m* thermique du processus || proceswarmte *f* || тепловой эффект *m* процесса

H77 **heat of reaction** || Wärmetönung *f* (der Reaktion), Reaktionswärme *f* || effet *m* thermique d'une réaction || reactiewarmte *f* || теплота *f* [тепловой эффект *m*] реакции

H78 **heat of solidification** || (latente) Erstarrungswärme *f* || chaleur *f* (latente) de solidification || (latente) stollingswarmte *f* || удельная [скрытая] теплота *f* затвердевания

H79 **heat of transition** || Umwandlungswärme *f* || chaleur *f* de transition [transformation] || omzettingswarmte *f* || теплота *f* (фазового) превращения

H80 **heat of vaporization** || Verdampfungswärme *f*, Verdunstungswärme *f* || chaleur *f* de vaporisation || verdampingswarmte *f* || (удельная) теплота *f* испарения, теплота *f* парообразования

H81 **heat power plant** || Wärmekraftwerk *n* || centrale *f* électrothermique || warmtekrachtcentrale *f* (*m*) || теплоэлектростанция *f*, тепловая электростанция *f*

H82 **heat producing** || wärme(ab)-gebend, wärmeerzeugend; exotherm || à dégagement calorifique || warmte afgevend || тепловыделяющий; экзотермический

H83 **heatproof** *see* **heat-resistant**

H84 **heat radiation** || Wärmestrahlung *f*, thermische Strahlung *f* || radiation *f* thermique || warmtestraling *f* || тепловое излучение *n*, тепловая радиация *f*

H85 **heat resistance** *see* **thermal resistance**

H86 **heat-resistant** || wärmebeständig, wärmefest; hitzebeständig, hitzefest || thermostable, résistant à la chaleur || warmtevast || теплостойкий, жаростойкий

H87 **heat-resisting alloy** || hitzebeständige Legierung *f* || alliage *m* refractaire [résistant à la chaleur] || vuurvaste legering *f* || жаростойкий [жароупорный] сплав *m*

H88 **heat stress** || Wärmespannung *f* || contrainte *f* thermique || warmtespanning *f* || тепловое [температурное] напряжение *n*

H89 **heat transfer** || Wärmeübertragung *f*, Wärmeübergang *m* || transmission *f* [transfert *m*] de chaleur || warmteoverdracht *f* (*m*) || перенос *m* тепла, теплопередача *f*

H90 **heat transfer medium** || Wärmeträger *m* || agent *m* de transfert de chaleur, véhicule *m* thermique || warmtedrager *m* || теплоноситель *m*

H91 **heat transmission** || Wärmeübertragung *f*; Wärmedurchgang *m* || transfert *m* de chaleur || warmteoverdracht *f* (*m*); warmtedoorgang *m* || теплопередача *f*

H92 **heat treamtent** || Wärmebehandlung *f*, thermische Behandlung *f* || traitement *m* thermique || warmtebehandeling *f*, thermische behandeling *f* || термическая обработка *f*

H93 **heavy-duty** || Schwerlast..., Hochleistungs... || de gros tonnage; à grande puissance, à grande capacité de charge || berekend op zware last || большой грузоподъёмности, большегрузный; большой мощности

H94 **heavy duty machine** || Hochleistungs(werkzeug)maschine *f* || machine-outil *f* à grande puissance || machine *f* met hoog rendement || станок *m* большой мощности

H95 **heavy-plate mill** || Grobblechwalzwerk *n* || laminoir *m* à tôles épaisses [fortes], train *m* a tôles fortes || walserij *f* voor zware plaat || толстолистовой прокатный стан *m*

H96 **heavy water** || schweres Wasser *n*, Deuteriumoxid *n* || eau *f* lourde || zwaar water *n* || тяжёлая вода *f*, окись *f* дейтерия

H97 **hecto...** || Hekto... || hecto... || hecto... || гекто... (*приставка в наименовании кратной единицы, соответствующая* 10^2)

H98 **hectopascal** || Hektopascal *n* || hectopascal *m* || hectopascal *n* || гектопаскаль (10^2 *Па*) *m*

H99 **height** || Höhe *f* || hauteur *f* || hoogte *f* || высота *f*

H100 **height above sea level** || Höhe *f* über dem Meeresspiegel || hauteur *f* au-dessus du niveau de la mer || hoogte *f* boven zeeniveau || высота *f* над уровнем моря

H101 **helical fluted reamer** || Schälreibahle *f* || alésoir *m* à dents hélicoïdales || gegroefde opruimer *m* met schroeftanden || развёртка *f* с винтовыми зубьями

H102 **helical gear** || Schrägzahnrad *n* || roue *f* hélicoïdale || schroeftandwiel *n* || косозубое колесо *n*

H103 **helical motion** || Schraubenbewegung *f*, schraubenförmige Bewegung *f* || mouvement *m* hélicoïdal || spiraalvormige beweging *f* || винтовое движение *n*

H104 **helical rack** || Zahnstange *f* mit Schrägverzahnung || crémaillère *f* à denture oblique || schroeftandstang *f* (*m*) || рейка *f* с косыми зубьями

H105 **helical spring** || Spiralfeder *f* || ressort *m* hélicoïdal [spiral] || schroefveer *f* (*m*) || спиральная пружина *f*

H106 **helical spur gear** || Schrägzahnstirnrad *n* || roue *f* cylindrique hélicoïdale || schroefwiel *n*, schroeflijnvormig tandwiel *n* || косозубое цилиндрическое колесо *n*

H107 **helicity** || Helizität *f*, Schraubensinn *m* || hélicité *f* || heliciteit *f* || спиральность *f* (*характеристика элементарной частицы со спином*)

H108 **helicoidal surface** || Schraubenfläche *f* || surface *f* hélicoïdale || spiraalvormige oppervlakte *f* || винтовая поверхность *f*

H109 **helium**, He || Helium *n* || hélium *m* || helium *n* || гелий *m*

H110 **helix angle** || Steigungswinkel *m* || angle *m* d'hélice || schroeflijnhoek *m* || угол *m* (подъёма) винтовой линии

H111 **henry**, H || Henry *n* || henry *m* || henry *f* (*m*) || генри *m*, Г

H112 **heptagon** || Heptagon *n* || heptagone *m* || heptagon *n* || семиугольник *m*

H113 **heptahedron** || Heptaeder *n* || heptaèdre *m* || zevenvlak *n* || гептаэдр *m*, семигранник *m*

H114 **Hermitian kernel** || hermitescher Kern *m* || noyau *m* hermitien || Hermite-kern *f* (*m*) || эрмитово ядро *n*

H115 **Hermitian matrix** || hermitische Matrix *f* || matrice *f* hermitienne || Hermite-matrix *f* || эрмитова матрица *f*

H116 **herringbone gear** || Pfeilrad *n* || roue *f* à chevrons || wiel *n* met pijltanden *m* (V-tanden) || шевронное зубчатое колесо *n*; шевронная шестерня *f*

H117 **hertz,** Hz || Hertz *n* || hertz *m* || hertz *m* || герц *m*, Гц

H118 **heterocyclic compound** || heterozyklische Verbindung *f*, Heterozyklus *m* || composé *m* hétérocyclique || heterocyclische verbinding *f* || гетероциклическое соединение *n*, гетероцикл *m*

H119 **heterodyne** || Überlagerer *m*, Überlagerungsoszillator *m*, Mischoszillator *m* || hétérodyne *m*, oscillateur *m* local || heterodyne *m* || гетеродин *m*

H120 **heterodyning** || Überlagerung *f* || hétérodynage *f* || heterodynering *f* || гетеродинирование *n*

H121 **heterogeneity** || Heterogenität *f*, Inhomogenität *f* || hétérogénéité *f* || heterogeniteit *f* || неоднородность *f*, гетерогенность *f*

H122 **heterogeneous** || heterogen, inhomogen || hétérogène || heterogeen || неоднородный, гетерогенный

H123 **heterogeneous catalysis** || heterogene Katalyse *f* || catalyse *f* hétérogène || heterogene katalyse *f* || гетерогенный катализ *m*

H124 **heterogeneous medium** || heterogenes [inhomogenes] Medium *n* || milieu *m* hétérogène || heterogeen medium *n* || неоднородная среда *f*

H125 **heterogeneous reaction** || heterogene Reaktion *f* || réaction *f* hétérogène || heterogene reactie *f* || гетерогенная реакция *f*

H126 **heterogeneous system** || heterogenes [inhomogenes] System *n* || système *m* hétérogène || heterogeen systeem *n* || гетерогенная [неоднородная] система *f*

H127 **heterojunction** || Heteroübergang *m* || jonction *f* hétérogène || hetero-overgang *m* || гетеропереход *m*

H128 **heterolytic reaction** || Heterolyse *f* || réaction *f* hétérolytique || heterolytische reactie *f* || гетеролитическая реакция *f*

H129 **heteropolar** || heteropolar || hétéropolaire || heteropolair || гетерополярный

H130 **heuristic programming** || heuristische Programmierung *f* || programmation *f* heuristique || heuristische programmering *f* || эвристическое программирование *n*

H131 **heuristics** || Heuristik *f* || heuristique *f* || heuristiek *f* || эвристика *f*

H132 **hexadecimal** || hexadezimal || hexadécimal || hexadecimaal || шестнадцатеричный

H133 **hexagone** || Hexagon *n* || hexagone *m* || hexagon *m* || шестиугольник *m*

H134 **hexagon head** || Sechskantkopf *m* || tête *f* hexagonale || zeskante kop *m* || шестигранная головка *f*

H135 **hexahedron** || Hexaeder *n* || hexaèdre *m* || hexaeder *m* || гексаэдр *m*, шестигранник *m*

H136 **hierarchical** || hierarchisch || hiérarchique || hiërarchisch || иерархический

H137 **hierarchy** || Hierarchie *f* || hiérarchie *f* || hiërarchie *f* || иерархия *f*

H138 **high-aperture** || lichtstark || lumineux || lichtsterk || светосильный

H139 **high-current** || Starkstrom... || à courant fort || sterkstroom... || сильноточный

H140 **high-energy** || energiereich, Hochenergie... || à haute énergie || hoogenergie... || высокоэнергичный, с большой [высокой] энергией

H141 **high-energy accelerator** || Hochenergiebeschleuniger *m* || accélérateur *m* à haute énergie || hoogenergieversneller *m* || ускоритель *m* на высокие энергии

H142 **high-energy physics** || Hochenergiephysik *f*, Physik *f* hoher

Energien || physique *f* des hautes énergies || hoogenergiefysica *f* || физика *f* высоких энергий

H143 **higher algebra** || höhere Algebra *f* || algèbre *f* supérieure || hogere algebra *f* (*m*) || высшая алгебра *f*

H144 **higher harmonic** || höhere Harmonische *f* || harmonique *m* supérieur [élevé] || hogere harmonische *f* || высшая гармоника *f*

H145 **higher mathematics** || höhere Mathematik *f* || mathématiques *f pl* supérieures || hogere mathematica *f* || высшая математика *f*

H146 **high-frequency** || hochfrequent, Hochfrequenz... || à haute fréquence || hoogfrequent... || высокочастотный

H147 **high-frequency discharge** || Hochfrequenzentladung *f* || décharge *f* de haute fréquence || hoogfrequentontlading *f* || высокочастотный разряд *m*

H148 **high-frequency generator** || Hochfrequenzgenerator *m*, HF-Generator *m* || générateur *m* à haute fréquence || hoogfrequentgenerator *m* || генератор *m* высокой частоты

H149 **high-gain** || Hochleistungs... || à haut gain || hoogvermogens... || с большим усилением

H150 **high-level language** || höhere Sprache *f* || langage *m* de haut niveau || hogere programmeertaal *f* || язык *m* (программирования) высокого уровня

H151 **high-melting** || schwerschmelzbar, hochschmelzbar || difficilement fusible || hoogsmeltend || тугоплавкий

H152 **high-pitch screw** || Grobgewindeschraube *f* || vis *f* à pas rapide || schroef *f* (*m*) met grote gang || винт *m* с резьбой крупного шага

H153 **high-power** || 1. leistungsstark, leistungsfähig 2. mit starker Vergrößerung *(Optik)* || 1. puissant 2. à grand grossissement || 1. van grote capaciteit 2. van grote vergroting || 1. мощный 2. с большим увеличением *(об оптике)*

H154 **high-precision machine** || Präzisionswerkzeugmaschine *f* || machine *f* de précision || precisie(werktuig)machine *f* || прецизионный станок *m*

H155 **high-pressure** || Hochdruck... || à haute pression || hogedruk... || высокого давления

H156 **high-pressure turbine** || Hochdruckturbine *f* || turbine *f* à haute pression || hoogdrukturbine *f* || турбина *f* высокого давления

H157 **high-resistance** || hochohmig, Hochohm..., mit hohem (elektrischen) Widerstand || à haute résistance || hoogohmig || с высоким сопротивлением, высокоомный

H158 **high-speed lathe** || Schnelldrehmaschine *f* || tour *f* à grande vitesse || sneldraaibank *f* (*m*) || быстроходный токарный станок *m*

H159 **high-speed steel** || Schnell(arbeits)stahl *m* || acier *m* rapide || sneldraaistaal *n* || быстрорежущая [инструментальная] сталь *f*

H160 **high-speed steel tool** || Schnellstahlwerkzeug *n* || outil *m* à coupe rapide, outil *m* en acier rapide || snelstaalwerktuig *n* || инструмент *m*, изготовленный из быстрорежущей стали

H161 **high-temperature** || von hoher Temperatur; Hochtemperatur... || à haute température || hoge temperatuur... || высокотемпературный

H163 **high-tension** || Hochspannungs..., Hochvolt... || à haute tension || hoogspannings... || высоковольтный

H164 **high-vacuum** || hochevakuiert, Hochvakuum... || à vide poussé [élevé] || hoogvacuum... || высоковакуумный

H165 **high-voltage** *see* **high-tension**

H166 **Hilbert space** || Hilbert-Raum *m* || espace *m* hilbertien || Hilbert-ruimte *f* || гильбертово пространство *n*

H167 **hinge** || Gelenk *n*; Drehachse *f*; Angel *f* || articulation *f*; charnière *f* || gewricht *n*, scharnier *n*; hengsel *n* || шарнир *m*; петля *f*

H168 **hinge(d) bolt** || Gelenkschraube *f*, Klappschraube *f* || boulon *m* articulé || scharnierbout *m*; scharnierpen *f* (*m*) || откидной болт *m*

H169 **hinge joint** || gelenkiger Knoten *m*, Gelenkknoten *m*; Gelenkverbindung *f* || articulation *f*, joint *m* à charnière || scharnierverbinding *f* || шарнирное соединение *n*

H170 **hob** || **1.** Wälzfräser *m* **2.** Prägestempel *m* || **1.** fraise-mère *f* **2.** poinçon *m* à froid || **1.** wormfrees *f* (*m*) **2.** persstempel *m* || **1.** червячная фреза *f* **2.** пуансон *m* для (холодного) выдавливания

H171 **hobbing** || Wälzfräsen *n* || taillage *m* par fraise-mère || frezen *n* met een wormfrees || фрезерование *n* червячной фрезой, обкатное фрезерование *n*

H172 **hob cutter** || Wälzfräser *m* || fraise-mère *f* || wormfrees *f* (*m*) || червячная фреза *f*

H173 **hodograph** || Hodograf *m*, Hodografenkurve *f* || (h)odographe *m*, lieu *m* des vecteurs || hodograaf *m* || годограф *m*

H174 **hoist** || Hebezeug *n*; Aufzug *m* || appareil *m* de levage; ascenseur *m*; élévateur *m*; treuil *m* || hefwerktuig *n*, hijswerktuig *n*; hijstoestel *n*; hijsinrichting *f* || подъёмный механизм *m*, подъёмное устройство *n*; подъёмник *m*; лебёдка *f*

H175 **hoist(ing) drum** || Seiltrommel *f* || tambour *m* de levage || hijstrommel *f* (*m*) || барабан *m* лебёдки; барабан *m* подъёмного устройства

H176 **hole** || **1.** Öffnung *f*, Loch *n*; Kanalöffnung *f*, Kanal *m* **2.** Loch *n* *(Halbleiter)* || **1.** ouverture *f* **2.** trou *m*, lacune *f* || **1.** opening *f*, kanaal *n* **2.** gat *n* || **1.** отверстие *n*; канал *m* **2.** дырка *f* *(в полупроводниках)*

H177 **hole basis limit system** || Einheitsbohrungssystem *n* || système *m* alésage || eenheidsboringssysteem *n* || система *f* отверстия *(в допусках)*

H178 **hole conduction** || Lochleitung *f*, Löcherleitung *f*, Defekt(elektronen)leitung *f*, p-Leitung *f* || conductibilité *f* par lacunes [par trous] || gatengeleiding *f* || дырочная проводимость *f*

H179 **hollow shaft** || Hohlwelle *f* || arbre *m* creux || holle as *f* (*m*) || полый вал *m*

H180 **hologram** || Hologramm *n* || hologramme *m* || hologram *n* || голограмма *f*

H181 **holography** || Holografie *f* || holographie *f* || holografie *f* || голография *f*

H182 **holomorphic** || holomorph || holomorphe || holomorf || голоморфный

H183 **holomorphic function** || holomorphe Funktion *f* || fonction *f* holomorphe || holomorfe functie *f* || голоморфная функция *f*

H184 **holomorphy** || Holomorphie *f* || holomorphie *f* || holomorfie *f* || голоморфность *f*

H185 **holonomic** || holonom || holonomique || holonoom || голономный

H186 **holonomic constraint** || holonome Bedingung *f*, holonome Bindung *f* || liaison *f* holonomique || holonome binding *f* || голономная связь *f*

H187 **holonomic system** || holonomes System *n* || système *m* holonomique || holonoom systeem *n* || голономная система *f*

H188 **holonomy** || Holonomie *f* || holonomie *f* || holonomie *f* || голономность *f*

H189 **homeomorphism** || Homöomorphie *f* || homéomorphisme *m* || homeomorfisme *n* || гомеоморфизм *m*

H190 **homocentric beam** || homozentrisches Bündel *m* || faisceau *m* homocentrique || homocentrische bundel *m* || гомоцентрический пучок *m*

H191 **homogeneity** || Homogenität *f*, Gleichartigkeit *f* || homogénéité *f* || homogeniteit *f* || однородность *f*, гомогенность *f*

H192 **homogeneous** || homogen || homogène || homogeen || гомогенный, однородный

H193 **homogeneous catalysis** || homogene Katalyse *f* || catalyse *f* homogène || homogene katalyse *f* || гомогенный катализ *m*

H194 **homogeneous field** || homogenes [gleichförmiges] Feld *n* || champ *m* homogène || homogeen veld *n* || однородное поле *n*

H195 **homogeneous medium** || homogenes Medium *n* || milieu *m* homogène || homogeen medium *n* || однородная среда *f*

H196 **homogeneous mixture** || homogenes Gemisch *n* || mélange *m* homogène || homogeen mengsel *n* || гомогенная [однородная] смесь *f*

H197 **homogeneous reaction** || homogene Reaktion *f* || réaction *f* homogène || homogene reactie *f* || гомогенная реакция *f*

H198 **homogeneous system** || homogenes System *n* || système *m* homogène || homogeen systeem *n* || гомогенная [однородная] система *f*

H199 **homologous** || homolog || homologique || homoloog || гомологичный; гомологический

H200 **homology** || Homologie *f* || homologie *f* || homologie *f* || гомология *f*

H201 **homolytic reaction** || Homolyse *f* || réaction *f* homolytique || homolytische reactie *f* || гомолитическая реакция *f*

H202 **homomorphism** || Homomorphie *f* || homomorphisme *m* || homomorfisme *n* || гомоморфизм *m*

H203 **homopolar** || homopolar; homöopolar || homopolaire; homéopolaire || homopolair || гомополярный; гомеополярный

H204 **homopolar bond** || homöopolare [kovalente] Bindung *f*, Kovalenz *f* || liaison *f* homopolaire || homopolaire binding *f* || гомеополярная [ковалентная] связь *f*

H205 **homothetic transformation** || Homothetie *f* || transformation *f* homothétique || homothetie *f* || гомотетия *f*

H206 **honing** || Honen *n*, Ziehschleifen *n* || honing *m* || honen *n* || хонингование *n*

H207 **honing machine** || Honmaschine *f*, Ziehschleifmaschine *f* || machine *f* à roder || hoonmachine *f* || хонинговальный станок *m*

H208 **honing tool** || Honahle *f*, Ziehschleifahle *f* || rodoir *m* || hoongereedschap *n* || хон *m*, инструмент *m* для хонингования

H209 **hook** || Haken *m* || crochet *m* || haak *m* || крюк *m*; крючок *m*

H210 **Hooke's joint** || Kardangelenk *n* || joint *m* universel, joint *m* cardan, articulation *f* à cardan || universele koppeling *f*, cardankoppeling *f* || шарнир *m* Гука, карданный [универсальный] шарнир *m*

H211 **Hooke's law** || Hookesches Gesetz *n* || loi *f* de Hooke || wet *f* (*m*) van Hooke || закон *m* Гука

H212 **horizontal boring machine** || Waagerechtbohrmaschine *f* || aléseuse *f* horizontale || horizontale boormachine *f* || горизонтально-расточный станок *m*

H213 **horizontal milling machine** || Waagerechtfräsmaschine *f* || fraiseuse *f* horizontale || horizontale freesmachine *f* || горизонтально-фрезерный станок *m*

H214 **horsepower** || Pferdestärke *f* || cheval-vapeur *m* || paardekracht *f* (*m*) || **1.** лошадиная сила *f* *(745,7 Вт)* **2.** мощность *f* в лошадиных силах

H215 **horseshoe magnet** || Hufeisenmagnet *m* || aimant *m* en fer à cheval || hoefijzermagneet *m* || подковообразный [U-образный] магнит *m*

H216 **hose** || Schlauch *m* || tuyau *m* (flexible) || slang *f* (*m*) || шланг *m*, гибкий рукав *m*

H217 **host crystal** || Wirtskristall *m*, Grundkristall *m* || cristal *m* principal || grondkristal *n* || основной кристалл *m*, кристалл-хозяин *m*

H218 **hot** || **1.** heiß **2.** hochaktiv, hochgradig radioaktiv || **1.** chaud **2.** hautement actif || **1.** heet **2.** sterk radioactief || **1.** горячий **2.** высоко(радио)активный

H219 **hot atom** || heißes [hochangeregtes] Atom *n* || atome *m* de recul || heet atoom *n* || атом *m* отдачи с большой энергией

H220 **hot forging** || Warmschmieden *n* || forgeage *m* à chaud || warm smeden *n* || горячая ковка *f*; горячая штамповка *f*

H221 **hot-rolling mill** || Warmwalzwerk *n* || laminoir *m* à chaud || warmwalserij *f* || стан *m* горячей прокатки

H222 **hot-wire instrument** || Hitzdrahtmeßgerät *n* || appareil *m* de mesure thermique || verhittingsdraadinstrument *m* || тепловой измерительный прибор *m*

H223 **hour** || Stunde *f* || heure *f* || uur *n* || час *m*

H224 **hour-glass worm** || Globoidschnecke *f* || vis *f* globique || globoïdaalworm *m* || глобоидальный червяк *m*

H225 **housing** || **1.** Gehäuse *n* **2.** Ständer *m*, Walzenständer *m* || **1.** carter *m*; boîte *f* **2.** montant *m* (de cage) || **1.** huis *n*; carter *n*, omhulsel *m*, *n* **2.** raam *n*, raamwerk *n* *(van walsenstoel)* || **1.** корпус *m*; кожух *m*; картер *m* 2. станина *f* *(клети прокатного стана)*

H226 **H-plane** || H-Ebene *f* || H-plan *m* || H-vlak *n* || плоскость *f* колебаний магнитного вектора

H227 **hub** || Nabe *f* || moyeu *m* || naaf *f* *(m)* || ступица *f*; втулка *f*

H228 **humidity** || Feuchtigkeit *f*, Feuchte *f* || humidité *f* || vochtigheid *f* || влажность *f*

H229 **Huygens-Fresnel principle** || Fresnelsches Prinzip *n* || principe *m* de Huygens-Fresnel || Huygens-Fresnelprincipe *n* || принцип *m* Гюйгенса — Френеля

H230 **H-wave** || H-Welle *f* || H-onde *f* || H-golf *f* *(m)* || волна *f* магнитного вектора (напряжённости)

H231 **hybrid computer** || Hybridrechner *m* || ordinateur *m* hybride || hybridecomputer *m* || гибридная вычислительная машина *f*

H232 **hybridization** || Hybridisierung *f* || hybridation *f* || hybridisatie *f* || гибридизация *f*

H233 **hydraulic analogy** || hydraulisches Analogon *n*, hydraulische Analogie *f* || analogie *f* hydraulique || hudraulische analogie *f* || гидравлическая аналогия *f*

H234 **hydraulic cylinder** || Hydraulikzylinder *m* || cylindre *m* hydraulique || hudraulische cilinder *f* *(m)* || гидроцилиндр *m*, гидравлический цилиндр *m*

H235 **hydraulic damper** || Flüssigkeitsdämpfer *m* || amortisseur *m* hydraulique [à huile] || hydraulische demper *m* || гидравлический амортизатор *m*

H236 **hydraulic drive** || hydraulisches Getriebe *n*, Flüssigkeitsgetriebe *n* || transmission *f* [entraînement *m*] hydraulique || hydraulische aandrijving *f* || гидравлический привод *m*, гидравлическая передача *f*

H237 **hydraulic engine** || Hydraulikmotor *m*, Flüssigkeitsmotor *m* || moteur *m* hydraulique || hydraulische motor *m* || гидравлический двигатель *m*

H238 **hydraulic fluid** || Druckflüssigkeit *f* || liquide *m* hydraulique || hydraulisch medium *n*; hydraulische vloeistof *f* *(m)* || гидравлическая жидкость *f*

H239 **hydraulic hoist** || hydraulisches Hebezeug *n*; hydraulischer Aufzug *m* || ascenseur *m* hydraulique; élévateur *m* hydraulique || hydraulische lift *m* || гидравлический подъёмник *m*

H240 **hydraulic impact** || Wasserschlag *m*, hydraulischer Schlag *m* || coup *m* de bélier || waterslag *m* || гидравлический удар *m*

H241 **hydraulic jump** || Wassersprung *m*, hydraulischer Sprung *m* || ressaut *m* hydraulique || watersprong *m* || гидравлический прыжок *m*

H242 **hydraulic lift** *see* **hydraulic hoist**

H243 **hydraulic loss** || hydraulische Verluste *m pl*; Druckhöhenverluste *m pl* || pertes *f pl* de charge || drukverliezen *n pl* || гидравлические потери *f pl*; потеря *f* напора

H244 **hydraulic press** || hydraulische Presse *f* || presse *f* hydraulique || hydraulische pers *f* *(m)* || гидравлический пресс *m*

H245 **hydraulic ram** || Druckwasserpresse *f*, hydraulischer Widder *m* || bélier *m* hydraulique || drukwaterpers *f* *(m)* || гидравлический таран *m*

H246 **hydraulics** || Hydraulik *f* || hydraulique *f* || hydraulica *f* || гидравлика *f*

H247 **hydraulic shock** || Wasserschlag *m*, hydraulischer Schlag *m* || coup *m* de bélier || waterslag *m* || гидравлический удар *m*

H248 **hydraulic turbine** || Wasserturbine *f* || turbine *f* hydraulique || hydraulische turbine *f* || гидравлическая турбина *f*, гидротурбина *f*

H249 **hydride** || Hydrid *n* || hydrure *m* || hydride *f* || гидрид *m*

H250 **hydrocarbon** || Kohlenwasserstoff *m* || hydrocarbure *m* || koolwaterstof *f* *(m)* || углеводород *m*

H251 **hydrochloric acid,** HCl || Chlorwasserstoffsäure *f*, Salzsäure *f* || acide *m* chlorhydrique || zoutzuur *n* || соляная кислота *f*

H252 **hydrochloride** || Hydrochlorid *n*, Chlorhydrat *n* || chlorhydrate *m*, hydrochlorure *m* || chloorhydraat *n*, hydrochloride *n* || хлоргидрат *m*, гидрохлорид *m*

H253 **hydrocyanic acid,** HCN || Zyanwasserstoffsäure *f*, Blausäure *f* || acide *m* cyanhydrique, nitrite *f* formique || cyaanwaterstof *f* *(m)*, blauwzuur *n* || цианистоводородная [синильная] кислота *f*

H254 **hydrodynamic drag** || hydrodynamischer Widerstand *m*, Strömungswiderstand *m* || résistance *f* hydrodynamique || hydrodynamische weerstand *m* || гидродинамическое сопротивление *n*

H255 **hydrodynamics** || Hydrodynamik *f* || hydrodynamique *f* || hydrodynamica *f* || гидродинамика *f*

H256 **hydrodynamic transmission** || Strömungsgetriebe *n* || transmission *f* hydrodynamique ||hydraulische overbrenging *f* || гидродинамическая передача *f*

H257 **hydro-electric plant** || Wasserkraftwerk *n* || centrale *f* hydro-électrique || waterkrachtcentrale *f* || гидроэлектростанция *f*

H258 **hydrofluoric acid,** HF || Fluorwasserstoffsäure *f* || acide *m* fluorhydrique || fluorwaterstofzuur *n* || фтористоводородная кислота *f*

H259 **hydrogen,** H || Wasserstoff *m* || hydrogène *m* || waterstof *f* *(m)* || водород *m*

H260 **hydrogenation** Hydrieren *n*, Hydrierung *f* || hydrogénation *f* || hydrogenering *f* || гидрогенизация *f*, гидрирование *n*

H261 **hydrogen bond** || Wasserstoff(brücken)bindung *f*, Wasserstoffbrücke *f* || liaison *f* d'hydrogène || waterstofbinding *f* || водородная связь *f*

H262 **hydrogen chloride,** HCl || Chlorwasserstoff *m* || hydrogène *m* chloré || chloorwaterstof *f* *(m)* || хлористый водород *m*

H263 **hydrogen electrode** || Wasserstoffelektrode *f* || électrode *f* à hydrogène || waterstofelektrode *f* || водородный электрод *m*

H264 **hydrogen fluoride,** HF || Fluorwasserstoff *m* || hydrogène *m* fluoré || fluorwaterstoff *f* *(m)* || фтористый водород *m*

H265 **hydrogen-like atom** || wasserstoffähnliches Atom *n* || atome *m* hydrogénoïde || waterstofachtig atoom *n* || водородоподобный атом *m*

H266 **hydrogen sulfide,** H_2S || Schwefelwasserstoff *m* || sulfure *m* d'hydrogène || zwavelwaterstof *f* *(m)* || сероводород *m*

H267 **hydrolysis** || Hydrolyse *f* || hydrolyse *f* || hydrolyse *f* || гидролиз *m*

H268 **hydromagnetic** || magnetohydrodynamisch, hydromagnetisch || magnéto-hydrodynamique || magnetohydrodynamisch || магнитогидродинамический

H269 **hydromagnetic generator** || magnetohydrodynamischer Generator *m*, MHD-Generator *m* || générateur *m* magnéto-hydrodynamique || magnetohydrodynamische generator *m* || магнитогидродинамический генератор *m*, МГД-генератор *m*

H270 **hydromagnetics** || Magnetohydrodynamik *f*, Hydromagnetik *f* || magnéto-hydrodynamique *f* || magnetohydrodynamica *f* || магнитная гидродинамика *f*, магнитогидродинамика *f*

H271 **hydromechanics** || Hydromechanik *f* || hydromécanique *f* || hydromechanica *f* || гидромеханика *f*

H272 **hydrometer** || Hydrometer *n* || hydromètre *m* || hydrometer *m* || **1.** ареометр *m*, плотномер *m* **2.** гидрометр *m*

H273 **hydroperoxide** || Hydroperoxid *n* || hydroperoxyde *m* || hydroperoxyde *n* || гидроперекись *f*, гидропероксид *m*

H274 **hydrostatic equilibrium** || hydrostatisches Gleichgewicht *n* || équilibre *m* hydrostatique || hydrostatisch evenwicht *n* || гидростатическое равновесие *n*

H275 **hydrostatic head, hydrostatic pressure** || hydrostatischer Druck *m*, hydrostatische Förderhöhe *f* || pression *f* [hauteur *f*] hydrostatique || hydrostatische druk *m*, hydrostatische stijghoogte *f* || гидростатическое давление *n*, напор *m*

H276 **hydrostatics** || Hydrostatik *f* || hydrostatique *f* || hydrostatica *f* || гидростатика *f*

H277 **hygrograph** || Hygrograf *m*, Feuchtigkeitsschreiber *m* || hygrographe *m* || hygrograaf *m* || гигрограф *m*, самопишущий гигрометр *m*

H278 **hygrometer** || Hygrometer *n*, Feuchtigkeitsmesser *m* || hygromètre *m* || hygrometer *m* || гигрометр *m*; психрометр *m*

H279 **hygroscopic** || hygroskopisch, wasseranziehend || hygroscopique || hygroscopisch || гигроскопический

H280 **hyperbola** || Hyperbel *f* || hyperbole *f* || hyperbool *f* (*m*) || гипербола *f*

H281 **hyperbolic functions** || Hyperbelfunktionen *f pl*, hyperbolische Funktionen *f pl* || fonctions *f pl* hyperboliques || hyperbolische functies *f pl* || гиперболические функции *f pl*

H282 **hyperbolic logarithm** || natürlicher Logarithmus *m* || logarithme *m* naturel || natuurlijke logaritme *f* (*m*) || натуральный логарифм *m*

H283 **hyperbolic point** || hyperbolischer Punkt *m* || point *m* hyperbolique || hyperbolische punt *f* (*m*) || гиперболическая точка *f*

H284 **hyperboloid** || Hyperboloid *n* || hyperboloïde *m* || hyperboloïde *f* (*m*) || гиперболоид *m*

H285 **hyperboloid of revolution** || Rotationshyperboloid *n* || hyperboloïde *m* de révolution || rotatiehyperboloïde *f* (*m*) || гиперболоид *m* вращения

H286 **hyperboloid of two sheets** || zweischaliges Hyperboloid *n* || hyperboloïde *m* à deux nappes || tweedelige hyperboloïde *f* (*m*) || двуполостный гиперболоид *m*

H287 **hypercharge** || Hyperladung *f* || hypercharge *f* || hyperlading *f* гиперзаряд *m*

H288 **hypercomplex** || hyperkomplex || hypercomplexe || hypercomplex || гиперкомплексный

H289 **hypercomplex numbers** || hyperkomplexe Zahlen *f pl* || nombres *m pl* hypercomplexes || hypercomplexe getallen *f* (*m*) || гиперкомплексные числа *n pl*

H290 **hyperconjugation** || Hyperkonjugation *f* || hyperconjugaison *f* hyperconjugatie *f* || сверхсопряжение *n*

H291 **hyperfine structure** || Hyperfeinstruktur *f* || structure *f* hyperfine hyperfijne structuur *f* || сверхтонкая структура *f* (*спектра*)

H292 **hypergeometric** || hypergeometrisch || hypergéométrique || hypergeometrisch || гипергеометрический

H293 **hypergeometric function** hypergeometrische Funktion *f* || fonction *f* hypergéométrique || hypergeometrische functie *f* || гипергеометрическая функция *f*

H294 **hypergeometric polynomials** hypergeometrische [Jacobische] Polynome *n pl* || polynômes *m pl* hypergéométriques || hypergeometrische polynomen *n pl* || многочлены *m pl* Якоби

H295 **hypergeometric series** || hypergeometrische Reihe *f* || séries *f pl* hypergéométriques || hypergeometrische reeks *f* (*m*) || гипергеометрический ряд *m*

H296 **hypernucleus** || Hyperfragment *n* || hypernoyau *m* || hyperkern *f* (*m*) || гипер-ядро *n*, гиперфрагмент *m*

H297 **hyperon** || Hyperon *n* || hypéron *m* || hyperon *n* || гиперон *m*

H298 **hyperplane** || Hyperebene *f* || hyperplan *m* || hypervlak *n* || гиперплоскость *f*

H299 **hypersensitization** || Übersensibilisierung *f*, Hypersensibilisierung *f* || hypersensibilisation *f* || hypersensibilisatie *f* || гиперсенсибилизация *f*

H300 **hypersonic** || hypersonisch, Hyperschall... || hypersonique || hypersonisch || гиперзвуковой

H301 **hypersonic aerodynamics** || Hyperschallaerodynamik *f* || aérodynamique *f* hypersonique || hypersonische aërodynamica *f* || аэродинамика *f* гиперзвуковых скоростей

H302 **hypersonic frequency** || Hyperschallfrequenz *f* || fréquence *f* hypersonique || hypersonische frequentie *f* || гиперзвуковая частота *f*

H303 **hypersound** || Hyperschall *m* || hyperson *m* || hypergeluid *n* || гиперзвук *m*

H304 **hypersphere** || Hypersphäre *f* || hypersphère *f* || hypersfeer *f* (*m*) || гиперсфера *f*

H305 **hyperstatic** || statisch unbestimmt || huperstatique || statisch onbepaalbaar || статически неопределимый

H306 **hypervelocity** || **1.** hypersonische Geschwindigkeit *f*, Hyperschallgeschwindigkeit *f* **2.** Überlichtgeschwindigkeit *f* || **1.** vitesse *f* hypersonique **2.** vitesse *f* surpassant la vitesse de la lumière || **1.** supersonische snelheid *f* **2.** overlichtsnelheid *f* || **1.** гиперзвуковая скорость *f* **2.** сверхсветовая скорость *f*

H307 **hypobromous acid,** HBrO || hypobromige Säure *f* || acide *m* hypobromeux || hypobromig zuur *n* || бромноватистая кислота *f*

H308 **hypochlorous acid,** HClO || hypochlorige Säure *f* || acide *m* hypochloreux || hypochlorig zuur *n* || хлорноватистая кислота *f*

H309 **hypocycloid** || Hypozykloide *f* || hypocycloïde *f* || hypocycloïde *f* || гипоциклоида *f*

H310 **hypoid (bevel) gear** || Hypoid(kegel)rad *n* || roue *f* dentée hypoïde || hypoïdrad *n* || гипоидное колесо *n*

H311 **hypoid gearing** || Hypoidgetriebe *n* || engrenage *m* hypoïde || hypoïdoverbrenging *f* || гипоидная передача *f*

H312 **hypophosphorous acid,** H_3PO_2 || hypophosphorige Säure *f* || acide *m* hypophosphoreux || hypofosforig zuur *n* || фосфорноватистая кислота *f*

H313 **hypothenuse** || Hypothenuse *f* || hypoténuse *f* || hypotenusa *f* (*m*) || гипотенуза *f*

H314 **hypothesis** || Hypothese *f* || hypothèse *f* || hypothese *f* || гипотеза *f*

H315 **hypsothermometer** || Hypsometer *n*, Siedebarometer *n*, Hypsothermometer *n* || thermobaromètre *m*, hypso(thermo)mètre *m* || barothermometer *m*, hypsothermometer *m* || термобарометр *m*

H316 **hysteresis** || Hysterese *f*, Hysteresis *f* || hystérésis *m* || hysteresis *f* || гистерезис *m*

H317 **hysteresis curve, hysteresis loop** || Hystereseschleife *f*, Hysteresisschleife *f* || boucle *f* d'hystérésis || hysteresislus *f* (*m*) || петля *f* гистерезиса

H318 **hysteresis loss** || Hysteresisverluste *m pl*, Hystereseverluste *m pl* || pertes *f pl* d'hystérésis || hysteresisverliezen *n pl* || потери *f pl* на гистерезис, гистерезисные потери *f pl*

H319 **hysteretic** || hystereseartig || de hystérésis || hysteresis... || гистерезисный

I

I1 **I beam** || Doppel-T-Träger *m*, I-Träger *m* || profil *m* [fer *m*] en double T || dubbel-T-balk *m* || двутавровый профиль *m*; двутавровая балка *f*

I2 **icosahedron** || Ikosaeder *n* || icosaèdre *m* || icosaeder *n* || икосаэдр *m*

I3 **ideal** || Ideal *n* || idéal *m* || ideaal *n* || идеал *m*

I4 **ideal cycle engine** || Motor *m* mit idealem Kreisprozeß || moteur *m* à cycle parfait || motor *m* met ideale cyclus || двигатель *m* с идеальным циклом

I5 **ideal efficiency** || idealer Wirkungsgrad *m* || rendement *m* parfait || ideaal rendement *n* || идеальный кпд

I6 **ideal gas** || ideales Gas *n* || gaz *m* parfait || ideaal gas *n* || идеальный газ *m*

I7 **ideal liquid** || ideale vollkommene Flüssigkeit *f* || fluide *m* parfait || ideale vloeistof *f* (*m*) || идеальная [невязкая и несжимаемая] жидкость *f*

I8 **ideal solution** || ideale Lösung *f* || solution *f* idéale || ideale oplossing *f* || идеальный раствор *m*

I9 **idempotent element** || Idempotent *n* || élément *m* idempotent || idempotent *n* || идемпотентный элемент *m*, идемпотент *m*

I10 **idempotent property** || Idempotenz *f* || idempotence *f* || idempotentie *f* || свойство *n* идемпотентности, идемпотентность *f*

I11 **identification** || Identifikation *f* || identification *f* || identificatie *f* || идентификация *f*

I12 **identifier** || Identifikator *m* || identificateur *m* || identificator *m* || идентификатор *m*

I13 **identity** || **1.** Identität *f* **2.** Einheit *f (in der Gruppentheorie)* || identité *f* || identiteit *f* || **1.** идентичность *f*; тождественность *f* **2.** тождество *n* **3.** мат. единица *f (в теории групп)*

I14 **identity matrix** || Einheitsmatrix *f* || matrice *f* unité || eenheidsmatrix *f* || единичная матрица *f*

I15 **identity relation** || Identitätsrelation *f* || relation *f* d'identité || identiteitsrelatie *f* || отношение *n* тождества

I16 **identity transformation** || identische Transformation *f* || transformation *f* identique || identieke transformatie *f* || тождественное преобразование *n*

I17 **idle pulley** || **1.** Leerscheibe *f*, Losscheibe *f* **2.** Leitrolle *f* || **1.** poulie *f* folle **2.** galet *m* guide [de guidage] || vrijlopende schijf *f* (*m*) || **1.** холостой шкив *m* **2.** направляющий ролик *m*; натяжной ролик *m*; натяжной шкив *m*

I18 **idler** *see* **idle pulley, idle wheel**

I19 **idle speed** || Leerlaufgeschwindigkeit *f* || vitesse *f* de rotation folle || vrijloopsnelheid *f* || частота *f* вращения холостого хода *(двигателя)*

I20 **idle wheel** || Zwischenrad *n* || roue *f* intermédiaire || tussenwiel *n* || промежуточное [паразитное] зубчатое колесо *n*

I21 **idling** || Leerlauf *m* || marche *f* [fonctionnement *m*] à vide || vrijloop *m (van motor)* || холостой ход *m*

I22 **ignition** || Zündung *f*; Entzündung *f* || allumage *m*; inflammation *f* || **1.** ontsteking *f* **2.** ontbranding *f* || **1.** зажигание *n* **2.** воспламенение *n*

I23 **ignition lag** || Zündverzug *m* || retard *m* à l'allumage || ontstekingsvertraging *f* || запаздывание *n* зажигания

I24 **ignition rectifier** || Ignitron *n*, Ignitronröhre *f*, Zündstiftröhre *f* || ignitron *m* || ignitron *n* || игнитронный выпрямитель *m*

I25 **ignition temperature** || Entzündungstemperatur *f*, Zündtemperatur *f* || température *f* d'allumage || ontbrandingstemperatuur *f* || температура *f* воспламенения

I26 **ignition timing** || Zünd(punkt)-verstellung *f* || temporisation *f* d'avance à l'allumage || ontstekingsverstelling *f* || установка *f* опережения зажигания

I27 **illuminance** || Beleuchtungsstärke *f*, Beleuchtungsgrad *m* || éclairement *m* || verlichtingssterkte *f* || освещённость *f*

I28 **illuminating lens** || Kondensorlinse *f*, Kondensor *m* || lentille *f* de condensateur, condenseur *m* || condensorlens *f* (*m*), condensor *m* || конденсорная линза *f*, конденсор *m*

I29 **illumination** || **1.** Beleuchtung *f* **2.** Bestrahlung *f (Licht)* **3.** Beleuchtungsstärke *f* || **1.** éclairage *m* **2.** irradiation *f* **3.** éclairement *m* || **1.** verlichting *f* **2.** bestraling *f* (*licht*) **3.** verlichtingssterkte *f* || **1.** освещение *n* **2.** облучение *n (светом)* **3.** освещённость *f*

I30 **illuminator** || **1.** Beleuchtungsgerät *n*, Beleuchtungsapparat *m*, Beleuchtungssystem *n* **2.** Kondensor *m* || **1.** illuminateur *m* **2.** condenseur *m* || **1.** verlichtingstoestel *n* **2.** condensor *m* || **1.** осветитель *m* **2.** конденсор *m*

I31 **image** || Abbildung *f*, Bild *n* || image *f* || beeld *n* || изображение *n*

I32 **image distance** || Bildweite *f* || distance *f* de l'image || beeldafstand *m* || расстояние *n* (до) изображения

I33 **image forces** || Bildkräfte *f pl*, Spiegelbildkräfte *f pl* || forces *f pl* d'image || beeldkrachten *f* (*m*) *pl* || силы *f pl* изображения, поляризационные силы *f pl*

I34 **image multiplier** || Bildverstärker *m* || intensificateur *m* [renforçateur *m*] d'image || beeldversterker *m* || усилитель *m* сигналов изображения, видеоусилитель *m*

I35 **image point** || Bildpunkt *m* || point *m* d'image || beeldpunt *n* || точка *f* изображения

I36 **image space** || Bildraum *m* || espace *m* des images || beeldruimte *f* || пространство *n* изображений

I37 **imaginary** || imaginär, Imaginär... || imaginaire || imaginair || мнимый

I38 **imaginary number** || imaginäre Zahl *f* || nombre *m* imaginaire || imaginair getal *n* || мнимое число *n*

I39 **imaginary part** || Imaginärteil *m* || partie *f* imaginaire || imaginair deel *n* || мнимая часть *f*

I40 **imaginary power** || **1.** Scheinleistung *f*, Gesamtleistung *f* **2.** Blindleistung *f* || **1.** puissance *f* apparente **2.** puissance *f* réactive || **1.** schijnbaar vermogen *n* **2.** blindvermogen *n* || **1.** кажущаяся [полная] мощность *f* **2.** реактивная мощность *f*

I41 **immersion** || **1.** Immersion *f* **2.** Eintauchen *n*, Tauchen *n* || **1.** immersion *f* **2.** plongement *m* || **1.** immersie *f* **2.** indompelen *n* || **1.** иммерсия *f* **2.** погружение *n*

I42 **immersion microscope** || Immersionsmikroskop *n* || microscope *m* à immersion || immersiemicroscoop *m* || иммерсионный микроскоп *m*

I43 **immersion objective** || Immersionsobjektiv *n* || objectif *m* à immersion || immersieobjectief *n* || иммерсионный объектив *m*

I44 **immersion thermometer** || Tauchthermometer *n* || thermomètre *m* à immersion || immersiethermometer *m* || погружной термометр *m*

I45 **immittance** || Immittanz *f* || immitance *f* || immittantie *f* || иммитанс *m*

I46 **impact** || Stoß *m*; Zusammenstoß *m*, Kollision *f* || choc *m*, coup *m*, impact *m* || slag *m*, stoot *m* || удар *m*

I47 **impact elasticity** || Stoßelastizität *f* || résilience *f*, résistance *f* au choc || stootviscositeit *f* || ударная вязкость *f*

I48 **impact energy** *see* **impact strength**

I49 **impact impulse** || Stoßimpuls *m* || impulsion *f* d'impact || stootimpuls *m* || ударный импульс *m*

I50 **impact load** || Stoßbelastung *f*, Stoßbeanspruchung *f* || charge *f* de choc || stootbelasting *f* || ударная нагрузка *f*

I51 **impact parameter** || Stoßparameter *m* || paramètre *m* d'impact || stootparameter *m* || прицельный параметр *m*, параметр *m* удара

I52 **impact strength** || Kerbschlagzähigkeit *f* || résilience *f* || slagsterkte *f* || ударная вязкость *f*

I53 **impact stress** || Spannung *f* bei stoßartiger Belastung || contrainte *f* par [due au] choc || stootspanning *f* || напряжение *n* при ударе

I54 **impact test** || Schlagversuch *m*; Kerbschlagversuch *m* || essai *m* de résistance au choc; essai *m* de résilience || slagproef *f* (*m*); kerfslagproef *f* (*m*) || испытание *n* на ударную прочность; определение *n* ударной вязкости

I55 **impedance** || Impedanz *f*, Scheinwiderstand *m* || impédance *f* || impedantie *f* || импеданс *m*, полное сопротивление *n*

I56 **impedance matching** || Impedanzanpassung *f* || adaptation *f* d'impédances || impedantieaanpassing *f* || согласование *n* импедансов

I57 **impeller** || Flügelrad *n*; Laufrad *n* || roue *f* à aubes; rotor *m* || vleugelrad *n*; rotor *m* || крыльчатка *f*; ротор *m*

I58 **imperfect crystal** || Realkristall *m*, realer [nichtidealer, fehlerhafter] Kristall *m* || cristal *m* imparfait || reëel kristal *n* || неидеальный [несовершенный] кристалл *m*

I59 **imperfect gas** || reales Gas *n*, Realgas *n* || gaz *m* imparfait || reëel gas *n* || реальный газ *m*

160 **imperfection** || Fehlstelle *f*, Defekt *m*, Fehlordnung *f* || défaut *m*, imperfection *f* || defect *n*, imperfectie *f* || дефект *m*, несовершенство *n* *(особ. кристалла)*

161 **impingement** || Aufschlag *m*, Aufprall *m* || chute *f* || val *m* || падение *n* *(на поверхность)*

162 **implication** || Implikation *f* || implication *f* || implicatie *f* || импликация *f*

163 **implicit function** || implizite Funktion *f* || fonction *f* implicite || impliciete functie *f* || неявная функция *f*

164 **implicit method** || implizite Methode *f* || méthode *f* implicite || impliciete methode *f* || неявный метод *m*

165 **implosion** || Implosion *f* || implosion *f* || implosie *f* || имплозия *f*, направленный внутрь взрыв *m*

166 **imponderability** || Schwerelosigkeit *f* || apesanteur *f*, impondérabilité *f* || gewichtloosheid *f* || невесомость *f*

167 **imponderable** || schwerelos || impondérable || gewichtloos || невесомый

168 **impossible event** || unmögliches Ereignis *n* || événement *m* impossible || onmogelijke gebeurtenis *f* || невозможное событие *n*

169 **improper** || uneigentlich || impropre || oneigenlijk || несобственный

170 **improper integral** || uneigentliches Integral *n* || intégrale *f* impropre || oneigenlijke integraal *f* *(m)* || несобственный интеграл *m*

171 **improper rotation** || **1.** unechte Drehung *f* **2.** *Math.* uneigentliche orthogonale Transformation *f*, uneigentliche orthogonale Abbildung *f*, Drehspiegelung *f* || **1.** rotation *f* impropre **2.** *math.* transformation *f* orthogonale impropre || **1.** oneigenlijke draaiing *f* **2.** *math.* oneigenlijke rotatie *f* || **1.** несобственное вращение *n* **2.** *мат.* несобственное вращение *n*, вращение *n* отражением

172 **improvement of convergence** || Konvergenzverbesserung *f* || amélioration *f* de convergence || convergentieverbetering *f* || улучшение *n* сходимости

173 **impulse** || Impuls *m*, Kraftstoß *m* || impulsion *f* || impuls *m* || импульс *m* *(механический)*

174 **impulse function** || Stoßfunktion *f*, Impulsfunktion *f* || fonction *f* impulsionnelle || impulsfunctie *f* || импульсная функция *f*

175 **impulse of force** || Kraftimpuls *m* || impulsion *f* de force || krachtimpuls *m* || импульс *m* силы

176 **impulse reaction turbine** || Gleichdruck-Überdruck-Turbine *f* || turbine *f* mixte à action et réaction || gecombineerde turbine *f* || активно-реактивная турбина *f*

177 **impulse response** || Sprungantwort *f* || réponse *f* impulsionnelle || impulsantwoord *n* || реакция *f* на единичное (ступенчатое) воздействие

178 **impulse turbine** || Gleichdruckturbine *f* || turbine *f* à action || actieturbine *f* || активная турбина *f*

178a **impulsive force** || **1.** Stoßkraft *f* **2.** Impulsstärke *f* || **1.** force *f* de choc [de frappe] **2.** force *f* d'impulsion || **1.** stootkracht *f* *(m)* **2.** impulssterkte *f* || **1.** ударная сила *f* **2.** импульсная сила *f*

179 **impurity atom** || Stör(stellen)-atom *n*, Fremdatom *n* || atome *m* d'impureté || vreemd atoom *n*, verontreinigingsatoom *n* || атом *m* примеси, примесный атом *m*

180 **impurity band** || Stör(stellen)-band *n* || bande *f* d'impureté || verontreinigingsband *m* || примесная энергетическая зона *f*

181 **impurity center** || Störstelle *f* || centre *m* d'impureté || verontreinigingscentrum *n* || примесный центр *m*

182 **impurity conduction** || Stör(stellen)leitung *f*, Fehlordnungsleitung *f* || conductibilité *f* extrinsèque d'impureté || verontreinigingsconductie *f* || несобственная примесная проводимость *f* *(в полупроводнике)*

183 **impurity level** || **1.** Stör(stellen)-gehalt *m*, Verunreinigungsgehalt *m* **2.** Stör(stellen)niveau *n*, Verunreinigungsniveau *n* || **1.** teneur *f* en impuretés **2.** niveau *m* d'impuretés || **1.** verontreinigingsgehalte *n* **2.** veront-

reinigingsniveau *n* || 1. содержание *n* примесей 2. примесный (энергетический) уровень *m*

I84 **impurity semiconductor** || Stör(stellen)halbleiter *m*, Verunreinigungshalbleiter *m* || semi-conducteur *m* dopé [à impuretés] || extrinsieke halfgeleider *m* || примесный полупроводник *m*

I85 **inaccuracy** || Fehler *m*; Ungenauigkeit *f*; Unsicherheit *f* || erreur *f* || fout *f* (*m*) || погрешность *f*; неточность *f*; ненадёжность *f* (*данных*)

I86 **incandescence** || (sichtbare) Wärmestrahlung *f*, Temperaturstrahlung *f*, Temperaturleuchten *n* || incandescence *f* || temperatuurstraling *f* || видимое тепловое излучение *n*, температурное свечение *n*

I87 **incandescent lamp** || Glühlampe *f* || lampe *f* à incandescence || gloeilamp *f* (*m*) || лампа *f* накаливания

I88 **incidence** || 1. Einfall *m*, Inzidenz *f*; Auftreffen *n* 2. Anstellwinkel *m* || 1. incidence *f* 2. angle *m* d'incidence [d'attaque] || 1. inval *m* 2. invalshoek *m* || 1. падение *n* (*напр. лучей*) 2. угол *m* атаки

I89 **incident** || einfallend || incident || invallend || падающий (*о лучах*); налетающий, бомбардирующий (*о частицах*)

I90 **incident beam** || auffallendes Bündel *n* || faisceau *m* incident || invallende bundel *m* || падающий пучок *m*

I91 **incident electron** || einfallendes Elektron *n* || électron *m* incident || invallend elektron *n* || налетающий [бомбардирующий] электрон *m*

I92 **incident light** || einfallendes [auftreffendes] Licht *n* || lumière *f* incidente || opvallend licht *m* || падающий свет *m*

I93 **incident wave** || einfallende [einlaufende] Welle *f* || onde *f* incidente || aankomende golf *f* (*m*) || падающая [набегающая] волна *f*

I94 **inclination** || 1. Inklination *f*, Neigung *f* 2. Neigung *f*, Neigungswinkel *m* || 1. inclinaison *f* 2. inclinaison *f*, angle *m* d'inclinaison || 1. inclinatie *f* 2. helling *f*; hellingshoek *m* || 1. наклонение *n* (магнитное) 2. наклон *m*

I95 **inclined plane** || schiefe Ebene *f* || plan *m* incliné || hellend vlak *n* || наклонная плоскость *f*

I96 **inclined trajectory** || geneigte Bahn *f* || trajectoire *f* inclinée || hellende baan *f* (*m*) || наклонная траектория *f*

I97 **inclusion** || Inklusion *f* || inclusion *f* || inclusie *f* || отношение *n* включения

I98 **inclusion relation** || Inklusionsbeziehung *f* || relation *f* d'inclusion || inclusierelatie *f* || отношение *n* включения

I99 **inclusive OR** || inklusives Oder *n* || OU inclusif || inclusieve OF || включающее ИЛИ, дизъюнкция *f*

I100 **incoherence** || Incohärenz *f*, Nichtkohärenz *f* || incohérence *f* || incoherentie *f* || некогерентность *f*

I101 **incoherent** || inkohärent || incohérent || incoherent || некогерентный

I102 **incommensurable** || inkommensurabel; unvergleichbar || incommensurable || incommensurabel || несоизмеримый

I103 **incommensurable quantities** || inkommensurable Größen *f pl* || quantités *f pl* incommensurables || incommensurabele grootheden *f pl* || несоизмеримые величины *f pl*

I104 **incomplete** || unvollständig || incomplet || onvolledig || неполный

I105 **incompressibility** || Inkompressibilität *f* || incompressibilité *f* || incompressibiliteit *f* || несжимаемость *f*

I106 **incompressible flow** || inkompressible Strömung *f*, inkompressibles Fließen *n* || courant *m* incompressible || incompressiebele stroming *f* || несжимаемое течение *n*, поток *m* несжимаемой жидкости

I107 **incompressible liquid** || inkompressible Flüssigkeit *f* || liquide *m* incompressible || incompressiebele vloeistof *f* (*m*) || несжимаемая жидкость *f*

I108 **inconsistency** || Inkonsistenz *f* || 1. contradiction *f* 2. incompatibilité *f* || 1. tegenstrijdigheid *f* 2. inconsistentie *f* || 1. противоречивость *f* 2. несовместимость *f* *(уравнений)*

I109 **inconsistent** || inkonsistent || 1. contradictoire 2. incompatible || 1. tegenstrijdig 2. inconsistent || 1. противоречивый 2. несовместный *(об уравнениях)*

I110 **inconsistent equations** || inkonsistente Gleichungen *f pl* || équations *f pl* incompatibles || inconsistente vergelijkingen *f pl* || несовместные уравнения *n pl*

I111 **increase** || Zunahme *f*, Vergrößerung *f*, Anstieg *m* || accroissement *m*, croissance *f*, augmentation *f* || toename *f* || увеличение *n*, возрастание *n* *(величины)*; увеличение *n*

I112 **increment** || Inkrement *n* || incrément *m*, accroissement *m* || increment *n* || инкремент *m*, приращение *n*

I113 **indefinable** || indefinierbar, unbestimmbar || indéterminable || onbepaalbaar || неопределимый, не поддающийся определению

I114 **indefinite** || undefiniert, unbestimmt || indéfini, indéterminé || onbepaald || неопределённый

I115 **indefinite form** || unbestimmte Form *f* || forme *f* indéfinie || onbepaalde vorm *m* || неопределённая форма *f*

I116 **indefinite integral** || unbestimmtes Integral *n* || intégrale *f* indéfinie || onbepaalde integraal *f* *(m)* || неопределённый интеграл *m*

I117 **indefinite metric** || indefinite Metrik *f* || métrique *f* indéfinie || onbepaalde metriek *f* || неопределённая метрика *f*

I118 **indentation hardness** || Eindruckhärte *f* || dureté *f* à la pénétration || indrukhardheid *f* || твёрдость *f*, определяемая по методу вдавливания, инденторная [индентометрическая] твёрдость *f*

I119 **independent** || unabhängig || indépendant || onafhankelijk || независимый

I120 **independent drive** || unabhängiger Antrieb *m* || commande *f* indépendante || afzonderlijke [onafhankelijke] aandrijving *f* || независимый привод *m*

I121 **independent experiments, independent trials** || unabhängige Versuche *m pl* || essais *m pl* indépendants || onafhankelijke proeven *f* *(m)* *pl* || независимые испытания *n pl*

I122 **independent variable** || unabhängige Variable *f* || variable *f* indépendante || onafhankelijke variabele *f* || независимая переменная *f*

I123 **indeterminacy principle** || (Heisenbergsche) Unbestimmtheitsrelation *f*, Unbestimmtheitsprinzip *n* || principe *m* d'incertitude [d'indétermination] || onbepaaldheidsprincipe *n* || принцип *m* неопределённости (Гейзенберга)

I124 **indeterminism** || Indeterminismus *m* || indéterminisme *m* || indeterminisme *m* || индетерминизм *m*

I125 **index** || Index *m* || indice *m*, index *m* || index *m* || индекс *m*; показатель *m*

I126 **indexing head** *see* **dividing head**

I127 **indexing table** || Drehtisch *m* mit Teileinrichtung || table *f* pivotante [tournante] avec un dispositif de division || indextafel *f* *(m)* || делительный стол *m* *(станка)*

I128 **index of refraction** || Brechungsindex *m*, Brech(ungs)zahl *f* || indice *m* de réfraction || brekingsindex *m* || показатель *m* преломления

I129 **indicated efficiency** || indizierter Wirkungsgrad *m* || rendement *m* indiqué || geïndiceerd rendement *n* || индикаторный кпд

I130 **indicated horsepower** || indizierte Pferdestärke *f*, indizierte Leistung *f* || puissance *f* indiquée || geïndiceerd vermogen *n* || индикаторная мощность *f*

I131 **indicated power** || indizierte Leistung *f*, Indikatorleistung *f* || puissance *f* indiquée || indicatorvermogen *n* || индикаторная мощность *f*

I132 **indicated pressure** || indizierter Druck *m* || pression *f* indiquée || geïndiceerde druk *m* || индикаторное давление *n*

I133 **indication** || Anzeige *f*, Indikation *f* || lecture *f*, indication *f* || indicatie *f* || индикация *f*, отсчёт *m*, показание *n* *(прибора)*

I134 **indicator** || Indikator *m*; Anzeiger *m* || indicateur *m* || indicator *m*, aanwijzer *m* || индикатор *m*; указатель *m*

I135 **indicator diagram** || Indikatordiagramm *n* || diagramme *m* d'indicateur || indicatordiagram *n* || индикаторная диаграмма *f*

I136 **indicator lamp** || **1.** Anzeigelampe *f* **2.** Indikatorröhre *f*, Anzeigeröhre *f* || lampe *f* témoin || indicatorbuis *f* (*m*) || индикаторная лампа *f*

I137 **indifferent equilibrium** || indifferentes [stetiges] Gleichgewicht *n* || équilibre *m* indifférent || onverschillig evenwicht *n* || безразличное равновесие *n*

I138 **indirect address** || indirekte Adresse *f* || adresse *f* indirecte || indirect adres *n* || косвенный адрес *m*

I139 **indirect measurement** || indirekte Messung *f*, Indirektmessung *f* || mesure *f* indirecte || indirecte meting *f* || косвенное [непрямое] измерение *n*

I140 **indistinguishability** || Ununterscheidbarkeit *f*, Nichtunterscheidbarkeit *f* || indiscernabilité *f* || identiteit *f* || неразличимость *f*, тождественность *f*

I141 **indoor temperature** || Zimmertemperatur *f*, Raumtemperatur *f* || température *f* ambiante || kamertemperatuur *f* || комнатная температура

I142 **induced charge** || Influenzladung *f*, induzierte Ladung *f* || charge *f* induite || geïnduceerde lading *f* || наведённый заряд *m*

I143 **induced current** || Induktionsstrom *m*, induzierter Strom *m*; Influenzstrom *m* || courant *m* induit || geïnduceerde stroom *m* || индуцированный [наведённый] ток *m*

I144 **induced emission** || induzierte [erzwungene, stimulierte] Emission *f* || émission *f* induite || geïnduceerde emissie *f* || индуцированное [стимулированное, вынужденное] излучение *n*

I145 **induced magnetism** || induzierter Magnetismus *m* || magnétisme *m* induit || geïnduceerd magnetisme *n* || наведённый [индуцированный] магнетизм *m*

I146 **induced transformations** || induzierte Transformationen *f pl* || transformations *f* induites || geïnduceerde transformaties *f pl* || индуцированные преобразования *n pl*

I147 **inductance** || **1.** Induktivität *f*, Selbstinduktionskoeffizient *m* **2.** Induktivitätsspule *f* || **1.** inductance *f* **2.** bobine *f* d'inductance || **1.** inductantie *f* **2.** inductantiespoel *f* (*m*) || **1.** индуктивность *f*, коэффициент *m* самоиндукции **2.** катушка *f* индуктивности

I148 **inductance voltage** || elektromotorische Kraft *f* der Selbstinduktion, Selbstinduktions-EMK *f* || force *f* électromotrice d'autoinduction || elektromotorische kracht *f* (*m*) van zelfinductie *f* || электродвижущая сила *f* самоиндукции

I149 **induction** || (elektrische) Induktion *f* || induction *f* || inductie *f* || (электрическая) индукция *f*

I150 **induction bridge** || Induktivitätsmeßbrücke *f* || pont *m* d'induction || inductiviteitsmeetbrug *f* (*m*) || мост *m* для измерения индуктивности

I151 **induction coil** || Induktionsspule *f*, Ruhmkorff-Induktor *m* || bobine *f* d'induction || inductiespoel *f* (*m*) || индукционная катушка *f*, катушка *f* Румкорфа

I152 **induction generator** || Induktionsgenerator *m* || générateur *m* à induction asynchrone || inductiegenerator *m* || индукционный [асинхронный] генератор *m*

I153 **induction heating** || Induktionsheizung *f*, induktive Heizung *f*; Induktionserhitzung *f*, induktive Erhitzung *f* || chauffage *m* par induction || ind uctieverwarming *f* || индукционный [высокочастотный] нагрев *m*

I154 **induction machine** || Induktionsmaschine *f* || machine *f* à induction || inductiemachine *f* || электрофорная индукционная машина *f*

I155 **induction manifold** || Saugleitung *f* || tuyau *m* [conduite *f*] d'aspiration || zuigbuis *f* (*m*) || впускной [всасывающий] трубопровод *m*

I156 **induction motor** || Asynchronmotor *m* || moteur *m* asynchrone ||

asynchrone motor *m* || асинхронный электродвигатель *m*

I157 **inductive** || induktiv, Induktiv...; Induktions... || inductif || inductief || индуктивный, индукционный

I158 **inductive component** || induktive Komponente *f*, inductiver Anteil *m* || composante *f* inductive || inductieve component *m* || индуктивная составляющая *f*

I159 **inductive coupling** || induktive Kopplung *f* || couplage *m* inductif || inductieve koppeling *f* || индуктивная трансформаторная связь *f*

I160 **inductive resistance** || induktiver Widerstand *m*, Induktanz *f*, induktive Reaktanz *f* || inductance *f*, réactance *f* inductive || inductieve weerstand *m* || индуктивное сопротивление *n*

I161 **inductivity** || Dielektrizitätskonstante *f* || constante *f* diélectrique, permittivité *f* || diëlektrische constante *f* (*m*) || диэлектрическая проницаемость *f*

I162 **inductor** || Induktor *m*; Induktionsspule *f*, Ruhmkorff-Induktor *m* || indcuteur *m*; bobine *f* à induction, bobine *f* de Ruhmkorff || inductor *m*; inductieklos *m*, *f* || индуктор *m*; индукционная катушка *f*, катушка *f* Румкорфа

I163 **inelastic** || unelastisch, inelastisch || non élastique, inélastique || inelastisch || неупругий

I164 **inelastic collision** || unelastischer [inelastischer] Stoß *m* || collision *f* inélastique || inelastische collisie *f* || неупругое столкновение *n*, неупругий удар *m*

I165 **inelastic impact** || unelastischer [inelastischer] Stoß *m* || choc *m* inélastique || inelastische stoot *m* || неупругий удар *m*

I166 **inelasticity** || Inelastizität *f* || inélasticité *f* || inelasticiteit *f* || неупругость *f*

I167 **inelastic scattering** || inelastische Streuung *f* || diffusion *f* inélastique || inelastische verspreiding *f* || неупругое рассеяние *n*

I168 **inequality** || Ungleichung *f* || inégalité *f* || ongelijkheid *f* || неравенство *n*

I169 **inert gas** || inertes Gas *n* Edelgas *n* || gaz *m* inerte; gaz *m* noble || inert gas *n*; edelgas *n* || инертный газ *m*; благородный газ *m*

I170 **inertia** || Trägheit *f*; Beharrungsvermögen *n*, Beharrung *f* || inertie *f*; force *f* d'inertie || traagheid *f*, inertie *f* || инерция *f*

I171 **inertia governor** || Beharrungsregler *m* || régulateur *m* d'inertie || traagheidsregelaar *m* || инерциальный регулятор *m*

I172 **inertial** || 1. Trägheits..., Inertial... 2. träge || 1. inertial 2. d'inertie || 1. inert 2. traag || 1. инерциальный (*о системе*) 2. инерционный (*о свойствах*)

I173 **inertialess** || trägheitslos, trägheitsfrei || sans inertie || traagheidsloos || безынерционный

I174 **inertial force** || Trägheitskraft *f* || force *f* d'inertie || traagheidskracht *f* (*m*), inertiekracht *f* (*m*) || сила *f* инерции

I175 **inertial frame** || Inertialsystem *n* || système *m* du centre de masses || traagheidssysteem *n* || инерциальная система *f*, система *f* (отсчёта) центра масс [центра инерции]

I176 **inertial mass** || Trägheitsmasse *f* || masse *f* d'inertie || inertiemassa *f* (*m*), traagheidsmassa *f* (*m*) || 1. инерционная масса *f* 2. инертная масса *f*

I177 **inertial motion** || Nachlauf *m* || mouvement *m* par inertie || naloop *m* || движение *n* по инерции

I178 **inertial system** || Inertialsystem *n* || système *m* à inertie || traagheidssysteem *n* || инерциальная система *f*

I179 **inertia pressure** || Trägheitsdruck *m* || pression d'inertie || traagheidsdruk *m* || инерционное давление *n*

I180 **inertia tensor** || Trägheitstensor *m* || tenseur *m* d'inertie || traagheidstensor *m* || тензор *m* инерции

I181 **inextensible filament** || undehnbarer Faden *m* || fil *m* inextensible || onrekbare draad *m* || нерастяжимая нить *f*

I182 **inference** || Schlußfolgerung *f*; Schluß *m*, Rückschluß *m* || inférence *f* || afleiding *f* || вывод *m* *(умозаключение)*

I183 **infinite-dimensional** || unendlichdimensional || à dimensions infinies || oneindig dimensionaal || бесконечномерный

I184 **infinite group** || unendliche Gruppe *f* || groupe *m* infini || oneindige groep *f* *(m)* || бесконечная группа *f*

I185 **infinitely far** || unendlich fern || éloigné à l'infini || oneindig ver || бесконечно удалённый

I186 **infinitely large** || unendlich groß || infiniment grand || oneindig groot || бесконечно большой

I187 **infinite matrix** || unendliche Matrix *f* || matrice *f* infinie || oneindige matrix *f* || бесконечная [бесконечномерная] матрица *f*

I188 **infinite product** || unendliches Produkt *n* || produit *m* infini || oneindig produkt *n* || бесконечное произведение *n*

I189 **infinite quantity** || unendlich große Größe *f* || quantité *f* infiniment grande || oneindig grote grootheid *f* || бесконечно большая величина *f*

I190 **infinite series** || unendliche Reihe *f* || série *f* infinie || oneindige reeks *f* *(m)* || бесконечный ряд *m*

I191 **infinite set** || unendliche Menge *f* || ensemble *m* infini || oneindige menigte *f* || бесконечное множество *n*

I192 I **infinitesimal** *see* **infinitesimal quantity**

I193 II **infinitesimal** || infinitesimal, unendlich klein || infinitésimal || oneindig klein || бесконечно малый

I194 **infinitesimal calculus** || Infinitesimalrechnung *f* || calcul *m* infinitésimal || infinitesimaalrekening *f* || исчисление *n* бесконечно малых

I195 **infinitesimal quantity** || unendlich kleine Größe *f* || quantité *f* infinitésimale || oneindig kleine grootheid *f* || бесконечно малая величина *f*

I196 **infinity** || Unendlichkeit *f* || infini *m*, infinité *f* || infiniteit *f* || бесконечность *f*

I197 **influence** || Einfluß *m*, Wirkung *f*, Einwirkung *f* || influence *f* || invloed *m* || влияние *n*, воздействие *n*

I198 **information** || Information *f* || information *f* || informatie *f* || информация *f*

I199 **information processing** || Informationsverarbeitung *f* || traitement *m* de l'information || informatieverwerking *f* || обработка *f* информации

I200 **information retrieval language** || Informationsrecherchesprache *f* || language *m* de recherche de l'information || informatierecherchetaal *f* || информационно-поисковый язык *m*, ИПЯ

I201 **information retrieval system** || Informationsrecherchesystem *n* || système *m* de recherche de l'information || informatie-recherchesysteem *n* || информационно-поисковая система *f*, ИПС

I202 **infrared** || **1.** Infrarotbereich *m*, Infrarotgebiet *n* **2.** Infrarotstrahlung *f*, infrarote Strahlung *f* || **1.** domaine *m* infrarouge **2.** radiation *f* infrarouge || **1.** infraroodgebied *n* **2.** infraroodstraling *f* || **1.** инфракрасная область *f* *(спектра)* **2.** инфракрасное излучение *n*

I203 **infrared radiation** || Infrarotstrahlung *f*, infrarote Strahlung *f* || radiation *f* infrarouge || infraroodstraling *f* || инфракрасное излучение *n*

I204 **infrared region** || Infrarotgebiet *n*, IR-Gebiet *n* || domaine *m* infrarouge || infraroodgebied *n* || инфракрасная область *f* *(спектра)*

I205 **infrared spectroscopy** || Infrarotspektroskopie *f* || spectroscopie *f* infrarouge || infrarode spectroscopie *f* || инфракрасная спектроскопия *f*

I206 **infrasonic** || Infraschall... || infra-sonore || infrasoon || инфразвуковой

I207 **infrasonic frequency** || Infraschallfrequenz *f* || fréquence *f* infrasonique || infrasone frequentie *f* || инфразвуковая частота *f*

I208 **infrasonic waves** || Infraschallwellen *f pl* || ondes *f pl* infrasoniques || infrasone golven *f* *(m)* *pl* || инфразвуковые волны *f pl*

I209 **infrasound** || Infraschall *m* || infra-son *m* || infrageluid *n* || инфразвук *m*

1210 **ingot** || Block *m*; Gußblock *m*; Barren *m*; Massel *f* || billette *f*; lingot *m* || gieteling *f*, baar *f* (*m*) || болванка *f*; слиток *m*; заготовка *f*

1211 **ingot iron** || Gußstahl *m* || fer *m* pur en lingot || gietstaal *n* || технически чистое железо *n* *(в слитках)*

1212 **inherent stress** || Eigenspannung *f* || contrainte *f* résiduelle || eigenspanning *f* || внутреннее остаточное напряжение *n* *(при отсутствии внешней силы)*

1213 **inhomogeneity** || Inhomogenität *f*, Heterogenität *f*; Ungleichförmigkeit *f* || inhomogénéité *f*, hétérogénéité *f* || inhomogeniteit *f*, heterogeniteit *f* || неоднородность *f*; неравномерность *f*

1214 **inhomogeneous** || inhomogen, heterogen || inhomogène, hétérogène || inhomogeen, heterogeen || неоднородный; неравномерный

1215 **inhomogeneous field** || inhomogenes Feld *n* || champ *m* inhomogène || inhomogeen veld *n* || неоднородное поле *n*

1216 **initial level** || Anfangspegel *m* || niveau *m* initial || aanvangsniveau *n* || начальный уровень *m* *(перехода)*

1217 **initial phase** || Anfangsphase *f* || phase *f* initiale || aanvangsfase *f* || начальная фаза *f*

1218 **initial pressure** || Anfangsdruck *m* || pression *f* initiale || aanvangsdruk *m* || начальное [исходное] давление *n*

1219 **initial state** ||Anfangszustand *m* || état *m* initial || aanvangstoestand *m* || начальное состояние *n*

1220 **initial time** || Anfangszeitmoment *m* || temps *m* initial || aanvangstijd *m* || начальный момент *m* *(времени)*

1221 **initial velocity** || Anfangsgeschwindigkeit *f* || vitesse *f* initiale || aanvangssnelheid *f* || начальная скорость *f*

1222 **injection** || Injektion *f*, Einspritzen *n* || injection *f* || injectie *f*, inspuiting *f* || инжекция *f*; впрыск *m*, впрыскивание *n*

1223 **injection lag** || Einspritzverzögerung *f* || retard *m* de l'injection || inspuitvertraging *f* || запаздывание *n* впрыска

1224 **injection pressure** || Einspritzdruck *m* || pression *f* d'injection || inspuitdruk *m* || давление *n* впрыска

1225 **injective function** || injektive Funktion *f*; Injektion *f* || fonction *f* injective || injectieve functie *f* || инъективная функция *f*; взаимнооднозначное соответствие *n*

1226 **injector** || Injektor *m* || injecteur *m* || injecteur *m*, straalpomp *f* (*m*) || инжектор *m*, струйный насос *m*

1227 **inlet stroke** || Ansaugtakt *m* || course *f* [temps *m*] d'aspiration || aanzuigslag *m* || ход *m* [такт *m*] всасывания [впуска]

1228 **inlet value** || Einlaßventil *n*, Einströmventil *n* || soupape *f* d'admission [d'arrivée] || inlaatklep *f* (*m*) || впускной [всасывающий] клапан *m*

1229 **in-line processing** || mitlaufende Verarbeitung *f*, In-line-Verarbeitung *f* || traitement *m* en ligne || in-line-verwerking *f* || оперативная обработка *f*, обработка *f* в линию

1230 **inner automorphism** || innere Automorphie *f* || automorphisme *m* intérieur || inwendig automorfisme *n* || внутренний автоморфизм *m*

1231 **inner dead center** || unterer Totpunkt *m* || point *m* mort inférieur [bas] || onderste dode punt *f* (*m*) || нижняя мёртвая точка *f*

1232 **inner diameter** || Innendurchmesser *m* || diamètre *m* intérieur || binnendiameter *m* || внутренний диаметр *m*

1233 **inner product (of two vectors)** || inneres Produkt *n*, Skalarprodukt *n* || produit *m* intérieur (de deux vecteurs) || inwendig produkt *n* || скалярное произведение *n* (двух векторов)

1234 **inner quantum number** || innere Quantenzahl *f*, Gesamtdrehimpulsquantenzahl *f* || nombre *m* quantique interne || inwendig quantumgetal *n* || внутреннее [полное] квантовое число *n*

1235 **inner race** || Innenring *m*, Innenlaufbahn *f* *(Lager)* || chemin *m* intérieur, bague *f* intérieure *(de roulement)* || binnenring *m (lager)* || внутреннее кольцо *n (подшипника качения)*

I236 **inner-shell electron** || Innenelektron *n* || électron *m* interne || elektron *n* van een binnenschil || электрон *m* внутренней оболочки *(атома)*

I237 **inorganic** || anorganisch, unorganisch || inorganique, anorganique, minéral || anorganisch || неорганический

I238 **inorganic chemistry** || anorganische Chemie *f* || chimie *f* inorganique [minérale] || anorganische chemie *f* || неорганическая химия *f*

I239 **in-parallel** || parallelgeschaltet, nebeneinandergeschaltet || en parallèle || parallelgeschakeld || включённый параллельно

I240 **inphase** || gleichphasig, phasengleich, synphas, in Phase (liegend) || concordant par phase || gelijkfasig || совпадающий по фазе, синфазный

I241 **input** || 1. Zufuhr *f (z. B. Leistung)* 2. Eingang *m (Gerät)* 3. Eingabe *f (Daten)* || 1. amenée *f*, alimentation *f* 2., 3. entrée *f* || 1. toevoer *m* 2.,3. ingang *m* || 1. ввод *m*; подвод *m (напр. мощности)* 2. вход *m (прибора)* 3. ввод *m (данных)*

I242 **input alphabet** || Eingabealphabet *n* || alphabet *m* d'entrée || ingangsalfabet *n* || входной алфавит *m*

I243 **input device** || Eingabegerät *n* || dispositif *m* d'entrée || ingangsinrichting *f* || устройство *n* ввода

I244 **input impedance** || Eingangsimpedanz *f*, Eingangsscheinwiderstand *m* || impédance *f* d'entrée || ingangsimpedantie *f* || входной импеданс *m*

I245 **input voltage** || Eingangsspannung *f* || tension *f* d'entrée || ingangsspanning *f* || входное напряжение *n*, напряжение *n* на входе

I246 **inscribed** || einbeschrieben || inscrit || ingeschreven || вписанный

I247 **inscribed circle** || einbeschriebener Kreis *m* || cercle *m* inscrit || ingeschreven cirkel *m* || вписанная окружность *f*

I248 **insensitive** || unempfindlich || insensible || ongevoelig || нечувствительный

I249 **in-series** || in Reihe [in Serie] geschaltet, hintereinandergeschaltet || en série || in serie geschakeld || включённый последовательно

I250 **inserted teeth milling cutter** || Fräser *m* mit eingesetzten Messern, Messerfräser *m*; Messerkopf *m* || fraise *f* à lames [à dents] rapportées || frees *f (m)* met ingezette tanden || фреза *f* со вставными зубьями

I251 **inside caliper** || Innentaster *m* || compas *m* d'intérieur || binnenpasser *m* || нутромер *m*

I252 **inside chaser** || Innengewindestrehler *m* || peigne *m* mâle || binnenschroefdraadbeitel *m* || гребёнка *f* для нарезания внутренней резьбы

I253 **inspection gauge** || Abnahmelehre *f* || jauge-étalon *m*; calibre *m* de contrôle || controlekaliber *n* || контрольный калибр *m*

I254 **instability** || Instabilität *f* || instabilité *f* || instabiliteit *f*, onstabiliteit *f* || неустойчивость *f*

I255 **instability mode** || Instabilitätsart *f* || mode *m* d'instabilité || instabiliteitsaard *m* || вид *m* [форма *f*, тип *m*] неустойчивости

I256 **installation** || Einrichtung *f*; Montage *f*; Einbau *m*; Ausrüstung *f* || installation *f* || inrichting *f*, installatie *f* || устройство *n*; установка *f*; оборудование *n*

I257 **instantaneous** || augenblicklich, momentan, instantan || instantané || ogenblikkelijk, momentaan || мгновенный, относящийся к данному моменту *(времени)*

I258 **instantaneous axis** || Momentanachse *f* (der Drehung), Momentandrehachse *f*, momentane Drehachse *f* || axe *m* (de rotation) instantané || momentane draaiingsas *f (m)* || мгновенная ось *f* (вращения)

I259 **instantaneous center** || Momentanzentrum *n* || point *m* [centre *m*] (de rotation) instantané || momentaan omloopcentrum *n* || мгновенный центр *m* (вращения)

1260 **instantaneous force** || 1. momentan wirkende Kraft *f*, Momentankraft *f* 2. Momentanwert *m* der Kraft || 1. force *f* instantanée 2. valeur *f* de la force instantanée || 1. momentkracht *f* (*m*) 2. momentwaarde *f* van de kracht || 1. мгновенная сила *f* 2. мгновенное значение *n* силы

1261 **instantaneous value** || Momentanwert *m*, Augenblickswert *m* || valeur *f* instantanée || momentwaarde *f* || мгновенное значение *n*

1262 **instantaneous velocity** || Momentangeschwindigkeit *f*, Momentanwert *m* der Geschwindigkeit *f* || vitesse *f* instantanée || momentsnelheid *f* || мгновенное значение *n* скорости

1263 **instant of time** || Zeitpunkt *m*, Zeitmoment *m*, Augenblick *m* || instant *m* || ogenblik *n* || момент *m* времени

1264 **instruction** || Instruktion *f*, Befehl *m* || instruction *f* || instructie *f* || инструкция *f*, команда *f*

1265 **instruction set** || Befehlsvorrat *m* || jeu *m* des instructions || instructieset *m* ||набор *m* команд; система *f* команд

1266 **instrument** || 1. Werkzeug *n* 2. Gerät *n*, Meßgerät *n* || 1. outil *m* 2. instrument *m (de mesure)* || instrument *n*; werktuig *n*; gereedschap *n* || 1. инструмент *m (как орудие для работы)* 2. *(контрольно-измерительный)* прибор *m*

1267 **instrument transformer** || Meßtransformator *m*, Meßwandler *m* || transformateur *m* de mesure || meettransformator *m* || измерительный трансформатор *m*

1268 **insulated conductor** || isolierter Draht *m*, isolierter Leiter *m* || conducteur *m* isolé || geïsoleerde leider *m* || изолированный провод *m*

1269 **insulation** || 1. Isolation *f*, Isolierung *f* 2. Isolierstoff *m*, Isoliermaterial *n*, Isolationsmaterial *n* || 1. isolation *f*, isolement *m* 2. isolant *m* || 1. isolatie *f* 2. isolatiemateriaal *n* || 1. изоляция *f*, изолирование *n* 2. изоляция *f*, изоляционный материал *m*

1270 **insulator** || Isolator *m*, Nichtleiter *m* || isolateur *m*, isolant *m* || isolator *m* || изолятор *m*

1271 **intake stroke** *see* **injection stroke**

1272 **integer** || ganze Zahl *f* || (nombre *m*) entier *m* || geheel getal *n* || целое (число) *n*

1273 **integer quantity** || ganzzahlige Größe *f* || quantité *f* entière || gehele grootheid *f* || целая величина *f*

1274 **integer spin** || ganzzahliger Spin *m* || spin *m* entier || gehele spin *m* || целый [целочисленный] спин *m*

1275 **integrability conditions** || Integrationsbedingungen *f pl* || conditions *f pl* d'intégrabilité || integratievoorwaarden *f pl* || условия *n pl* интегрируемости

1276 I **integral** || Integral *n* || intégrale *f* || integraal *f* (*m*) || интеграл *m*

1277 II **integral** || integral, Integral... || intégral || integraal || интегральный

1278 **integral calculus** || Integralrechnung *f* || calcul *m* intégral || integraalrekening *f* || интегральное исчисление *n*

1279 **integral curvature** || Integralkrümmung *f* || courbure *f* intégrale || integraalkromming *f* || полная кривизна *f*

1280 **integral equation** || Integralgleichung *f* || équation *f* intégrale || integraalvergelijking *f* || интегральное уравнение *n*

1281 **integral function** || Integralfunktion *f* || fonction *f* intégrale || integraalfunctie *f* || интегральная функция *f*

1282 **integral of motion** || Integral *n* der Bewegung, Bewegungsintegral *n* || intégrale *f* du mouvement || bewegingsintegraal *f* || интеграл *m* движения

1283 **integral transform, integral transformation** || Integraltransformation *f* || transformation *f* intégrale || integraaltransformatie *f* || интегральное преобразование *n*

1284 **integrated automation** || komplexe Automatisierung *f* || automatisation *f* intégrale || complexe automatisering *f* || комплексная автоматизация *f*

I285 **integrated circuit** || integrierte Schaltung *f* || circuit *m* intégral [intégré] || geïntegreerd circuit *n* || интегральная схема *f*

I286 **integrating factor** || integrierender Faktor *m* || facteur *m* intégrant || integrerende factor *m* || интегрирующий множитель *m*

I287 **integration** || Integration *f*, Integrieren *n* || intégration *f* || integratie *f* || интегрирование *n*

I288 **integration by parts** || partielle Integration *f* || intégration *f* par parties || partiële integratie *f* || интегрирование *n* по частям

I289 **integrator** || Integrator *m*, Integriergerät *n* || intégrateur *m* || integrator *m* || интегратор *m*

I290 **integrity** || Ganzheit *f*, Integrität *f* || intégrité *f* || integriteit *f* || целостность *f*

I291 **intense radiation** || intensive Strahlung *f* || rayonnement *n* intensif || intensieve straling *f* || интенсивное излучение *n*

I292 **intensification** || Verstärkung *f* *(fotografisches Bild)* || intensification *f* || versterking *f* || усиление *n* *(изображения)*

I293 **intensity** || 1. Intensität *f*, Stärke *f* *(Strahlung)* 2. Feldstärke *f*, Feldintensität *f* || intensité *f* || intensiteit *f* || 1. интенсивность *f* *(излучения)* 2. напряжённость *f* *(поля)*

I294 **intensity of gravity** || Gravitationsfeldstärke *f* || intensité *f* de la pesanteur || gravitatieveldsterkte *f* || напряжённость *f* поля тяжести, ускорение *n* силы тяжести

I295 **intensity of spectral line** || Intensität *f* der Spektrallinie, Spektrallinienintensität *f* || intensité *f* de la ligne [de la raie] spectrale || intensiteit *f* van de spectraallijn || яркость *f* [интенсивность *f*] спектральной линии

I296 **intensive variable** || intensive Variable *f*, intensiver Parameter *m* || paramètre *m* intensif || intensieve parameter *m* || интенсивный параметр *m*

I297 **interaction** || Wechselwirkung *f* || interaction *f* || wisselwerking *f* || взаимодействие *n*

I298 **interaction energy** || Wechselwirkungsenergie *f* || énergie *f* d'interaction || wisselwerkingsenergie *f* || энергия *f* взаимодействия

I299 **interatomic distance** || interatomarer Abstand *m*, Atomabstand *m* || distance *f* interatomique || interatomaire afstand *m* || межатомное расстояние *n*

I300 **interatomic interaction** || interatomare Wechselwirkung *f* || interaction *f* interatomique || interatomaire wisselwerking *f* || межатомное взаимодействие *n*

I301 **intercircuit coupling** || Stromkreiskopplung *f* || couplage *m* des circuits || circuitkoppeling *f* || связь *f* между (электрическими) контурами

I302 **interdiffusion** || Interdiffusion *f* || interdiffusion *f* || interdiffusie *f* || взаимная диффузия *f*

I303 **interface** || Grenzfläche *f*, Trenngrenze *f*, Phasengrenze *f* || interface *f* || grensvlak *n* || граница *f* раздела *(между фазами)*; поверхность *f* раздела

I304 **interference** || 1. Interferenz *f* 2. Störung *f* || 1. interférence *f* 2. interaction *f* || 1. interferentie *f* 2. storing *f* || 1. интерференция *f* 2. помехи *f pl*; взаимное влияние *n*

I305 **interference band** || Interferenzstreifen *m* || frange *f* d'interférence || interferentiestreep *f* *(m)* || интерференционная полоса *f*

I306 **interference colors** || Interferenzfarben *f pl* || couleurs *f pl* d'interférence || interferentiekleuren *m, n pl* || интерференционные цвета *m pl*

I307 **interference figure** *see* **interference pattern**

I308 **interference fit** || Preßpassung *f* || ajustement *m* forcé; ajustement *m* sans jeu || vaste passing *f* || посадка *f* с натягом; неподвижная посадка *f*

I309 **interference fringe** || Interferenzstreifen *m* || frange *f* d'interférence || interferentiestreep *f* *(m)* || интерференционная полоса *f*

I310 **interference microscope** || Interferenzmikroskop *n* || microscope *m* interférentiel || interferentiemicroscoop *m* || интерференционный микроскоп *m*

I311 **interference of light** || Lichtinterferenz *f*, Interferenz *f* des Lichtes || interférence *f* de lumière || lichtinterferentie *f* || интерференция *f* света

I312 **interference of polarized light** || Interferenz *f* polarisierter Lichtstrahlen || interférence *f* des rayons polarisés || interferentie *f* van gepolariseerde lichtstralen || интерференция *f* поляризованных (световых) лучей

I313 **interference pattern interferogram** || Interferenzbild *n*; Interferogramm *n* || image *f* d'interférence, interférogramme *m* || interferentiepatroon *n* || интерференционная картина *f*; интерферограмма *f*

I314 **interferometer** || Interferometer *n*, Interferenzgerät *n* || interféromètre *m* || interferometer *m* || интерферометр *m*

I315 **interferometry** || Interferometrie *f* || interférométrie *f* || interferometrie *f* || интерферометрия *f*

I316 **interior measure** || inneres Maß *n* || mesure *f* intérieure || inwendige maat *f* (*m*) || внутренняя мера *f*

I317 **interlock** || Blockierung *f*; Sperrung *f*; Verriegelung *f* || verrouillage *m* || blokkering *f*, grendeling *f* || (внутренняя) блокировка *f* (цепи)

I318 **intermediate frequency** || Zwischenfrequenz *f* || fréquence *f* intermédiaire || tussenfrequentie *f* || промежуточная частота *f*

I319 **intermediate image** || Zwischenbild *n*, Zwischenabbildung *f* || image *f* intermédiaire || tussenbeeld *n* || промежуточное изображение *n*

I320 **intermittent** || intermittierend, aussetzend, oszillierend || intermittent || intermitterend || прерывистый; пульсирующий

I321 **intermittent friction** || intermittierende Reibung *f* || frottement *m* intermittent || intermitterende wrijving *f* || прерывистое трение *n*, трение *n* с проскальзыванием

I322 **intermittent motion** || aussetzende Bewegung *f* || mouvement *m* intermittent || intermitterende beweging *f*, intermitterende loop *m* || прерывистое движение *n*

I323 **intermittent oiling** || unterbrochene Schmierung *f* || graissage *m* périodique || onderbroken smering *f* || периодическая смазка *f*

I324 **intermittent weld** || unterbrochene Schweißnaht *f* || soudure *f* discontinue [entrecoupée] || onderbroken lasnaat *m* || прерывистый сварной шов *m*

I325 **intermolecular interaction** || intermolekulare [zwischenmolekulare] Wechselwirkung *f* || interaction *f* intermoléculaire || intermoleculaire wisselwerking *f* || межмолекулярное взаимодействие *n*

I326 **internal combustion engine** || Verbrennungsmotor *m* || moteur *m* à combustion interne, moteur *m* à explosion || verbrandingsmotor *m* || двигатель *m* внутреннего сгорания, ДВС

I327 **internal conversion** || innere Konversion *f* *(Gamma-Strahlung)* || conversion *f* interne || inwendige conversie *f* || внутренняя конверсия *f* *(гамма-излучения)*

I328 **internal cylindrical grinding** || Innenrundschleifen *n* || rectification *f* cylindrique intérieure || binnenrondslijpen *n* || круглое внутреннее шлифование *n*

I329 **internal energy** || innere Energie *f* || énergie *f* interne || binnenenergie *f* || внутренняя энергия *f*

I330 **internal expanding brake** || Innenbackenbremse *f* || frein *m* à sabots [à mâchoires] articulé[e]s || rem *f* (*m*) met remblokken binnen de remtrommel || тормоз *m* с разжимными колодками внутри тормозного барабана

I331 **internal force** || innere Kraft *f* || force *f* interne || inwendige kracht *f* (*m*) || внутренняя сила *f*

I332 **internal friction** || innere Reibung *f*; Viskosität *f* || frottement *m* intérieur; viscosité *f* || inwendige wrijving *f*; viscositeit *f* || внутреннее трение *n*; вязкость *f*

I333 **internal grinding machine** || Innenrundschleifmaschine *f* || machine *f* à rectifier les intérieurs, rectifieuse *f* intérieure || binnenslijpmachine *f* || внутришлифовальный станок *m*

I334 **internal lapping** || Innenläppen *n* || rodage *m* intérieur || binnenlappen *n* || лапингование *n* [доводка *f*] внутренней поверхности

I335 **internal pressure** || Innendruck *m* || pression *f* intérieure || binnendruk *m* || внутреннее давление *n*

I336 **internal resistance** || innerer Widerstand *m*, Innenwiderstand *m* || résistance *f* interne || binnenweerstand *m*, inwendige weerstand *m* || внутреннее сопротивление *n*

I337 **internal splines** || Keilnabenprofile *n pl* || cannelures *f pl* intérieures || inwendige sleuven *f (m)* || шлицы *m pl* в отверстии, внутренние шлицы *m pl*

I338 **internal stress** || Eigenspannung *f*; innere Spannung *f* || contrainte *f* interne || inwendige spanning *f* || внутреннее напряжение *n*

I339 **internal thread** *see* **female thread**

I340 **internal voltage** || **1.** elektromotorische Kraft *f* des Generators, Generator-EMK *f* **2.** gegenelektromotorische Kraft *f* des Motors || **1.** force *f* électromotrice d'un générateur **2.** force *f* contre-électromotrice d'un moteur || **1.** elektromotorische kracht *f (m)* van een generator **2.** tegenelektromotorische kracht *f (m)* van een motor || **1.** электродвижущая сила *f* генератора **2.** противоэлектродвижущая сила *f* двигателя

I341 **international screw thread** || metrisches Gewinde *n (ISO-Gewinde mit einem Gewindeprofilwinkel von 60°)* || filet *m* métrique *(filetage normalisé avec un angle de profil de 60°)* || metrische schroefdraad *m* met profielhoek van 60° || (стандартная) метрическая резьба *f (с углом профиля 60°)*

I342 **interpolation** || Interpolation *f*, Interpolierung *f* || interpolation *f* || interpolatie *f* || интерполяция *f*, интерполирование *n*

I343 **interpolation formula** || Interpolationsformel *f* || formule *f* d'interpolation || interpolatieformule *f (m)* || интерполяционная формула *f*

I344 **interpolation polynomial** || Interpolationspolynom *n* || polynôme *m* d'interpolation || interpolatiepolynoom *n* || интерполяционный полином *m*

I345 **interpretation** || Interpretation *f*, Interpretierung *f* || interprétation *f* || interpretatie *f* || интерпретация *f*

I346 **interquartile range** || Interquartilintervall *n* || intervalle *m* interquartial || interquartiel interval *n* || интерквартильная широта *f*

I347 **interruption** || Unterbrechung *f* || interruption *f* || interruptie *f* || прерывание *n*

I348 **intersection** || Schneiden *n*; Durchdringung *f* || intersection *f* || doorsne(d)e *f (m)* || пересечение *n*

I349 **intersection of curves** || Schneiden *n* von Kurven || intersection *f* des courbes || krommenkruising *f* || пересечение *n* кривых

I350 **intersection of events** || Koinzidenz *f* von Ereignissen, Konjunktion *f* || intersection *f* des événements || samenvallen *n* van gebeurtenissen || совмещение *n* событий, логическое произведение *n (событий)*

I351 **intersection of sets** || Vereinigung *f* von Mengen, Produktmenge *f* || intersection *f* des ensembles || menigtenkruising *f* || пересечение *n* множеств, логическое произведение *n (множеств)*

I352 **intersection of surfaces** || Schneiden *n* von Oberflächen || intersection *f* des surfaces || vlakkensnijding *f* || пересечение *n* поверхностей

I353 **interstitial(atom)** || Zwischengitteratom *n*, Lückenatom *n* || atome *m* interstitiel || interstitiëel atoom *n* || атом *m* внедрения, меж(до)узельный атом *m*

I354 **interturn capacity** || Windungskapazität *f* || capacité *f* entre spires || windingscapaciteit *f* || меж(ду)витковая ёмкость *f*

I355 **interval** || Intervall *n*, Abstand *m* || intervalle *m* || interval *n* || интервал *m*; промежуток *m*

I356 **intramolecular** || intramolekular || intramoléculaire || intramoleculair || внутримолекулярный

I357 **intranuclear** || intranuklear, innernuklear || intranucléaire || intranucleair || внутриядерный

I358 **intranuclear forces** || innernukleare [intranukleare] Kräfte *f pl* || forces *f pl* intranucléaires || intranucleaire krachten *f (m) pl* || внутриядерные силы *f pl*

I359 **intrinsic conduction** || Eigenleitfähigkeit *f*, Eigen(halb)leitung *f*, Intrinsic-Leitfähigkeit *f* || conductibilité *f* intrinsèque || eigenconductiviteit *f* || собственная проводимость *f (в полупроводнике)*

I360 **intrinsic derivative** || absolute Ableitung *f* || dérivée *f* intrinsèque || absolute afgeleide *f* || абсолютная производная *f*

I361 **intrinsic differencial geometry** || Differentialgeometrie *f* || géométrie *f* différentielle intrinsèque || differentiaalgeometrie *f* || дифференциальная геометрия *f*

I362 **intrinsic equation of a curve** || natürliche Kurvengleichung *f* || équation *f* intrinsèque d'une courbe || natuurlijke krommenvergelijking *f* || натуральное уравнение *n* кривой

I363 **intrinsic geometry of a surface** || innere Flächengeometrie *f* || géométrie *f* intrinsèque d'une surface || intrinsieke vlakgeometrie *f* || внутренняя геометрия *f* поверхности

I364 **intrinsic photoeffect** || innerer Photoeffekt *m*, Photoleitungseffekt *m* || effet *m* photorésistif [photoélectrique interne] photoconduction *f* || inwendig fotoelektrisch effect *n* || внутренний фотоэффект *m*, фотопроводимость *f*

I365 **intrinsic semiconductor** || Eigenhalbleiter *m*, Intrinsic-Halbleiter *m* || semi-conducteur *m* intrinsèque || intrinsieke halfgeleider *m* || собственный [чистый, беспримесный] полупроводник *m*

I366 **intrinsic viscosity** || Grenzviskosität *f*, Grundviskosität *f* || viscosité *f* intrinsèque || intrinsieke viscositeit *f* || характеристическая вязкость *f*

I367 **invariable address** || konstante Adresse *f* || adresse *f* invariable || constant adres *n* || постоянный адрес *m*

I368 **invariance** || Invarianz *f* || invariance *f* || invariance *f* || инвариантность *f*

I369 I **invariant** || Invariante *f* || invariant *m* || invariante *f* || инвариант *m*

I370 II **invariant** || invariant || invariant || invariant || инвариантный

I371 **invariant manifold** || invariante Mannigfaltigkeit *f* || variété *f* invariante || invariante menigvoudigheid *f* || инвариантное многообразие *n*

I372 **inverse** || invers, reziprok || inverse || invert || обратный, взаимно-обратный

I373 **inverse Fourier transform(ation)** || inverse Fourier-Transformation *f* || transformation *f* de Fourier inverse || inverte Fourier-transformatie *f* || обратное преобразование *m* Фурье

I374 **inverse function** || Umkehrfunktion *f*, inverse Funktion *f* || fonction *f* inverse || omgekeerde functie *f* || обратная функция *f*

I375 **inverse homomorphism** || inverser Homomorphismus *m* || homomorphisme *m* réciproque || invert homomorfisme *n* || обращение *n* гомоморфизма

I376 **inverse hyperbolic functions** || Areafunktionen *f pl* || fonction *f* hyperbolique inverse || inverte hyperbolische functies *f pl* || обратные гиперболические функции *f pl*

I377 **inverse logarithm** || Antilogarithmus *m* || logarithme *m* inverse || inverte logaritme *f (m)* || антилогарифм *m*

I378 **inversely proportional** || umgekehrt proportional || inversement proportionnel || omgekeerd evenredig || обратно пропорциональный

I379 **inverse operator** || inverser Operator *m* || opérateur *m* inverse || inverte operator *m* || обратный оператор *m*

I380 **inverse relation** || Umkehrrelation *f* || relation *f* inverse || inverte relatie *f* || обратное отношение *n*

I381 **inverse-square-law** || dem Abstandsquadrat umgekehrt proportional || inversement proportionnel au carré de la distance || omgekeerd evenredig op afstandskwadraat || обратно пропорциональный квадрату расстояния

I382 **inverse transformation** || inverse Transformation *f*; Umkehrabbildung *f* || transformation *f* inverse || inverte transformatie *f* || обратное преобразование *n*

I383 **inverse trigonometric(al) functions** || Arkusfunktionen *f pl* || fonctions *f pl* trigonométriques inverses || inverte trigonometrische functies *f pl* || обратные тригонометрические функции *f pl*

I384 **inversion** || **1.** Inversion *f*, Invertierung *f*, Umkehr(ung) *f* **2.** Wechselrichtung *f* *(Strom)* || inversion *f* || inversie *f* || **1.** инверсия *f*, обращение *n* **2.** инвертирование *n* *(тока, напряжения)*

I385 **inversion of image** || Bildumkehr *f* || inversion *f* de l'image || beeldinversie *f* || переворачивание *n* изображения

I386 **inversion symmetry** || Inversionssymmetrie *f* || symétrie *f* de l'inversion || inversiesymmetrie *f* || симметрия *f* инверсии, инверсионная симметрия *f*

I387 **inversion theorem** || Umkehrsatz *m* || théorème *m* d'inversion || inversietheorema *n* || теорема *f* об инверсии

I388 **inverted image** || kopfstehendes [höhenverkehrtes] Bild *n* || image *f* retournée || omgekeerd beeld *n* || перевёрнутое изображение *n*

I389 **inverter** || Wechselrichter *m* || inverseur *m*, convertisseur *m* || invertor *m* || инвертор *m*

I390 **investment casting** || Ausschmelzverfahren *n*, Modellausschmelzgießverfahren *n* || moulage *m* [fonderie *f*, coulée *f*] à cire perdue || verloren-wasmethode *f* || литьё *n* по выплавляемым моделям

I391 **invisible radiation** || unsichtbare Strahlung *f* || radiation *f* invisible || onzichtbare straling *f* || невидимое излучение *n*

I392 **involute** || Evolute *f*, Involute *f* || involute *f* || involute *f* || эволюта *f*

I393 **involute gear** || Evolventenrad *n* || engrenage *m* à développante de cercle || evolvente vertanding *f* || зубчатое колесо *n* с эвольвентным зацеплением

I394 **involute surface** || Evolventenfläche *f* || surface *f* développante || evolvente oppervlakte *f* || эвольвентная поверхность *f*

I395 **involute tooth** || Evolventenzahn *m* || dent *m* de profil en développante || evolventvormige tand *m* || зуб *m* эвольвентного профиля

I396 **involution** || Involution *f* || involution *f* || involutie *f* || **1.** инволюция *f* **2.** возведение *n* в степень

I397 **inward-flow turbine** || Zentripetalturbine *f* || turbine *f* centripète || centripetaalturbine *f* || центростремительная [радиальная] турбина *f*

I398 **iodine,** I || Jod *n* || iode *m* || jood *n*, jodium *n* || йод *m*

I399 **ion** || Ion *n* || ion *m* || ion *n* || ион *m*

I400 **ion-bombarded** || ionenbeschossen, ionenbombardiert || bombardé par ions || ionengebombardeerd || бомбардируемый ионами

I401 **ion emission** || Ionenemission *f* || émission *f* d'ions || ionenemissie *f* || ионная эмиссия *f*

I402 **ionic binding, ionic bond** || Ionenbindkng *f* || liaison *f* ionique || ionaire binding *f* || ионная связь *f*

I403 **ionic conduction** || Ionenleitung *f*, Ionenleitfähigkeit *f* || conduction *f* ionique || ionengeleiding *f* || ионная проводимость *f*

I404 **ionic crystal** || Ionenkristall *m* || cristal *m* d'ions || ionenkristal *m* || ионный кристалл *m*

I405 **ion implantation** || Ionenimplantation *f* || implantation *f* d'ions || ionenimplantatie *f* || ионное внедрение *n*, ионная имплантация *f*

I406 **ionization** || Ionisation *f*, Ionisierung *f* || ionisation *f* || ionisatie *f* || ионизация *f*

1407 **ionization by collision** || Stoßionisation *f*, Stoßionisierung *f* || ionisation *f* par choc [par impact] || botsingsionisatie *f* || ударная ионизация *f*

1408 **ionization by electron impact** || Elektronenstoßionisation *f* || ionisation *f* par impact d'électrons || elektronenbotsingsionisatie *f* || ионизация *f* электронным ударом

1409 **ionization by radiation** || Strahlungsionisation *f* || ionisation *f* par radiation || stralingsionisatie *f* || ионизация *f* излучением

1410 **ionization chamber** || Ionisationskammer *f* || chambre *f* d'ionisation || ionisatiekamer *f* (*m*) || ионизационная камера *f*

1411 **ionization energy** || Ionisationsenergie *f* || énergie *f* d'ionisation || ionisatieënergie *f* || энергия *f* ионизации

1412 **ionization gauge** || Ionisationsmanometer *n* || jauge *f* [manomètre *m*] à ionisation || ionisatiemanometer *m* || ионизационный манометр *m*

1413 **ionization potential** || Ionisationspotential *f* || potentiel *m* d'ionisation || ionisatiepotentiaal *m* || потенциал *m* ионизации, ионизационный потенциал *m*

1414 **ionized atom** || ionisiertes Atom *n* || atome *m* ionisé || geioniseerd atoom *n* || иониз(ир)ованный атом *m*

1415 **ionized gas** || ionisiertes Gas *n* || gaz *m* ionisé || geioniseerd gas *n* || иониз(ир)ованный газ *m*

1416 **ionizing radiation** || ionisierte Strahlung *f* || radiation *f* ionisante || ionisatiestraling *f* || иониз(ир)ующее излучение *n*

1417 **ionosphere** || Ionosphäre *f* || ionosphère *f* || ionosfeer *f* (*m*) || ионосфера *f*

1418 **ion temperature** || Ionentemperatur *f* || température *f* ionique || ionentemperatuur *f* || ионная температура *f*

1419 **iris (diaphragm)** || Irisblende *f* || (diaphragme-)iris *f* || irisdiafragma *n* || ирисовая диафрагма *f*

1420 **iron,** Fe || **1.** Eisen *n*; Roheisen *n*; Gußeisen *n* **2.** *Chem.* Eisen *n* || **1.** fer *m*; fonte *f* **2.** *chim.* fer *m* || **1.** ijzer *n*; gietijzer *n* **2.** *chem.* ijzer *n* || **1.** железо *n*; чугун *m* **2.** *хим.* железо *n*

1421 **iron core** || Eisenkern *m* || noyau *m* en fer || ijzerkern *f* (*m*) || (магнитный) сердечник *m*, магнитопровод *m*

1422 **irradiance** || Bestrahlungsstärke *f* || éclairement *m* énergétique || stralingssterkte *f* || облучённость *f*, энергетическая освещённость *f*

1423 **irradiation** || **1.** Einstrahlung *f* **2.** Irradiation *f*, Überstrahlung *f* || irradiation *f* || irradiatie *f* || **1.** облучение *m* **2.** иррадиация *f*

1424 **irradiator** || Strahler *m* || irradiateur *m* || straler *m* || облучатель *m*; излучатель *m*

1425 **irrational** || irrational || irrationnel || irrationeel || иррациональный

1426 **irrational numbers** || irrationale Zahlen *f pl* || nombres *m pl* irrationnels || irrationele getallen *n pl* || иррациональные числа *n pl*

1427 **irrational value** || irrationale Größe *f* || valeur *f* irrationnelle || irrationele grootheid *f* || иррациональная величина *f*

1428 **irreducible representation** || irreduzible Darstellung *f* || représentation *f* irréductible || irreducibele voorstelling *f* || неприводимое представление *n*

1429 **irregular wear** || ungleichmäßiger Verschleiß *m* || usure *f* non uniforme || onregelmatige slijtage *f*, ongelijkmatige afslijting *f* || неравномерный износ *m*

1430 **irreversibility** || Irreversibilität *f*, Nichtumkehrbarkeit *f* || irréversibilité *f* || onomkeerbaarheid *f* || необратимость *f* (*напр. в термодинамике*)

1431 **irreversible** || irreversibel, nicht umkehrbar || irréversible || onomkeerbaar || необратимый (*напр. процесс*)

1432 **irreversible cycle** || irreversibler [nicht umkehrbarer] Kreisprozeß *m* || cycle *m* irréversible || onomkeerbare cyclus *m* || необратимый цикл *m*

1433 **irreversible process** || irreversibler [nicht umkehrbarer] Prozeß *m* || procédé *m* irréversible || irreversibel proces *n* || необратимый процесс *m*

1434 **irreversible transformation** || irreversible Umwandlung *f* || transformation *f* irréversible || onomkeerbare transformatie *f* || необратимое превращение *n*

1435 **irrotational flow** || wirbelfreie Strömung *f*, Potentialströmung *f* || écoulement *m* potentiel [sans tourbillons] || wervelvrije stroming *f* || безвихревое [потенциальное] течение *n*

1436 **irrotational vector field** || wirbelfreies Vektorfeld *n* || champ *m* vectoriel irrotationnel || wervelvrij vectorveld *n* || безвихревое векторное поле *n*

1437 **isentropic process** || isentrop(isch)er [adiabatischer] Prozeß *m* || procédé *m* [transformation *f*] isentropique [adiabatique] || isentropisch [adiabatisch] proces *n* || изэнтропический [адиабатический] процесс *m*

1438 **isobar** || **1.** Isobare *f*, Isobarenlinie *f* **2.** Isobar *n* *(Nuklied)* || isobare *f* || isobaar *m* || **1.** изобара *f* **2.** изобар *m* *(нуклид)*

1439 **isobaric** || isobar || isobare || isobaar || изобарный, изобарический

1440 **isobaric nucleus** || isobarer Kern *m* || noyau *m* isobare || isobare kern *f* || изобарное ядро *n*

1441 **isobaric process** || isobarer Prozeß *m*, isobarer Vorgang *m* || procédé *m* [processus *m*] isobare || isobaar proces *n* || изобарический [изобарный] процесс *m*

1442 **isochore** || Isochore *f* || isochore *f* || isochoor *m* || изохора *f*

1443 **isochoric** || isochor || isochore || isochoor || изохорный, изохорический

1444 **isochoric process** || isochorer Prozeß *m*, isochorer Vorgang *m* || procédé *m* [transformation *f*] isochore || isochoor proces *n* || изохорический [изохорный] процесс *m*

1445 **isochrone pendulum** || isochrones Pendel *n* || pendule *m* isochrone || isochroon pendel *n* || изохронный маятник *m*

1446 **isochronism** || Isochronie *f*, Isochronismus *m* || isochronisme *m* || isochronisme *n* || изохронность *f*

1447 **isochronous vibrations** || isochrone Schwingungen *f pl* || oscillations *f pl* isochrones || isochrone trillingen *f pl* || изохронные колебания *n pl*

1448 **isocline** || Isokline *f* || isocline *f* || isocline *f* || изоклина *f*

1449 **isoelectronic** || isoelektronisch || isoélectronique || iso-elektronisch || изоэлектронный

1450 **isoelectronic sequence** || isoeletkronische Reihe *f* || série *f* isoélectronique || iso-elektronische reeks *f* (*m*) || изоэлектрический ряд *m*, изоэлектронная последовательность *f*

1451 **isoenergetic** || isoenergetisch || isoénergétique || iso-energetisch || изоэнергетический

1452 **isogonal mapping** || isogonale [winkeltreue] Abbildung *f*, konforme Abbildung *f* **2.** Art *f* || application *f* isogonale || isogone [conforme] afbeelding *f* || изогональное отображение *n*, конформное отображение *n* второго рода

1453 **isogonal trajectory** || isogonale Bahn *f* || trajectoire *f* isogonale || isogone baan *f* (*m*) || изогональная траектория *f*

1454 **isolated point** || isolierter Punkt *m* || point *m* isolé || geïsoleerd punt *n* || изолированная точка *f*

1455 **isolated set** || isolierte Menge *f* || ensemble *m* isolé || geïsoleerde menigte *f* || изолированное множество *n*

1456 **isolated singularity** || isolierte Singularität *f* || singularité *f* isolée, point *m* singulier isolé || geïsoleerd singulier punt *n* || изолированная особая точка *f*

1457 **isolation** || Isolation *f*, Isolierung *f* || isolation *f* || isolatie *f* || изоляция *f*, изолирование *n*

1458 **isolator** || Isolator *m*, Nichtleiter *m* || isolateur *m* || isolator *m* || (электрический) изолятор *m*

1459 **isoline** || Isolinie *f*, Linie *f* gleicher Werte || isoligne *f*, ligne *f* d'égale valeur || lijn *f* (*m*) van gelijke waarde || изолиния *f*

1460 **isomeric nucleus** || isomerer Kern *m* || noyau *m* isomérique || isomere kern *f* || изомерное ядро *n*

1461 **isomeric transition** || isomerer Übergang *m*, Isomerenübergang *m* || transition *f* isomérique || isomere overgang *m* || изомерный переход *m*

1462 **isomerism** || Isomerie *f* || isomérie *f* || isomerie *f* || изомерия *f*

1463 **isomerization** || Isomerisierung *f*, Isomerisation *f* || isomérisation *f* || isomerisatie *f* || изомеризация *f*

1464 **isometric mapping** || isometrische Abbildung *f* || application *f* isométrique || isometrische afbeelding *f* || изометрическое отображение *n*

1465 **isometric spaces** || isometrische Räume *m pl* || espaces *m pl* isométriques || isometrische ruimten *f pl* || изометричные пространства *n pl*

1466 **isometric surfaces** || isometrische Oberflächen *f pl* || surfaces *f* isométriques || isometrische vlakken *n pl* || изометричные поверхности *f pl*

1467 **isomorphic** || isomorph || isomorphe || isomorf || изоморфный

1468 **isomorphism** || Isomorphie *f*, Isomorphismus *m* || isomorphisme *m* || isomorfisme *n* || изоморфизм *m*

1469 **isoperimetric problem** || isoperimetrisches Problem *n* || problème *m* isopérimétrique || isoperimetrisch probleem *n* || изопериметрическая задача *f*

1470 **isosceles triangle** || gleichschenkliges Dreieck *n* || triangle *m* isocèle || gelijkbenige driehoek *m* || равнобедренный треугольник *m*

1471 **isospin** || Iso(topen)spin *m*, isotoper Spin *m* || spin *m* isotopique || isotopenspin *m* || изоспин *m*, изотопический спин *m*

1472 **isostasy** || Isostasie *f*, Massenkompensation *f* || isostasie *f* || isostasie *f* || изостазия *f*

1473 **isotactic polymer** || isotaktisches Polymer(es) *n* || polymère *m* isotactique || isotactisch polymeer *n* || изотактический полимер *m*

1474 **isotherm** || Isotherme *f*, isothermische Kurve *f* || isotherme *f* || isotherm *m* || изотерма *f*

1475 **isothermal** || isotherm || isotherme || isotherm(isch) || изотермический

1476 **isothermal compression** || isotherme Kompression *f* || compression *f* isothermique || isotherme compressie *f* || изотермическое сжатие *n*

1477 **isothermal expansion** || isotherme Expansion *f* || expansion *f* isotherme || isotherme expansie *f* || изотермическое расширение *n*

1478 **isothermal process** || isotherme Zustandsänderung *f* || procédé *m* [transformation *f*] isotherme || isotherme toestandsverandering *f*, isotherm proces *n* || изотермический процесс *m*

1479 **isothermic** *see* **isothermal**

1480 **isotone** || Isoton *n* || isotone *m* || isotoon *n* || изотон *m*

1481 **isotope** || Isotop *n* || isotope *m* || isotoop *n* || изотоп *m*

1482 **isotope separation** || Isotopentrennung *f* || séparation *f* d'isotopes || isotopenscheiding *f* || разделение *n* изотопов

1483 **isotope shift** || Isotopieverschiebung *f*, Isotopenverschiebung *f* || déplacement *m* isotope || isotopenverschuiving *f* || изотопическое смещение *n*

1484 **isotopic spin** || Iso(topen)-spin *m* || spin *m* isotope || isotopenspin *m* || изотопический спин *m*, изоспин *m*

1485 **isotopy** || Isotopie *f* || isotopie *f* || isotopie *f* || изотопия *f*

1486 **isotropic** || isotrop || isotrope || isotroop || изотропный

1487 **isotropic surface** || isotrope Oberfläche *f* || surface *f* isotrope || isotroop vlak *n* || изотропная поверхность *f*

1488 **isotropy** || Isotropie *f* || isotropie *f* || isotropie *f* || изотропия *f*

1489 **isovolumic process** || isochore Zustandsänderung *f* || procédé *m* [transformation *f*] isochore || isochore toestandsverandering *f*, isochoor proces *n* || изохорический [изохорный] процесс *m*

I490 **iterated-interpolation method** || iteratives Interpolationsverfahren *n* || méthode *f* d'interpolation itérative || iteratieve interpolatiemethode *f* || итерационно-интерполяционный метод *m*

I491 **iterated kernel** || iterierter Kern *m* || noyau *m* itéré || geïtereerde kern *f* || итерированное ядро *n*

I492 **iteration** || Iteration *f* || itération *f* || iteratie *f* || итерация *f*

I493 **iteration methods** || Iterationsverfahren *n pl*, Iterationsmethoden *f pl* || méthodes *f pl* par itérations || iteratiemethodes *f pl* || итерационные методы *m pl*

I494 **Izod test** || (Izod-)Kerbschlagprüfung *f* || essai *m* d'Isod || kerfslagproef *f* (*m*) van Izod || определение *n* ударной вязкости по Изоду, испытание *n* на ударную вязкость

J

J1 **jack** || Winde *f*, Hebebock *m* || cric *m*, vérin *m* || dommekracht *f* (*m*); windas *n* || домкрат *m*

J2 **jacket** || Mantel *m* || enveloppe *f*, chemise *f* || mantel *n* || кожух *m*; чехол *m*; рубашка *f* (*цилиндра*)

J3 **jacketed cylinder** || Wassermantelzylinder *m* || cylindre *m* à chemise d'eau || watermantelcilinder *m*, *f* || цилиндр *m* с водяной рубашкой

J4 **jackscrew** || 1. Schraubenwinde *f* 2. Hebespindel *f* || 1. cric *m* [vérin *m*] à vis 2. vis *f* de commande du cric du vérin || schroefvijzel *f* (*m*) || 1. винтовой домкрат *m* 2. ходовой винт *m* домкрата

J5 **Jacobian** || Funktionaldeterminante *f*, Jakobische Determinante *f* || jacobien *m* || Jacob-determinant *m* || якобиан *m*

J6 **jaw** || 1. Kloben *m*; Backe *f*, Klobenbacke *f* 2. Klaue *f* || 1. étau *m*; mâchoire *f* 2. came *f* || 1. bek *f*, klauwplaat *f* (*m*) 2. klauw *m*, *f* || 1. тиски *pl*; щека *f* (*тисков*); зажимное приспособление *n* 2. кулачок *m*

J7 **jaw chuck** || Backenfutter *n* || mandrin *m* à mâchoires || klauwplaat *f* (*m*) || кулачковый патрон *m*

J8 **jaw clutch** || Backenkupplung *f* || embrayage *m* à mâchoires || klauwkoppeling *f* || кулачковая муфта *f*

J9 **jet** || 1. Düse *f*; Brennstoffdüse *f*; Vergaserdüse *f* 2. Strahl *m* 3. Strahltriebwerk *n* || 1., 2. jet *m* 3. moteur *m* à réaction, réacteur *m* || 1. sproeier *m*; broes *f* (*m*) 2. straal *n* 3. straalandrijving *f* || 1. форсунка *f*; сопло *n*; жиклёр *m* 2. струя *f*; реактивная струя *f* 3. реактивный двигатель *m*

J10 **jet engine** || Strahltriebwerk *n* || moteur *m* à réaction, réacteur *m* || straalmotor *m* || реактивный двигатель *m*

J11 **jet flow** || Strahlströmung *f* || jet *m* d'air, jet-stream *m* || straalstroom *m* || струйное течение *n*

J12 **jet-propelled** || strahlgetrieben, Strahl..., Düsen... || à réaction || straal... || с реактивной тягой, реактивный

J13 **jet propulsion** || 1. Strahlantrieb *m* 2. Schub *m* || 1. propulsion *f* par réaction [à jet] 2. poussée *f* de réaction || 1. straalbeweging *f* 2. straalvoortstuwingskracht *f* (*m*) || 1. реактивное движение *n* 2. реактивная тяга *f*

J14 **jet pump** || Strahlpumpe *f* || pompe *f* à jet || straalpomp *f* (*m*) || струйный насос *m*

J15 **jet stream** || 1. Strahlströmung *f*, Strahlstrom *m* 2. Schubstrahl *m* || 1. jet-stream *m* 2. jet *m* de propulsion || 1. straalstroming *f* 2. straalstroom *m* || 1. струйное течение *n* 2. реактивная струя *f*

J16 **jet thrust** || Schub *m*; Impuls *m* || jet *m* de propulsion || straaldrijfkracht *f* || реактивная сила *f*, реактивная тяга *f*

J17 **jib crane** || Auslegerkran *m*, Schwenkkran *m* || grue *f* à flèche || kraan met arm *m*, giekkraan *m* || (самоходный) стреловой кран *m*

J18 **jig** || Spannvorrichtung *f*; Bohrschablone *f* || guide-outil *m*;

dispositif *m* de serrage || boorsjabloon *f* (*m*) || зажимное приспособление *n*; кондуктор *m*

J19 **jig boring machine** || Lehrenbohrmaschine *f* || machine *f* à aléser par coordonnées || sjablonenboormachine *f*; matrijsboormachine *f* || координатно-расточный станок *m*

J20 **job** || Job *m*, Auftrag *m* || travail *m* || opgave *f* || задание *n*; работа *f*

J21 **job control language** || Jobsteuersprache *f* || langage *m* de commande de travaux || jobbesturingstaal *f* || язык *m* управления заданиями

J22 **jockey pulley** || Leitrolle *f* || poulie *f* de tension || spanrol *f* (*m*) || натяжной ролик *m*

J23 **joint** || **1.** Verbindung *f*, Anschluß *m*; Knoten(punkt) *m* **2.** Gelenk *n* || **1.** joint *m*; raccord *m* **2.** articulation *f* || **1.** verbinding *f*; koppeling *f*; voeg *f* (*m*) **2.** scharnier *n* || **1.** соединение *n*; шов *m*, стык *m*; узел *m* **2.** шарнир *m*

J24 **joint distribution** || gemeinsame mehrdimensionale Verteilung *f* || distribution *f* jointe || gemeenschappelijke verdeling *f* || совместное распределение *n*

J25 **joint pin** || Gelenkbolzen *m* || essieu *m* du joint articulé; essieu *m* d'articulation || scharnierbout *m* || палец *m* шарнира; шарнирный болт *m*

J26 **Jordan curve** || Jordan-Kurve *f* || courbe *f* de Jordan || Jordankromme *f* (*m*) || кривая *f* Жордана, Жорданова кривая

J27 **Jordan's test** || Jordansches Konvergenzkriterium *n* || test *m* de Jordan, Jordan-test *m* || признак *m* Жордана *(сходимости рядов Фурье)*

J28 **joule,** J || Joule *n* || joule *m* || joule *m* || джоуль *m*, Дж

J29 **Joule heat** || Joulesche Wärme *f* || chaleur *f* (de) Joule || joule-warmte *f* || джоулево тепло *n*

J30 **Joule heating** || ohmsche Heizung *f*, Joulesche Aufheizung *f* || chauffage *m* ohmique [par résistance] || ohmse verwarming *f* || омический нагрев *m*

J31 **Joule's equivalent** || mechanisches Wärmeäquivalent *n* || équivalent *m* mécanique de chaleur || mechanisch warmteëquivalent *n* || механический эквивалент *m* тепла

J32 **journal** || Zapfen *m*, Wellenzapfen *m* || portée *f* (d'arbre); tourillon *m* || ashals *m*; astap *m* || цапфа *f*, шейка *f*, шип *m (вала)*

J33 **journal bearing** || Gleitlager *n* || palier *m* lisse [de glissement] || glijlager *n* || подшипник *m* скольжения

J34 **journal box** || Buchse *f*; Lagerbuchse *f* || boîte *f* (d'essieu); manchon *m*; bague *f* (de coussinet) || bus *f* (*m*); lagerbus *f* (*m*); lager *n* || букса *f*; втулка *f*; подшипник *m*

J35 **jump** || **1.** Sprung *m* **2.** sprunghafte Änderung *f* || saut *m* || sprong *m* || **1.** скачок *m* (величины) **2.** переход *m*, передача *f* управления

J36 **jump function** || Sprungfunktion *f* || fonction *f* de saut || sprongfunctie *f* || функция *f* скачков

J37 **jump instruction** || Sprungbefehl *m* || instruction *f* de saut || spronginstructie *f* || команда *f* перехода, команда *f* передачи управления

J38 **junction** || Übergang *m*, Halbleiterübergang *m* || jonction *f* || overgang *m* || (полупроводниковый) переход *m*

K

K1 **Kantor section** || Kantorscher Schnitt *m* || section *f* de Cantor || Kantor-snede *f* (*m*) || канторово сечение *n*

K2 **kaon** || Kaon *n*, K-Meson *n* || kaon *m* || kaon *n* || каон *m*, K-мезон *m*

K3 **Kater's pendulum** || Katersches Pendel *n*, Reversionspendel *n* || pendule *m* réversible || Kater-pendel *n* || оборотный маятник *m*

K4 **kayser** || Kayser *n* || kayser *m* || kayser *n* || кайзер *m* ($см^{-1}$)

K5 **kelvin** || Kelvin *n* || kelvin *m* || kelvin *n* || кельвин *m (единица температуры в СИ)*

K6 **Kelvin balance** || Stromwaage *f* nach Kelvin, Kelvin-Stromwaage *f* || balance *f* électrodynamique || Kelvin-balans *f* (*m*) || токовые весы *pl*, весы *pl* Кельвина

K7 **Kelvin temperature** || absolute Temperatur *f*, Kelvin-Temperatur *f* || température *f* absolue thermodynamique || absolute temperatuur *f* || абсолютная термодинамическая температура *f*

K8 **kernel math** || Kern *m* || noyau *m* || kern *f* (*m*) || ядро *n*

K9 **Kerr cell** || Kerr-Zelle *f* || cellule *f* de Kerr || Kerr-cel *f* (*m*) || ячейка *f* Керра *(оптический затвор)*

K10 **ketone** || Keton *n* || cétone *f* || keton *n* || кетон *m*

K11 **key** || **1.** Keil *m*; Feder *f* **2.** Schlüssel *m* || **1.** clavette *f*; cale *f* **2.** clef *f*, dé *m* || **1.** spie *f* (*m*); **2.** sleutel *m* || **1.** шпонка *f*; клин *m* **2.** ключ *m*

K12 **keyboard computer** || Tastenrechner *m* || calculateur *m* à clavier || toetsenbordcomputer *m* || клавишная вычислительная машина *f*

K13 **key diagram** || Prinzipschema *n*, Prinzipschaltung *f* || schéma *m* de principe || principeschema *n*, principeschakeling *f* || принципиальная схема *f*

K14 **keyseat(ing)** || Keilnut *f* || rainure *f* [cannelure *f*] de clavetage || spiebaan *f* (*m*) || шпоночный паз *m*

K15 **keyseating milling machine** || Langlochfräsmaschine *f*, Keilnutenfräsmaschine *f* || machine *f* à fraiser les rainures (de clavetage) || spiesleuffreesmachine *f* || шпоночно-фрезерный станок *m*

K16 **keyway** *see* **keyseat(ing)**

K17 **keyway cutting machine** || Keilnutenstoßmaschine *f* || mortaiseuse *f* à rainer [à rainurer], mortaiseuse-raineuse *f* || spiesleufsteekmachine *f* || шпоночно-долбёжный станок *m*

K18 **keyway milling machine, keyway seating machine** *see* **keyseating milling machine**

K19 **keyway stotting machine** *see* **keyway cutting machine**

K20 **keyword** || Schlüsselwort *n* || mot *m* de clé || sleutelwoord *n* || ключевое слово *n*

K21 **Khintchine's theorem** || Chintschinscher Satz *m* || théorème *m* de Khintchine || Khintchine-theorema *n* || теорема *f* Хинчина

K22 **kilo...** || Kilo... || kilo... || kilo... || кило... *(приставка в наименовании кратной единицы, соответствующая* 10^3*)*

K23 **kilogram** || Kilogramm *n* || kilogramme *m* || kilogram *n* || килограмм *m*

K24 **kilogram-force** || Kilopond *n* || kilogramme-force *m* || kilogramkracht *f* (*m*) || килограмм-сила *f*, кгс *(9,80665 Н)*

K25 **kilowatt-hour** || Kilowattstunde *f* || kilowatt-heure *m* || kilowattuur *n* || киловатт-час *m*, квт·ч *(*$3{,}6\cdot10^6$ *Дж)*

K26 **kinematic chain** || kinematische Kette *f* || chaîne *f* cinématique || kinematische keten *f* (*m*) || кинематическая цепь *f*

K27 **kinematic pair** || kinematisches Paar *n* || paire *f* cinématique || kinematisch paar *n* || кинематическая пара *f*

K28 **kinematics** || Kinematik *f*, Bewegungslehre *f* || cinématique *f* || kinematica *f* || кинематика *f*

K29 **kinematic viscosity** || kinematische Viskosität *f* || viscosité *f* cinématique || kinematische viscositeit *f* || кинематическая вязкость *f*

K30 **kinetic** || kinetisch || cinétique || kinetisch || кинетический

K31 **kinetic energy** || kinetische Energie *f*, Bewegungsenergie *f* || énergie *f* cinétique || kinetische energie *f* || кинетическая энергия *f*

K32 **kinetic friction** || Bewegungsreibung *f* || frottement *m* cinétique || kinetische wrijving *f*, bewegingswrijving *f* || кинетическое трение *n*

K33 **kinetic gas theory** || kinetische Theorie *f* der Gase || théorie *f* cinétique des gaz || kinetische gastheorie *f* || кинетическая теория *f* газов

K34 **kinetic head** || Geschwindigkeitshöhe *f*, Stauhöhe *f* || pression *f* cinétique || stuwdruk *m* || скоростной [динамический] напор *m*

K35 **kinetic heating** || aerodynamische Erwärmung *f* || réchauffement *m* aérodynamique || aërodynamische verwarming *f* || аэродинамический нагрев *m*

K36 **kinetic moment(um)** || kinetischer Impuls *m*, Schwung *m* || moment *m* cinétique || kinetisch moment *n* || кинетический момент *m*, количество *n* движения, импульс *m*

K37 **kinetics** || Kinetik *f* || cinétique *f* || kinetica *f* || кинетика *f*

K38 **kinetostatics** || Kinetostatik *f* || cinétostatique *f* || kinetostatica *f* || кинетостатика *f*

K39 **Kjeldahl's method** || Kjeldahl-Bestimmung *f* || méthode *f* de Kjeldahl || Kjeldahl-methode *f* || способ *m* Кельдаля

K40 **knee type milling machine** || Konsolfräsmaschine *f* || fraiseuse *f* à console || consolefreesmachine *f* || консольно-фрезерный станок *m*

K41 **knife-edge diffraction** || Kantenbeugung *f*, Beugung *f* an der Kante || diffraction *f* par un bord || breking *f* aan de rand || дифракция *f* на (остром) крае

K42 **knock-on collision** || Frontalzusammenstoß *m* || collision *f* frontale || frontale collisie *f* || прямое [лобовое] столкновение *n*

K43 **knock-on electron** || Anstoßelektron *n* || électron *m* d'éjection || uitgestoten elektron *n* || выбитый электрон *m*

K44 **knockout** Herausschlagen *n* *(Teilchen)* || dégagement *m* || uitstoting *f* || выбивание *n* *(частицы)*

K45 **knuckle joint** || Gabelgelenk *n* || articulation *f* à chape || gaffelgewricht *n* || шарнирное соединение *n* вилкой, вильчатый шарнир *m*

K46 **knuckle thread** || Rundgewinde *n* || filet *m* demi-rond || ronde *m* schroefdraad *m* || резьба *f* полукруглого профиля

K47 **knurling** || Rändeln *n*; Kordeln *n* || moletage *m* || kartelen *n* || накатывание *n*

K48 **knurling tool** || Rändelrolle *f*; Kordel *f* || molette *f* || kartelrol *f* (*m*) || (резьбо)накатный ролик *m*

K49 **Kolmogorov's test** || Kolmogorowsches Kriterium *n* || test *m* de Kolmogorov || Kolmogorov-test *m* || критерий *m* Колмогорова

K50 **Kotelnikov sampling theorem** || Abtasttheorem *n* || théorème *m* d'échantillonnage de Kotelnikov || Kotelnikov-theorema *n* || теорема *f* о дискретном представлении, теорема *f* Котельникова

L

L1 **label** || Etikett *n*, Marke *f* *(in einem Programm, einem Graphen)* || label *m* || etiket *n* || метка *f* *(в программе, графе)*

L2 **labeling** || Markierung *f* *(eines Graphen)* || étiquetage *m* || markering *f* || разметка *f* *(графа)*

L3 **labelled atom** || markiertes Atom *n*, Tracer *m* || atome *m* marqué || gemerkt atoom *n* || меченый атом *m*

L4 **labelled graph** || markierter Graph *m* || graphe *m* étiqueté || gemarkeerde graaf *f* (*m*) || помеченный граф *m*

L5 **laboratory frame** *see* **laboratory system**

L6 **laboratory system** || Labor(koordinaten)system *n*, L-System *n* || système *m* de référence au laboratoire || laboratorium(referentie)systeem *n* || лабораторная система *f* отсчёта

L7 **laboratory test** || Laborversuch *m* || essai *m* de laboratoire || laboratoriumtoets *m* || laboratoriumproef *f* (*m*) || лабораторное испытание *n*

L8 **labyrinth seal** || Labyrinthdichtung *f* || joint *m* en labyrinthe || labyrint(af)dichting *f* || лабиринтное уплотнение *n*

L9 **lack of balance** || Unbalance *f*; Unwucht *f* || déséquilibre *m* || onbalans *f* (*m*) || неуравновешенность *f*

L10 **lag** || zeitliche Verzögerung *f*, Zeitverzögerung *f*, Verzug *m*, (zeitliches) Nacheilen *n* || retard *m* || vertraging *f* || отставание *n*, запаздывание *n*; инерционность *f*

L11 **Lagrange's equations** || Lagrangesche Gleichung *f*, Lagrangesche Schlüsselgleichung *f* || équations *f pl* lagrangiennes || Lagrange-vergelijkingen *f pl* || уравнения *n pl* Лагранжа

L12 **Lagrangian function** || Lagrange-Funktion *f* || fonction *f* lagrangienne || Lagrange-functie *f* || функция *f* Лагранжа

L13 **laminar flow** || laminare Strömung *f*, Laminarströmung *f*, Laminarbewegung *f* || écoulement *m* laminaire || laminaire stroming *f* || ламинарное течение *n*

L14 **laminarity** || laminares Verhalten *n* || laminarité *f* || laminariteit *f* || ламинарность *f* *(течения)*

L15 **lantern pinion** || Triebstockrad *n* || roue *f* à la lanterne [à fuseaux] || lantaarnwiel *n* || цевочное колесо *n*

L16 **lanthanum, La** || Lanthan *n* || lanthane *m* || lanthaan *n*, lanthanium *n* || лантан *m*

L17 **lanthanum compound** || Lanthanid *n*, Lanthanoid *n* || lanthanide *m* || lanthanide *n* || лантанид *m*, лантаноид *m*

L18 **lap joint** || Überlappstoß *m* || joint à recouvrement || lapverbinding *f* || соединение *n* внахлёстку

L19 **Laplace transform(ation)** || Laplace-Transformation *f* || transformation *f* laplacienne || Laplace-transformatie *f* || преобразование *n* Лапласа

L20 **Laplacian operator** || Laplace-Operator *m*, Laplacescher Operator *m* || opérateur *m* laplacien, laplacien *m* || Laplace-operator *m* || лапласиан *m*, оператор *m* Лапласа

L21 **lapper** *see* **lapping machine**

L22 **lapping** || Läppen *n* || lapping *m*; rodage *m* (manuel) || lappen *n* || лаппингование *n*, доводка *f*

L23 **lapping compound** || Läppaste *f* || pâte *f* à roder || slijppasta *m* || притир *m*, абразивная паста *f*

L24 **lapping machine** || Läppmaschine *f* || rodeuse *f* || lapmachine *f* || доводочный [притирочный] станок *m*

L25 **lap weld** || Überlapp(schweiß)-naht *f* || soudure *f* à clin [à recouvrement] || overlapnaad *m* || нахлёсточный сварной шов *m*; нахлёсточное сварное соединение *n*

L26 **large calorie** || Kilokalorie *f* || kilocalorie *f*, grande calorie *f* || kilocalorie *f* || килокалория *f* *(4,1868 кДж)*

L27 **large-scale integration circuit** || Großintegrationsschaltung *f* || circuit *m* intégré à large échelle || LSI-schakeling *f* || схема *f* с большой степенью интеграции, большая интегральная схема *f*, БИС

L28 **Larmor precession** || Larmor-Präzession *f*, Larmorsche Präzession *f* || précession *f* de Larmor || Larmor-precessie *f* || ларморова прецессия *f*

L29 **laser** || Laser *m*, optischer Maser *m* || laser *m*, maser *m* optique || laser *m* || лазер *m*

L30 **laser beam** || Laserstrahl *m* || faisceau *m* de laser || laserbundel *m* || лазерный луч *m*

L31 **latent** || latent || latent || latent || латентный, скрытый

L32 **latent heat** || latente Wärme *f*, Umwandlungswärme *f* || chaleur *f* latente || latente warmte *f* || удельная скрытая теплота *f*

L33 **latent image** || latentes Bild *n* || image *f* latente || latent beeld *n* || скрытое изображение *n*

L34 **lateral axis** || Querachse *f* || axe *m* transversal || dwarsas *f* (*m*) || поперечная ось *f*

L35 **lateral contraction** || Querzusammenziehung *f*, Querkontraktion *f* || contraction *f* latérale [transversale] || dwarscontractie *f* || поперечное сжатие *n*

L36 **lateral force** || Querkraft *f* || force *f* transversale || dwarskracht *f* (*m*) || поперечная сила *f*

L37 **lateral inversion** || Seitenvertauschung *f* des Bildes || inversion *f* latérale || zijïnversie *f* || зеркальное горизонтальное обращение *n* изображения

L38 **lateral oscillations** || Transversalschwingungen *f pl*, Querschwingungen *f pl* || oscillations *f pl* transversales || dwarstrillingen *f pl* || поперечные колебания *n pl*

L39 **lateral strain** || Querdeformation *f* || déformation *f* transversale || dwarsvervorming *f* || поперечная деформация *f*

L40 **lateral wave** || Seitenwelle *f* || onde *f* transversale || dwarsgolf *f* (*m*) || поперечная волна *f*

L41 **lathe** || Drehmaschine *f* || tour *m* || draaibank *f* (*m*) || токарный станок *m*

L42 **lattice** || Gitter *n*; Kristallgitter *n* || **1.** grille *f*; treillis *m* **2.** réseau *m* cristallin *n* || **1.** rooster *n* **2.** kristalrooster *n* || **1.** решётка *f* **2.** (кристаллическая) решётка *f*

L43 **lattice cell** || Gitterzelle *f* || cellule *f* cristallographique || roostercel *f* (*m*) || элементарная ячейка *f* *(кристалла)*

L44 **lattice defect** || Gitter(bau)fehler *m*, Gitterfehlstelle *f* || défaut *m* de réseau || roosterdefect *n* || дефект *m* кристаллической решётки

L45 **lattice energy** || Gitterenergie *f* || énergie *f* réticulaire || roosterenergie *f* || энергия *f* (кристаллической) решётки

L46 **lattice point, lattice site** || Gitterpunkt *m* || point *m* de réseau || roosterpunt *f* (*m*) || узел *m* (кристаллической) решётки

L47 **lattice structure** || Gitterstruktur *f*, Gitter(auf)bau *m* || structure *f* réticulaire du réseau || roosterstructuur *f* || структура *f* (кристаллической) решётки

L48 **lattice vibrations** || Gitterschwingungen *f pl* || oscillations *f pl* du réseau || roostertrillingen *f pl* || колебания *n pl* (кристаллической) решётки

L49 **Laue pattern** || Laue-Diagramm *n* || diagramme *m* de Laue || Laue-diagram *n* || диаграмма *f* Лауэ, лауэграмма *f*

L50 **Laurent series** || Laurent-Reihe *f* || série *f* de Laurent || reeks *f* (*m*) van Laurent || ряд *m* Лорана

L51 **law of assertion** || Behauptungssatz *m* || loi *f* d'assertion || beweringswet *f* (*m*) || закон *m* утверждения

L52 **law of conservation** || Erhaltungssatz *m*, Erhaltungsgesetz *n* || loi *f* de conservation || behoudswet *f* (*m*) || закон *m* сохранения

L53 **law of cosines** || Kosinussatz *m* || théorème *m* des cosinus || cosinustheorema *n* || теорема *f* косинусов

L54 **law of large numbers** || Gesetz *n* der großen Zahlen || loi *f* des grands nombres || wet *f* (*m*) van de grote getallen || закон *m* больших чисел

L55 **law of mass action** || Massenwirkungsgesetz *n* || loi *f* d'action de masse || wet *f* (*m*) van de massawerking || закон *m* действующих масс

L56 **law of motion** || Bewegungsgesetz *n* || loi *f* de mouvement || bewegingswet *f* (*m*) || закон *m* движения, закон *m* Ньютона

L57 **law of sines** || Sinussatz *m* || théorème *m* des sinus || sinustheorema *n* || теорема *f* синусов

L58 **law of the excluded middle** || Satz *m* vom ausgeschlossenen Dritten || loi *f* du tiers exclu || wet *f* (*m*) van de uitgesloten derde || закон *m* исключённого третьего

L59 **layer** || Schicht *f*, Überzugsschicht *f*; Beschichtung *f* || couche *f* || laag *f* || слой *m*

L60 **lay shaft** || Zwischenwelle *f* || arbre *m* intermédiaire || hulpas *f* (*m*), tussenas *f* (*m*) || промежуточный вал *m*

L61 **lead** || Anschluß *m*; Anschlußdraht *m* || fil *m* de connexion || aansluitdraad *m* || ввод *m* или вывод *m* *(схемы)*; провод *m*

L62 **lead, Pb** || Blei *n* || plomb *m* || lood *n* || свинец *m*

L63 **lead angle** *see* **helix angle**

L64 **lead bronze** Bleibronze *f* || bronze *f* au plomb || loodbrons *n* || свинцовая бронза *f*

L65 **leader** || Leader *m* || leader *m* || leader *m* || лидер *m* (разряда)

L66 **lead screw** *see* **feed screw**

L67 **lead wire** || Leitung *f* unter Last || conducteur *m* sous charge || leider *m* onder belasting || провод *m* под нагрузкой

L68 **leaf** || Blatt *n* || **1.** feuille *f* **2.** *math.* folium *m* || blad *n* || **1.** листок *m* *(электроскопа)* **2.** *мат.* лист *m*

L69 **leaf of Deskartes** || Cartesisches [Descartessches] Blatt *n* || folium *m* de Descartes || Descartes-blad *n* || декартов лист *m*

L70 **leaf spring** || Blattfeder *f* || ressort *m* à feuilles || bladveer *f* (*m*) || листовая рессора *f*

L71 **leak** || **1.** Leck *n*, Leckage *f* **2.** Leckstelle *f*, Undichtigkeit(sstelle) *f* || fuite *f* || lek *n* || **1.** утечка *f* **2.** негерметичность *f*

L72 **leakage** || Leckage *f*; Verlust *m* || fuite *f* || lek *n* || утечка *f*

L73 **leakage current** || Verluststrom *m*, Fehlstrom *m*, Leckstrom *m*, Ableit(ungs)strom *m* || courant *m* de fuite || lekstroom *m* || ток *m* утечки

L74 **leak-free** || hermetisch, (hermetisch) dicht, luftdicht || étanche, hermétique || hermetisch, dicht || герметический, герметичный

L75 **least squares approximation** || Approximation *f* nach der Methode der kleinsten Quadrate || approximation *f* par moindres carrés || benadering *f* volgens de methode van de kleinste kwadraten || приближение *n* по методу наименьших квадратов

L76 **least squares method** || Methode *f* der kleinsten Quadrate || méthode *f* des moindres carrés || methode *f* van de kleinste kwadraten || метод *m* наименьших квадратов

L77 **least time principle** || Prinzip *n* der kürzesten Zeit || principe *m* du moindre temps || principe *n* van de kortste tijd || принцип *m* наименьшего времени (Ферма)

L78 **least upper bound** || Supremum *n* || borne *f* supérieure (la plus petite) || kleinste bovenkant *m* || верхняя точная грань *f*

L79 **least work principle** || Prinzip *n* der kleinsten Arbeit || principe *m* de travail minimal || principe *n* van de kleinste arbeid || принцип *m* наименьшей работы

L80 **Lebesgue convergence theorem** || Lebesguescher Konvergenzsatz *m* || théorème *m* de convergence de Lebesgue || convergentietheorema *n* van Lebesgue || теорема *f* сходимости Лебега

L81 **Lebesgue integral** || Lebesguesches Integral *n* || intégrale *f* de Lebesgue || integraal *f* (*m*) van Lebesgue || интеграл *m* Лебега

L82 **Lebesgue integration** || Integration *f* im Lebesgueschen Sinne || intégration *f* lebesguienne || Lebesgue-integratie *f* || интегрирование *n* по Лебегу

L83 **Lebesgue measure** || Lebesguesches Maß *n* || mesure *f* lebesguienne || Lebesgue-maat *f* (*m*) || лебегова мера *f*

L84 **Lebesgue-Stieltjes integral** || Lebesgue-Stieltjessches Integral *n* || intégrale *f* de Stieltjes-Lebesgue || integraal *f* (*m*) van Lebesgue-Stieltjes || интеграл *m* Лебега—Стилтьеса

L85 **left-hand derivative** || linksseitige Ableitung *f* || dérivée *f* à gauche || linkerzijde afgeleide *f* || производная *f* слева, левая производная *f*

L86 **left-handed coordinate system** || Linkssystem *n*, linkshändiges Koordinatensystem *n* || système *m* de coordonnées à gauche || linker coördinatenstelsel *n* || левая система *f* координат

L87 **left-hand rule** || Linke-Hand-Regel *f*, Linkehandregel *f* || règle *f* de la main gauche || linkerhandregel *m* || правило *n* левой руки, правило *n* «мотора»

L88 **left-hand screw** || Linksgewindeschraube *f* || vis *f* sinistrorsum; vis *f* à filet gauche || linkse schroef *f* (*m*) || винт *m* с левой резьбой

L89 **left-hand thread** || Linksgewinde *n* || filet *m* à gauche || linkse schroefdraad *m* || левая резьба *f*

L90 **Legendre function** || Legendresche Funktion *f* || fonction *f* de Legendre || Legendre-functie *f* || функция *f* Лежандра

L91 **Legendre polynomials** || Legendresche Polynome *n pl* || polynômes *m pl* de Legendre || polynomen *n pl* van Legendre || многочлены *m pl* Лежандра

L92 **lemma** || Lemma *n* || lemme *m* || lemma *n* || лемма *f*

L93 **lemmiscate** || Lemniskate *f* || lemniscate *f* || lemniscaat *f* (*m*) || лемниската *f*

L94 **length** || Länge *f* || longueur *f* || lengte *f* || длина *f*

L95 **length standard** || Längennormal *n* || étalon *m* de longueur || lengtenormaal *f* (*m*) || эталон *m* длины

L96 **lens** || Linse *f* || lentille *f* || lens *f* (*m*) || линза *f*

L97 **lens combination** *see* **lens system**

L98 **lens power** || Brechkraft *f* einer Linse || puissance *f* optique d'une lentille || brekingsvermogen *n* (van een lens) || оптическая сила *f* линзы

L99 **lens system** || Linsensystem *n*, Linsenkombination *f* || système *m* de lentilles || lenzenstelsel *n* || система *f* линз

L100 **lepton** || Lepton *n* || lepton *m* || lepton *n* || лептон *m*

L101 **level** || Niveau *n*, Stand *m*, Höhe(nlage) *f*; Pegel *m* || niveau *m* || niveau *n* || уровень *m*

L102 **level of significance** || Konfidenzzahl *f* || niveau *m* de signification || belangrijkheidsniveau *n* || уровень *m* значимости

L103 **level population** || Besetzung *f* des Niveaus || population *f* des niveaux || bezetting *f* van het (energie)-niveau || заселённость *f* (энергетического) уровня

L104 **lever** || 1. Hebel *m* 2. Griff *m* || levier *m* || 1. hefboom *m* 2. handgreep *f* (*m*) || 1. рычаг *m* 2. рукоятка *f*

L105 **lever arm** Hebelarm *m* || bras *m* de levier || hefboomarm *m* || плечо *n* рычага

L106 **lever jack** || Lastwinde *f*, Hebelwinde *f* || cric *m*, vérin *m* à levier || dommekracht *f* (*m*) || рычажный домкрат *m*

L107 **levitation** || Levitation *f*, (freies) Schweben *n* || lévitation *f* || levitatie *f* || левитация *f*, парение *n*

L108 **Lewis key** || Tangentkeil *m* || clavette *f* tangentielle || tangentiaalwig *f* (*m*) || тангенциальная шпонка *f*

L109 **library (sub)routine** || Bibliotheksprogramm *n* || sous-programme *m* de bibliothèque || bibliotheekprogramma *n* || библиотечная (под)программа *f*

L110 **libration** || Libration *f* || libration *f* || libratie *f* || либрация *f*

L111 **Lie's group** || Liesche Gruppe *f* || groupe *m* de Lie || Lie-groep *f* (*m*) || группа *f* Ли

L112 **life test** || Lebensdauerprüfung *f* || essai *m* de durée (de vie) || levensduurtoets *m* || испытание *n* на долговечность, испытание *n* для определения срока службы

L113 **lifetime** || 1. Lebensdauer *f* 2. Betriebszeit *f*, Betriebsdauer *f* || 1. durée *f* de vie 2. durée *f* de service || levensduur *m* || 1. время *n* [продолжительность *f*] жизни *(частиц, изотопов)* 2. срок *m* службы; долговечность *f*

L114 **lift** *see* **elevator**

L115 **lifting force** || Auftriebskraft *f*, Auftrieb *m* || portance *f*; force *f* de levage || opstuwende kracht *f* (*m*) || подъёмная сила *f*

L116 **lifting pump** || Wasserheberpumpe *f*; Saugpump *f* || pompe *f* élévatoire || hefpomp *f* (*m*) || водоподъёмный насос *m*; всасывающий насос *m*

L117 **lifting tackles** || Hebezug *m*, Flaschenzug *m* || palan *m* || hijsblok *n* || таль *f*, полиспаст *m*

L118 **ligand** || Ligand *m* || ligand *m* || ligand *n* || лиганд *m*

L119 **light absorption** || Absorption *f* des Lichtes, Lichtabsorption *f* || absorption *f* de la lumière || lichtabsorptie *f* || поглощение *n* света, светопоглощение *n*

L120 **light adaptation** || Helladaptation *f*, Hellanpassung *f* || adaptation *f* à la lumière || lichtadaptatie *f* || световая адаптация *f*, адаптация *f* к свету

L121 **light alloy** || Leichtmetall *n*, Leichtmetallegierung *f* || alliage *m* léger || lichte legering *n* || лёгкий сплав *m*

L122 **light beam** || Lichtstrahl *m* || rayon *m* lumineux || lichtstraal *m*, *f* || световой луч *m*

L123 **light breakdown** || Lichtdurchschlag *m*, optischer Durchbruch *m* || claquage *m* lumineux || lichtdoorslag *m* || световой пробой *m*

L124 **light cone** || Lichtkegel *m* || cône *m* lumineux [de lumière] || lichtkegel *m* || световой конус *m* *(в теории относительности)*

L125 **light diode** || Leuchtdiode *f* || diode *f* lumineuse || lichtdiode *f* (*m*) || светодиод *m*

L126 **light efficiency** || Lichtausbeute *f*, Lichtabgabe *f* || efficacité *f* lumineuse || lichtopbrengst *f* || световая отдача *f*, светоотдача *f*

L127 **light filter** || Lichtfilter *n* || filtre *m* de lumière || lichtfilter *n*, *m* || светофильтр *m*

L128 **light guide** || Lichtleiter *m* || conduit *m* [guide *m*] de lumière || lichtleider *m* || световод *m*

L129 **light intensity** || Lichtintensität *f*, Lichtstärke *f* || intensité *f* lumineuse || lichtintensiteit *f* || интенсивность *f* света

L130 **light microscope** || Lichtmikroskop *n*, optisches Mikroskop *n* || microscope *m* optique [à lumière] || lichtmicroscoop *m* || оптический [световой] микроскоп *m*

L131 **light modulation** || Modulation *f* des Lichtes, Lichtmodulation *f* || modulation *f* lumineuse [de lumière] || lichtmodulatie *f* || **1.** модуляция *f* света **2.** модуляция *f* светом

L132 **lightning arrester** || Überspannungsableiter *m*, Blitzableiter *m* || parafoudre *m*, paratonnerre *m* || bliksemafleider *m* || грозовой разрядник *m*, молниеотвод *m*

L133 **lightning protection** || Blitzschutz *m* || protection *f* contre la foudre || bliksembeveiliging *f* || грозозащита *f*, молниезащита *f*

L134 **light output** *see* **light efficiency**

L135 **light pen** || Lichtstift *m* || crayon *m* électronique || lichtpen *f* (*m*) || световое перо *n*, световой карандаш *m*

L136 **light pulse** || Lichtimpuls *m* || éclat *m* lumineux, impulsion *f* lumineuse || lichtimpuls *m* || световой импульс *m*

L137 **light quantum** || Lichtquant *n*, Photon *n* || quantum *m* de lumière, photon *m* || lichtquantum *n*, foton *n* || световой квант *m*, фотон *m*

L138 **light running** || leichter Lauf *m* || allure *f* légère || lichte loop *m*, lichte gang *m* || лёгкий ход *m*

L139 **light scattering** || Licht(zer)-streuung *f*, Streuung *f* des Lichtes || diffusion *f* de la lumière || lichtdiffusie *f* || рассеяние *n* света

L140 **light sensitivity** || Lichtempfindlichkeit *f* || photosensibilité *f*, sensibilité *f* à la lumière || lichtgevoeligheid *f* || светочувствительность *f*

L141 **light source** || Lichtquelle *f* || source *f* lumineuse || lichtbron *f* (*m*) || источник *m* света

L142 **light wave** || Lichtwelle *f* || onde *f* lumineuse || lichtgolf *f* (*m*) || световая волна *f*

L143 **light year** || Lichtjahr *n* || année-lumière *f* || lichtjaar *n* || световой год *m* (*9,46 · 10^{15} м*)

L144 **like-charged** || gleichnamig geladen || de même signe || gelijknamig geladen || одноимённо заряженный

L145 **like charges** || gleichnamige Ladungen *f pl* || charges *f pl* de même signe || gelijknamige ladingen *f pl* || одноимённые заряды *m pl*

L146 **likelihood function** || Likelihood-Funktion *f* || fonction *f* de vraisemblance || likelihood-functie *f* || функция *f* правдоподобия

L147 **like poles** || gleichnamige Pole *m pl* || pôles *m pl* semblables [de même nom] || gelijknamige polen *m pl* || одноимённые полюса *m pl*

L148 **lime** || Kalk *m* || chaux *f* || kalk *m* || известь *f*

L149 **lime-water** || Kalkwasser *n* || eau *f* calcique [de chaux] || kalkwater *n* || известковая вода *f*

L150 **limit** || Grenzwert *m*, Limes *m* || limite *f* || limit *f* (*m*), grens *f* (*m*); grenswaarde *f* || предел *m*

L151 **limitation** || Einschränkung *f*, Beschränkung *f*, Begrenzung *f* || limitation *f* || begrenzing *f* || ограничение *n*

L152 **limit cycle** || Grenzzykel *m* || cycle *m* limite || grenscyclus *m* || предельный цикл *m*

L153 **limiter** || Begrenzer *m* || limiteur *m* || begrenzer *m* || ограничитель *m*

L154 **limit gauge** || Grenzlehre *f* || calibre *m* à limites || tolerantiekaliber *n* || предельный калибр *m*

L155 **limiting speed** || Grenzgeschwindigkeit *f* || vitesse *f* limite || grenssnelheid *f* || предельная скорость *f*

L156 **limiting stress** *see* **limit stress**

L157 **limiting value** || Grenzwert *m* || valeur *f* limite || grenswaarde *f* || предельная величина *f*, предельное значение *n*

L158 **limit-in-mean** || Limes *m* in medio || convergence *f* en moyenne || convergentie *f* in middelwaarde || предел *m* [сходимость *f*] в среднем, среднеквадратичная сильная сходимость *f*

L159 **limit load** || Grenzbelastung *f*, Grenzbeanspruchung *f* || charge *f* limite || belastingsgrens *f* (*m*) || предельная нагрузка *f*

L160 **limit of elasticity** *see* **elastic limit**

L161 **limit of proportionality** *see* **proportional limit**

L162 **limit point** || Grenzpunkt *m*, Häufungspunkt *m* || point *m* limite || grenspunt *f* (*m*) || предельная точка *f*, точка *f* накопления [конденсации]

L163 **limit speed** *see* **limiting speed**

L164 **limit stress** || Grenzspannung *f* || contrainte *f* limite || grensspanning *f* || предельное напряжение *n*

L165 **limit system** || Toleranzsystem *n* || tolérances *f pl* limites || tolerantiesystem *n* || система *f* (предельных) допусков

L166 **limit theorem** || Grenzwertsatz *m* || théorème *m* limite || grenstheorema *n* || предельная теорема *f*

L167 **line** || **1.** Linie *f* **2.** Leitung *f* || ligne *f* || lijn *f* || линия *f*

L168 **linear accelerator** || Linearbeschleuniger *m*, linearer Beschleuniger *m* || accélérateur *m* linéaire || lineaire versneller *m* || линейный ускоритель *m*

L169 **linear algebra** || lineare Algebra *f* || algèbre *f* linéaire || lineaire algebra *f* (*m*) || линейная алгебра *f*

L170 **linear characteristic** || lineare Kennlinie *f*, lineare Charakteristik *f* || caractéristique *f* linéaire || lineaire karakteristiek *f* || линейная характеристика *f*

L171 **linear dependence** || lineare Abhängigkeit *f* || dépendance *f* linéaire || lineaire afhankelijkheid *f* || линейная зависимость *f*

L172 **linear differential equation** || lineare Differentialgleichung *f* || équation *f* différentielle linéaire || lineaire differentiaalvergelijking *f* || линейное дифференциальное уравнение *n*

L173 **linear dimension** || lineare Dimension *f* || dimension *f* linéaire || lineaire dimensie *f* || линейная размерность *f*

L174 **linear dispersion** || Lineardispersion *f*, lineare Dispersion *f* || dispersion *f* linéaire || lineaire dispersie *f* || линейная дисперсия *f*

L175 **linear equation** || lineare Gleichung *f* || équation *f* linéaire || lineaire vergelijking *f* || линейное уравнение *n*

L176 **linear expansion** || lineare Ausdehnung *f*, Längenausdehnung *f* || dilatation *f* linéaire || lineaire uitzetting *f* || линейное (тепловое) расширение *n*

L177 **linear form** || Linearform *f* || forme *f* linéaire || lineaire vorm *m* || линейная форма *f*

L178 **linear function** || lineare Funktion *f* || fonction *f* linéaire || lineaire functie *f* || линейная функция *f*

L179 **linearization** || Linearisierung *f* || linéarisation *f* || linearisatie *f* || линеаризация *f*

L180 **linearly polarized wave** || linear polarisierte Welle *f* || onde *f* polarisée rectilignement || lineair gepolariseerde golf *f* (*m*) || линейно-поляризованная [плоскополяризованная] волна *f*

L181 **linear manifold** || lineare Mannigfaltigkeit *f* || variété *f* linéaire || lineaire meervoudigheid *f* || линейное многообразие *n*

L182 **linear momentum** || Impuls *m*, Bewegungsgröße *f* || quantité *f* de mouvement, impulsion *f* || bewegingshoeveelheid *f*, impuls *m* || количество *n* движения, импульс *m*

L183 **linear motor** || Linearmotor *m* || moteur *m* linéaire || lineaire motor *m* || линейный электродвигатель *m*

L184 **linear operator** || linearer Operator *m* || opérateur *m* linéaire || lineaire operator *m* || линейный оператор *m*

L185 **linear order** || lineare Ordnung *f* || ordre *m* linéaire || lineaire orde *f* (*m*) || линейный порядок *m*

L186 **linear polarization** || lineare Polarisation *f*, Linearpolarisation *f* || polarisation *f* linéaire || lineaire polarisatie *f* || линейная плоская поляризация *f*

L187 **linear programming** || lineare Programmierung *f* || programmation *f* linéaire || lineaire programmering *f* || линейное программирование *n*

L188 **linear programming problem** || Aufgabe *f* der linearen Programmierung || problème *m* de programmation linéaire || opgave *f* van lineaire programmering || задача *f* линейного программирования

L189 **linear response** || lineare Charakteristik *f*, lineare Kennlinie *f* || caractéristique *f* [réponse *f*] linéaire || lineaire karakteristiek *f*, lineair antwoord *n* || линейная характеристика *f*, линейный отклик *m*

L190 **linear set** || lineare Menge *f* || ensemble *m* linéaire || lineaire menigte *f* || линейное множество *n*

L191 **linear system** || lineares System *n* || système *m* linéaire || lineair system *n* || линейная система *f*

L192 **linear transformation** || lineare Transformation *f* || transformation *f* linéaire || lineaire transformatie *f* || линейное преобразование *n*

L193 **linear velocity** || lineare Geschwindigkeit *f* || vitesse *f* linéaire || lineaire snelheid *f* || линейная скорость *f*

L194 **line coordinates** || Linienkoordinaten *f pl* || coordonnées *f pl* tangentielles || tangentiaalcoördinaten *m pl* || тангенциальные координаты *f pl* прямой

L195 **line element** || lineares Element *n* || élément *m* linéaire || lineair element *n* || линейный элемент *m*

L196 **line integral** || Linienintegral *n*, Kurvenintegral *n* || intégrale *f* curviligne || lijnintegraal *f* (*m*) || (криво)линейный интеграл *m*

L197 **line intensity** || Intensität *f* der Spektrallinie, Spektrallinienintensität *f* || brillance *f* de la ligne du spectre || spectraal lijnintensiteit *f* || яркость *f* спектральной линии

L198 **line of action** || Angriffslinie *f*, Wirkungslinie *f* *(Kraft)* || ligne *f* d'action || actielijn *f* *(m)* || линия *f* действия *(силы)*

L199 **line of centres** || Zentrallinie *f* *(Kräftesystem)* || ligne *f* des centres || centerlijn *f* (*m*) || линия *f* центров

L200 **line of curvature** || Krümmungslinie *f* || ligne *f* de courbure || krommingslijn *f* (*m*) || линия *f* кривизны

L201 **line of flux** || Flußlinie *f*, Kraftflußlinie *f* || ligne *f* d'écoulement [de courant] || stromingslijn *f* (*m*) || линия *f* тока

L202 **line of force** || Feldlinie *f*, Kraftlinie *f*, Feldstärkelinie *f* || ligne *f* de force || krachtlijn *f* (*m*) || силовая линия *f*

L203 **line of induction** || Induktionslinie *f* || ligne *f* d'induction || inductielijn *f* (*m*) || линия *f* индукции

L204 **line of sight** || Gesichtslinie *f*, Sichtlinie *f* || rayon *m* visuel || gezichtslijn *f* (*m*) || луч *m* зрения

L205 **line printer** || Zeilendrucker *m* || imprimante *f* par ligne || regeldrukker *m* || построчно-печатающее устройство *n*

L206 **line segment** || Geradenstück *n* || segment *m* de droite || segment *n* van een rechte ljn || отрезок *m* прямой

L207 **line spectrum** || Linienspektrum *n* || spectre *m* de lignes [de raies] || lijnenspectrum *n* || линейчатый [дискретный] спектр *m*

L208 **line width** || Linienbreite *f*, Spektrallinienbreite *f* || largeur *f* de ligne [de raie] || lijnbreedte *f* || ширина *f* (спектральной) линии

L209 **link** || **1.** Glied *n* *(Kette)* **2.** Gelenk *n* **3.** Verbindungsglied *n* || maillon *m*; anneau *m* || **1.** schakel *m* **2.** gewricht *n*, scharnier *n* **3.** verbinding *f*; koppeling *f* || **1.** звено *n* *(цепи)*; трак *m* *(гусеницы)* **2.** шарнир *m* **3.** соединение *n*; связь *f*

L210 **link bolt** *see* **hinge(d) bolt**

L211 **link mechanism** || Gelenkgetriebe *n* || système *m* d'articulations || gewrichtmechanisme *f* || шарнирный механизм *m*

L212 **lip angle** || Keilwinkel *m*, Meißelwinkel *m* || angle *m* d'affûtage || snijhoek *m* || угол *m* заострения *(режущей части инструмента)*

L213 **liquefaction** || Verflüssigung *f* || liquéfaction *f* || vloeibaarmaking *f* || ожижение *n*, сжижение *n* *(газа)*

L214 **liquefied gas** || verflüssigtes [flüssiges] Gas *n* || gaz *m* liquéfié || vloeibaargemaakt gas *n* || ожижённый [сжиженный] газ *m*

L215 **liquefier** || Verflüssiger *m* || appareil *m* de liquéfaction || vloeibaarmaker *m* || ожижитель *m*

L216 I **liquid** || Flüssigkeit *f* || liquide *m* || vloeistof *f* *(m)* || жидкость *f*

L217 II **liquid** || flüssig || liquide || vloeibaar || жидкий

L218 **liquid-column manometer** || Flüssigkeitsmanometer *n* || manomètre *m* à liquide || vloeistofmanometer *m* || жидкостный манометр *m*

L219 **liquid-cooled** || flüssigkeitsgekühlt || à refroidissement par liquide || vloeistofgekoeld || с жидкостным охлаждением

L220 **liquid crystal** || flüssiger Kristall *m* || cristal *m* liquide || vloeibaar kristal *n* || жидкий кристалл *m*

L221 **liquid-filled thermometer** || Flüssigkeitsthermometer *n* || thermomètre *m* à liquide || vloeistofthermometer *m* || жидкостный термометр *m*

L222 **liquid-gas transition** || Flüssigkeit-Gas-Übergang *m* || transformation *f* liquide-gaz || vloeistof-gasovergang *m* || (фазовый) переход *m* жидкость — газ

L223 **liquid phase** || flüssige Phase *f*, Flüssigphase *f* || phase *f* liquide [fluide] || vloeibare fase *f* || жидкая фаза *f*

L224 **liquid state** || flüssiger Zustand *m* || état *m* liquide || vloeibare toestand *m* || жидкое состояние *n*

L225 **liquidus (curve)** || Liquiduskurve *f*, Liquiduslinie *f* || liquidus *m* || liquidus *n* || кривая *f* ликвидуса, ликвидус *m*

L226 **Lissajous figures** || Lissajous-Figuren *f pl* || figures *f pl* de Lissajous || Lissajous-figuren *f* *(m)* *pl* || фигуры *f pl* Лиссажу

L227 **list** || Liste *f* || liste *f* || lijst *f* || список *m*

L228 **liter** || Liter *n* || litre *m* || liter *m* || литр *m* *(10^{-3} м^3)*

L229 **lithium,** Li || Lithium *n* || lithium *m* || lithium *n* || литий *m*

L230 **litmus** || Lackmus *n* || tournesol *m*, papier *m* de tournesol || lakmoes *n* || лакмус *m*

L231 **live axle** *see* **driving axle**

L232 **live contact** || geschlossener Kontakt *m* || contact *m* fermé à ouverture || gesloten contact *n* || замкнутый контакт *m*

L233 **live head** *see* **headstock**

L234 **live load** || vorübergehende Last *f*; Verkehrslast *f* || surcharge *f*; charge *f* passagère; charge *f* utile || mobiele belasting *f* || временная нагрузка *f*; подвижная нагрузка *f*; динамическая нагрузка *f*

L235 **load** || Belastung *f*, Last *f* || charge *f* || belasting *f*; last *f* *(m)* || нагрузка *f*; груз *m*

L236 **load circuit** || Belastungs(strom)kreis *m* || circuit *m* utilisateur de charge || belastingskring *m* || цепь *f* нагрузки, нагрузочная цепь *f*

L237 **load current** || Laststrom *m* || courant *m* de charge [de fonctionnement] || laststroom *m* || ток *m* через нагрузку, нагрузочный ток *m*

L238 **load diagram** || Belastungsdiagramm *n* || schéma *m* de chargement; diagramme *m* de chargement || belastingsdiagram *n* || эпюра *f* нагрузки

L239 **loader** || **1.** Lader *m* **2.** Ladeprogramm *n* || **1.** chargeur *m* **2.** chargeur *m*, programme *m* chargeur || **1.** laadmachine *f* **2.** laadprogramma *n* || **1.** погрузчик *m*, погрузочная машина *f* **2.** загрузчик *m*, программа-загрузчик *m*

L240 **loading** || 1. Belastung *f*; Beanspruchung *f* 2. Laden *n*, Aufladen *n* || chargement *m* || 1. belasting *f* 2. lading *f* || 1. приложение *n* нагрузки, нагружение *n* 2. зарядка *f*

L241 **loading program** || Ladeprogramm *n* || chargeur *m*, programme *m* chargeur || laadprogramma *n* || загрузчик *m*, программа-загрузчик *m*

L242 **loading test** *see* **load test**

L243 **load module** || Lademodul *m* || module *m* chargeable || laadmodule *n* || загрузочный модуль *m*

L244 **load resistor** || Lastwiderstand *m*, Belastungswiderstand *m* || résistance *f* de charge || belastingsweerstand *m* || нагрузочный резистор *m*

L245 **load speed** || Lastdrehzahl *f* || vitesse *f* (de rotation) en charge || toerental *n* onder belasting || частота *f* вращения под нагрузкой

L246 **load test** || Belastungsversuch *m*; Lastprobe *f* || essai *m* de charge(ment) || belastingsproef *f* (*m*) || испытание *n* (под) нагрузкой

L247 **load voltage** || Lastspannung *f* || tension *f* de charge || belastingsspanning *f* || напряжение *n* на нагрузке

L248 **Lobachevskis' geometry** || Lobatschewskische Geometrie *f* || géométrie *f* hyperbolique || Lobatchevski-geometrie *f* || геометрия *f* Лобачевского

L249 **lobe** || Keule *f*, Zipfel *m* || lobe *m* || lob *f* (*m*) || лепесток *m* *(диаграммы направленности)*

L250 **local extremum** || lokales Extremum *n* || extrémum *m* local || lokaal extremum *n* || локальный экстремум *m*

L251 **localization** || Lokalisierung *f*, Ortsbestimmung *f*, Eingrenzung *f* || localisation *f* || lokalisatie *f* || локализация *f*

L252 **localized mass** || Punktmasse *f*, konzentrierte Masse *f* || masse *f* concentrée || geconcentreerde massa *f* (*m*) || сосредоточенная масса *f*

L253 **local oscillator** || Überlagerer *m*, Überlagereroszillator *m* || oscillateur *m* local || heterodyne *m* || гетеродин *m*

L254 **local stability** || lokale [örtliche] Stabilität *f* || stabilité *f* locale || lokale stabiliteit *f* || местная [локальная] устойчивость *f*

L255 **local system** || lokales Koordinatensystem *n* || système *m* de coordonnées local || lokaal coördinatenstelsel *n* || местная система *f* координат

L256 **local variable** || lokale Variable *f* || variable *f* locale || lokale variabele *f* || локальная переменная *f*

L257 **location** || Speicherplatz *m* || location *f* || cel *f* || ячейка *f* памяти, ячейка *f* ЗУ

L258 **location number** || Speicherplatzadresse *f* || numéro *m* de location || geheugenadres *n* || адрес *m* ячейки памяти [ячейки ЗУ]

L259 **locator** || Ortungsgerät *n* || détecteur *m*, localisateur *m* || lokator *m* || локатор *m*

L260 **lock** || Verschluß *m*; Sperre *f* || serrure *f*; verrou *m* || slot *n*; sluitingsorgaan *n*; grendel *m* || замок *m*; стопор *m*; фиксатор *m*

L261 **locknut** || Sicherungsmutter *f* || contre-écrou *m* || borgmoer *f* (*m*), tegenmoer *f* (*m*) || контргайка *f*; стопорная гайка *f*

L262 **lockout** || Blockierung *f*, Sperre *f* || blocage *m* || blokkering *f*, grendeling *f* || блокировка *f*, запирание *n*

L263 **lock screw** || Sicherungsschraube *f* || vis *f* d'about [d'arrêt] || grendelschroef *f* (*m*); drukschroef *f* (*m*) || упорный винт *m*

L264 **lock washer** || Federring *m* || rondelle *f* Grower; rondelle *f* à ressort || veerring *m* || (разрезная) пружинная шайба *f*

L265 **logarithm** || Logarithmus *m* || logarithme *m* || logaritme *f* (*m*) || логарифм *m*

L266 **logarithmic decrement** || logarithmisches Dekrement *n*, logarithmisches Dämpfungsdekrement *n* || décrément *m* logarithmique || logaritmische decrement *m* || логарифмический декремент *m* *(затухания)*

L267 **logarithmic function** || Logarithmusfunktion *f* || fonction *f* logarithmique || logaritmische functie *f* || логарифмическая функция *f*

L268 **logarithmic potential** || logarithmisches Potential *n* || potentiel *m* logarithmique || logaritmische potentiaal *m* || логарифмический потенциал *m*

L269 **logarithmic property** || Haupteigenschaft *f* des Logarithmus || propriété *f* logarithmique || hoofdeigenschap *f* van logaritme || основное свойство *n* логарифма

L270 **logical add(ition)** || Disjunktion *f* || addition *f* logique || logische additie *f* || логическое сложение *n*

L271 **logical connective** || logische Verknüpfung *f* || couple *f* logique || logische koppeling *f* || логическая связка *f*

L272 **logical inclusion** || Inklusion *f* || inclusion *f* logique || logische insluiting *f* || (логическое) включение *n*

L273 **logical multiplication** || Konjunktion *f* || multiplication *f* logique || logische vermenigvuldiging *f* || логическое умножение *n*

L274 **logical product** || logisches Produkt *n* || produit *m* logique || logisch produkt *n* || логическое произведение *n*

L275 **logical record** || Datensatz *m*, logischer Satz *m* || enregistrement *m* logique || logische inschrijving *f* || логическая запись *f*

L276 **logical sum** || logische Summe *f* || somme *f* logique || logische som *f* (*m*) || логическая сумма *f*

L277 **log-log characteristic** || logarithmische Charakteristik *f* || caractéristique *f* d'un logarithme || logaritmische karakteristiek *f* || логарифмическая характеристика *f*

L278 **logometer** || Qiotient(en)-messer *m* || logomètre *m*, quotientmètre *m* || quotiëntmeter *m* || логометр *m*

L279 **lone** || Einzel..., einzeln || célibataire || enkel || одиночный

L280 **long-distance** *see* **long-range**

L281 **long division** || schriftliche Division *f* || division *f* d'angle || schriftelijke deling *f* || деление *n* углом

L282 **long-focus lens** || Langfokuslinse *f*, langbrennweitige Linse *f* || lentille *f* de grande focale || langfocuslens *f* (*m*) || длиннофокусная линза *f*

L283 **longitudinal** || Längs..., Lang..., longitudinal || longitudinal || langs..., longitudinaal || продольный

L284 **longitudinal displacement** || Längsverschiebung *f* || déplacement *m* longitudinal || langsverschuiving *f* || продольное смещение *n*

L285 **longitudinal force** || Längskraft *f* || force *f* longitudinale || longitudinale kracht *f* (*m*) || продольная сила *f*

L286 **longitudinal oscillations** || Longitudinalschwingungen *f pl*, Längsschwingungen *f pl* || oscillations *f pl* longitudinales || longitudinale trillingen *f pl* || продольные колебания *n pl*

L287 **longitudinal wave** || Longitudinalwelle *f*, Längswelle *f* || onde *f* longitudinale || longitudinale golf *f* (*m*) || продольная волна *f*

L288 **long-lived isotope** || langlebiges Isotop *n* || isotope *m* à longue période || langlevende isotoop *m* || долгоживущий изотоп *m*

L289 **long-range** || Fernwirk(ungs)...; weitreichend || à long parcours || verreikend || дальнодействующий, дальнего действия

L290 **long-range force** || Fernwirkungskraft *f*, fernwirkende [weitreichende] Kraft *f* || force *f* agissante à longue distance || ver-werkende kracht *f* (*m*) || дальнодействующая сила *f*

L291 **long-range order** || Fernordnung *f* (*Kristall*) || ordre *m* étalé à grande distance || verre ordening *f* || дальний порядок *m* (*в кристалле*)

L292 **long-term strength, long-time strength** || Dauer(stand)festigkeit *f* || résistance *f* au temps [dans le temps] || duurstevigheid *f* || длительная прочность *f* (*при продолжительном нагружении*)

L293 **long-wavelength** || langwellig,

Langwellen... || de longues ondes || langegolf... || длинноволновый

L294 **loop** || 1. (stromdurchflossene) Windung *f*, Schleife *f*, Stromschleife *f* 2. Schleife *f* || 1. spire *f*; boucle *f* 2. cycle *m* || 1. winding *f*, lus *f* (*m*) 2. cyclus *m* || 1. виток *m*, рамка *f*, петля *f* (с током) 2. цикл *m*

L295 **loose axoid** || Polhodiekegel *m*, Gangpolkegel *m*, Polkegel *m* || axoïde *f* mobile, surface *f* mobile des axes instantanés || mobiele axoïde *n* || подвижный аксоид *m*

L296 **loose packing** || lockere Pakkung *f* || assemblage *m* meuble || losse pakking *f* || рыхлая упаковка *f*

L297 **loose pulley** || Losscheibe *f* || poulie *f* folle || loose schijf *f* (*m*); losse katrol *f* (*m*) || холостой шкив *m*

L298 **Lorentz contraction** || Lorentz-Kontraktion *f*, Lorentz-Verkürzung *f* || contraction *f* de Lorentz || Lorentz-contractie *f* || лоренцово сокращение *n* (*в теории относительности*)

L299 **lorry** || 1. Lastkraftwagen *m* 2. Hunt *m*; Lore *f* 3. Flachwagen *m* || 1. camion *m* 2. berline *f* 3. wagon *m* || 1. vrachtauto *m* 2. lorrie *f*; wagentje *m* 3. platte wagen *m* || 1. грузовой автомобиль *m* 2. вагонетка *f* 3. железнодорожная платформа *f*

L300 **Loschmidt's number** || Loschmidt-Konstante *f* || nombre *m* de Loschmidt || Loschmidt-getal *n* || число *n* Лошмидта

L301 **loss** || Verlust *m*; Einbuße *f* || perte *f* || verlies *n* || потеря *f*, потери *f pl*; проигрыш *m*

L302 **loss angle** || dielektrischer Verlustwinkel *m* || angle *m* de pertes diélectriques || diëlektrische verlieshoek *m* || угол *m* (диэлектрических) потерь

L303 **loss factor** || (dielektrischer) Verlustfaktor *m* || facteur *m* de pertes diélectriques || (diëlektrische) verliesfactor *m* || коэффициент *m* потерь (*в диэлектрике*)

L304 **lossless** || verlustlos, verlustfrei || sans pertes || verliesvrij || без потерь

L305 **loss of head** || Druckhöhenverlust *m* || perte *f* de charge || drukverlies *n* || потеря *f* напора

L306 **loss tangent** || Verlustfaktor *m* || facteur *m* de dissipation [de pertes diélectriques] || diëlektrische verlies-factor *m* || тангенс *m* угла (диэлектрических) потерь

L307 **lossy** || verlustbehaftet || à pertes considérables || met verliezen || с (большими) потерями

L308 **lost motion** || Leerlauf *m* || marche *f* [course *f*, allure *f*] à vide || vrijloop *m* || холостой ход *m*

L309 **loudness** || Lautstärke *f* || intensité *f* de son, volume *m* sonore || geluidsterkte *f* || громкость *f*

L310 **loudness level** || Lautstärkepegel *m* || niveau *m* de l'intensité sonore || geluidsterkteniveau *n* || уровень *m* громкости

L311 **low-carbon steel** || kohlenstoffarmer Stahl *m* || acier *m* doux || koolstofarm staal *n* || низкоуглеродистая [малоуглеродистая] сталь *f*

L312 **low-energy** || energiearm, niederenergetisch, Niederenergie... || à faible énergie || laagenergetisch || малой энергии, медленный

L313 **lower hound** || untere Schranke *f* || limite *f* inférieure || ondergrens *f* || нижняя граница *f*; нижний предел *m*

L314 **lowering of indices** || Herunterziehen *n* von Indexen || abaissement *m* des indices || daling *f* van indexen || опускание *n* индексов

L315 **low-expansion** || mit geringem Wärme(aus)dehnungsvermögen || à faible coefficient de dilatation thermique || laag warmteuitzettingsvermogen *n* || с малым коэффициентом теплового расширения

L316 **low-frequency** || niederfrequent, Niederfrequenz... || de basse fréquence || laagfrequent || низкочастотный

L317 **low-level language** || niedere Sprache *f* || langage *m* de bas niveau || lagere programmeertaal *f* || язык *m* низкого уровня

L318 **low-noise** || rauscharm; geräuscharm || à bas niveau de bruits || geruisarm || с низким уровнем шумов, малошумящий

L319 **low-power** || (von) kleiner Leistung, leistungsschwach, leistungsarm || à basse puissance || zwak, krachteloos || малой мощности, маломощный

L320 **low-pressure** || Niederdruck... || à basse pression || lagedruk... || низкого давления

L321 **low-pressure turbine** || Niederdruckturbine *f* || turbine *f* à basse pression || lage-druk-turbine *f* || турбина *f* низкого давления, низконапорная турбина *f*

L322 **low-resistance** || niederohmig, Niederohm... || de faible résistance || laagohms... || с малым сопротивлением, низкоомный

L323 **low-temperature** || Niedertemperatur... || de basse température || lagetemperatuur... || низкотемпературный

L324 **low-tension** || Niedervolt..., Niederspannungs... || à basse tension || laagspannings... || низковольтный, низкого напряжения

L325 **low-viscosity** || schwachviskos, niedriger Viskosität || faiblement visqueux || zwak visceus || малой вязкости, маловязкий

L326 **low-voltage** || Niedervolt..., Niederspannungs... || à basse tension || laagspannings... || низковольтный

L327 **lubricant** || Schmiermittel *n*; Schmierstoff *m* || lubrifiant *m*; graisse *f* || smeermiddel *n*, smeerstof *f* (*m*) || смазка *f*; смазочный материал *m*

L328 **lubricating grease** || Schmierfett *n* || graisse *f* (consistante) || smeervet *n* || консистентная смазка *f*

L329 **lubricating gun** || Schmierpresse *f* || pompe *f* à graisse || smeerpistool *n*, *m* || смазочный шприц *m*

L330 **lubricating oil** || Schmieröl *n* || huile *f* de graissage || smeerolie *f* (*m*) || смазочное масло *n*

L331 **lubricating pump** *see* **oil pump**

L332 **lubrication** || Schmierung *f* || lubrification *f*; graissage *m* || smeren *n*, smering *f*; oliën *n* || смазка *f*, смазывание *n*

L333 **lubricator** || Öler *m* || lubrificateur *m*; graisseur *m* || smeerinrichting *f*, smeertoestel *n* || смазочное устройство *n*

L334 **lubricator box** || Schmierbüchse *f*; Öler *m* || boîte *f* à graissage || smeerpot *m*, smeerdop *m* || смазочная коробка *f*; маслёнка *f*

L335 **lumen**, lm || Lumen *n* || lumen *m* || lumen *n* || люмен *m*

L336 **luminance** || Leuchtdichte *f*; Helligkeit *f* || brillance *f*, luminance *f* || helderheid *f*, luminantie *f* || яркость

L337 **luminescence** || Lumineszenz *f*, kaltes Leuchten *n* || luminescence *f* || luminescentie *f* || люминесценция *f*

L338 **luminescent lamp** || Lumineszenzlampe *f*, Lumineszenzröhre *f* || lampe *f* fluorescente à luminescence || luminescentielamp *f* (*m*) || люминесцентная лампа *f*

L339 **luminophor** || Luminophor *m*, Lumineszenzstoff *m*, Leuchtstoff *m* || luminophore *m*, composé *m* luminescent || luminofoor *m* || люминофор *m*

L340 **luminosity** || **1.** Helligkeit *f*, Leuchtdichte *f* **2.** Lichtausbeute *f*, Lichtleistung *f* || **1.** luminosité *f* **2.** efficacité *f* lumineuse || **1.** helderheid *f* **2.** lichtopbrengst *f* || **1.** яркость *f*, светимость *f* **2.** видность *f*, световая отдача *f*

L341 **luminous emittance** || spezifische Lichtausstrahlung *f*, spezifische Leuchtstärke *f* || émittance *f* lumineuse || specifieke lichtsterkte *f* || светимость *f*

L342 **luminous energy** || Lichtenergie *f*; Strahlungsenergie *f* || énergie *f* rayonnante; énergie *f* lumineuse || lichtenergie *f* || лучистая энергия *f*; световая энергия *f*

L343 **luminous flux** || Lichtstrom *m*, Lichtfluß *m* || flux *m* lumineux || lichtstroom *m* || световой поток *m*

L344 **luminous intensity** || Strahlstärke *f* || intensité *f* lumineuse || lichtintensiteit *f* || энергетическая сила *f* света, сила *f* излучения

L346 **lux**, lx || Lux *n* || lux *m* || lux *f* || люкс *m*, лк

L347 **Lyapunov's stability theory** || Lyapunowsche Stabilitätstheorie *f* || théorie *f* de stabilité au sens de Lyapunov || stabiliteitstheorie *f* van

Lyapunov || теория *f* устойчивости по Ляпунову

L348 **lyophilic** || lyophil || lyophile || lyofiel || лиофильный

L349 **lyophobic** || lyophob || lyophobe || lyofoob || лиофобный

M

M1 **Mach cone** || Machscher Kegel *m* || cône *m* de Mach || Mach-kegel *m* || конус *m* Маха

M2 **machinability** || Spanbarkeit *f*, Zerspanbarkeit *f* || usinabilité *f* || bewerkbaarheid *f* || обрабатываемость *f* *(на станках)*

M3 **machine** || **1.** Maschine *f* ; Mechanismus *m* **2.** Werkzeugmaschine *f* **3.** Transportmittel *n* || machine *f* || **1.** machine *f*; mechanisme *n*; mechanische inrichting *f* **2.** werktuigmachine *f* **3.** transportmiddel *n* || **1.** машина *f*; механизм *m*; механическое устройство *n* **2.** станок *m* **3.** транспортное средство *n*

M4 **machine code** || Rechnerkode *m*, maschineninterne Kodierung *f* || code *m* de machine || machinecode *m* || машинный код *m*

M5 **machine language** || Maschinensprache *f* || langage-machine *m*, langage *m* de machine || machinetaal *f* || машинный язык *m*

M6 **machine mo(u)lding** || Maschinenformerei *f* || moulage *m* à la machine || machinaal vormen *n* || машинная формовка *f*

M7 **machine oil** || Maschinenöl *n* || huile *f* à machines || machineolie *f* (*m*) || машинное масло *n*

M8 **machine-oriented language** || rechnerorientierte [maschinenorientierte] Sprache *f* || langage *m* orienté vers le calculateur || machinegeoriënteerde taal *f* || машино-ориентированный язык *m*

M9 **machinery** || Maschinen *f pl*; Maschinenpark *m*; Mechanismen *m pl* || machines *f pl*; machinerie *f* || machinerie(ën) *n* || машины *f pl*; механизмы *m pl*; машинное оборудование *n*

M10 **machinery steel** || Maschinenbaustahl *m* || acier *m* de construction || machinestaal *n* || конструкционная сталь *f*

M11 **machine screw** || Maschinenschraube *f* || vis *f* à métaux || kolomschroef *f* (*m*); machine(span)schroef *f* (*m*) || винт *m* для металла

M12 **machine tool** || Werkzeugmaschine *f* || machine *f* à tailler les métaux || gereedschapwerktuig *n*, werktuigmachine *f* || металлорежущий станок *m*

M13 **machining** || Maschinenarbeit *f*; spanende Bearbeitung *f* || usinage *m* (sur machine) || mechanische [machinale] bewerking *f* || механическая обработка *f*

M14 **machining allowance** || Bearbeitungszugabe *f* || surépaisseur *m* d'usinage || machinale bijslag *m*, machinale speling *m* || припуск *m* на обработку

M15 **Mach number** || Mach-Zahl *f*, Machsche Zahl *f*, Mach *n* || nombre *m* de Mach || Mach-getal *n* || число *n* Маха

M16 **macrodefinition** || Makrodefinition *f* || macrodéfinition *f* || macrodefinitie *f* || макроопределение *n*

M17 **macroexpansion** || Makroauflösung *f* || macro-expansion *f* || macro-expansie *f* || макрорасширение *n*

M18 **macrolanguage** || Makrosprache *f* || macrolangage *m* || macrotaal *f* || макроязык *m*

M19 **macromolecule** || Makromolekül *n* || macromolécule *f* || macromolecule *f* (*m*) || макромолекула *f*

M20 **macroscopic** || makroskopisch || macroscopique || macroscopisch || макроскопический

M21 **macroscopic structure, macrostructure** || Makrostruktur *f*, Makrogefüge *n* || structure *f* macroscopique || macroscopische structuur, macrostructuur *f* || макроскопическая структура *f*, макростуктура *f*

M22 **magenta** || Fuchsin *n* || fuchsine *f*, magenta *m* || fuchsine *f* (*m*) || фуксин *m*

M23 **magic nucleus** || magischer Kern *m* || noyau *m* magique || magische kern *f* || магическое ядро *n*

M24 **magic number** || magische Zahl *f* || nombre *m* magique || magisch getal *n* || магическое число *n*

M25 **magnesia,** MgO || Magnesiumoxid *n*, Magnesia *f* || magnésie *f*, hydroxyde *m* de magnésium || magnesia *f*, magnesiumoxyde *n* || магнезия *f*, окись *f* магния

M26 **magnesium,** Mg || Magnesium *n* || magnésium *m* || magnesium *n* || магний *m*

M27 **magnesium alloy** || Magnesiumlegierung *f* || alliage *m* à magnésium || magnesiumlegering *f* || магниевый сплав *m*

M28 **magnesium carbonate,** $MgCO_3$ || Magnesiumkarbonat *n* || carbonate *m* de magnésium || magnesiumcarbonaat *n* || карбонат *m* магния, углекислый магний *m*

M29 **magnesium oxide,** MgO || Magnesiumoxid *n*, Magnesia *f* || oxyde *m* de magnésium, magnésie *f* || magnesiumoxyde *n* || окись *f* магния, магнезия *f*

M30 **magnet** || Magnet *m* || aimant *m* || magneet *m* || магнит *m*

M31 **magnet gap** || Luftspalt *m* *(Magnet)* || écartement *m* des pôles || luchtspleet *f* (*m*) *(magneet)* || зазор *m* между полюсами магнита

M32 I **magnetic** || Magnetikum *n*, magnetischer Werkstoff *m* || magnétique *m* || magneetstof *f* (*m*), magnetische stof *f* (*m*) || магнетик *m*, магнитное вещество *n*

M33 II **magnetic** || magnetisch || magnétique || magnetisch || магнитный

M34 **magnetic bar** || Stabmagnet *m* || barreau *m* aimanté, aimant *m* droit || staafmagneet *m* || полосовой магнит *m*, стержневой магнит

M35 **magnetic brake** || Magnetbremse *f* || frein *m* magnétique || magneetrem *f* (*m*) || (электро)магнитный тормоз *m*

M36 **magnetic chuck** || Magnetspannfutter *n* || mandrin *m* magnétique || magnetische opspanplaat *f* (*m*) || электромагнитный патрон *m* *(для зажима обрабатываемых деталей)*

M37 **magnetic circuit** || magnetischer Kreis *m* || circuit *m* magnétique || magneetcircuit *n*, magnetisch circuit *n* || магнитная цепь *f*, магнитопровод *m*

M38 **magnetic clutch** || elektromagnetische Kupplung *f* || embrayage *m* magnétique || elektromagnetische koppeling *f* || электромагнитная муфта *n*

M39 **magnetic-core storage** || Ferrit(kern)speicher *m* || mémoire *f* à tores magnétiques || magneetkerngeheugen *n* || память *f* на магнитных сердечниках

M40 **magnetic deflection** || magnetische Ablenkung *f* || déflexion magnétique || magnetische deflectie *f* || магнитное отклонение *n* *(пучка)*

M41 **magnetic dip** *see* **magnetic inclination**

M42 **magnetic disk** || Magnetplatte *f* || disque *m* magnétique || magneetschijf *f* (*m*) || магнитный диск *m*

M43 **magnetic-disk storage** || Magnetplattenspeicher *m* || mémoire *f* à disques magnétiques || magneetschijvengeheugen *n* || память *f* на магнитных дисках

M44 **magnetic displacement** || magnetische Verschiebung *f*; Vormagnetisierung *f* || déplacement *m* magnétique || magnetische verschuiving *f* || магнитное смещение *n*, магнитная индукция *f*

M45 **magnetic domain** || Weißscher Bezirk *m*, ferromagnetischer Bereich *m* || domain *m* magnétique || magnetisch domein *n* || магнитный домен *m*

M46 **magnetic drum** || Magnettrommel *f* || tambour *m* magnétique || magneettrommel *f*, *m* || магнитный барабан *m*

M47 **magnetic-drum storage** || Magnettrommelspeicher *m* || mémoire *f* à tambour magnétique || magneettrommelgeheugen *n* || память *f* на магнитном барабане

M48 **magnetic field** || Magnetfeld *n*, magnetisches Feld *n* || champ *m* magnétique || magneetveld *n* || магнитное поле *n*

M49 **magnetic flux** || magnetischer Fluß *m*, Magnetfluß *m* || flux *m* magnétique || magneetstroom *m* || магнитный поток *m*

M50 **magnetic hydrodynamics** *see* **magnetohydrodynamics**

M51 **magnetic inclination** || (magnetische) Inklination *f* || inclinaison *f* magnétique || magnetische inclinatie *f* || магнитное наклонение *n*

M52 **magnetic induction** || magnetische Induktion *f* || induction *f* magnétique || magnetische inductie *f* || магнитная индукция *f*

M53 **magnetic insulation** || magnetische Abschirmung *f* || blindage *m* magnétique || magnetische afscherming *f* || 1. магнитное экранирование *n* 2. магнитный экран *m*

M54 **magnetic lens** || magnetische Linse *f* || lentille *f* magnétique || magneetlens *f* || магнитная линза *f*

M55 **magnetic levitation** *see* **magnetic suspension**

M56 **magnetic linkage** || Flußverkettung *f* || flux *m* magnétique local || stromingscohesie *f* || потокосцепление *n*

M57 **magnetic mirror** || magnetischer Pfropfen *m*; magnetischer Spiegel *m* || bouchon *m* à aimant, obturateur *m* magnétique; miroir *m* magnétique || magnetisch kurk *n*, *m*; magnetische spiegel *m* || магнитная пробка *f*; магнитное зеркало *n*

M58 **magnetic moment** || magnetisches Moment *n* || moment *m* magnétique || magnetisch moment *n* || магнитный момент *m*

M59 **magnetic permeability** || (magnetische) Permeabilität *f* || perméabilité *f* (magnetique) || magnetische permeabiliteit *f* || магнитная проницаемость *f*

M60 **magnetic pole** || Magnetpol *m* || pôle *m* magnétique || magneetpool *m* || магнитный полюс *m*

M61 **magnetic quantum number** || magnetische Quantenzahl *f* || nombre *m* quantique magnétique || magnetisch quantengetal *n* || магнитное квантовое число *n*

M62 **magnetic resistance** || magnetischer Widerstand *m*, Reluktanz *f* || résistence *f* magnétique || magnetische weerstand *m* || магнитное сопротивление *n*, магнитосопротивление *n*

M63 **magnetic rotation** || magnetische Drehung *f* (der Polarisationsebene) ||effet *m* Faraday, rotation *f* de Faraday, rotation *f* magnétique du plan de la polarisation || magnetische draaiing *f* || вращение *n* плоскости поляризации света в магнитном поле, эффект *m* Фарадея

M64 **magnetic shield** || magnetische Abschirmung *f* || écran *m* magnétique || magneetscherm *n* || магнитный экран *m*, магнитная защита *f*

M65 **magnetic storm** || magnetischer Sturm *m*, magnetisches Gewitter *n* || tempête *f* magnétique || magnetische storm *m* || магнитная буря *f*

M66 **magnetic susceptibility** || magnetische Suszeptibilität *f* || susceptibilité *f* magnétique || magnetische susceptibiliteit *f* || магнитная восприимчивость *f*

M67 **magnetic suspension** || magnetische Aufhängung *f* || suspension *f* magnétique || magnetische ophanging *f* || магнитный подвес *m*, магнитная подвеска *f*

M68 **magnetic transition** || magnetische Umwandlung *f* || transition *f* magnétique || magnetische omvorming *f* || магнитное превращение *n*

M69 **magnetic variation** || magnetische Variation *f* || variation *f* magnétique || magnetische variatie *f* || магнитная вариация *f*

M70 **magnetic vector** || magnetischer Vektor *m* || vecteur *m* magnétique || magnetische vector *m* || вектор *m* напряжённости магнитного поля, магнитный вектор *m*

M71 **magnetic viscosity** || 1. magnetische Viskosität *f* 2. magnetische Hysterese *f* || 1. viscosité *f* magnétique 2. posteffet *m* magnétique || 1. magnetische viscositeit *f* 2. magnetische hysteresis *f* || 1. магнитная вязкость *f* 2. магнитное последействие *n*

M72 **magnetic wave** || magnetische Vektorwelle *f* || onde *f* de vecteur magnétique || magnetische vectorgolf *f* (*m*) || волна *f* магнитного вектора (напряжённости)

M73 **magnetism** || Magnetismus *m* || magnétisme *m* || magnetisme *n* || магнетизм *m*

M74 **magnetization** || **1.** Magnetisierung *f* **2.** Magnetisierungsstärke *f*, Magnetisierungskraft *f* || **1.** aimantation *f*, magnétisation *f* **2.** intensité *f* d'aimantation || **1.** magnetisatie || **2.** magnetiseringskracht *f* (*m*) *f* **1.** намагничивание *n* **2.** намагниченность *f*

M75 **magnetization current** || Magnetisierungsstrom *m* || courant *m* d'aimantation || magnetiseringsstroom *m* || намагничивающий ток *m*

M76 **magnetization curve** || Magnetisierungskurve *f*, Magnetisierungskennlinie *f* || courbe *f* d'aimantation || magnetiseringskromme *f* (*m*) || кривая *f* намагничивания

M77 **magnetization reversal** || Ummagnetisierung *f* || renversement *m* [inversion *f*] d'aimantation || magnetisatieompoling *f* || перемагничивание *n*

M78 **magnetization wave** || Magnetisierungswelle *f* || onde *f* de renversement d'aimantation || magnetisatieompolingsgolf *f* (*m*) || волна *f* перемагничивания

M79 **magnetizing field** || Magnetisier(ungs)feld *n* || champ *m* d'aimantation || magnetiseringsveld *n* || намагничивающее поле *n*

M80 **magneto** || magnetelektrische Maschine *f*, magnetelektrischer Generator *m* || magnéto *f*, machine *f* magnéto-électrique || magneto *m* || магнитоэлектрическая машина *f*, магнето *n*

M81 **magnetodielectric** || Magnetodielektrikum *n* || magnétodiélectrique *m* || magnetodiëlektricum *n* || магнитодиэлектрик *m*

M82 **magnetoelectric** || Magnetoelektrikum *n* || matériau *m* ferro-électrique, ferro-électrique *m* || ferro-elektricum *n* || сегнетомагнетик *m*

M83 **magnetoelectric generator** || magnetoelektrischer Generator *m* || dynamo *f* magnéto-électrique || magneto-elektrische generator *m* || (магнитный) индуктор *m*, магнето *n*

M84 **magnetohydrodynamic generator** || magnetohydrodynamischer Generator *m*, MHD-Generator *m* || générateur *m* magnéto-hydrodynamique || magnetohydrodynamische generator *m* || магнитогидродинамический генератор *m*, МГД генератор *m*

M85 **magnetohydrodynamics** || Magnetohydrodynamik *f* || magnéto-hydrodynamique *f* || magnetohydrodynamica *f* || магнитная гидродинамика *f*, магнитогидродинамика *f*

M86 **magnetohydrodynamic wave** || magnetohydrodynamische Welle *f* || onde *f* magnéto-hydrodynamique || magnetohydrodynamische golf *f* (*m*) || магнитогидродинамическая волна *f*

M87 **magnetomechanical factor** || magnetomechanisches Verhältnis *n*, magnetomechanischer Faktor *m* || rapport *m* gyromagnétique || magnetomechanische factor *m* || гиромагнитное отношение *n*

M88 **magnetometer** || Magnetometer *n* || magnétomètre *m*, magnétimètre *m* || magnetometer *m* || магнитометр *m*

M89 **magnetomotive force** || magnetomotorische Kraft *f* || force *f* magnétomotrice || magnetomotorische kracht *f* (*m*) || магнитодвижущая сила *f*

M90 **magneton** || Magneton *n* || magnéton *m* || magneton *n* || магнетон *m*

M91 **magneto-optics** || Magnetooptik *f* || magnéto-optique *f* || magneto-optica *f* || магнитооптика *f*

M92 **magnetosphere** || Magnetosphäre *f* || magnétosphère *f* || magnetosfeer *f* (*m*) || магнитосфера *f*

M93 **magnetostatics** || Magnetostatik *f* || magnétostatique *f* || magnetostatica *f* || магнитостатика *f*

M94 **magnetostriction** || Magnetostriktion *f* || magnétostriction *f*, striction *f* magnétique || magnetostrictie *f* || магнитострикция *f*

M95 **magnetostriction vibrator** || magnetostriktiver Schwinger *m* || vibrateur *m* magnétostrictif || magnetostrictieve triller *m* || магнитострикционный вибратор *m*

M96 **magnet pole** || Magnetpol *m* || pôle *m* d'un aimant || magneetpool *m* || 1. полюс *m* магнита 2. магнитный полюс *m*

M97 **magnet winding** || Magnetwicklung *f* || bobinage *m* d'électroaimant || magneetwikkeling *f* || обмотка *f* электромагнита

M98 **magnification** || 1. (optische) Vergrößerung *f* 2. Verstärkung *f* (Spannung *f*) || 1. grossissement *m* 2. amplification *f* || 1. vergroting *f* 2. versterking *f* || 1. увеличение *n* *(изображения)* 2. усиление *n* *(напряжения)*

M99 **magnifier, magnifying glass** || Vergrößerungsglas *n*, Lupe *f* || loupe *f*, verre *m* grossissant || vergrootglas *n*, loupe *f* || лупа *f*, увеличительное стекло *n*

M100 **magnitude** || (absoluter) Betrag *m* || magnitude *f*, module *m* || absoluut bedrag *n* || абсолютная величина *f*, модуль *m*

M101 **main bearing** || Grundlager *n*, Kurbelwellenlager *n* || palier *m* principal || krukas(kussen)blok *n*; krukaslager *n* || подшипник *m* коленчатого вала, коренной подшипник *m*

M102 **main storage** || Arbeitsspeicher *m*; Hauptspeicher *m* || mémoire *f* centrale || hoofdgeheugen *n* || оперативная память *f*; основная память *f*

M103 **majorant** || Majorante *f* || majorant *m* || majorante *f* || мажоранта *f*

M104 **majorant of a function** || Majorante *f* einer Funktion || majorant *m* d'une fonction || majorante *f* van een functie || мажоранта *f* функции

M105 **majorant of a series** || Majorante *f* einer Reihe || majorant *m* d'une série || majorante *f* van een reeks || мажоранта *f* ряда

M106 **major axis** || große Achse *f*; Hauptachse *f* || grand axe *m*; axe *m* principal || hoofdas *f* (*m*) || большая ось *f*; главная ось *f*

M107 **major cutting edge** || Hauptschneide *f* || arête *f* principale de coupe || hoofdsnijkant *m* || главная режущая кромка *f*

M108 **major diameter** || Außendurchmesser *m* || diamètre *m* extérieur || buitenmiddellijn *f* (*m*) || наружный диаметр *m*

M109 **major flank** || Hauptfreifläche *f* *(Meißel)* || flanc *m* d'appui *(d'un outil)* || hoofdflank *f* (*m*) *(beitel)* || главная задняя поверхность *f* *(резца)*

M110 **majority carrier** || Majoritäts(ladungs)träger *m*, Hauptladungsträger *m* || porteur *m* des charges majoritaire || hoofdladingsdrager *m* || основной носитель *m* заряда *(в полупроводнике)*

M111 **majorizing** || Majorisieren *n* || majoration *f* || majoratie *f* || мажорирование *n*

M112 **major premise** || Hauptprämisse *f* || grande prémisse *f* || hoofdpremisse *f* || бóльшая посылка *f* *(в математической логике)*

M113 **male thread** || Außengewinde *n* || filet *m* mâle || buitendraad *m* || наружная резьба *f*

M114 **malleability** || Schmiedbarkeit *f* || forgeabilité *f* || smeedbaarheid *f* || ковкость *f*

M115 **malleable cast iron** || Temperguß *m* || fonte *f* forgeable || smeedbaar gietijzer *n* || ковкий чугун *m*

M116 **Maltese cross (mechanism)** || Malteserkreuz *n* || croix *f* de Malte [d'arrêtage] || maltezer kruis *n* || мальтийский механизм *m*, мальтийский крест *m*

M117 **mandrel** || Dorn *m* || mandrin *m* || doorn *m* || оправка *f*

M118 **manganese,** Mn || Mangan *n* || manganèse *m* || mangaan *n* || марганец *m*

M119 **manganese bronze** || Manganbronze *f* || bronze *m* au manganèse || mangaanbrons *n* || марганцовистая бронза *f*

M120 **manganese steel** || Manganstahl *m* || acier *m* au manganèse || mangaanstaal *n* || марганцовистая сталь *f*

M121 **manifold** || 1. Rohrverzweigung *f* 2. Sammelleitung *f* 3. *Math.* Mannigfaltigkeit *f* || 1. tuyau *m* à branchements 2. collecteur *m* 3. variété *f* || 1. pijp *f* (*m*) met aftakkingen 2. verzamelleiding *f* 3. *math.* verscheidenheid *f* || 1. труба *f* с отводами 2. коллектор *m* 3. *мат.* многообразие *n*

M122 **manipulator** || 1. Manipulator *m*; Greifarm *m* 2. Taster *m* || 1. manipulateur *m* 2. clé *f* Morse || seinsleutel *m* || 1. манипулятор *m* 2. телеграфный ключ *m*

M123 **man-made interference** || industrielle Störungen *f pl*, Industriestörungen *f pl* || brouillages *m pl* industriels || industrie-storingen *f pl* || промышленные помехи *f pl* *(радиоприёму)*

M124 **manometer** || Manometer *n*, Druckmesser *m* || manomètre *m* || manometer *m* || (жидкостный) манометр *m*

M125 **mantissa** || Mantisse *f* || mantisse *f* || mantisse *f* || мантисса *f*

M126 **manual control** || manuelle Steuerung *f*; Handbetätigung *f* || commande *f* à main [manuelle] || handbesturing *f* || ручное управление *n*

M127 **manufacturing lathe** || Produktionsdrehmaschine *f* || machine *f* de production || draaibank *f* (*m*) voor massaproduktie || токарный станок *m* для массового [серийного] производства

M128 **manufacturing process** || Fertigungsverfahren *n* || procédé *m* technologique (de production) || technologisch procédé *n* || технологический процесс *m*

M129 **many-body problem** || Mehrkörperproblem *n* || problème *m* de corps multiples || veellichaamprobleem *n* || задача *f* многих тел

M130 **many-body system** || Mehrkörpersystem *n*; Mehrteilchensystem *n* || système *m* de plusieurs corps || veeldeeltjessysteem *n* || система *f* многих тел; система *f* многих частиц

M131 **many-electron atom** || Mehrelektronenatom *n* || atome *m* à plusieurs électrons || veel-elektronenatoom *n* || многоэлектронный атом *m*

M132 **many-particle system** || Mehrteilchensystem *n* || système *m* à plusieurs particules || veeldeeltjessysteem *n* || система *f* многих частиц

M133 **mapping** *math.* || Abbildung *f* || application *f* || afbeelding *f* || отображение *n*

M134 **marginal distribution** || Marginalverteilung *f* || distribution *f* marginale || marginale distributie *f* || маргинальное распределение *n*

M135 **Markov chain** || Markowsche Kette *f* || chaîne *f* markovienne || Markovsche ketting *m*, *f* || марковская цепь *f*

M136 **Markownikoff's rule** || Markownikowsche Regel *f* || règle *f* de Marcovnikoff || regel *m* van Markovnikov || правило *n* Марковникова

M137 **marsh gas**, CH_4 || Sumpfgas *n*, Methan *n* || gaz *m* des marais, méthane *m* || methaan *n* || болотный газ *m*, метан *m*

M138 **maser** || Maser *m* || maser *m* || maser *m* || мазер *m*, микроволновый квантовый генератор *m* и усилитель *m*

M139 **mass** || Masse *f* || masse *f* || massa *f* (*m*) || масса *f*

M140 **mass-analyzing magnet** || magnetischer Massenanalysator *m* || analyseur *m* de masses magnétique || magnetische massa-analysator *m* || магнитный анализатор *m* *(частиц по массам)*

M141 **mass at rest** || Ruh(e)masse *f* || masse *f* au repos || rustmassa *f* (*m*) || масса *f* покоя

M142 **mass-center integral** || Schwerpunktintegral *n* || intégrale *f* du centre des masses || zwaartepuntintegraal *f* (*m*) || интеграл *m* движения центра масс [центра инерции]

M143 **mass defect** || Massendefekt *m*, Kernschwund *m* || défaut *m* de masse || massadefect *n* || дефект *m* массы

M144 **mass density** || Massendichte *f* || densité *f* de matière || massadensiteit *f* || плотность *f* *(вещества)*

M145 **mass-energy relation** || Masse-Energie-Beziehung *f*, Energie-Masse-Beziehung *f* || relation *f* masse-énergie || massa-energieverhouding *f* || взаимосвязь *f* массы и энергии

M146 **mass force** || Massenkraft *f* || force *f* volumique || massakracht *f* (*m*) || массовая [объёмная] сила *f*

M147 **massive** || Massiv..., massiv; schwer || massif; lourd || massief || массивный; тяжёлый

M148 **massless** || **1.** trägheitslos, trägheitsfrei; schwerelos **2.** masselos || **1.** sans inertie **2.** sans masse || massaloos || **1.** безынерционный; невесомый *(о теле)* **2.** безмассовый *(о частице)*

M149 **mass number** || Massenzahl *f* || nombre *m* de masse || massa-getal *n* || массовое число *n*

M150 **mass percent** || Masseprozent *n* || pourcentage *m* en masse || massaprocent *n* || массовый процент *m*, процент *m* по массе

M151 **mass point** || Massepunkt *m*, materieller Punkt *m* || point *m* matériel || massapunt *n* || материальная точка *f*; точечная масса *f*

M152 **mass-spectrograph** || Massenspektrograf *m* || spectrographe *m* de masse || massaspectrograaf *m* || масс-спектрограф *m*

M153 **mass spectrometer** || Massenspektrometer *n* || spectromètre *m* de masse || massaspectrometer *m* || масс-спектрометр *m*

M154 **mass spectroscopy** || Massenspektroskopie *f* || spectroscopie *f* de masse || massaspectroscopie *f* || масс-спектроскопия *f*

M155 **mass-spectrum** || Massenspektrum *n* || spectre *m* de masse || massaspectrum *n* || спектр *m* масс, масс-спектр *m*

M156 **mass storage** || Großspeicher *m* || mémoire *f* de masse || massageheugen *n* || память *f* большой ёмкости

M157 **mass transfer, mass transport** || Stofftransport *m*, Massentransport *m* || transport *m* de la masse || massatransport *n* || перенос *m* массы, массоперенос *m*

M158 **master gauge** || Prüflehre *f* || calibre-étalon *m*, jauge-étalon *m* || proefkaliber *n* || эталонный калибр *m*

M159 **master gear** || Meisterrad *n* || roue-étalon *f* || sjabloontandwiel *n* || эталонная шестерня *f*

M160 **master oscillator** || Muttergenerator *m*, Steueroszillator *m* || maître *m* oscillateur, oscillateur *m* d'excitation || moederoscillator *m* || задающий генератор *m*

M161 **master pattern** || **1.** Modell *n*, Vergleichsmodell *n*; Schablone *f* **2.** Kopierschablone *f* || modèle-maître *m*, modèle-type *m*; gabarit *m* || **1.** monster *n*; model *n*; sjabloon *f* (*m*) **2.** kopiëersjabloon *f* (*m*) || **1.** модель-эталон *m*; мастер-модель *f*; шаблон *m* **2.** копир *m*

M162 **master workpiece** || Muster *n*; Meisterstück *n* || échantillon *m* témoin; échantillon *m* type || monster *n*; model *n*; patroon *n* || образец *m* [эталон *m*] продукции; эталонная деталь *f*

M163 **matched** || angepaßt *(Belastung)* || adapté, accordé || aangepast || согласованный *(напр. о нагрузке)*

M164 **matched impedances** || angepaßte [reflexionsfreie] Widerstände *m pl* || impédances *f pl* adaptées || aangepaste impedanties *f pl* || согласованные сопротивления *n pl*

M165 **matched load** || angepaßte Belastung *f* || charge *f* adaptée || aangepaste belading *f* || согласованная нагрузка *f*

M166 **matching** || Anpassung *f* *(Belastung)* || adaptation *f* || aanpassing *f* || согласование *n* *(напр. нагрузки)*

M167 **material** || Material *n*, Stoff *m*, Werkstoff *m*; Substanz *f* || matériau *m*, matière *f* || stof *f* (*m*) || материал *m*; вещество *n*

M168 **materialization** || Materialisation *f* || matérialisation *f* || materialisatie *f* || **1.** превращение *n* квантов в частицы вещества **2.** материализация *f*

M169 **mathematical expectation** || Erwartungswert *m* || espérance *f* mathématique || mathematische [wiskundige] verwachting *f* || математическое ожидание *n*

M170 **mathematical induction** || mathematische Induktion *f* || induction *f* mathématique || mathematische [wiskundige] inductie *f* || математическая индукция *f*

M171 **mathematical logic** || mathematische Logik *f* || logique *f* mathématique || mathematische logica *f* || математическая логика *f*

M172 **mathematical model** || mathematisches Modell *n* || modèle *m* mathématique || mathematisch model *n* || математическая модель *f*

M173 **mathematical pendulum** || mathematisches Pendel *n* || pendule *m* mathématique [simple] || mathematisch pendel *n* || математический маятник *m*

M174 **mathematical programming** || mathematische Programmierung *f* || programmation *f* mathématique || mathematische programmering *f* || математическое программирование *n*

M175 **mathematization** Mathematisierung *f* || mathématisation *f* || mathematisatie *f* || математизация *f*

M176 **matrix** || Matrix *f* || matrice *f* || matrix *f* || матрица *f*

M177 **matrix algebra** || Matrixalgebra *f*, Matrizenalgebra *f* || algèbre *f* matricielle || matrixalgebra *f* (*m*) || матричная алгебра *f*

M178 **matrix differential equation** || Matrix-Differentialgleichung *f* || équation *f* différentielle matricielle || matrix-differentiaalvergelijking *f* || матричное дифференциальное уравнение *n*

M179 **matrix inversion** || Matrizeninversion *f*, Matrizenumkehrung *f* || inversion *f* de matrices || matrices-inversie *f* || обращение *n* матриц

M180 **matrix norm** || Matrixnorm *f* || norme *f* d'une matrice || matrixnorm *f* (*m*) || норма *f* матрицы

M181 **matrix notation** || Matrizenschreibweise *f* || notation *f* matricielle || matrixnotitie *f* || матричная запись *f*

M182 **matrix operation** || Matrixoperation *f* || opération *f* matricielle || matrixoperatie *f* || матричная операция *f*

M183 **matrix representation** || Matrizendarstellung *f* || représentation matricielle || matrixvoorstelling *f* || матричное представление *n*

M184 **matrix ring** || Matrizenring *m* || anneau *m* des matrices || matrixring *m* || кольцо *n* матриц

M185 **matter** || **1.** Materie *f* **2.** Stoff *m*, Substanz *f* || matière *f* || materie *f* stof *f* (*m*) || **1.** материя *f* **2.** вещество *n*

M186 **matter waves** || Materiewellen *f pl*, De-Broglie-Wellen *f pl* || ondes *f pl* de matière || De-Brogliegolven *f* (*m*) *pl* || волны *f pl* вещества [материи], волны *f pl* де Бройля

M187 **maxima** || Maxima *n pl* || maximums *m pl*, maxima *m pl* || maxima *m pl* || максимумы *m pl*

M188 **maximizing** || Maximierung *f* || maximisation *f* || maximisatie *f* || максимизирование *n*

M189 **maximum** || Maximum *n* || maximum *m* || maximum *n* || максимум *m*

M190 **maximum-likelyhood estimate** || Maximum-Likelyhood-Schätzung *f*, plausibelste Schätzung *f* || estimation *f* à plus vraisemblable || maximaal waarschijnlijke waardering *f* || наиболее правдоподобная оценка *f*

M191 **maximum likelyhood method** || Maximum-Likelyhood-Methode *f* || méthode *f* de la vraisemblance maximale || maximaal waarschijnlijke methode *f* || метод *m* максимального правдоподобия

M192 **maximum load** || Höchstbelastung *f*; Grenzbelastung *f* || charge *f* maximum [maximale] || maximale belasting *f* || максимальная нагрузка *f*

M193 **maximum modulus theorem** || Satz *m* vom Modulmaximum || théorème *m* du maximum d'un module || theorema *n* van het modulusmaximum || теорема *f* о максимуме модуля

M194 **maximum speed** || Höchstgeschwindigkeit *f* || vitesse *f* maximum [maximale] || maximale snelheid *f* || максимальная скорость *f*

M195 **maxwell** || Maxwell *n* || maxwell *m* || maxwell *m* || максвелл *m* (10^{-8}] *Вб*)

M196 **Maxwell equations** || Maxwellsche Beziehungen *f pl*, Maxwellsche Relationen *f pl* || équations *f pl* de Maxwell || Maxwellvergelijkingen *f pl* || уравнения *n pl* Максвелла

M197 **mean count rate** || mittlere Zählgeschwindigkeit *f* *(im Poissonschen Prozeß)* || vitesse *f* moyenne de comptage || middelbare aftelsnelheid *f* || средняя скорость *f* отсчётов *(в процессе Пуансона)*

M198 **mean curvature** || mittlere Krümmung *f* || courbure *f* moyenne || middelbare [gemiddelde] kromming *f* || средняя кривизна *f*

M199 **mean cycle stress** || Lastspielmittelspannung *f* || contrainte *f* moyenne du cycle || gemiddelde spanning *f* van de cyclus || среднее напряжение *n* цикла

M200 **mean deviation** || mittlere Abweichung *f* || déviation *f* moyenne || gemiddelde afwijking *f* || среднее отклонение *n*

M201 **mean effective pressure** || mittlerer Effektivdruck *m* || pression *f* effective moyenne || gemiddelde effectieve druk *m* || среднее эффективное давление *n*

M202 **mean error** || mittlerer Fehler *m* || erreur *f* moyenne || gemiddelde fout *f* || средняя ошибка *f*, средняя погрешность *f*

M203 **mean fatigue stress** || mittlere Ermüdungsspannung *f* || contrainte *f* moyenne à la fatigue || gemiddelde vermoeiingsspanning *f* || среднее напряжение *n* усталости *(за цикл)*

M204 **mean free path** || mittlere freie Weglänge *f*, mittlere freie Wegstrecke *f* || libre parcours *m* moyen || gemiddelde vrije weglengte *f* || средний свободный пробег *m* *(частицы)*

M205 **mean free time** || mittlere Zeit *f* für die freie Wegstrecke || libre temps *m* moyen || gemiddelde tijd *m* van de vrije weglengte || среднее время *n* свободного пробега

M206 **mean radial deviation, mean radial error** || mittlere radiale Abweichung *f* || déviation *f* [erreur *f*] moyenne radiale || gemiddelde radiale afwijking *f* || средняя радиальная ошибка *f*

M207 **mean speed** *see* **average speed**

M208 **mean square** || quadratisches Mittel *n* || moyenne *f* quadratique || kwadratisch middel *n* || среднее квадратичное *n*

M209 **mean-square deviation** || mittlere quadratische Abweichung *f* || déviation *f* moyenne quadratique || gemiddelde kwadratische alwijking *f* || среднее квадратичное отклонение *n*

M210 **mean-square error** || mittlerer quadratischer Fehler *m* || erreur *f* quadratique moyenne || gemiddelde kwadratische fout *f* (*m*) || средняя квадратическая погрешность *f*, средняя квадратичная ошибка *f*

M211 **mean stress** || mittlere Spannung *f*, Mittelspannung *f* || contrainte *f* moyenne || gemiddelde spanning *f* || среднее напряжение *n*

M212 **mean value** || Mittelwert *m*, Mittel *n* || valeur *f* moyenne || gemiddelde waarde *f* || среднее (значение) *n*

M213 **mean-value theorem** || Mittelwertsatz *m* || théorème *m* de la moyenne || theorema *f* over de gemiddelde waarde || теорема *f* о среднем (значении)

M214 **mean velocity** || mittlere Geschwindigkeit *f* || vitesse *f* moyenne || gemiddelde snelheid *f* || средняя скорость *f*

M215 **measurable function** || meßbare Funktion *f* || fonction *f* mesurable || meetbare functie *f* || измеримая функция *f*

M216 **measurable set** || meßbare Menge *f* || ensemble *m* mesurable || meetbare menigte *f* || измеримое множество *n*

M217 **measure** || Maß *n* || mesure *f* || maat *f* (*m*) || мера *f*

M218 **measure algebra** || Maßalgebra *f*; Maßring *m*, Maßverband *m* || algèbre *f* à mesure || maatalbgebra *f* (*m*) || алгебра *f* с мерой

M219 **measurement** || Messung *f*, Ausmessung *f*, Vermessung *f* || mesure *f*, mesurage *m* || meting *f* || измерение *n*

M220 **measure of dispersion** || Streuungsmaß *n* || mesure *f* de dispersion || dispersiemaat *f* (*m*), strooiingsmaat *f* (*m*) || мера *f* дисперсии

M221 **measure theory** || Maßtheorie *f* || théorie *f* de la mesure || maattheorie *f* || теория *f* меры

M222 **measuring** || Durchführung *f* von Messungen, Messung *f* || mesurage *m* || meting *f* || проведение *n* измерений, измерения *n pl*

M223 **measuring apparatus** || Meßapparatur *f*, Meßgeräte *n pl* || appareillage *m* de mesure || meetapparatuur *f* || измерительная аппаратура *f*

M224 **measuring instrument** || Meßgerät *n*, Meßinstrument *n* || instrument *m* de mesure || meetinstrument *n* || измерительный прибор *m*

M225 **measuring system** || Meßsystem *n* || système *m* de mesure || meetsysteem *n* || измерительная система *f*

M226 **mechanical** || mechanisch || mécanique || mechanisch || механический

M227 **mechanical advantage** || Kraftgewinn *m* || gain *m* de force || krachtwinst *f* || выигрыш *m* в силе

M228 **mechanical efficiency** || mechanischer Wirkungsgrad *m* || rendement *m* mécanique || mechanisch rendement *n* || механический коэффициент *m* полезного действия, механический кпд

M229 **mechanical energy** || mechanische Energie *f* || énergie *f* mécanique || mechanische energie *f* || механическая энергия *f*

M230 **mechanical equivalent of heat** || mechanisches Wärmeäquivalent *n* || équivalent *m* mécanique de chaleur || mechanisch warmteëquivalent *n* || механический эквивалент *m* тепла

M231 **mechanical fatigue** || mechanische Ermüdung *f* || fatigue *f* mécanique || mechanische moeheid *f* || механическая усталость *f*

M232 **mechanical fatigue test** || Prüfung *f* auf mechanische Ermüdung *f* || éssai *m* de fatigue mécanique || test *m* op mechanische moeheid || испытание *n* на механическую усталость

M233 **mechanical load** || mechanische Belastung *f* || charge *f* mécanique || mechanische belasting *f* || механическая нагрузка *f*

M234 **mechanical power** || mechanische Energie *f* || énergie *f* mécanique || mechanische energie *f* || механическая энергия *f*

M235 **mechanical properties** || mechanische Eigenschaften *f pl* || propriétés *f pl* mécaniques || mechanische eigenschappen *n pl* || механические свойства *n pl*

M236 **mechanical strength** || mechanische Festigkeit *f* || résistance *f* mécanique || mechanische sterkte *f* || механическая прочность *f*

M237 **mechanical system** || mechanisches System *n* || système *m* mécanique || mechanisch systeem *n* || механическая система *f*

M238 **mechanical testing** || mechanisches Prüfverfahren *n pl* || éssai *m* mécanique || mechanische proefmethode *f* || механические испытания *n pl*

M239 **mechanical trajectory** || Bewegungsbahn *f* || trajectoire *f* de mouvement || bewegingsbaan *f* (*m*) || траектория *f* движения

M240 **mechanical treatment** || mechanische Bearbeitung *f* || traitement *m* mécanique || mechanische [koude] bewerking *f* || механическая обработка *f*

M241 **mechanical vibrations** || mechanische Schwingungen *f pl* || vibrations *f pl* mécaniques || mechanische trillingen *f pl* || механические колебания *n pl*

M242 **mechanics** || Mechanik *f* || mécanique *f* || mechanica *f* || механика *f*

M243 **mechanics of continua** || Kontinuumsmechanik *f*, Mechanik *f* der Kontinua || mécanique *f* du milieu continu || continuïteitsmechanica *f* || механика *f* сплошных сред

M244 **mechanics of fluids** || Strömungsmechanik *f*, Mechanik *f* der Flüssigkeiten und Gase || mécanique *f* des fluides || hydromechanica *f* || механика *f* жидкостей и газов, гидроаэродинамика *f*

M245 **mechanics of rigid bodies** || Mechanik *f* des absolut starren Körpers || mécanique *f* des solides || mechanica *f* van vaste lichamen || механика *f* (абсолютно) твёрдого тела

M246 **mechanism of a reaction** || Reaktionsmechanismus *m* || mecanisme *m* de réaction || reactiemechanisme *n* || механизм *m* реакции

M247 **mechanostriction** || Mechanostriktion *f* || mécanostriction *f* || mechanostrictie *f* || механострикция *f*

M248 **median** || **1.** Median *m*, Medianwert *m* **2.** Seitenhalbierende *f*, Mediane *f* || médiane *f* || mediaan *f* || медиана *f*

M249 **median of a triangle** || Seitenhalbierende *f* [Mediane *f*] eines Dreiecks || médiane *f* d'un triangle || mediaan *f* van een driehoek || медиана *f* треугольника

M250 **medium** || Medium *n*, Mittel *n* || milieu *m* || medium *n* || среда *f*

M251 **mega...** || Mega... || méga... || mega... || мега... *(приставка в наименовании кратной единицы, соответствующая 10^6)*

M252 **megajoule** || Megajoule *n* || mégajoule *m* || megajoule *m* || мегаджоуль *m* *(10^6 Дж)*

M253 **megger** || Megger *m*, Megohmmeter *n* || mégohmmètre *m*, ohmmètre *m* à magnéto || megohmmeter *m* || мегомметр *m*, меггер *m*

M254 **melting** || Schmelzen *n*, Schmelzung *f* || fusion *f*, fonte *f* || smelting *f* || плавление *n*; расплавление *n*

M255 **melting point, melting temperature** || Schmelztemperatur *f*, Schmelzpunkt *m* || température *f* [point *m*] de fusion || smelttemperatuur *f*, smeltpunt *f m* || температура *f* [точка *f*] плавления

M256 **member** || Glied *n*; Term *m* || membre *m* || term *m* || член *m*

M257 **member of a class** || Klassenelement *n* || membre *m* d'une classe || term *m* van een klas || элемент *m* класса

M258 **membrane** || **1.** Membran *f* **2.** Trennwand *f*, Zwischenwand *f* || membrane *f* || **1.** membraan *n*, *f* (*m*) **2.** tussenschot *n* || **1.** мембрана *f* **2.** перегородка *f*

M259 **membrane equilibrium** || Membrangleichgewicht *n* || équilibre *m* de membrane || membraanevenwicht *n* || мембранное равновесие *n*

M260 **memory** || Speicher *m* || mémoire *f* || geheugen *n* || память *f*, запоминающее устройство *n*, ЗУ

M261 **mendelevium**, Md || Mendelevium *n* || mendélévium *m* || mendelevium *n* || менделевий *m*

M262 **meniscus** || Meniskus *m*, Meniskuslinse *f* || lentille *f* à ménisque, ménisque *m* || hol-bolle lens *f* (*m*), meniscus *n* || мениск *m*, менисковая линза *f*

M263 **meniscus correction** || Meniskuskorrektion *f* || correction *f* de ménisque || meniscuscorrectie *f* || поправка *f* на мениск *(жидкости)*

M264 **meniscus lens** *see* **meniscus**

M265 **mercuric chloride,** $HgCl_2$ || Quecksilberchlorid *n*, Sublimat *n* || chlorure *m* mercurique, sublimé *m* corrosif || mercurochloride *n*, sublimaat *n* || хлорная ртуть *f*, сулема *f*

M266 **mercury,** Hg || Quecksilber *n* || mercure *m* || kwik(zilver) *n* || ртуть *f*

M267 **mercury barometer** || Quecksilberbarometer *n* || baromètre *m* à mercure || kwikzilverbarometer *m* || ртутный барометр *m*

M268 **mercury column** || Quecksilbersäule *f* || colonne *f* de mercure || kwikzuil *f* || ртутный столб *m*

M269 **mercury-discharge lamp** || Quecksilber(dampf)lampe *f*; Quecksilberdampfröhre *f* || lampe *f* à (vapeur de) mercure || kwik(damp)lamp *f* (*m*) || ртутная лампа *f*

M270 **mercury manometer** || Quecksilbermanometer *n* || manomètre *m* à mercure || kwik(zilver)manometer *m* || ртутный манометр *m*

M271 **mercury thermometer** || Quecksilberthermometer *n* || thermomètre *m* à mercure || kwik(zilver)thermometer *m* || ртутный термометр *m*

M272 **meromorphic function** || meromorphe Funktion *f* || fonction *f* méromorphe || meromorfe functie *f* || мероморфная функция *f*

M273 **mesh** || 1. Eingriff *m (Zahnräder)* 2. Masche *f (Sieb)* || 1. engrènement *m* 2. maille *f (de tamis)* || 1. ineengrijpen *n (tandwielen)* 2. maas *f (m)* || 1. зацепление *n (зубчатых колёс)* 2. отверстие *n (решета, сита)*

M274 **mesh-connection** || Vieleckschaltung *f*, Polygonschaltung *f* || couplage *m* en polygone || veelhoekschakeling *f*, polygoonschakeling *f* || соединение *n* многоугольником

M275 **mesomerism** || Mesomerie *f*, Strukturresonanz *f* || mésomérie *f* || mesomerie *f* || мезомерия *f*

M276 **mesomorphic state** || mesomorpher Zustand *m* || état *m* mésomorphe || mesomorfe toestand *m* || мезоморфное состояние *n*

M277 **meson** || Meson *n* || méson *m* || meson *n* || мезон *m*

M278 **mesonic atom** || Mesoatom *n* || mésoatome *m*, atome *m* mésique || mesonisch atoom *m* || мезоатом *m*

M279 **metacentre** || Metazentrum *n* || métacentre *m* || metacentrum *n* || метацентр *m*

M280 **metal** || Metall *n* || métal *m* || metaal *n* || металл *m*

M281 **metalanguage** || Metasprache *f* || métalangage *m* || meta-taal *f* || метаязык *m*

M282 **metallic binding** || Metallbindung *f*, metallische Bindung *f* || liaison *f* métallique || metallische binding *f* || металлическая связь *f*

M283 **metallic conduction** || metallische Leitfähigkeit *f*, elektrische Leitung *f* der Metalle || conductibilité *f* métallique || metallische leiding *f* || металлическая проводимость *f*

M284 **metallic crystal** || metallischer Kristall *m* || cristal *m* métallique || metallisch kristal *n* || металлический кристалл *m*

M285 **metallography** || Metallografie *f* || métallographie *f* || metallografie *f* || металлография *f*

M286 **metal mo(u)ld** || Dauergießform *f* || moule *m* à couler métallique || metallische gietvorm *m* || металлическая (литейная) форма *f*

M287 **metal optics** || Metalloptik *f* || optique *f* des métaux || metaaloptiek *f* || металлооптика *f*, оптика *f* металлов

M288 **metal-working machine** || spanabhebende Werkzeugmaschine *f* || machine *f* de coupe || metaalbewerkingsmachine *f* || металлорежущий [металлообрабатывающий] станок *m*

M289 **metamagnetic** || Metamagnetikum *n*, metamagnetischer Stoff *m* || métamagnétique *m* || metamagneticum *n* || метамагнетик *m*

M290 **metamathematics** || Metamathematik *f* || métamathématique *f* || metamathematica *f* || метаматематика *f*

M291 **metamerism** || Metamerie *f* || métamérie *f* || metamerie *f* || метамерия *f*

M292 **metastability** || Metastabilität *f* || métastabilité *f* || metastabiliteit *f* || метастабильность *f*

M293 **metastable** || metastabil || métastable || metastabiel || метастабильный

M294 **metastable atom** || metastabiles Atom *n* || atome *m* métastable || metastabiel atoom *n* || метастабильный атом *m*

M295 **metastable equilibrium** || metastabiles Gleichgewicht *n* || équilibre *m* métastable || metastabiel evenwicht *n* || метастабильное равновесие *n*

M296 **metastable level** || metastabiles Niveau *n* || niveau *m* métastable || metastabiel niveau *n* || метастабильный (энергетический) уровень *m*

M297 **metastable state** || metastabiler Zustand *m* || état *m* métastable || metastabiele toestand *m* || метастабильное состояние *n*

M298 **metathetical reaction** || Austauschreaktion *f* || réaction *f* d'échange || omzettingsreactie *f* || реакция *f* обмена

M299 **meter** || Meßgerät *n*, Meßinstrument *n* || appareil *m* de mesure, mesureur *m* || meetinstrument *n* || измерительный прибор *m*

M300 **metering** || Messung *f*; Registrierung *f* || mesure *f*, mesurage *m* || meten *n* || измерение *n*, регистрация *f*

M301 **methane,** CH_4 || Methan *n* || méthane *m* || methaan *n* || метан *m*

M302 **methanol, methyl alcohol,** CH_3OH || Methylalkohol *m*, Methanol *n* || alcool *m* méthylique, méthanol *m* || methylalcohol *m*, methanol *n* || метиловый спирт *n*, метанол *m*

M303 **methylation** || Methylieren *n*, Methylierung *f* || méthylation *f* || methylering *f* || метилирование *n*

M304 **metre** || Meter *n* || mètre *m* || meter *m* || метр *m*

M305 **metre-candle** || Lux *n* || lux *m* || lux *f* || люкс *m* *(1 лм/м²)*

M306 **metric** || Metrik *f* || métrique *f* || metriek *f* || метрика *f*

M307 **metric invariant** || metrische Invariante *f* || invariant *m* métrique || metrieke invariant *m* || метрический инвариант *m*

M308 **metric screw thread** || metrisches Gewinde *n* || filet *m* métrique || metrische schroefdraad *m* || метрическая резьба *f*

M309 **metric space** || metrischer Raum *m* || espace *m* métrique || metrieke ruimte *f* || метрическое пространство *n*

M310 **metric system** || metrisches System *n*, metrisches Einheitensystem *n* || système *m* métrique || metrisch systeem *n* [stelsel *n*] || метрическая система *f* (мер)

M311 **mho** || Kehrwert *m* des Ohms, mho || siemens *m* || siemens *m* || обратный ом *m*, сименс *m*, сим *m*

M312 **micelle** || Mizelle *f*, Mizell *n* || micelle *f* || micel *f* (*m*) || мицелла *f*

M313 **Michelson echleon** || Michelsonsches Staffelgitter *n*, Echelon *n* || échelon *m* de Michelson || Michelson-echelon *m* || эшелон *m* Майкельсона

M314 **micro...** || Mikro... || micro... || micro... || микро... *(приставка в наименовании дольной единицы, соответствующая 10⁻⁶)*

M315 **microammeter** || Mikroamperemeter *n* || microampèremètre *m* || microampèremeter *m* || микроамперметр *m*

M316 **microampere** || Mikroampere *n* || microampère *m* || microampère *m* || микроампер *m* *(10⁻⁶ A)*

M317 **microcircuit module** || Mikromodul *m*, Mikromodulbaustein *m* || module *m* micro-miniature, micromodule *m* || micromodule *n* || микромодуль *m* *(электронной схемы)*

M318 **microelectronics** || Mikroelektronik *f* || microélectronique *f* || micro-elektronica *f* || микроэлектроника *f*

M319 **micrometer (gauge)** || Mikrometer *n*, Feinmeßschraube *f* || micromètre *m* || micrometer *m* || микрóметр *m*

M320 **micrometer microscope** || Meßmikroskop *n* || microscope *m* de mesure || meetmicroscoop *m* || измерительный микроскоп *m*

M321 **micromicro...** *see* **pico...**

M322 **micron** || Mikrometer *n* || micromètre *m*, micron *m* || micrometer *m*, micron *n*, *m* || микромéтр *m* *(10⁻⁶ м)*, микрон *m*

M323 **microphone** || Mikrofon *n* || micro(phone) *m* || microfoon *m* || микрофон *m*

M324 **microphonics** || Mikrofoneffekt *m* || effet *m* microphonique, microphonicité *f*, microphonie *f* || microfooneffect *n*, microfonisch effect *n* || микрофонный эффект *m*

M325 **microphotography** || Mikrophotographie *f* || microphoto(graphie) *f*, photomicrographie *f* || microfotografie *f* || микрофотография *f*, микрофотографирование *n*

M326 **microphysics** || Mikrophysik *f* || microphysique *f* || microfysica *f* || физика *f* микромира

M327 **microprocessor** || Mikroprozessor *m* || microprocesseur *m* || microprocessor *m* || микропроцессор *m*

M328 **microscope** || Mikroskop *n* || microscope *m* || microscoop *m* || микроскоп *m*

M329 **microscopic** || mikroskopisch || microscopique || microscopisch || микроскопический

M330 **microscopic structure** || Mikrostruktur *f*, Feinstruktur *f*, Feingefüge *n* || microstructure *f* || microstructuur *f* || микроскопическая структура *f*, микроструктура *f*

M331 **microscopic theory** || mikroskopische Theorie *f* || théorie *f* microscopique || microscopische theorie *f* || микроскопическая теория *f*

M332 **microscopy** || Mikroskopie *f* || microscopie *f* || microscopie *f* || микроскопия *f*

M333 **microstructure** *see* **microscopic structure**

M334 **microwave** || Mikrowelle *f*, Höchstfrequenzwelle *f* || micro-onde *f* || microgolf *f* (*m*) || ультравысокочастотная волна *f*

M335 **microwave imaging** || Funkvision *f* || radiovision *f* || radiovisie *f* || радиовидение *n*

M336 **microwave radiation** || Mikrowellenstrahlung *f* || radiation *f* à micro-ondes || microgolvenstraling *f* || микроволновое излучение *n*

M337 **microwaves** || Mikrowellen *f pl* || micro-ondes *f pl* || microgolven *f* (*m*) *pl* || микроволны *f pl*

M338 **microwave spectroscopy** || Mikrowellenspektroskopie *f*, Radiofrequenzspektroskopie *f* || radiospectrographie *f* || microgolfspectroscopie *f* || микроволновая спектроскопия *f*, радиоспектроскопия *f*

M339 **midpoint of a segment** || Segmentmitte *f* || mi-point *m* d'un segment || middelpunt *n* van een lijn || середина *f* отрезка

M340 **midship section** || Mittelquerschnitt *m* || maître *m* couple || middeldoorsne(d)e *f* (*m*) || миделево(е) сечение *n*

M341 **mild steel** || Flußstahl *m*, niedriggekohlter Stahl *m* || acier *m* doux || zachtstaal *n* || мягкая [низкоуглеродистая] сталь *f*

M342 **mill** || **1.** Werk *n*; Fabrik *f* **2.** Walzwerk *n* **3.** Fräser *m* || **1.** usine *f* **2.** laminoir *m*, train *n* **3.** fraise *f* || **1.** fabriek *f* **2.** walswerk *n* **3.** frees *f* (*m*) || **1.** завод *m*; фабрика *f* **2.** прокатный стан *m* **3.** фреза *f*

M343 **milli...** || Milli... || milli... || milli... || милли... *(приставка в наименовании дольной единицы, соответствующая* 10^{-3}*)*

M344 **milliammeter** || Milliamperemeter *n* || milliampèremètre *m* || milliampèremeter *m* || миллиамперметр *m*

M345 **milliard** || Milliarde *f* || billion *m*, milliard *m* || miljard *n* || миллиард *m*, биллион *m*

M346 **millimicron** || Nanometer *n* || nanomètre *m* || nanometer *m* || нанометр *m* *(*10^{-9} *м)*

M347 **milling** || Fräsen *n* || fraisage *m* || frezen *n* || фрезерование *n*

M348 **milling cutter** || Fräser *m* || fraise *f* || frees *f* (*m*) || фреза *f*

M349 **milling head** || Fräskopf *m* || tête *f* de fraisage || freeskop *m* || фрезерная головка *f*

M350 **milling machine** || Fräsmaschine *f* || fraiseuse *f*; machine *f* à fraiser || freesbank *f* (*m*), freesmaschine *f* || фрезерный станок *m*

M351 **milling machine arbor** || Fräsdorn *m* || mandrin *m* de fraisage || freesdoorn *m* || фрезерная оправка *f*

M352 **milling unit** *see* **milling head**

M353 **millivolt** || Millivolt *n* || millivolt *m* || millivolt *m* || милливольт *m* *(*10^{-3} *В)*

M354 **millivoltmeter** || Millivoltmeter *n* || millivoltmètre *m* || millivoltmeter *m* || милливольтметр *m*

M355 **mill rolls** || Walzen *f pl* || cylindres *m pl* de laminage || walsen *m pl* || прокатные валки *m pl*

M356 **mineral oil** || Mineralöl *n* || huile *f* minérale || minerale olie *f* (*m*) || минеральное масло *n*

M357 **minicomputer** || Minirechner *m* || mini-ordinateur *m* || minicomputer *m* || мини-ЭВМ *f*

M358 **minima** || Minima *n pl* || minimums *m pl*, minima *m pl* || minima *n pl* || минимумы *m pl*

M359 **minimal surface** || Minimalfläche *f* || surface *f* minimale || minimaalvlak *n* (*m*) || минимальная поверхность *f*

M360 **minimax** || Minimax *n* || minimax *m* || minimax *n* || минимакс *m*

M361 **minimax test** || Minimax-Kriterium *n* || critère *m* de minimax || minimax-criterium *n* || минимаксный критерий *m*

M362 **minimax theorem** || Minimax-Satz *m* || théorème *m* de minimax || minimax-theorema *n* || теорема *f* минимакса

M363 I **minimizing** || Minimierung *f* || minimisation *f* || minimisatie *f* || минимизация *f*
M364 II **minimizing** || minimierend || minimisant || minimierend || минимизирующий
M365 **minimum** || Minimum *n* || minimum *m* || minimum *n* || минимум *m*
M366 **minium** || Mennige *f*, Minium *n* || minium *m* || meme *f* (*m*) || сурик *m*
M367 **Minkowski's geometry** || Minkowskische Geometrie *f* || géométrie *f* de Minkowski || Minkowskische geometrie *f* || геометрия *f* Минковского
M368 **minor** || Minor *m* || mineur *m* || minor *m* || минор *m*
M369 **minor axis** || Nebenachse *f* || petit axe *m* || kleine as *f* (*m*) || малая ось *f* *(орбиты, эллипса)*
M370 **minor diameter** || Innendurchmesser *m* || diamètre *m* intérieur du filet || binnendiameter *m* || внутренний диаметр *m*
M372 **minority carrier** || Minoritäts(ladungs)träger *m* *(im Halbleiter)* || porteur *m* des charges minoritaire || minoriteits(ladings)drager *m* || неосновной носитель *m* заряда *(в полупроводнике)*
M373 **minor premise** || Nebenprämisse *f* || prémisse *f* mineure || kleine premisse *f* || малая посылка *f* *(в математической логике)*
M373a **minor cutting edge** *see* **front cutting edge**
M374 **mirror** || Spiegel *m*; Reflektor *m*, Rückstrahler *m* || miroir *m*; réflecteur *m* || spiegel *m*; reflector *m* || зеркало *n*; рефлектор *m*, отражатель *m*
M375 **mirror image** || Spiegelbild *n* || image *f* reflétée || spiegelbeeld *n* || зеркальное изображение *n*
M376 **mirror surface** || Spiegelfläche *f* || surface *f* à polissage brillant || spiegelvlak *n* || зеркальная поверхность *f*
M377 **mirror symmetry** || Spiegelsymmetrie *f* || symétrie *f* par réflexion || spiegelsymmetrie *f* || зеркальная симметрия *f*
M378 **misalignment** || Fluchtungsfehler *m*; Verlagerung *f* || non-coaxialité *f*; non-alignement *m* || niet-coaxiaal aanleg *m* || несоосность *f*; смещение *n*; нарушение *n* центрирования
M379 **miscibility** || Mischbarkeit *f*; Löslichkeit *f* || miscibilité *f* || mengbaarheid *f*; oplosbaarheid *f* || смешиваемость *f*; растворимость *f*
M380 **mismatch** || Fehlanpassung *f* || désadaptation *f* || foutieve aanpassing *f* || рассогласование *n*
M381 **misphasing** || Phasenverschiebung *f*, Phasendrehung *f* || déphasage *m* || faseverschuiving *f* || расфазирование *n*, расфазировка *f*
M382 **miter** || Gehrung *f*; Winkel *m* von 45° || biais *m*, angle *m* oblique (à 45 degrés) || verstek *n*; hoek *m* van 45° || скос *m*; угол *m* 45°
M383 **miter valve** || Kegelventil *n* || soupape *f* à siège conique à 45 degrés || kegelklep *f* (*m*) || клапан *m* с коническим седлом 45°
M384 **miter wheel** || Kegelrad *n* mit einem Teilkegelwinkel von 45° || roue *f* d'angle à 45°; pignon *m* conique à 45° || kegelwiel *n* met deelkegelhoek van 45° || коническое зубчатое колесо *n* с углом делительного конуса 45°
M385 **mitre** *see* **miter**
M386 **mixed friction** || Mischreibung *f*, gemischte Reibung *f* || frottement *m* mixte || gemengde wrijving *f* || смешанное трение *n*
M387 **mixed gas** || Gasgemisch *n* || mélange *m* gazeux || gasmengsel *n* || газовая смесь *f*, смесь *f* газов
M388 **mixed tensor** || gemischter Tensor *m* || tenseur *m* mixte || gemengde tensor *m* || смешанный тензор *m*
M389 **mixer** || Mischstufe *f*, Mischer *m* || mélangeur *m* || menginrichting *f* || смеситель *m*
M390 **mixing chamber** || Mischkammer *f* || chambre *f* de mélange [de malaxage] || mengkamer *f* (*m*) || смесительная камера *f*
M391 **mixture** || Gemisch *n*, Mischung *f* || mélange *m* || mengsel *n* || смесь *f*
M392 **mobile load** || bewegliche [wandernde] Belastung *f* || charge *f*

mobile || mobele belasting *f* || подвижная нагрузка *f*

M393 **mobility** || Beweglichkeit *f* || mobilité *f* || mobiliteit *f* || подвижность *f*

M394 **modal column** || Eigenvektor *m (einer Matrix)* || colonne *f* modale || modale kolom *f* (*m*) || модальный столбец *m*, собственный вектор *m*

M395 **modal logic** || modale Logik *f* || logique *f* modale || modale logica *f* || модальная логика *f*

M396 **modal matrix** || Modalmatrix *f* || matrice *f* modale || modale matrix *f* || модальная матрица *f*

M397 **mode** || **1.** Art *f*, Gattung *f*, Form *f*, Typ *m* **2.** Schwingungsart *f*, Schwingungsform *f* **3.** Betriebsart *f*, Betriebsweise *f* **4.** *Math.* Mode *f*, Modalwert *m* || **1.** aspect *m*, forme *f*, genre *m*, mode *m* **2.** mode *m* de vibrations **3.** régime *m* **4.** *math.* mode *f* || **1.** aard *m*, type *n*, grondvorm *m* **2.** trillingsmodus *m* **3.** regime *n* **4.** *math.* mode *f* || **1.** вид *m*, тип *m*, форма *f* **2.** вид *m* [форма *f*] колебаний **3.** режим *n* **4.** *мат.* мода *f*

M398 **model** || Modell *n* || modèle *m* || model *n* || модель *f*

M399 **model(l)ing** || Modellierung *f*, Nachbildung *f*, Simulation *f* || simulation *f* || simulatie *f*, modelleren *n* || моделирование *n (явления)*

M400 **mode of decay** || **1.** Zerfallsart *f* **2.** Abklingscharakter *m* || **1.** mode *m* de désintégration **2.** nature *f* de baisse || **1.** vervals-aard *m* **2.** dalingskarakter *n* || **1.** вид *m* распада **2.** характер *m* спада

M401 **mode of operation** || Betriebsweise *f*, Betriebsart *f*, Betrieb *m* || régime *m* de fonctionnement || werkingsregime *n* || режим *m* работы

M402 **mode of vibration** || Schwingungsart *f*, Schwingungsform *f* || mode *m* de vibrations [d'oscillations] || trillingsmodus *m* || вид *m* [форма *f*] колебаний

M403 **moderator** || Moderator *m*, Bremsmittel *n*, Bremsstoff *m*, Neutronenbremse *f* || modérateur *m*, substance *f* modératrice || moderator *m* || замедлитель *m (нейтронов)*

M404 **modification** || Modifikation *f* || modification *f* || modificatie *f* || модификация *f*

M405 **modified function** || modifizierte Funktion *f* || fonction *f* modifiée || gemodificeerde functie *f* || модифицированная функция *f*

M406 **modifier** || Modifikator *m*; Modifikation *f* || modifieuse *f*; modificateur *m* || modificator *m*; modificatie *f* || модификатор *m*; модификация *f*

M407 **modular** || modular, Modul... Baukasten... || modulaire || modulair || агрегатный; модульный

M408 **modular angle** || Modularwinkel *m* || angle *m* modulaire || modulaire hoek *m* || модулярный угол *m*

M409 **modulated oscillations** || modulierte Schwingungen *f pl* || oscillations *f pl* modulées || gemoduleerde trillingen *f pl* || модулированные колебания *n pl*

M410 **modulated wave** || modulierte Welle *f* || onde *f* modulée || gemoduleerde golf *f* (*m*) || модулированная волна *f*

M411 **modulation** || Modulation *f*, Modelung *f* || modulation *f* || modulatie *f* || модуляция *f*

M412 **modulation factor** || Modulationsgrad *m*, Modulationstiefe *f* || profondeur *f* [taux *m*] de modulation || modulatiegraad *m*, modulatiediepte *f*, modulatiefactor *m* || глубина *f* [коэффициент *m*] модуляции

M413 **modulation frequency** || Modulationsfrequenz *f* || fréquence *f* de modulation || modulatiefrequentie *f* || частота *f* модуляции

M414 **modulator** || Modulator *m*, Modler *m* || modulateur *m* || modulator *m* || модулятор *m*

M415 **module** || Modul *m* || module *m* || modulus *m* || модуль *m*

M416 **modulus** || Modul *m*; Betrag *m* || module *m* || modulus *m* || модуль *m*

M417 **modulus of elasticity** || E-Modul *m*, Elastizitätsmodul *m* || module *m* d'élasticité || elasticiteitsmodulus *m* || модуль *m* упругости

M418 **modulus of elasticity in shear, modulus of rigidity** || Schub(elastizitäts)modul *m* || module *m* de cisaillement; module *m* de rigidité || elasticiteits- en afschuivingsmodulus *m* || модуль *m* упругости при сдвиге

M419 **modulus of section** *see* **section modulus**

M420 **Moebius strip** || Möbiussches Blatt *n* || bande *f* de Möbius || band *n* van Möbius || лист *m* Мёбиуса

M421 **moisture** || Feuchtigkeit *f*, Feuchte *f* || humidité *f* || vochtigheid *f* || влажность *f*; влага *f*

M422 **moisture content** || Feuchtigkeitsgehalt *m*, Feuchtegehalt *m* || teneur *f* en humidité || vochtigheidsgehalte *n* || влагосодержание *n*, влагоёмкость *f*; (удельная) влажность *f*

M423 **molality** || Molalität *f* || molalité *f* || molaliteit *f* || моляльность *f*

M424 **molar fraction** || Molenbruch *m* || fraction *f* molaire || molaire fractie *f* || мольная [молярная] доля *f*, мольная долевая концентрация *f*

M425 **molarity** || Molarität *f* || molarité *f* || molariteit *f* || молярность *f*

M426 **mold** *see* **mould**

M427 **mole** || Mol *n*, Grammolekül *n* || mole *f*, gramme-molécule *m*, molécule-gramme *f* || mol *f*, grammolecule *f* (*m*) || моль *m*, грамм-молекула *f*

M428 **molecular** || molekular, Molekular... || moléculaire || moleculair || молекулярный

M429 **molecular beam** || Molekularstrahl *m*, Molekülstrahl *m* || faisceau *m* moléculaire || moleculaire bundel *m* || молекулярный пучок *m*

M430 **molecular conductivity** || molare Leitfähigkeit *f*, Molarleitfähigkeit *f* || conductibilité *f* électrique molaire || molair elektrisch geleidingsvermogen *n* || мольная [молярная] электропроводность *f*

M431 **molecular crystal** || Molekülkristall *m*, molekularer Kristall *m* || cristal *m* moléculaire || moleculairkristal *n* || молекулярный кристалл *m*

M432 **molecular distilation** || Molekulardestillation *f* || distillation *f* moléculaire || moleculaire distillatie *f* || молекулярная перегонка *f*

M433 **molecular electronics** || Molekularelektronik *f*, Molelektronik *f* || électronique *f* moléculaire || moleculaire elektronica *f* || функциональная [молекулярная] электроника *f*

M434 **molecular heat** || molare Wärmekapazität *f* || capacité *f* calorifique mol(écul)aire || mol(ecul)aire warmtecapaciteit *f* || мол(екул)ярная теплоёмкость *f*

M435 **molecular ion** || Molekülion *n* || ion *m* moléculaire || moleculairion *n* || молекулярный ион *m*

M436 **molecular-kinetic theory** || molekularkinetische Theorie *f* || théorie *f* moléculaire-cinétique || moleculairkinetische theorie *f* || молекулярно-кинетическая теория *f*

M437 **molecular mass** || Molekülmasse *f*, Molekularmasse *f* || masse *f* moléculaire || molecuulmassa *f* (*m*) || масса *f* молекулы, молекулярная масса *f*

M438 **molecular oscillator** || Molekulargenerator *m*, Molekularoszillator *m* || oscillateur *m* moléculaire || moleculairgenerator *m* || молекулярный генератор *m*

M439 **molecular physics** || Molekularphysik *f*, Molekülphysik *f* || physique *f* moléculaire || moleculairfysica *f* || молекулярная физика *f*

M440 **molecular refraction, molecular refractivity** || Mol(ekular)refraktion *f* || réfraction *f* mol(écul)aire || mol(ecul)aire refractie *f* || мол(екул)ярная рефракция *f*

M441 **molecular scattering** || Molekularstreuung *f*, molekulare Streuung *f* || diffusion *f* moléculaire || moleculaire diffusie *f* || молекулярное рассеяние *n*

M442 **molecular sieve** || Molekularsieb *n* || tamis *m* moléculaire || molecuulzeef *f* (*m*) || молекулярное сито *n*

M443 **molecular spectroscopy** || Molekülspektroskopie *f* || spectroscopie *f* moléculaire || moleculairspectroscopie *f*, molecuulspectroscopie *f* || молекулярная спектроскопия *f*

M444 **molecular spectrum** || Molekülspektrum *n* || spectre *m* moléculaire || moleculairspectrum *n*, molecuulspectrum *n* || молекулярный спектр *m*

M445 **molecular structure** || 1. molekularer Aufbau *m*, molekulare Struktur *f*, Molekulargefüge *n* 2. Molekülbau *m*, Molekülstruktur *f* || 1. structure *f* moléculaire 2. structure *f* de molécule || moleculaire structuur *f*, molecuulstructuur *f*, molecuulopbouw *m* 1. молекулярное строение *n* 2. структура *f* молекулы

M446 **molecular weight** || Molekularmasse *f*, (relative) Molekülmasse *f* || masse *f* moléculaire || moleculairgewicht *n*, molecuulgewicht *n* || молекулярная масса *f*

M447 **molecule** || Molekül *n* || molécule *f* || molecule *f* (*m*) || молекула *f*

M448 **molybdenum,** Mo || Molybdän *n* || molybdène *m* || molybdeen *n* || молибден *m*

M449 **moment** || 1. Moment *n* 2. Moment *m*, Zeitmoment *m*, Augenblick *m* || moment *m* || moment *n* || 1. момент *m* 2. момент *m* (времени)

M450 **moment arm** || Kraftarm *m*, Arm *m* des Kräftepaares || bras *m* de levier du couple || krachtenpaar *n* || плечо *n* пары сил

M451 **momentary load** || kurzzeitige Belastung *f* || charge *f* momentanée || momentane belasting *f* || кратковременная нагрузка *f*

M452 **moment curve** || Momentenkurve *f*, Momentenlinie *f* || épure *f* [ligne *f*] des moments || momentenkromme *f* (*m*) || эпюра *f* (изгибающих) моментов

M453 **moment diagram** || Momentendiagramm *n* || diagramme *m* des moments (fléchissants) || momentenkromme *f* (*m*) || диаграмма *f* моментов; эпюра *f* (изгибающих) моментов

M454 **moment-generating function** || Momentenerzeugungsfunktion *f* || fonction *f* génératrice des moments || functie *f* van de momentenverwekking || производящая функция *f* моментов

M455 **moment matrix** || Momentmatrix *f* || matrice *f* des moments || momentenmatrix *f* || матрица *f* моментов

M456 **moment method of estimation** || Momentenmethode *f* || méthode *f* des moments pour l'estimation || momentenmethode *f* || метод *m* моментов *(для нахождения оценок)*

M457 **moment of couple** || Moment *n* eines Kräftepaares || moment *m* d'un couple || krachtenpaarmoment *n* || момент *m* (пары) сил

M458 **moment of force** || Kraftmoment *n* || moment *m* d'une force || krachtmoment *n* || момент *m* силы

M459 **moment of inertia** || Trägheitsmoment *n* || moment *m* d'inertie || traagheidsmoment *n* || момент *m* инерции

M460 **moment of momentum** || Impulsmoment *n*, Drehimpuls *m* || moment *m* des quantités de mouvement || impulsmoment *n* || момент *m* количества движения, кинетический момент *m*

M461 **moment of resistance** || axiales Widerstandsmoment *n* || moment *m* des forces intérieures || weerstandsmoment *n* || момент *m* внутренней пары, момент *m* внутренних сил в сечении (при изгибе)

M462 **moment problem** || Momentenproblem *n* || problème *m* des moments || momentenprobleem *n* || проблема *f* моментов

M463 **momentum** || Impuls *m*, Bewegungsgröße *f* || quantité *f* de mouvement || bewegingshoeveelheid *f*, impuls *m* || импульс *m*, количество *n* движения

M464 **momentum-energy tensor** || Energie-Impuls-Tensor *m* || tenseur *m* d'énergie-impulsion || tensor *m* van de energie-impuls || тензор *m* энергии импульса

M465 **momentum transport** || Impulstransport *m*, Impulsübertragung *f* || transport *m* de l'impulsion || impulsoverbrenging *f* || перенос *m* импульса

M466 **monadic** *math.* || monadisch || monadique || monadisch || одноместный

M467 **Monge axis** || Mongesche Achse *f* || axe *m* de Monge || Mongesche as *f* (*m*) || ось *f* Монжа

M468 **Monge cone** || Mongescher Kegel *m* || cône *m* de Monge || Mongesche conus *m* || конус *m* Монжа

M469 **monkey** || Fallbär *m*; Fallhammer *m* || mouton *m* || heiblok *n*, beer *m* || баба *f* (*свайного копра*)

M470 **monochromatic** || monochromatisch, einfarbig || monochromatique || monochromatisch || монохроматический

M471 **monochromatic beam** || monochromatisches Bündel *n* || faisceau *m* monochromatique || monochromatische bundel *m* || монохроматический пучок *m*

M472 **monochromatic light** || monochromatisches Licht *n* || lumière *f* monochromatique || monochromatisch licht *n* || монохроматический свет *m*

M473 **monochromatization** || Monochromatisierung *f*, Lichtzerlegung *f* || monochromatisation *f* || monochromatisatie *f* || монохроматизация *f*

M474 **monochromator** || Monochromator *m* || monochromateur *m* || monochromator *m* || монохроматор *m*

M475 **monoclinic** || monoklin || monoclinique || monoklien || моноклинный

M476 **monocrystal** || Monokristall *m*, Einkristall *m* || monocristal || monokristal *n* || монокристалл *m*

M477 **monoenergetic** || monoenergetisch (*Teilchen*); monochromatisch (*Photon*) || monoénergétique; monochromatique || mono-energetisch; monochromatisch || моноэнергетический (*о частицах*); монохроматический (*о фотонах*)

M478 **monomer** || Monomer(es) *n*, monomere Substanz *f* || monomère *m* || monomeer *n* || мономер *m*

M479 **monomial matrix** || monomiale Matrix *f* || matrice *f* monomiale || monomiale matrix *f* || мономиальная матрица *f*

M480 **monomolecular layer** || monomolekulare Schicht *f* || couche *f* monomoléculaire || monomoleculaire laag *f* (*m*) || мономолекулярный слой *m*

M481 **monopole** || Monopol *m* || monopôle *m* || monopool *f* (*m*) || уединённый магнитный полюс *m*, монополь *m* (Дирака)

M482 **monotone increasing function** || monoton steigende [monoton wachsende] Funktion *f* || fonction *f* monotone croissante || monotoon [constant] toenemende functie *f* || монотонно возрастающая функция *f*

M483 **monotonic function** || monotone Funktion *f* || fonction *f* monotone || monotone functie *f* || монотонная функция *f*

M484 **monotropy** || Monotropie *f* || monotropie *f* || monotropie *f* || монотропия *f*

M485 **most powerful test** || trennschärfster Test *m* || test *m* le plus puissant || meest-vermogende test *m* || наиболее мощный критерий *m*

M486 **motion** || Bewegung *f*, Lauf *m*; Verschiebung *f* || mouvement *m*; déplacement *m* || beweging *f*; verplaatsing *f* || движение *n*; перемещение *n*

M487 **motive force** *see* **moving force**

M488 **motor** || **1.** Motor *m*; Triebwerk *n* **2.** Elektromotor *m* || **1.** moteur *m* **2.** électromoteur *m* || **1.** motor *m* **2.** elektromotor *m* || **1.** двигатель *m* **2.** электродвигатель *m*

M489 **mould** || Form *f* || moule *m* || vorm *m* || форма *f* || (*напр. литейная*)

M490 **moulder tool** || Formwerkzeug *n* || outillage *m* de mouleur || vormersgereedschap *n* || формовочный инструмент *m*

M491 **moulding** || Formerei *f* || moulage *m*; formage *m* || vormen *n* || формовка *f*

M492 **moulding machine** || Formmaschine *f* || machine *f* à mouler || vormmachine || формовочная машина *f*

M493 **mounting** || Montage *f*; Befestigung *f* || montage *m*; installation *f* || montage *f*; opstelling *f* || монтаж *m*; установка *f*; крепление *n*

M494 **movable rest** || mitlaufender Setzstock *m* || lunette *f* à suivre || beweegbaar lunet *f* (*m*) *(werktuigmachine)* || подвижный люнет *m* *(станка)*

M495 **movement** || Fortbewegung *f*, Bewegung *f*, Verschiebung *f* || mouvement *m*, déplacement *m* || beweging *f*; verplaatsing *f* || движение *n*; перемещение *n*

M496 **mover** || Bewegungsantrieb *m*, Vortriebsorgan *n* || groupe *m* moteur || motor *m* || двигатель *m*; движитель *m*

M497 **moving axes** || bewegliche Achsen *f pl* || axes *m pl* mobiles || losse assen *f* (*m*) *pl* || подвижные оси *f pl*

M498 **moving coordinate system** || bewegtes Koordinatensystem *n* || système *m* de coordonnées mobile || beweeglijk coördinatenstelsel *n* || движущаяся система *f* координат

M499 **moving force** || Antriebskraft *f*, treibende Kraft *f* || force *f* motrice; force *f* mouvante; force *f* propulsive || beweegkracht *f* (*m*) || движущая сила *f*

M500 **moving trihedron** || begleitendes Dreibein *n* || trièdre *m* mobile || begeleide driekant *m* || подвижной трёхгранник *m*

M501 **muffle furnace** || Muffelofen *m* || four *m* à moufle || moffeloven *m* || муфельная печь *f*

M502 **muffler** || Dämpfer *m*, Schalldämpfer *m* || silencieux *m* || demper *m* || глушитель *m*

M503 **multibeam interference** || Mehrstrahlinterferenz *f* || interférence *f* multifaisceau || meerbundelinterferentie *f* || многолучевая интерференция *f*

M504 **multichannel** || Mehrkanal... || multicanal || multikanaal... || многоканальный

M505 **multielectrode** || Mehrelektroden... || à plusieurs électrodes || multi-elektroden... || многоэлектродный

M506 **multimeter** || Mehrfachmeßgerät *n* || multimètre *m* || multimeter *m* || универсальный (электро)-измерительный прибор *m*

M507 **multimodal distribution** || multimodale Verteilung *f* || distribution *f* multimodale || multimodale distributie *f* || мультимодальное распределение *n*

M508 **multimolecular layer** || Mehrfachschicht *f*, polymolekulare Schicht *f* || couche *f* multimoléculaire || meermoleculenlaag *f* (*m*) || полимолекулярный слой *m*

M509 **multinomial coefficient** || multinomialer Koeffizient *m* || coefficient *m* multinomial || multinomiale coëfficiënt *m* || мультиномиальный [полиномиальный] коэффициент *m*

M510 **multinomial distribution** || multinomiale Verteilung *f* || distribution *f* multinomiale || multinomiale distributie *f* || полиномиальное распределение *n*

M511 **multiple automatic (lathe)** || Mehrspindeldrehautomat *m* || tour *m* automatique à plusieurs broches || veelspindeldraaiautomaat *m* || многошпиндельный токарный автомат *m*

M512 **multiple correlation coefficient** || multipler Korrelationskoeffizient *m* || coefficient *m* de corrélation multiple || multipele correlatiecoëfficiënt *m* || сводный коэффициент *m* корреляции

M513 **multiple drilling machine** || Mehrspindelbohrmaschine *f* || perceuse *f* multiple; perceuse *f* à broches multiples || veelspindelboormachine *f* || многошпиндельный сверлильный станок *m*

M514 **multiple integral** || mehrfaches Integral *n* || intégrale *f* multiple || multipele integraal *m* || кратный интеграл *m*

M515 **multiple ionization** || Mehrfachionisation *f* || ionisation *f* multiple || meervoudige ionisatie *f* || многократная ионизация *f*

M516 **multiple limit** || multipler Grenzwert *m* || limite *f* multiple || multipele grenswaarde *f* || предел *m* по совокупности

M517 **multiple roots** || mehrfache Wurzeln *f pl*; mehrfache Nullstellen *f pl* || racines *f pl* multiples ||

multipele wortels *m pl* || кратные корни *m pl*

M518 **multiplet** || Multiplett *n* || multiplet *m* || multiplet *n* || мультиплет *m*

M519 **multiple thread** *see* **multistart screw thread**

M520 **multiple-threaded tap** || mehrgängiger Gewindebohrer *m* || taraud *m* à denture multiple || meervoudige tap *m* || метчик *m* с многозаходной резьбой

M521 **multiple tool block** || Mehrfachmeißelhalter *m* || porte-outil *m* multiple || veelbeitelhouder *m* || резцовый блок *m*, многорезцовая державка *f*

M522 **multiple-valued function** || mehrwertige Funktion *f* || fonction *f* multivalente || meerwaardige functie *f* || многозначная функция *f*

M523 I **multiplex** || Multiplex *m* || multiplex *m* || multiplex *m* || мультиплекс *m*

M524 II **multiplex** || Multiplex... || multiplex || multiplex... || мультиплексный

M525 **multiplex channel** || Multiplexkanal *m* || canal *m* multiplex || multiplexkanaal *n* || мультиплексный канал *m*

M526 **multiplication** || **1.** Multiplikation *f*, Vermehrung *f* *(Neutronen)* **2.** Verstärkung *f* **3.** *Math.* Multiplikation *f* || **1.** multiplication *f* **2.** amplification *f* **3.** *math.* multiplication *f* || **1.** vermeerdering *f* **2.** versterking *f* **3.** *math.* vermenigvuldiging *f* || **1.** размножение *n* *(нейтронов)* **2.** усиление *n* **3.** *мат.* умножение *n*

M527 **multiplication factor** || **1.** Multiplikationsfaktor *m*, Vermehrungsfaktor *m* **2.** Verstärkungsfaktor *m* || **1.** facteur *m* de multiplication **2.** facteur *m* d'amplification || **1.** vermeerderingsfactor *m* **2.** versterkingsfactor *m* || **1.** коэффициент *m* размножения *(нейтронов)* **2.** коэффициент *m* усиления

M528 **multiplier** || Vervielfacher *m* || multiplicateur *m* || vermenigvuldiger *m* || умножитель *m*

M529 **multiply charged ion** || mehrfach geladenes Ion *n* || ion *m* à charges multiples || meervoudig geladen ion *n* || многозарядный ион *m*

M530 **multipoint tool** || mehrschneidiges Werkzeug *n* || outil *m* à pointes multiples || veelsnedig gereedschap *n* || многолезвийный инструмент *m*

M531 **multipolar** || Mehrpol... mehrpolig || multipolaire || meerpolig || многополюсный

M532 **multipolarity** || Multipolarität *f*, Multipolordnung *f* || multipolarité *f* || multipolariteit *f* || мультипольность *f*

M533 **multipole** || Multipol *m* || multipôle *m* || multipool *f* *(m)* || мультиполь *m*

M534 **multipole expansion** || Multipolentwicklung *f* || développement *m* en multipôles || ontbinding *f* in multipolen || разложение *n* по мультипо́лям

M535 **multiprogramming** || Multiprogrammierung *f* || multiprogrammation *f* || multiprogrammering *f* || мультипрограммирование *n*

M536 **multispeed gearbox** || Mehrganggetriebe *n* || boîte *f* de vitesses à rapports multiples || meervoudige versnellingsbak *m* || многоступенчатая коробка *f* передач

M537 **multispindle head** || Mehrspindelkopf *m* || tête *f* à broches multiples || veelspindelkop *m* || многошпиндельная головка *f*

M538 **multispindle machine** || Mehrspindler *m* || machine *f* à broches multiples || veelspindelmachine *f* || многошпиндельный станок *m*

M539 **multistage compressor** || Mehrstufenverdichter *m* || compresseur *m* à plusieurs étages || meertrapscompressor *m* || многоступенчатый компрессор *m*

M540 **multistage turbine** || Mehrstufenturbine *f* || turbine *f* à plusieurs étages || meertrapsturbine *f* || многоступенчатая турбина *f*

M541 **multistart screw thread** || mehrgängiges Gewinde *n* || filetage *m* à plusieurs filets || meervoudige schroefdraad *m* || многозаходная резьба *f*

M542 **multistart worm** || mehrgängige Schnecke *f* || vis *f* sans fin à plusieurs filets || meervoudige worm *m* || многоходовой червяк *m*

M543 **multistep method** || Mehrschrittverfahren *n* || méthode *f* de plusieurs pas || meertrapsmethode *f* || многошаговый метод *m*

M544 **multiterminal network** || Mehrpol *m* || multipôle *m*, circuit *m* multipôle || meeroolp *f* (*m*) || многополюсник *m*

M545 **multitool** || Mehrspindel... || (tour) multiples || veelbeitel... || многорезцовый

M546 **multivariante distribution** || mehrdimensionale Verteilung *f* || distribution *f* multidimensionnelle || multidimensionale distributie *f* || многомерное распределение *n*

M547 **multivariante sample** || multivariable Stichprobe *f* || échantillon *m* multidimensionnel || multivariante steekproef *f* (*m*) || многомерная выборка *f*

M548 **multivibrator** || Multivibrator *m* || multivibrateur *m* || multivibrator *m* || мультивибратор *m*

M549 **muon** || Myon *n* || muon *m* || muon *n* || мюон *m*

M550 **muonium** || Myonium *n* || muonium *m* || muonium *n* || мюоний *m*

M551 **mushroom valve** *see* **disk valve**

M552 **mutual attraction** || gegenseitige Anziehung *f* || attraction *f* mutuelle || onderlinge aantrekking *f* || (взаимное) притяжение *n*

M553 **mutual diffusion** || gegenseitige Diffusion *f* || diffusion *f* mutuelle || onderlinge diffusie *f* || взаимодиффузия *f*

M554 **mutual inductance** || Gegeninduktivität *f* || inductance *f* mutuelle || onderlinge inductantie *f* || взаимная индуктивность *f*, коэффициент *m* взаимоиндукции

M555 **mutuality** || Gegenseitigkeit *f* || mutualité *f* || mutualiteit *f* || взаимность *f*

M556 **mutually complementary** *math.* || paarweise komplementär || mutuellement complémentaire || paarsgewijs complementair || взаимнодополнительный

M557 **mutually disjoint** || paarweise disjunkt || mutuellement disjoint || paarsgewijs disjunctief || попарно-непересекающийся

M558 **mutually exclusive** *math.* || inkompatibel || mutuellement exclusif || elkaar uitsluitend || взаимноисключающий

M559 **mutual repulsion** || gegenseitige Abstoßung *f* || répulsion *f* mutuelle || onderling afstoten *n* || (взаимное) отталкивание *n*

N

N1 **nabla** || Nablaoperator *m*, Hamiltonoperator *m* || nabla *m* || Hamilton-operator *m* || набла(-оператор) *m*, оператор *m* Гамильтона

N2 **NAND** || NAND *n* || ET NON || EN NIET || НЕ-И

N3 **nano...** || Nano... || nano... || nano... || нано... *(приставка в наименовании дольной единицы, соответствующая 10^{-9})*

N4 **nanosecond** || Nanosekunde *f* || nanoseconde *f* || nanoseconde *f* (*m*) || наносекунда *f* *(10^{-9} с)*

N5 **narrow-band** || Schmalband..., schmalbandig || à bande étroite || smalbandig || **1.** узкополосный **2.** узкозонный *(о полупроводнике)*

N6 **narrow-band filter** || Schmalbandsperre *f* || filtre *m* à bande étroite || smalle-bandfilter *n*, *m* || узкополосный фильтр *m*

N7 **natural** || Natur..., natürlich || naturel || natuurlijk || естественный; природный

N8 **natural boundary** || natürliche Grenzen *f pl* || bornes *f pl* [limites *f pl*, frontières *f pl*] naturelles || natuurlijke grenzen *f* (*m*) *pl* || естественные границы *f pl*

N9 **natural boundary condition** || natürliche Grenzbedingung *f* || condition *f* aux limites naturelle || natuurlijke grensvoorwaarde *f* || естественное граничное условие *n*

N10 **natural cooling** || natürliche Kühlung *f*, Selbstkühlung *f* || refroidissement *m* naturel || natuurlijke koeling *f* || естественное охлаждение *n*

N11 **natural frequency** || Eigenfrequenz *f* || fréquence *f* propre || eigenfrequentie *f* || собственная частота *f*, частота *f* собственных колебаний

N12 **natural gas** || Erdgas *n* || gaz *m* naturel || aardgas *n* || природный газ *m*

N13 **natural logarithm** || natürlicher Logarithmus *m* || logarithme *m* naturel || natuurlijke logaritme *f* (*m*) || натуральный логарифм *m*

N14 **natural mode of vibration** || natürliche Schwingungsform *f* || mode *m* d'oscillations naturel || natuurlijke trillingsmodus *m* || собственный [нормальный] вид *m* колебаний; собственная мода *f* колебаний

N15 **natural number** || natürliche Zahl *f* || nombre *m* naturel || natuurlijk getal *n* || натуральное число *n*

N16 **natural occurrence** || Häufigkeit *f* in der Natur || abondance *f* naturelle || verspreiding *f* in de natuur || распространённость *f* *(напр. элемента)* в природе

N17 **natural oscillation** || Eigenschwingungen *f pl* || oscillations *f pl* propres || eigentrillingen *f pl* || собственные [свободные] колебания *n pl*

N18 **natural radioactivity** || natürliche Radioaktivität *f* || radioactivité *f* naturelle || natuurlijke radioactiviteit *f* || естественная радиоактивность *f*

N19 **navel point** || Nabelpunkt *m* || ombilic *m*, point *m* ombilical || ombiliek *f* || омбилическая точка *f*, омбилик *m*

N20 **near-equilibrium** || Fastgleichgewichts... || presque-équipondérant || bijna balanceer... || почти равновесный

N21 **near infrared** || mittleres Infrarot(gebiet) *n* || domaine *m* infrarouge proche || nabijgelegen infraroodgebied *n* || ближняя инфракрасная область *f* *(спектра)*

N22 **near ultraviolet** || Schwarzlichtbereich *m* || domaine *m* proche || nabijgelegen ultravioletgebied *n* || ближняя ультрафиолетовая область *f* *(спектра)*, ближний ультрафиолет *m*

N23 **needle lubricator** || Nadelöler *m* || graisseur *m* à aiguille || naaldsmeerpot *m* || игольчатая масленка *f*

N24 **needle roller bearing** || Nadelrollenlager *n* || roulement *m* à aiguilles; palier *m* à aiguilles || naaldrollager *n* || игольчатый подшипник *m*

N25 **needle valve** || Nadelventil *n* || soupape *f* à aiguilles || naaldklep *f* (*m*) || игольчатый клапан *m*

N26 **negation** || Negation *f* || négation *f* || negatie *f* || отрицание *n*

N27 I **negative** || Negativ *f* || négatif *m* || negatief *n* || негатив *m*, негативное изображение *n*

N28 II **negative** || negativ || négatif || negatief || отрицательный; негативный

N29 **negative acceleration** || negative Beschleunigung *f*, Verlangsamung *f*, Bremsung *f* || accélération *f* négative, décélération *f* || negatieve versnelling *f*, vertraging *f* || отрицательное ускорение *n*, замедление *n*

N30 **negative charge** || negative Ladung *f* || charge *f* négative || negative lading *f* || отрицательный заряд *m*

N31 **negative electricity** || negative Elektrizität *f* || électricité *f* négative || negatieve electriciteit *f* || отрицательное электричество *n*

N32 **negative electrode** || negative Elektrode *f*, Katode *f* || électrode *f* négative, cathode *f* || negatieve elektrode *f*, kathode *f* || отрицательный электрод *m*, катод *m*

N33 **negative image** || Negativbild *n* || image *f* renversée [négative] || negatief *n* || негативное изображение *n*, негатив *m*

N34 **negative ion** || negatives Ion *n*, Anion *n* || ion *m* négatif, anion *m* || negatief ion *n*, anion *n* || отрицательный ион *m*, анион *m*

N35 **negative rake** || negativer Spanwinkel *m* || angle *m* de coupe négatif || negatieve spaanhoek *m* || отрицательный угол *m* *(режущего инструмента)*

N36 **negentropy** || negative Entropie *f*, Negentropie *f* || entropie *f* négative, néguentropie *f* || negatieve entropie *f* || отрицательная энтропия *f*, негэнтропия *f*

N37 **neighbo(u)rhood** || Umgebung *f* || voisinage *m* || omgeving *f* || окрестность *f*; близость *f*

N38 **Neil's parabola** || Neilsche [semikubische] Parabel *f* || parabole *f* de Neil [semi-cubique] || parabool *f* *(m)* van Neil || парабола *f* Нейля, полукубическая парабола *f*

N39 **neon,** Ne || Neon *n* || néon *m* || neon *n* || неон *m*

N40 **neoprene** || Neopren *n* || néoprène *m* || neopreen *n* || неопрен *m*

N41 **neper** || Neper *n* || néper *m* || neper *m* || непер *m* *(8,68 дБ)*

N42 **nephelometry** || Nephelometrie *f*; Trübungsmessung *f* || néphélométrie *f* || nefelometrie *f* || нефелометрия *f*

N43 **Nernst heat theorem** || Nernstsches Wärmetheorem *n*, Nernstscher Wärmesatz *m* || théorème *m* thermique de Nernst || warmtetheorema *n* van Nernst || тепловая теорема *f* Нернста

N44 **Nernst principle** || Nernstscher Wärmesatz *m*, dritter Hauptsatz *m* der Thermodynamik || principe *m* de Nernst || Nernst-principe *n* || принцип *m* недостижимости абсолютного нуля (температуры), третье начало *n* термодинамики, теорема *f* Нернста

N45 I **net** || **1.** Eigengewicht *n*; Eigenmasse *f*; Nettogewicht *n*; Nettomasse *f* **2.** Gitter *n*; Netz *n* || **1.** poids *m* propre; masse *f* propre; poids *m* net; masse *f* nette **2.** grille *f*, filet *m*; crépine *f* || **1.** eigengewicht *n*, eigenmassa *f* *(m)*; netto-gewicht *n*, netto-massa *f* *(m)* **2.** net *n* || **1.** собственный вес *m*; собственная масса *f*; чистый вес *m*, вес *m* нетто; масса *f* нетто **2.** сетка *f*

N46 II **net** || **1.** Netto... **2.** resultierend || **1.** netto **2.** résultant || **1.** netto **2.** resulterend || **1.** нетто **2.** результирующий

N47 **net force** || resultierende Kraft *f*, Resultierende *f*, Resultante *f* || force *f* résultante || resulterende kracht *f* *(m)* || равнодействующая [результирующая] сила *f*

N48 **net power** || Nutzleistung *f* || puissance *f* utile || nuttig vermogen *n* || полезная мощность *f*

N49 **net weight** || Nettogewicht *n*; Nettomasse *f* || poids *m* net || nettogewicht *n* || чистый вес *m*, вес *m* нетто; чистая масса *f*

N50 **network** || Netz *n*; Netzwerk *n* || réseau *m*; circuit *m* || netwerk *n* || сеть *f*; схема *f*; разветвлённая цепь *f*

N51 **network analysis** || Netzwerkanalyse *f* || analyse *f* de réseau || netwerkanalyse *f* || схемный анализ *m*

N52 I **neutral** || **1.** Nulleiter *m* **2.** neutrales Teilchen *n*, Neutralteilchen *n* || **1.** conducteur *m* neutre **2.** particule *f* non chargée || **1.** nulleider *m* **2.** neutraal deeltje *n* || **1.** нейтраль *f*, нейтральный провод *m* **2.** нейтрал *m*, незаряженная частица *f*

N53 II **neutral** || neutral || neutre || neutraal || нейтральный

N54 **neutral atom** || neutrales Atom *n* || atome *m* neutre || neutraal atoom *n* || нейтральный атом *m*

N55 **neutral axis** || neutrale Achse *f*, Nullachse *f* || axe *m* neutre || neutrale as *f* *(m)* || нейтральная ось *f*

N56 **neutral equilibrium** || indifferentes Gleichgewicht *n* || équilibre *m* indifférent || indifferent evenwicht *n* || безразличное равновесие *n*

N57 **neutral filter** || Neutralfilter *n* || filtre *m* neutre || neutraal filter *n* || нейтральный (свето)фильтр *m*

N59 **neutrality** || Neutralität *f* || neutralité *f* || neutraliteit *f* || нейтральность *f*

N60 **neutralization** || Neutralisation *f*, Neutralisieren *n* || neutralisation *f* || neutralisatie *f* || нейтрализация *f*

N61 **neutral axis** || neutrale Linie *f*, Nullinie *f* *(bei der elastischen Biegung)* || axe *m* neutre || neutrale as *f* *(m)* || нейтральная ось *f* *(при изгибе)*

N62 **neutral layer** || neutrale Schicht *f* *(bei der elastischen Biegung)* || couche *f* neutre || neutrale laag *f* || нейтральный слой *m* *(при изгибе)*

N63 **neutral particle** || neutrales Teilchen *n*, Neutralteilchen *n* || particule *f* neutre || neutraal deeltje *n* || нейтральная [незаряженная] частица *f*

N64 **neutrino** || Neutrino *n* || neutrino *m* || neutrino *n* || нейтрино *n*

N65 **neutron** || Neutron *n* || neutron *m* || neutron *n* || нейтрон *m*

N66 **neutron-deficient nucleus** || neutronenarmer Kern *m* || noyau *m* déficient en neutrons || neutronenarme kern *f* *(m)* || нейтронодефицитное ядро *n*

N67 **neutron deficit** || Neutronendefizit *n* || déficit *m* de neutrons || neutronendeficit *n* || дефицит *m* нейтронов *(в нуклиде)*

N68 **neutron diffraction analysis** || neutronendiffraktometrische Untersuchung *f*, Neutronenstrukturuntersuchung *f* || diffractométrie *f* neutronique || neutronendiffractieanalyse *f* || нейтронография *f*

N69 **neutron-rich nucleus** || neutronenreicher Kern *m* || noyau *m* riche en neutrons || neutronenrijke kern *f* *(m)* || нейтроноизбыточное ядро *n*

N70 **neutron star** || Neutronenstern *m* || étoile *f* neutronique [de neutrons] || neutronenster *f* *(m)* || нейтронная звезда *f*

N71 **newton,** N || Newton *n* || newton *m* || newton *m* || ньютон *m*, Н

N72 **Newtonian mechanics** || klassische [Newtonsche] Mechanik *f* || mécanique *f* classique || klassieke mechanica *f* || классическая [ньютоновская] механика *f*

N73 **Newtonian viscosity** || Newtonsche Viskosität *f* || viscosité *f* de Newton || Newton-viscositeit *f* || ньютоновская [нормальная] вязкость *f*

N74 **Newton's fringes** || Newtonsche Ringe *m pl* || anneaux *m pl* de Newton || Newton-ringen *m pl* || кольца *n pl* Ньютона

N75 **N-gon, n-gon** || Polygon *n*, Vieleck *n* || polygone *m* || polygoon *m*, veelhoek *m* || многоугольник *m*

N76 **nickel,** Ni || Nickel *n* || nickel *m* || nikkel *n* || никель *m*

N77 **nickel alloy** || Nickellegierung *f* || alliage *m* de nickel || nikkellegering *f* || никелевый сплав *m*

N78 **Nicol (prism)** || Nicol *n*, Nicolsches Prisma *n* || prisme *m* de Nicol, nicol *m* || Nicol-prisma *n*, nicol *n* || призма *f* Николя, николь *m*

N79 **nilpotent group** || nilpotente Gruppe || groupe *m* nilpotent || nilpotente groep *f* *(m)* || нильпотентная группа *f*

N80 **nilpotent operator** || nilpotenter Operator *m* || opérateur *m* nilpotent || nilpotente operator *m* || нильпотентный оператор *m*

N81 **niobium,** Nb || Niob *n* || niobium *m* || niobium *n* || ниобий *m*

N82 **nit,** cd/m² || Nit *n* || nit *m* || nit *n* || нит, кд/м²

N83 **nitrate** || Nitrat *n* || nitrate *m* || nitraat *n* || нитрат *m*

N84 **nitration** || Nitrierung *f*, Nitrieren *n* || nitration *f* || nitratie *f* || нитрование *n*

N85 **nitric acid,** HNO_3 || Salpetersäure *f* || acide *m* nitrique || salpeterzuur *n* || азотная кислота *f*

N86 **nitric oxide,** NO || Stickstoff(mon)oxid *n*, Stickoxid *n* || oxyde *m* azotique [nitrique] || stikstof(mon)oxyde *n* || окись *f* азота

N87 **nitride** || Nitrid *n* || nitrure *m* || nitride *n* || нитрид *m*, соединение *n* азота с металлом

N88 **nitrided steel** || nitrierter Stahl *m* || acier *m* nitruré || genitreerd staal *n* || азотированная сталь *f*

N89 **nitriding** || Nitrieren *n*, Nitrierung *f* || nitruration *f* || nitreren *n* || азотирование *n*; нитрирование *n*

N90 **nitrite** || Nitrit *n* || nitrite *m* || nitrite *n* || нитрит *m*, соль *f* *или* эфир *m* азотистой кислоты

N91 **nitrogen,** N || Stickstoff *m* || azote *m* || stikstof *f* (*m*) || азот *m*

N92 **nitrogen dioxide,** NO_2 || Stickstoffdioxid *n* || dioxyde *m* d'azote, hypoazotide *m* || stikstofdioxyde *n* || двуокись *f* азота

N93 **nitrogen oxide** || Stick(stoff)-oxid *n* || oxyde *m* azotique [nitrique] || stikstof(mon)oxyde *n* || окись *f* азота

N94 **nitrogen pentoxide,** N_2O_5 || Stickstoffpentoxid *n* || pentoxyde *m* d'azote || stikstofpentoxyde *n* || пятиокись *f* азота

N95 **nitrogen trioxide,** N_2O_3 || Stickstofftrioxid *n* || trioxyde *m* d'azote || stikstoftrioxyde *n* || трёхокись *f* азота

N96 **nitroso dye** || Nitrosofarbstoff *m* || colorant *m* nitrosé || nitrosokleurstof *f* (*m*) || нитрозокраситель *m*

N97 **nitrous acid,** HNO_2 || salpetrige Säure *f* || acide *m* nitreux || salpeterigzuur *m* || азотистая кислота *f*

N98 **nitrous oxide,** N_2O || Distickstoff(mon)oxid *n* || oxyde *m* azoteux || stikstofdioxyde *n* || закись *f* азота

N99 **nodal line** || Knotenlinie *f* || ligne *f* des nœuds || knooplijn *f* (*m*) || узловая линия *f*, линия *f* узлов

N100 **nodal point** || Knoten(punkt) *m* || nœud *m*, point *m* nodal || knooppunt *n* || узел *m*; узловая точка *f*

N101 **node** || Knoten(punkt) *m* || nœud *m* || knoop *m* || узел *m*; вершина *f* *(графа)*

N102 **no-go gauge** || Ausschußlehre *f* || calibre *m* n'entre pas || uitschotkaliber *n* || непроходной калибр *m*

N103 **noise** || Rauschen *n* || bruit *m* || lawaai *n* || шум *m*

N104 **noise factor** || Rauschzahl *f*, Rauschfaktor *m* || coefficient *m* [facteur *m*] de bruit || lawaaifactor *m* || коэффициент *m* шумов, шумфактор *m*

N105 **noise-free** || rauschfrei || sans bruits || lawaaivrij || свободный от помех, беспомеховый

N106 **noiseless running** || geräuschloser Lauf *m* || allure *f* silencieuse || geruisloze gang *m* || бесшумный ход *m*

N107 **noise evel** || Rauschpegel *m*; Geräuschpegel *m* || niveau *m* de bruit || lawaainiveau *n* || уровень *m* шума

N108 **noise suppression** || Rauschunterdrückung *f*, Störungsunterdrükkung *f* || suppression *f* des brouillages, déparasitage *m* || lawaaibestrijding *f* || подавление *n* помех

N109 **noise temperature** || Rauschtemperatur *f* || température *f* de bruits || lawaaitemperatuur *f* || эквивалентная температура *f* шумов; шумовая температура *f*

N110 **no-load** || unbelastet; Leerlauf... || non chargé; à vide || onbelast || без нагрузки; холостой

N111 **no-load speed** *see* **idle speed**

N112 **nomenclature** || Nomenklatur *f* || nomenclature *f* || nomenclatuur *f* || номенклатура *f*

N113 **nominal load** || Nennlast *f* || charge *f* nominale || nominale belasting *f* || номинальная нагрузка *f*

N114 **nominal power** || Nennleistung *f* || puissance *f* nominale || nominaal vermogen *n* || номинальная мощность *f*

N115 **nominal size** || Nennmaß *n* || dimension *f* nominale || nominale maat *f* (*m*) || номинальный размер *m*

N116 **nominal speed** || **1.** Nenngeschwindigkeit *f* **2.** Nenndrehzahl *f* || vitesse *f* nominale || nominale snelheid *f* || **1.** номинальная скорость *f* **2.** номинальная частота *f* вращения

N117 **nominal stress** || Nennspannung *f* || contrainte *f* nominale || nominale spanning *f* || расчётное напряжение *n* *(вычисленное в предположении упругой работы тела)*

N118 **nominal value** || Nennwert *m* || valeur *f* nominale || nominale waarde *f* || номинальное [расчётное] значение *n*

N119 **non-aqueous solution** || nichtwäßrige Lösung *f* || solution *f* non aqueuse || niet-waterige oplossing *f* || неводный раствор *m*

N120 **nonautonomous system** || nichtautonomes System *n* || système *m* non autonome || niet-autonoom systeem *n* || неавтономная система *f*

N121 **non-central force** || nichtzentrale Kraft *f* || force *f* non centrale || niet-centrale kracht *f* (*m*) || нецентральная сила *f*

N122 **non-central load** *see* **eccentric load**

N123 **noncommutative field** || nichtkommutatives Feld *n* || corps *m* non commutatif || niet-commutatief veld *n* || некоммутативное поле *n*

N124 **noncommutative group** || nichtkommutative Gruppe *f* || groupe *m* non commutatif || niet-commutatieve groep *f* (*m*) || некоммутативная группа *f*

N125 **non-commutativity** || Nichtkommutativität *f*, Nichtvertauschbarkeit *f* || non-commutativité *f* || niet-commutativiteit *f* || некоммутативность *f*

N126 **nonconducting** || nichtleitend || non conducteur || niet-geleidend || непроводящий

N127 **non-conjunction** || NAND-Verknüpfung *f* || non-conjonction *f* || niet-conjunctie *f* || антиконъюнкция *f*

N128 **non-constructive definition** || nichtkonstruktive Definition *f* || définition *f* non constructive || niet-constructieve definitie *f* || неконструктивное определение *n*

N129 **nondecreasing function** || nichtabnehmende [nichtfallende] Funktion *f* || fonction *f* non décroissante || niet-afnemende functie *f* || неубывающая функция *f*

N130 **nondegenerate level** || nichtentartetes Energieniveau *n* || niveau *m* non dégénéré || niet-ontaard (energie) niveau *n* || невырожденный (энергетический) уровень *m*

N131 **nondeterministic automaton** || nichtdeterminierter Automat *m* || automate *m* non déterministe || niet-gedetermineerde automaat *m* || недетерминированный автомат *m*

N132 **nondimensional** || dimensionslos || non dimensionnel || dimensieloos || безразмерный

N133 **non-disjunction** || NOR-Verknüpfung *f* || non-disjonction *f* || niet-disjunctie *f* || антидизъюнкция *f*

N134 **nonequilibrium process** || Nichtgleichgewichtsprozeß *m* || processus *m* déséquilibré || niet-evenwichtsproces *n* || неравновесный процесс *m*

N135 **nonequilibrium state** || Nichtgleichgewichtszustand *m* || état *m* déséquilibré || niet-evenwichtstoestand *m* || неравновесное состояние *n*

N136 **non-Euclidean geometry** || nichteuklidische Geometrie *f* || géométrie *f* non euclidienne || niet-euclidische meetkunde *f* || неевклидова геометрия *f*

N137 **non-excited atom** || nichtangeregtes Atom *n* || atome *m* non excité || niet-opgewekt atoom *n* || невозбуждённый атом *m*

N138 **non-ferrous metals** || NE-Metalle *n pl*, Buntmetalle *n pl* || métaux *m* non ferreux || nonferrometalen *n pl* || цветные металлы *m pl*

N139 **nonhomogeneous differential equation** || nichthomogene Differentialgleichung *f* || équation *f* différentielle non homogène || niet-homogene differentiaalvergelijking *f* || неоднородное дифференциальное уравнение *n*

N140 **non-increasing function** || nichtwachsende Funktion *f* || fonction *f* non croissante || niet-opklimmende functie *f* || невозрастающая функция *f*

N141 **non-inertial frame** || Nichtinertialsystem *n* || système *m* de référence non-inertial || niet-inertiaal stelsel *n* || неинерциальная система *f* отсчёта

N142 **nonlinear operations** || nichtlineare Operationen *f pl* || opérations *f pl* non linéaires || niet-lineaire operaties *f pl* || нелинейные операции *f pl*

N143 **nonlinear optics** || nichtlineare Optik *f* || optique *f* non linéaire || niet-lineaire optica *f* || нелинейная оптика *f*

N144 **nonlinear oscillations** || nichtlineare Schwingungen *f pl* || oscillations *f pl* non linéaires || niet-lineaire trillingen *f pl* || нелинейные колебания *n pl*

N145 **nonlinear programming** || nichtlineare Programmierung *f* || programmation *f* non linéaire || niet-lineaire programmering *f* || нелинейное программирование *n*

N146 **nonmetal** || Nichtmetall *n* || non-métal *m* || niet-metaal *n* || неметалл *m*

N147 **nonnegative form** || nicht-negative Form *f* || forme *f* non négative || niet-negatieve form *m* || неотрицательная форма *f*

N148 **nonnegative integral form** || nichtnegative Integralform *f* || forme *f* intégrale non négative || niet-negatieve integraalvorm *m* || неотрицательная интегральная форма *f*

N149 **nonnegative matrix** || nicht-negative Matrix *f* || matrice *f* non négative || niet-negatieve matrix *f* || неотрицательная матрица *f*

N150 **non-Newtonian viscosity** || nicht-Newtonsche [anomale] Viskosität *f* || viscosité *f* non newtonienne || niet-Newtonse viscositeit *f* || неньютоновская [аномальная] вязкость *f*

N151 **nonparametric test** || parameterfreier Test *m* || test *m* non paramétrique || parametervrije test *m* || непараметрический критерий *m*

N152 **non-periodic** || aperiodisch || apériodique || aperiodisch || апериодический

N153 **nonpolar** || nichtpolar, apolar || non polaire || niet-polair || неполярный

N154 **nonpolar molecule** || unpolares Molekül *n* || molécule *f* non polaire || niet-polaire molecule *f* (*m*) || неполярная молекула *f*

N155 **nonpositive form** || nicht-positive Form *f* || forme *f* non positive || niet-positieve vorm *m* || неположительная форма *f*

N156 **nonpositive integral form** || nichtpositive Integralform *f* || forme *f* intégrale non positive || niet-positieve integraalvorm *m* || неположительная интегральная форма *f*

N157 **nonpositive matrix** || nicht-positive Matrix *f* || matrice *f* non positive || niet-positieve matrix *f* || неположительная матрица *f*

158 **nonradiative transition** || strahlungsloser Übergang *m*, Auger-Übergang *m* || transition *f* sans radiation || stralingsloze overgang *m* || безызлучательный [безрадиационный] переход *m*, оже-переход *m*

N159 **non-return valve** || Rückschlagventil *n* || soupape *f* de retenue; clapet *m* de retour || terugslagklep *f* (*m*) || обратный клапан *m*

N160 **non-self-maintained discharge** || unselbständige Entladung *f* || décharge *f* semi-autonome || onzelfstandige ontlading *f* || несамостоятельный разряд *m*

N161 **nonsingular matrix** || nicht-singuläre Matrix *f* || matrice *f* non singulière || niet-singulaire matrix *f* || невырожденная [неособенная] матрица *f*

N162 **nonsingular operator** || nicht-singulärer Operator *m* || opérateur *m* non singulier || niet-singulaire operator *m* || невырожденный оператор *m*

N163 **non-spherical** || nichtsphärisch || non sphérique || niet-sferisch || асферический

N164 **nonstationary notion** || nicht-stationäre Bewegung *f* || mouvement *m* non stationnaire || niet-stationaire beweging *f* || неустановившееся [переменное] движение *n*

N165 **nonuniform motion** || ungleichförmige Bewegung *f* || mouvement *m* irrégulier || ongelijkmatige beweging *f* || неравномерное движение *n*

N166 **nonviscous liquid** || reibungsfreie Flüssigkeit *f* || liquide *m* parfait || niet-taaie vloeistof *f* || невязкая жидкость *f*

N167 **NOR** || NOR *n* || OU, OU-NON || NIET-OF || НЕ-ИЛИ

N168 **norm** || Norm *f* || norme *f* || norm *f* (*m*) || норма *f*

N169 I **normal** || Normale *f* || normale *f* || normaal *f* (*m*) || нормаль *f*; эталон *m*

N170 II **normal** || normal, Normal... || normal || normaal || нормальный

N171 **normal acceleration** || Normalbeschleunigung *f* || accélération *f* normale || normale versnelling *f* || нормальное ускорение *n*

N172 **normal algorithm** || Normalalgorithmus *m* || algorithme *m* normal || normaalalgoritme *n* || нормальный алгоритм *m*

N173 **normal atom** || normales Atom *n* || atome *m* normal || normaal atoom *n* || атом *m* в основном состоянии

N174 **normal coordinates** || Normalkoordinaten *f pl* || coordonnées *f pl* normales || normale coördinaten *m pl* || нормальные координаты *f pl*

N175 **normal curvature** || Normalkrümmung *f* || courbure *f* normale || normaalkromming *f* || нормальная кривизна *f*

N176 **normal derivative** || Normalableitung *f* || dérivée *f* normale || normale afgeleide *f* || нормальная производная *f*

N177 **normal deviation** || Normalabweichung *f* || déviation *f* normale || normale afwijking *f* || нормальное отклонение *n*

N178 **normal dispersion** || normale Dispersion *f* || dispersion *f* normale || normale dispersie *f* || нормальная дисперсия *f*

N179 **normal distribution** || Normalverteilung *f* || distribution *f* normale || normale verdeling *f* || нормальное [гауссово] распределение *n*

N180 **normal distribution function** || Normalverteilungsfunktion *f* || fonction *f* de distribution *f* normale [de distribution *f* gaussienne] || normale verdelingsfunctie *f* || функция *f* нормального распределения

N181 **normal divisor** || Normalteiler *m* || deviseur *m* normal || normale deler *m* || нормальный делитель

N182 **normal elliptic integral** || elliptisches Integral *n* in per Legendreschen Normalform || intégrale *f* elliptique normale || normale elliptische integraal *f* (*m*) || нормальный эллиптический интеграл *m*

N183 **normal error integral** || Gaußsches Fehlerintegral *n* || intégrale *f* d'erreur normale || foutenintegraal *f* (*m*) van Gauss || интеграл *m* ошибок, интеграл *m* вероятности

N184 **normal force** || Normalkraft *f* || force *f* normale || normale kracht *f* (*m*) || нормальная составляющая *f* силы, нормальная сила *f*

N185 **normal form** || Normalform *f* || forme *f* normale || normale vorm *m* || нормальная форма *f*

N186 **normal frequency** || Normalfrequenz *f* || fréquence *f* normale || normale frequentie *f* || нормальная частота *f*

N187 **normal frequency function** || Normalverteilungsdichte *f* || fonction *f* de fréquence normale || normaleverdelingsdichtheid *f* || плотность *f* (нормального) распределения

N188 **normal incidence** || normaler [senkrechter] Einfall *m* || incidence *f* normale || normale inval *m* || падение *n* по нормали, нормальное падение *n*

N189 **normalizable function** || normierbare Funktion *f* || fonction *f* normable [normalisable] || normeerbare functie *f* || нормируемая функция *f*

N190 **normalizable kernel** || normierbarer Kern *m* || noyau *m* normable [normalisable] || normeerbare kern *f* (*m*) || нормируемое ядро *n*

N191 **normalization** || Normierung *f*; Normalisierung *f*; Normung *f*; Standardisation *f* || normalisation *f*; standardisation *f* || normalisatie *f* || нормирование *n*; стандартизация *f*; нормализация *f*

N192 **normalization factor** || Normierungsfaktor *m* || facteur *m* de normalisation || normalisatiefactor *m* || нормирующий множитель *m*

N193 **normalized form** || normalisierte Form *f* || forme *f* normalisée || genormaliseerde vorm *m* || нормализованная форма *f*

N194 **normalizer** || Normalisator *m* || normalisateur *m* || normalisator *m* || нормализатор *m*

N195 **normal matrix** || Normalmatrix *f* || matrice *f* normale || normale matrix *f* || нормальная матрица *f*

N196 **normal mode** || Normalschwingung *f* || oscillation *f* normale || normale trilling *f* || нормальное [основное] колебание *n*

N197 **normal of plane** || Normale *f* zu einer Ebene || normale *f* à un plan || normaal *f* (*m*) van een vlak || нормаль *f* к плоскости

N198 **normal of surface** || Normale *f* zu einer Fläche, Flächennormale *f* || normale *f* à une surface || normaal *f* (*m*) van een oppervlakte || нормаль *f* к поверхности

N199 **normal operator** || normaler Operator *m* || opérateur *m* normal || normale operator *m* || нормальный оператор *m*

N200 **normal pitch** || Normalteilung *f* || pas *m* normal || normale stap *m*; normale spoed *m*; normale steek *m* || нормальный шаг *m*

N201 **normal plane** || Normalebene *f* || plan *m* normal || normaalvlak *n* || нормальная плоскость *f*

N202 **normal population** || normalverteilte Gesamtheit *f* || population *f* normale || populatie *f* met normale verdeling || нормальная совокупность *f*

N203 **normal pressure** *see* **standard pressure**

N204 **normal random numbers** || normalverteilte Zufallszahlen *f pl* || nombres *m pl* aléatoires normaux || normaalverdeelde toevalsgetallen *n pl* || нормально распределённые случайные числа *n pl*

N205 **normal random process** || Gaußscher Zufallsprozeß *m* || processus *m* aléatoire normal || normaal toevalsproces *n* || нормальный случайный [гауссовский случайный] процесс *m*

N206 **normal random variable** || normale Zufallsvariable *f* || variable *f* aléatoire normale || normale toevalsvariabele *f* || нормальная случайная переменная *f*

N207 **normal response** || Normalreaktion *f* || réponse *f* normale || normale reactie *f* || нормальная реакция *f* (*на внешнюю нагрузку*)

N208 **normal sample** || normale Stichprobe *f* || échantillon *m* normal || normaal monster *n* || нормальная выборка *f*

N209 **normal section** || Normalschnitt *m* || section *f* normale || normale snede *f* (*m*) || нормальное сечение *n*

N210 **normal series** || Normalreihe *f* || série *f* normale || normale reeks *f* (*m*) || нормальный ряд *m*

N211 **normal solution** || Normallösung *f* || solution *f* normale || normaaloplossing *f* || нормальный раствор *m*

N212 **normal speed** || Normalgeschwindigkeit *f* || vitesse *f* normale || normale snelheid *f* || нормальная скорость *f*

N213 **normal state** || Normalzustand *m* || état *m* normal || normale toestand *m* || нормальное состояние

N214 **normal stress** || Normalspannung *f* || contrainte *f* normale || normale spanning *f* || нормальное напряжение *n*

N215 **normal subgroup** || normale Untergruppe *f* || sous-groupe *m* normal || normale ondergroep *f* (*m*) || нормальная подгруппа *f*

N216 **normal vector** || Normalenvektor *m*, Normale *f* || vecteur *m* normal || normale vector *m* || нормальный вектор *m* (*к поверхности*), нормаль *f*

N217 **normal velocity** || Normalgeschwindigkeit *f* || vitesse *f* normale || normale snelheid *f* || нормальная скорость *f*, нормальная составляющая *f* скорости

N218 **normal vibration** || Normalschwingungen *f pl* || vibrations *f pl* [oscillations *f pl*] normales || normale trillingen *f pl* || нормальные колебания *n pl*

N219 **norm of a complex number** || absoluter Betrag *m* einer komplexen Zahl || norme *f* d'un nombre complexe || absoluut bedrag *n* van een complex getal || модуль *m* комплексного числа

N220 **norm of a function** || Norm *f* einer Funktion || norme *f* d'une fonction || norm *f* (*m*) van een functie || норма *f* функции

N221 **norm of a matrix** || Matrixnorm *f* || norme *f* d'une matrice || matrixnorm *f* (*m*) || норма *f* матрицы

N222 **norm of a vector** || Vektornorm *f* || norme *f* d'un vecteur || norm *f* (*m*) van een vector || норма *f* вектора

N225 **NOT** || Nicht *n* (*logische Funktion bzw. Operation*) || NON (*fonction ou opération logique*) || NIET (*logische functie of operatie*) || НЕ (*логическая функция или операция*)

N226 **notch-bar test** || Kerbschlagversuch *m* || essai *m* de résilience sur barreau entaillé || kerfslagproef *f* (*m*) || испытание *n* образца (с надрезом) на ударную вязкость

N227 **notch brittleness** || Kerbsprödigkeit *f* || fragilité *f* au choc || kerfbrosheid *f* || ударная хрупкость *f*

N228 **notched specimen** || gekerbter Probestab *m* || éprouvette *f* entaillée || gekerfde proefstaaf *f* (*m*) || образец *m* с надрезом

N229 **notching** || **1.** Kerben *n* **2.** Kerbe *f* || entaille *f*, entaillage *m*; encoche *f*; rainure *f* || kerf *f* (*m*) || **1.** надрез *m*, надрезание *n* **2.** пропил *m*, прорезь *f*

N230 **notch strength** || Kerbschlagzähigkeit *f* || résilience *f*, résistance *f* au choc d'une éprouvette entaillée || kerfslagtaaiheid *f*, kerfslagwaarde *f* || ударная вязкость *f* (*образца с надрезом*)

N231 **nozzle** || Düse *f*; Mundstück *n* || tuyère *f*; buse *f*; bec *m* (d'injecteur) || mondstuk *n*; sproeier *m*; broes *m* || сопло *n*; форсунка *f*

N232 **n-sided polygon** *see* **N-gon**

N233 **n-type semiconductor** || Überschußhalbleiter *m*, n-Halbleiter *m* || semi-conducteur *m* type *n* || n-halfgeleider *m* || полупроводник *m* n-типа, электронный полупроводник *m*

N234 **nuclear** || Kern..., nuklear || nucléaire || kern... || ядерный

N235 **nuclear energy** || Kernenergie *f*, Atomenergie *f*, nukleare Energie *f* || énérgie *f* nucléaire || kernenergie *f* || ядерная [атомная] энергия *f*

N236 **nuclear engineering** || Kerntechnik *f* || technique *f* nucléaire || kerntechniek *f* || ядерная техника *f*

N237 **nuclear explosion** || Kernexplosion *f* || explosion *f* nucléaire || kernexplosie *f* || ядерный взрыв *m*

N238 **nuclear field** || Kernfeld *n*, Kraftfeld *n* des Atomkernes || champ *m* nucléaire || kernveld *n* || поле *n* (атомного) ядра, ядерное поле *n*

N239 **nuclear fission** || Kernspaltung *f* || fission *f* nûcléaire || kernsplijting *f* || деление *n* (атомного) ядра, ядерное деление *n*

N240 **nuclear forces** || Kernkräfte *f pl* || forces *f pl* nucléaires || kernkrachten *f* (*m*) *pl* || ядерные силы *f pl*

N241 **nuclear fragment** || Kernbruchstück *n*, Kernfragment *n* || fragment *m* nucléaire || kernfragment *n* || осколок *m* (расщепления) ядра

N242 **nuclear fuel** || Kernbrennstoff *m* || combustible *m* nucléaire || kernbrandstof *f* (*m*) || ядерное топливо *n*

N243 **nuclear fusion** || Kernfusion *f*, Fusion *f*, Kernsynthese *f* || fusion *f* nucléaire || kernfusie *f*, kernversmelting *f* || слияние *n* ядер, ядерный синтез *m*

N244 **nuclear interaction** || Kernwechselwirkung *f* || interaction *f* nucléaire || kernwisselwerking *f* || ядерное взаимодействие *n*

N245 **nuclear isomerism** || Kernisomerie *f* || isomérie *f* nucléaire || kernisomerie *f* || ядерная изомерия *f*

N246 **nuclear magnetic resonance** || magnetische Kernresonanz *f* || résonance *f* nucléaire magnétique || magnetische kernresonantie *f*, kernmagnetische resonantie *f* || ядерный магнитный резонанс *m*, ЯМР

N247 **nuclear magneton** || Kernmagneton *n* || magnéton *m* nucléaire || kernmagneton *n* || ядерный магнетон *m*

N248 **nuclear physics** || Kernphysik *f* || physique *f* nucléaire ||

kernfysica *f* || ядерная физика *f*, физика *f* атомного ядра

N249 **nuclear power** || Kernenergie *f* || énergie *f* nucléaire || kernenergie *f* || атомная [ядерная] энергия *f*

N250 **nuclear power engineering** || Kernenergetik *f* || énergétique *f* nucléaire || kernenergetica *f* || ядерная энергетика *f*

N251 **nuclear power plant** || **1.** Kernkraftwerk *n*; Kernenergieanlage *f* || centrale *f* nucléaire; installation *f* nucléaire || kerncentrale *f* (*m*); kerninstallatie *f* || атомная [ядерная] электростанция *f*; атомная [ядерная] силовая установка *f*

N252 **nuclear reaction** || Kernreaktion *f* || réaction *f* nucléaire || kernreactie *f* || ядерная реакция *f*

N253 **nuclear reactor** || Kernreaktor *m* || réacteur *m* nucléaire || kernreactor *m* || ядерный [атомный] реактор *m*

N254 **nuclear spin** || Kernspin *m* || spin *m* nucléaire || kernspin *m* || спин *m* (атомного) ядра

N255 **nuclear transformation** || Kernumwandlung *f* || transformation *f* nucléaire || kerntransformatie *f* || ядерное превращение *n*, ядерная реакция *f*

N256 **nucleation** || Keimbildung *f* || germination *f*; nucléation *f* || kiemvorming *f* || образование *n* зародышей, зародышеобразование *n*

N257 **nucleic acid** || Nukleinsäure *f* || acide *m* nucléique || nucleïnezuur *n* || нуклеиновая кислота *f*

N258 **nucleon** || Nukleon *n* || nucléon *m* || nucleon *n* || нуклон *m*

N259 **nucleoside** || Nukleosid *n* || nucléoside *m* || nucléoside *n* || нуклеозид *m*

N260 **nucleotide** || Nukleotid *n* || nucléotide *m* || nucleotide *n* || нуклеотид *m*

N261 **nucleus** || Kern *m* || noyau *m* || kern *f* (*m*) || ядро *n*

N262 **nuclide** || Nuklid *n* || nucl(é)ide *m* || nuclide *n* || нуклид *m*

N263 **nuisance vibration** || störende Vibration *f* || vibration *f* nuisible || schadelijke trilling *f* || вредная [нежелательная] вибрация *f*

N264 **null curve** || Nullkurve *f* || courbe *f* nulle || nulkromme *f* (*m*) || кривая *f* нулевой длины

N265 **null direction** || isotrope Richtung *f* *(einer Geodätischen)* || direction *f* nulle [isotrope] || isotrope richting *f* || изотропное направление *n* *(геодезической линии)*

N266 **null displacement** || isotrope Verschiebung *f* || déplacement *m* nul || isotrope verplaatsing *f* || изотропное перемещение *n*

N267 **null geodesic** || Nullgeodätische *f* || géodésique *f* nulle || geodetische lijn *f* (*m*) van nullengte *f* || геодезическая линия *f* нулевой длины

N268 **null gravity** || Schwerelosigkeit *f* || apesanteur *f*, non-pesanteur *f* || gewichtsloosheid *f* || невесомость *f*

N269 **nullity** || Nullität *f* || nullité *f* || nulliteit *f* || размерность *f* *(линейного преобразования)*

N270 **null matrix** || Nullmatrix *f* || matrice *f* nulle || nulmatrix *f* || нулевая матрица *f*

N271 **null set** || Nullmenge *f*, Menge *f* vom Maß Null || ensemble *m* de mesure nulle || nulmenigte *f* || множество *n* меры нуль, пустое множество *n*

N272 **null space** *math.* || Nullraum *m* || espace *m* nul || nulruimte *f* || пространство *n* нулей, ядро *n* *(линейного преобразователя)*

N273 **null tensor** || Nulltensor *m* || tenseur *m* nul || nultensor *m* || нуль-тензор *m*

N274 **null transformation** || Nulltransformation *f* || transformation *f* nulle || nultransformatie *f* || нулевое преобразование *n*

N275 **null vector** || Nullvektor *m* || vecteur *m* nul || nulvector *m* || нуль-вектор *m*

N276 **number** || Zahl *f* || nombre *m* || getal *n* || число *n*

N277 **number of addresses** || Adressenkapazität *f* || nombre *m* d'adresses || adrescapaciteit *f* || адресность *f*

N278 **number of revolutions** || Drehzahl *f* || nombre *m* de tours [de révolutions] || toerental *n* || число *n* оборотов; частота *f* вращения

N279 **numeral** || Numeral *n* || numéral *m* || numeral *n* || нумерал *m*, числовое поле *n*

N280 **numerical analysis** || numerische Analyse *f* || analyse *f* numérique || numerieke analyse *f* || численный анализ *m*

N281 **numerical aperture** || numerische Apertur *f* || ouverture *f* numérique || numerieke apertuur *f* || числовая апертура *f*

N282 **numerical differentiation** || numerische Differentiation *f* || différentiation *f* numérique || numerieke differentiatie *f* || численное дифференцирование *n*

N283 **numerical integration** || numerische Integration *f* || intégration *f* numérique || numerieke integratie *f* || численное интегрирование *n*

N284 **numerically controlled machine** || numerisch gesteuerte Werkzeugmaschine *f*, NC-gesteuerte Werkzeugmaschine *f* || machine-outil *f* à commande numérique || NC-werktuigmachine *f* || станок *m* с числовым программным управлением [с ЧПУ]

N285 **numerical solution** || numerische Lösung *f* || solution *f* numérique || numerieke oplossing *f* || численное решение *n*

N286 **numeric control** || numerische Steuerung *f*, NC-Steuerung *f* || commande *f* numérique || numerieke besturing *f* || числовое программное управление *n*, ЧПУ

N287 **nut** || Schraubenmutter *f*, Mutter *f* || écrou *m* || moer *f* (*m*) || гайка *f*

N288 **nutation** || Nutation *f* || nutation *f* || nutatie *f* || нутация *f*

N289 **nut tapping machine** || Muttergewindeschneidmaschine *f* || machine *f* à fileter [tailler] les écrous || moer(en)tapautomaat *m* || гайконарезной станок *m*

N290 **Nyquist criterion** || Nyquist-Kriterium *n* || critère *m* de Nyquist || criterium *m* van Nyquist || критерий *m* Найквиста

O

O1 **object** || Objekt *n* || objet *m* || object *n* || объект *m*; предмет *m*, объект *m* (*в оптике*)

O2 **object distance** || Dingweite *f* || distance *f* à l'objet || objectafstand *m* || расстояние *n* (до) предмета

O3 **objective (lens)** || Objektiv *n* || objectif *m* || objectief *n* || объектив *m*

O4 **object language** || Objektsprache *f* || langage *m* objet || objecttaal *f* || объектный язык *m*

O5 **object point** || Dingpunkt *m* || point *m* de l'objet || objectpunt *f* (*m*) || точка *f* предмета

O6 **object program** || Objektprogramm *n* || programme *m* objet || objectprogramma *n* || объектная программа *f*

O7 **object space** || Dingraum *m* || espace *m* d'objets || objectruimte *f* || пространство *n* предметов

O8 **oblate ellipsoid** || abgeplattetes Ellipsoid *n* || ellipsoïde *m* aplati || platgedrukte ellipsoïde *f* || сплющенный эллипсоид

O9 **oblique impact** || schiefer Stoß *m* || choc *m* indirect [oblique] || schuine stoot *m* || косой удар *m*

O10 **oblique incidence** || schräger Einfall *m* || incidence *f* oblique || schuine inval *m* || косое [наклонное] падение *n*

O11 **oblique parallelepiped** || schiefes Parallelepipedon *n* || parallélépipède *m* oblique || schuin parallelepipedum *n* || наклонный параллелепипед *m*

O12 **oblique wave** || Schrägwelle *f* || onde *f* oblique || schuine golf *f* (*m*) || косая волна *f*

O13 **observation** || Beobachtung *f* || observation *f* || waarneming *f*, observatie *f* || наблюдение *n*

O14 **obturator ring** || Verdichtungsring *m*, Kompressionsring *m* || segment *m* étanche [d'étanchéité] || compressiering *m* || компрессионное кольцо *n*

O15 **obtuse triangle** || stumpfwinkliges Dreieck *n* || triangle *m* obtusangle || stomphoekige driehoek *m* || тупоугольный треугольник *m*

O16 **occlusion** || Okklusion *f* || occlusion *f* || occlusie *f* || окклюзия *f*

O17 **occupied** || besetzt || occupé || bezet || заполненный, занятый *(напр. уровень энергии)*

O18 **octagon** || Oktagon *n*, Achteck *n* || octagon *m* || achthoek *m* || восьмиугольник *m*

O19 **octahedron** || Oktaeder *n*, Achtflächner *m* || octaèdre *m* || achtvlak *n*, octaëder *m* || восьмигранник *m*, октаэдр *m*

O20 **octal** || oktal || octal || octal || восьмеричный

O21 **octane**, CH_3 $(CH_2)_6$ CH_3 || Oktan *n* || octane *m* || octaan *n* || октан *m*

O22 **octet** || Oktett *n* || octet *m* || octet *n* || октет *m*

O23 I **octupole** || Oktupol *m* || octopôle *m* || octupool *f (m)* || октуполь *m*

O24 II **octupole** || achtfach || octuple || achtvoudig || восьмикратный

O25 **ocular (lens)** || Okularlinse *f* || oculaire *m*, lentille *f* oculaire || oculair *n*, oculairlens *f (m)* || окуляр *m*

O26 **odd-even nucleus** || ug-Kern *m*, Ungerade-gerade-Kern *m* || noyau *m* impair-pair || oneven-even-kern *f* || нечётно-чётное ядро *n*

O27 **odd function** || ungerade Funktion *f* || fonction *f* impaire || oneven functie *f* || нечётная функция *f*

O28 **oddness** || Ungeradzahligkeit *f* || imparité *f* || onevenheid *f* || нечётность *f*

O29 **odd-odd nucleus** || uu-Kern *m*, Ungerade-ungerade-Kern *m* || noyau *m* impair-impair || oneven-oneven-kern *f* || нечётно-нечётное ядро *n*

O30 **oersted**, Oe, O || Oersted *n* || œrsted *m* || oerstedt *m* || эрстед *m*, э

O31 **off-axis** || außeraxial || extra-axial || buiten-axiaal || не лежащий на оси; внеосевой

O32 **off-center collision** || nichtzentraler Stoß *m* || choc *m* non central || niet-centrale stoot *m* || нецентральный удар *m*

O33 **off-line** || Off-line- || autonome, off-line || off-line, autonoom || автономный, «оф-лайн»

O34 **off-line processing** || Off-line-Datenverarbeitung *f* || traitement *m* autonome, traitement *m* off-line || off-line databewerking *f*, autonome databewerking *f* || автономная обработка *f* данных, обработка *f* «оф-лайн»

O35 **off-resonance** || verstimmt || sans résonance || buiten resonantie || нерезонансный; далёкий от резонанса

O36 **offset tool** || gekröpfter Meißel *m* || outil *m* coudé [de côté] || gekromde beitel *m* || изогнутый резец *m*

O37 **ohm**, Ω || Ohm *n* || ohm *m* || ohm *n* || ом *m*

O38 **ohmic** || ohmsch || ohmique || ohms || омический, активный

O39 **ohmic heating** || ohmsche Aufheizung *f* || chauffage *m* ohmique || ohmse verhitting *f* || омический нагрев *m*

O40 **ohmic loss** || ohmscher Verlust *m* || pertes *f pl* ohmiques || ohmse verliezen *n pl* || омические потери *f pl*

O41 **ohmic resistance** || ohmscher Widerstand *m* || résistance *f* ohmique || ohmse weerstand *m* || омическое сопротивление *n*

O42 **ohmmeter** || Ohmmeter *n* || ohmmètre *m* || ohmmeter *m* || омметр *m*

O43 **oil** || 1. Öl *n*, Schmieröl *n* 2. Erdöl *n* || huile *f* || 1. olie *f (m)*; smeerolie *f (m)* 2. aardolie *f (m)* || 1. смазочное масло *n*, жидкая смазка *f* 2. нефть *f*

O44 **oil bath** || Ölbad *n* || bain *m* d'huile || oliebad *n* || масляная ванна *f*

O45 **oil container** || Ölbehälter *m* || réservoir *m* d'huile || oliereservoir *n* || резервуар *m* для масла

O46 **oiler** || Öler *m* || graisseur *m*; lubrificateur *m* || oliedop *m* || маслёнка *f*

O47 **oil-filled transformer** *see* **oil transformer**

O48 **oil film** || Ölfilm *m*, Ölschicht *f* || pellicule *f* d'huile || oliefilm *m* || масляная плёнка *f*

O49 **oil filter** || Ölfilter *n* || filtre *m* à huile || oliefilter *f* (*m*) || масляный фильтр *m*

O50 **oil groove** || Schmiernut *f* || rainure *f* de graissage; patte *f* d'araignée || oliegroef *f* (*m*) || смазочная канавка *f*

O51 **oil hole** || Schmierloch *n* || orifice *m* [trou *m*] de graissage || smeergaatje *n* || смазочное отверстие *n*

O52 **oiling** || Schmieren *n*, Ölen *n* || graissage *m*; lubrification *f* || smering *f* || смазывание *n*, смазка *f*

O53 **oil pressure gauge** || Ölmanometer *n* || manomètre *m* d'huile; indicateur *m* de pression d'huile || oliemanometer *m* || масляный манометр *m*

O54 **oil pumpe** || Ölpumpe *f* || pompe *f* à huile || oliepomp *f* (*m*) || масляный насос *m*

O55 **oil tank** || Ölbehälter *m* || réservoir *m* d'huile || olietank *m*, oliereservoir *n* || масляный бак *m*; резервуар *m* для масла

O56 **oil transformer** || Öltransformator *m* || transformateur *m* à huile || olietransformator *m* || масляный трансформатор *m*

O57 **olefin** || Olefin *n*, Alken *n* || oléfine *f* || olefine *f* (*m*) || олефин *m*, алкен *m*

O58 **omnidirectional radiation** || ungerichtete Strahlung *f* || radiation *f* omnidirectionnelle || ongerichte straling *f* || ненаправленное излучение *n*

O59 **on-and-off switch** || Ein-Aus-Schalter *m* || interrupteur *m* à deux positions || tweestandenschakelaar *m* || двухпозиционный выключатель *m*

O60 **one-dimensional** || eindimensional, linear || unidimensionnel || eendimensionaal || одномерный, линейный

O61 **one-half period** || Halbperiode *f* || demi-période *f* || halfperiode *f* || полупериод *m*

O62 **one-phase** || Einphasen..., einphasig || monophasé || eenfasig || однофазный

O63 **one-sided limit** || einseitiger Limes *m* || limite *f* unilatérale || eenzijdige limiet *f* (*m*) || односторонний предел *m*

O64 **one-to-one** || eineindeutig || un à un, biunivoque || one-to-one || взаимно-однозначный

O65 **one-to-one correspondence, one-to-one-mapping** || eineindeutige Abbildung *f* || correspondance *f* biunivoque, correspondance *f* un à un || one-to-one-afbeelding *f* || взаимно-однозначное соответствие *n*, взаимно-однозначное отображение *n*

O66 **on-line** || On-line- || on-line, en ligne || on-line-... || неавтономный; «он-лайн»; в темпе поступления данных

O67 **on-line processing** || On-line-Datenverarbeitung *f* || traitement *m* on-line [en ligne] || on-line-bewerking *f* || неавтономная [управляемая] обработка *f* данных, обработка *f* «он-лайн»

O68 **opacity** || Undurchsichtigkeit *f* || opacité *f* || opaciteit *f* || непрозрачность *f*

O69 **opalescence** || Opaleszenz *f* || opalescence *f* || opalescentie *f* || опалесценция *f*

O70 **opaque** || undurchsichtig || opaque || opaak || непрозрачный

O71 **open circuit** || offener Stromkreis *m* || circuit *m* ouvert || open circuit *m* || незамкнутая [разомкнутая] цепь *f*

O72 **open-hearth furnace** || Siemens-Martin-Ofen *m* || four *f* (Siemens-)Martin || martinoven *m* || мартеновская печь *f*

O73 **open-hearth steel** || Siemens-Martin-Stahl *m* || acier *m* Martin || martinstaal *n* || мартеновская сталь *f*

O74 **open interval** || offenes Intervall *n* || intervalle *m* ouvert || open interval *n* || открытый интервал *m*

O75 **open-loop** || Open-loop-, mit aufgeschnittenem [offenem] Regelkreis || à cycle ouvert || met open cyclus || без обратной связи, с разомкнутым циклом

O76 **open set** || offene Menge *f* || ensemble *m* ouvert || open menigte *f* || открытое множество *n*

O77 **open system** || offenes System *n* || système *m* ouvert || open systeem *n* || незамкнутая [открытая] система *f* *(в термодинамике)*

O78 **operand** || Operand *m* || opérande *f* || operand *f* (*m*) || операнд *m*

O79 **operand address** || Operandenadresse *f* || adresse *f* d'opérande || operand-adres *n* || адрес *m* операнда

O80 **operating characteristic** || 1. Operationscharakteristik *f*, O-C-Funktion *f* 2. Betriebsverhalten *n* || caractéristique *f* opérationnelle || 1. operatiekarakteristiek *f* 2. arbeidskarakteristiek *f* || 1. оперативная характеристика *f* *(статистического критерия)* 2. эксплуатационная [рабочая] характеристика

O81 **operating conditions** || Betriebsbedingungen *f pl*; Betriebszustand *m*, Betriebsart *f* || conditions *f pl* de travail [de fonctionnement] || arbeidsvoorwaarden *f pl*, bedrijfsvoorwaarden *f pl* || рабочие условия *n pl*; режим *m* работы

O82 **operating system** || Betriebssystem *n* || système *n* d'exploitation, système *m* opérationnel || bedrijfssysteem *n* || операционная система *f*

O83 **operating temperature** || Betriebstemperatur *f* || température *f* de travail || bedrijfstemperatuur *f* || рабочая температура *f*

O84 **operating voltage** || Betriebsspannung *f* || tension *f* de travail || bedrijfsspanning *f* || рабочее напряжение *n*

O85 **operation** || Operation *f* || opération *f* || operatie *f* || 1. операция *f* 2. работа *f*

O86 **operational calculus** || Operatorenkalkül *m* || calcul *m* opérationnel || operatorenberekening *f* || операционное исчисление *n*

O87 **operator** || Operator *m* || opérateur *m* || operator *m* || оператор *m*

O88 **opposing force** || Gegenkraft *f* || force *f* opposée || tegenkracht *f* (*m*) || противодействующая сила *f*

O89 **opposite** || Gegen...; gegenüberliegend opposé || tegen... || противоположный

O90 **opposite charges** || ungleichnamige Ladungen *f pl* || charges *f pl* de noms [de signes] contraires || ongelijknamige ladingen *f pl* || разноимённые заряды *m pl*

O91 **opposite in phase** || gegenphasig || opposé en phase || tegenfasig || противоположный по фазе, в противофазе

O92 **opposite poles** || ungleichnamige Pole *m pl* || pôles *m pl* opposables || ongelijknamige polen *m pl* || разноимённые полюса *m pl*

O93 **opposite spins** || antiparallele Spine *m pl* || spins *m pl* antiparallèles || antiparallele spinen *m pl* || противоположные [антипараллельные] спины *m pl*

O94 **optical** || optisch || optique || optisch || оптический; видимый

O95 **optical activity** || optische Aktivität *f*, optisches Dreh(ungs)-vermögen *n* || activité *f* optique || optische activiteit *f* || оптическая активность *f*

O96 **optical axis** || optische Achse *f* || axe *m* optique || optische as *f* (*m*) || оптическая ось *f*

O97 **optical density** || 1. Extinktion *f* 2. Schwärzung *f* *(Photomaterialien)* || densité *f* optique || optische dichtheid *f* || оптическая плотность *f*

O98 **optical length** || optische Länge *f* || longueur *f* optique || optische lengte *f* || оптическая длина *f* пути

O99 **optically flat** || optisch flach || optiquement plan || optisch effen || оптически плоский

O100 **optical maser** || Laser *m*, optischer Maser *m* || laser *m*, maser *m* optique || laser *m*, optische maser *m* || лазер *m*, оптический квантовый генератор *m*

O101 **optical-mechanical analogy** || optisch-mechanische Analogie *f* || analogie *f* optique-mécanique || optisch-mechanische analogie *f* || оптико-механическая аналогия *f*

O102 **optical path** *see* **optical length**

O103 **optical profile grinding machine** || optische Kopierschleifmaschine *f* || machine *f* à rectifier les profils à projection optique || optische kopiëerslijpmachine *f* || копировально-шлифовальный станок *m* с оптическим устройством

O104 **optical pumping** || optisches Pumpen *n* || pompage *m* optique || optisch pompen *n* || оптическая накачка *f*

O105 **optical rotation** || Drehung *f* der Polarisationsebene || rotation *f* du plan de polarisation de la lumière || draaiing *f* van polarisatievlak || вращение *n* плоскости поляризации (света)

O106 **optical system** || optisches System *n* || système *m* optique || optisch systeem *n* || оптическая система *f*

O107 **optical thickness** || optische Dicke *f* || épaisseur *f* optique || optische dikte *f* || оптическая толщина *f*

O108 **optical transmission** || Lichtdurchlässigkeit *f* || transparence *f*, translucidité *f* || doorschijnendheid *f* || прозрачность *f* *(для света)*, светопроницаемость *f*

O109 **optics** || Optik *f* || optique *f* || optica *f* || оптика *f*

O110 **optimal** || optimal || optimal, optimum || optimaal || оптимальный

O111 **optimal control** || Optimalregelung *f* || contrôle *m* optimal || optimale besturing *f* || оптимальное управление *n*

O112 **optimality principle** || Optimalitätsprinzip *n* || principe *m* d'optimalité || optimaliteitsprincipe *n* || принцип *m* оптимальности

O113 **optimal policy** || optimale Strategie *f* || stratégie *f* optimale || optimale strategie *f* || оптимальная стратегия *f*

O114 **optimeter** || Optimeter *n* || optimètre *m* || optimeter *m* || оптиметр *m*

O115 **optimization** || Optimie(sie)rung *f* || optimisation *f* || optimalisering *f* || оптимизация *f*

O116 **optimum** || Optimum *n* || optimum *m* || optimum *n* || оптимум *m*

O117 **optoelectronics** || Optoelektronik *f* || optoélectronique *f* || opto-elektronica *f* || оптоэлектроника *f*

O118 **OR** || OR *n* || OU || OF || ИЛИ, дизъюнкция *f*, логическое сложение *n*

O119 **orbit** || Bahn *f*, Umlaufbahn *f*, Orbit *m* || orbite *f* || baan *f* *(m)* || орбита *f*

O120 I **orbital** || Orbital *n* || orbitale *f* || orbital *n* || орбиталь *f*

O121 II **orbital** || Bahn..., Orbital... || orbitaire, orbital, planétaire || baan... || орбитальный

O122 **orbital flight** || Orbitalflug *m* || vol *m* orbital || baanvlucht *f* *(m)* || орбитальный полёт *m*

O123 **orbital motion** *see* **orbiting**

O124 **orbital plane** || Bahnebene *f* || plan *m* orbital || baanvlak *n* || плоскость *f* орбиты

O125 **orbital quantum number** || Bahnquantenzahl *f*, Nebenquantenzahl *f* || nombre *m* quantique orbital || baanquantengetal *n* || орбитальное [азимутальное] квантовое число *n*

O126 **orbital velocity** || Orbitalgeschwindigkeit *f*, erste kosmische Geschwindigkeit *f* || vitesse *f* orbitale || orbitaalsnelheid *f* || первая космическая скорость *f*, орбитальная скорость *f*

O127 **orbiter** || Orbiter *m* || orbiter *m* || orbiter *m* || орбитальный (искусственный) спутник *m*

O128 **orbiting** || Bahnbewegung *f*; Umlauf *m* || mouvement *m* orbital || orbitaalbeweging *f*, baanbeweging *f* || движение *n* по орбите, орбитальное движение *n*

O129 **order** || **1.** Ordnung *f* **2.** Befehl *m*, Instruktion *f* || ordre *m* || **1.** orde *f* *(m)* **2.** instructie *f*, commando *n* || **1.** порядок *m*; упорядоченность *f* **2.** команда *f*, инструкция *f*

O130 **ordered** || geordnet || ordonné || geordend geregeld || упорядоченный; регулярный

O131 **ordered set** || geordnete Menge *f* || ensemble *m* ordonné || geordende menigte *f* || упорядоченное множество *n*

O132 **ordering** || **1.** Ordnen *n* **2.** Ordnung *f* || ordre *m* || ordening *f* || **1.** упорядочение *n* **2.** порядок *m*; упорядоченность *f*

O133 **order of a curve** || Kurvenordnung *f* || ordre *m* d'une courbe || krommenorde *f* (*m*) || порядок *m* кривой

O134 **order of a differential equation** || Ordnung *f* einer Differentialgleichung || ordre *m* d'une équation différentielle || orde *f* (*m*) van een differentiaalvergelijking || порядок *m* дифференциального уравнения

O135 **order of a function** || Ordnung *f* einer Funktion || ordre *m* d'une fonction || orde *f* (*m*) van een functie || порядок *m* функции

O136 **order of a matrix** || Matrixordnung *f* || rang *m* d'une matrice || matrixorde *f* (*m*) || порядок *m* матрицы

O137 **order of magnitude** || Größenordnung *f* || ordre *m* de grandeur || grootheidsorde *f* (*m*) || порядок *m* величины

O138 **order of reaction** || Reaktionsordnung *f*, Ordnung *f* der Reaktion || ordre *m* d'une réaction || reactieorde *f* (*m*) || порядок *m* реакции

O139 **ordinal** || Ordinal *n* || ordinal *m* || ordinal *n* || ординал *m*

O140 **ordinary differential equation** || gewöhnliche Differentialgleichung *f* || équation *f* différentielle ordinaire || gewone differentiaalvergelijking *f* || обыкновенное дифференциальное уравнение *n*

O141 **ordinary ray** || ordentlicher [ordinärer] Strahl *m* || rayon *m* ordinaire || gewone straal *m*, *f* || обыкновенный луч *m*

O142 **ordinary wave** || ordentliche [ordinäre] Welle *f* || onde *f* ordinaire || gewone golf *f* (*m*) || обыкновенная волна *f*

O143 **ordinate** || Ordinate *f* || ordonnée *f* || ordinaat *f* || ордината *f*

O144 **organic compound** || organische Verbindung *f* || composé *m* [combinaison *f*] organique || organische verbinding *f* || органическое соединение *n*

O145 **organo-metallic compound** || metallorganische Verbindung *f*, Organometallverbindung *f* || composé *m* organométallique || organometallieke verbinding *f* || металл(о)органическое соединение *n*

O146 **orientation** || Orientierung *f* || orientation *f* || oriëntering *f*, oriëntatie *f* || ориентация *f*; ориентировка *f*

O147 **oriented** || orientiert; ausgerichtet || orienté || georiënteerd || ориентированный

O148 **oriented graph** || orientierter Graph *m* || graphe *m* orienté || georiënteerde graaf *f* (*m*) || ориентированный граф *m*

O149 **origin (of co-ordinates)** || Koordinatenursprung *m* || origine *f* (des coordonnées) || coördinatenoorsprong *m* || начало *n* координат

O150 **origin of force** || Kraftangriffspunkt *m* || point *m* d'application de la force || krachtingrijppunt *f* (*m*) || точка *f* приложения силы

O151 **orpiment**, As_2S_3 || Auripigment *n* || auripigment *m*, orpiment *m* || auripigment *n* || аурипигмент *m*

O152 **orthogonal basis** || orthogonale Basis *f* || base *f* orthogonale || orthogonale basis *f* || ортогональный базис *m*

O153 **orthogonal coordinates** || orthogonale [rechtwinklige] Koordinaten *f pl* || coordonnées *f pl* orthogonales || orthogonale [rechthoekige] coördinaten *m pl* || ортогональные координаты *f pl*

O154 **orthogonal functions** || orthogonale Funktionen *f pl* || fonctions *f pl* orthogonales || orthogonale functies *f pl* || ортогональные функции *f pl*

O155 **orthogonality** || Orthogonalität *f* || orthogonalité *f* || orthogonaliteit *f* || ортогональность *f*

O156 **orthogonalization** || Orthogonalisierung *f* || orthogonalisation *f* || orthogonalisatie *f* || ортогонализация *f*

O157 **orthogonal matrix** || orthogonale Matrix *f* || matrice *f* orthogonale || orthogonale matrix *f* || ортогональная матрица *f*

O158 **orthogonal polynomials** || orthogonale Polynome *n pl* || polynômes *m pl* orthogonaux || orthogonale polynomen *n pl* || ортогональные многочлены *m pl*

O159 **orthogonal projection** || orthogonale Projektion *f* || projection *f* orthogonale || orthogonale projectie *f* || ортогональная проекция *f*

O160 **orthogonal representation** || orthogonale Darstellung *f* || représentation *f* orthogonale || orthogonale voorstelling *f* || ортогональное представление *n*

O161 **orthogonal transformation** || orthogonale Transformation *f* || transformation *f* orthogonale || orthogonale transformatie *f* || ортогональное преобразование *n*

O162 **orthogonal vectors** || orthogonale Vektoren *m pl* || vecteurs *m pl* orthogonaux || orthogonale vectoren *m pl* || ортогональные векторы *m pl*

O163 **orthonormal basis** || orthonormale Basis *f* || base *f* orthonormale || orthonormale basis *f* || ортонормальный базис *m*

O164 **orthonormal functions** || orthonormale Funktionen *f pl* || fonctions *f pl* orthonormales || orthonormale functies *f pl* || ортонормальные функции *f pl*

O165 **oscillating** || schwingend, oszillierend || oscillatoire || oscillatie..., trillings... || колеблющийся; осциллирующий; колебательный

O166 **oscillating superfinishing machine** || Schwingschleifmaschine *f* || machine *f* de superfinition [à superfinir] || superfinishmachine *f*, slingerslijpmachine *f* || суперфинишный станок *m*

O167 **oscillating system** || Schwing(ungs)system *n* || système *m* oscillatoire || oscillatiesysteem *n* || колебательная система *f*

O168 **oscillation** || Schwingung(en) *f* (*pl*), Oszillation *f* || oscillation *f* || oscillatie *f*; trilling *f* || колебание *n*, колебания *n pl*

O169 **oscillation frequency** || Schwingungsfrequenz *f* || fréquence *f* d'oscillations || oscillatiefrequentie *f*, trillingsfrequentie *f* || частота *f* колебаний

O170 **oscillation mode** || Schwingungsart *f*, Schwingungsform *f*, Mode *f* || mode *m* de vibrations [d'oscillations] || trillingsmodus *m* || вид *m* [форма *f*] колебаний, мода *f*

O171 **oscillation period** || Schwingungsperiode *f* || période *f* d'oscillations || trillingstijd *m* || период *m* колебаний

O172 **oscillator** || **1.** Oszillator *m* **2.** Schwingungserzeuger *m* || **1.** oscillateur *m* **2.** générateur *m* d'oscillations || oscillator *m* || **1.** осциллятор *m* **2.** генератор *m* (электромагнитных) колебаний

O173 **oscillatory motion** || schwingende Bewegung *f*, Schwing(ungs)-bewegung *f* || mouvement *m* oscillatoire || oscillatie *f*; slingerbeweging *f* || колебательное движение *n*

O174 **oscillogram** || Oszillogramm *n* || oscillogramme *m* || oscillogram *n* || осциллограмма *f*

O175 **oscillograph** || Oszillograph *m* || oscillographe *m* || oscillograaf *m* || осциллограф *m*

O176 **oscillography** || Oszillographie *f* || oscillographie *f* || oscillografie *f* || осциллография *f*, осциллографирование *n*

O177 **oscilloscope** || Oszilloskop *n*; Oszillograph *m* || oscilloscope *m*, ondoscope *m* || oscilloscoop *m* || осциллоскоп *m*; осциллограф *m*

O178 **osculating circle** || Schmieg(ungs)kreis *m*, Krümmungskreis *m* || cercle *m* osculateur [de courbure] || krommingscirkel *m* || соприкасающаяся окружность *f*, круг *m* кривизны

O179 **osculating plane** || Schmieg(ungs)ebene *f* || plan *m* osculateur ||

aanrakingsvlak *n*, raakvlak *n* || соприкасающаяся плоскость *f*

O180 **osculating sphere** || Schmieg(ungs)kugel *f* || sphère *f* osculatrice || aanrakingssfeer *f* (*m*) || соприкасающаяся сфера *f*

O181 **osculation** || Berührung *f*; *Math.* Schmiegung *f* || osculation *f* || aanraking *f* || касание *n*

O182 **osmium,** Os || Osmium *n* || osmium *m* || osmium *n* || осмий *m*

O183 **osmosis** || Osmose *f* || osmose *f* || osmose *f* || осмос *m*

O184 **osmotic coefficient** || osmotischer Koeffizient *m* || coefficient *m* osmotique || osmotische coëfficiënt *m* || осмотический коэффициент *m*

O185 **osmotic pressure** || osmotischer Druck *m* || pression *f* osmotique || osmotische druk *m* || осмотическое давление *n*

O186 **outdoor** temperature || Außentemperatur *f* || température *f* ambiante || buitentemperatuur *f* || температура *f* наружного воздуха, внешняя температура *f*

O187 **outer dead center** || oberer Totpunkt *m* || point *m* mort avant [haut] || bovenste dode punt *f* (*m*) || верхняя мёртвая точка *f*

O188 **outer diameter** *see* **outside diameter**

O189 **outer product** || äußeres Produkt *n*; Vektorprodukt *n* || produit *m* extérieur || uitwendig produkt *n*, vectorprodukt *n* || внешнее [прямое, векторное] произведение *n*

O190 **outer product of matrices** || direktes Matrizenprodukt *n* || produit *m* extérieur des matrices || direct matricesprodukt *n* || произведение *n* матриц

O191 **outer product of tensors** || Tensorprodukt *n* || produit *m* extérieur des tenseurs || uitwendig tensorenprodukt *n* || (внешнее) произведение *n* тензоров

O192 **outer race** || Außenring *m* (*Kugellager*) || chemin *m* [anneau *m*] de roulement extérieur || buitenring *m* (*kogellager*) || наружное кольцо *n* (*подшипника*)

O193 **outer-shell electron** || Außenelektron *n*, Valenzelektron *n* || électron *m* périphérique || elektron *n* van de buitenschil || внешний [валентный] электрон *m*

O194 **out-of-balance** || unbalanciert, nicht ausbalanciert; unausgewuchtet; unabgeglichen (*Brücke*) || non équilibré, déséquilibré || onbalans, onevenwichtig || неуравновешенный

O195 **out-of-focus** || nichtfokussiert || non focalisé || niet-gefocusseerd || несфокусированный; не в фокусе

O196 **out-of-gear** || außer Eingriff gebracht || désengrené || buiten ingrijping gebracht || выведенный из зацепления

O197 **out-of-phase** || phasenverschoben, außer Phase || déphasé || buiten fase || сдвинутый по фазе, не в фазе

O198 **output** || **1.** Leistung(sfähigkeit) *f*; Nutzeffekt *m*; Abgabe *f*; Ausbeute *f*; erzeugte Leistung *f*; abgegebene Leistung *f* **2.** Ausgang *m* (*Gerät*) **3.** Ausgabe *f* (*Daten*) || **1.** rendement *m*; efficacité *f*; débit *m*; puissance *f* de sortie; **2.** sortie *f* **3.** prélèvement *m* || **1.** produktiviteit *f*; efficiëntie *f*; debiet *n*; uitgangsvermogen *n* **2.** uitgang *m* **3.** output *m* || **1.** производительность *f*; отдача *f*; выход *m*; генерируемая мощность *f*; выходная мощность *f* **2.** выход *m* (*прибора*) **3.** вывод *m* (*данных*)

O199 **output alphabet** || Ausgabealphabet *n* || alphabet *m* de sortie || uitgangsalfabet *n* || выходной алфавит *m*

O200 **output impedance** || Ausgangsimpedanz *f* || impédance *f* de sortie || uitgangsimpedantie *f* || выходной импеданс *m*

O201 **output power** || Ausgangsleistung *f* || puissance *f* motrice || uitgangsvermogen *n* || выходная мощность *f*

O202 **output speed** || Ausgangsdrehzahl *f* || vitesse *f* à la sortie || uitgangssnelheid *f* || скорость *f* на выходе; частота *f* вращения ведомого вала

O203 **output torque** || Ausgangsmoment *n* || couple *m* (de rotation) secondaire || uitgangstorsiemoment *n* || крутящий момент *m* ведомого вала, крутящий момент *m* на выходном валу

O204 **output transformer** || Ausgangstransformator *m* || transformateur *m* de sortie || uitgangstransformator *m* || выходной трансформатор *m*

O205 **output voltage** || Ausgangsspannung *f* || tension *f* de sortie || uitgangsspanning *f* || выходное напряжение *n*, напряжение *n* на выходе

O206 **outside diameter** || **1.** Außendurchmesser *m* **2.** Kopfkreisdurchmesser *m (Zahnrad)* || diamètre *m* extérieur || **1.** buitendiameter *m* **1.** buitenmiddellijn *f* (*m*) **2.** kopomtrekdiameter *m (tandwiel)* || **1.** наружный [внешний] диаметр *m* **2.** диаметр *m* вершин зубьев *(зубчатого колеса)*

O207 **outward-flow turbine** || Zentrifugalturbine *f* || turbine *f* centrifuge || centrifugaalturbine *f* || центробежная турбина *f*

O208 **oval head** || Flachrundkopf *m* || tête *f* ovale [mi-fraisée] || halfverzonken kop *m* || полупотайная [овальная] головка *f (напр. винта)*

O209 **overall gain** || Gesamtverstärkung *f (Signal)*; Gesamtverstärkungsfaktor *m* || amplification *f* totale || totale versterking *f* || общее усиление *n (сигнала)*

O210 **overcurrent** || Überstrom *m* || surintensité *f* de courant || overstroom *m* || сверхток *m*

O211 **overflow** || Überlaufrohr *n* || trop-plein *m* || overlooppijp *f* (*m*) || перепускная [сливная] труба *f*

O212 **overheat** || Überhitzung *f* || surchauffage *m* || oververhitting *f* || перегрев *m*

O213 **overlap** || Überlappung *f* || recouvrement *m* || overlapping *f* || нахлёстка *f*; перекрытие *n*

O214 **overlay** || Overlay *n*, Overlaying *n*, Überlagerung *f* || recouvrement *m*, overlay *m* || overlay *n* || перекрытие *n*, оверлей *m*

O215 **overload** || Überlastung *f* || surcharge *f* || overbelasting *f* || чрезмерная нагрузка *f*, перегрузка *f*

O216 **overload protection** || Überlastungsschutz *m* || protection *f* contre les surcharges || overbelastingsbescherming *f* || защита *f* от перегрузок

O217 **overload test** || Überlastungsprüfung *f* || essai *m* de surcharge || overbelastingsproef *f* (*m*) || испытание *n* на перегрузку

O218 **overpressure** || Überdruck *m* || surpression *f* || overdruk *m* || избыточное давление *n*

O219 **overrelaxation** || Überrelaxation *f* || superrelaxation *f* || overrelaxatie *f* || сверхрелаксация *f*

O220 **overspeed** || übernormale Geschwindigkeit *f* || survitesse *f* || overnormale snelheid *f* || скорость *f* выше нормальной

O221 **overstrain** || Überbeanspruchung *f* || surtension *f* || overspanning *f* || перенапряжение *n*; перегрузка *f*

O222 **overtone** || Oberton *m* || son *m* concomitant; harmonique *f* || boventoon *m*, bijtoon *m* || обертон *m*, высшая гармоника *f*

O223 **overvoltage** || Überspannung *f* || surtension *f* || overspanning *f* || (электрическое) перенапряжение *n*

O224 **oxidation** || Oxydation *f*, Oxydierung *f* || oxydation *f* || oxydatie *f* || окисление *n*

O225 **oxidation-reduction potential** || Oxydations-Reduktions-Potential *n*, Redoxpotential *n* || potentiel *m* d'oxydo-réduction || oxydatie-reductiepotentiaal *m* || окислительно-восстановительный потенциал *m*, редокс-потенциал *m*

O226 **oxide** || Oxid *n* || oxyde *m* || oxyde *n* || окись *f*, оксид *m*; окисел *m*

O227 **oxyacetylene welding** || Azetylen-Sauerstoff-Schweißen *n*; Gasschweißen *n* || soudage *m* oxyacétylénique [aux gaz] || lassen *n* met acetyleen en zuurstof; gaslassen *n* || ацетилено-кислородная сварка *f*; газовая сварка *f*

O228 **oxyazo dye** || Hydroxyazofarbstoff *m* || oxyazocolorant *m* || oxyazokleurstof *f* (*m*) || оксиазокраситель *m*

O229 **oxygen**, O || Sauerstoff *m* || oxygène *m* || zuurstof *f* (*m*) || кислород *m*

O230 **oxyhydrogen welding** || Sauerstoff-Wasserstoff-Schweißen *n*; Gasschweißen *n* || soudage *m* à flamme oxhydrique; soudage *m* à l'hydrogène-oxygène || knalgaslassen *n*; gaslassen *n* || кислородно-водородная сварка *f*; газовая сварка *f*

O231 **ozone,** O_3 || Ozon *n* || ozone *m* || ozon *n* || озон *m*

P

P1 **packing** || **1.** Packung *f*; Liderung *f*; Dichtung *f* **2.** *Phys., Chem.* Packung *f* || **1.** garniture *f*; joint *m* **2.** *phys., chim.* (em)paquetage *m* || **1.** pakking *f*; verpakking *f*; dichting *f* **2.** *phys., chem.* pakking *f* || **1.** (уплотняющая) прокладка *f*; уплотнение *n* **2.** *физ., хим.* упаковка *f*

P2 **packing density** || Packungsdichte *f* || densité *f* de paquetage || pakkingsdichtheid *f* || плотность *f* упаковки

P3 **packing factor** || Packungsfaktor *m (für Atomkern)* || facteur *m* de compression || pakkingsfactor *m* || упаковочный множитель *m (для ядра)*

P4 **paddle wheel** || Schaufelrad *n* || roue *f* à palettes; roue *f* à aubes || schoepenrad *n* || лопастное колесо *n*

P5 **pair** || Paar *n* || couple *m*, paire *f* || paar *n* || пара *f*

P6 **pair creation** || Paarerzeugung *f* || création *f* des paires || paarvorming *f* || рождение *n* [образование *n*] (электрон-позитронных) пар

P7 **pairing** || Paarung *f* || appariement *m* || paring *f* || спаривание *n (электронов)*

P8 **pair production** *see* **pair creation**

P9 **palladium,** Pd || Palladium *n* || palladium *m* || palladium *n* || палладий *m*

P10 **panel** || **1.** Platte *f*; Tafel *f* **2.** Schalttafel *f* || **1.** panneau *m* **2.** tableau *m* de commande, tableau *m* de distribution || **1.** plaat *f (m)* **2.** schakelpaneel *n* || **1.** панель *f* **2.** щит *m* управления, распределительный щит *m*

P11 **parabola** || Parabel *f* || parabole *f* || parabool *f (m)* || парабола *f*

P12 **parabolic** || parabolisch || parabolique || parabolisch || параболический

P13 **parabolic aerial** || Parabolantenne *f* || antenne *f* parabolique || paraboolantenne *f (m)* || параболическая антенна *f*

P14 **parabolic mirror** || Parabolspiegel *m* || miroir *m* parabolique || paraboolspiegel *m* || параболическое зеркало *n*

P15 **paraboloid** || Paraboloid *n* || paraboloïde *m* || paraboloïde *f* || параболоид *m*

P16 **paraboloid of revolution** || Rotationsparaboloid *n* || paraboloïde *m* de révolution || rotatieparaboloïde *f* || параболоид *m* вращения

P17 **parachor** || Parachor *m* || parachor *m* || parachoor *m* || парахор *m*

P18 **paradox** || Paradoxon *n* || paradoxe *m* || paradox *m* || парадокс *m*

P19 **paraelectric** || Paraelektrikum *n* || paraélectrique *m* || **para-elektricum** *n* || параэлектрик *m*

P20 **parallel axiom** || Parallelenaxiom *n* || axiome *m* de parallélisme || parallelenaxioma *n* || аксиома *f* параллельности

P21 **parallel beam** || Parallelstrahlenbündel *n* || faisceau *m* parallèle || parallelbundel *m* || пучок *m* параллельных лучей

P22 **parallel circuit** || Parallelkreis *m*; Spannungskreis *m* || circuit *m* parallèle || parallelschakeling *f* || параллельная [шунтовая] цепь *f*; цепь *f* напряжения

P23 **parallel displacement** || Parallelverschiebung *f* || déplacement *m* parallel || parallelverschuiving *f* || **1.** *мат.* параллельный перенос *m* **2.** параллельное смещение *n*

P24 **parallelepiped(on)** || Parallelepiped(on) *n*, Parallelflach *n*, Spat *m* || parallélépipède *m* || parallelepipedum *n* || параллелепипед *m*

P25 **paralleling** || Parallelschaltung *f*, Nebenschluß *m* || couplage *m* [mise *f*, connexion *f*] en parallèle || parallelschakeling *f* || параллельное включение *n*, параллельное соединение *n*

P26 **parallelism** || **1.** Parallelität *f* **2.** Parallelismus *m* || parallélisme *m* ||

parallelisme *m* || 1. параллельность *f* 2. параллелизм *m*

P27 **parallelogram** || Parallelogramm *n* || parallélogramme *m* || parallelogram *n* || параллелограмм *m*

P28 **parallelogram law** || Parallelogrammregel *f* || loi *f* de parallélogramme || parallelogramregel *m* || правило *n* параллелограмма

P29 **parallelogram of forces** || Kraftparallelogramm *n* || parallélogramme *m* des forces || krachtenparallelogram *n*, parallelogram *n* van krachten || параллелограмм *m* сил

P30 **parallelogram of velocities** || Geschwindigkeitsparallelogramm *n* || parallélogramme *m* des vitesses || snelheidsparallelogram *n*, parallelogram *n* van snelheden || параллелограмм *m* скоростей

P31 **parallel-plane** || planparallel || plan-parallèle || planparallel || плоско-параллельный

P32 **parallel screw thread** || zylindrisches Gewinde *n* || filet *m* cylindrique || cilindervormige schroefdraad *m* || цилиндрическая резьба *f*

P33 **parallel-series** || Parallelreihen..., Parallelserien... || série-parallèle || parallel-serie... || параллельно-последовательный, смешанный *(о включении в цепь)*

P34 **parallel winding** || Parallelwicklung *f* || enroulement *m* shunt [parallèle, en dérivation] || parallelwikkeling *f* || параллельная [шунтовая] обмотка *f*

P35 **paramagnet** *see* **paramagnetic I**

P36 I **paramagnetic** || Paramagnetikum *n* || paramagnétique *m* || paramagneet *n* || парамагнетик *m*

P37 II **paramagnetic** || paramagnetisch || paramagnétique || paramagnetisch || парамагнитный

P38 **paramagnetism** || Paramagnetismus *m* || paramagnétisme *m* || paramagnetisme *n* || парамагнетизм *m*

P39 **parameter** || Parameter *m* || paramètre *m* || parameter *m* || параметр *m*

P40 **parameter space** || Parameterraum *m* || espace *m* des paramètres || parameterruimte *f* || параметрическое пространство *m*

P41 **parametric representation** || Parameterdarstellung *f* || représentation *f* paramétrique || parametervoorstelling *f* || параметрическое представление *n*

P42 **parasitic** || parasitär; Stör... || parasite || parasitair || паразитный

P43 **paraxial** || paraxial || paraxial || paraxiaal || параксиальный, приосевой

P44 **parent nucleus** || Mutterkern *m*, Ausgangskern *m* || noyau *m* père || moederkern *f* || материнское ядро *n*

P45 **parity** || Parität *f* || parité *f* || pariteit *f* || чётность *f*

P46 **parity violation** || Nichterhaltung *f* der Parität || non-conservation *f* de la parité || niet-behoud *n* van pariteit || несохранение *n* чётности

P47 **partial** || partiell, Partial...; Teil..., partiell || partiel || partiëel || парциальный *(напр. о давлении)*; частичный

P48 **partial derivative** || partielle Ableitung *f* || dérivée *f* partielle || partiële afgeleide *f* || частная производная *f*

P49 **partial difference equation** || partielle Differentialgleichung *f* || équation *f* différentielle aux dérivées partielles || partiële differentiaalvergelijking *f* || уравнение *n* в частных производных

P50 **partially filled shell** || teilweise gefüllte Elektronenschale *f* || couche *f* partiellement remplie || gedeeltelijk gevulde schil *m* || частично заполненная (электронная) оболочка *f*

P51 **partially occupied band** || teilweise besetztes Energieband *n* || bande *f* partiellement occupée || gedeeltelijk bezette band *m* || частично заполненная энергетическая зона *f*

P52 **partially ordered set** || teilweise geordnete Menge *f* || ensemble *m* partiellement ordonné || partiëel geregelde menigte *f* || частично упорядоченное множество *n*

P53 **partially polarized light** || teilweise polarisiertes Licht *n* || lumière *f* partiellement polarisée || gedeeltelijk gepolariseerd licht *n* || частично поляризованный свет *m*

P54 **partial order** || partielle Ordnung *f* || ordre *m* partiel || partiële orde *f* (*m*) || частичный порядок *m*

P55 **partial polarisation** || Teilpolarisation *f*, teilweise Polarisation *f* || polarisation *f* partielle || gedeeltelijke polarisatie *f* || частичная [неполная] поляризация *f*

P56 **partial pressure** || Partialdruck *m* || pression *f* partielle || partiële druk *m* || парциальное давление *n*

P57 **partial valency** || Partialvalenz *f*, Restvalenz *f* || valence *f* partielle || partiële valentie *f* || парциальная валентность *f*

P58 **partial volume** || Partialvolumen *n* || volume *m* partiel || partiëel volume *n* || парциальный объём *m*

P59 **particle** || **1.** Teilchen *n* **2.** Massenpunkt *m* || **1.** particule *f* **2.** point *m* matériel || **1.** deeltje *n*, partikel *n* **2.** massapunt *f* (*m*) || **1.** частица *f* **2.** материальная точка *f*

P60 **particle detector** || Teilchendetektor *m* || détecteur *m* des particules || deeltjesdetector *m* || детектор *m* частиц

P61 **particle flux** || Teilchenfluß *m* || flux *m* de particules || deeltjesflux *m* || поток *m* частиц

P62 **particle mechanics** || Massenpunktmechanik *f* || mécanique *f* du point matériel || deeltjesmechanica *f* || механика *f* (материальной) точки

P63 **particle physics** || Physik *f* der Elementarteilchen || physique *f* des particules fondamentales [élémentaires] || fysica *f* van elementaire deeltjes || физика *f* элементарных [фундаментальных] частиц

P64 **particle scattering** || Teilchenstreuung *f* || diffusion *f* des particules || deeltjesdispersie *f*, deeltjesstrooiing *f* || рассеяние *n* частиц

P65 **particular integral** || partikuläre Lösung *f* *(einer Differentialgleichung)* || intégrale *f* particulière || particulaire oplossing *f* || частное решение *n* *(дифференциального уравнения)*

P66 **particular solution** || partikuläre Lösung *f* || solution *f* particulière || particulaire oplossing *f* || частное решение *n*

P67 **parting-off tool** || Stechmeißel *m* || outil *m* à tronçonner || steekbeitel *m* || отрезной резец *m*

P68 **partition of energy** || Energieverteilung *f* || distribution *f* de l'énergie || energieverdeling *f* || распределение *n* энергии

P69 **parton** || Parton *n* || parton *m* || parton *n* || партон *m*

P70 **pascal**, Pa || Paskal *n* || pascal *m* || pascal *n* || паскаль *m*, Па ($1 Н/м^2$)

P71 **Pascals's triangle** || Pascalsches Zahlendreieck *n* || triangle *m* pascalien || driehoek *m* van Pascal || треугольник *m* Паскаля

P72 **passband** || Durchlaßbereich *m* || bande *f* passante || doorlaatband *m* || полоса *f* пропускания

P73 **passive** || passiv, inert || passif, inerte || passief, inert || пассивный, инертный

P74 **passivity** || Passivität *f*, **Metall**passivität *f* || passivité *f* || passiviteit *f* || пассивность *f*, инертность *f*

P75 **path** || **1.** Weg *m*, Pfad *m*; Bahn *f* **2.** Weg *m* || **1.** chemin *m*; trajectoire *f* **2.** chemin *m* || **1.** weg *m*; baan *f* (*m*) **2.** weg *m* || **1.** путь *m*, траектория *f* **2.** путь *m* *(напр. в графе)*

P76 **path difference** || Gangunterschied *m*, Gangdifferenz *f* || différence *f* de propagation (des faisceaux) || gangonderscheid *n* || разность *f* хода *(лучей)*

P77 **path equation** || Bahngleichung *f* || équation *f* de la trajectoire || baanvergelijking *f* || уравнение *n* траектории

P78 **path length** || Weglänge *f* || longueur *f* de parcours || weglengte *f* || длина *f* пути

P79 **path velocity** || Bahngeschwindigkeit *f* || vitesse *f* de trajectoire || baansnelheid *f* || скорость *f* движения по траектории; путевая скорость *f*

P80 **pattern** || **1.** Muster *n*; Schablone *f*; Modell *n* **2.** Bild *n*; Muster *n*, Gestalt *f* || **1.** modèle *m*; échantillon *m* **2.** image *f*, figure *f* || **1.** monster *n*; sjabloon *f* (*m*); model *n* **2.** patroon *f* (*m*) || **1.** образец *m*; шаблон *m*; модель *f* **2.** фигура *f*, картина *f*; образ *m*

P81 **pattern recognition** || Mustererkennung *f*, Gestalterkennung *f* || reconnaissance *f* des images || afbeeldingsherkenning *f* || распознавание *n* образов

P82 **pawl** || Sperrklinke *f* || cliquet *m* *(d'arrêt)* || pal *m* || собачка *f* *(храпового механизма)*

P83 **payoff matrix** || Gewinnmatrix *f* || matrice *f* de gain || winstmatrix *f* || матрица *f* выигрыша

P84 **peak current** || Stromspitzenwert *m*, Spitzenstrom *m* || courant *m* maximal [de pointe] || piekstroom *m* || максимальный ток *m*

P85 **peak load** || Spitzenlast *f* || charge *f* de pointe || piekbelasting *f* || пиковая нагрузка *f*

P86 **peak-to-peak amplitude** || Spitze-zu-Spitze-Amplitude *f* || amplitude *f* totale [crête à crête] || spits-tot-spits-amplitude *f* || двойная амплитуда *f*, размах *m*

P87 **peaking transformer** || Impulstransformator *m* || transformateur *m* de crête || spitstransformator *m* || пик-трансформатор *m*

P88 **peak value** || Spitzenwert *m*; Scheitelwert *m* || valeur *f* de crête [de pointe] || topwaarde *f* || максимальное [пиковое, амплитудное] значение *n*

P89 **peak voltage** || Spannungsspitzenwert *m*, Spitzenspannung *f* || tension *f* de crête [de pointe] || piekspanning *f* || пиковое [амплитудное] напряжение *n*

P90 **peak voltmeter** || Spitzenspannungsmesser *m* || voltmètre *m* de crête || spitsspanningsmeter *m* || пиковый [амплитудный] вольтметр *m*

P91 **pedal curve** || Fußpunktkurve *f*, Podoide *f* || courbe *f* de pédale || voetpuntskromme *f* *(m)* || подэра *f*

P92 **peening** || Kaltverfestigung *f*; Hämmern *n* || écrouissage *m* || koudversteviging *f*; hameren *n* || наклёп *m*; проковка *f*

P93 **pendulum** || Pendel *n* || pendule *m* || pendel *n* || маятник *m*

P94 **pendulum governor** || Pendelregler *m* || régulateur *m* à pendule || pendelregelaar *m* || маятниковый регулятор *m*

P95 **pendulum suspension** || Pendelaufhängung *f* || suspension *f* à pendule || pendelophanging *f* || маятниковый подвес *m*

P96 **penetrability** || Durchdringungsfähigkeit *f*; Durchlässigkeit *f* || pénétrabilité *f* || doorlaatbaarheid *f* || проницаемость *f*; прозрачность *f*

P97 **penetrating radiation** || durchdringende [harte] Strahlung *f* || radiation *f* pénétrante || harde straling *f* || проникающее [жёсткое] излучение *n*; проникающая радиация *f*

P98 **penetration** || Durchdringung *f* || pénétration *f* || doordringing *f*, penetratie *f* || проникновение *n*, проникание *n*, пенетрация *f*

P99 **pentagon** || Pentagon *n*, Fünfeck *n* || pentagone *m* || vijfhoek *m* || пятиугольник *m*

P100 **peptization** || Peptisation *f*, Peptisierung *f* || peptisation *f* || peptisatie *f* || пептизация *f*

P101 **per-acid** || Peroxosäure *f* || peracide *m* || perzuur *n* || перкислота *f*, надкислота *f*

P102 **percentage** || Prozentgehalt *m* || pourcentage *m* || percentage *n* || процентное содержание *n*, процент *m*

P103 **percentage modulation** || Aussteuerungsgrad *m* || profondeur *f* [taux *m*] de modulation || modulatiefactor *m* || глубина *f* модуляции; коэффициент *m* модуляции

P104 **percent by mass** || Massenprozent *m* || pourcentage *m* en masse || massapercentage *n* || процент *m* по массе, массовый процент *m*

P105 **percent by volume** || Volum(en)prozent *n* || pourcentage *m* en volume || volumepercentage *n* || процент *m* по объёму, объёмный процент

P106 **percentile** || Perzentil *n*, Prozentil *n* || percentile *m* || percentiel *n* || процентиль *m*

P107 **perchloric acid,** $HClO_4$ || Perchlorsäure *f*, Chlor (VII)-Säure *f* || acide *m* perchlorique [oxychlorique] || perchloorzuur *n* || хлорная кислота *f*

P108 **perfect absorber** || schwarzer Strahler *m* || corps *m* noir || zwarte straler *m* || абсолютно чёрное тело *n*

P109 **perfect conductor** || idealer Leiter *m* || conducteur *m* parfait || ideale leider *m* || идеальный проводник *m*

P110 **perfect crystal** || Idealkristall *m*, fehlerfreier Kristall *m* || cristal *m* parfait [idéal] || ideaal kristal *n* || идеальный [совершенный] кристалл *m*

P111 **perfect gas** || ideales Gas *n* || gaz *m* parfait || ideaal gas *n* || идеальный газ *m*

P112 **perfect number** || vollkommene Zahl *f* || nombre *m* parfait || volkomen getal *n* || совершенное число *n*

P113 **perfect polarization** || vollkommene [ideale] Polarisation *f* || polarisation *f* parfaite || ideale polarisatie *f* || полная [идеальная] поляризация *f*

P114 **perfect radiator** || schwarzer Strahler *m* || corps *m* noir || zwarte straler *m* || абсолютно чёрное тело *n*

P115 **performance** || **1.** Betriebsverhalten *n*; Leistungsfähigkeit *f* **2.** Wirkungsgrad *m* || rendement *m* || **1.** arbeidskarakteristiek *f*; prestatie *f* **2.** rendement *n* || **1.** (рабочая) характеристика *f*; производительность *f* **2.** коэффициент *m* полезного действия, кпд

P116 **period** || Periode *f* || période *f* || periode *f* || период *m*

P117 **periodic acid** || Perjodsäure *f* || acide *m* périodique || perjodiumzuur *n* || йодная кислота *f*

P118 **periodic force** || periodische Kraft *f*; periodisch wirkende Kraft *f* || force *f* périodique || periodische kracht *f* (*m*) || периодическая сила *f*, периодически действующая сила *f*

P119 **periodic function** || periodische Funktion *f* || fonction *f* périodique || periodieke functie *f* || периодическая функция *f*

P120 **periodicity** || Periodizität *f* || périodicité *f* || periodiciteit *f* || периодичность *f*

P121 **periodic law** || Gesetz *n* der Periodizität || loi *f* périodique de Mendéléev || periodieke wet *f* (*m*) || периодический закон *m* (химических элементов Менделеева)

P122 **periodic motion** || periodische Bewegung *f* || mouvement *m* périodique || periodieke beweging *f* || периодическое движение *n*

P123 **periodic quantity** || periodische Größe *f* || quantité *f* périodique || periodieke grootheid *f* || периодическая величина *f*

P124 **periodic random process** || periodischer Zufallsprozeß *m* || processus *m* aléatoire périodique || periodiek toevalsproces *n* || периодический случайный процесс *m*

P125 **periodic sampling** || periodische Stichprobe *f* || échantillon *m* périodique || periodieke steekproef *f* (*m*) || периодическая выборка *f*

P126 **periodic table** || Periodensystem *n* (der Elemente) || table *f* périodique (des éléments) || periodiek systeem *n* || периодическая таблица *f* (химических элементов)

P127 **peripheral speed** *see* **circumferential speed**

P128 **peripheral stress** || Umfangsspannung *f* || effort *m* circonférenciel || omtrekspanning *f* || окружное усилие *n*

P129 **peripheral unit** || peripheres Gerät *n*, Anschlußgerät *n* || unité *f* périphérique || perifere eenheid *f* || периферийное [внешнее] устройство *n*

P130 **periphery grinding** || Außenrundschleifen *n* || meulage *m* circulaire périphérique || buitenrondslijpen *n* || круглое наружное шлифование *n*

P131 **permanent deformation** || bleibende Verformung *f*, bleibende Deformation *f* || déformation *f* permanente || blijvende deformatie *f* || остаточная деформация *f*

P132 **permanent extension** || bleibende Dehnung *f* || extension *f* permanente || blijvende verlenging *f* || остаточное удлинение *n*

P133 **permanent load** || ständige Last *f*; Dauerbelastung *f* || charge *f* permanente || duurzame [permanente] belasting *f* || постоянная нагрузка *f*; статическая нагрузка *f*

P134 **permanent magnet** || Dauermagnet *m*, Permanentmagnet *m* ||

aimant *m* permanent || permanente magneet *m* || постоянный магнит *m*

P135 **permanent-magnet ammeter** || magnetelektrisches Amperemeter *n*, Strommesser *m* mit Dauermagnetmeßwerk || ampèremètre *m* magnétoélectrique || magneetelektrische ampèremeter *m* || магнитоэлектрический амперметр *m*

P136 **permanganic acid,** $HMnO_4$ || Permangansäure *f*, Mangan (VII)-säure *f* || acide *m* permanganique || permangaanzuur *n* || марганцовая кислота *f*

P137 **permeability** || Permeabilität *f* || perméabilité *f* || permeabiliteit *f* || (магнитная) проницаемость *f*

P138 **permeance** || magnetischer Leitwert *m*, Permeanz *f* || conductivité *f* magnétique, perméance *f* || permeantie *f* || магнитная проводимость *f*

P139 **permissible load** *see* **allowable load**

P140 **permissible stress** *see* **allowable stress**

P141 **permissible tolerance** *see* **accepted tolerance**

P142 **permittivity** || Dielektrizitätskonstante *f* || constante *f* diélectrique, permittivité *f* || diëlektrische permeabiliteit *f* || диэлектрическая проницаемость *f*

P143 **permutation** || Permutation *f* || permutation *f* || permutatie *f*; verschikking *f* || перестановка *f*; подстановка *f*

P144 **permutation group** || Permutationsgruppe *f* || groupe *m* des permutations || permutatiegroep *f* (*m*) || группа *f* подстановок

P145 **permutation matrix** || Permutationsmatrix *f* || matrice *f* des permutations || permutatiematrix *f* || матрица *f* перестановок

P146 **peroxide** || Peroxid *n* || peroxyde *m* || peroxyde *n* || перекись *f*, пероксид *m*

P147 **perpetual motion machine, perpetuum mobile** || Perpetuum *n* mobile || moteur *m* à mouvement perpétuel || perpetuum *n* mobile || вечный двигатель *m*

P148 **persistent waves** || ungedämpfte Wellen *f pl* || ondes *f pl* entretenues || ongedempte golven *f* (*m*) *pl* || незатухающие волны *f pl*

P149 **perspective geometry** || perspektive Geometrie *f* || géométrie *f* perspective || perspectieve [beschrijvende] meetkunde *f* || начертательная геометрия *f*; аксонометрия *f*

P150 **persulphuric acid,** $H_2S_2O_8$ || Peroxodischwefelsäure *f* || acide *m* persulfurique || perzwavelzuur *n* || надсерная кислота *f*

P151 **perturbation** || Störung *f* || perturbation *f* || storing *f*, perturbatie *f* || возмущение *n*

P152 **perturbation theory** || Störungstheorie *f* || théorie *f* de perturbations || storingstheorie *f* || теория *f* возмущений

P153 **per unit mass, per unit weight** || je Masseneinheit, spezifisch || par unité de masse, spécifique || per massaeenheid, specifiek || удельный, на единицу массы

P154 **peta...** || Peta... || péta... || peta... || пета... *(приставка в наименовании кратной единицы, соответствующая* 10^{15}*)*

P155 **petrol** || Benzin *n* || essence *f*, benzine *f* || benzine *f* (*m*) || бензин *m*

P156 **petrol-air mixture** || Kraftstoff-Luft-Gemisch *n* || mélange *m* air-combustible || brandstof-luchtmengsel *n* || топливовоздушная смесь *f*

P157 **petrolatum** || Petrolat(um) *n*, Rohvaseline *f*, Vaseline *f* || pétrolatum *m*; vaseline *f* || petrolatum *n*; vaseline *f* (*m*) || петролатум *m*; вазелин *m*

P158 **petrol engine** || Benzinmotor *m* || moteur *m* à essence || benzinemotor *m* || бензиновый двигатель *m*

P159 **petroleum** || Erdöl *n* || pétrole *m* || aardolie *f* (*m*) || нефть *f*

P160 **petroleum coke** || Erdölkoks *m*, Petrolkoks *m* || coke *m* de pétrole || aardoliecokes *f* (*m*) || нефтяной кокс *m*

P161 **petroleum ether** || Petroläther *m* || gazolène *m*, éther *m* de pétrole || petroleumether *m* || петролейный эфир *m*

P162 **petroleum jelly** *see* **petrolatum**

P163 I **phase** || Phase *f* || phase *f* || fase *f* || фаза *f*

P164 II **phase** || Phasen... || de phase || fase... || фазовый

P165 **phase angle** || Phasenwinkel *m* || angle *m* de phase || fasehoek *m* || фазовый угол *m*

P166 **phase cell** || Phasenraumzelle *f* || cellule *f* de phase || fasecel *f* (*m*) || ячейка *f* фазового пространства

P167 **phase change** *see* **phase transformation**

P168 **phase compensation** || Phasenausgleich *m*, Phasenkompensation *f* || compensation *f* du déphasage || fasencompensatie *f* || компенсация *f* сдвига фаз, фазовая компенсация *f*

P169 **phase-contrast microscope** || Phasenkontrastmikroskop *n* || microscope *m* à contraste de phase || fase(n)contrastmicroscoop *m* || фазоконтрастный микроскоп *m*

P170 **phase diagram** || Zustandsdiagramm *n* || diagramme *m* d'état [de phases] || toestandsdiagram *n* || диаграмма *f* состояния, фазовая диаграмма *f*

P171 **phase difference** || Phasenendifferenz *f*, Phasenverschiebung *f* || déphasage *m* || fasedifferentie *f* || разность *f* [сдвиг *m*] фаз

P172 **phase equilibrium** || Phasengleichgewicht *n* || équilibre *m* de phase || fase-evenwicht *n* || фазовое равновесие *n*

P173 **phase factor** || Phasenfaktor *m* || facteur *m* de phase || fasefactor *m* || фазовый множитель *m*

P174 **phase integral** || Phasenintegral *n* || intégrale *f* de phase || fase-integraal *f* (*m*) || фазовый интеграл *m*, интеграл *m* движения

P175 **phase modulation** || Phasenmodulation *f* || modulation *f* de phase || fasemodulatie *f* || фазовая модуляция *f*

P176 **phase plane** || Phasenebene *f* || plan *m* de phase || fasevlak *n* || фазовая плоскость *f*

P177 **phase plate** || Phasenplatte *f* || plaque *f* de phase || faseplaat *f* || фазовая пластинка *f*

P178 **phase reversal** || Phasenumkehr *f* || inversion *f* de phase || fase-inversie *f* || опрокидывание *n* фазы

P179 **phase rule** || (Gibbssche) Phasenregel *f* || règle *f* des phases || fasenregel *m* || правило *n* фаз (Гиббса)

P180 **phase shift** || Phasenverschiebung *f* || déphasage *m* || faseverschuiving *f* || смещение *n* по фазе, фазовый сдвиг *m*

P181 **phase space** || Phasenraum *m* || espace *m* de phase || faseruimte *f* || фазовое пространство *n*

P182 **phase state** || Phasenzustand *m* || état *m* d'agrégation || fasetoestand *m* || фазовое состояние *n*; агрегатное состояние *n*

P183 **phase transformation, phase transition** || Phasenübergang *m*, Phasenumwandlung *f* || transformation *f* de phase || faseovergang *m* || фазовый переход *m*, фазовое превращение *n*

P184 **phase velocity** || Phasengeschwindigkeit *f* || vitesse *f* de phase || fasesnelheid *f* || фазовая скорость *f*

P185 **phase voltage** || Phasenspannung *f* || tension *f* de phase || fasespanning *f* || фазное напряжение *n*

P186 **phasing** || In-Phase-Bringen *n*, Phaseneinstellung *f* || mise *f* en phase || fase-instelling *f* || фазирование *n*, фазировка *f*

P187 **phasor** || Phasor *m* || phasor *m* || fasor *m* || фазор *m*

P188 **phenomenological theory** || phänomenologische Theorie *f* || théorie *f* phénoménologique || fenomenologische theorie *f* || феноменологическая теория *f*

P189 **pH-meter** || pH-Meter *n*, PH-(Wert-)Meßgerät *n*, pH-(Wert-)Messer *m* || pH-mètre *m* || pH-meter *m* || pH-метр *m*

P190 **phonon** || Phonon *n*, Schallquant *n* || phonon *m* || fonon *n* || фонон *m*

P191 **phosphide** || Phosphid *n* || phosphure *m* || fosfide *n* || фосфид *m*

P192 **phosphite** || Phosphit *n* || phosphite *m* || fosfiet *n* || фосфит *m*

P193 **phosphor** || Luminophor *m*, Leuchtstoff *m*, Phosphor *m* || phosphore *m* || fosfor *m* || люминофор *m*, фосфóр *m*

P194 **phosphorescence** || Phosphoreszenz *f* || phosphorescence *f* || fosforescentie *f* || фосфоресценция *f*

P195 **phosphoric acid,** H_3PO_4 || Phosphorsäure *f* || acide *m* phosphorique || fosforzuur *n* || фосфорная кислота *f*

P196 **phosphorous acid,** H_3PO_3 || phosphorige Säure *f* || acide *m* phosphoreux || fosforigzuur *n* || фосфористая кислота *f*

P197 **phosphorus,** P || Phosphor *m* || phosphore *m* || fosfor *m* || фосфор *m*

P198 **phosphorus pentoxide,** P_4O_{10} || Phosphorpentoxid *n*, Phosphorsäureanhydrid *n* || anhydride *m* phosphorique || fosforpentoxyde || пятиокись *f* фосфора, фосфорный ангидрид *m*

P199 **phosphorus trioxide,** P_4O_6 || Phosphortrioxid *n*, Phosphorigsäureanhydrid *n* || anhydride *m* phosphoreux || fosforig-anhydride *n* || трёхокись *f* фосфора, фосфористый ангидрид *m*

P200 **photocathode** || Fotokatode *f* || photocathode *f* || fotokathode *f* || фотокатод *m*

P201 **photocell** || Fotoelement *n*; Fotozelle *f* || cellule *f* photo-électrique || fotocel *f* (*m*) || фотоэлемент *m*

P202 **photochemical reaction** || fotochemische Reaktion *f* || réaction *f* photochimique || fotochemische reactie *f* || фотохимическая реакция *f*

P203 **photochemistry** || Fotochemie *f* || photochimie *f* || fotochemie *f* || фотохимия *f*

P204 **photochromism** || Photochromismus *m* || photochromisme *m* || fotochromisme *n* || фотохромизм *m*

P205 **photoconductivity** || Fotoleitfähigkeit *f* || photoconductibilité *f* || fotogeleidbaarheid *f* || фотопроводимость *f*, внутренний фотоэффект *m*

P206 **photoconductor** || Fotoleiter *m* || photoconducteur *m* || fotogeleider *m* || фотопроводник *m*

P207 **photocurrent** || Fotostrom *m* || courant *m* photo-électrique, photocourant *m* || fotostroom *m* || фототок *m*

P208 **photodiode** || Fotodiode *f* || photodiode *f* || fotodiode *f* (*m*) || фотодиод *m*

P209 **photodissociation** || Fotodissoziation *f* || photodissociation *f* || fotodissociatie *f* || фотодиссоциация *f*

P210 **photoeffect** || Fotoeffekt *m*, lichtelektrischer [fotoelektrischer] Effekt *m* || effet *m* photo-électrique, photo-effect *m* || foto-elektrisch effect *n* || фотоэлектрический эффект *m*, фотоэффект *m*

P211 **photoelasticity** || Fotoelastizität *f*, Spannungsoptik *f* || photoélasticité *f* || foto-elasticiteit *f* || фотоупругость *f*

P212 **photoelectric cell** || Fotozelle *f* || cellule *f* photo-électrique || fotocel *f* (*m*) || фотоэлемент *m*

P213 **photoelectric current** || Fotostrom *m* || courant *m* photo-électrique, photocourant *m* || fotostroom *m* || фототок *m*

P214 **photoelectric effect** *see* **photoeffect**

P215 **photoelectric multiplier** *see* **photomultiplier**

P216 **photoelectromotive force** || Foto-EMK *f* || force *f* photo-électromotrice || foto-elektromotorische kracht *f* (*m*) || фотоэлектродвижущая сила *f*, фото-эдс

P217 **photoelectron** || Fotoelektron *n* || photo-électron *m* || foto-elektron *n* || фотоэлектрон *m*

P218 **photoelectron emission** || lichtelektrische Emission *f*; äußerer Fotoeffekt *m*, Fotoemissionseffekt *m* || émission *f* photo-électronique; photo-effet *m* extérieur || lichtelektrische emissie *f*, (uitwendig) foto-elektrisch effect *n* || фотоэлектронная эмиссия *f*; (внешний) фотоэлектрический эффект *m*, фотоэффект *m*

P219 **photoelectron spectroscopy** || Fotoelektronenspektroskopie *f* || spectroscopie *f* photo-électronique || foto-elektronenspectroscopie *f* || фотоэлектронная спектроскопия *f*

P220 **photoemission** || Fotoemission *f*, äußerer Fotoeffekt *m* || photoémission *f* || foto-emissie *f* || фотоэмиссия *f*, внешний фотоэффект *m*, внешний фотоэлектрический эффект *m*

P221 **photoemulsion** || fotografische Emulsion *f*, Fotoemulsion *f* || photoémulsion *f* || foto-emulsie *f* || фотографическая эмульсия *f*, фотоэмульсия *f*

P222 **photographic blackening** || fotografische Schwärzung *f* || noircissement *m* photographique || fotografische zwarting *f* || фотографическое почернение *n*

P223 **photographic density** || fotografische Dichte *f* || densité *f* photographique || fotografische densiteit *f* || плотность *f* почернения, фотографическая плотность *f*

P224 **photographic emulsion** || fotografische Emulsion *f*, Fotoemulsion *f* || émulsion *f* photographique || foto-emulsie *f* || фотоэмульсия, фотографическая эмульсия *f*

P225 **photographic film** || fotografischer Film *m* || pellicule *f* [film *m*] photographique || fotografische film *m* || фотоплёнка *f*

P226 **photographic plate** || fotografische Platte *f*, Fotoplatte *f* || plaque *f* photographique || fotoplaat *f* (*m*) || фотографическая пластинка *f*, фотопластинка *f*

P227 **photography** || Fotografie *f* || photographie *f* || fotografie *f* || фотографирование *n*, фотография *f*

P228 **photoionization** || Fotoionisation *f* || photo-ionisation *f* || foto-ionisatie *f* || фотоионизация *f*

P229 **photoluminescence** || Fotolumineszenz *f* || photoluminescence *f* || fotoluminescentie *f* || фотолюминесценция *f*

P230 **photolysis** || Fotolyse *f* || photolyse *f* || fotolyse *f* || фотолиз *m*

P231 **photometer** || Fotometer *n* || photomètre *m* || fotometer *m* || фотометр *m*

P232 **photometer wedge** || Fotometerprismenkeil *m* || coin *m* optique [photométrique] || fotometerwig *f* (*m*) || фотометрический клин *m*

P233 **photometry** || Fotometrie *f* || photométrie *f* || fotometrie *f* || фотометрия *f*

P234 **photomultiplier** || Fotovervielfacher *m*, Sekundärelektronenvervielfacher *m*, SEV || photomultiplicateur *m* || fotovermenigvuldiger *m* || фотоэлектронный умножитель *m*, фотоумножитель *m*

P235 **photon** || Photon *n*, Lichtquant *n* || photon *m* || foton *n* || фотон *m*, световой квант *m*

P236 **photonics** || Fotonik *f* || photonique *f* || fotoniek *f* || фотоника *f*, корпускулярные действия *n pl* света

P237 **photoresist** || Fotoresist *m*, Fotolack *m* || photorésiste *m* || fotoresist *m* || фоторезист *m*

P238 **photoresistance** || Fotowiderstand *m* || photorésistance *f* || fotoweerstand *m* || фотосопротивление *n*

P239 **photoresistor** || Fotowiderstand *m* || résistance *f* photosensible || fotoweerstand *m* || фоторезистор *m*

P240 **photosynthesis** || Fotosynthese *f* || photosynthèse *f* || fotosynthese *f* || фотохимический синтез *m*, фотосинтез *m*

P241 **phototransistor** || Fototransistor *m* || transistor *m* à effet photoélectrique, phototransistor *m* || fototransistor *m* || фототранзистор *m*

P242 **phototropy** || Fototropie *f* || phototropie *f* || fototropie *f* || фототропия *f*

P243 **phototube** || Vakuum(foto)zelle *f* || tube *m* photo-électrique || vacuüm(foto)cel *f* (*m*) || вакуумный фотоэлемент *m*

P244 **physical** || physikalisch || physique || fysisch, fysiek... || физический

P245 **physical meaning** || physikalische Deutung *f* || sens *m* physique || fysische zin *m* || физический смысл *m*

P246 **physical pendulum** || physikalisches Pendel *n* || pendule *m* physique || fysisch pendel *n* || физический маятник *m*

P247 **physical properties** || physikalische Eigenschaften *f pl* || propriétés *f pl* physiques || fysische eigenschappen *n pl* || физические свойства *f pl*

P248 **physical significance** *see* **physical meaning**

P249 **physics** || Physik *f* || physique *f* || fysica *f* || физика *f*

P250 **pickup** || **1.** Fühler *m*; Geber *m* **2.** Tonabnehmer(kopf) *m* || **1.** détecteur *m* **2.** tête *f* de pick-up || **1.** gever *m*, voeler *m* **2.** pick-up *m* || **1.** датчик *m*, чувствительный элемент *m* 2. звукосниматель *m*; головка *f* звукоснимателя

P251 **pico...** || Piko... || pico... || pico... || пико... *(приставка в наименовании дольной единицы, соответствующая 10^{-12})*

P252 **picofarad** || Pikofarad *n* || picofarad *m* || picofarad *m* || пикофарад *m* (10^{-12} Ф)

P253 **picture** || Bild *n*; Aufnahme *f* || image *f* || beeld *n* || изображение *n*; картина *f*; снимок *m*

P254 **piecewise continuous function** || stückweise stetige Funktion *f* || fonction *f* continue par morceaux || stuksgewijze onophoudelijke functie *f* || кусочно-непрерывная функция *f*

P255 **piercing** || Lochen *n* || poinçonnage *m*; perçage *m* (de trous) || doorsteken *n*; doorboren *n* || пробивание *n* отверстий, перфорирование *n*, перфорация *f*

P256 **piezocrystal** || Piezokristall *m* || cristal *m* piézo-électrique || piëzokristal *n* || пьезокристалл *m*, пьезоэлектрический кристалл *m*

P257 **piezoelectric** || piezoelektrisch || piézo-électrique || piëzo-elektrisch || пьезоэлектрический

P258 **piezoelectric cell** || Piezoelement *n* || cellule *f* piézo-électrique || piëzocel *f* *(m)* || пьезоэлемент *m*

P259 **piezoelectric crystal** *see* **piezocrystal**

P260 **piezoelectric effect** || piezoelektrischer Effekt *m*, Piezoeffekt *m* || effet *m* piézo-électrique || piëzoelektrisch effect *n* || пьезоэлектрический эффект *m*

P261 **piezoelectricity** || Piezoelektrizität *f* || piézo-électricité *f* || piëzo-elektriciteit *f* || пьезоэлектричество *n* *(пьезоэлектрический эффект)*

P262 **piezoelectric vibrator** || piezoelektrischer Schwinger *m* || vibrateur *m* piézo-électrique || piëzo-elektrische triller *m* || пьезоэлектрический вибратор *m*

P263 **piezomagnetism** || piezomagnetischer Effekt *m* || piézo-magnétisme *m* || piëzomagnetisme *n* || пьезомагнетизм *m*, пьезомагнитный эффект *m*

P264 **piezoquartz** || Piezoquarz *n*, Schwingquarz *n* || quartz *m* piézoélectrique || piëzokwarts *n* || пьезокварц *m*

P265 **piezosemiconductor** || Piezohalbleiter *m* || semi-conducteur *m* piézo-électrique || piëzohalfgeleider *m* || пьезополупроводник *m*

P266 **pile** || 1. Kernreaktor *m* 2. (elektrische) Batterie *f* || 1. pile *f*, réacteur *m* 2. batterie *f* || 1. kernreactor *m* 2. batterij *f* || 1. (ядерный) реактор *m* 2. (электрическая) батарея *f*

P267 **pillar drill** || Säulenbohrmaschine *f* || perceuse *f* à colonne [à montant] || zuilenboormachine *f* || вертикально-сверлильный станок *m* на колонне

P268 **pilot bearing** || Führungslager *n* || palier *m* de guidage || leikussenblok *n* || направляющий подшипник *m*

P269 **pilot valve** || Vorsteuerventil *m* || clapet *m* de réglage || stuurventiel *n* || клапан *m* управления

P270 **pin** || 1. Stift *m*; Bolzen *m* 2. Zapfen *m* 3. Stift *m* || goujon *m*; goupille *f*; cheville *f*; axe *m* || 1. stift *f* *(m)*; pin *f* *(m)*, pen *f* *(m)* 2. pen *f* *(m)* 3. stift *f* *(m)* || 1. шпилька *f*; штырь *m*; палец *m* 2. цапфа *f* 3. штифт *m*

P271 **pinch, pinch-effect** || Pincheffekt *m*, Selbsteinschnürung *f* || effet *m* de pincement, rhéostriction *f* || pinch-effect *n*, insnoeringseffect *n* || пинч-эффект *m*, самосжатие *n* разряда

P272 **pinhole camera** || Lochkammer *f*, Camera *f* obskura || chambre *f* obscure [noire], sténope *m* || camera-obscura *f* *(m)* || камера-обскура *f*

P273 **pinion** || Ritzel *n* || roue *f* [pignon *m*] de commande || rondsel *n* || шестерня *f*; ведущее зубчатое колесо *n*

P274 **pinion-shaped (gear) cutter** || Schneidrad *n* || outil-pignon *m* || snijwiel *n* || (зуборезный) долбяк *m*

P275 **pinning (down)** || Auflaufen *n*, Blockierung *f* *(Versetzungen)* || pinning *m* || pinning *n* || пиннинг *m*, зацепление *n* *(дислокаций)*

P276 **pintle** || Gelenkstift *m*; Gelenkachse *f* || pivot *m* || gewrichtstift *f* (*m*); scharnierpen *f* (*m*) || (поворотный) шкворень *m*; ось *f* шарнира

P277 **pin wheel** || Triebstockrad *m*, Zapfenrad *n* || roue *f* à la lanterne; roue *f* à fuseaux; roue *f* à chevilles || penrad *n* || цевочное зубчатое колесо *n*, штифтовое колесо *n*

P278 **pion** || Pion *n*, Pi-Meson *n* || pion *m* || pion *n* || пион *m*, пи-мезон *m*

P279 **pipe** || Rohr *n*; Röhre *f* || tube *m*; tuyau *m*; conduite *f* || pijp *f* (*m*), buis *f* (*m*) || труба *f*; трубка *f*

P280 **pipe cutter** || Rohr(ab)schneider *m* || coupe-tube *m*; coupe-tuyau *m* || buissnijder *m*, pijpsnijder *m* || труборез *m*

P281 **pipe mill** || Rohrwalzwerk *n* || laminoir *m* à tubes; train *m* à tubes || pijpwalswerk *n* || трубопрокатный стан *m*

P282 **pipe thread** || Rohrgewinde *n* || filet *m* au pas du gaz || pijpschroefdraad *m* || трубная резьба *f*

P283 **pipe threading machine** || Rohrgewindeschneidmaschine *f* || tour *m* [machine *f*] à fileter les tubes || pijpschroefdraadsnijmachine *f* || трубонарезной станок *m*

P284 **piping** || Rohranlage *f*; Rohrnetz *n*, Leitungssystem *n* || tuyauterie *f* || buisleiding *f*, pijpleiding *f* || трубопровод *m*; система *f* труб

P285 **piston** || Kolben *m* || piston *m* || zuiger *m*; plunjer *m* || поршень *m*; плунжер *m*

P286 **piston compressor** || Kolbenverdichter *m* || compresseur *m* à piston || zuigercompressor *m* || поршневой компрессор *m*

P287 **piston engine** || Kolbenmotor *m* || moteur *m* à piston || zuigermotor *m*, zuigermachine *f* || поршневой двигатель *m*

P288 **piston pin** || Kolbenbolzen *m* || tourillon *m* de piston || zuigerpen *m* || поршневой палец *m*

P289 **piston pump** || Kolbenpumpe *f* || pompe *f* à piston || zuigerpomp *f* (*m*) || поршневой насос *m*

P290 **piston ring** || Kolbenring *m* || segment *m* du piston || zuigerring *m* || поршневое кольцо *n*

P291 **piston rod** || Kolbenstange *f*; Pleuelstange *f* || tige *f* de piston || zuigerstang *f* (*m*), plunjerstang *f* (*m*) || шток *m* поршня, поршневой шток *m*; шатун *m*

P292 **piston stroke, piston travel** || Kolbenhub *m* || course *f* du piston || zuigerslag *m* || ход *m* [длина *f* хода] поршня

P293 **piston valve** || Kolbenschieber *m* || tiroir *m* à piston || bosschuif *f* (*m*) || поршневой [цилиндрический] золотник *m*

P294 **pitch** || **1.** Steigung *f* *(Gewinde)*; Teilung *f*; Ganghöhe *f* **2.** Gefälle *n* || **1.** pas *m* *(p. ex. du filet)* **2.** pente *f*, angle *m* de pente || **1.** spoed *m* *(schroefdraad)* **2.** helling *f* || **1.** шаг *m* *(резьбы, зубьев)*; питч *m* **2.** уклон *m*; наклон *m*

P295 **pitch angle** || Teilkegelwinkel *m* *(Zahnrad)* || angle *m* du cône primitif d'engrenage || hoek *m* van de deelkegel *(kegelwiel)* || угол *m* делительного конуса *(конического зубчатого колеса)*

P296 **pitch chain** || kalibrierte Kette *f* || chaîne *f* à maillons calibrés || kaliberketting *m*, *f* || калиброванная цепь *f*

P297 **pitch circle** || Teilkreis *m* *(Zahnrad)* || cercle *m* primitif de taillage || steekcirkel *m* *(tandwiel)* || делительная окружность *f* *(зубчатого колеса)*

P298 **pitch diameter** || Teilkreisdurchmesser *m*; Flankendurchmesser *m* *(Gewinde)* || diamètre *m* à flancs de filet, diamètre *m* primitif || steekcirceldiameter *m*; gemiddelde diameter *m* *(schroefdraad)*; diameter *m* van de deelkegel || диаметр *m* делительной окружности; средний диаметр *m* *(резьбы)*; диаметр *m* делительного конуса

P299 **pitch of sound** || Tonhöhe *f* || hauteur *f* de son || geluidhoogte *f* || высота *f* звука

P300 **pitch point** || Wälzpunkt *m* *(Verzahnung)* || point *m* de contact [d'approche] des dents engagées || walspunt *f* (*m*) *(vertanding)* || полюс *m* зацепления

P301 **pivot** || Zapfen *m*, Drehzapfen *m*, Tragzapfen *m*; Drehachse *f*; Drehpunkt *m* || pivot *m* || tap *m*;

draaias *f* (*m*); steunstuk *n*; gewricht *n* || цапфа *f*; ось *f* (вращения); пята *f*; шарнир *m*

P302 **pivot reamer** *see* **floating reamer**

P303 **plain-carbon steel** || unlegierter Stahl *m* || acier *m* à carbon || ongelegeerd staal *n* || углеродистая [нелегированная] сталь *f*

P304 **plain grinding machine** *see* **cylindrical grinding machine**

P306 **planar** || planar || planaire || planair || планарный

P307 **Planck's constant** || Plancksche Konstante *f*, Plancksches Wirkungsquantum *n* || constante *f* de Planck || constante *f* (*m*) van Planck || постоянная *f* Планка

P308 **plane mirror** || Planspiegel *m*, ebener Spiegel *m* || miroir *m* plan || vlakke spiegel *m* || плоское зеркало *n*

P309 **plane of incidence** || Einfallsebene *f* || plan *m* d'incidence || invalsvlak *n* || плоскость *f* падения (луча)

P310 **plane of symmetry** || Symmetrieebene *f* || plan *m* de symétrie || symmetrievlak *n* || плоскость *f* симметрии

P311 **plane-parallel** || planparallel || plan-parallèle || planparallel || плоскопараллельный

P312 **plane-parallel plate** || planparallele Platte *f* || plaque *f* plan-parallèle || planparallele plaat *f* || плоскопараллельная пластинка *f*

P313 **plane polarization** || lineare [geradlinige] Polarisation *f*, Linearpolarisation *f* || polarisation *f* linéaire [rectiligne], polarisation *f* dans un plan || lineaire polarisatie *f* || линейная [плоская] поляризация *f*

P314 **plane-polarized wave** || linear polarisierte Welle *f* || onde *f* polarisée dans un plan || lineair gepolariseerde golf *f* (*m*) || плоскополяризованная [линейнополяризованная] волна *f*

P315 **plane** || **1.** Hobelmaschine *f* **2.** Hobel *m* || **1.** raboteuse *f* **2.** rabot *m* || **1.** schaafmachine *f* **2.** schaaf *f* (*m*) || **1.** строгальный станок *m* **2.** рубанок *m*

P316 **plane stress** || ebener [zweiachsiger] Spannungszustand *m* || état *m* de contrainte à deux dimensions || vlakke spanningstoestand *m* || плоское напряжённое состояние *n*

P317 **plane surface** || ebene Oberfläche *f* || surface *f* plane || vlak [plat] oppervlak *n* || плоская поверхность *f*

P318 **panlet** || Planet *m* || planète *f* || planeet *f* (*m*) планета *f*

P319 **planetary gear** || Planetenrad *n* || roue *f* [pignon *m*] planétaire || planeetwiel *n* || планетарное зубчатое колесо *n*, планетарная шестерня *f*, сателлит *m*

P320 **planetary gearing** || Planetengetriebe *n* || engrenage *m* planétaire || planeetdrijfwerk *n* || планетарный механизм *m*; планетарная передача *f*

P321 **planetary pinion** *see* **planetary gear**

P322 **planet carrier** || Steg *m* (*Planetengetriebe*) || porte-satellite *m* || planeetdrager *m* || водило *n* (*планетарной передачи*)

P323 **planet wheel** *see* **planetary gear**

P324 **plane wave** || ebene Welle *f* || onde *f* plane || vlakke golf *f* (*m*) || плоская волна *f*

P325 **planing** || Hobeln *n* || rabotage *m*, rabotement *m* || schaafbewerking *f* || строгание *n*

P326 **planing machine** *see* **planer 1.**

P327 **planing tool** || Hobelmeißel *m* || outil *m* de rabotage || schaafbeitel *m* || строгальный резец *m*

P328 **planishing** || **1.** Glattdrücken *n* **2.** Richten *n*, Abrichten *n* || **1.** planage *m* **2.** rectification *f*, ponçage *m* || **1.** vlakken *n*; pletten *n* **2.** richten *n* || **1.** правка *f*, рихтовка *f* **2.** шлифование *n*, шлифовка *f*, полирование *n*, полировка *f*

P329 **plano-type milling machine** || Planfräsmaschine *f* || fraiseuse-raboteuse *f* || langfreesbank *f* (*m*) || продольно-фрезерный станок *m*

P330 **plasma** || Plasma *n* || plasma *m* || plasma *n* || плазма *f*

P331 **plasma-arc welding** || Plasmabogenschweißen *n* || soudage *m* au plasma || plasmabooglassen *n* || плазменная сварка *f*

P332 **plasma confinement** || Plasmaeinschließung *f* || confinement *m* de plasma || plasma-insluiting *f* || удержание *n* плазмы *(в установке термоядерного синтеза)*

P333 **plasma generator** || Plasmatron *n*, Plasmabrenner *m* || plasmatron *m* || plasmabrander *m* || плазмотрон *m*, плазматрон *m*, плазменный генератор *m*, плазменная горелка *f*

P334 **plasma instability** || Plasmainstabilität *f* || instabilité *f* du plasma || plasmaonevenwicht *n* || плазменная неустойчивость *f*

P335 **plasma jet** || Plasmastrahl *m* || jet *m* de plasma || plasmastraal *m*, *f* || плазменная струя *f*, струя *f* плазмы

P336 **plasma physics** || Plasmaphysik *f*, Physik *f* des Plasmas || physique *f* du plasma || plasmafysica *f* || физика *f* плазмы

P337 **plasma state** || Plasmazustand *m* || état *m* de plasma || plasmatoestand *m* || плазменное состояние *n*

P338 **plasma wave** || Plasmawelle *f* || onde *f* de plasma || plasmagolf *f* (*m*) || плазменная волна *f*

P339 I **plastic** || Plast *m*, Kunststoff *m* || plastique *m*; matière *f* plastique || kunststof *f* (*m*); plastic *n* || пластмасса *f*, пластик *m*

P340 II **plastic** || plastisch || plastique || plastisch || пластичный; пластический

P341 **plastic deformation** || plastische Verformung *f*, plastische Deformation *f* || déformation *f* plastique || plastische vervorming *f* || пластическая деформация *f*

P342 **plasticity** || Plastizität *f* || plasticité *f* || plasticiteit *f* || пластичность *f*

P343 **plasticizer** || Weichmacher *m*, Plastizier(ungs)mittel *n*, Plastifikator *m* || plastifiant *m*, amollissant *m* || plastificeermiddel *n* || пластификатор *m*

P344 **plate** || **1.** Platte *f* **2.** Blech *n* **3.** Anode *f* || **1.** plaque *f* **2.** tôle *m* **3.** anode *f* || **1.** plaat *f* (*m*) **2.** plaatmateriaal *n* **3.** anode *f* || **1.** пластина *f*; плита *f* **2.** лист *m*; листовой металл *m*; **3.** анод *m*

P345 **plateau** || Plateau *n*, horizontaler Abschnitt *m* *(Kurve)* || plateau *m* || plateau *n* || плато *n* *(на характеристике)*

P346 **plate cutch** *see* **disk clutch**

P347 **plate edge planing machine** || Kantenhobelmaschine *f* || chanfreineuse *f* || kantschaafmachine *f* || кромкострогальный станок *m*

P348 **plate-flattening machine** || Blechrichtmaschine *f* || machine *f* à dresser les tôles || plaatrichtmachine *f* || листоправильная машина *f*

P349 **plate load** || Anodenbelastung *f* || charge *f* d'anode || anodebelasting *f* || анодная нагрузка *f*

P350 **plate mill** || Blechwalzwerk *n* || laminoir *m* [train *m*] à tôles || plaatwalswerk *n* || листовой прокатный стан *m*

P351 **plate-shearing machine** || Blechtafelschere *f*, Tafelschere *f* || cisaille(s) *f* (*pl*) à tôles || plaatschaar *f* (*m*) || листовые ножницы *pl*

P352 **plate voltage** || Anodenspannung *f* || tension *f* d'anode || anodespanning *f* || анодное напряжение *n*

P353 **plating** || **1.** Blechummantelung *f* **2.** Metallisierung *f*; Plattierung *f*; Galvanisieren *n*; galvanischer Überzug *m* || revêtement *m*; recouvrement *m*; métallisation *f* électrolytique, plaquage *m* || **1.** plaatmantel *m* **2.** metallisatie *f*, galvanisering *f*; plateren *n* || **1.** листовая обшивка *f* **2.** металлизация *f*; плакирование *n*; гальванопокрытие *n*

P354 **platinum,** Pt || Platin *n* || platine *m* || platina *n* || платина *f*

P355 **play** || Spiel *n*, Spielraum *m* || jeu *m* || speling *f* (*m*) || зазор *m*

P356 **pleochroism** || Pleochroismus *m* || pléochroïsme *m* || pleochroïsme *n* || плеохроизм *m*

P357 **pliers** || Zange *f*; Flachzange *f*; Kombizange *f* || pince *f* (universelle) || tang *f* (*m*); buigtang *f* (*m*); combinatietang *f* (*m*) || клещи *pl*; плоскогубцы *pl*; пассатижи *pl*

P358 **plot** || graphische Darstellung *f*; Kurvendiagramm *n*, Kurvenbild *n* || graphique *m*, courbe *f*, diagramme *m*

|| grafische voorstelling *f*, diagram *n* || график *m*

P359 **plug** || 1. Pfropfen *m*; Stopfen *m*; Spund *m* 2. Zündkerze *f* 3. Stekker *m* || 1. bouchon *m*, tampon *m* 2. bougie *f* d'allumage 3. fiche *f* || 1. stop *m*; prop *f* (*m*) 2. bougie *f* *(auto)* 3. stekker *m* || 1. пробка *f*; заглушка *f* 2. свеча *f* зажигания 3. штепсель *m*; штепсельная вилка *f*; штекер *m*

P360 **plug gauge** || Lehrdorn *m* || tampon *m*, calibre *m* à tampon || kernkaliber *n* || калибр-пробка *m*

P361 **plumbate** || Plumbat *n* || plombate *m* || loodzout *n* || плумбат *m*, соль *f* свинцовой кислоты

P362 **plunger** || 1. Tauchkolben *m* 2. Stempel *m* 3. Tauchkern *m (Elektromagnet)* || 1. plongeur *m*; piston *m* plein [plongeur] 2. poinçon *m* 3. noyau *m* || 1. plunjer *m*; dompelaar *m* 2. plunjer *m* 3. kern *f* (*m*) || 1. плунжер *m*; поршень *m* 2. пуансон *m* 3. сердечник *m (электромагнита)*

P363 **plunger pump** || Tauchkolbenpumpe *f* || pompe *f* à piston plongeur || plunjerpomp *f* (*m*) || плунжерный насос *m*

P364 **plunging trajectory** || steile Flugbahn *f*, Steilbahn *f* || trajectoire *f* courbe || kromlijnige baan *f* (*m*) || навесная [крутая] траектория *f*

P365 **plutonium,** Pu || Plutonium *n* || plutonium *m* || plutonium *n* || плутоний *m*

P366 **pneumatic hoist** || Drucklufthebezeug *n* || élévateur *m* pneumatique: treuil *m* pneumatique || pneumatisch hefwerktuig *n*, pneumatisch hijswerktuig *n* || пневматический подъёмник *m*, пневмоподъёмник *m*, пневматическая лебёдка *f*

P367 **pneumatic tool** || Preßluftwerkzeug *n* || outil *m* pneumatique || pneumatisch werktuig *n* || пневматический инструмент *m*, пневматическая ручная машина *f*

P368 **p-n junction** || pn-Übergang *m* || jonction *f* électron-trou, jonction *f* *p-n* || *p-n*-overgang *m* || *p-n*-переход *m*, электронно-дырочный переход *m*

P369 **point** || 1. Punkt *m* 2. Punkt *m*, Komma *n (Stellenwertzeichen)* 3. Schneidenkopf *m*; Spitze *f*, Meißelspitze *f* || 1. point *m* 2. virgule *f* 3. pointe *f*; bec *m* d'outil || 1. punt *n* 2. punt *n*, komma *f* 3. snijkop *m* (gereedschap); beitelpunt *f* (*m*) || 1. точка *f* 2. точка *f*, запятая *f (в позиционном представлении числа)* 3. режущая часть *f (инструмента)*; вершина *f (резца)*

P370 **point charge** || Punktladung *f* || charge *f* ponctuelle || puntlading *f* || точечный заряд *m*

P371 **point mass** || Massenpunkt *m* || point *m* matériel || puntmassa *f* (*m*), massapunt *f* (*m*) || материальная точка *f*, точечная масса *f*

P372 **point of application** || Angriffspunkt *m* || point *m* d'application || aangrijpingspunt *n* || точка *f* приложения силы

P373 **point of inflexion** || Wendepunkt *m* || point *m* d'inflexion || buigingspunt *f* (*m*) || точка *f* перегиба

P374 **point of support** || Stützpunkt *m*; Auflagepunkt *m*; Auflagerpunkt *m* || point *m* d'appui || steunpunt *f* (*m*) || точка *f* опоры

P376 **point source** || punktförmige Quelle *f*, Punktquelle *f* || source *f* ponctuelle || puntbron *f* (*m*) || точечный источник *m*

P377 **point spectrum** || diskretes Spektrum *n* || spectre *m* ponctuel || puntspectrum *n*, discontinu spectrum *n* || дискретный спектр *m*

P378 **Poisson distribution** || Poisson-Verteilung *f* || distribution *f* de Poisson || Poisson-verdeling *f* || распределение *n* Пуассона

P379 **Poisson integral** || Poissonsches Integral *n* || intégrale *f* de Poisson || integraal *f* (*m*) van Poisson, Poisson-integraal *f* (*m*) || интеграл *m* Пуассона

P380 I **polar** || Polare *f* || polaire *f* || polair *f* || поляра *f*

P381 II **polar** || polar, Polar... || polaire || polair || полярный

P382 **polar bond** || polare Bindung *f* || liaison *f* polaire || polaire binding *f* || полярная связь *f*

P383 **polar coordinates** || Polarkoordinaten *f pl* || coordonnées *f pl* polaires || poolcoördinaten *m pl* || полярные координаты *f pl*

P384 **polar curve** || Pol(ar)kurve *f* || courbe *f* polaire || poolcurve *f (m)* || полярная кривая *f*

P385 **polarimeter** || Polarimeter *n* || polarimètre *m* || polarimeter *n* || поляриметр *m*

P386 **polarity** || Polarität *f* || polarité *f* || polariteit *f* || **1.** полярность *f* **2.** полярное соответствие *n*, полярная корреляция *f*

P387 **polarity reversal** || Polwechsel *m*; Umpolung *f* || inversion *f* de polarité || poolwisseling *f*; ompoling *f* || обращение *n* [перемена *f*] полярности; переполюсовка *f*, перемагничивание *n*

P388 **polarizability** || Polarisierbarkeit *f* || polarisabilité *f* || polariseerbaarheid *f* || поляризуемость *f*

P389 **polarizability ellipsoid** || Polarisierbarkeitsellipsoid *n* || ellipsoïde *m* de polarisabilité || polariseerbaarheidsellipsoïde *f* || эллипсоид *m* поляризуемости

P390 **polarization** || Polarisation *f* || polarisation *f* || polarisatie *f* || поляризация *f*

P391 **polarization by reflection** || Reflexionspolarisation *f* || polarisation *f* à la réflexion || reflectiepolarisatie *f* || поляризация *f* при отражении *(волны)*

P392 **polarization plane** || Polarisationsebene *f* || plan *m* de polarisation || polarisatievlak *n* || плоскость *f* поляризации

P393 **polarized beam** || polarisiertes Bündel *n* || faisceau *m* polarisé || gepolariseerde bundel *m* || поляризованный пучок *m*

P394 **polarized light** || polarisiertes Licht *n* || lumière *f* polarisée || gepolariseerd licht *n* || поляризованный свет *m*

P395 **polarizer** || Polarisator *m* || polarisateur *m* || polarisator *m* || поляризатор *m*

P396 **polar molecule** || polares Molekül *n* || molécule *f* polaire || polaire molecule *f (m)* || полярная молекула *f*

P397 **polarography** || Polarografie *f* || polarographie *f* || polarografie *f* || полярография *f*

P398 **polar plane** || Polarebene *f* || plan *m* polaire || polair vlak *n* || полярная плоскость *f*

P399 **polar triangle** || Poldreieck *n* || triangle *m* polaire || polaire driehoek *m* || полярный треугольник *m*

P400 **polar vector** || polarer Vektor *m* || vecteur *m* polaire || polaire vector *m* || полярный вектор *m*

P401 **pole** || Pol *m* || pôle *m* || pool *f (m)* || полюс *m*

P402 **pole core** || (ferro)magnetischer Kern *m*, Magnetkern *m* || noyau *m* magnétique || magneetkern *f (m)* || магнитный сердечник *m*

P403 **pole face** *see* **pole piece**

P404 **pole gap** || Pollücke *f* || entrefer *m* || poolafstand *m* || зазор *m* между полюсами магнита

P405 **pole piece** || Polschuh *m* || épanoissement *m* [pièce *f*] polaire || poolpunt *m* || полюсный наконечник *m (магнита)*

P406 **polhode** || Polhodie *f* || polhodie *f* || polhodie *f* || полодия *f*, полоида *f*

P407 **polishing** || Polieren *n* || (po)lissage *m* || polijsten *n* || полирование *n*

P408 **polishing wheel** || Polierscheibe *f* || disque *m* à polir || polijstschijf *f (m)* || полировальный круг *m*

P409 **polonium,** Po || Polonium *n* || polonium *m* || polonium *n* || полоний *m*

P410 **polychromatic beam** || polychromatisches Bündel *n* || faisceau *m* polychromatique || polychromatische bundel *m* || немонохроматический [полиэнергетический] пучок *m*

P411 **polycrystal** || Polykristall *m* || polycristal *m* || polykristal *n* || поликристалл *m*

P412 **polyester** || Polyäther *m* || polyester *m* || polyester *m* || полиэфир *m*

P413 **polygon of forces** || Kräftepolygon *n* || polygone *m* des forces || krachtenveelhoek *m* || многоугольник *m* сил, силовой многоугольник *m*

P414 **polymer** || Polymer(es) *n*, Polymerisat *n* || polymère *m*, polymérisat *m* || polymeer *n* || полимер *m*

P415 **polymerization** || Polymerisation *f*, Polymerisierung *f* || polymérisation *f* || polymerisatie *f* || полимеризация *f*

P416 **polymorphic transformation** || polymorphe Umwandlung *f* || transformation *f* polymorphyque || polymorfe omvorming *f* || полиморфное превращение *n*

P417 **polymorphism** || Polymorphie *f* || polymorphisme *m* || polymorfie *f* || полиморфизм *m*, полиморфия *f*

P418 I **polynomial** || Polynom *n* || polynôme *m* || veelterm *m*, polynoom *m* || многочлен *m*, полином *m*

P419 II **polynomial** || polynomial || polynomial || veeltermig || полиномиальный

P420 **polynomial approximation** || polynomiale Approximation *f* || approximation *f* polynomiale || veeltermige approximatie *f* || полиномиальное приближение *n*

P421 **polypeptide** || Polypeptid *n* || polypeptide *m* || polypeptide *n* || полипептид *m*

P422 **polyphase system** || Mehrphasensystem *n* || système *m* polyphasé || veelfasensysteem *n* || многофазная система *f*

P423 **polypropylene** || Polypropylen *n* || polypropylène *m* || polypropeen *n* || полипропилен *m*

P424 **polystyrene** || Polystyrol *n* || polystyrolène *m*, polystyrol *m* || polystyreen *n* || полистирол *m*

P425 **polythene** || Polyäthylen *n*, Polyäthen *n* || polyéthylène *m*, polythène *m* || polyeth(yl)een *n* || полиэтилен *m*

P426 **polytrope, polytropic curve** || Polytrope *f* || courbe *f* polytropique || polytroop *m* || политропа *f*

P427 **polytropic process** || polytroper Prozeß *m* || transformation *f* polytropique [polytrope] || polytroop proces *n* || политропический [политропный] процесс *m*

P428 **polyurethane** || Polyurethan *n* || polyurétha(n)ne *m* || polyurethaan *n* || полиуретан *m*

P429 **polyvinyl chloride** || Polyvinylchlorid *n* || polyvinylchlorure *m* || polyvinylchloride *n* || поливинилхлорид *m*

P430 **ponderomotive force** || ponderomotorische Kraft *f* || force *f* pondéromotrice || ponderomotorische kracht *f* (*m*) || пондеромоторная сила *f*

P431 **pooled sample** || zusammengefaßte Stichprobe *f* || échantillon *m* combiné || verbonden steekproef *f* (*m*) || объединённая выборка *f*

P432 **pooled variance** || Gesamtstreuung *f* || variance *f* combinée || summarische variantie *f* || суммарная дисперсия *f*

P433 **poor conductor** || schlechter Leiter *m* || mauvais conducteur *m* || slechte geleider *m* || плохой проводник *m*

P434 **poppet head** *see* **tailstock**

P435 **poppet valve** *see* **disk valve**

P436 **population** || **1.** Gesamtheit *f* *(in der Statistik)* **2.** Besetzung *f* *(eines Energieniveaus)* || population *f* || **1.** populatie *f* *(statistiek)* **2.** bezetting *f* || **1.** совокупность *f* *(в статистике)* **2.** заселённость *f* *(энергетического уровня)*

P437 **population inversion** || Besetzungsinversion *f* || inversion *f* de population || bezettingsinversie *f* || инверсная заселённость *f*

P438 **porosity** || Porosität *f*, Porigkeit *f* || porosité *f* || poreusheid *f* || пористость *f*

P439 **portal (jib) crane** || Portalkran *m* || grue *f* sur portique; grue *f* à portique || portaalkraan *f* (*m*) || портальный кран *m*

P440 **portion** || Teil *m* || portion *f* || deel *n* || часть *f*; участок *m*; область *f*

P441 **portion of a curve** || Kurvenabschnitt *m* || portion *f* d'une courbe || kurvensegment *n* || отрезок *m* кривой

P442 **position** || **1.** Stellung *f*, Lage *f* **2.** Position *f*, Stellung *f* || position *f*, disposition *f* || **1.** ligging *f*, plaatsing *f* **2.** positie *f* || **1.** положение *n*, расположение *n* **2.** позиция *f*

P443 **positional tolerance** || Lagetoleranz *f* || tolérance *f* sur les positions

|| positietolerantie *f* || допуск *m* на точность расположения *(напр. центров отверстий)*

P444 **position modulation** || Pulsphasenmodulation *f* || modulation *f* d'espacement d'impulsions || impulsfasemodulatie *f* || фазово-импульсная модуляция *f*

P445 **position vector** || Radiusvektor *m* || rayon-vecteur *m* || radiusvector *m* || радиус-вектор *m*

P446 I **positive** || Positiv(bild) *n* || positif *m* || positief *n* || позитив(ное изображение *n*) *m*

P447 II **positive** || positiv, Positiv... || positif || positief || положительный; позитивный

P448 **positive charge** || positive Ladung *f* || charge *f* positive || positieve lading *f* || положительный заряд *m*

P449 **positive definite form** || positiv definierte Form *f* || forme *f* définie positive || positief gedefiniëerde vorm *m* || положительно определённая форма *f*

P450 **positive definite matrix** || positiv definierte Matrix *f* || matrice *f* définie positive || positief gedefiniëerde matrix *f* || положительно определённая матрица *f*

P451 **positive direction** || positive Richtung *f* || direction *f* positive || positieve richting *f* || положительное направление *n*

P452 **positive electricity** || positive Elektrizität *f* || électricité *f* positive || positieve elektriciteit *f* || положительное электричество *n*

P453 **positive electrode** || positive Elektrode *f*, Anode *f* || électrode *f* positive, anode *f* || positieve elektrode *f*, anode *f* || положительный электрод *m*, анод *m*

P454 **positive image** || Positivbild *n* || image *f* positive || positief beeld *n* || позитивное изображение *n*, позитив *m*

P455 **positive ion** || positives Ion *n*, Kation *n* || ion *m* positif, cation *m* || positief ion *n*, kation *n* || положительный ион *m*, катион *m*

P456 **positron** || Positron *n* || positron *m* || positron *n* || позитрон *m*

P457 **positronium** || Positronium *n* || positronium *m* || positronium *n* || позитроний *m*

P458 **postulate** || Postulat *n* || postulat *m* || postulaat *n* || постулат *m*

P459 **potassium,** K || Kalium *n* || potassium *m* || kalium *n* || калий *m*

P460 I **potential** || **1.** Potential *n* **2.** Potentialfunktion *f* || **1.** potentiel *m* **2.** fonction *f* potentielle || **1.** potentiaal *m* **2.** potentiële functie *f*, potentiaalfunctie *f* || **1.** потенциал *m* **2.** потенциальная функция *f*, потенциал *m*

P461 II **potential** || Potential... || potentiel || potentiëel, potentiaal... || потенциальный

P462 **potential barrier** || Potentialbarriere *f*, Potentialschwelle *f* || barrière *f* [montagne *f*] de potentiel || potentiaalbarrière *f* || потенциальный [энергетический] барьер *m*

P463 **potential difference** || Potentialunterschied *m*, Potentialdifferenz *f* || différence *f* de potentiel || potentiaalverschil *n* || разность *f* потенциалов, напряжение *n*

P464 **potential energy** || potentielle Energie *f* || énergie *f* potentielle || potentiaalenergie *f* || потенциальная энергия *f*

P465 **potential field** || potentiales [wirbelfreies] Feld *n* [Vektorfeld *n*] || champ *f* potentiel || potentiaalveld *n* || потенциальное [безвихревое] поле *n*

P466 **potential function** || Potentialfunktion *f* || fonction *f* potentielle || potentiële functie *f*, potentiaalfunctie *f* || потенциальная функция *f*, потенциал *m*

P467 **potential gradient** || Potentialgradient *n*, Potentialgefälle *n* || gradient *m* de potentiel || potentiaalgradiënt *m* || градиент *m* потенциала, напряжённость *f* поля

P468 **potential well** || Potentialtopf *m*, Potentialsenke *f* || puits *m* [cuvette *f*] de potentiel || potentiaalput *m* || потенциальная яма *f*

P469 **potentiometer** || Potentiometer *n* || potentiomètre *m* || potentiometer *m* || потенциометр *m*

P470 **potentiometric titration** || potentiometrische [elektrometrische] Titration *f* || titrage *m* potentiométrique || potentiometrische titratie *f* || потенциометрическое титрование *n*

P471 **powder metal** || Sinterlegierung *f*; Sintermetall *n* || métal *m* fritté || sinter *m*; sintermetaal *n* || спечённый сплав *m*; (металлический) спечённый материал *m*

P472 **powder metallurgy** || Pulvermetallurgie *f* || métallurgie *f* des poudres || sintermetallurgie *f* || порошковая металлургия *f*

P473 **power** || 1. Leistung *f* 2. Energie *f* 3. Fähigkeit *f*, Vermögen *n* 4. *Math.* Grad *m* || 1. puissance *f* 2. énergie *f* 3. capacité *f* 4. *math.* puissance *f* || 1. vermogen *n* 2. energie *f* 3. capaciteit *f* 4. *math.* graad *m* || 1. мощность *f* 2. энергия *f* 3. способность *f* 4. *мат.* степень *f*; порядок *m*

P474 **power consumption** || Leistungsverbrauch *m* || puissance *f* consommée || vermogensopname *f* (*m*) || 1. расход *m* мощности; потребление *n* энергии 2. потребляемая мощность *f*

P475 **power dissipation** || Leistungsstreuung *f* || dissipation *f* de puissance || vermogensverstrooiing *f* || рассеяние *n* мощности

P476 **power drive** || Kraftantrieb *m* || commande *f* par moteur [mécanique]; entraînement *m* par moteur [mécanique] || krachtaandrijving *f* || механический привод *m*; силовой привод *m*; привод *m* от двигателя

P477 **powered gyroscope** || nichtfreies Gyroskop *n* || plate-forme *f* gyroscopique || niet-vrije gyroscoop *m* || несвободный [силовой] гироскоп *m*

P478 **power engineering** || Energetik *f* || énergétique *f* || energetica *f* || энергетика *f*

P479 **power factor** || Leistunsfaktor *m* || coefficient *m* [facteur *m*] de puissance || vermogensfactor *m* || коэффициент *m* (электрической) мощности

P480 **powerful** || leistungsfähig, leistungsstark, Groß... || puissant || van grote capaciteit || мощный

P481 **power gain** || Leistungsverstärkung *f* || gain *m* en puissance || vermogensversterking *f* || усиление *n* по мощности

P482 **power hammer** || Maschinenhammer *m* || marteau *m* mécanique || machinehamer *m* || механический молот *m*

P483 **power input** || zugeführte Leistung *f* || puissance *f* appliquée || toegevoerd vermogen *n* || подводимая мощность *f*

P484 **power level** || Leistungspegel *m*; Leistungshöhe *f* || niveau *m* de puissance || vermogensniveau *n* || уровень *m* мощности

P485 **power loss** || Leistungsverlust *m* || perte *f* de puissance || vermogensverlies *n* || потеря *f* мощности

P486 **power of test** || Trennschärfe *f* eines Tests, Teststärke *f* || puissance *f* d'un test || testvermogen *n* || мощность *f* критерия

P487 **power-operated feed** || Kraftvorschub *m* || avance *f* mécanique || mechanische toevoer *m* || механическая подача *f*

P488 **power-operated tool** || Kraftwerkzeug *n* || outils *m pl* mécaniques || machinegereedschap *n* || механизированный инструмент *m*

P489 **power-operated wrench** || Schrauber *m* || clef *f* [clé *f*] mécanique; visseuse *f* || moeraanzetter *m* || гайковёрт *m*

P490 **power output** || Ausgansleistung *f*, abgegebene Leistung *f* || puissance *f* de sortie || uitgangsvermogen *n* || генерируемая [выходная] мощность *f*

P491 **power plant** || 1. Kraftwerk *n* 2. Energie(erzeugungs)anlage *f* || 1. centrale *f* électrique 2. groupe *m* électrogène || 1. elektrische centrale *f* 2. energie-installatie *f* || 1. электростанция *f* 2. энергетическая [силовая] установка *f*

P492 **power rating** || Nennleistung *f* || puissance *f* nominale || nominaal vermogen *n* || номинальная мощность *f*

P493 **power reactor** || Leistungsreaktor *m* || réacteur *m* énergétique || energiereactor *m* || энергетический (ядерный) реактор *m*

P494 **power series** || Potenzreihe *f* || série *f* de puissances || machtreeks *f* (*m*) || степенной ряд *m*

P495 **power shovel** || Bagger *m* || pelle *f* mécanique || baggermolen *m*; motorlaadschop *f* (*m*) || механическая лопата *f*; экскаватор *m*

P496 **power station** || Kraftwerk *n* || centrale *f* électrique || elektrische centrale *f* || электростанция *f*

P497 **power stroke** || Arbeitshub *m*; Arbeitstakt *m* || course *f* active [effective, de travail] || hefhoogte *f*; arbeidstakt *m* || рабочий ход *m* (*поршня*); рабочий такт *m*

P498 **power supply** || 1. Stromversorgung *f* 2. Stromquelle *f* || 1. alimentation *f* 2. source *f* d'alimentation || 1. voeding *f* 2. voedingsbron *f* (*m*) || 1. питание *n* 2. источник *m* питания

P499 **power system** || Energiesystem *n* || système *m* [réseau *m*] énergétique || energiesysteem *n* || энергетическая система *f*, энергосистема *f*

P500 **power transformer** || Leistungstransformator *m*, Krafttransformator *m* || transformateur *m* de puissance || krachttransformator *m* || силовой трансформатор *m*

P501 **power transmission** || Energieübertragung *f* || transmission *f* de (la) puissance || energieoverbrenging *f* || силовая передача *f*; передача *f* электроэнергии

P502 **power unit** || Kraftaggregat *f*; Kraftanlage *f*; Antriebsgruppe *f* || groupe *m* moteur; installation *f* motrice || krachtaggregaat *n*, krachtinstallatie *f* || силовой агрегат *m*; силовая установка *f*

P503 **Poynting vector** || Poyntingscher Vektor *m*, Energiestromdichte *f* || vecteur *m* de propagation [de Poynting] || vector *m* van Poynting || вектор *m* Умова—Пойнтинга, вектор *m* потока электромагнитной энергии

P504 **pragmatics** || Pragmatik *f* || pragmatique *f* || pragmatiek *f* || прагматика *f*

P505 **precedence** || 1. Präzedenz *f* 2. Priorität *f* || 1. précédence *f* 2. priorité *f* || 1. precedentie *f* 2. prioriteit *f* || 1. предшествование *n* 2. приоритет *m*

P506 **precedence matrix** || Präzedenzmatrix *f* || matrice *f* de précédence || precedentiematrix *f* || матрица *f* предшествования

P507 **precedence table** || Präzedenztabelle *f* || table *f* de précédence || precedentietabel *f* || таблица *f* предшествования

P508 **precission** || Präzession *f* || précession *f* || precessie *f* || прецессия *f*

P509 **prechamber** || Vorkammer *f* || chambre *f* de précombustion; antichambre *f*; préchambre *f* (d'allumage) || voorkamer *f* (*m*) || предкамера *f*

P510 **precipitation hardening** || Aus(scheidungs)härtung *f* || durcissement *m* par précipitation || uitscheidingsharding *f* || дисперсионное твердение *n*

P511 **precision** || Genauigkeit *f*, Präzision *f* || précision *f* || precisie *f*, nauwkeurigheid *f* || точность *f*; прецизионность *f*

P512 **precision boring machine** *see* **fine boring machine**

P513 **precision turning** *see* **high precision turning**

P514 **precombustion chamber** *see* **prechamber**

P515 **precompact set** || kompakte Menge *f* || ensemble *m* précompact || compacte menigte *f* || компактное множество *n*

P516 **predicate** || Prädikat *n* || prédicat *m* || predikaat *n* || предикат *m*

P517 **predicate calculus** || Prädikatenkalkül *n* || calcul *m* des prédicats || predikatenberekening *f* || исчисление *n* предикатов

P518 **predissociation** || Prädissoziation *f* || prédissociation *f* || predissociatie *f* || предиссоциация 1

P519 **preferred orientation** || bevorzugte Orientierung *f*, Vorzugsorientierung *f* || orientation *f* préférentielle [privilégiée] || overwegende oriëntatie *f* || преимущественная [предпочтительная, избирательная] ориентация *f*

P520 **preheating** || Vorwärmung *f* || préchauff(ag)e *m* || voorverwarming *f* || предварительный нагрев *m* (предварительный) подогрев *m*

P521 **premiss** || Prämisse *f* || prémisse *f* || premisse *f* || посылка *f*; предпосылка *f*

P522 **prescribed value** || vorgeschriebener Wert *m*, Sollwert *m* || valeur *f* prescrite [spécifiée] || voorgeschreven waarde *f* || заданное значение *n*

P523 **press** || Presse *f* || presse *f* || pers *f* (*m*) || пресс *m*

P524 **press forging** || Gesenkschmieden *n* || forgeage *m* à la presse; matriçage *m* || perssmeden *n* || ковка *f* в штампах

P525 **pressing** || 1. Pressen *n* 2. Gesenkschmieden *n* || 1. pressage *m* 2. matriçage *m*, estampage *m*; emboutissage *m* || 1. persen *n*; samendrukking *f* 2. perssmeden *n*; persing || 1. прессование *n* 2. ковка *f* в штампах; штамповка *f*

P526 **pressing tool** || Preßwerkzeug *n* || éstampe *f*; outils *m pl* d'estampage [de matriçage] || perswerktuig *n* || штамп *m*, штамповочный инструмен *m*

P527 **pressure** || Druck *m* || pression *f* || druk *m* || давление *n*

P528 **pressure blower** || Druckgebläse *n*; Luftgebläse *n* || soufflerie *f* [soufflante] à (haute) pression || blaasmachine *f*, blazer *m* || воздуходувная машина *f*; воздуходувка *f* высокого давления

P529 **pressure casting** || Spritzgießen *n*, Spritzguß *m* || coulée *f* sous pression; moulage *m* par injection || persgieten *n* || литьё *n* под давлением

P530 **pressure differential** || Druckgefälle *n* || pression *f* différentielle; différence *f* de pression || differentiaaldruk *m* || перепад *m* давления, разность *f* давления

P531 **pressure drop** || Druckgefälle *n*; Druckabfall *m* || chute *f* de pression || drukverschil *n* || перепад *m* давления; падение *n* давления

P532 **pressure force** || Druckkraft *f* || force *f* de pression || drukkracht *f* (*m*) || сила *f* давления

P533 **pressure forging** *see* **die forging**

P534 **pressure gauge** || Manometer *n*, Druckmesser *m* || manomètre *m* || manometer *m*, drukmeter *m* || манометр *m*

P535 **pressure gradient** || Druckgradient *m* || gradient *m* de pression || drukgradiënt *m* || градиент *m* давления

P536 **pressure jump** || Drucksprung *m* || saut *m* de pression || druksprong *m* || скачок *m* давления

P537 **pressure loss** || Druckverlust *m* || perte *f* de pression || drukverlies *n* || гидравлическое сопротивление *n*; падение *n* [потеря *f*] давления [напора]

P538 **pressure profile** || Druckprofil *n* || épure *f* de pressions || drukprofiel *n* || эпюра *f* [профиль *m*] давлений

P539 **pressure ratio** || Kompressionsgrad *m* || taux *m* de compression || compressiegraad *m* || степень *f* сжатия; степень *f* повышения давления

P540 **pressure spring** || Druckfeder *f* || ressort *m* de pression || drukveer *f* (*m*) || пружина *f* сжатия

P541 **pressure test** || Druckversuch *m*; Druckprobe *f* || essai *m* [épreuve *f*] sous [par] pression || drukproef *f* (*m*) || 1. испытание *n* на сжатие 2. испытание *n* под давлением; опрессовка *f*

P542 **pressure-tight** || druckdicht || étanche [imperméable] à l'air *(ou au gaz sous pression)* || hermetisch, gasdicht || герметический, газонепроницаемый

P543 **pressure-tight joint** || druckdichte Verbindung *f* || joint *m* [raccord *m*] étanche [imperméable] || hermetische verbinding *f* || герметичный шов *m*; герметичный стык *m*, герметичное соединение *n*

P544 **pressure valve** || Druckventil *n* || soupape *f* [clapet *m*] de refoulement || drukventiel *n* || нагнетательный [напорный] клапан *m*

P545 **pressure wave** || Druckwelle *f* || onde *f* de pression || drukgolf *f* (*m*) || волна *f* сжатия

P546 **pressurized** || 1. Druck... 2. Lade..., mit Aufladung *(Verbrennungsmotor)* || 1. sous pression 2. sous surpression, sous pressurisation || 1. onder druk gebracht 2. met oplading || 1. находящийся под давлением *(выше атмосферного)* 2. с наддувом *(о двигателе)*

P547 I **primary** || 1. Primärwicklung *f* 2. Primärteilchen *n* || 1. enroulement *m* primaire 2. particule *f* primaire || 1. primaire wikkeling *f* 2. primair deeltje *n* || 1. первичная обмотка *f* 2. первичная частица *f*

P548 II **primary** || primär, Primär... || primaire || primair || первичный

P549 **primary colour** || Grundfarbe *f*, Primärfarbe *f* || couleur *f* primaire || hoofdkleur *f* || основной [первичный] цвет *m*

P550 **primary face** *see* **face 2.**

P551 **primary ion** || Primärion *n* || ion *m* primaire || primair ion *n* || первичный ион *m*

P552 **primary particle** || Primärteilchen *n* || particule *f* primaire || primair deeltje *n* || первичная частица *f*

P553 **primary voltage** || Primärspannung *f* || tension *f* primaire || primaire spanning *f* || напряжение *n* в первичной обмотке

P554 **primary winding** || Primärwicklung *f* || enroulement *m* primaire || primaire wikkeling *f* || первичная обмотка *f*

P555 **prime mover** || Kraftmaschine *f* || source *f* d'énergie motrice; source *f* de force motrice || primaire machine *f*; primaire motor *m* || первичный двигатель *m*

P556 **primitive character** || primitiver Charakter *m (Darstellung)* || caractère *m* primitif || primitief karakter *n* || простой [примитивный] характер *m (представления)*

P557 **primitive recursive function** || primitiv-rekursive Funktion *f* || fonction *f* récursive primitive || primitief-recursieve functie *f* || примитивно-рекурсивная функция *f*

P558 **principal axis** || Hauptachse *f* || axe *m* principal || hoofdas *f (m)* || главная ось *f*

P559 **principal curvature** || Hauptkrümmung *f* || courbure *f* principale || hoofdkromming *f* || главная кривизна *f*

P560 **principal focus** || Hauptbrennpunkt *m* || foyer *m* principal || hoofdbrandpunt *n* || главный фокус *m*

P561 **principal normal** || Hauptnormale *f* || normale *f* principale || hoofdnormaal *f (m)* || главная нормаль *f*

P562 **principal normal section** || Hauptnormalschnitt *m* || section *f* normale principale || hoofdnormaalsnede *f (m)* || главное нормальное сечение *n*

P563 **principal quantum number** || Hauptquantenzahl *f* || nombre *m* quantique principal || hoofdquantengetal *n* || главное квантовое число *n*

P564 **principal value** || Hauptwert *m* || valeur *f* principale || hoofdwaarde *f* || главное значение *n*

P565 **principle** || Prinzip *n*; Gesetz *n* || principe *m* || principe *n* || принцип *m*; начало *n*; закон *m*

P566 **principle of action and reaction** || Wechselwirkungsgesetz *n*, Prinzip *n* von Wirkung und Gegenwirkung || principe *m* de l'action et de la réaction || principe *n* van werking en tegenwerking || закон *m* равенства (сил) действия и противодействия

P567 **principle of least action** || (Hamiltonsches) Prinzip *n* der kleinsten Wirkung || principe *m* de la moindre action || principe *n* van de kleinste werking || принцип *m* наименьшего действия (Гамильтона)

P568 **principle of least constraint** || (Gaußsches) Prinzip *n* des kleinsten Zwanges || principe *m* de la moindre contrainte || principe *n* van de kleinste dwang || принцип *m* наименьшего принуждения (Гаусса)

P569 **principle of least curvature** || (Hertzsches) Prinzip *n* der kleinsten Krümmung, Prinzip *n* der geradesten Bahn || principe *m* de la moindre courbure || principe *n* van de kleinste kromming || принцип *m* кратчайшего пути [наименьшей кривизны] (Герца)

P570 **principle of optimality** || Optimalitätsprinzip *n* || principe *m* d'optimalité || optimaliteitsprincipe *n* || принцип *m* оптимальности

P571 **principle of relativity** || Relativitätsprinzip *n* || principe *m* de relativité || relativiteitsprincipe *n* || принцип *m* относительности

P572 **printed circuit** || gedruckte Schaltung *f* || circuit *m* imprimé || gedrukte schakeling *f* печатная схема *f*

P573 **printer** || Drucker *m* || imprimante *f* || drukker *m* || печатающее устройство *n*

P574 **prism** || Prisma *n* || prisme *m* || prisma *n* || призма *f*

P575 **prism spectrograph** || Prismenspektrograph *m* || spectrographe *m* à prisme || prismaspectrograaf *m* || призменный спектрограф *m*

P576 **probabilistic** || probabilistisch, Wahrscheinlichkeits... || probabilistique || waarschijnlijkheids... || вероятностный

P577 **probabilistic automaton** || Wahrscheinlichkeitsautomat *m* || automate *m* probabilistique || waarschijnlijkheidsautomaat *m* || вероятностный автомат *m*

P578 **probability** || Wahrscheinlichkeit *f* || probabilité *f* || waarschijnlijkheid *f* || вероятность *f*

P579 **probability density** || Wahrscheinlichkeitsdichte *f*, Verteilungsdichte *f*, Dichtefunktion *f* || densité *f* de probabilité || waarschijnlijkheidsdichtheid *f* || плотность *f* вероятности

P580 **probability function** || Wahrscheinlichkeitsfunktion *f* || fonction *f* de probabilité || waarschijnlijkheidsfunctie *f* || вероятностная функция *f*

P581 **probability integral** || Wahrscheinlichkeitsintegral *n* || intégrale *f* de probabilité || waarschijnlijkheidsintegraal *f* (*m*) || интеграл *m* вероятности

P582 **probability theory** || Wahrscheinlichkeitstheorie *f* || théorie *f* des probabilités || waarschijnlijkheidstheorie *f* || теория *f* вероятностей

P583 **probable deviation** || wahrscheinliche Abweichung *f* || déviation *f* probable || waarschijnlijke afwijking *f* || вероятное отклонение *n*

P584 **probe** || Sonde *f* || sonde *f* || sonde *f* (*m*) || зонд *m*; пробник *m*; щуп *m*; датчик *m*

P585 **problem** || Problem *n* || problème *m* || probleem *n* || проблема *f*; задача *f*

P586 **problem-oriented language** || problemorientierte Sprache *f* || langage *m* orienté vers le problème || probleemgeoriënteerde taal *f* || проблемно-ориентированный язык *m*

P587 **procedure** || Prozedur *f* || procédure *f* || werkwijze *f* || процедура *f*

P588 **process** || Prozeß *m*, Vorgang *m* || processus *m* || proces *n*; procédé *n* || процесс *m*

P589 **process control** || Prozeßsteuerung *f* || contrôle *m* du processus de fabrication || procesbesturing *f* || управление *n* (технологическим) процессом

P590 **producer gas** || Generatorgas *n* || gaz *m* de gazogène [de générateur] || generatorgas *n* || генераторный газ *m*

P591 **production** || **1.** Produktion *f* **2.** Erzeugung *f* || production *f* || **1.** produktie *f* **2.** vorming *f* || **1.** производство *n* **2.** образование *n*, генерация *f*

P592 **production lathe** *see* **manufacturing lathe**

P593 **product space** || Produkt-Raum *m* || produit *m* des espaces || produktruimte *f* || топологическое произведение *n*

P594 **profile** || Profil *n* || profil *m* || profiel *n* || профиль *m*

P595 **profile cutter** || Formfräser *m* || fraise *f* à dents profilées || profielfrees *f* (*m*) || фасонная фреза *f*

P596 **profile flow** || Profilumströmung *f* || écoulement *m* d'un profil || profielomstroming *f* || обтекание *n* профиля

P597 **profile grinding** || Profilschleifen *n* || rectification *f* de forme || profielslijpen *n* || фасонное [профильное] шлифование *n*

P598 **profile grinding machine** || Profilschleifmaschine *f* || rectifieuse *f* de profils complexes || profielslijpmachine *f* || профилешлифовальный станок *m*

P599 **profile lathe** *see* **copying lathe**

P600 **profile milling** || Formfräsen *n* || fraisage *m* profiling [par reproduction, par profilage] || profielfrezen *n*, kopiëerfrezen *n* || фрезерование *n* по копиру

P601 **profile milling machine** *see* **copying milling machine**

P602 **profiling** || Kopierarbeit *f*, Kopieren *n* || profilage *m* par copiage; gabariage *m* || profileren *n*; kopiëren *n* || профилирование *n*, обработка *f* по копиру

P603 **profilometer** || Profilmesser *m* || profilomètre *m* || profielmeter *m* || профилометр *m*

P604 **program** || Programm *n* || programme *m* || programma *n* программа *f*

P605 **programming language** || Programmiersprache *f* || langage *m* de programmation || programmertaal *f* (*m*) || язык *m* программирования

P606 **progressive motion** || translatorische Bewegung *f* || mouvement *m* de translation || translatorische beweging *f* || поступательное движение *n*

P607 **progressive transmission** *see* **stepless transmission**

P608 **progressive wave** || Wanderwelle *f* || onde *f* progressive || lopende golf *f* (*m*) || бегущая волна *f*

P609 **projectile** || **1.** auftreffendes Teilchen *n*, Beschußteilchen *n* **2.** Geschoß *n* (*Mechanik*) || **1.** particule *f* bombardante **2.** projectile *m*, obus *m* || **1.** bombarderend deeltje *n* **2.** projectiel *n* || **1.** налетающая [бомбардирующая] частица *f* **2.** снаряд *m*; летящее тело *n*

P610 **projection** || Projektion *f* || projection *f* || projectie *f* || проекция *f*

P611 **projection of a curve** || Kurvenprojektion *f* || projection *f* d'une courbe || curvenprojectie *f* || проекция *f* кривой

P612 **projection theorem** || Projektionssatz *m* || théorème *m* de projection || projectietheorema *n* || теорема *f* о проекциях

P613 **projective geometry** || projektive Geometrie *f* || géométrie *f* projective || projectieve meetkunde *f* || проективная геометрия *f*

P614 **projector** || Bildwerfer *m* || projecteur *m*, appareil *m* de projection || projectieapparaat *n* || проектор *m*, проекционный аппарат *m*

P615 **prolate ellipsoid** || verlängertes [gestrecktes] Ellipsoid *n* || ellipsoïde *m* allongé || verlengde ellipsoïde *f* || вытянутый эллипсоид *m*

P616 **prompt neutron** || promptes Neutron *n*, Spaltneutron *n* || neutron *m* immédiat [instantané] || onmiddellijk neutron *n* || мгновенный нейтрон *m* (*деления*)

P617 **proof** || Beweis *m* || démonstration *f* || bewijs *n* || доказательство *n*

P618 **proof load** || **1.** Prüflast *f* **2.** Versuchslast *f* || **1.** charge *f* d'essai [d'épreuve] **2.** charge *f* (à la) limite d'élasticité; charge *f* élastique limite || proefbelasting *f* || **1.** пробная [контрольная] нагрузка *f* **2.** нагрузка *f*, не вызывающая остаточных деформаций; максимально допустимая нагрузка *f*

P619 **proof stress** || **1.** Dehngrenze *f* **2.** Prüfbeanspruchung *f* || **1.** limite *f* d'écoulement [d'élasticité] conventionnelle **2.** contrainte *f* d'essai [d'épreuve] || **1.** rekgrens *f* (*m*), vloeigrens *f* (*m*) **2.** proefspanning *f* || **1.** условный предел *m* текучести (*соответствующий установленной величине остаточной деформации*) **2.** испытательное напряжение *n*; испытание *n*, установленное методикой испытаний

P620 **propagation** || Ausbreitung *f*, Fortpflanzung *f* || propagation *f* || verspreiding *f* || распространение *n*

P621 **propagation length** || Fortpflanzungsweite *f* || distance *f* de propagation || verspreidingslengte *f* || дальность *f* распространения

P622 **propagation of disturbance** || Störungsfortpflanzung *f* || propagation *f* de perturbation || storingsverspreiding *f* || распространение *n* возмущений

P623 **propagation of errors** || Fehlerfortpflanzung *f* || propagation *f* des erreurs || foutenverspreiding *f* || распространение *n* ошибок

P624 **propagation vector** || Wellenvektor *m* || vecteur *m* d'onde || golfvector *m* || волновой вектор *m*

P625 **propeller** || Propeller *m*, Luftschraube *f*; Schiffsschraube *f* || hélice *f* || luchtschroef *f* (*m*); propeller *m*, scheepsschroef *f* (*m*) || воздушный винт *m*, пропеллер *m*; гребной винт *m*

P626 **propeller fan** || Flügelradgebläse *n* || ventilateur *m* à palettes || schroefbladventilator *m* || лопастный вентилятор *m*

P627 **propeller shaft** || Gelenkwelle *f*, Kardanwelle *f* || arbre *m* à cardan || drijfas *f* (*m*); cardanas *f* (*m*) || карданный вал *m*

P628 **propene,** $CH_3CH : CH_2$ || Propen *n*, Propylen *n* || propène *m*, propylène *n* || propeen *n* || пропен *m*, пропилен *m*

P629 **proper absorption** || echte [absolute, wahre] Absorption *f* || absorption *f* vraie || echte absorptie *f* || истинное поглощение *n*

P630 **proper function** || Eigenfunktion *f* || fonction *f* propre || eigenfunctie *f* || собственная функция *f*

P631 **proper rotation** || Eigendrehung *f* || rotation *f* propre || eigendraaiing *f* || собственное вращение *n*

P632 **proper subset** || Eigen-Untermenge *f* || sous-ensemble *m* propre || eigen-ondermenigte *f* || собственное подмножество *n*

P633 **proper time** || Eigenzeit *f* || temps *m* propre || eigentijd *m* || собственное время *n*

P634 **property** || Eigenschaft *f* || propriété *f* || eigenschap *f* || свойство *n*

P635 **proper value** || Eigenwert *m* || valeur *f* propre || eigenwaarde *f* || собственное значение *n*

P636 **proper vector** || Eigenvektor *m* || vecteur *m* propre || eigenvector *m* || собственный вектор *m*

P637 **proportion** || Proportion *f*, Verhältnis *n* || proportion *f* || proportie *f* || пропорция *f*

P638 **proportional limit** || P-Grenze *f*, Proportionalitätsgrenze *f* || limite *f* proportionnelle || proportionaliteitsgrens *f* (*m*) || предел *m* пропорциональности

P639 **proposition** || Aussage *f*, || proposition *f* || propositie *f*, mening *f* || суждение *n*, высказывание *n*

P640 **propositional calculus** || Aussagenkalkül *n* || calcul *m* propositionnel || propositionele berekening *f* || исчисление *n* высказываний, пропозициональное исчисление *n*

P641 **propulsion** || **1.** Antrieb *m* **2.** Traktion *f*; Triebkraft *f* || **1.** entraînement *m* **2.** traction *f* || **1.** aandrijving *f* **2.** tractie *f*; drijfkracht *f* || **1.** приведение *n* в движение **2.** тяга *f*; движущая сила *f*

P642 **propulsion by reaction** || reaktive Bewegung *f*; Rückstoßantrieb *m* || propulsion *f* par réaction || straalbeweging *f* || реактивное движение *n*

P643 **propulsion nozzle** || Strahlantriebsdüse *f*; Schubdüse *f* || tuyère *f* à réaction || straalmotordoes *m* || сопло *n* реактивного двигателя; реактивное сопло *n*

P644 **propulsive force** || Zugkraft *f*; Antriebskraft *f*; Traktionskraft *f* || force *f* propulsive || tractiekracht *f* (*m*) || тяговое [толкающее] усилие *n*; сила *f* тяги

P645 **propylene** *see* **propene**

P646 **protective colloid** || Schutzkolloid *n* || colloïde *m* protecteur || beschermingscolloïde *f* (*m*) || защитный коллоид *m*

P647 **protein** || Protein *n* || protéine *f* || proteïne *n* || протеин *m*, белок *m*

P648 **proton** || Proton *n* || proton *m* || proton *n* || протон *m*

P649 **prototype workpiece** *see* **master workpiece**

P650 **proximity** || Nähe *f* || proximité *f* || nabijheid *f*, proximiteit *f* || близость *f*

P651 **proximity of zero order** || Nähe *f* nullter Ordnung || proximité *f* de l'ordre zéro || nabijheid *f* van de nulorde || близость *f* нулевого порядка

P652 **prussic acid**, HCN || Zyanwasserstoff(säure *f*) *m*, Blausäure *f* || acide *m* cyanhydrique, nitrile *m* formique || cyaanwaterstof *f* (*m*) ||

цианистоводородная [синильная] кислота *f*

P653 **pseudoperiodic quantity** || pseudoperiodische Größe *f* || quantité *f* pseudo-périodique || pseudoperiodische grootheid *f* || псевдопериодическая величина *f*

P654 **pseudo-random** || Pseudozufalls... || pseudo-aléatoire || pseudotoevallig || псевдослучайный

P655 **pseudosphere** || Pseudosphäre *f* || pseudo-sphère *f* || pseudosfeer *f* (*m*) || псевдосфера *f*

P656 **pseudotensor** || Pseudotensor *m* || pseudo-tenseur *m* || pseudotensor *m* || псевдотензор *m*

P657 **psi-function** || Psi-Funktion *f* || fonction *f* psi || psi-functie *f* || пси-функция *f*

P658 **psychrometer** || Psychrometer *n* || psychromètre *m* || psychrometer *m* || психрометр *m*

P659 **p-type conductivity** || p-Leitung *f*, Defektelektronenleitung *f* || conductibilité *f* par lacunes [par trous] || defectelektronleiding *f* || дырочная проводимость *f*

P660 **p-type semiconductor** || p-Halbleiter *m*, Löcherhalbleiter *m* || semi-conducteur *m* type p || p-halfgeleider *m* || полупроводник *m* *p*-типа, дырочный полупроводник *m*

P661 **pull** || Zug *m*; Zugkraft *f* || poussée *f* || trekkracht *f* (*m*) || тяга *f*; сила *f* тяги

P662 **pulley** || Scheibe *f*; Flasche *f*, Rolle *f* || poulie *f* || schijf *f* (*m*); rol *f* (*m*) || шкив *m*; блок *m*

P663 **pulley block** || Flaschenzug *m*, Rollenzug *m* || palan *m* || talie *f* || **1.** полиспаст *m*; таль *m* **2.** обойма *f* блока

P664 **pulsar** || Pulsar *m* || pulsar *m* || pulsar *m* || пульсар *m*

P665 **pulsating current** || pulsierender Strom *m* || courant *m* pulsatoire [pulsé] || pulserende stroom *m* || пульсирующий ток *m*

P666 **pulsating load** || pulsierende Belastung *f* || charge *f* pulsatoire || pulserende belasting *f* || пульсирующая нагрузка *f*

P667 **pulsating voltage** || Welligkeitsspannung *f* || tension *f* pulsatoire || pulserende spanning *f* || пульсирующее напряжение *n*

P668 **pulsation** || Pulsation *f* || pulsation *f* || pulsatie *f* || пульсация *f*

P669 **pulse** || Impuls *m*, Puls *m* || impulsion *f* || (im)puls *m* || импульс *m*

P670 **pulse amplitude** || Impulsamplitude *f*, Impulshöhe *f* || amplitude *f* de l'impulsion || impulsamplitude *f* || амплитуда *f* импульса

P671 **pulse-code modulation** || Pulskodemodulation *f* || modulation *f* en code d'impulsions, modulation *f* d'impulsions codées || (im)pulscodemodulatie *f* || импульсно-кодовая [кодово-импульсная] модуляция *f*

P672 **pulsed** || Impuls..., Puls...; pulsierend || d'impulsion; pulsatoire || impuls...; pulserend || импульсный; пульсирующий

P673 **pulsed laser** || Impulslaser *m* || laser *m* pulsé || impulslaser *m*, pulserende laser *m* || импульсный лазер *m*

P674 **pulse generator** || Impulserzeuger *m*, Impulsgenerator *m* || générateur *m* d'impulsions || impulsgenerator *m* || импульсный генератор *m*, генератор *m* импульсов

P675 **pulse height** *see* **pulse amplitude**

P676 **pulse-height analyzer** || Amplitudenanalysator *m* || analyseur *m* d'amplitudes || amplitudeanalysator *m* || амплитудный анализатор *m* (импульсов)

P677 **pulse length** || Impulsdauer *f*, Impulsbreite *f* || largeur *f* [durée *f*] d'impulsion || impulsduur *m* || длительность *f* импульса

P678 **pulse mode** || Impulsbetrieb *m* || régime *m* impulsionnel [pulsatoire] || impulsmodus *m* || импульсный режим *m*

P679 **pulse modulation** || Pulsmodulation *f* || modulation *f* d'impulsions || impulsmodulatie *f* || импульсная модуляция *f*

P680 **pulse time** || Pulsperiode *f*, Impulsfolgeperiode *f* || période *f* (de répétition) des impulsions || impulsperiode *f* || период *m* (повторения) импульсов

P681 **pulse train** || Impulsfolge *f*, Impulsserie *f* || train *m* d'impulsions || impulsserie *f*, impulsreeks *f* (*m*) || серия *f* импульсов

P682 **pulse width** *see* **pulse length**

P683 **pulse-width modulation** || Pulsbreitenmodulation *f*, Pulslängenmodulation *f*, Pulsdauermodulation *f* || modulation *f* par impulsions à largeur variable [d'impulsions en durée] || impulsduurmodulatie *f* || широтно-импульсная модуляция *f*

P684 **pump** || Pumpe *f* || pompe *f* || pomp *f* (*m*) || насос *m*

P685 **pumping** || **1.** Pumpen *n*; Umwälzen *n* **2.** Auspumpen *n* **3.** Pumpen *n* (*Laser*) || pompage *m* || **1.** pompen *n* **2.** uitpompen *n* **3.** pompen *n* || **1.** качание *n*; накачивание *n*; нагнетание *n* **2.** откачка *f*; откачивание *n* **3.** накачка *f* (*лазера*)

P686 **punch** || **1.** Locheisen *n* **2.** Locher *m* || **1.** poinçon *m* **2.** perforateur *m*, perforatrice *f* || **1.** drevel *m*; doorslag *m* **2.** perforator *m* || **1.** пуансон *m*; пробойник *m*, бородок *m*; прошивень *m* **2.** перфоратор *m*

P687 **punch(ed) card** || Lochkarte *f* || carte *f* perforée || ponskaart *f* (*m*) || перфокарта *f*

P688 **punching** || **1.** Ausschneiden *n*, Stanzen *n* **2.** Lochen *n* || poinçonnage *m* || **1.** ponsen *n*, ponsing *f*, uithakking *f*, uitsnijding *f* **2.** perforeren *n*; uitsnijding *f*; stansen *n* || **1.** пробивка *f* (*дыр*), продавливание *n* **2.** штамповка *f* **3.** прошивка *f*

P689 **punching bear** || Handlochpresse *f* || poinçonneuse *f* à main || handgatpers *f* (*m*); handponsmachine *f* || ручной дыропробивной пресс *m*

P690 **punching machine** || Perforiermaschine *f*; Perforierpresse *f*, Lochpresse *f* || poinçonneuse *f* || perforeermachine *f*; drevel *m*; doorslag *m* || дыропробивной станок *m*, прошивной пресс *m*

P691 **pupil** || Pupille *f* (*des optischen Systems*) || pupille *f* || pupil *f* || зрачок *m* (*оптической системы*)

P692 **purely random process** || reiner Zufallsprozeß *m* || processus *m* purement aléatoire || zuiver toevalsproces *n* || чисто случайный процесс *m*

P693 **pure mathematics** || reine Mathematik *f* || mathématiques *f pl* pures || zuivere mathematica *f* || чистая математика *f*

P694 **pure physics** || reine [fundamentale] Physik *f* || physique *f* pure || zuivere fysica *f* || чистая [фундаментальная] физика *f*

P695 **pure rotation** || reine Rotation *f* || rotation *f* pure || zuivere rotatie *f* || чистое вращение *n*

P696 **pure strategy** || reine Strategie *f* || stratégie *f* pure || reine strategie *f* || чистая стратегия *f*

P697 **push button control** || Druckknopfsteuerung *f* || commande *f* par boutons-poussoirs || drukknopbesturing *f* || кнопочное управление *n*

P698 **push-down store** || Verschiebespeicher *m*, Push-down-Speicher *m* || mémoire hiérarchisé || push-down geheugen *n* || магазинная память *f*

P699 **pushing force** || Druckkraft *f* || force *f* de poussage || drukkracht *f* (*m*) || толкающее усилие *n*

P700 **push rod** || Stoßstange *f* || tige *f* de poussoir || stootstang *f* (*m*) || штанга *f*, толкатель *m*; шток *m* [штанга *f*] толкателя клапана

P701 **pycnometer** || Pyknometer *n* || pycnomètre *m* || pyknometer *m* || пикнометр *m*

P702 **pyramid** || Pyramide *f* || pyramide *f* || piramide *f* || пирамида *f*

P703 **pyroelectric** || pyroelektrisch || pyro-électrique || pyro-elektrisch || пироэлектрический

P704 **pyroelectricity** || Pyroelektrizität *f* || pyro-électricité *f* || pyroelektriciteit *f* || пироэлектричество *n*

P705 **pyrometer** || Pyrometer *n* || pyromètre *m* || pyrometer *m* || пирометр *m*

Q

Q1 **quadrant** || **1.** Quadrant *m* **2.** Wechselräderschere *f* **3.** Zahnsektor *m* || **1.** quadrant *m* **2.** lyre *f*; tête de cheval **3.** secteur *m* denté || **1.** kwadrant *n* 2. kwadrant *n*, schaar *f* (*m*) (*draaimachine*) 3. kwadrant *n*; tandsektor *m* || 1. квадрант *m* 2. гитара *f* (*станка*) 3. (зубчатый) сектор *m*

Q2 **quadrantal triangle** || Quadrantdreieck *n* || triangle *m* quadrantal || kwadrantdriehoek *m* || квадрантный треугольник *m*

Q3 **quadrant plate** *see* **quadrant 2.**

Q4 **quadratically integrable function** || quadratisch integrierbare Funktion *f* || fonction *f* quadratiquement intégrable || kwadratisch integreerbare functie *f* || квадратично интегрирующая [интегрирующая с квадратом] функция *f*

Q5 **quadratic equation** || quadratische Gleichung *f* || équation *f* quadratique || kwadratische vergelijking *f* || квадратное уравнение *n*

Q6 **quadratic form** || quadratische Form *f* || forme *f* quadratique || kwadratische vorm *m* || квадратичная форма *f*

Q7 **quadratic-variation theorem** || Satz *m* von der quadratischen Abweichung || théorème *m* de variation quadratique || theorema *n* van de kwadratische afwijking || теорема *f* о квадратичном отклонении

Q8 **quadrature formula** || Quadraturformel *f* || formule *f* de quadrature || kwadratuurformule *f* (*m*) || формула *f* квадратуры

Q9 **quadric (surface)** || Quadrik *f* || quadrique *f*, surface *f* du second ordre || kwadriek *f* || поверхность *f* второго порядка

Q10 **quadripole** || Vierpol *m* || quadripôle *m* || vierpool *f* (*m*) || четырёхполюсник *m*

Q11 **quadrupole** || Quadrupol *m* || quadripôle *m* || quadrupool *f* (*m*) || квадруполь *m*

Q12 **quadrupole moment** || Quadrupolmoment *n* || moment *m* quadripôle || quadrupoolmoment *n* || квадрупольный момент *m*

Q13 **qualitative analysis** || qualitative Analyse *f* || analyse *f* qualitative || kwalitatieve analyse *f* || качественный анализ *m*

Q14 **quality factor** || Gütefaktor *m* (*Hohlraumresonator, Verstärker*) || facteur *m* de qualité || kwaliteitsfactor *m* || добротность *f* (*резонатора*)

Q15 **quantification** || Quantifizierung *f* || quantification *f* || quantificatie *f* || квантификация *f*

Q16 **quantitative analysis** || quantitative Analyse *f* || analyse *f* quantitative || quantitatieve analyse *f* || количественный анализ *m*

Q17 **quantity** || 1. Menge *f* 2. Größe *f* || 1. quantité *f* 2. grandeur *f* || 1. quantiteit *f* 2. grootheid *f* || 1. количество *n* 2. величина *f*

Q18 **quantity of elektricity** || Elektrizitätsmenge *f*, elektrische Ladung *f* || quantité *f* d'électricité || elektriciteitshoeveelheid *f* || количество *n* электричества, (электрический) заряд *m*

Q19 **quantity of heat** || Wärmemenge *f* || quantité *f* de chaleur || warmtehoeveelheid *f* || количество *n* теплоты [тепла]

Q20 **quantity production** || Massenproduktion *f* || production *f* en masse [en série] || massaproduktie *f* || массовое [серийное] производство *n*

Q21 **quantization** || Quantelung *f*, Quantisierung *f* || quantification *f* || quantisering *f* || квантование *n*

Q22 **quantized** || gequantelt || quantifié || gequantiseerd || квантованный

Q23 **quantized system** || gequanteltes System *n* || système *m* quantifié || gequantiseerd systeem *n* || квантованная система *f*

Q24 **quantum** || Quant *n* || quantum *m* || quantum *n* || квант *m*

Q25 **quantum efficiency** *see* **quantum yield**

Q26 **quantum energy** || Quantenenergie *f* || énergie *f* de quantum || quantumenergie *f* || энергия *f* кванта

Q27 **quantum mechanics** || Quantenmechanik *f* || mécanique *f* des quanta [quantique] || quantummechanica *f* || квантовая механика *f*

Q28 **quantum number** || Quantenzahl *f* || nombre *m* quantique || quantumgetal *n* || квантовое число *n*

Q29 **quantum of action** || Wirkungsquant(um) *n*, Plancksche Konstante *f* || quantum *m* d'action || werkingsquantum *n* || квант *m* действия, постоянная *f* Планка

Q30 **quantum oscillator** || Quantenoszillator *m* || oscillateur *m* quantique || quantengenerator *m* || квантовый осциллятор *m*

Q31 **quantum statistics** || Quantenstatistik *f* || statistique *f* quantique || quantenstatistiek *f* || квантовая статистика *f*

Q32 **quantum theory** || Quantentheorie *f* || théorie *f* quantique [des quanta] || quantumtheorie *f* || квантовая теория *f*

Q33 **quantum transition** || Quantenübergang *m*, quantenmechanischer Übergang *m* || transition *f* [saut *m*] quantique || quantenovergang *m* || квантовый переход *m*

Q34 **quantum yield** || Quantenausbeute *f* || rendement *m* quantique || quantenopbrengst *f* || квантовый выход *m*

Q35 **quark** || Quark *n* || quark *m* || quark *n* || кварк *m*

Q36 **quarter wavelength** || Viertelwelle(nlänge) *f* || quart *m* d'onde || kwartgolflengte *f* || четверть *f* длины волны, четвертьволна *f*

Q37 **quarter-wave plate** || Viertelwellenplättchen *n* || plaque *f* de quart d'onde || kwartgolflengteplaat *f* (*m*) || пластинка *f* в четверть волны, четвертьволновая пластинка *f*

Q38 **quartic equation** || biquadratische Gleichung *f*, Gleichung *f* vierten Grades || équation *f* quartique || vergelijking *f* van de vierde graad || уравнение *n* четвёртой степени

Q39 **quartile** || Quartil *n* || quartile *m* || kwartiel *n* || квартиль *m*

Q40 **quartz (crystal) clock** || Quarzuhr *f* || horloge *f* à quartz || kwartsklok *f* (*m*) || кварцевые часы *pl*

Q41 **quartz stabilization** || Quarzstabilisation *f* || stabilisation *f* à quartz || kwartsstabilisatie *f* || (пьезо-)кварцевая стабилизация *f* (частоты)

Q42 **quasar** || Quasar *m* || quasar *m* || quasar *m* || квазар *m*

Q43 **quasi-classical** || quasiklassisch || quasi classique || quasi-klassiek || квазиклассический

Q44 **quasi-elastic** || quasielastisch || quasi élastique || quasi-elastisch || квазиупругий

Q45 **quasilinear** || quasilinear || quasi linéaire || quasi-lineair || квазилинейный

Q46 **quasilinearization** || Quasilinearisation *f* || quasi-linéarisation *f* || quasi-linearisatie *f* || квазилинеаризация *f*

Q47 **quasi-particle** || Quasiteilchen *n* || quasi-particule *f* || quasideeltje *n* || квазичастица *f*

Q48 **quasi-stationary current** || quasistationärer Strom *m* || courant *m* quasi stationnaire || quasi-stationaire stroom *m* || квазистационарный ток *m*

Q49 **quasi-steady flow** || quasistationäre Strömung *f* || écoulement *m* quasi stationnaire || quasi-stationaire stroming *f* || квазистационарное течение *n*

Q50 **quaternion** || Quaternion *f* || quaternion *m* || quaternion *n* || кватернион *m*

Q51 **quaternion algebra** || Quaternionenalgebra *f* || algèbre *f* de quaternions || algebra *f* (*m*) van quaternionen || алгебра *f* кватернионов

Q52 **quenching of luminescence** || Lumineszenzlöschung *f* || extinction *f* de la luminescence || luminescentiedoving *f* || тушение *n* люминесценции

Q53 **queued access method** || erweiterte Zugriffsmethode *f* || méthode *f* d'accès à queues || toegangsmethode *f* met rijen || метод *m* доступа с очередями

Q54 **queueing theory** || Bedienungstheorie *f*, Warteschlangentheorie *f* || théorie *f* des queues [des files d'attente] || bedieningstheorie *f* || теория *f* массового обслуживания, теория *f* очередей

Q55 **quick-operating** || schnellwirkend, Schnell... || à action rapide || snelwerkend || быстродействующий

Q56 **quick release** || Schnellauslösung *f* || déclenchement *m* rapide [instantané] || sneluitschakeling *f*, sneluitkoppeling *f* || быстрое [мгновенное] выключение *n*; быстрое [мгновенное] расцепление *n*

Q57 **quick release gear** || schnell ausrückendes Getriebe *n* || mécanisme *m* de déclenchement à action rapide [instantanée] || mechanisme *n* met snelontkoppeling || механизм *m* с мгновенным расцеплением

Q58 **quicksilver,** Hg || Quecksilber *n* || mercure *m* || kwik *n*, kwikzilver *n* || ртуть *f*

Q59 **quick test** || Kurzzeitversuch *m* || essai *m* rapide || snelproef *f* (*m*) || ускоренное испытание *n*

Q60 **quill** || **1.** Hülse *f* *(z. B. Spindelhülse)* **2.** Rolle *f* *(Nadellager)* **3.** Hohlwelle *f* || **1.** fourreau *m* (de broche) **2.** roulette *f* *(d'un palier à aiguilles)* **3.** arbre *m* creux || **1.** huls *f* (*m*) *(bijv. spindelhuls)* **2.** rol *f* (*m*) *(naaldlager)* **3.** holle as *f* (*m*) || **1.** гильза *f* *(напр. шпинделя)* **2.** ролик *m* *(игольчатого подшипника)* **3.** полый вал *m*

Q61 **quill drive** || Hohlwellenantrieb *m* || transmission *f* par arbre creux || aandrijving *f* met holle as || передача *f* с полым валом

Q62 **quotient-difference algorithm** || Algorithmus *m* dividierter Differenzen || algorithme *m* de différences (de) quotients || algoritme *n* van gedeelde verschillen || алгоритм *m* разделённых разностей

Q63 **quotient group** || Faktorgruppe *f* || groupe *m* quotient, groupe *m* facteur || factorgroep *f* (*m*) || факторгруппа *f*

Q64 **Q value** || **1.** Gütefaktor *m* **2.** Q-Wert *m* *(Kernreaktor)* || **1.** facteur *m* de qualité **2.** Q-valeur *f* || **1.** kwaliteitsfactor *m* **2.** Q-waarde *f* || **1.** добротность *f* *(контура, конденсатора)* **2.** энергия *f* (ядерной) реакции

R

R1 **race** || **1.** Ring *m* *(Wälzlager)*; Laufbahn *f* *(am Lagerring)* **2.** Einlaufkanal *m* || **1.** chemin *m* de roulement **2.** canal *m* d'amenée || **1.** lagerring *m* *(walslager)*; loopbaan *n* **2.** toevoer- of afvoerkanaal *m* || **1.** кольцо *n*, обойма *f* *(подшипника качения)*; дорожка *f* качения *(на кольце подшипника)* **2.** подводящий канал *m*; отводящий канал *m*

R2 **racemic compound** || razemische Molekülverbindung *f*, Razemverbindung *f*, Razemat *n* || composé *m* racémique, racémate *m* || racemische verbinding *f* || рацемическое соединение *n*, рацемат *m*

R3 **rack** || Zahnstange *f* || crémaillère *f* || tandstang *f* (*m*) || (зубчатая) рейка *f*

R4 **rack jack** || Zahnstangenwinde *f* || cric *m* à crémaillère || dommekracht *f* (*m*) || реечный домкрат *m*

R5 **radar** || Radargerät *n* || radar *m* || radar *m* || радиолокационная станция *f*, РЛС, радиолокатор *m*, радар *m*

R6 **radial bearing** || Radiallager *n* || palier *m* [roulement *m*] radial || radiaallager *n* || радиальный подшипник *m*

R7 **radial clearance** || Radialspalt *m* || jeu *m* radial || radiale spleet *f* (*m*) || радиальный зазор *m*

R8 **radial deviation** || radiale Abweichung *f* || déviation *f* radiale || radiale afwijking *f* || радиальное отклонение *n*

R9 **radial drilling machine** || Radialbohrmaschine *f* || perseuse *f* radiale || radiaalboormachine *f* || радиально-сверлильный станок *m*

R10 **radial error** || radialer Fehler *m* || erreur *f* radiale || radiale fout *f* (*m*) || радиальная ошибка *f*

R11 **radial load** || Radialbelastung *f*, Radiallast *f* || charge *f* radiale || radiale belasting *f* || радиальная нагрузка *f*

R12 **radial-thrust bearing** || Schräglager *n* || palier *m* radial et de butée || radiaaldrukkogellager *n* || радиально-упорный подшипник *m*

R13 **radial velocity** || Radialgeschwindigkeit *f* || vitesse *f* radiale || radiale snelheid *f* || радиальная скорость *f*

R14 **radiance** || Strahlendichte *f* || luminance *f* énergétique || stralendichtheid *f* || энергетическая яркость *f*, лучистость *f*

R15 **radiant emittance** || spezifische Ausstrahlung *f* || radiance *f* énergétique || specifieke uitstraling *f* || энергетическая светимость *f*

R16 **radiant energy** || Strahlungsenergie *f* || énergie *f* de rayonnement; énergie *f* rayonnante || stralingsenergie *f* || энергия *f* излучения; лучистая энергия *f*

R17 **radiant heat** || Strahlungswärme *f* || chauleur *f* radiante [rayonnante] || stralingswarmte *f* || лучистая теплота *f*, теплота *f* лучеиспускания

R18 **radiation** || Strahlung *f*, Ausstrahlung *f* || radiation *f*; rayonnement *m* || straling *f*, uitstraling *f* || излучение *n*; радиация *f*

R19 **radiation belt** || Strahlungsgürtel *m* || ceinture *f* de radiations || stralingsgordel *m* || радиационный пояс *m*

R20 **radiation cooling** || Strahlungsabkühlung *f* || refroidissement *m* par radiation || stralingsafkoeling *f* || охлаждение *n* за счёт излучения; охлаждение *n* излучением

R21 **radiation detector** || Strahlungsdetektor *m* || détecteur *m* de rayonnement || stralingsdetector *m* || детектор *m* излучения

R22 **radiation dose** || Strahlungsdosis *f* || dose *f* de radiation || stralingsdosis *f* || 1. доза *f* излучения 2. доза *f* облучения

R23 **radiation field** || Strahlungsfeld *n* || champ *m* de radiation || stralingsveld *n* || поле *n* излучения

R24 **radiation flux** || Strahlungsfluß *m* || flux *m* de rayonnement [de radiation], flux *m* énergétique || stralingsstroom *m* || поток *m* излучения, лучистый поток *m*

R25 **radiation heating** || Strahlungserwärmung *f* || chauffage *m* par radiation || stralingsverwarming *f*, stralingsverhitting *f* || радиационный нагрев *m*

R26 **radiation-induced** || strahlungsinduziert || induit par radiation || door straling geïnduceerd || радиационно-стимулированный, вызванный облучением

R27 **radiationless transition** || strahlungsloser Übergang *m*, Auger-Übergang *m* || transition *f* sans radiation || stralingsloze overgang *m* || безызлучательный переход *m*, оже-эффект *m*, оже-переход *m*

R28 **radiation loss** || Strahlungsverlust *m* || pertes *f pl* par radiation || stralingsverliezen *n pl* || потери *f pl* *(энергии)* на излучение; радиационные потери *f pl*

R29 **radiation pressure** || Strahlungsdruck *m* || pression *f* de radiation || stralingsdruk *m* || радиационное давление *n*, давление *n* излучения

R30 **radiation protection** || Strahlenschutz *m* || protection *f* contre les radiations || stralingsbescherming *f* || радиационная защита *f*, защита *f* от излучения

R31 **radiation receiver** || Strahlungsempfänger *m* || récepteur *m* de rayonnement || stralingopvanger *m* || приёмник *m* излучения

R32 **radiation shield** || Strahlungsabschirmung *f*; Strahlenabschirmung *f*; Strahlenschutz *m* || protection *f* contre les radiations || bescherming *f* tegen straling || радиационная защита *f*

R33 **radiation-stimulated** *see* **radiation-induced**

R34 **radiation temperature** || Strahlungstemperatur *f* || température *f* de radiation || stralingstemperatuur *f* || радиационная температура *f*

R35 **radiation transfer** || Strahlungstransport *m* || transfert *m* par radiation || stralingstransport *n* || радиационный перенос *m*, перенос *m* излучением

R36 **radiative capture** || Strahlungseinfang *m* *(Neutroneneinfang)* || capture *f* radiative || stralingsvangst *f* || радиационный захват *m* *(нейтрона)*

R37 **radiative transition** || Strahlungsübergang *m* || transition *f* radiative || stralingsovergang *m* || радиационный [излучательный] переход *m*

R38 **radiator** || Strahler *m* || radiateur *m* || straler *m* || излучатель *m*; радиатор *m*

R39 I **radical** || Radikal *n* || radical *m* || radicaal *n* || радикал *m*

R40 II **radical** || radikal || radical || radicaal || радикальный

R41 **radicand** || Radikand *m* || radicande *m* || wortelgrootheid *f* || подкоренное выражение *n*

R42 **radioactive** || radioaktiv || radio-actif || radioactief || радиоактивный

R43 **radioactive chain** || radioaktive Zerfallsreihe *f* || série *f* radioactive || radioactieve reeks *f* (*m*) || цепочка *f* радиоактивных превращений, цепочка *f* распада

R44 **radioactive change** || radioaktive Umwandlung *f* || transformation *f* radio-active || radioactieve omzetting *f* || радиоактивное превращение *n*

R45 **radioactive decay** || radioaktiver Zerfall *m* || désintégration *f* || desintegratie *f*, radioactief verval *n* || радиоактивный распад *m*

R46 **radioactive family** || radioaktive Zerfallsreihe *f* || famille *f* radioactive || radioactieve reeks *f* (*m*) || радиоактивное семейство *f*, радиоактивный ряд *m*

R47 **radioactive isotope** || Radioisotop *n*, radioaktives Isotop *n* || isotope *m* radio-actif || radio-isotoop *m*, radioactieve isotoop *m* || радиоактивный изотоп *m*, радиоизотоп *m*

R48 **radioactive nucleus** || radioaktiver Kern *m* || noyau *m* radioactif || radioactieve kern *f* (*m*) || радиоактивное ядро *n*

R49 **radioactive series** *see* **radioactive family**

R50 **radioactive source** || Aktivitätsquelle *f* || source *f* de radio-activité || bron *f* van radioactiviteit || радиоактивный источник *m*

R51 **radioactivity** || Radioaktivität || radio-activité *f* || radioactiviteit *f* || радиоактивность *f*

R52 **radioactivity heat** || radiogene Wärme *f* || chaleur *f* de la désintégration radio-active || radioactieve warmte *f* || теплота *f* радиоактивного распада

R53 **radio astronomy** || Radioastronomie *f* || radio-astronomie *f* || radiosterrenkunde *f* || радиоастрономия *f*

R54 **radio emission** || Funkwellenausstrahlung *f* || émission *f* radioélectrique || radiogolvenstraling *f* || радиоизлучение *n*, излучение *n* радиоволн

R55 **radio engineering** || Radiotechnik *f* || radiotechnique *f* || radiotechniek *f* || радиотехника *f*

R56 **radio frequency** || Funkfrequenz *f*, Radiofrequenz *f*; Hochfrequenz *f* || radiofréquence *f*, fréquence *f* radio-électrique || radiofrequentie *f* || высокая частота *f*, радиочастота *f*

R57 **radio-frequency** || Hochfrequenz..., HF- || à haute fréquence || hoogfrequent, radiofrequent || высокочастотный, радиочастотный

R58 **radio-frequency generator** || Hochfrequenzgenerator *m* || générateur *m* à haute fréquence || hoogfrequentgenerator *m* || генератор *m* высокой частоты

R59 **radio-frequency radiation** || Hochfrequenzstrahlung *f*, Radiofrequenzstrahlung *f*, Radiostrahlung *f* || rayonnement *m* [émission *f*] radioélectrique || radio(frequentie)straling *f* || радиоизлучение *n*, радиочастотное [высокочастотное] излучение *n*

R60 **radio-frequency range** || Funkfrequenzbereich *m*, Radiofrequenzbereich *m* || gamme *f* de radiofréquences || radiofrequentiebe-reik *n* || диапазон *m* радиочастот, радиодиапазон *n*

R61 **radiography** || Radiographie *f* || radiographie *f* || radiografie *f* || радиография *f*

R62 **radio interferometer** || Radiointerferometer *n* || radio-interféromètre *m* || radio-interferometer *m* || радиоинтерферометр *m*

R63 **radioisotop** || Radioisotop *n*, radioaktives Isotop *n* || radio-isotope *m* || radio-isotoop *m* || радиоактивный изотоп *m*, радиоизотоп *m*

R64 **radio location** || Funkortung *f* || radiodétection *f*, radiorepérage *m* || radiolocatie *f* || радиолокация *f*

R65 **radioluminescence** || Radiolumineszenz *f* || radioluminescence *f* || radioluminescentie *f* || радиолюминесценция *f*

R66 **radiometer** || Radiometer *n*, Strahlenmesser *m* || radiomètre *m* || radiometer *m* || радиометр *m*

R67 **radio noise** || Funkstörungen *f pl* || bruit *m* radio-électrique, parasites *m pl* || radioruis *m*; radiostoringen *f pl* || радиошум *m (космический)*; радиопомехи *f pl*; помехи *f pl* радиоприёму [радиосвязи]

R68 **radionuclide** || Radionuklid *n* || radionucl(é)ide *m* || radionuclide *f* || радиоактивный нуклид *m*, радионуклид *m*

R69 **radiophysics** || Radiophysik *f*, Hochfrequenzphysik *f* || radiophysique *f* || radiofysica *f* || радиофизика *f*

R70 **radiosity** || Strahlungsflußdichte *f* || radiance *f* || stralingsstroomdichtheid *f* || энергетическая светимость *f*

R71 **radiospectroscopy** || Radiospektroskopie *f* || radiospectroscopie *f* || radiospectroscopie *f* || радиоспектроскопия *f*

R72 **radiotelemetry** || Funkfernmessung *f*, Hochfrequenzfernmessung *f* || télémétrie *f*, radiotélémesure *f* || radiotelemetrie *f* || радиотелеметрия *f*

R73 **radiotelescope** || Radioteleskop *n* || radiotélescope *m* || radiotelescoop *m* || радиотелескоп *m*

R74 **radio wave** || Funkwelle *f* || onde *f* radio-électrique || radiogolf *f (m)* || радиоволна *f*

R75 **radium,** Ra || Radium *n* || radium *m* || radium *n* || радий *m*

R76 **radius** || Radius *m*, Halbmesser *m* || rayon *m* || radius *m* || радиус *m*

R77 **radius of convergence** || Konvergenzradius *m* || rayon *m* de convergence || convergentieradius *m* || радиус *m* сходимости

R78 **radius of curvature** || Krümmungshalbmesser *m* || rayon *m* de courbure || krommingsradius *m* || радиус *m* кривизны

R79 **radius of gyration** || Trägheitsradius *m*, Trägheitshalbmesser *m* || rayon *m* de giration || traagheidsradius *m* || радиус *m* инерции

R80 **radius of torsion** || Torsionsradius *m*, Windungsradius *m* || rayon *m* de torsion || torsieradius *m* || радиус *m* кручения

R81 **radius vector** || Radiusvektor *m* || rayon-vecteur *m* || radiusvector *m* || радиус-вектор *m*, вектор *m* положения точки

R82 **radon,** Rn || Radon *n* || radon *m* || radon *n* || радон *m*

R83 **rail mill** || Schienen- und Trägerwalzwerk *n* || train *m* à rails; laminoir *m* à rails || rail- en balkwalserij *f* || рельсобалочный стан *m*

R84 **raising to the... power** || Potenzieren *n* zur n-ten Potenz || élévation *f* à la puissance... || machtsverheffing *f* || возведение *f* в ... степень

R85 **rake** || **1.** Neigungswinkel *m*; Anzugswinkel *m (Keil)*; Spanwinkel *m (Schneidwerkzeug)* **2.** Sägezahnteilung *f* || **1.** pente *f*; angle *m* de pente *(d'un outil coupant)* **2.** pas *m* de la scie || **1.** hellingshoek *m*; spaanhoek *m* **2.** zaagtandsteek *m* || **1.** угол *m* наклона; угол *m* уклона; передний угол *m (режущего инструмента)* **2.** шаг *m* зубьев пилы

R86 **rake angle** || Spanwinkel *m* der Hauptschneide *(Meißel)*; Spanwinkel *m (Schneidwerkzeug)* || angle *m* de dégagement *(d'un outil coupant)* || hoofdspaanhoek *m (bijv. beitel)*; spaanhoek *m* || главный передний угол *m (резца)*; передний угол *m (режущего инструмента)*

R87 **ram** || **1.** Tauchkolben *m* **2.** Supportschlitten *m* **3.** Rammbär *m* || **1.** plongeur *m* **2.** support *m (à chariot)* **3.** dame *f* || **1.** plunger *m* **2.** werktuigdrager *m* **3.** ram *m* || **1.** плунжер *m* **2.** суппорт *m* на салазках **3.** трамбовка *f*

R88 **Raman effect, Raman scattering** || Raman-Effekt *m*, Raman-Streuung *f* || effet *m* Raman || Raman-effect *n*, Raman strooing *n* || эффект *m* Рамана, эффект *m* Ландсберга — Мандельштама — Рамана, комбинационное рассеяние *n* света

R89 **Raman spectrum** || Raman-Spektrum *n* || spectre *m* de Raman || Raman-spectrum *n* || спектр *m* Рамана, спектр *m* комбинационного рассеяния *(света)*

R90 **ramification** || Verzweigung *f* || ramification *f* || vertakking *f* || (раз)ветвление *n (напр. программы)*

R91 **random** || **1.** zufällig **2.** beliebig || aléatoire || **1.** wisselvallig **2.** willekeurig || **1.** случайный **2.** произвольный; непостоянный; изменчивый

R92 **random access** || direkter [wahlfreier] Zugriff *m* || accès *m* aléatoire || willekeurige toegang *m* || произвольный доступ *m*

R93 **random access memory** || Direktzugriffsspeicher *m* || mémoire *f* à accès aléatoire || vrij toegankelijk geheugen *n* || память *f* с произвольным доступом

R94 **random error** || zufälliger Fehler *m*, Zufallsfehler *m* || erreur *f* accidentelle || toevallige fout *f* (*m*) || случайная погрешность *f*

R95 **randomizing** || Randomisierung *f* || randomisation *f*, arrangement *m* au hasard || randomisatie *f* || рандомизация *f*

R96 **random number generator** || Zufallszahlengenerator *m* || générateur *m* de nombres aléatoires || toevalsgetallengenerator *m* || генератор *m* случайных чисел

R97 **random numbers** || Zufallszahlen *f pl* || nombres *m pl* aléatoires || toevalsgetallen *n pl* || случайные числа *n pl*

R98 **random-perturbation optimization** || Optimierung *f* nach der Methode der zufälligen Störungen || optimisation *f* des perturbations aléatoires || optimalisering *f* van toevalsstoringen || метод *m* случайных возмущений

R99 **random process** || Zufallsprozeß *m* || évolution *f* aléatoire || toevalsproces *n* || случайный [стохастический] процесс *m*

R100 **random quantity** || Zufallsgröße *f* || quantité *f* aléatoire || toevalsgrootheid *f* || случайная величина *f*

R101 **random sample** || zufällige Stichprobenentnahme *f* || échantillon *m* aléatoire || toevallige steekproef *f* (*m*) || случайная выборка *f*

R102 **random series** || Zufallsfolge *f* || suite *f* aléatoire || toevalsvolgorde *f* (*m*) || случайная последовательность *f*

R103 **random value** || Zufallswert *m* || valeur *f* aléatoire || toevalswaarde *f* || случайное значение *n*

R104 **random variable** || Zufallsvariable *f* || variable *f* aléatoire || toevalsvariabele *f* || случайная переменная *f*

R105 **random walk** || (stochastische) Irrfahrt *f* || errance *f* || toevalszwerven *n* || случайное блуждание *n*

R106 **range** || **1.** Reichweite *f* **2.** Bereich *m* **3.** Intervall *n* **4.** Gebiet *n* || **1.** rayon *m* d'action **2.** parcours *m* **3.** gamme *f*; intervalle *m* **4.** domaine *m* || **1.** draagwijdte *f* **2.** traject *n* **3.** bereik *n*; interval *n* **4.** gebied *n* || **1.** дальность *f* (*напр. действия*) **2.** длина *f* пробега, пробег *m* **3.** диапазон *m*, интервал *m* **4.** область *f*

R107 **range finder** || Entfernungsmesser *m* || télémètre *m*, distancemètre *m* || afstandsmeter *m* || дальномер *m*

R108 **range of a function** || Definitionsbereich *m* einer Funktion || domaine *m* de détermination d'une fonction || definitiebereik *n* van een functie || область *f* определения функции

R109 **range of a sample** || Stichprobenweite *f* || étendue *f*, amplitude *f* (d'échantillon) || monsteromvang *m* || размах *m* выборки

R110 **range of a variable** || Wertebereich *m* einer Variablen || domaine *m* d'une variable || waardebereik *n* van een variabele || область *f* значений переменной

R111 **range of vision** || Sichtreichweite *f* || profondeur *f* du champ visuel || gezichtsvelddiepte *f* || глубина *f* поля зрения

R112 **rank** || Rang *m* || rang *m* || rang *m* || ранг *m*

R113 **rank of a linear operator** || Rang *m* eines linearen Operators || rang *m* d'un opérateur linéaire || rang *m* van een lineaire operator || ранг *m* линейного оператора

R114 **rank of a matrix** || Rang *m* einer Matrix || rang *m* d'une matrice || rang *m* van een matrix || ранг *m* матрицы

R115 **rank of a representation** || Rang *m* einer Darstellung || rang *m* d'une représentation || rang *m* van een voorstelling || ранг *m* представления

R116 **rank of a tensor** || Rang *m* eines Tensors || rang *m* d'un tenseur || rang *m* van een tensor || ранг *m* тензора

R117 **rare-earth element** || Seltenerdmetall *n* || élément *m* des terres rares || zeldzame-aardmetaal *n* || редкоземельный элемент *m*

R118 **rarefied gas** || verdünntes Gas *n* || gaz *m* raréfié || verdund gas *n* || разрежённый газ *m*

R119 **rare gas** || Edelgas *n*, Inertgas *n* || gaz *m* inerte [noble, rare] || inert [indifferent] gas *n* || инертный [благородный] газ *m*

R120 **ratchet** || Gesperre *n*, Knarre *f*, Ratsche *f* || rochet *m* || palwerk *n* || храповой механизм *m*; стопор *m*

R120a **ratchet wheel** || Sperrad *n*, Klinkenrad *n* || roue *f* à rochet || palrad *n* || храповое колесо *n*

R121 **rate** || Geschwindigkeit *f*; Rate *f* || taux *m*; vitesse *f*, rythme *m* || snelheid *f*, vlugheid *f*, vaardigheid *f* || скорость *f*, быстрота *f*, темп *m* *(изменения, процесса)*

R122 **rated** || Soll..., Nenn... || nominal || berekenings.., nominaal || расчётный; номинальный

R123 **rated capacity** || **1.** Nennkapazität *f* **2.** Nennleistung *f* || **1.** capacité *f* nominale **2.** puissance *f* nominale || **1.** nominale capaciteit *f* **2.** nominaal vermogen *n* || **1.** номинальная ёмкость *f* **2.** номинальная мощность *f*

R124 **rated speed** *see* **nominal speed**

R125 **rate of cutting** || Schnittgeschwindigkeit *f* || vitesse *f* de coupe || snijsnelheid *f* || скорость *f* резания

R126 **rate of decay** || Zerfallsgeschwindigkeit *f* || taux *m* de désintégration || vervalsnelheid *f* || скорость *f* распада

R127 **rate of feed** || Vorschubgeschwindigkeit *f* || vitesse *f* d'avance || aanzet *m*; toevoersnelheid *f* || скорость *f* подачи

R128 **rate of heating** || Erwärmungsgeschwindigkeit *f*; Wärmestufe *f*, Heizstufe *f* || taux *m* de chauffage || verhittingssnelheid *f*; verhittingsgraad *m* || скорость *f* нагрева; степень *f* нагрева

R129 **rating** || Nenndaten *pl*; Entwurfsdaten *pl*; Nennwert *m* || valeur *f* nominale; taux *m*, caractéristique *f* || nominale karakteristiek *f*; nominale grootheid *f* || номинальная характеристика *f*; проектная характеристика *f*; номинальная величина *f*

R130 **ratio** || **1.** Übersetzungsverhältnis *n* **2.** Verhältnis *n*, Proportion *f* **3.** Grad *m*; Koeffizient *m* || **1.** rapport *m* **2.** rapport *m*, rélation *f* **3.** degré *m*; facteur *m* || **1.** overbreng(ings)getal *n* **2.** verhouding *f* **3.** graad *m*; coëfficiënt *m* || **1.** передаточное число *n* **2.** пропорция *f*, соотношение *n* **3.** степень *f*; коэффициент *m*

R131 **ratio of compression** *see* **compression ratio**

R132 **ratio of expansion** *see* **expansion ratio**

R133 **rational** || rational || rationnel || rationeel || рациональный

R134 **rational function** || rationale Funktion *f* || fonction *f* rationnelle || rationele functie *f* || рациональная функция *f*

R135 **rational number** || rationale Zahl *f* || nombre *m* rationnel || rationeel getal *n* || рациональное число *n*

R136 **ray** || Strahl *m* || rayon *m* || straal *m*, *f* || луч *m*

R137 **ray optics** || geometrische Optik *f*, Strahlenoptik *f* || optique *f* des rayons || stralenoptica *f* || лучевая [геометрическая] оптика *f*

R138 **reactance** || Blindwiderstand *m*, Reaktanz *f* || réactance *f* || reactantie *f* || реактивное (электрическое) сопротивление *n*, реактанс *m*

R139 **reaction** || Reaktion *f* || réaction *f* || reactie *f* || реакция *f*

R140 **reaction force** || Reaktion(skraft) *f*, Gegendruck *m* || force *f* de réaction || reactiekracht *f* (*m*) || сила *f* реакции, реакция *f* *(напр. связи, опоры)*

R141 **reaction pressure** || Gegendruck *m* || pression *f* réactive || **1.** tegendruk *m* **2.** reactiedruk *m* || реакгивное давление *n*; противодавление *n*

R142 **reaction turbine** || Überdruckturbine *f*, Reaktionsturbine *f* || turbine *f* à réaction || reactieturbine *f* || реактивная турбина *f*

R143 **reaction velocity** || Reaktionsgeschwindigkeit *f* || vitesse *f* de réaction (chimique) || reactiesnelheid *f* || скорость *f* реакции

R144 **reactive component** || Blindanteil *m*, Blindkomponente *f*, wattlose Komponente *f* || composante *f* réactive || wattloze component *m* || реактивная составляющая *f*

R145 **reactive force** *see* **reaction force**

R146 **reactive power** || Blindleistung *f*, wattlose Leistung *f* || puissance *f* réactive || wattloos vermogen *n*, reactievermogen *n* || реактивная мощность *f*

R147 **reactive torque** || Rückdrehmoment *n* || moment *m* de torsion réactif || reactief draaimoment *n* || реактивный крутящий момент *m*

R148 **reactive voltage** || Blindspannung *f*, wattlose Spannung *f* || tension *f* réactive || wattloze spanning *f* || реактивное напряжение *n*

R149 **reactive volt-ampere** || Var *n*, Blindwatt *n* || var *m* || blindwatt *m* || реактивный вольт-ампер *m*, вар *m*

R150 **reactivity** || Reaktivität *f* || réactivité *f* || reactiviteit *f* || реактивность *f*

R151 **reactor** || 1. Reaktor *m*, Kernreaktor *m* 2. *El.* Reaktanzspule *f*, Reaktanz *f* || 1. réacteur *m* nucléaire 2. *el.* inductance *f* d'arrêt, bobine *f* de choc || 1. kernreactor *m* 2. *el.* reactantiespoel *f* (*m*) || 1. (ядерный) реактор *m* 2. *эл.* реактивная катушка *f*, реактор *m*

R152 **reading** || Lesen *n*, Ablesen *n*, Abfühlen *n* || lecture *f* || lezen *n* || считывание *n*

R153 **read-only storage** || Fest(wert)speicher *m* || mémoire *f* fixe [permanente] || permanent geheugen *n* || постоянная память *f*, память *f* с нестираемой записью

R154 **real axis** || reelle Achse *f* || axe *m* réel || reële as *f* (*m*) || действительная ось *f*

R155 **real focus** || wirklicher Brennpunkt *m* || foyer *m* réel || werkelijk brandpunt *n* || действительный фокус *m*

R156 **real gas** || reales Gas *n*, Realgas *n* || gaz *m* réel || reëel gas *n* || реальный газ *m*

R157 **real image** || wirkliches Bild *n* || image *f* réelle || werkelijk beeld *n* || действительное изображение *n*

R158 **real number** || reelle Zahl *f* || nombre *m* réel || reëel getal *n* || действительное [вещественное] число *n*

R159 **real part** Realteil *m* || partie *f* réelle || reëel deel *n* || действительная часть *f*

R160 **real root** || reelle Wurzel *f* || racine *f* réelle || reële wortel *m* || вещественный корень *m*

R161 **real-time processing** || Echtzeitverarbeitung *f* || fonctionnement *m* [traitement *m*] en temps réel || echttijdbewerking *f* || обработка *f* в реальном [истинном] масштабе времени

R162 **real vector space** || reeller Vektorraum *m* || espace *m* vectoriel réel || reële vectorruimte *f* || действительное векторное пространство *n*

R163 **reamed bolt** || Paßbolzen *m*, Paßschraube *f* || boulon *m* ajusté || pasbout *m* || конусный [призонный] болт *m*

R164 **reamer** || Reibahle *f* || alésoir *m* || (op)ruimer *m* || развёртка *f*

R165 **reaming** || Aufreiben *n*, Aufreibarbeit *f* || alésage *m* || opruimen *n* || развёртывание *n*

R166 **rear focus** || hinterer Brennpunkt *m* || foyer *m* image || achterbrandpunt *n* || задний [второй] фокус *m*

R167 **rearrangement** || Umlagerung *f*, Umgruppierung *f* || réarrangement *m*, transposition *f* || hergroepering *f* || перегруппировка *f*

R168 **Réaumur scale** || Réaumur-Skale *f* || échelle *f* Réaumur || Réaumur-schaal *f* (*m*) || (температурная) шкала *f* Реомюра

R169 **rebound hardness** || Rückprallhärte *f* || dureté au rebondissement ||

terugslaghardheid *f* || твёрдость *f* по отскоку

R170 **receiver** || Empfänger *m* || récepteur *m* || ontvanger *m* || приёмник *m*

R171 **receiving** || Empfang *m* || réception *f* || ontvangst *m* || приём *m*

R172 **receiving loop** || Rahmenantenne *f* || antenne *f* à cadre || raamantenne *f* (*m*) || рамочная антенна *f*

R173 **recessed head** || Kopf *m* (*z. B. einer Schraube*) mit innerer Ausdrehung für Aufsteckschlüssel || tête *f* (*p. ex. du boulon*) à encoche pour clef à douille || kop *m* (*bijv. schroef*) met inwendige uitdraaiing voor imbussleutel *m* || головка *f* (*напр. винта*) с внутренней расточкой под торцевой ключ

R174 **reciprocal** || reziprok || réciproque || reciprook || взаимный; обратный

R175 **reciprocal basis** || reziproke Basis *f* || base réciproque || reciproke basis *f* || обратный базис *m*

R176 **reciprocal differences** || reziproke Differenzen *f pl* || différences réciproques || reciproke differentiën *f pl* || обратные разности *f pl*

R177 **reciprocal milling** || Pendelfräsen *n* || fraisage *m* réciproque [alterné] || pendelfrezen *n* || маятниковое фрезерование *n*

R178 **reciprocal one-to-one correspondence** || eineindeutige Abbildung *f* || correspondance *f* [application *f*] biunivoque || reciproke one-to-one afbeelding *f* || взаимно-однозначное соответствие *n*, взаимно-однозначное отображение *n*

R179 **reciprocal (value)** || Kehrwert *m* || valeur *f* réciproque [inverse] || reciproke waarde *f* || обратное значение *n*, обратная величина *f*

R180 **reciprocal wavelength** *see* **wave number**

R181 **reciprocating engine** *see* **piston engine**

R182 **reciprocating motion** || hin- und hergehende Bewegung *f* || mouvement *m* de va-et-vient [alternatif] || heen-en-weergaande beweging *f* || возвратно-поступательное движение *n*

R183 **reciprocity** || Reziprozität *f* || réciprocité *f* || reciprociteit *f*, wederkerigheid *f* || взаимность *f*

R184 **recoil** || Rückstoß *m*; Rückprall *m* || recul *m* || terugslag *m* || отдача *f*; отскок *m*

R185 **recoil atom** || Rückstoßatom *n* || atome *m* de recul || terugslagatoom *n* || атом *m* отдачи

R186 **recoil energy** || Rückstoßenergie *f* || énergie *f* de recul || terugslagenergie *f* || энергия *f* (частицы) отдачи

R187 **recoil momentum** || Rückstoßimpuls *m* || impulsion *f* de recul || terugslagimpuls *m* || импульс *m* отдачи

R188 **recoil nucleus** || Rückstoßkern *m* || noyau *m* de recul || terugslagkern *f* || ядро *n* отдачи

R189 **recombination** || Rekombination *f* || recombinaison *f* || recombinatie *f* || рекомбинация *f*

R190 **record** || **1.** Aufzeichnung *f* **2.** Satz *m*, Datensatz *m* || enregistrement *m* || registratie *f* || запись *f*

R191 **recrystallization** || Umkristallisieren *n*, Umkristallisation *f*, Rekristallisation *f* || recristallisation *f* || herkristallisatie *f* || перекристаллизация *f*

R192 **rectangular cartesian coordinates** || rechtwinklige [cartesische] Koordinaten *f pl* || coordonnées *f pl* cartésiennes [rectangulaires] || rechthoekige [cartesiaanse] coördinaten *m pl* || прямоугольные [декартовы] координаты *f pl*

R193 **rectangular distribution** || Gleichverteilung *f*, Rechteckverteilung *f* || distribution *f* rectangulaire [uniforme] || rechthoekige verdeling *f* || равномерное [прямоугольное] распределение *n*

R194 **rectangular hyperbola** || gleichseitige Hyperbel *f* || hyperbole *f* équilatère || gelijkzijdige hyperbool *f* (*m*) || равносторонняя гипербола *f*

R195 **rectangular parallelepiped** || Quader *m*, rechtwinkliges Parallelepiped(on) *n* || parallélépipède *m* rectangle || rechthoekig parallelepipedum *n* || прямоугольный параллелепипед *m*

R196 **rectangular pulse** || Rechteckimpuls *m* || impulsion *f* en créneau || rechthoekige impuls *m* || прямоугольный импульс *m*

R197 **rectifiable curve** || rektifizierbare Kurve *f* || courbe *f* rectifiable || rectificeerbare kromme *f* (*m*) || спрямляемая кривая *f*

R198 **rectification** || **1.** Gleichrichtung *f* (*Strom*) **2.** *Math.* Rektifikation *f*, Rektifizierung *f* **3.** *Chem.* Rektifikation *f* || rectification *f*; redressement *m* || **1.** gelijkrichting *f* **2.** *math.* rectificatie *f* **3.** *chem.* rectificatie *f* || **1.** выпрямление *n*, детектирование (*тока*) **2.** *мат.* спрямление *n* (*кривой*) **3.** *хим.* ректификация *f*

R199 **rectification circuit** || Gleichrichterschaltung *f* || circuit *m* redresseur || gelijkrichterschakeling *f* || выпрямительная схема *f*

R200 **rectified current** || gleichgerichteter Strom *m* || courant *m* redressé || gelijkgerichte stroom *m* || выпрямительный ток *m*

R201 **rectifier** || Gleichrichter *m*; Demodulator *m* || redresseur *m* || gelijkrichter *m* || выпрямитель *m*; (электрический) вентиль *m*; детектор *m*

R202 **rectifying barrier** || Sperrschicht *f* (*Halbleiter*) || couche *f* redresseuse || gelijkrichtlaag *f* || выпрямляющий (барьерный) слой *m* (*в полупроводнике*)

R203 **rectifying column** || Rektifizierkolonne *f*, Rektifikationskolonne *f* || colonne *f* de rectification || rectificeerkolom *f* (*m*) || ректификационная колонна *f*

R204 **rectifying plane** || rektifizierende Ebene *f* || plan *m* rectifiant || rectificerend vlak *n* || спрямляющая плоскость *f*

R205 **rectilinear translation** || geradlinige translatorische Bewegung *f* || mouvement *m* de translation rectiligne || rechtlijnige voorwaartsbeweging *f* || прямолинейное поступательное движение *n*

R206 **recurrence relation** || rekurrente Beziehung *f* || relation *f* de récurrence || recurrente relatie *f* || рекуррентное соотношение *n*, рекуррентная формула *f*

R207 **recursion** || Rekursion *f* || récursion *f* || recurrentie *f* || рекурсия *f*

R208 **recursion formula** || Rekursionsformel *f* || formule *f* de récurrence || recurrentieformule *f* (*m*) || рекуррентная [рекурсивная] формула *f*

R209 **recursive** || rekursiv, rekurrent || récursif || recursief || рекурсивный

R210 **recursive algorithm** || rekursiver Algorithmus *m* || algorithme *m* récursif || recursief algoritme *n* || рекурсивный алгоритм *m*

R211 **recursive function** || rekursive Funktion *f* || fonction *f* récursive || recursieve functie *f* || рекурсивная функция *f*

R212 **recursively enumerable set** || rekursiv aufzählbare Menge *f* || ensemble *m* récursivement dénombrable || recursief optelbare menigte *f* || рекурсивно перечислимое множество *n*

R213 **recursively presented** || rekursiv definiert || récursivement présenté || recursief gedefiniëerd || рекурсивно определённый

R214 **recursive partial function** || partiell-rekursive Funktion *f* || fonction *f* partiellement récursive || partiëel-recursieve functie *f* || частично-рекурсивная функция *f*

R215 **recursive set** || rekursive Menge *f* || ensemble *m* récursif || recursieve menigte *f* || рекурсивное множество *n*

R216 **red magnetism** || Nordmagnetismus *m* || magnétisme *m* de nord || noordmagnetisme *n* || северный магнетизм *m*

R217 **redox** || Oxydations-Reduktions-Reaktion *f*, Reduktions-Oxydations-Reaktion *f*, Redoxreaktion *f* || oxydo-réduction *f* || oxydatie-reductiereactie *f* || окисление-восстановление *n*

R218 **redox-potential** || Oxydations-Reduktions-Potential *n*, Redoxpotential *n* || potentiel *m* d'oxydoréduction || redoxpotentiaal *m* ||

редокс-потенциал *m*, окислительно-восстановительный потенциал *m*

R219 **red shift** || Rotverschiebung *f* || déplacement *m* [décalage *m*] vers le rouge || roodverschuiving *f* || красное смещение *n* *(спектра)*

R220 **reduced equation** || reduzierte Gleichung *f* || équation *f* réduite || gereduceerde vergelijking *f* || приведённое уравнение *n*

R221 **reduced length** || reduzierte Länge *f* || longueur *f* réduite || gereduceerde lengte *f* || приведённая длина *f*

R222 **reduced mass** || reduzierte Masse *f* || masse *f* réduite || gereduceerde massa *f* (*m*) || приведённая масса *f*

R223 **reduced temperature** || reduzierte Temperatur *f* || température *f* réduite || gereduceerde temperatuur *f* || приведённая температура *f*

R224 **reducibility** || Reduzierbarkeit *f* || réductibilité *f* || reduceerbaarheid *f* || приводимость *f*

R225 **reducible operator** || reduzibler Operator *m* || opérateur *m* réductible || reducibele operator *m* || приводимый оператор *m*

R226 **reducible representation** || reduzible Darstellung *f* || représentation *f* réductible || reducibele voorstelling *f* || приводимое представление *n*

R227 **reducing valve** || Reduzierventil *n*, Druckminderer *m* || soupape *f* de réduction || reduceerklep *f* (*m*) || редукционный клапан *m*

R228 **reduction** || **1.** Verringerung *f*, Verminderung *f*, Reduzierung *f* **2.** *Phys., Math.* Reduktion *f*, Reduzierung *f* **3.** Reduktion *f* **4.** Untersetzung *f*, Reduktion *f* || réduction *f* || reductie *f* || **1.** понижение *n*, уменьшение *n* **2.** *физ., мат.* приведение *n* **3.** обжатие *n* **4.** понижение *n*, замедление *n* при передаче **5.** *хим.* восстановление *n*; раскисление *n*

R229 **reduction of forces** || Kräftereduktion *f* || réduction *f* des forces || krachtenreductie *f* || приведение *n* сил

R230 **reductor** || Reduziergetriebe *n*, Untersetzungsgetriebe *n*, Getriebe *n*; Geschwindigkeitswandler *m* || réducteur *m*; réducteur *m* de vitesse || versnellingsbak *m*; reductiemechanisme *f* || редуктор *m*; преобразователь *m* скорости

R231 **redundancy** || Redundanz *f* || redondance *f* || redundantie *f* || **1.** избыточность *f* **2.** резервирование *n*; дублирование *n*

R231a **reference** || Referenz *f*, Bezugnahme *f* || référence *f* || referentie *f* || ссылка *f*; обращение *n*

R232 **reference frame** || Bezugssystem *n* || système *m* de référence || referentiesysteem *n* || система *f* отсчёта

R233 **reference-frame velocity** || Bezugssystemgeschwindigkeit *f* || vitesse *f* du système de référence || referentiesysteemsnelheid *f* || переносная скорость *f*

R234 **reference gauge** || Normallehre *f*, Prüflehre *f* || calibre *m* de contrôle; jauge-étalon *m* || sjablonenkaliber *n* || эталонный [контрольный] калибр *m*

R235 **reference plane** *see* **datum plane**

R236 **reference voltage** || Bezugsspannung *f* || tension *f* de référence || referentiespanning *f* || опорное напряжение *n*

R237 **refined oil** || raffiniertes Öl *n* || huile *f* raffinée || geraffineerde olie *f* (*m*) || очищенное масло *n*

R238 **refining** || Raffinieren *n*, Raffination *f*, Reinigung *f* || raffinage *m*, affinage *m* || raffinage *f*, raffinering *f* || рафинирование *n*, очищение *n*

R239 **reflectance** || Reflexionskoeffizient *m*, Reflexionsfaktor *m*; Reflexionsvermögen *n* || coefficient *m* de réflexion, réflectance *f*; pouvoir *m* réfléchissant [de réflexion] || reflectiecoëfficient *m*; reflectievermogen *n* || коэффициент *m* отражения; отражательная способность *f*

R240 **reflected beam** || reflektierter Strahl *m* || faisceau *m* reflecté || gereflecteerde bundel *m* || отражённый пучок *m*, отражённый луч *m*

R241 **reflected impulse** || Echoimpuls *m* || impulsion *f* réfléchie [d'écho]

|| echo-impuls *m* || отражённый импульс *m*

R242 **reflected wave** || reflektierte Welle *f* || onde *f* réfléchie || gereflecteerde golf *f* (*m*) || отражённая волна *f*

R243 **reflecting telescope** || Spiegelteleskop *n*, Reflektor *m* || télescope *m* réflecteur [à miroir] || spiegeltelescoop *m* || телескоп-рефлектор *m*, зеркальный телескоп *m*

R244 **reflection** || Reflexion *f* || réflexion *f* || reflectie *f* || отражение *n*

R245 **reflectivity** || Reflexionsvermögen *n*; Reflexionskoeffizient *m*, Reflexionsfaktor *m* || pouvoir *m* réfléchissant [de réflexion]; coefficient *m* de réflexion, réflectance *f* || reflectievermogen *n*; reflectiecoëfficiënt *m* || отражательная способность *f*; коэффициент *m* отражения

R246 **reflector** || 1. Reflektor *m*, Rückstrahler *m* 2. Reflektor *m*, Spiegelteleskop *n* || 1. réflecteur *m* 2. télescope *m* à miroir [à réflecteur] || 1. reflector *m* 2. spiegeltelescoop *m* || 1. отражатель *m*, рефлектор *m* 2. телескоп-рефлектор *m*, зеркальный телескоп *m*

R247 **reflexive relation** || reflexives Verhaltnis *n* || relation *f* réflexive || reflexieve verhouding *f* || рефлексивное отношение *n*

R248 **reflexivity** || Reflexivität *f* || réflexivité *f* || reflexiviteit *f* || рефлексивность *f*

R249 **reflux** || Rücklauf *m*, Rückfluß *m* || reflux *m* || reflux *m* || 1. флегма *f*; конденсат *m* 2. обратный ток *m*; отток *m*

R249a **reflux ratio** || Rücklaufverhältnis *n*, Rücklaufzahl *f* || rapport *m* [taux *m*] de reflux || refluxcoëfficiënt *m* || 1. коэффициент *m* дефлегмации, флегмовое число *n* 2. коэффициент *m* противотока

R250 **reflux valve** *see* **non-return valve**

R251 **reforming** || Reform(ier)en *n*, Reforming *n* || reforming *m* || reforming *f* || риформинг *m*

R252 **refracted beam** || gebrochener Strahl *m* || faisceau *m* réfracté || gebroken bundel *m* || преломлённый пучок *m*, преломлённый луч *m*

R253 **refracted light** || gebrochenes Licht *n* || lumière *f* réfractée || gebroken licht *n* || преломлённый свет *m*

R254 **refracted wave** || gebrochene Welle *f* || onde *f* réfractée || gebroken golf *f* (*m*) || преломлённая волна *f*

R255 **refracting telescope** || Linsenteleskop *n*, Refraktor *m* || lunette *f*, réfracteur *m* || refractor *m* || телескоп-рефрактор *m*, линзовый телескоп *m*

R256 **refraction** || Brechung *f*; Strahlenbrechung *f*; Refraktion *f* || réfraction *f* || refractie *f* || преломление *n*; рефракция *f*

R257 **refraction angle** || Brechungswinkel *m* || angle *m* de réfraction || brekingshoek *m* || угол *m* преломления

R258 **refraction of extremals** || Extremalenreflexion *f* || réfraction *f* des extrémales || extremalenrefractie *f* || преломление *n* экстремалей

R259 **refractive index** || Brechungsindex *m*, Brechungsfaktor *m*, Brechzahl *f* || indice *m* de réfraction || brekingsindex *m*, refra || (абсолютный) показатель *m* преломления

R260 **refractivity** || Refraktivität *f* || pouvoir *m* réfractif [réfringent] || refractiviteit *f* || преломляющая способность *f*, преломляемость *f*

R261 **refractor** *see* **refracting telescope**

R263 **refrigeration** || 1. Kühlung *f*, Abkühlung *f* 2. Kälteerzeugung *f* || 1. réfrigération *f* 2. production *f* du froid || 1. (af)koeling *f* 2. koudeopwekking *f* || 1. охлаждение *n* 2. производство *n* холода

R264 **refrigerator** || 1. Kältemaschine *f* 2. Refrigerator *m* 3. Kühlschrank *m* || 1. machine *f* frigorifique 2. appareil *m* frigorifique, réfrigérateur *m* 3. réfrigérateur *m* || 1. koelapparaat *n*, koelmachine *f* 2. vriesaggregaat *n* 3. koelkast *f* || 1. холодильная машина *f* 2. рефрижератор *m* 3. холодильник *m*

R265 **regeneration** || Regeneration *f*; Regenerierung *f* || régénération *f* || regeneratie *f* || 1. обратная связь *f* 2. регенерация *f*

R266 **region** || Gebiet *n*, Bereich *m*, Zone *f* || région *f*; zone *f*; domaine *m* || gebied *n*, bereik *n*, zone *f* || область *f*, зона *f*

R267 **register** || Register *n* || registre *m* || register *n* || регистр *m*

R268 **regression** || Regression *f* || régression *f* || regressie *f* || регрессия *f*

R269 **regression coefficient** || Regressionskoeffizient *m* || coefficient *m* de régression || regressiecoëfficiënt *m* || коэффициент *m* регрессии

R270 **regrinding** || Nachschleifen *n*, Nachschliff *m* || raffûtage *m* || bijslijpen *n*; nadraaiing *f* (snijgereedschap) || перешлифовка *f*; перезаточка *f* *(режущего инструмента)*

R272 **regula falsi** || Regula *f* falsi || règle *f* de fausse position, regula *f* falsi || regula *f* falsi || способ *m* ложного положения

R273 **regular** || regulär || régulier || regulair || регулярный; правильный

R274 **regular curve** || reguläre Kurve *f* || courbe *f* régulière || regulaire kromme *f* *(m)* || регулярная кривая *f*

R275 **regular expression** || regulärer Ausdruck *m* || expression *f* régulière || regulaire uitdrukking *f* || регулярное выражение *n*

R276 **regular function** || reguläre [holomorphe] Funktion *f* || fonction *f* régulière || regulaire functie *f* || (аналитическая) регулярная функция *f*, голоморфная функция *f*

R277 **regular point** || regulärer Punkt *m* || point *m* régulier || regulair punt *n* || обыкновенная [правильная, неособая] точка *f*

R278 **regular point of a differential equation** || regulärer Punkt *m* einer Differentialgleichung || point *m* régulier d'une équation différentielle || regulaire punt *f* *(m)* van een differentiaalvergelijking || правильная [неособая] точка *f* дифференциального уравнения

R279 **regular polygon** || regelmäßiges Vieleck *n* || polygone *m* régulier || regulaire veelhoek *m* || правильный многоугольник *m*

R280 **regular polyhedron** || regelmäßiges Polyeder *n* || polyèdre *m* régulier || regulaire polyeder *m* || правильный многогранник *m*

R281 **regular representation** || reguläre Darstellung *f* || représentation *f* régulière || regulaire voorstelling *f* || регулярное представление *n*

R282 **regular set** || reguläre Menge *f* || ensemble *m* régulier || regulaire menigte *f* || регулярное множество *n*

R283 **regular singular point** || regulärer singulärer Punkt *m* || point *m* singulier régulier || regulair singulier punt *n* || регулярная особая точка *f*

R284 **regular surface** || reguläre Fläche *f* || surface *f* régulière || regulair vlak *n* || регулярная поверхность *f*

R285 **regular transformation** || reguläre Transformation *f* || transformation *f* régulière || regulaire transformatie *f* || невырожденное преобразование *n*

R286 **regualtor** || Regler *m*, Regel(ungs)einrichtung *f* || régulateur *m* || regelaar *m* || регулятор *m*, регулирующее устройство *n*

R287 **reheat** || **1.** Nachverbrennung *f*, Nachbrennen *n* **2.** Anwärmen *n* || postcombustion *f* *(du carburant)* **2.** réchauffage *m* || **1.** naverbranding *f* **2.** verhitting *f* || **1.** дожигание *n* *(топлива)*; форсаж *m* **2.** подогрев *m*

R288 **reinforced plastic** || verstärkter Plast *m* || plastique *m* armé || gewapende kunststof *f* *(m)* || армированная пластмасса *f*

R289 **rejection filter** || Sperrfilter *n* || filtre *m* d'arrêt [de blocage] || sperfilter *n* || заграждающий фильтр *m*

R290 **rejection limit** || Ablehnungsgrenze *f* || limite *f* de rejet || verwerpingsgrens *f* *(m)*, rejectiegrens *f* *(m)* || предел *m* отбраковки

R291 **rejection region** || Ablehnungsbereich *m* || région *f* de rejet || rejectiebereik *n* || критическая область *f*

R292 **rejector** *see* **rejection filter**
R293 **relation** || Beziehung *f*, Relation *f* || relation *f* || relatie *f* || отношение *n*
R294 **relational algebra** || Relationsalgebra *f* || algèbre *f* de relations || algebra *f* van relaties || алгебра *f* отношений
R295 **relative** || relativ || relatif || relatief || относительный
R296 **relative address** || relative Adresse *f* || adresse *f* relative || relatief adres *n* || относительный адрес *m*
R297 **relative aperture** || relative Apertur *f* || ouverture *f* relative || relatieve apertuur *f* || относительное отверстие *n*
R298 **relative efficiency** || relativer Wirkunsgrad *m* || efficacité *f* relative, rendement *m* relatif || relatieve werkinsgraad *m* || относительный кпд
R299 **relative error** || relativer Fehler *m* || erreur *f* relative || relatieve fout *f* (*m*) || относительная погрешность *f*; относительная ошибка *f*
R300 **relative frequency** || relative Häufigkeit *f* || fréquence *f* relative || relatieve frequentie *f* || относительная частота *f*
R301 **relative humidity** || relative Feuchtigkeit *f* || humidité *f* relative || relatieve vochtigheid *f* || относительная влажность *f*
R302 **relative index of refraction** || Brechungsverhältnis *n*, relativer Brechungsindex *m* || indice *m* de réfraction relatif || relatieve brekingsindex *m* || относительный показатель *m* преломления
R303 **relative intensity** || relative Intensität *f* || intensité *f* relative || relatieve intensiteit *f* || относительная интенсивность *f*
R304 **relative motion** || Relativbewegung *f*, relative Bewegung *f* || mouvement *m* relatif || relatieve beweging *f* || относительное движение *n*
R305 **relative speed** || relative Geschwindigkeit *f* || vitesse *f* relative || relatieve snelheid *f* || относительная скорость *f*
R306 **relative stability** || relative Stabilität *f* || stabilité *f* relative || relatieve stabiliteit *f* || относительная устойчивость *f*
R307 **relative velocity** *see* **relative speed**
R308 **relativistic correction** || relativistische Korrektion *f* || correction *f* relativiste || relativistische correctie *f* || релятивистская поправка *f*
R309 **relativistic mass equation** || Masse-Geschwindigkeit-Beziehung *f* || relation *f* masse-vitesse || massasnelheidsrelatie *f* || зависимость *f* массы от скорости
R310 **relativistic mechanics** || relativistische Mechanik *f*, Relativitätsmechanik *f* || mécanique *f* relativiste [de relativité] || relativiteitsmechanica *f* || релятивистская механика *f*
R311 **relativistic velocity** || relativistische Geschwindigkeit *f* || vitesse *f* relativiste || relativistische snelheid *f* || релятивистская скорость *f*
R312 **relativity** || **1.** Relativität *f* **2.** Relativitätstheorie *f* || **1.** relativité *f* **2.** théorie *f* de relativité || **1.** relativiteit *f* **2.** relativiteitstheorie *f* || **1.** относительность *f* (*движения*) **2.** теория *f* относительности
R313 **relativity of simultaneity** || Relativität *f* der Gleichzeitigkeit *f* || relativité *f* de la simultanéité || relativiteit *f* van gelijktijdigheid || относительность *f* одновременности
R314 **relativity principle** || Relativitätsprinzip *n* || principe *m* de relativité || relativiteitsprincipe *n* || принцип *m* относительности (*движения*)
R315 **relaxation** || Relaxation *f* || relaxation *f* || relaxatie *f* || релаксация *f*
R316 **relaxation method** || Relaxationsverfahren *n* || méthode *f* de relaxation || relaxatiemethode *f* || релаксационный метод *m*, метод *m* релаксации
R317 **relaxation oscillation** || Relaxationsschwingungen *f pl*, Kippschwingungen *f pl* || oscillations *f pl* de relaxation || relaxatietrillingen *f pl* || релаксационные колебания *n pl*
R318 **relaxation time** || Relaxationszeit *f* || temps *m* de relaxation || rela-

xatietijd *m* || время *n* [период *m*] релаксации

R319 **relay** || Relais *n* || relais *m* || relais *n* || реле *n*

R319a **relaying** || Relaisschutz *m* || protection *f* par relais || relaisbeveiliging *f* || релейная защита *f*

R320 **release** || Auslösung *f* || libération *f* || vrijlating *f* || освобождение *n*, высвобождение *n*

R321 **relief-slotting machine** || Hinterstoßmaschine *f* || mortaiseuse *f* à détalonner || achterstootmachine *f* || долбёжно-затыловочный станок *m*

R322 **relief valve** *see* **safety valve**

R323 **reluctance** || magnetischer Widerstand *m*, Reluktanz *f* || réluctance *f*, résistance *f* magnétique || magnetische weerstand *m*, reluctantie *f* || магнитное сопротивление *n*

R324 **reluctivity** || spezifischer magnetischer Widerstand *m* || réluctance *f* spécifique || soortelijke reluctantie *f*, reluctiviteit *f* || удельное магнитное сопротивление *n*

R325 **remainder** || Rest *m*, Restglied *n* || reste *m*; terme *m* résiduel || rest *f* (*m*), restlid *n* || остаток *m*; остаточный член *m*

R326 **remainder theorem** || Restsatz *m* || théorème *m* des restes || resttheorema *n* || теорема *f* об остатках

R327 **remanence** || Remanenz *f*, remanente Induktion *f* || induction *f* magnétique rémanente [résiduelle] || remanentie *f* || остаточная магнитная индукция *f*

R328 **remanent induction** *see* **remanence**

R329 **remanent magnetism** || Restmagnetismus *m* || magnétisme *m* résiduel [rémanent] || restmagnetisme *n* || остаточный магнетизм *m*, остаточная намагниченность *f*

R330 **remanent magnetization** || Restmagnetisierung *f* || aimantation *f* rémanente [résiduelle] || restmagnetisering *f* || остаточная намагниченность *f*, остаточный магнетизм *m*

R331 **remote control** || Fernsteuerung *f* || télécommande *f* || besturing *f* op afstand || дистанционное управление *n*, телеуправление *n*

R332 **removable singularity** || hebbare Singularität *f* || fausse singularité *f* || ophefbare singulariteit *f* || устранимая особенность *f*

R333 **repetend** || Periode *f* (*eines periodischen Bruches*) || période *f* (*d'une fraction périodique*) || periode *f* (*van een repeterende breuk*) *f* || период *m* (*периодической дроби*)

R334 **repetition** || Wiederholung *f* || répétition *f* || herhaling *f* || повторение *n*

R335 **repetition rate** || Impulsfolgefrequenz *f*, Folgefrequenz *f* || fréquence *f* [taux *m*] de répétition (*des impulsions*) || impulsfrequentie *f* || частота *f* повторения импульсов

R336 **repetition time** || Zyklusdauer *f* || durée *f* du cycle || cyclustijd *m* || период *m*, продолжительность *f* цикла

R337 **representation** || **1.** Darstellung *f* **2.** Abbildung *f* || **1.** représentation *f* **2.** application *f* || **1.** voorstelling *f* **2.** afbeelding *f* || **1.** представление *n* **2.** отображение *n*

R338 **representation of a group** || Darstellung *f* einer Gruppe || représentation *f* d'un groupe || voorstelling *f* van een groep || представление *n* группы

R339 **repulsion** || Abstoßung *f* || répulsion *f* || repulsie *f* || отталкивание *n*

R340 **repulsive force** || Abstoßungskraft *f* || force *f* répulsive || repulsiekracht *f* (*m*) || сила *f* отталкивания

R341 **repulsive interaction** || gegenseitige Abstoßung *f* || répulsion *f* mutuelle || wederzijdse repulsie *f* || (взаимное) отталкивание *n*

R342 **repulsive potential** || Abstoßungspotential *n* || potentiel *m* de répulsion || repulsieve potentiaal *m* || потенциал *m* отталкивания, отталкивательный потенциал *m*

R343 **rerun** || Wiederholung *f*, Wiederanlauf *m* || réexécution *f* || herhaling *f* || повторный прогон *m* (*программы*)

R343a **reseating** || Einschleifen *n* || rodage *m* || inslijpen *n* || притирка *f* (*клапанов*)

R344 **resetting** || Rücksetzen *n* || remise *f* à zéro || terugstelling *f* ||

сброс *m* (на нуль), установка *f* в исходное положение

R345 I **residual** || Rest *m* || résidu *m*, reste *m* || rest *f* (*m*) || остаток *m*

R346 II **residual** || Rest... || résiduel || rest... || остаточный

R347 **residual magnetization** || Restmagnetisierung *f* || aimantation *f* résiduelle [rémanente] || restmagnetisering *f* || остаточная намагниченность *f*

R348 **residual stress** || bleibende Spannung *f*, Restspannung *f* || contrainte *f* résiduelle || restspanning *f* || остаточное напряжение *n*

R349 **residue** || **1.** Rest *m*; Niederschlag *m* **2.** *Math.* Residuum *n* || résidu *m* || residu *n* || **1.** остаток *m*; осадок *m*, отстой *m* **2.** *мат.* вычет *m*

R350 **residue class** || Restklasse *f* || classe *f* des résidus || residu-klasse *f* || класс *m* вычетов

R351 **residue theorem** || Residuensatz *m* || théorème *m* des résidus || residu-theorema *n* || теорема *f* о вычетах

R352 **resistance** || **1.** Widerstand *m* **2.** Wirkwiderstand *m* || **1.** résistance *f* **2.** résistance *f* active [ohmique, pure] || **1.** weerstand *m* **2.** actieve weerstand *m* || **1.** сопротивление *n* **2.** активное [омическое] сопротивление *n*

R353 **resistance box** || Widerstandskasten *m* || boîte *f* de résistances || weerstandsbank *f* (*m*) || магазин *m* резисторов [сопротивлений]

R354 **resistance thermometer** || Widerstandsthermometer *n* || thermomètre *m* à résistance || weerstandsthermometer *m* || термометр *m* сопротивления

R355 **resistance to abrasion** *see* **abrasion resistance**

R356 **resistant welding** || Widerstandsschweißen *n* || soudage *m* par résistance || elektrisch stomplassen *n* || контактная сварка *f*

R357 **resistive component** || Wirkstromkomponente *f*, ohmsche Komponente *f* || composante *f* active [wattée] || ohmse component *m* || активная [омическая] составляющая *f*

R358 **resistivity** || spezifischer Widerstand *m* || résistance *f* spécifique; résistivité *f* || soortelijke weerstand *m* || удельное сопротивление *n*

R359 **resistor** || Widerstand *m* || résistance *f* || weerstand *m* || резистор *m*

R360 **resolution** || **1.** Auflösungsvermögen *n*, Auflösung *f* **2.** Zerlegung *f* (*z. B. eines Vektors in seine Komponenten*) || **1.** pouvoir *m* résolvant [de résolution] **2.** décomposition *f* (*p. ex. d'un vecteur*) || **1.** oplossingsvermogen *n* **2.** ontleding *f* || **1.** разрешающая способность *f*; разрешающая сила *f*; разрешение *n* **2.** разложение *n* (*напр. вектора*)

R361 **resolution time** || Auflösungszeit *f* || temps *m* de résolution || oplossingstijd *m* || разрешение *n* во времени, временнóе разрешение *n*

R362 **resolvent** || Resolvente *f* || résolvante *f* || resolvente *f* || резольвента *f*

R363 **resolvent kernel** || lösender Kern *m* || noyau *m* résolvant || oploskern *f* (*m*) || резольвентное ядро *n*

R364 **resolvent operator** || lösender Operator *m* || opérateur *m* résolvant || oplos-operator *m* || резольвентный оператор *m*

R365 **resolvent set** || Resolventenmenge *f* || ensemble *m* résolvant || resolventenmenigte *f* || резольвентное множество *n*

R366 **resolving power** || Auflösungsvermögen *n* || pouvoir *m* résolvant [de résolution] || oplossingsvermogen *n* || **1.** разрешающая способность *f*, разрешающая сила *f* **2.** *хим.* растворяющая способность *f*

R367 **resonance** || Resonanz *f* || résonance *f* || resonantie *f* || резонанс *m*

R368 **resonance absorption** || Resonanzabsorption *f* || absorption *f* résonnante || resonantieabsorptie *f* || резонансное поглощение *n*

R369 **resonance amplitude** || Resonanzamplitude *f* || amplitude *f* de résonance || resonantieamplitude *f* || амплитуда *f* в резонансе, резонансная амплитуда *f*

R370 **resonance curve** || Resonanzkurve *f* || courbe *f* de résonance ||

resonantiekromme *f (m)* || резонансная кривая *f*

R371 **resonance frequency** || Resonanzfrequenz *f* || fréquence *f* de résonance || resonantiefrequentie *f* || резонансная частота *f*

R372 **resonance oscillation** || Resonanzschwingungen *f pl* || oscillations *f pl* de résonance || resonantietrillingen *f pl* || резонансные колебания *n pl*

R373 **resonance particle** || Resonanzteilchen *n* || particule *f* de résonance || resonantiedeeltje *n* || резонансная частица *f*, резонон *m*

R374 **resonator** || Resonator *m* || résonateur *m* || resonator *m* || резонатор *m*

R375 **response** || **1.** Antwort *f*; Reaktion *f* **2.** Ansprechen *n* **3.** Frequenzgang *m*; Kennlinie *f*, Kurve *f* || réponse *f* || **1.** antwoord *m*; reactie *f* **2.** karakteristiek *f* || **1.** отклик *m*; срабатывание *n* **2.** характеристика *f*

R376 **response time** || Ansprechzeit *f* || temps de réponse || aanspreektijd *m* || время *n* отклика [срабатывания] *(прибора)*

R377 **rest** || **1.** Auflage *f*, Stütze *f*; Anschlag *m* **2.** Support *m (Werkzeugmaschine)* **3.** Ruhe *f*, Ruhezustand *m* || **1.** appui *m*; butée *f* **2.** support *m* **3.** repos *m*, état *m* de repos || steun *m*; steunpunt *n* **2.** support *n* **3.** rust *f (m)* rusttoestand *m* || **1.** опора *f*; упор *m* **2.** суппорт *m* **3.** покой *m*, состояние *n* покоя

R378 **restart** || Wiederanlauf *m* || reprise *f* || opnieuw aanzetten *n* || рестарт *m*, перезапуск *m (программы)*

R379 **rest energy** || Ruh(e)energie *f* || énergie *f* de repos || rustenergie *f* || энергия *f* покоя

R380 **rest mass** || Ruhemasse *f* || masse *f* de repos || rustmassa *f (m)* || масса *f* покоя

R381 **restoring force** || Richtkraft *f*, Richtgröße *f*, Rückstellgröße *f* || force *f* de rappel || terugkeerkracht *f (m)* || возвращающая сила *f*

R382 **rest position** || Ruhestellung *f* || position *f* de repos || ruststelling *f* || положение *n* покоя

R383 **restraint, restriction** || Einschränkung *f* || restriction *f* || beperking *f* || ограничение *n*

R384 **result** || Resultat *n*, Ergebnis *n* || résultat *m* || resultaat *n*, uitkomst *f* || результат *m*

R385 **resultant** *math.* || Resultante *f* || résultant *m* || resultante *f (m)* || результант *m*

R386 **resultant, resultant force** || Resultierende *f*, Resultante *f*, resultierende Kraft *f* || résultante *f* || resultante *f (m)* || результирующая [равнодействующая] сила *f*, равнодействующая *f*

R387 **resultant velocity** || resultierende Geschwindigkeit *f* || vitesse *f* résultante || resulterende snelheid *f* || результирующая скорость *f*

R 388 **retardation** || Verzögerung *f* || retardation *f* || vertraging *f* || замедление *n*

R389 **retarded potential** || retardiertes Potential *n* || potentiel *m* retardé || geretardeerde potentiaal *m* || запаздывающий потенциал *m*

R390 **retarding field** || Bremsfeld *n* || champ *m* retardateur || remveld *n* || тормозящее [замедляющее] поле *n*

R391 **retarding force** || Bremskraft *f* || effort *m* de freinage || remkracht *f (m)* || тормозящая [замедляющая] сила *f*

R392 **retrieval** || Wiederauffindung *f*, Recherche *f* || recherche *f* || recherche *f (m)*, naspeuring *f* || поиск *m (информации)*

R393 **retroreflection** || katadioptrische Reflexion *f*, Reflexreflexion *f* || retroréflexion *f* || reflexreflectie *f* || обратное отражение *n*, отражение *n* назад

R394 **return** || Rückkehr *f* || retour *m* || terugkeer *m* || возврат *m*

R395 **return travel** || Rück(wärts)gang *m*, Rück(wärts)lauf *m*, Rückwärtsbewegung *f* || course *f* retour [rétrograde] || teruggang *m* || обратный ход *m*

R396 **return valve** || Rückschlagventil *n* || soupape *f* de retour || terugslagklep *f (m)* || обратный клапан *m*

R397 **reverberation** || Nachhall *m* || réverbération *f* || reverberatie *f*, terugkaatsing *f* || реверберация *f*

R398 **reversal load** || alternierende Belastung *f* || charge *f* alternative || wisselende belasting *f* || (знако)переменная нагрузка *f*

R399 **reversal of charge** || Ladungsumkehr *f* || recharge *f* || ladingsomkeer *m* || перезарядка *f*, перемена *f* знака заряда

R400 **reversal of spectrum lines** || Linienumkehr *f*, Umkehr *f* der Spektrallinien || renversement *m* des raies || lijnenomkeer *m* || обращение *n* спектральных линий

R401 **reversed image** || 1. umgekehrtes Bild *n* 2. Negativ *n* || 1. image *f* renversée 2. négatif *m* || 1. omgekeerd beeld *n* 2. negatief *n* || 1. перевёрнутое изображение *n* 2. негативное изображение *n*, негатив *n*

R402 **reverse gear** || Umsteuergetriebe *n*, Reversiervorrichtung *f* || mécanisme *m* inverseur [de renversement de marche], engrenage *m* réversible || omkeermechanisme *n* || реверс *m*, реверсивный механизм *m*

R403 **reverse motion** || Rückwärtsbewegung *f* || mouvement *m* de retour; mouvement inverse || teruggaande beweging *f* || обратное [возвратное] движение *n*

R404 **reverser** *see* **reverse gear**

R405 **reverse voltage** || Sperrspannung *f* || tension *f* inverse || tegenspanning *f* || обратное напряжение *n*, противоэлектродвижущая сила *f*, противоэдс *f*

R406 **reversibility** || Umkehrbarkeit *f* || réversibilité *f* || omkeerbaarheid *f* || обратимость *f*

R407 **reversible cycle** || reversibler Kreisprozeß *m* || cycle *m* réversible || omkeerproces *n* || обратимый цикл *m*

R408 **reversible pendulum** || Reversionspendel *n*, Umkehrpendel *n* || pendule *m* réversible || omkeerpendel *n* || оборотный маятник *m*

R409 **reversible process** || reversibler [umkehrbarer] Prozeß *m* || processus *m* réversible || omkeerproces *n* || обратимый процесс *m*

R410 **reversible transformation** || reversible [umkehrbare] Umwandlung *f* || transformation *f* réversible || reversibele [omkeerbare] omvorming *f* || обратимое превращение *n*

R411 **reversing** || Bewegungsumkehr *f*, Umsteuerung *f*, Reversierung *f* || renversement *m* (*de marche*) || verandering *f* van de bewegingsrichting, omkering *f* || изменение *n* направления (движения), реверсирование *n*

R412 **reversing engine** || Umsteuermotor *m*, Reversiermotor *m* || moteur *m* réversible || omkeermotor *m* || реверсивный (электро)двигатель *m*

R413 **reversing shaft** || Rücklaufwelle *f* || arbre *m* de renversement || as *f* (*m*) van achteruit || вал *m* заднего хода

R414 **reversion** || 1. Umsteuerung *f* 2. *Math.* Umkehrung *f*, Umkehr *f* || réversion *f* || omkering *f* || 1. реверсирование *n* 2. *мат.* обращение *n*

R415 **reversion of series** || Reihenumkehrung *f* || réversion *f* de série || reeksomkering *f* || обращение *n* ряда

R416 **revolution** || 1. Drehung *f*, Rotation *f*; Umlauf *m* 2. Umdrehung *f* || rotation *f*, révolution *f* || draaiing *f* || 1. вращение *n*; обращение *n* 2. оборот *m*

R417 **revolutions per minute** || Umdrehungszahl *f* je Minute || (nombre de) révolutions par minute || omwentelingen *f pl* per minuut || частота *f* вращения, число *n* оборотов в минуту

R418 **rheology** || Rheologie *f*, Fließkunde *f* || rhéologie *f* || reologie *f* || реология *f*

R419 **rheonomic constraint** || rheonome Bedingung *f* || liaison *f* rhéonome || reonome verbinding *f* || реономная [нестационарная] связь *f*

R420 **rheostat** || Rheostat *m*, Regelwiderstand *m* || rhéostat *m* || reostaat *m* || реостат *m*

R421 **Riemannian geometry** || Riemannsche Geometrie *f* || géométrie *f* riemannienne || meetkunde *f* van Riemann || риманова геометрия *f*

R422 **Riemann space** || Riemannscher Raum *m* || espace *m* riemannien || ruimte *f* van Riemann || пространство *n* Римана, риманово пространство *n*

R423 **right continuous function** ||

rechtsseitig stetige Funktion *f* || fonction *f* continue à droite || rechtszijdige repeterende functie *f* || непрерывная справа функция *f*

R424 **right-hand derivative** || rechtsseitige Ableitung *f* || dérivée *f* à droite || rechtszijdige afleiding *f* || правая производная *f*, производная *f* справа

R425 **right-handed coordinate system** || Rechtssystem *n*, rechtshändiges Koordinatensystem *n* || système *m* de coordonnées à droite || rechtse coördinatenstelsel *n* || правая система *f* координат

R426 **right-hand rule** || Rechte-Hand-Regel *f*, Dreifingerregel *f* der rechten Hand || règle *f* de la main droite || rechterhandregel *m* || правило *n* правой руки, «правило *n* буравчика»

R426a **right-hand screw** || rechtsgängige Schraube *f* || vis *f* à filet droit || schroef *f* (*m*) met rechtse draad || винт *m* с правой резьбой

R427 **right-hand thread** || Rechtsgewinde *n* || filet *m* [filetage *m*] à droite || rechtse schroefdraad *m* || правая резьба *f*

R428 **righting moment** || aufrichtendes [rückdrehendes] Moment *n*, Aufrichtmoment *n*, Rückdrehmoment *n* || moment *m* de redressement || terugdraaiend moment *n* || выпрямляющий [восстанавливающий] момент *m*

R429 **right parallelepiped** || Quader *m*, rechtwinkliges Parallelepiped(on) *n* || parallélépipède *m* droit || rechthoekig parallelepipedum *n* || прямой параллелепипед *m*

R430 **right triangle** || rechtswinkliges Dreieck *n* || triangle *m* rectangle || rechthoekige driehoek *m* || прямоугольный треугольник *m*

R431 **rigid** || starr || rigide || stijf || жёсткий, недеформируемый

R432 **rigid body** || starrer Körper *m* || corps *m* rigide, solide *m* || stijf lichaam *n* || абсолютно жёсткое тело *n*

R433 **rigid-body dynamics** || Dynamik *f* starrer Körper || dynamique *f* du corps rigide || dynamica *f* van stijf lichaam || динамика *f* твёрдого тела, динамика *f* неизменяемых систем

R434 **rigidity** || **1.** Steifheit *f*, Steifigkeit *f* **2.** Formelastizität *f* || **1.** rigidité *f* **2.** élasticité *f* de forme || **1.** hardheid *f* **2.** vormelasticiteit *f* || **1.** жёсткость *f* **2.** упругость *f* на сдвиг, упругость *f* второго рода

R435 **rigidity modulus** *see* **modulus of elasticity in shear**

R436 **rigid rotator** || starrer Rotator *m* || rotateur *m* rigide || stijve rotator *m* || жёсткий ротатор *m*

R437 **rigid top** || starrer Kreisel *m* || toupie *f* rigide || stijve tol *m* || жёсткий волчок *m*

R438 **rim** || **1.** Felge *f*; Rand *m*; Bord *m* **2.** Zahnkranz *m* || **1.** jante *f*; rebord *m*, couronne *f* **2.** couronne *f* dentée [à denture] || **1.** krans *f* (*m*); rand *m* **2.** tandkrans *f* (*m*) || **1.** обод *m*; закраина *f*; реборда *f* **2.** зубчатый венец *m*

R439 **ring** || Ring *m*; Bügel *m* || anneau *m*, bague *f* || ring *m*; krans *f* (*m*); beugel *m* || кольцо *n*; обод *m*; хомут *m*

R440 **ring gauge** || Lehrring *m*, Ringlehre *f* || bague-étalon *f* || kaliberring *m* || кольцевой калибр *m*, измерительное кольцо *n*

R441 **ring lubrication** || Ringschmierung *f* || lubrification *f* [graissage *m*] à bague || ringsmering *f* || кольцевая смазка *f*

R442 **ring spring** || Ringfeder *f* || ressort *m* en anneau || ringveer *f* (*m*) || кольцевая пружина *f*

R443 **ring with identity, ring with unity** || Ring *m* mit Einheit || anneau *m* unitaire || ring *m* met eenheid || кольцо *n* с единицей

R444 **ripple** || **1.** Kapillarwellen *f pl*, Kräuselwellen *f pl* **2.** Pulsation *f*, Pulsierung *f* (*Strom u. ä.*) || **1.** rides *f pl*, ondes *f pl* [rides] capillaires **2.** pulsation(s) *f* (*pl*) || **1.** rimpeling *f* **2.** pulseren *n* || **1.** рябь *f*, капиллярные волны *f pl* **2.** пульсация *f* (*напр. тока*)

R445 **ripple tank** || Wellenküvette *f* || cuvette *f* à ondes || golvencuvet *n* || волновая кювета *f*

R446 **ripple voltage** || Kräuselspannung *f*, pulsierende Spannung *f* || tension *f* pulsée || pulserende spanning *f* || пульсирующее напряжение *n*

R447 **rivet** || Niet *m* || rivet *m* || klinknagel *m* || заклёпка *f*

R448 **riveter** || Nietmaschine *f*; Niethammer *m* || riv(et)euse *f*; marteau-riveur *m* || klinkmachine *f*; klinkhamer *m* || клепальная машина *f*; клепальный молоток *m*

R449 **rivet head** || Nietkopf *m* || tête *f* de rivet || klinknagelkop *m* || головка *f* заклёпки

R450 **riveting** || Nieten *n*, Nietung *f* || rivetage *m* || klinken *n* || клёпка *f*

R451 **rocker** || Schwinghebel *m*, Kipphebel *m*; Schwinge *f*, Wippe *f* || basculeur *m* || hefboom *m*; balansarm *m*; tuimelaar *m* || коромысло *n*; балансир *m*

R452 **rocker arm** || Ventilkipphebel *m* || levier *m* de soupape || kleptuimelaar *m* || коромысло *n* клапана

R453 **rocket dynamics** || Raketendynamik *f* || balistique *f* des fusées || raketdynamica *f* || ракетодинамика *f*

R454 **rocket engine, rocket motor** || Raketentriebwerk *n* || moteur-fusée *m* || raketmotor *m* || ракетный двигатель *m*

R455 **rocket-powered** || raketengetrieben || propulsé par moteur-fusée || met raketmotor || с ракетным двигателем

R456 **Rockwell hardness test** || Härteprüfung *f* nach Rockwell, Rockwell-Härteprüfung *f* || essai *m* de dureté Rockwell || Rockwell-test *m* || определение *n* твёрдости по Роквеллу

R457 **rod** || Zugstange *f*; Stange *f*; Stock *m*; Stab *m*, Schaft *m* || tige *f*, tringle *f*, bielle *f*; barre *f* || stang *f* (*m*); trekstang *f* (*m*); staaf *f* (*m*) || тяга *f*; шток *m*; штанга *f*; стержень *m*

R458 **roentgen** || Röntgen *n*, Röntgeneinheit *f* || rœntgen *m* || röntgen *f* (*m*) || рентген *m* *(2,58 10^{-4} Кл/кг)*

R459 **roentgenoluminescence** || Röntgenlumineszenz *f* || radioluminescence *f*, luminescence *f* X || röntgenluminescentie *f* || рентгенолюминесценция *f*

R460 **roll** || Rolle *f*; Welle *f*; Walze *f* || galet *m*, rouleau *m* || rol *f* (*m*); wals *f* (*m*) || ролик *m*; валик *m*; валок *m*

R461 **rolled steel sections** || Formstahl *m*, Profilstahl *m*; Walzstahl *m* || acier *m* laminé; profilés *m pl* en acier || gewalst profielstaal *n*; gewalste staalprofielen *n pl* || прокатная фасонная сталь *f*; стальной прокат *m*

R462 **rolled thread** || gewalztes [gerolltes] Gewinde *n* || filet *m* roulé [laminé] || gewalste schroefdraad *m* || накатанная резьба *f*

R463 **roller** || Rolle *f*; Walze *f*; Trommel *f*, Zylinder *m* || galet *m*, rouleau *m*, cylindre *m* || rol *f* (*m*); wals *f* (*m*); trommel *f* (*m*) || ролик *m*; валик *m*; барабан *m*

R464 **roller bearing** || Rollenlager *n* || palier *m* à rouleaux || rollager *n* || роликоподшипник *m*, роликовый подшипник *m*

R465 **roller chain** || Rollenkette *f* || chaîne *f* à rouleaux || rollenketting *m* (*f*) || втулочно-роликовая цепь *f*

R466 **rolling** || **1.** Rollen *n*; Wälzen *n* **2.** Walzen *n* *(bei Druckumformen von Metall)* || **1.** roulement *m* **2.** laminage *m* || **1.** rollen *n* **2.** walsen *n* || **1.** качение *n* **2.** прокатка *f* *(металла)*

R467 **rolling bearing** || Wälzlager *n* || roulement *m*. palier *m* à roulement || walslager *n* || подшипник *m* качения

R468 **rolling friction** || Rollreibung *f* || frottement *m* de roulement || rolwrijving *f*, rollende wrijving *f* || трение *n* качения

R469 **rolling mill** || Walzwerk *n*, Walzstraße || laminoir *m*, train *m* || walserij *f* || прокатный стан *m*

R470 **rolling surface** || Wälzfläche *f* || surface *f* de roulement || walsoppervlakte *f* || поверхность *f* качения

R471 **roll squeezers** || Vorwalzen *f pl* || cylindre *m* dégrossisseur [ébaucheur] || samendrukwals *f* (*m*), samenperswals *f* (*m*) || обжимные вальцы *pl*

R472 **roll turning lathe** || Walzendrehmaschine *f* || tour *m* à cylindres ||

walsdraaibank *f* *(m)* || вальцетокарный станок *m*

R473 **ROM** *see* **read-only storage**

R474 **room acoustics** || Raumakustik *f* || acoustique *f* de salle || ruimteakoestiek *f* || геометрическая акустика *f*, акустика *f* помещений

R475 **room temperature** || Raumtemperatur *f* || température *f* ambiante || ruimtetemperatuur *f* || комнатная температура *f*

R476 **root** || **1.** Lücke *f*, Rille *f* *(Gewindeprofil)* **2.** Grund *m*, Fuß *m* *(eines Zahns)* **3.** Schweißnahtwurzel *f* || **1.** creux *m* *(du filet)* **2.** pied *m* [racine *f*] *(de la dent)* **3.** racine *f* de la soudure || **1.** gleuf *f* *(m)* (schroefdraadprofiel) **2.** tandvoet *m* *(tandwiel)* **3.** naadwortel *m* || **1.** впадина *f* *(профиля резьбы)* **2.** основание *n* *(зуба шестерни)* **3.** корень *m* (сварного) шва

R477 **root-mean-square deviation, root-mean-square error** || mittlere quadratische Abweichung *f*, mittlerer quadratischer Fehler *m*, Standardfehler *m*, Standardabweichung *f* || erreur *f* moyenne quadratique || standaardafwijking *f* *(m)*, standaardfout *f* *(m)* || среднеквадратичное [стандартное] отклонение *n*

R478 **root-mean-square value** || quadratischer Mittelwert *m* || valeur *f* quadratique moyenne || kwadratisch gemiddelde *n* || среднее квадратическое (значение) *n*

R479 **root of an equation** || Wurzel *f*, Nullstelle *f* *(einer Gleichung)* || racine *f* d'une équation || wortel *m*, nulpunt *n* *(van een vergelijking)* || корень *m* уравнения

R480 **rotary casting** *see* **centrifugal casting**

R481 **rotary compressor** || Rotationskompressor *m*, Drehkolbenverdichter *m* || compresseur *m* rotatif || roterende compressor *m* || ротационный компрессор *m*

R482 **rotary engine** || Umlauf(kolben)motor *m* || moteur *m* rotatif || roterende motor *m* || роторный двигатель *m*

R483 **rotary pump** || Rotationspumpe *f*, Drehkolbenpumpe *f* || pompe *f* rotative || rotatiepomp *f* *(m)* || роторный насос *m*

R484 **rotary table** || Drehtisch *m*, Schalttisch *m* *(Werkzeugmaschine)* || table *f* pivotante || draaitafel *f* *(m)* *(werktuigmachine)* || поворотный стол *m* *(станка)*

R485 **rotary transformer** || **1.** Gleichstromumformer *m* **2.** Drehtransformator *m* || **1.** convertisseur *m* alternatif-continu **2.** transformateur *m* rotatif || **1.** gelijkstroom-wisselstroomomvormer *m* **2.** draaitransformator *m* || **1.** двигатель-генератор *m*, умформер *m* **2.** вращающийся трансформатор *m*; фазорегулятор *m*

R486 **rotated-planewave** || Welle *f* mit rotierender Polarisationsebene || onde *f* à plan de polarisation rotatif || golf *f* *(m)* met roterend polarisatievlak || вращающаяся волна *f*, волна *f* с вращающейся плоскостью поляризации

R487 **rotating coordinate system** || rotierendes Koordinatensystem *n* || système *m* de coordonnées rotatif || roterend coördinatenstelsel *n* || вращающаяся система *f* координат

R488 **rotation** || **1.** Drehung *f*, Drehbewegung *f*, Rotation *f*, Drehung *f* **2.** *Math.* Rotation *f* || **1.** rotation *f* **2.** *math.* rotationnel *m* || **1.** draaiing *f* **2.** *math.* rotatie *f* || **1.** вращение *n*, вращательное движение *n*; поворот *m* **2.** *мат.* ротор *m*

R489 **rotational** || Rotations..., Dreh... || rotatif, giratoire || rotatie..., draai... || вращательный

R490 **rotational energy** || Rotationsenergie *f* || énergie *f* de rotation || rotatie-energie *f* || энергия *f* вращательного движения

R491 **rotational flow** || Wirbelströmung *f* || écoulement *m* tourbillonnaire || wervelstroming *f* || вихревое течение *n*

R492 **rotational moment** || Drehmoment *n*, Rotationsmoment *n* || couple *m* moteur, moment *m* de rotation || draaimoment *n*, rotatiemoment *n* || вращающий момент *m*

R493 **rotational motion** *see* **rotation 1.**

R494 **rotational spectrum** || Rotationsspektrum *n* || spectre *m* de rotation || rotatiespectrum *n* || вращательный спектр *m*

R495 **rotational speed** || Drehzahl *f* || vitesse *f* de rotation || omwentelingssnelheid *f* || частота *f* вращения

R496 **rotational symmetry** || Rotationssymmetrie *f* || symétrie *f* de rotation || rotatiesymmetrie *f* || вращательная симметрия *f*

R497 **rotational velocity** *see* **rotational speed**

R498 **rotation axis** || Drehachse *f*, Rotationsachse *f* || axe *m* de rotation || rotatieas *f* (*m*) || ось *f* вращения

R499 **rotation group** || Drehgruppe *f* || groupe *m* de rotations || rotatiegroep *f* (*m*) || группа *f* вращений

R500 **rotation of plane of polarization** || Polarisationsdrehung *f*, Drehung *f* der Polarisationsebene || rotation *f* de plan de la polarisation || draaiing *f* van polarisatievlak *n* || вращение *n* плоскости поляризации

R501 **rotation-reflection group** || Dreh-Reflexionsgruppe *f* || groupe *m* de rotations-réflections || draaireflectiegroep *f* (*m*) || группа *f* вращений-отражений

R502 **rotator** || Rotator *m* || rotateur *m* || rotator *m* || ротатор *m*

R503 **rotor** || 1. Laufrad *n*, Läufer *m* *(Turbine)* 2. Tragschraube *f* || rotor *m* || 1. rotor *m*; looprad *n*, loopwiel *n* *(turbine)* 2. rotor *m* || 1. ротор *m*; рабочее колесо *n* *(турбины)* 2. несущий винт *m* *(вертолёта)*

R504 **rotor blade** || 1. Lauf(rad)-schaufel *f* 2. Tragschraubenschaufel *f* || aile *f* du rotor || rotorblad *n*; loopschoep *f* (*m*) || 1. (рабочая) лопатка *f* *(ротора, рабочего колеса)* 2. лопасть *f* несущего винта

R505 **rough grinding machine** || Grobschleifmaschine *f* || machine *f* à ébarber à meule || ruwslijpmachine *f*, grofslijpmachine *f* || обдирочно-шлифовальный станок *m*

R506 **roughing cutter** || Schruppfräser *m*, Vorfräser *m*; Zahnradvorfräser *m* || fraise *f* d'ébauche || voordraaifrees *f* (*m*); voorfrees *f* (*m*); tandradvoorfrees *f* (*m*) || обдирочная фреза *f*; фреза *f* для чернового нарезания зубчатых колёс

R507 **roughing lathe** || Schrupp-drehmaschine *f* || tour *m* à ébaucher || voorschaafdraaibank *f* (*m*) || обдирочный токарный станок *m*

R508 **roughing-out** || Schruppen *n*, Grobbearbeitung *f* || ébauchage *m*, ébauche *f* || grove verwerking *f* || черновая обработка *f*. обработка *f* начерно

R509 **roughing tool** || Schruppmeißel *m* || outil *m* dégrossisseur [de dégrossissage]; lame *f* à ébaucher || voordraaibeitel *m*, voorschaafbeitel *m* || обдирочный резец *m*

R510 **roughness** || Rauh(ig)keit *f* *(Oberfläche)* || rugosité *f* *(de la surface)* || ruwheid *f* *(oppervlakte)* || шероховатость *f* *(поверхности)*

R511 **rough surface** || rauhe Oberfläche *f* || surface *f* rugueuse || ruw oppervlak *n* || шероховатая поверхность *f*

R512 **rough vacuum** || Grobvakuum *n* || vide *m* faible [bas] || laag vacuüm *n* || низкий вакуум *m*

R513 **round bar** || Rundstab *m* || rond *m*, barre *f* ronde || rondstaaf *f* (*m*) || брус(ок) *m* круглого сечения

R514 **round head** || Halbrundkopf *m* || tête *f* ronde || halfronde kop *m* || полукруглая головка *f* *(напр. винта)*

R515 **rounding** || Runden *n*, Rundung *f* || arrondissement *m* || ronding *f*, afronding *f* || 1. округление *n* 2. закругление *n*

R516 **round-off error** || Rundungsfehler *m* || erreur *f* d'arrondissement || afrondingsfout *f* (*m*) || ошибка *f* округления

R517 **round thread** || Rundgewinde *n* || filet *m* rond || ronde schroefdraad *m* || круглая резьба *f*

R518 **routine** (program) || Routine *f* || routine *f*, programme *m* || routine *(programma)* || (стандартная) программа *f*

R519 **row matrix** || Zeilenmatrix *f* || matrice *f* ligne || regelmatrijs *f* (*m*) || матрица-строка *f*

R520 **rubber washer** || Gummischeibe *f* || rondelle *f* en caoutchouc ||

rubberonderlegplaat *f* *(m)* || резиновая шайба *f*

R521 **rubbing surface** *see* **friction surface**

R522 **Ruhmkorff coil** || Ruhmkorffscher Funkeninduktor *m*, Ruhmkorff-Induktor *m* || bobine *f* de Ruhmkorff || Ruhmkorff-spoel *f* *(m)* || катушка *f* Румкорфа, индукционная катушка *f*

R523 **ruled surface** || Regelfläche *f* || surface *f* réglée || regelvlak *n*, regeloppervlak *n* || линейчатая поверхность *f*

R524 **rule of correspondence** || Korrespondenzregel *f* || règle *f* de correspondance || overeenkomstregel *m* || правило *n* соответствия

R525 **rule of inference** || Ableitungsregel *f* *(in einem formalen System)* || règle *f* d'inférence || afleidingsregel *m* || правило *n* вывода *(в формальной системе)*

R526 **run** || Durchlauf *m*, Programmdurchlauf *m* || exécution *f* || programmadoorloop *m* || прогон *m* *(программы)*

R527 **running block** || lose Rolle *f* || poulie *f* baladeuse [courante] || beweeglijke rol *f* *(m)* || подвижной блок *m* *(полиспаста)*

R528 **running fit** || Laufsitz *m*, Spielpassung *f* || ajustement *m* avec jeu || looppassing *f*, gangpassing *f*; bewegingspassing *f* || подвижная посадка *f*

R529 **rupture** || Zerreißen *n*, Reißen *n*, Riß *m*; Bruch *m* || rupture *f* || breuk *f* *(m)* || разрыв *m*; разрушение *n*; излом *m*

R530 **rusting** || Korrosion *f*, Rostung *f*, Verrosten *n* || enrouillement *m*, rouillage *m*; corrosion *f* || roesten *n*; corrosie *f*; aantasting *f* || ржавление *n*, коррозия *f*

R531 **rydberg** || Ridberg *n* || rydberg *m* || rydberg *m* || ридберг *m* *(13,6 эВ)*

S

S1 **saddle** || **1.** Schlitten *m*; Support *m* 2. Ventilsitz *m* || **1.** chariot *m*, porte-outil *m*, support *m* **2.** siège *m* *(de soupape)* || **1.** slede *f* *(m)*; support *n* **2.** zadel *m*, *n* || **1.** салазки *pl*; суппорт *m* **2.** седло *n* *(клапана)*

S2 **saddle point** || Sattelpunkt *m* || point *m* selle, point (de) col || zadelpunt *n* || седловая точка *f*, седло *n*

S3 **saddle-point method** || Sattelpunktmethode *f* || méthode *f* du point (de) col || zadelpuntmethode *f* || метод *m* перевала

S4 **safety bolt** || Sicherungsschraube *f* || boulon *m* de sécurité || veiligheidsbout *m*, borgbout *m* || стопорный болт *m*

S5 **safety clutch** || Sicherheitskupplung *f*, Überlastungskupplung *f* || accouplement *m* à friction de sécurité, accouplement de sûreté || veiligheidsmof *f* *(m)* || предохранительная муфта *f*

S6 **safety factor** || Sicherheitsfaktor *m*, Sicherheitsgrad *m* || coefficient *m* de sécurité || veiligheidscoëfficiënt *m* || коэффициент *m* запаса (прочности)

S7 **safety interlock** || selbsttätige Sperre *f* || blocage *m* de sécurité || zelfblokkering *f*, automatische blokkering *f* || автоблокировка *f*

S8 **safety stop** || Sicherungssperre *f*; Sicherheitsanschlag *m* || butée *f* fixante, arrêt *m* de sûreté || veiligheidsstop *m* || предохранительный останов *m*; предохранительный упор *m*

S9 **safety valve** || Sicherheitsventil *n* || soupape *f* de sûreté || veiligheidsklep *f* *(m)* || предохранительный клапан *m*; аварийный вентиль *m*

S10 **salt** || Salz *n* || sel *m* || zout *n* || соль *f*

S11 **sample** || **1.** Muster *n*; Probe *f*, Probestück *n* **2.** Stichprobe *f* **3.** Schablone *f* || **1.** échantillon *m*, témoin *m* **2.** gabarit *m*, modèle *m* || **1.** monster *m*; proef *f* *(m)*, staal *n* **2.** steekproef *f* *(m)* **3.** sjabloon *f* *(m)* || **1.** образец *m*; проба *f* **2.** выборка *f* **3.** шаблон *m*

S12 **sample average** || Stichprobenmittel *n* || moyenne *f* d'échantillon || gemiddelde steekproefwaarde *f* || среднее выборочное *n*

S13 **sample dispersion** || Stichprobenstreuung *f* || variance *f* d'échantillonnage || steekproefafwijking *f* || выборочное стандартное отклонение *n*

S14 **sample distribution** || Stichprobenverteilung *f* || distribution *f* d'échantillonnage || steekproefverdeling ||выборочное распределение *n*

S15 **sample function** || Stichprobenfunktion *f* || fonction *f* d'échantillon || steekproeffunctie *f* || выборочная функция *f*, функция *f* выборки

S16 **sample mean** *see* **sample average**

S17 **sample quantile** || Stichprobenquantil *n* || quantile *m* d'échantillon || steekproefquantile *n* || выборочная квантиль *f*

S18 **sample range** || Stichprobenweite *f* || amplitude *f* d'échantillon || steekproefwijdte *f* || размах *m* выборки

S19 **sample size** || Stichprobenumfang *m* || taille *f* d'échantillon || steekproefomvang *m* || объём *m* выборки

S20 **sample space** || Stichprobenraum *m* || espace *m* d'échantillons || steekproefruimte *f* || пространство *n* выборок

S21 **sample value** || Stichprobenwert || valeur *f* d'échantillon || steekproefwaarde *f* || выборочное значение *n*

S22 **sample variance** || Stichprobenvarianz *f* || variance *f* d'échantillon || steekproefvariantie *f* || выборочная дисперсия *f*

S23 **sampling** || **1.** Probenahme *f* **2.** Stichprobenerhebung *f* || échantillonnage *m* || steekproefkeuze *f* || **1.** отбор *m* [взятие *n*] проб **2.** выбор *m*; взятие *n* выборки

S24 **sampling theorem** || Abtasttheorem *n* || théorème *m* d'échantillonnage || aftasttheorema *n* || теорема *f* о дискретном представлении, теорема *f* Котельникова

S25 **sand blasting** || Sandstrahlen *n* || sablage *m*, grésage *m* || zandstraling *f* || пескоструйная обработка *f*

S26 **sand casting** || Sand(form)guß *m* || moulage *m* en sable || zandgieting *f* || литьё *n* в песчаные формы

S27 **sanding** || **1.** Sandreinigung *f* **2.** Sandstrahlen *n* || **1.** décapage *m* par émeri, ponçage *m* **2.** sablage *m*, grésage *m* || **1.** schuren *(met schuurpapier)* **2.** zandstralen *n* || **1.** очистка *f (напр. шкуркой)* **2.** пескоструйная очистка *f*

S28 **satellite** || Satellit *m*; Sputnik *m* || satellite *m* || satelliet *m* || (искусственный) спутник *m*

S29 **saturated vapour** || Sattdampf *m* || vapeur *f* saturée || verzadigde stoom *m* || насыщенный пар *m*

S30 **saturation** || Sättigung *f* || saturation *f* || verzadiging *f* || насыщение *n*

S31 **saturation current** || Sättigungsstrom *m* || courant *m* de saturation || verzadigingsstroom *m* || ток *m* насыщения

S32 **saturation induction** || Sättigungsinduktion *f* || induction *f* de saturation || verzadigingsinductie *f* || (магнитная) индукция *f* насыщения

S33 **saturation magnetization** || Sättigungsmagnetisierung *f* || aimantation *f* de saturation || verzadigingsmagnetisering *f* || намагниченность *f* насыщения

S34 **sawing machine** || mechanische Säge *f*, Sägemaschine *f* || scie *f* mécanique, machine *f* à scier || zaagmachine *f* || механическая [приводная] пила *f*, пильный станок *m*

S35 I **scalar** || Skalar *m* || scalaire *m* || scalair *n* || скаляр *m*

S36 II **scalar** || skalar || scalaire || scalair || скалярный

S37 **scalar field** || Skalarfeld *n*, skalares Feld *n* || champ *m* scalaire || scalair veld *n* || скалярное поле *n*

S38 **scalar potential** || skalares Potential *n* || potentiel *m* scalaire || scalaire potentiaal *m* || скалярный потенциал *m*

S39 **scalar product** || Skalarprodukt *n* || produit *m* scalaire || scalair produkt *n* || скалярное произведение *n*

S40 **scale factor** || Skalenfaktor *m* || facteur *m* d'échelle || schaalfactor *m* || масштабный множитель *m*

S41 **scanning** || Abtasten *n* || balayage *m* || scanning *m*, zwaai *m* || сканирование *n*; развёртка *f*

S42 **scanning beam** || Abtast(elektronen)strahl *m* || faisceau *m* explorateur || zwaaistraal *m*, *f* || сканирующий [развёртывающий] луч *m*

S43 **scanning electron microscope** || Rasterelektronenmikroskop *n*, Abtastmikroskop *n* || microscope *m* électronique à trame || rasterelektronenmicroscoop *m* || растровый электронный микроскоп *m*

S44 **scattered beam** || Streustrahl *m* || faisceau *m* diffusé || verstrooide bundel *m* || рассеянный пучок *m*

S45 **scattered light** || Streulicht *n* || lumière *f* diffuse || verstrooid licht *n* || рассеянный свет *m*

S46 **scattered wave** || Streuwelle *f* || onde *f* diffusée || verstrooide golf *f* (*m*) || рассеянная волна *f*

S47 **scattering** *f* || Streuung *f* || diffusion *f* || verstrooiing *f* || **1.** рассеяние *n* **2.** разброс *m*

S48 **scattering factor** || Streufaktor *m* || coefficient *m* de dispersion || verstrooiingsfactor *m* || коэффициент *m* рассеяния

S49 **scattering medium** || Streumedium *n* || milieu *m* diffusant [dispersif, dissipatif] || verstrooiingsmedium *n* || рассеивающая среда *f*

S50 **scattering power** || Streuvermögen *n* || pouvoir *m* diffusant || verstrooiingsvermogen *n* || рассеивающая способность *f*

S51 **scavenging pressure** || Spüldruck *m* || pression *f* de soufflage || uitblaasdruk *m* || давление *n* продувки

S52 **schematic diagram** || Prinzipschaltbild *n*, Prinzipschema *n* || schéma *m* de principe || principeschema *n* || принципиальная схема *f*

S53 **scintillation** || Szintillation *f* || scintillation *f* || scintillatie *f* || сцинтилляция *f*

S54 **scintillation counter** || Szintillationszähler *m* || compteur *m* à scintillation || scintillatieteller *m* || сцинтилляционный счётчик *m*

S55 **sclerometer** || Sklerometer *n*, Härtemesser *m* || scléromètre *m* || sclerometer *m* || склерометр *m*

S56 **scleronomic constraint** || skleronome Bindung *f* || liaison *f* scléronomiqke || scleronome binding *f* || склерономная [стационарная] связь *f*

S57 **scleroscope** || Skleroskop *n*, Rückprallhärteprüfer *m* || scléroscope *m* || scleroskoop *m* || склероскоп *m*

S58 **scleroscope hardness test** || Skleroskophärteprüfung *f* || essai *m* de dureté au scléroscope || scleroskoophardheid(s)proef *f* (*m*) || определение *n* твёрдости склероскопом

S59 **scratch hardness** || Ritzhärte *f* || dureté *f* à la rayure [par striage] || krashardheid *f* || твёрдость *f* по методу царапания

S60 **screen** || Schirm *m* || écran *m* || scherm *n* || экран *m*

S61 **screening** || Abschirmung *f* || blindage *m* || afscherming *f* || экранирование *n*

S62 **screw** || Schraube *f*; Holzschraube *f* || vis *f* || schroef *f* (*m*) || винт *m*; шуруп *m*

S64 **screw coupling** || Verschraubung *f*, Schraubenkupplung *f* || tendeur *m* à vis || schroefkoppeling *f* || винтовая стяжка *f*

S65 **screw cutting** || Gewindeschneiden *n* || filetage *m* || schroefdraadsnijden *n* || нарезание *n* (винтовой) резьбы

S66 **screw-cutting lathe** || Zeit- und Zugspindeldrehmaschine *f* || tour *m* à fileter || leischroefdraaimachine *f* (*m*) || токарно-винторезный станок *m*

S67 **screw die** || Gewindekluppe *f*; Gewindeschneidkopf *m*; Schneidbakke *f* || filière *f*, cage *f* de filière || (schroef-)draadsnijblok *n* || клупп *m*; винтонарезная головка *f*; нарезная плашка *f*

S68 **screw gear** || Schraubenrad *n* || roue *f* hélicoïdale || schroefwiel *n* || косозубое колесо *n*

S69 **screw gearing** || Schraubenrädergetriebe *n* || engrenage *m* hélicoïdale || schroefwieloverbrenging *f* || винтовая зубчатая передача *f*, червячная передача *f*

S70 **screw jack** || Schraubenwinde *f* || cric *m* à vis || vijzel *f* (*m*), vijzeldommekracht *f* (*m*) || винтовой домкрат *m*

S71 **screw machine** *see* **screw-cutting lathe**

S72 **screw pitch** || Gewindesteigung *f* || pas *m* du filet [de filetage] || schroefdraadspoed (*m*) || шаг *m* (винтовой) резьбы

S73 **screw press** || Spindelpresse *f*, Schraubenpresse *f* || presse *f* à vis || schroefpers *f* (*m*) || винтовой пресс *m*

S74 **screw shackle** *see* **turnbuckle**

S75 **screw tap** || Gewindebohrer *m*, Schneidbohrer *m* || taraud *m* || snijtap *m*, schroeftap *m* || метчик *m*

S76 **screw thread** || Schraubengewinde *n* || filet *m*; filetage *m* || schroefdraad *m* || (винтовая) резьба *f*

S77 **screw thread cutting machine** || Gewindeschneidmaschine *f* || fileteuse *f*, machine *f* à fileter || schroefdraadsnijmachine *f* (*m*) || резьбонарезной станок *m*

S78 **screw thread milling machine** || Gewindefräsmaschine *f* || machine *f* à fraiser les filets || schroefdraadfreesmachine *f* || резьбофрезерный станок *m*

S79 **screw thread rolling machine** || Gewinderollmaschine *f*, Gewindewalzmaschine *f* || machine *f* à rouler les filets [à fileter à la molette ou par roulage] || schroefdraadwalsmachine *f* || резьбонакатный станок *m*

S80 **screw tool** || Gewindeschneidwerkzeug *n* || outil *m* à fileter [à tarauder] || schoefdraadsnijwerktuig *n* || резьбонарезной инструмент *m*

S81 **screw vice** || Schraubstock *m* || étau *m* à vis || schroefklem *f* (*m*) || винтовые тиски *pl*

S82 **scroll chuck** || Universal-Spannfutter *n* || mandrin *m* (de serrage) universel, douille *f* de serrage universelle || universele klauwplaat *f* (*m*) || универсальный зажимной патрон *m*

S83 **scuffing wear** || Verschleiß *m* durch Gleiten || usure *f* des surfaces en contact glissant || slijtage *f* bij glijding || износ *m* при скольжении

S84 **seal** || Dichtung *f*; Stopfbuchse *f* || garniture *f* d'étanchéité || pakking *f*; pakkingbus *f* (*m*) || уплотнение *n*; сальник *m*

S85 **sealing** || Abdichtung *f*; Hermetisierung *f* || étanchement *m*, hermétisation *f* || afdichting *f* || уплотнение *n*; герметизация *f*

S86 **seamless pipe** || nahtloses [ganzgezogenes] Rohr *n* || tube *m* sans soudure || naadloze buis *f* (*m*) || бесшовная [цельнотянутая] труба *f*

S87 **seam welding** || Rollennahtschweißen *n* || soudage *m* à la molette || naadlassen *n* || шовная [роликовая] контактная сварка *f*

S88 **search** || Suche *f*; Recherche *f* || recherche *f* || recherche *f* (*m*) || поиск *m*

S89 **seat** || Sitz *m* (*Ventil*); Auflagefläche *f*, Anlagefläche *f* || siège *m*, assise *f* || zadel *m*, *n*; klepzitting *f*; steunvlakte *f*; kussen *n* || седло *n*, гнездо *n* (*клапана*); опорная поверхность *f*

S90 **second** || Sekunde *f* || seconde *f* || seconde *f* (*m*) || секунда *f*

S91 I **secondary** || **1.** Sekundärwicklung *f* **2.** Sekundärteilchen *n* || **1.** enroulement *m* secondaire **2.** particule *f* secondaire || **1.** secondaire wikkeling *f* **2.** secondair deeltje *n* || **1.** вторичная обмотка *f* **2.** вторичная частица *f*

S92 II **secondary** || Sekundär..., sekundär || secondaire || secondair, sekundair || вторичный

S93 **secondary electron** || Sekundärelektron *n* || électron *m* secondaire || secondair elektron *n* || вторичный электрон *m*

S94 **secondary electron emission** || Sekundärelektronenemission *f* || émission *f* (d'électrons) secondaire || secondaire-elektronenemissie *f*, emissie *f* van secondaire elektronen || вторично-электронная эмиссия *f*

S95 **secondary electron multiplier** || Sekundärelektronenvervielfacher *m*, SEV || multiplicateur *m* (d'émission secondaire) électronique || secondaire-elektronenvermenigvuldiger *m* || вторично-электронный умножитель *m*

S96 **secondary emission** *see* **secondary electron emission**

S97 **secondary particle** || Sekunderteilchen *n* || particule *f* secondaire ||

secondair deeltje *n* || вторичная частица *f*

S98 **secondary voltage** || Sekundärspannung *f* || tension *f* secondaire || secondaire spanning *f* || напряжение *n* во вторичной обмотке

S99 **secondary winding** || Sekundärwicklung *f* || enroulement *m* secondaire || secondaire wikkeling *f* || вторичная обмотка *f*

S100 **section** || **1.** Querschnitt *m*; Profil *n*; Schnitt *m* **2.** Sektion *f*; Abschnitt *m* || section *f*, coupe *f*, profil *m* || doorsnede *f* (*m*); snede *f* (*m*); profiel *n* **2.** sectie *f* || **1.** сечение *n*; разрез *m*, профиль *m* **2.** секция *f*; участок *m*

S101 **section modulus** || Widerstandsmoment *n* || module *m* d'inertie; module *m* de résistance || weerstandsmoment *n* || момент *m* сопротивления

S102 **sector** || Sektor *m* || secteur *m* || sector *m* || сектор *m*

S103 **sector of a circle** || Kreissektor *m*, Kreisausschnitt *m* || secteur *m* circulaire || cirkelsector *m* || круговой сектор *m*

S104 **sector of a sphere** || sphärischer Sektor *m* || secteur *m* sphérique || bolsector *m*, sferische sector *m* || сферический сектор *m*

S105 **secular equation** || sekuläre Gleichung *f* || équation *f* séculaire || seculaire vergelijking *f* || вековое уравнение *n*

S106 **seed crystal** || Kristallkeim *m*, Imfkristall *m* || germe *m* de cristallisation || kristallisatiekern *f* (*m*) || затравочный кристалл, затравка

S107 **segment** || **1.** Segment *n*, Abschnitt *m* **2.** Strecke *f* || segment *m* || segment *n* || **1.** сегмент *m* **2.** отрезок *m*, сегмент *m*

S108 **segmental chip** *see* **discontinuous chip**

S110 **segment of a curve** || Kurvensegment *n* || segment *m* d'une courbe || segment *n* van een kromme || отрезок *m* кривой

S111 **segregation** || Segregation *f*, Trennung *f*, Scheidung *f* || séparation *f* || segregatie *f* || разделение *n*; сегрегация *f*

S112 **selection rule** || Auswahlregel *f* || règle *f* de sélection || keuzeregel *m* || правило *n* отбора

S113 **selective reflection** || selektive Reflexion *f* || réflexion *f* sélective || selectieve reflectie *f* || избирательное [селективное] отражение *n*

S114 **selectivity** || Selektivität *f* || sélectivité *f* || selectiviteit *f* || селективность *f*, избирательность *f*

S115 **selenium,** Se || Selen *n* || sélénium *m* || seleen *n* || селен *m*

S116 **self-acting** || automatisch || automatique || zelfwerkend, automatisch || автоматический

S117 **self-adjoint operator** || selbstadjungierter Operator *m* || opérateur *m* auto-adjoint || zelf-geadjungeerde operator *m* || самосопряжённый оператор *m*

S118 **self-adjusting** || selbstregelnd, selbstregulierend; selbsteinstellend || à réglage automatique, à autoréglage, à autorégulation || zelf-instellend || саморегулирующийся, самоустанавливающийся; самонастраивающийся

S119 **self-aligning bearing** || Pendellager *n* || palier *m* à réglage automatique || zelf-instellend kussenblok *n* || самоустанавливающийся подшипник *m*

S120 **self-centering chuck** || selbstzentrierendes Futter *n* || mandrin *m* à serrage automatique || zelfcentrerende klauwplaat *f* (*m*) || самоцентрирующий (зажимной) патрон *m*

S121 **self-centring** || Selbstzentrierung *f* || à cintrage *m* automatique || zelfcentrering *f* || самоцентрирование *n*

S122 **self-conjugate operator** *see* **self-adjoint operator**

S123 **self-cooled transformer** || Trockentransformator *m* || transformateur *m* refroidi par air [à refroidissement par air] || luchtgekoelde transformator *m* || трансформатор *m* с воздушным охлаждением

S124 **self-diffusion** || Selbstdiffusion *f* || autodiffusion *f* || zelfdiffusie *f* || самодиффузия *f*

S125 **self-energizing** || Selbsterregung *f* || auto-excitation *f* || zelf-opwekking *f* || самовозбуждение *n*

S126 **self-excited oscillation** || selbsterregte Schwingungen *pl* || oscillations *pl* auto-excitées || self-geïnduceerde trillingen *pl* || автоколебания *pl*

S127 **self-excited vibration** *see* **self-excited oscillation**

S128 **self-focusing** || Selbstfokussierung *f* || autofocalisation *f* || zelffocussering *f* || самофокусировка *f*

S129 **self-hardening steel** *see* **air-hardening steel**

S130 **self-induced vibrations** *see* **self-excited oscillation**

S131 **self-inductance** || Selbstinduktion *f*, Selbstinduktivität *f* || self-inductance *f* || zelfinductie *f* || индуктивность *f* *(количественная характеристика явления самоиндукции)*

S132 **self-induction** || **1.** Selbstinduktion *f* 2. Selbstinduktivität *f* || **1.** auto-induction *f* 2. self-inductance *f* || zelfinductie *f* || **1.** самоиндукция *f* 2. индуктивность

S133 **self-locking bolt** || selbstsperrende Schraube *f* || boulon *m* à verrouillage automatique, boulon *m* à autoserrage || zelfklemmende [zelfstoppende] bout *m* || самостопорящийся болт *m*

S134 **self-maintained discharge** || selbständige Entladung *f* || décharge *f* auto-entretenue [autonome] || zelfstandige ontlading *f* || самостоятельный разряд *m*

S135 **self-modulation** || Selbstmodulation *f* || automodulation *f* || zelfmodulatie *f* || автомодуляция *f*

S136 **self-oscillations** *see* **self-excited oscillation**

S137 **self-osculation point** || Rückkehrpunkt *m* || point *m* d'auto-osculation || terugkeerpunt *f* *(m)* || точка *f* возврата

S138 I **self-programming** || Selbstprogrammierung *f* || autoprogrammation *f* || zelfprogrammering *f* || автоматическое программирование

S139 II **self-programming** || selbstprogrammierend || autoprogrammeur *m* || zelfprogrammerend || самопрограммирующийся

S140 **self-regulation** || Selbstregelung *f*, automatische Regelung *f* || réglage *m* automatique || zelfregeling *f* || автоматическое регулирование *n*

S141 **self-sealing** || selbstdichtend || à étanchement automatique || zelfdichtend || самоуплотняющийся

S142 **self-sharpening** || Selbstschärfungs... || à auto-affûtage; à affûtage automatique || zelfscherpend || самозатачивающийся

S143 **self-similarity** || Selbstähnlichkeit *f* || self-similarité *f* || zelfgelijksoortigheid *f* || автомодельность *f*

S144 **self-sustaining** || selbständig || auto-entretenu, autonome || zelfstandig || самоподдерживающийся, самостоятельный *(напр. о разряде)*

S145 **self-tapping screw** || selbstschneidende Schraube *f* || vis *f* d'autotaraudage (assurant taraudage du trou pendant le vissage) || zelftappende schroef *f* *(m)*|| самонарезающий винт *m*

S146 **Sellers screw thread** || Sellers-Gewinde *n* || filet *m* Sellers || schroefdraad *m* van Sellers, amerikaanse standaardschroefdraad *m* || резьба *f* Селлерса, американская нормальная резьба *f* *(с углом профиля 60°)*

S147 **selsyn** || Drehmelder *m*, Drehfeldgeber *m* || selsyn *m* || selsyn *n* || сельсин *m*

S148 **semiconductor** || Halbleiter *m* || semi-conducteur *m* || halfgeleider *m* || полупроводник *m*

S149 **semiconductor rectifier** || Halfleitergleichrichter *m* || redresseur *m* à semi-conducteur || halfgeleidergelijkrichter *m* || полупроводниковый выпрямитель *m*

S150 **semiconvergence** || Semikonvergenz *f* || semi-convergence *f* || halfconvergentie *f* || полусходимость *f*

S151 **semidefinite matrix** || semidefinite Matrix *f* || matrice *f* semi-définie || halfgedefiniëerde matrix *f* || полуопределённая матрица *f*

S152 **semi-elliptical spring** *see* **cantilever spring**

S153 **semi-invariant** || Semiinvariante *f* || semi-invariant *m* || halfinvariante *f* || семиинвариант *m*

S154 **semikilled steel** || halbberuhigter Stahl *m* || acier *m* semi-calmé || halfgedesoxigeerd staal *n* || полуспокойная сталь *f*

S155 **semi-linear set** || halblineare Menge *f* || ensemble *m* semi-linéaire || halflineaire menigte *f* || полулинейное множество *n*

S156 **semimetal** || Halbmetall *n* || semi-métal *m* || halfmetaal *n* || полуметалл *m*

S157 **semi-polar bond** || semipolare Bindung *f* || liaison *f* semi-polaire || semipolaire binding *f* || семиполярная связь *f*

S158 **semitransparent** || halbdurchsichtig; halbdurchlässig || semi-transparent || halfdoorzichtig || полупрозрачный

S159 **sense of rotation** || Drehsinn *m*, Drehrichtung *f* || sens *m* de rotation || draairichting *f* || направление *n* вращения

S160 **sensitive** || empfindlich || sensible || sensibel, gevoelig || чувствительный

S161 **sensitivity** || Empfindlichkeit *f* || sensibilité *f* || sensibiliteit *f*, gevoeligheid *f* || чувствительность *f*

S162 **sensitivity coefficient** || Empfindlichkeitszahl *f*, Empfindlichkeitsfaktor *m* || coefficient *m* de sensibilité || gevoeligheidscoëfficiënt *m* || коэффициент *m* чувствительности

S163 **sensitization** || Sensibilisierung *f* || sensibilisation *f* || sensibilisatie *f* || сенсибилизация *f*

S164 **sensitometry** || Sensitometrie *f* || sensitométrie *f* || sensitometrie *f* || сенситометрия *f*

S165 **sensor** || Sensor *m* || élément *m* sensible [de perception]; capteur *m* || sensor *m*, gevoelig element *n* || чувствительный элемент *m*, сенсор *m*

S166 **separable kernel** || separabler Kern *m* || noyau *m* séparable || scheidbare kern *f* (*m*) || сепарабельное ядро *n*

S167 **separable space** || separabler Raum *m* || espace *m* séparable || scheidbare ruimte *f* || сепарабельное пространство *n*

S168 **separated flow** || abgelöste Strömung *f* || courant *m* séparé || afgescheiden stroming *f* || отрывное течение *n*

S169 **separate oiling** || Einzelschmierung *f* || graissage *m* séparé || lokale smering *f* || местная смазка *f*

S170 **separation** || 1. Trennung *f*; Separation *f* 2. Intervall *n*, Abstand *m* 3. Anreicherung *f*; Aufbereitung *f* || 1. séparation *f* 2. intervalle *m*, distance *f* || 1. scheiding *f*; separatie *f* 2. interval *n*, afstand *m* 3. veredeling *f* || 1. разделение *n*; отделение *n*; сепарация *f* 2. отрыв 3. интервал *m*; расстояние *n*; разнос *m* 4. обогащение *n*

S171 **separation of variables** || Trennung *f* von Variablen || séparation *f* des variables || scheiding *f* van variabelen || разделение *n* переменных

S172 **sequence** || Folge *f* || séquence *f* || volgorde *f* (*m*) || последовательность *f*

S173 **sequential test** || Sequentialtest *m* || test *m* séquentiel || sequentiële test *m* || последовательный критерий *m*

S174 **series** || 1. Serie *f* 2. Reihe *f* || série *f* || 1. serie *f* 2. reeks *f* (*m*) || 1. серия *f* 2. ряд *m*

S175 **series circuit** || Serien(strom)kreis *m* || circuit *m* en série || serieschakeling *f* || последовательная цепь *f*, последовательный контур *m*

S176 **series-parallel** || Reihenparallel..., Serienparallel ... || sérieparallèle || serie-parallel || последовательно-параллельный, смешанный (*о включении*)

S177 **series-parallel winding** || Reihenparallelwicklung *f*, Serienparallelwicklung *f* || enroulement *m* série-parallèle || serie-parallelwikkeling *f* || последовательно-параллельная обмотка *f*

S178 **series winding** || Reihen(schluß)wicklung *f* || enroulement *m* en série || seriewikkeling *f* || последовательная обмотка *f*

S178a **series-wound motor** || Reihenschlußmotor *m* || moteur *m* en série || seriemotor || (электро)двигатель с последовательным возбуждением, сериесный (электро)двигатель *m*

S179 **service life** || Betriebsdauer *f*, Lebensdauer *f*; Nutzungsdauer *f* || durée *f* de vie [d'emploi, de fonctionnement] || bedrijfsduur *m*, gebruiksduur *m* || эксплуатационный срок *m* службы; (технический) ресурс *m*

S180 **servo** *see* **servomechanism**

S181 **servocontrol** || Servosteuerung *f* || servo-contrôle *m* || servobediening *f* || сервоуправление *n*

S182 **servomechanism** || Servomechanismus *m* || servo-mécanisme *m* || servomechanisme *n* || сервомеханизм *m*

S183 **servo-motor** || Servomotor *m*, Stellmotor *m* || servo-moteur *m* || servomotor *m* || сервомотор *m*, серводвигатель *m*

S184 **set** || Menge *f* || ensemble *m* || menigte *f* || множество *n*

S185 **set screw** || Stellschraube *f*; Justierschraube *f* || vis *f* de réglage; vis *f* d'arrêt || stelschroef *f* (*m*); justeerschroef *f* (*m*) || установочный винт *m*; юстировочный винт *m*; регулировочный винт *m*

S186 **setting** || Einstellung *f*; Regelung *f* || pose *f*, posage *m*; ajustage *m*, réglage *m* || instelling; regeling *f* || установка *f*; регулирование *n*, регулировка *f*

S187 **shadow** || 1. Schatten *m* 2. (empfangs)tote Zone *f* || 1. ombre *f* 2. zone *f* morte || 1. schaduw *f* (*m*) 2. dode zone *f* || 1. тень *f* 2. мёртвая зона *f*, зона *f* молчания

S188 **shaft** || Welle *f*; Achswelle *f*, Achse *f*; Spindel *f*; Schaft *m* || arbre *m*, axe *m*, broche *f* || as, spindel *m* || вал *m*; ось *f*; шпиндель *m*

S189 **shaft basis limit system** || Einheitswellensystem *n*, System *n* der Einheitswelle *f* || système *m* d'arbre (normal) || systeem *n* van de eenheidsas || система *f* вала (*в допусках*)

S190 **shafting** || Kraftübertragung *f*, Transmission *f*; Wellenleitung *f* || transmission *f* générale [principale] (par courroie) || asleiding *f*, asoverbrenging *f* || трансмиссионная передача *f*, трансмиссия *f*; линия *f* валов, валопровод *m*

S191 **shaft torque** || Drehmoment *n* an der Welle || couple *m* de torsion le long de l'arbre || draaimoment *n* aan de as || крутящий момент *m* на валу

S192 **shank** || 1. Stab *m*, Stange *f* 2. Schaft *m* || tige *f*; queue *f* || 1. staaf *f* (*m*) 2. schacht *f* (*m*) || 1. стержень *m* 2. хвостовик *m* (*напр. сверла*)

S193 **shank end mill** || Schaftfräser *m* || fraise *f* à queue || schachtfrees *f* (*m*) || концевая [хвостовая] фреза *f*

S194 **shaped casting** || Formstück *n* || moulage *m* façonné [profolé] || profielgietstuk *n* || фасонная отливка *f*

S195 **shaped tool** || Formmeißel *m* || outil *m* profilé || profielbeitel *m* || фасонный [профильный] резец *m*

S196 **shape elasticity** || Formelastizität *f* || élasticité *f* de forme || vormelasticiteit *f* || упругость *f* на сдвиг, упругость *f* второго рода

S197 **shaper** *see* **shaping machine**

S198 **shaping** || 1. Profilieren *n*; Formgebung *f*, Gestaltung *f* 2. Druckumformen *n* || 1. profilage *m* 2. formage *m* || 1. vormen *f*, vormgeving *f*, profilering 2. verwerking *f* onder druk || 1. профилирование *n*, придание *n* формы 2. обработка *f* давлением

S199 **shaping machine** || Kurzhobler *m*, Shapingsmaschine *f* || étau-limeur *m* || dwarsschaafmachine *f* || поперечно-строгальный станок *m*

S200 **sharpening** || Schärfen *n*, Scharfschleifen *n* (*Schneidwerkzeug*) || affûtage *m*, meulage *m* || scherpen *n*, scherpslijpen *n* || затачивание *n*, заточка *f* (*режущего инструмента*)

S201 **sharpness of image** || Bildschärfe *f* || netteté *f* de l'image || beeldscherpte *f* || резкость *f* изображения

S202 **sharpness of resonance** || Resonanzschärfe *f* || acuité *f* de résonance || resonantiescherpte *f* || острота *f* резонанса

S203 **shaving** || 1. Beschneiden *n*, Entgraten *n* (*nach dem Ausschneiden*) 2. Schaben *n*, Schabwälzverfahren *n* || 1. ébarbage *m* 2. procédé *m* shaving, rasage *m*, (rasage-)shaving *m* || 1. besnoeiing *f* 2. fijnschaven *n* || 1. обрезка *f*, снятие *n* заусенцев (*после вырубки*) 2. шевингование *n*

S204 **shaving cutter** || Schabwerkzeug *n* || outil *m* pour rasage, couteau *m* de shaving || schaafwerktuig *n* || шевер *m*

S205 **shaving tool** || Schabewerkzeug *n* || outil *m* à dresser [de finition]; couteau *m* de shaving || 1. gereedschap *n* voor fijnbewerking 2. schaafwerktuig *n* || 1. отделочный инструмент *m* 2. шевер *m*

S206 **shear angle** || Scherwinkel *m* || angle *m* de cisaillement || schuifhoek *m* || угол *m* сдвига [среза]

S207 **shearing** || Abscheren *n*, Scherung *f* || cisaillement *m* || afschuiving *f* || срез *m*; сдвиг *m*; скалывание *n*

S208 **shearing force** || Scherkraft *f* || effort *m* tranchant; force *f* transversale; effort *m* de cisaillement || afschuifkracht *f* (*m*) || срезывающее усилие *n*; скалывающее усилие *n*

S209 **shearing machine** || Schermaschine *f*; Maschinenschere *f* || cisaille *f* à tôles, machine *f* à cisailler || schaar *f* (*m*), knipmachine *f* || машина *f* для резки листового металла; механические ножницы *pl*

S210 **shearing steel** || Werkzeugstahl *m* || acier *m* à outillage [à outils] || gereedschapsstaal *n* || инструментальная сталь *f*

S211 **shearing strain** || Schubverformung *f* || déformation *f* de cisaillement || schuifvervorming *f* || деформация *f* сдвига

S212 **shearing strength** || Schubwiderstand *m*, Scherwiderstand *m* || résistance *f* au cisaillement || schuifweerstand *m* || сопротивление *n* сдвигу [срезу]; прочность *f* при сдвиге

S213 **shearing stress** || Schubspannung *f* || contrainte *f* de cisaillement || (af)schuifspanning *f* || срезывающее напряжение *n*; скалывающее напряжение *n*

S214 **schear modulus** || Schubmodul *m*, Schubelastizitätsmodul *m* || module *m* d'élasticité transversal, module de Coulomb || schuifmodulus *m*, elasticiteitsmodulus bij schuifkrachten || модуль *m* сдвига, модуль *m* упругости при сдвиге

S215 **shear plane** || Scherebene *f* || plan *m* de cisaillement || schuifvlakte *f* || плоскость *f* среза; плоскость *f* сдвига

S216 **sheave (wheel)** || Rillenscheibe *f*; Rillenrolle *f* || poulie *f* || kabelschijf *f* (*m*); katrol(schijf) *f* (*m*) || (желобочный) шкив *m*; (желобочный) блок *m*

S217 **sheet metal** || Feinblech *n* || tôle *f* mince, feuille *f* métallique || (dun) plaatijzer *n*; (тонкий) листовой металл *m*

S218 **sheet mill** || Blechwalzwerk *n* || laminoir *m* [train *m*] à tôles || plaatwalserij *f* || (листо)прокатный стан *m*

S219 **shell** || Hülle *f*; Schale *f* || voile *m*, enveloppe *f*, coque *f* || schaal *f* (*m*), schil *f* || оболочка *f*

S220 **shell-end mill** || walzenförmiger Aufsteckfräser *m* || fraise *f* à trou [creuse] cylindrique || cilinderformige opzetfrees *f* (*m*) || цилиндрическая насадная фреза *f*

S221 **shell structure** *f* || Schalenstruktur *f* (*Atom*) || structure *f* quantique (*atome*) || schilstructuur *f* (*atoom*) || оболочечная структура *f* (*атома*)

S222 **shield** || Abschirmung *f*; Schutzmantel *m*, Schutzgehäuse *m*; Sohild *m* || capot *m*; tableau *m*; écran *m* || bescherming *f*, beschermingsmantel *m*; scherm *n*; schild *n* || защита *f*, экран *m*; (защитный) кожух *m*; щит(ок) *m*

S223 **shielding** || Abschirmen *n* || mise *f* sous écran || beschermen *n* || экранирование *n*

S224 **shift** || Verschiebung *f*; Verlagerung *f* || déplacement *m*, écart *m* || verschuiving || *f* смещение *n*, сдвиг *m*

S225 **shifting** || Gangschaltung *f*, Getriebeschaltung *f*; Riemenumlegung *f*, Riemenwechsel *m* || changement *m* (de vitesse), déplacement *m* (de courroie); décalage *m* || overschakeling *f* (snelheid); riemoverschakeling *f* || переключение *n* (скоростей); перевод *m* (ремня)

S226 **shift operator** || Verschiebungsoperator *m* || opérateur *m* de décalage

|| schuifoperator *m* || оператор *m* сдвига [смещения]

S227 **shift register** || Verschiebungsregister *n* || registre *m* de décalage || schuifregister *n* || регистр *m* сдвига

S228 **shim** || Beilage *f* || cale *f* || latoentje *n*, tussenlaag *f* (*m*) || прокладка *f*

S229 **shock** || 1. Stoß *m*; Schlag *m* || choc *m* || slag *m* || удар *m*

S230 **shock absorber, shock damper** || Stoßdämpfer *m* || amortisseur *m* (de chocs) || schokopnemer *m*; schokbreker *m*; schokdemper *m* || амортизатор *m*

S231 **shock-stall** || Verdichtungsstoßabreißen *n* || formation *f* des ondes de choc || schokgolfverschijnsel *n* || волновой кризис *m*

S232 **shock stress** || Spannung *f* bei dynamischen Beanspruchungen, Schlagbeanspruchung *f* || contrainte *f* due au choc || stootspanning *f*; dynamische spanning *f* || напряжение *n* от удара, ударное напряжение *n*

S233 **shock test** || Schlagprobe *f*, Schlagversuch *m* || essai *m* de choc [au choc] || slagproef *f* (*m*) || ударное [динамическое] испытание *n*

S234 **shock wave** || Stoßwelle *f* || onde *f* de choc || schokgolf *f* (*m*) || ударная волна *f*

S235 **shop test** || Betriebsprobe *f*, Betriebsprüfung *f* || essai *m* en usine || bedrijfsproef *f* (*m*) || заводское испытание *n*

S236 **short circuit** || Kurzschluß *m* || court-circuit *m* || kortsluiting *f* || короткое замыкание *n*

S237 **short-circuited** || kurzgeschlossen || court-circuité || kortgesloten || короткозамкнутый, замкнутый накоротко

S238 **short-focus lens** || Kurzlinse *f* || lentille *f* à courte focale || kortlens *f* (*m*) || короткофокусная линза *f*

S239 **shorting** || Kurzschließen *n* || court-circuit *m* || kortsluiting *f* || замыкание *n* накоротко

S240 **shorting voltage** || Kurzschlußspannung *f* || tension *f* de court-circuit || kortsluit(ings)spanning *f* || напряжение *f* короткоге замыкания

S241 **short-lived isotope** || kurzlebiges Isotop *n* || isotope *m* à vie courte || isotoop *n* met korte levensduur || короткоживущий изотоп *n*

S242 **short-range forces** || Nahwirkungskräfte *f pl* || forces *f pl* à rayon d'action court || nabijwerkende krachten *f* (*m*) *pl* || короткодействующие [близкодействующие] силы *f pl*

S243 **short-range order** || Nahordnung *f* || ordre *m* local [à petite distance] || nabijheidsorde *f* (*m*) || ближний порядок *m*

S244 **short-wavelength** || Kurzwellen... || à ondes courtes || korte-golven..., kortgolvig || коротковолновый

S245 **shot blasting** || Abstrahlen *n*, Strahlreinigen *n*; Kugelstrahlen *n*; Sandstrahlen *n* || grenaillage *m* || kogelstralen *n* || дробеструйная очистка *f*, дробеструйная обработка *f*; пескоструйная очистка *f*, пескоструйная обработка *f*

S246 **shot noise** || Schroteffekt *m*, Schrotrauschen *n* || bruit *m* de grenaille || schrootgeruis *n* || дробовой шум *m*, дробовой эффект *m*

S247 **shot peening** || Kugelstrahlen *n* || grenaillage *m* || kogelstralen *n* || дробеструйная очистка *f*, дробеструйная обработка *f*

S248 **shrinkage allowance** || Schwind(maß)zugable *f* || tolérance *f* de retrait || toegave op de krimp || допуск *m* на усадку

S249 **shrinkage stress** || Schwindspannung *f*, Schrumpfspannung *f* || contrainte *f* de retrait || krimpspanning *f* || напряжение *n* усадки, усадочное напряжение *n*

S250 **shunt** || Nebenschluß *m*, Nebenwiderstand *m*, Shunt *m* || shunt *m* || shunt *m* || шунт *m*

S251 **shutter** || 1. Verschluß *m*; Riegel *m*; Klappe *f* 2. Blende *f* || 1. volet *m* 2. obturateur *m* || 1. schreif *f* (*m*), klep *f* (*m*) 2. sluiter *m* || 1. затвор *m*; заслонка *f* 2. обтюратор *m*

S252 **shuttle** || Teil *n* mit hin- und hergehender Bewegung; Schützen *m* || navette *f*, détail *m* à mouvement alternatif [va-et-vient] || detail *n* met heen-en-weergaande beweging; schietspoel *f* (*m*) || деталь *f* с возвратно-поступательным движением; челнок *m*

S253 **side cutting edge** *see* **major cutting edge**

S254 **side cutting edge angle** *see* **plan approach angle**

S255 **side milling cutter** || dreiseitig schneidender Fräser *m* || fraise *f* à trois tailles [faces] || driezijdig snijdende frees *f* (*m*) || трёхсторонняя фреза *f*

S256 **side tool** || Seitenmeißel *m*; Plandrehmeißel *m* || outil *m* de coupe à saigner || beitel *m* met zijdelingse snede; uitdraaibeitel *m* || упорный проходной резец *m*; подрезной резец *m*

S257 **siemens,** S || Siemens *n* || siemens *m* || siemens *m* || сименс *m*, См, обратный ом *m*

S258 **sight feed oiler** || Tropföler *m* || graisseur *m* à gouttes || druppelssmeertoestel *n* || капельная маслёнка *f*

S259 **sigma function** || Sigmafunktion *f* || fonction *f* sigma || sigmafunctie *f* || сигма-функция *f*

S260 **sign** || Zeichen *n* || signe *m* || teken *n* || знак *m*

S261 **signal** || Signal *m* || signal *m* || signaal *n* || сигнал *m*; импульс *m*

S262 **signal generator** || Signalgenerator *m* || générateur *m* de signaux || signaalgenerator *m* || генератор *m* сигналов

S263 **signal output** || Ausgangssignal *n* || signal *m* de sortie || uitgangssignaal *n* || выходной сигнал *m*

S264 **signal-to-noise ratio** || Signal-Rausch-Verhältnis *n* || rapport *m* signal-bruit || signaal-ruisverhouding *f* || отношение *n* сигнал/шум

S265 **sign convention** || Zeichenregel *f* || règle *f* des signes || tekenregel *m* || правило *n* знаков *(напр. зарядов)*

S266 **signed minor** || Adjunkte *f*, algebraisches Komplement *n* || mineur *m* signé || algebraïsch complement *n* || алгебраическое дополнение *n*

S267 **significance** || Gewicht *n*, Signifikanz *f* || signification *f* || betekenis *f*, gewicht *n* || значимость *f*, вес *m*

S268 **significant figure** || gültige Ziffer *f* || chiffre *m* significatif || geldig cijfer *n* || значащая цифра *f* *(в числе)*

S269 **sign of rotation** || Drehsinn *m*, Drehrichtung *f* || sens *m* de rotation || draairichting *f* || направление *n* вращения

S270 **sign position** || Vorzeichenstelle *f* || position *f* de signe || tekenstelling *f* || знаковый разряд *m*

S271 **sign test** || Zeichentest *m* || test *m* de signes || tekentest *m* || критерий *m* знаков

S272 **signum function** || Signum-Funktion *f* || fonction *f* signe, sgn || signumfunctie *f* || сигнум-функция *f*

S273 **silent chain** || geräuschlose Kette *f* || chaîne *f* silencieuse || geruisloze ketting *f* || бесшумная цепь *f*

S274 **silent discharge** || stille Entladung *f* || décharge *f* silencieuse || stille ontlading *f* || тихий разряд *m*

S275 **silent running** *see* **noiseless running**

S276 **silica,** SiO_2 || Siliziumdioxid *n* || bioxyde *m* de silicium || siliciumdioxyde *n* || двуокись *f* кремния, ангидрид *m* кремнёвой кислоты

S277 **silicic acid** || Kieselsäure *f* || acide *m* silicique || kiezelzuur *n* || кремнёвая кислота *f*

S278 **silico-manganese steel** || Silizium-Manganstahl *m* || acier *m* au silicium-manganèse || silicium-mangaanstaal *n* || кремнемарганцовистая сталь *f*

S279 **silicon,** Si || Silizium *n* || silicium *m* || silicium *n* || кремний *m*

S280 **silicon steel** || Siliziumstahl *m*, siliziumlegierter Stahl *m* || acier *m* au silicium || siliciumstaal *n* || кремнистая сталь *f*

S281 **silver,** Ag || Silber *n* || argent *m* || zilver *n* || серебро *n*

S282 **similarity** || Ähnlichkeit *f*; Analogie *f* || similitude *f*; analogie *f* || gelijkenis *f*, analogie *f* || подобие *n*; аналогия *f*

S283 **similarity relation** *see* **similitude parameter**

S284 **similarity theory** || Ähnlichkeitstheorie *f* || théorie *f* de la similitude || gelijkenistheorie *f* || теория *f* подобия

S285 **similarity transformation** || Ähnlichkeitstransformation *f* || transformation *f* homothétique || gelijkvormigheids-transformatie *f* || преобразование *n* подобия

S286 **similar matrices** || ähnliche Matrizen *f pl* || matrices *f pl* semblables || gelijkvormige matrices *f* (*m*) *pl* || подобные матрицы *f pl*

S287 **similar representations** || ähnliche [äquivalente] Darstellungen *f pl* || représentation *f* semblable || gelijksoortige voorstellingen *f pl* || подобные [эквивалентные] представления *pl*

S288 **similitude** || Ähnlichkeit *f* || similitude *f* || gelijkvormigheid *f* || подобие *n*

S289 **similitude parameter** || Ähnlichkeitskennzahl *f* || argument *m* de similitude || gelijkeniscriterium *n* || критерий *m* подобия

S290 **simple** || einfach; primitiv || simple || eenvoudig, simpel || простой; примитивный

S291 **simple event** || Elementarereignis *n* || événement *m* simple || eenvoudige gebeurtenis *f* || элементарное событие *n*

S292 **simplex channel** || Simplexkanal *m* || canal *m* simplex || simplexkanaal *n*, simultaankanaal *n* || симплексный канал *m*

S293 **simplex method** || Simplexmethode *f* || méthode du simplexe || simplexmethode *f* || симплекс-метод *m*

S294 **simply ordered set** || einfach geordnete Menge *f* || ensemble *m* totalement ordonné || eenvoudig geregelde menigte *f* || линейно упорядоченное множество *n*

S295 **simulation** || Simulieren *n*, Nachbilden *n* || simulation *f* || simulatie *f* || моделирование *n*

S296 **simulator** || Simulator *m* || simulateur *m* || simulator *m* || моделирующее устройство *n*; тренажёр *m*

S297 **simultaneity** || Gleichzeitigkeit *f* || simultanéité *f* || gelijktijdigheid *f* || одновременность *f*

S298 **simultaneous** || simultan, gleichzeitig || simultané || simultaan..., gelijktijdig || одновременный; параллельный; параллельного действия *(ЭВМ)*

S299 **simultaneous equations** || simultane Gleichungen *f pl* || équations *f pl* simultanées || simultaanvergelijkingen *f pl* || совместные уравнения *pl*; система *f* уравнений

S300 **sine** || Sinus *m* || sinus *m* || sinus *m* || синус *m*

S301 **sine curve** || Sinuskurve *f*, Sinusoide *f* || courbe *f* sinusoïdale, sinusoïde *f* || sinuskromme *f* (*m*) || синусоида

S302 **sine integral** || Integralsinus *m* || sinus *m* intégral || integraalsinus *m* || интегральный синус *m*

S303 **sine law** || **1.** Sinusgesetz *n* **2.** Sinussatz *m* || **1.** loi *f* sinusoïdale **2.** théorème *m* des sinus || **1.** sinuswet *f* (*m*) **2.** sinustheorema *n* || **1.** синусоидальный закон *m* **2.** теорема *f* синусов

S304 **sine transformation** || Sinustransformation *f* || transformation *f* sinus || sinustransformatie *f* || синус-преобразование *n*

S305 **single-acting** || einfachwirkend || à simple effet || enkelvoudig werkend || одностороннего действия

S306 **single bond** || Einfachbindung *f*, einfache Bindung *f* || liaison *f* simple || enkelvoudige binding *f* || ординарная [простая] связь

S307 **single-row ball bearing** || einreihiges Kugellager *n* || roulement *m* [palier *m*] simple à billes || eenrijig kogellager *n* || однорядный шариковый подшипник *m*

S308 **single-stage compressor** || Einstufenverdichter *m* || compresseur *m* à un seul étage || enkeltrapscompressor *m* || одноступенчатый компрессор *m*

S309 **single-start screw thread, single thread** || eingängiges Gewinde *n* || filet *m* simple, filetage *m* à un filet || eengangige schroefdraad *m* || однозаходная [одноходовая] резьба *f*

S310 **single-valued function** || eindeutige Funktion *f* || fonction *f* univoque || functie *f* met een cijfer || однозначная функция *f*

S311 **single vortex** || Einzelwirbel *m* || tourbillon *m* solitaire || eenzame wervel *m* || уединённый [изолированный, дискретный] вихрь *m*

S312 **singly charged ion** || einfach geladenes Jon *n* || ion *m* simple || eenmaal geladen ion *n* || однозарядный ион *m*

S313 **singular** || singulär || singulier || singulier || особый; сингулярный; вырожденный

S314 **singular integral** || singuläres Integral *n* || intégrale *f* singulière || singuliere integraal *f* (*m*) || сингулярный интеграл *m*

S315 **singular matrix** || singuläre Matrix *f* || matrice *f* singulière || singuliere matrix *f* || сингулярная матрица *f*

S316 **singular operator** || singulärer Operator *m* || opérateur *m* singulier || singuliere operator *m* || вырожденный оператор *m*

S317 **singular point** || singulärer Punkt *m*, Singularität *f* || point *m* singulier || singuliere pun *f* (*m*) || особая точка *f* *(кривой, поверхности)*

S318 **singular transformation** || singuläre Transformation *f* || transformation *f* singulière || singuliere transformatie *f* || вырожденное [особенное] преобразование *n*

S319 **sintering** || Sintern *n*, Sinterung *f* || agglomération *f*, frittage *m* *(en métallurgie de poudres)* || sintering *f* || спекание *n*, агломерация *f*

S320 **sinusoid, sinusoidal curve** *see* **sine curve**

S321 **sinusoidal oscillations** || sinusförmige Schwingungen *f pl*, Sinusschwingungen *f pl* || oscillations *f pl* sinusoïdales || sinustrillingen *f pl* || синусоидальные колебания *pl*

S322 **SI system** || Internationales Einheitensystem *n*, SI || système *m* international d'unités, SI || internationaal eenhedenstelsel *n*, SI || Международная система *f* единиц, СИ

S323 **SI unit** || SI-Einheit *f* || unité *f* SI || SI-eenheid *f* || единица *f* Международной системы, единица *f* СИ

S324 **size** || Größe *f*, Abmessung *f*, Maß *n* || taille *f*; dimension *f* || grootheid *f* || величина *f*, размер *m*

S325 **sizing** || **1.** Dimensionierung *f*, Bemessung *f* **2.** Eichung *f*; Kalibrierung *f* || **1.** mesure *f*, mesurage *m* **2.** calibrage *m* || **1.** meting *f* **2.** kalibreren *n*, kalibrering *f* || **1.** измерение *n*; расчёт *m* размеров **2.** калибрование *n*, калибровка *f*

S326 **skew field** || Schiefkörper *m* || corps *m* gauche || scheeflichaam *n* || некоммутативное поле *n*

S327 **skew-hermitian matrix** || schiefhermitische Matrix || matrice *f* antihermitienne || scheefhermitische matrix *f* || косоэрмитова [альтернирующая] матрица *f*

S328 **skew-hermitian operator** || schiefhermitischer Operator *m* || opérateur *m* antihermitien || scheefhermitische operator *m* || косоэрмитов оператор *m*

S329 **skew lines** || windschiefe Geraden *f pl* || droites *f pl* gauches || gekruiste rechten *f pl* || скрещивающиеся прямые *pl*

S330 **skewness** || Schiefe, Schiefheitsmaß *n*, Asymmetrie *f* *(Stichprobe)* || asymétrie *f* || asymmetrie *f* || асимметрия *f* *(выборки)*

S331 **skew-symmetric matrix** || schiefsymmetrische Matrix *f* || matrice *f* symétrique gauche || scheefsymmetrische matrix *f* || кососимметрическая матрица *f*

S332 **skew-symmetric operator** || schiefsymmetrischer Operator *m* || opérateur *m* symétrique gauche || scheefsymmetrische operator *m* || кососимметричный оператор *m*

S333 **skew-symmetry** || Schiefsymmetrie *f*, Antisymmetrie *f* || antisymétrie *f* || scheefsymmetrie *f* || антисимметрия *f*

S334 **skin effect** || Skineffekt *m* || effet *m* pelliculaire [de peau] || skineffect *n* || скин-эффект *m*

S335 **sky wave** || Raumwelle *f* || onde *f* spatiale || ruimtegolf *f* (*m*) || пространственная [небесная] волна

S336 **slabbing mill** || Brammenwalzwerk *n*, Brammenstraße *f* || slabbing *m*, laminoir *m* [train *m*]

à brames || slabwalserij *f* || слябинг *m*

S337 **sleeve** || 1. Büchse *f*, Buchse *f*; Hülse *f* 2. Rohr *m* || manchon *m*, douille *f* || **1.** bus *f* (*m*); huls *f* (*m*) **2.** buis *f* (*m*), pijp *f* (*m*) || **1.** втулка *f*; гильза *f* 2. трубка *f*

S338 **slide** || Schlitten *m*; Wagen *m* || glissière *f*; coulisse *f*, coulisseau *m* || slede *f* (*m*) || салазки *pl*; каретка *f*; суппорт *m*, ползун *m*

S339 **slide bar** || Führung *f*, Führungsstab *m* || guide *m*, guidage *m*, glissière *f* || leibaan *f* (*m*), leistang *f* (*m*) || направляющая *f*

S340 **slide block** || Gleitstein *m*, Gleitstück *n*, Kreuzkopf *m*; Schlitten *m* || coulisseau *m*, coulisse *f*; crosse *f*, patin *m* || leiblok *n*, kruishoofd *n*, kruiskop *m*; slede *f* (*m*) || ползун *m*; крейцкопф *m*; салазки *pl*

S341 **slide rest** || Kreuzsupport *m*, Kreuzschieber *m*; Drehteilsupport *m* || support *m* à chariot croisé || kruissupport *n*; support *n* met draaibare slede || крестовый суппорт *m*; суппорт *m* с поворотными салазками

S342 **slide valve** || Längsschieber *m*, Ventilschieber *m* || tiroir *m* (à piston) || stoomschuif *f* (*m*) || золотник *m*

S343 **slide-wire** || Schleifdraht *m* || rhéostat *m* à fil || glijdraad *m* || реохорд *m*; струна *f* реохорда

S344 **sliding** || Gleiten *n* || glissement *m* || glijding || скольжение *n*

S345 **sliding friction** || Gleitreibung *f* || frottement *m* de glissement || glijdende wrijving *f* || трение *n* скольжения

S347 **slipper (block)** *see* **slide block**

S348 **slit** || **1.** Spalt *m* **2.** Einschnitt *m*; Schlitz *m* || fente *f*; échancrure *f*, cannelure *f* || **1.** spleet *f*; sleuf *f* (*m*), gleuf *f* (*m*); spleet *f* (*m*) || **1.** (оптическая) щель **2.** прорезь *f*; шлиц *m*

S349 **slitting** || Schlitzen *n* || cannelage *m*; taillage *m* des fentes || sleuven *n* || шлицевание *n*

S350 **slope of a tangent** || Tangentenneigung *f* || pente *f* d'une tangente || tangentneiging *f*, *n* || наклон *m* касательной

S351 **slot** || Nut *f*; Spalt *m*; Schlitz *m* || rainure *f*, mortaise *f* || gleuf *f* (*m*); spleet *f* (*m*); sleuf *f* (*m*) || паз *m*; щель *f*; шлиц *m*

S352 **slot-milling machine** || Langlochfräsmaschine *f*, Nutenfräsmaschine *f* || machine *f* à fraiser les rainures || sleuffreesmachine *f* || шпоночно-фрезерный станок *m*

S353 **slotted head** || geschlitzter Kopf *m* || tête *f* fendue || sleufkop *m* || головка *f* с прорезью *(под отвёртку)*

S354 **slotter** *see* **slotting machine**

S355 **slotting cutter** || Einstechfräser *m* || fraise *f* à tailler les cannelures || insteekfrees *f* (*m*) || прорезная фреза *f*

S356 **slotting end mill** || Nutenfräser *m*; Stirnfräser *m* für Nuten || fraise *f* en bout pour rainures || steeksleuffrees *f* (*m*) || торцевая пазовая фреза *f*

S357 **slotting machine** || Stoßmaschine *f* || mortaiseuse *f*, machine *f* à mortaiser || (af)steekmachine *f* || долбёжный станок *m*

S358 **slotting tool** || Keilnutenstoßmeißel *m* || outil *m* à mortaiser [de mortaisage] || steekbeitel *m* || долбёжный резец *m*, резец *m* для долбления канавок

S359 **slot washer** || geschlitzte Scheibe *f* || rondelle *f* à fente [fendue] || sleufschijf *f* (*m*) || разрезная шайба *f*

S360 **slowing-down** || Bremsung *f* || décélération *f*, freinage *m* || vertraging *f*, remming *f* || замедление *n*, торможение *n*

S361 **small calorie** || kleine Kalorie *f*, Grammkalorie *f* || (petite) calorie *f* || kleine calorie *f* || калория *f* *(4, 1868 Дж)*

S362 **small tool** *see* **hand tool**

S363 **smoothing** || Glätten *n*, Glättung *f* || lissage *m* || gladmaken *n* || сглаживание *n*

S364 **smoothing filter** || Glättungsfilter *n* || filtre *m* de lissage || gladfilter *m* || сглаживающий фильтр *m*

S365 **snap** || 1. Döpper *m*, Kopfsetzer *m* 2. Sperre *f*, Sperrklinke *f*; Spannvorrichtung *f* || 1. bouterolle *f* 2. cliquet *m*, griffe *f* || 1. snapper *m* 2. klink *f*, klem *f* (*m*) || 1. обжимка *f (для получения заклёпочной головки)* 2. защёлка *f*, зажим *m*

S366 **snap gauge** || Rachenlehre *f* || calibre *m* à entaille, calibre mâchoire, calibre ouvert || kalibrekram *f* (*m*) || калибр-скоба *m*

S367 **socket** || 1. Muffe *f*, Verbindungsmuffe *f*; Hülse *f* 2. Rohrstutzen *m*; Rohrmuffe *f* || 1. douille *f* 2. emboîtement *m* || mof *f* (*m*); huls *f* (*m*); buisaansluiting *f* || 1. муфта *f*; гильза *f*; патрубок *m* 2. раструб *m*; переходной конус *m*

S368 **socket wrench** || Aufsteckschlüssel *m* || clef *f* [clé *f*] à douille || steeksleutel *m* || торцевой гаечный ключ *m*

S369 **soda ash,** Na_2CO_3 || kalzinierte Soda *f*, Natriumkarbonat *n* || carbonate *m* de soude anhydre, sel *m* de soude || gecalcineerde soda *m* || кальцинированная сода *f*, карбонат *m* натрия

S370 **sodium,** Na || Natrium *n* || sodium *m* || natrium *n* || натрий *m*

S371 **soft-magnetic** || weichmagnetisch || magnétique doux || zachtmagnetisch || магнитно-мягкий

S372 **soft radiation** || weiche Strahlung *f* || radiation *f* douce, rayonnement *m* mou || zachte straling *f* || мягкое [слабопроникающее] излучение *n*

S373 **soft steel** || niedriggekohlter Stahl *m* || acier *m* à bas carbone [peu carburé] || zacht staal *n* || низкоуглеродистая [малоуглеродистая] сталь *f*

S374 **software** || Software *f* || logiciel *m* || software *f* || программное [математическое] обеспечение *n*

S375 **soft X-rays** || weiche Röntgenstrahlung *f* || radiation *f* X douce, rayonnement *m* X mou || zachte röntgenstraling *f* || мягкое рентгеновское излучение *n*

S376 **sol** || Sol *n* || sol *m* || sol *n* || золь *m*

S377 **solar radiation** || Sonnenstrahlung *f* || rayonnement *m* solaire || zonnestraling *f* || солнечная радиация *f*

S378 **solar system** || Sonnensystem *n* || système *m* solaire || zonnestelsel *n* || солнечная система *f*

S379 **solar wind** || Sonnenwind *m* || vent *m* solaire || zonnewind *m* || солнечный ветер *m*

S380 **solder** || Lot *n*, Weichlot *n* || brasure *f* (tendre) || soldeer *n*, *m* || припой *m*

S381 **solenoid** || Solenoid *n* || solénoïde *m* || solenoïde *f* || соленоид *m*

S382 **solenoidal vector field** || solenoidales Vektorfeld *n*, quellenfreies Feld *n* || champ *m* vectoriel solénoïdal || solenoïdaal vectorveld *n* || соленоидальное векторное поле *n*

S383 I **solid** *see* **solid body**

S384 II **solid** || fest || solide || vast || твёрдый

S385 **solid body** || Festkörper *m* || corps *m* solide || vast lichaam *n* || твёрдое тело *n*

S386 **solid fuel** || fester Brennstoff *m*, fester Treibstoff *m*, Festtreibstoff *m* || combustible *m* solide || vaste brandstof *f* (*m*) || твёрдое топливо *n*

S387 **solidification** *f* || Erstarren *n* || solidification *f* || verharding *f* || затвердевание *n*

S388 **solidification point** || Erstarrungspunkt *m*, Erstarrungstemperatur *f* || point *m* de solidification || verhardingspunt *f* (*m*) || точка *f* [температура *f*] затвердевания

S389 **solid phase** || feste Phase *f* || phase *f* solide || vaste fase *f* || твёрдая фаза *f*

S390 **solid solution** || feste Lösung *f* || solution *f* solide [cristallisée] || harde oplossing *f* || твёрдый раствор *m*

S391 **solid state** || fester Zustand *m* || état *m* solide || vaste toestand *m* || твёрдое состояние *n*

S392 **solid state physics** || Festkörperphysik *f* || physique *f* de l'état solide || vastestof-fysica *f* || физика *f* твёрдого тела

S393 **solidus (curve)** || Solidulinie *f*, Soliduskurve *f* || solidus *m* || solidus *m* || солидус *m*, линия *f* солидуса

S394 **solitary** || Einzel... || solitaire || enkel || одиночный, уединённый

S395 **solitary wave, soliton** || Einzelwelle *f* || onde *f* solitaire || enkele golf *f* (*m*) || уединённая волна *f*, солитон *m*

S396 **solubility** || Löslichkeit *f*, Lösbarkeit *f* || (dis)solubilité *f* || oplosbaarheid *f* || растворимость *f*

S397 **solubility product** || Löslichkeitsprodukt *n* || produit *m* de solubilité || oplosbaarheidsprodukt *n* || произведение *n* растворимости

S398 **solution** || **1.** *Math.* Lösung *f* **2.** *Chem.* Lösung *f* **3.** *Chem.* Lösen *n*, Auflösen *n* || solution *f* || **1.** *math.* oplossing *f* **2.** *chem.* solutie *f* **3.** *chem.* oplossing *f* || **1.** *мат.* решение *n* **2.** *хим.* раствор *m* **3.** *хим.* растворение *n*

S399 **solvable group** || auflösbare Gruppe *f* || groupe *m* soluble || oplosbare groep *f* (*m*) || разрешимая группа *f*

S400 **solvation** || Solvatation *f*, Solvatisierung *f* || solvatation *f* || solvatatie *f* || сольватация *f*

S401 **solvent** || Lösungsmittel *n*, Lösemittel *n*, Löser *m*, Solvens *n* || (dis)solvant *m* || oplosmiddel *n* || растворитель *m*

S402 **sonar** || Unterwasserortungsgerät *n*, Sonar *m* || détecteur *m* [sondeur *m*] sous-marin; sonar *m* || sonar *m* || гидролокатор *m*, сонар *m*

S403 **sonde** || Sonde *f* || sonde *f* || sonde *f* (*m*) || зонд *m*; пробник *m*; щуп *m*

S404 **sonic** || sonisch || sonique || geluid... || звуковой, акустический

S405 **sonic barrier** || Schallmauer *f* || barrière *f* du son, barrière *f* [mur *m*] sonique || geluidbarriere *f* (*m*) || звуковой барьер *m*

S406 **sonic boom** || Überschallknall *m* || choc *m* sonique || geluidslag *m* || звуковой удар *m*

S407 **sonic frequency** || akustische Frequenz *f*, Tonfrequenz *f* || fréquence *f* sonore, audiofréquence *f* || akoestische frequentie *f* || звуковая частота *f*, частота *f* звука

S408 **sonics** || Sonik *f*, technische Akustik *f* || acoustique *f* technique || technische akoestik *f*, technische geluidleer *f*, geluidtechniek *f* || соника *f*, техническая акустика *f*

S409 **sonic wave** || Schallwelle *f* || onde *f* sonore || geluidgolf *f* (*m*) || звуковая волна *f*

S410 **sorption** || Sorption *f* || sorption *f* || sorptie *f* || сорбция *f*

S411 **sound** || **1.** Schall *m* **2.** Sonde *f* || **1.** son *m* **2.** sonde *f* || **1.** geluid *n* **2.** sonde *f* || **1.** звук *m* **2.** зонд *m*

S412 **sound-absorbing** || schallschluckend || ... absorbeur acoustique || geluidabsorberend || звукопоглощающий

S413 **sound-conducting** || schalleitend || ... transmetteur acoustique || geluidgeleidend || звукопроводящий, звукопроводный

S414 **sounding** || Sondierung *f* || sondage *m* || sonderen *n* || зондирование *n*

S415 **sound-insulating** || schalldämmend || à isolation [à protection] phonique, d'insonorisation || geluidisolerend || звукоизоляционный

S416 **sound insulation** || Schalldämmung *f*, Schallisolierung *f*, Schallisolation *f* || isolation *f* [protection *f*] sonore, insonorisation *f* || geluidisolatie *f* || звукоизоляция *f*

S417 **sound intensity** || Schallstärke *f* || intensité *f* acoustique [sonore] || geluidsterkte *f* (*m*) || сила *f* звука

S418 **sound locator** *see* **sonar**

S419 **sound proofing** *see* **sound insulation**

S420 **sound propagation** || Schallausbreitung *f* || propagation *f* du son || geluidvoortplanting *f* || распространение *n* звука

S421 **sound velocity** || Schall(ausbreitungs)geschwindigkeit *f* || vitesse *f* de propagation du son || geluidsnelheid *f* || скорость *f* (распространения) звука

S422 **sound wave** || Schallwelle *f* || onde *f* sonore || geluidgolf *f* (*m*) || звуковая волна *f*

S423 **source** || Quelle *f* || source *f* || bron *f* (*m*) || источник *m*

S424 **source intensity** || Quellenleistung *f* || intensité *f* de source || bronvermogen *n* || мощность *f* [сила *f*] источника

S425 **source of field** || Feldquelle *f* || source *f* de champ || veldbron *f* (*m*) || источник *m* поля

S426 **source program** || Quellenprogramm *n* || programme *m* source || uitgangsprogramma *n* || исходная программа *f*

S427 **source resistance** || Quellenwiderstand *m* || résistance *f* de source || eigenweerstand *m* || внутреннее сопротивление *n* источника тока

S428 **space** || Raum *m* || espace *m* || ruimte *f* || пространство *n*; космическое пространство *n*, космос *m*

S429 **space charge** || Raumladung *f* || charge *f* d'espace || ruimtelading *f* || объёмный [пространственный] заряд *m*

S430 **space flight** || Raumflug *m* || vol *m* cosmique [spatial] || ruimtevaart *f* (*m*) || космический полёт *m*

S431 **space lattice** || Raumgitter *n* (*Kristall*) || réseau *m* spatial || ruimterooster *m*, *n* || пространственная решётка *f*. (*кристалла*)

S432 **space-time** || Raum-Zeit *f* || espace-temps *m* || ruimte-tijd *m* || пространство-время *n*

S433 **spanner** || Schraubenschlüssel *m* || clef *f* [clé *f*] à écrou || moersleutel *m*, schroefsleutel *m*; Engelse sleutel *m* || гаечный ключ *m*

S434 **spanner wrench** || Ringmutternschlüssel *m* || clef *f* [clé *f*] à ergot, clef *f* [clé *f*] crochet || ringmoersleutel *m* || ключ *m* для круглых гаек

S435 **spark discharge** || Funkenentladung *f* || décharge *f* à étincelles || vonkontlading *f* || искровой разряд *m*

S436 **spark eroding machine** || Funkenabtrag(ungs)maschine *f* || machine *f* d'usinage par l'étincellement [par électroérosion] || vonkverspaningsmachine *f* || электроимпульсный станок *m*

S437 **spark gap** || Funkenstrecke *f* || distance *f* explosive [de contournement] || vonkbrug *f* (*m*) || искровой промежуток *m*

S438 **sparking** || Funkenbildung *f* || étincellement *m*, formation *f* d'étincelles || vonkvorming *f* || искрение *n*, искрообразование *n*

S439 **sparking coil** || Induktionsspule *f* || bobine *f* à induction || inductieklos *m*, *f* || индукционная катушка *f*, катушка *f* Румкорфа, индуктор *m*

S440 **sparking plug** || Zündkerze *f* || bougie *f* d'allumage || bougie *f* || свеча *f* зажигания

S441 **spark machining** || Elektrofunkenbearbeitung *f*, Funkenerosion *f*, Funkenabtragen *n* || usinage *m* par étincelage [par électroérosion] || vonkverspaning *f* || электроимпульсная обработка *f*

S442 **sparkover** || Funkenüberschlag *m* || chevauchement *m* d'étincelles || vonkoverslag *m* || перекрытие *n* искрой, искровое перекрытие *n*

S443 **spark-over voltage** || Überschlagsspannung *f* || tension *f* de contournement [d'amorçage] || overslagspanning *f* || перекрывающее [разрядное] напряжение *n*

S444 **special relativity** || spezielle Relativitätstheorie *f* || théorie *f* de relativité restreinte [spéciale] || speciale relativiteitstheorie *f* || специальная теория *f* относительности

S445 **specific** || spezifisch || spécifique || specifiek || удельный

S446 **specific activity** || spezifische Aktivität *f* || activité *f* spécifique || specifieke activiteit *f* || удельная активность *f*

S447 **specific charge** || spezifische Ladung *f*, Ladung-Masse-Verhältnis *n* || charge *f* spécifique, rapport *m* charge-masse || specifieke lading *f*, lading-massa-verhouding *f* || удельный заряд *m*, отношение *n* заряда к массе

S448 **specific energy** || spezifische Energie *f* || énergie *f* spécifique || specifieke energie *f* || удельная энергия *f*

S449 **specific gravity** || spezifisches Gewicht *n* || poids *m* spécifique || specifiek [soortelijk] gewicht *n* || удельный вес *m*

S450 **specific heat** || spezifische Wärme *f* || chaleur *f* spécifique || specifieke [soortelijke] warmte *f* || удельная теплоёмкость *f*

S451 **specific heat at constant pressure** || spezifische Wärme *f* bei konstantem Druck || chaleur *f* spécifique à la pression constante || soortelijke warmte *f* bij constante druk || удельная теплоёмкость *f* при постоянном давлении, изобарная теплоёмкость *f*

S452 **specific heat at constant volume** || spezifische Wärme *f* bei konstantem Volumen || chaleur *f* spécifique au volume constant || soortelijke warmte *f* bij constant volume || удельная теплоёмкость *f* при постоянном объёме, изохорная теплоёмкость *f*

S453 **specific heat per mole** || molare spezifische Wärme *f* || chaleur *f* [capacité *f* calorifique] molaire || molaire soortelijke warmte *f* || молярная удельная теплоёмкость *f*

S454 **specific heat ratio** || Verhältnis *n* der spezifischen Wärmen || rapport *m* des chaleurs spécifiques || verhouding *f* van soortelijke warmtes || отношение *n* изобарной теплоёмкости к изохорной теплоёмкости, показатель *m* адиабаты

S455 **specific humidity** || spezifische Feuchtigkeit *f* || humidité *f* spécifique || specifieke vochtigheid *f* || удельная влажность *f*

S456 **specific impulse** || spezifischer Schub *m* (des Raketentreibstoffs) || impulsion *f* spécifique || soortelijke impuls *m* || удельный импульс *m* (топлива)

S457 **specific resistance** || spezifischer Widerstand *m* || résistance *f* spécifique || soortelijke weerstand *m* || удельное сопротивление *n*

S458 **specific thrust** || spezifischer Schub *m* || poussée *f* spécifique || specifieke sleekracht *f* (*m*) || удельная тяга *f*

S459 **specific volume** || spezifisches Volumen *n* || volume *m* spécifique || soortelijk volume *n* || удельный объём *m*

S460 **specified conditions** || Betriebsbedingungen *f pl (gemäß technischen Anforderungen)* || conditions *f pl* spécifiées par les normes || bedrijfsvoorwaarden *f pl (in overeenstemming met vastgestelde technische eisen)* || условия *pl*, установленные техническими нормами

S461 **specified load** || rechnerische Belastung *f*, Lastannahme *f*; Nennlast *f* || charge *f* normalisée; charge *f* nominale || aangenomen [nominale] belasting *f* || нормативная нагрузка *f*; номинальная нагрузка *f*

S462 **specified value** || vorgeschriebene Größe *f* || valeur *f* spécifiée || voorgeschreven grootheid *f* || заданная величина *f*

S463 **specimen** || Probekörper *m*, Probestück *n*, Prüfling *m* || éprouvette *f*, épreuve *f* || proefstuk *n* || образец *m* (для испытаний), испытуемый образец *m*

S464 **spectral** || spektral, Spektral... || spectral || spectraal || спектральный

S465 **spectral analysis** || Spektralanalyse *f* || analyse *f* spectrale || spectraalanalyse *f* || спектральный анализ *m*

S466 **spectral band** || Spektralbande *f* || bande *f* spectrale [de spectre] || spectraalband *m* || спектральная полоса *f*

S467 **spectral line** || Spektrallinie *f* || raie *f* [ligne *f*] du spectre, raie spectrale || spectraallijn *f* (*m*) || спектральная линия *f*, линия спектра *f*

S468 **spectral sensitivity** || Spektralempfindlichkeit *f* || sensibilité *f* spectrale || spectraalgevoeligheid *f* || спектральная чувствительность *f*

S469 **spectral series** || Spektralserie *f* || série *f* spectrale || spectraalserie *f* || спектральная серия *f*

S470 **spectral term** || Spektralterm *m* || terme *m* spectral || spectraalterm *m* || спектральный терм *m*

S471 **spectrogram** || Spektrogramm *n* || spectrogramme *m* || spectrogram *n* || спектрограмма *f*

S472 **spectrograph** || Spektrograph *m* || spectrographe *m* || spectrograaf *m* || спектрограф *m*

S473 **spectrometer** || Spektrometer *n* || spectromètre *m* || spectrometer *m* || спектрометр *m*

S474 **spectrometer transmission** || Spektrometertransmission *f* || transmission *f* de spectromètre || spectrometertransmissie *f* || светосила *f* спектрометра

S475 **spectrometry** || Spektrometrie *f* || spectrométrie *f* || spectrometrie *f* || спектрометрия *f*

S476 **spectrophotometer** || Spektrophotometer *n* || spectrophotomètre *m* || spectrofotometer *m* || спектрофотометр *m*

S477 **spectroscope** || Spektroskop *n* || spectroscope *m* || spectroscoop *m* || спектроскоп *m*

S478 **spectroscopy** || Spektroskopie *f* || spectroscopie *f* || spectroscopie *f* || спектроскопия *f*

S479 **spectrum** || Spektrum *n* || spectre *m* || spectrum *n* || спектр *m*

S480 **spectrum analyzer** || Spektralanalysator *m* || analyseur *m* de spectre || spectrumanalysator *m* || спектроанализатор *m*, анализатор *m* спектра

S481 **specular reflection** || gerichtete Reflexion *f*, Spiegelreflexion *f* || réflexion *f* spéculaire [régulière] || spiegelreflex *m* || зеркальное отражение *n*

S482 **speed** || Geschwindigkeit *f* || vitesse *f*; rapidité *f* || snelheid *f* || скорость *f*

S483 **speed change gear** *see* **gear box**

S484 **speed lathe** *see* **high-speed lathe**

S485 **speed of revolution, speed of rotation** || Drehzahl *f*, Drehgeschwindigkeit *f* || vitesse *f* de rotation || omwentelingssnelheid *f*, rotatiesnelheid *f* || частота *f* вращения

S487 **speed variator** || stufenloses Getriebe *n*, Variator *m* || variateur *m* (continu de vitesse) || traploze overbrenging *f*; variator *m* || бесступенчатая передача *f*, вариатор *m*

S488 **sphere** || Sphäre *f* || sphère *f* || sfeer *f* (*m*) || сфера *f*, шар *m*

S489 **sphere-gap voltmeter** || Kugelvoltmeter *n* || voltmètre *m* à sphères || kogelvoltmeter *m* || измерительный шаровой разрядник *m*

S490 **spherical aberration** || sphärische Aberration *f* || aberration *f* sphérique || sferische aberratie *f* || сферическая аберрация *f*

S491 **spherical coordinates** || Kugelkoordinaten *f pl*, sphärische Koordinaten *f pl* || coordonnées *f pl* sphériques || ruimtelijke coördinaten *m pl*, bolcoördinaten *m pl* || сферические координаты *pl*

S492 **spherical mirror** || sphärischer Spiegel *m* || miroir *m* sphérique || bolvormige spiegel *m* || сферическое зеркало *n*

S493 **spherical roller bearing** || Pendelrollenlager *n* || roulement *m* à deux rangées de rouleaux-tonneaux; palier *m* double à rouleaux-tonneaux || rollager *n* met bolvormig buitenloopvlak || роликовый сферический подшипник *m*

S494 **spherical surface** || Kugelfläche *f* || surface *f* sphérique || bolvormige oppervlakte *f* || сферическая поверхность *f*

S495 **spherical triangle** || sphärisches Dreieck *n* || triangle *m* sphérique || bolvormige [sferische] driehoek *m* || сферический треугольник *m*

S496 **spherical wave** || Kugelwelle *f* || onde *f* sphérique || bolvormige [sferische] golf *f* (*m*) || сферическая волна *f*

S497 **spherics** || **1.** atmosphärische Störungen *f pl* **2.** sphärische Geometrie *f*; sphärische Trigonometrie *f* || **1.** atmosphériques *m pl*, perturbations *f pl* [parasites *m pl*] atmosphériques **2.** géométrie sphérique; trigonométrie *f* sphérique || **1.** atmosferische storingen *f pl* **2.** bolvormige geometrie *f*, bolvormige trigonometrie *f* || **1.** атмосферные помехи *pl*, (атмо)сферики *pl* **2.** сферическая геометрия *f*; сферическая тригонометрия *f*

S498 **spheroid** || Sphäroid *n*, Rotationsellipsoid *n* || sphéroïde *m* || sferoïde *f* || сфероид *m*
S499 **spherometer** || Sphärometer *n*, Wölbungsmesser *m* || sphéromètre *m* || sferometer *m* || сферометр *m*
S500 **spin** || Spin *m* || spin *m* || spin *m* || спин *m*, спиновый момент *m* количества движения
S501 **spindle** || Spindel *f*; Welle *f*; Achse *f*; Stab *m* || broche *f*; arbre *m*, axe *m* || spindel *m*; as *f* (*m*); staaf *f* (*m*) || шпиндель *m*; вал *m*; ось *f*; стержень *m*
S502 **spin echo** || Spinecho *n* || écho *m* de spin || spin-echo *m* || спиновое эхо *n*
S503 **spin moment** || Spinmoment *n* || (moment *m* de) spin *m* || spinmoment *n* || **1.** спиновый (механический) момент *m*, спин *m* **2.** спиновый магнитный момент *m*
S504 **spinning** || **1.** Eigendrehung *f* **2.** Drücken *n* *(auf der Drückmaschine)* **3.** Spinnen *n* || **1.** rotation *f* propre **2.** repoussage *m* *(au tour)* **3.** filature *f*, filage *m* || **1.** eigendraaiing *f* **2.** uitdrukking *f* *(aan de vloeidraaibank)* **3.** spinnen *n* || **1.** собственное вращение *n* **2.** выдавливание *n* *(на токарно-давильном станке)* **3.** прядение *n*
S505 **spinning machine** || **1.** Drückmaschine *f*, Fließdrückmaschine *f* **2.** Spinnmaschine *f* || **1.** tour *m* à repousser **2.** machine *f* de filature, fileuse *f* || **1.** vloeidraaibank *f* **2.** spinmachine *f* || **1.** токарнодавильный станок *m* **2.** прядильная машина *f*
S506 **spinning top** || rotierender Kreisel *m* || toupie *f* tournante || draaiende tol *m* || (вращающийся) волчок *m*
S507 **spino** || Spinor *m* || spineur *m* || spinor *m* || спинор *m*
S508 **spin-orbit** || Spin-Bahn-... || à orbite de spin || spin-baan-... || спин-орбитальный
S509 **spin-spin** || Spin-Spin- || (à) spin-spin || spin-spin-... || спин-спиновый
S510 **spin wave** || Spinwelle *f* || onde *f* de spin || spingolf *f* (*m*) || спиновая волна *f*, магнон *m*
S511 I **spiral** || Spirale *f* || spirale *f* || spiraal *f* (*m*) || спираль *f*
S512 II **spiral** || spiral, Spiral... || spiral || spiraal... || спиральный
S513 **spiral angle** *see* **helix angle**
S514 **spiral fluted reamer** *see* **helical fluted reamer**
S515 **spiral gear** || spiralverzahntes Zahnrad *n* || roue *f* à denture hélicoïdale || tandwiel *n* met spiraaltanden || колесо *n* со спиральными зубьями
S516 **spiral of Archimedes** || Archimedische Spirale *f* || spirale *f* d'Archimède || spiraal *f* (*m*) van Archimedes || спираль *f* Архимеда
S517 **spiral roller bearing** || Federrollenlager *n* || roulement *m* [palier *m*] à rouleaux en hélice || rollager *n* met spiraalrollen || роликовый подшипник *m* с витыми роликами
S518 **spiral spring** *see* **helical spring**
S519 **spline** || **1.** Schlitz *m*, Nut *f* **2.** Gleitfeder *f*; Schiebekeil *m* || **1.** cannelure *f*, mortaise *f* **2.** clavette *f* *(coulissante)* || **1.** sleuf *f* (*m*) **2.** glijspie *f* (*m*) **1.** шлиц *m*, паз *m* **2.** скользящая шпонка *f*
S520 **spline broaching machine** || Nutenziehräummaschine *f* || machine *f* à brocher les fentes || spiesleufschaafmachine *f* || шлицепротяжный станок *m*
S521 **splined shaft** || Kiel(nuten)-welle *f* || arbre *n* cannelé || as *f* (*m*) met sleuven || зубчатый [шлицевый] вал *m*
S522 **split bearing** || geteiltes Lager *n* || palier *m* démontable [ouvert] || gedeeld kogellager *n* *(kussenblok)* || разъёмный подшипник *m*
S523 **split gear** || geteiltes Zahnrad *n* || roue *f* *(dentée)* en deux pièces || gedeeld tandwiel *n* || разъёмное зубчатое колесо *n*
S524 **split key** || Drahtsplint *m* || clavette *f* [goupille *f*] fendue || splitpen *f* (*m*) || разводная чека *f*, разводной шплинт *m*

S525 **split elns** || Spaltlinse *f* || bilentille *f* || bilens *f* (*m*) ||. билинза *f*
S526 **split nut** || Schloßmutter *f*, Leitspindelmutter *f* || écrou *m* débrayable || gedeelde moer *f* (*m*) || разъёмная гайка *f*
S527 **split pin** || Splint *m*; Vorstecker *m* || goupille *f*, clavette *f* *(fendue)* || splijtpen *f* (*m*) || шплинт *m*; чека *f*
S528 **splitting of spectral lines** || Linienaufspaltung *f*, Aufspaltung *f* von Spektrallinien || décomposition *f* des raies spectrales || splijting *f* van de spectraallijnen || расщепление *n* спектральных линий
S529 **spontaneous** || spontan || spontané || spontaan || спонтанный, самопроизвольный
S530 **spontaneous emission** || spontane Emission *f*, Spontanemission *f* || émission *f* spontanée || spontane emissie *f* || спонтанное [самопроизвольное] излучение *n*
S531 **spontaneous fission** || Spontanspaltung *f*, spontane Kernspaltung *f* || fission *f* spontanée || spontane (kern)splijting *f* || спонтанное деление *n* *(ядра)*
S532 **spontaneous radiation** || spontane Strahlung *f* || radiation *f* spontanée || spontane straling *f* || спонтанное [самопроизвольное] излучение *n*
S533 **spontaneous transformation** || spontaner Zerfall *m*, spontane Kernumwandlung *f* || transformation *f* spontanée || spontane omzetting *f* || самопроизвольное превращение *n*
S534 **spontaneous transition** || spontaner Übergang *m* || transition *f* spontanée || spontane overgang *m* || спонтанный [самопроизвольный] переход *m*
S535 **spot welding** || Punktschweißen *n* || soudage *f* par points || puntlassen *n* || точечная сварка *f*
S536 **spread of points** || Streuung *f* von Punkten *(in einem Diagramm)* || dispersion *f* des points || puntenspreiding *f* || разброс *m* точек *(на графике)*
S537 **spring** || Feder *f* || ressort *m* || veer *f* (*m*) || пружина *f*
S538 **spring balance** || Federwaage *f* || balance *f* à ressort || veerbalans *f* (*m*) || пружинные весы *pl*
S539 **spring chuck** *see* **collet chuck**
S540 **spring dynamometer** || Federdynamometer *n* || dynamomètre *m* à ressort || veerdynamometer *m* || пружинный динамометр *m*
S541 **spring-loaded governor** || Feder(kraft)regler *m* || régulateur *m* centrifuge à ressort || centrifugaal veerregelaar *m* || пружинный центробежный регулятор *m*
S542 **spring rate** || Federsteife *f*; Federsteifigkeitszahl *f* || rigidité *f* du ressort || veerstijfheid *f*; veerstijfheids-coëfficiënt *m* || жёсткость *f* пружины; коэффициент *m* жёсткости пружины
S543 **spring ring** || Federring *m* || segment *m* élastique [fendu] || veerring *m* || (разрезная) пружинная шайба *f*
S544 **spring stiffness** *see* **spring rate**
S545 **spring valve** || federbelastetes Ventil *n* || soupape *f* à ressort || ventiel *n* met veerbelasting *n* || пружинный клапан *m*
S546 **spring washer** *see* **elastic washe**
S547 **sprocket** || Kettenrad *n*, Sternrad *n* || roue *f* dentée pour chaîne || kettingschijf *f* (*m*), kettingwiel *n* || (цепная) звёздочка *f*; цепное колесо *n*
S548 **sprocket chain** || Gelenkkette *f* || chaîne *f* à rouleaux || scharnierketting *f* || шарнирная цепь *f*
S549 **sprocket wheel** *see* **sprocket**
S550 **spur** || Spur *f* || trace *f* || spoor *n* || след *m*
S551 **spur gear** || geradzahniges Stirnrad *n*, Geradzahnstirnrad *n* || pignon *m* droit; engrenage *m* droit || rechttandwiel *n* || цилиндрическое прямозубое колесо *n*
S552 **spurious pulse** || unechter Impuls *m* || fausse impulsion *f* || onechte impuls *m* || ложный импульс *m*
S553 **spur rack** || Zahnstange *f* mit Geradverzahnung || crémaillère *f* à denture droite [à denture rectili-

gnel || rechte tandstang *f* (*m*) || прямозубая рейка *f*

S554 **square** || Quadrat *n* || carré *m* || kwadraat *n* || квадрат *m*

S555 **square error** || quadratischer Fehler *m* || erreur *f* quadratique || kwadratische fout *f* (*m*) || квадратическая ошибка *f*

S556 **square-law detection** || quadratische Gleichrichtung *f* || détection *f* quadratique || kwadratische gelijkrichting *f* || квадратичное детектирование *n*

S557 **square root** || Quadratwurzel *f* || racine *f* carrée || kwadraatwortel *m* || квадратный корень *m*

S558 **square thread** || flachgängiges Gewinde *n*, Flachgewinde *n* || filet *m* carré || vierkante (platte)schroefdraad *m* || прямоугольная резьба *f*

S559 **stability** || Stabilität *f* || stabilité *f* || stabiliteit *f* || устойчивость *f*; стабильность *f*

S560 **stability in the large** || Gesamtstabilität *f* || stabilité *f* globale || totaalstabiliteit *f* || глобальная устойчивость *f*, устойчивость *f* в целом

S561 **stabilization** || Stabilisierung *f*, Stabilisation *f* || stabilisation *f* || stabilisatie *f* || стабилизация *f*

S562 **stabilizer** || Stabilisator *m* || stabilisateur *m* || stabilisator *m* || стабилизатор *m*

S563 **stable** || stabil || stable || stabiel || устойчивый; стабильный

S564 **stable equilibrium** || stabiles Gleichgewicht *n* || équilibre *m* stable || stabiel evenwicht *n* || устойчивое равновесие *n*

S565 **stable isotope** || stabiles Isotop *n* || isotope *m* stable || stabiel isotoop *n* || устойчивый [нерадиоактивный] изотоп *m*

S566 **stable motion** || stabile Bewegung *f* || mouvement *m* stable || stabiele beweging *f* || устойчивое движение *n*

S567 **stable nucleus** || stabiler Kern *m* || noyau *m* stable || stabiele kern *f* || устойчивое [нерадиоактивное] ядро *n*

S568 **stable state** || stabiler Zustand *m* || état *m* stable || stabiele toestand *m* || устойчивое состояние *n*

S569 **stable system** || stabiles System *n* || système *m* stable || stabiel systeem *n* || устойчивая система *f*

S570 **stage** || 1. Stufe *f* (*z. B. Turbine*) 2. Stadium *n*; Phase *f* 3. *El.* Stufe *f* || 1. étage *m* (*p. ex. d'une turbine*) 2. stade *m* 3. *él.* étage *m* || 1. trap *m* (*bijv. turbine*) 2. stadium *n*, fase *f* 3. *el.* trap *f* || 1. ступень *f* (*напр. турбины*) 2. стадия *f*, фаза *f* 3. каскад *m* (*электрической схемы*)

S571 **stainless steel** || nichtrostender [rostfreier] Stahl *m* || acier *m* inoxydable || roestvrij staal *n* || нержавеющая сталь *f*

S572 **stall** || Abreißen *n* (*Strömung*) || décollement *m* [décrochage *m*] de la veine || afbreken *n* || срыв *m* потока

S573 **stamping** || Stanzen *n*; Gesenkschmieden *n* || estampage *m* || smeden *n* || штамповка *f*; ковка *f* в штампах

S574 **stamping press** || Gesenk(schmiede)presse *f*; Stanzpresse *f* || presse *f* à estamper || matrijspers *f* (*m*) || штамповочный пресс *m*

S575 **stand** || 1. Gestell *n*; Ständer *m* 2. Prüfeinrichtung *f*, Stand *m* || 1. bâti *m* 2. stand *m* (d'épreuve) || 1. gestel *n* 2. proefstand *m*; stand *m* || 1. станина *f* 2. испытательная установка *f*; стенд *m*

S576 I **standard** || Standard *m*; Norm *f*; Normal *n* || standard *m*, étalon *m* || **standaard** *m*, etalon *m* || стандарт *m*; эталон *m*

S577 II **standard** || Standard...; Normal... || standard, normalisé || standaard..., normaal... || стандартный; эталонный; нормальный

S578 **standard atmosphere** || Normalatmosphäre *f*, physikalische Atmosphäre *f* || atmosphère *f* normale || normaalatmosfeer *f* (*m*) || нормальная [физическая] атмосфера *f* (*101325 Па*)

S579 **standard deviation** || Standardabweichung *f* || déviation *f* standard || standaardafwijking *f* ||

среднеквадратичное [стандартное] отклонение *n*

S580 **standard pressure** || Norm(al)druck *m* || pression *f* normale || normaaldruk *m* || нормальное давление *n*

S581 **standard program** || Standardprogramm *n* || programme *m* standard || standaardprogramma *n* || стандартная программа *f*

S582 **standard spectrum** || Bezugsspektrum *n*, Vergleichsspektrum *n* || spectre *m* de comparaison || vergelijkingsspectrum *n* || спектр *m* сравнения

S583 **standard temperature and pressure** || Normalbedingungen *f pl* || conditions *f pl* normales || normaalvoorwaarden *f pl* || нормальные условия *pl* *(1,013·10⁵ Па и 273,15 К)*

S584 **standard volume** || Normvolumen *n* || volume *m* standard || normaalvolume *n* || объём *m* моля газа при нормальных условиях

S585 **stand-by** || Reserve..., Stand-by- || réservé, de réserve || reserve... || резервный

S586 **standing wave** || Stehwelle *f* || onde *f* stationnaire [immobile] || staande golf *f* (*m*) || стоячая волна *f*

S587 **standing-wave ratio** || Stehwellenverhältnis *n* || taux *m* d'ondes stationnaires || staande-golfverhouding *f* || коэффициент *m* стоячей волны

S588 **star connection** || Sternschaltung *f* || couplage *m* en étoile || sterschakeling *f* || соединение *n* звездой

S589 **starting drill** *see* **center drill**

S590 **starting torque** || Anlaufdrehmoment *n*, Startdrehmoment *n*; Anfahr(dreh)moment *n* || couple *m* de démarrage || aanloopkoppel *n*; || пусковой (крутящий) момент *m*; начальный момент *m*; момент *m* трогания

S591 **starting voltage** || Einsatzspannung *f* || tension *f* de démarrage || aanloopspanning *f* || пусковое напряжение *n*

S592 **star voltage** || Phasenspannung *f*, Sternspannung *f* || tension *f* de phase || fasespanning *f* || напряжение *n* между фазой и нейтралью; фазовое напряжение *n*

S593 **star wheel** || **1.** Handrad *n* **2.** Sperrad *n*; Klinkenrad *n* || **1.** volant *m* de manœuvre à main **2.** roue *f* étoilée; roue *f* à cliquet || **1.** handwiel *n* **2.** sterrenwiel *n* || **1.** (ручной) маховичок *m* **2.** звёздочка *f*; зубчатый диск *m*; храповое колесо *n*; храповик *m*

S594 **state** || Zustand *m* || état *m* || toestand *m* || состояние *n*

S595 **statement** || **1.** Statement *n*, Vereinbarung *f*, Anweisung *f* **2.** *Math.* Ansatz *m* || **1.** proposition *f* **2.** conception *f* mathématique || **1.** voorstel *n*; aanwijzing *f* **2.** aanpak *m* || **1.** предложение *n*; высказывание *n* **2.** *мат.* подход *m* к решению

S596 **state of rest** || Ruhezustand *m* || état *m* de repos || rusttoestand *m* || состояние *n* покоя

S597 **state of stress** || Spannungszustand *m*, gespannter Zustand *m* || état *m* de contrainte || spanningstoestand *m* || напряжённое состояние *n*

S598 **state parameter, state variable** || Zustandsgröße *f*, Zustandsparameter *m* || paramètre *m* d'état || toestandsparameter *m* || параметр *m* состояния

S599 **statically determinate system** || statisch bestimmtes System *n* || système *m* isostatique [statiquement déterminé] || statisch bepaalbaar systeem *n* || статически определимая система *f*

S600 **statically indeterminate system** || statisch unbestimmtes System *n* || système *m* hyperstatique [statiquement indéterminé] || statisch onbepaalbaar systeem *n* || статически неопределимая система *f*

S601 **statically stable** || statisch stabil || statiquement stable || statisch stabiel || статически устойчивый

S602 **static balance** *see* **static equilibrium**

S603 **static characteristic** || statische Kennlinie *f*, statische Charakteristik *f* || caractéristique *f* statique || statische karakteristiek *f* || статическая характеристика *f*

S604 **static charge** || statische Ladung *f* || charge *f* statique || statische lading *f* || статический заряд *m*

S605 **static electricity** || statische Elektrizität *f* || électricité *f* statique || statische elektriciteit *f* || статическое электричество *n*

S606 **static equilibrium** || statisches Gleichgewicht *n* || équilibre *m* statique || statisch evenwicht *n* || статическое равновесие *n*

S607 **static force** || statische Kraft *f* || force *f* statique || statische kracht *f* (*m*) || статическая сила *f*

S608 **static friction** *see* **friction of rest**

S609 **static head** || statische Druckhöhe *f* || pression *f* (hydro)statique || statische drukhoogte *f* || (гидро)-статический напор *m*

S610 **static lift** || statischer Auftrieb *m* || effort *m* de levée statique || statische opstijgkracht *f* (*m*) || статическая выталкивающая сила *f*

S611 **static load** || ruhende [statische] Belastung *f*, Ruhelast *f* || charge *f* [sollicitation *f*] statique || statische (rustende) belasting *f* || статическая нагрузка *f*

S612 **static machine** || elektrostatische Maschine *f* || machine *f* électrostatique || elektrostatische machine *f* || электростатическая машина *f*

S613 **static moment (of area)** || statisches Flächenmoment *n* || moment *m* statique d'une aire || statisch moment *n* van het vlak || статический момент *m* площади

S614 **statics** || Statik *f* || statique *f* || statica *f* || статика *f*

S615 **static test** || statische Erprobung *f* || essai *m* [épreuve *f*] statique || statische proef *f* (*m*) || статическое испытание *n*, испытание *n* статической нагрузкой

S616 **stationary** || stationär || stationnaire || stationair || стационарный; установившийся

S617 **stationary blade** *see* **stator blade**

S618 **stationary motion** || stationäre Bewegung *f* || mouvement *m* stationnaire || stationaire beweging *f* || установившееся [стационарное] движение *n*

S619 **stationar random process** || stationärer Zufallsprozeß *m* || processus *m* aléatoire stationnaire || stationair toevalsproces *n* || стационарный случайный процесс *m*

S620 **stationary state** || stationärer Zustand *m* || état *m* stationnaire || stationaire toestand *m* || стационарное состояние *n*; установившееся состояние *n*

S621 **statistical dependence** || statistische Abhängigkeit *f* || dépendence *f* statistique || statistische afhankelijkheid *f* || статистическая зависимость *f*

S622 **statistical hypothesis** || statistische Hypothese *f* || hypothèse *f* statistique || statistische hypothese *f* || статистическая гипотеза *f*

S623 **statistical thermodynamics** || statistische Thermodynamik *f* || thermodynamique *f* statistique || statistische thermodynamica *f* || статистическая термодинамика *f*

S624 **statistics** || Statistik *f* || statistique *f* || statistiek *f* || статистика *f*

S625 **stator** || Stator *m*, Ständer *m* || stator *m* || stator *m* || статор *m*

S626 **stator blade** || Leitschaufel *f*, Statorschaufel *f* || aile *f* du stator || vaste schoep *f* (*m*) || лопатка *f* статора, неподвижная лопатка *f*

S627 **stay** || **1.** Untersatz *m*, Bock *m*; Stütze *f* **2.** Abspannseil *n*, Ankerseil *n* **3.** Verbindungsstange *f* **4.** Setzstock *m*, Lünette *f* || **1.** étai *m* **2.** hauban *m* **3.** tige *f* d'ancrage **4.** lunette *f* || **1.** steun *m*, stut *m*; schoor *m* **2.** afspankabel *m*; *f* (*m*) **3.** verbindingsstand *f* (*m*) **4.** bril *m* || **1.** подставка *f*; опора *f* **2.** оттяжка *f*, ванта *f* **3.** соединительная штанга *f* **4.** люнет *m*

S628 **I steady** || **1.** Stütze *f*, Abstützung *f* **2.** Setzstock *m*, Lünette *f* || **1.** appui *m*; étai *m* **2.** lunette *f* || **1.** steun *m*; stut *m* **2.** bril *m* || **1.** опора *f*; подпорка *f* **2.** люнет *m*

S629 II **steady** || stabil, konstant; stationär || stable || stabiel || устойчивый; установившийся

S630 **steady pin** || Einstellstift *m*, Paßstift *m*, Stellstift *m* || cheville *f* d'appui [d'accouplement] || paspen *f* (*m*) || установочный [зажимной] штифт *m*

S631 **steady pull** || gleichförmige Zugkraft *f* || traction *f* uniforme || gelijkmatige trekkracht *f* (*m*) || равномерная сила *f* тяги

S632 **steady rotation** || gleichmäßige Drehung *f* *(mit konstanter Winkelgeschwindigkeit)* || rotation *f* uniforme || gelijkmatige draaiing *f* || равномерное вращение *n* *(с постоянной угловой скоростью)*

S633 **steady-state** || stationär || stationnaire || stationair || установившийся; стационарный

S634 **steady-state conditions** || stationärer Zustand *m* || état *m* stationnaire || stationaire toestand *m* || установившийся [стационарный] режим *m*

S635 **steady-state oscillation** || stationäre Schwingungen *f pl* || oscillations *f pl* stationnaires || stationaire trillingen *f pl* || установившиеся колебания *n pl*

S636 **steady-state vibration** || stationäre Schwingung *f*, Beharrungsschwingung *f* || vibrations *f pl* équilibrées [stabilisées] || stationaire schommeling *f* || установившиеся механические колебания *n pl*

S637 **steam distillation** || Wasserdampfdestillation *f* || distillation *f* à vapeur d'eau || waterdampdestillatie *f* || перегонка *f* с водяным паром

S638 **steam engine** || Dampfkraftanlage *f*; Dampfmaschine *f* || machine *f* à vapeur || stoommachine *f* || паровая машина *f*

S639 **steam power plant** || **1.** Wärmekraftwerk *n* **2.** Dampfkraftanlage *f* || **1.** centrale *f* électrothermique **2.** installation *f* à vapeur || stoomkrachtinstallatie *f* || **1.** теплоэлектростанция *f* **2.** паросиловая установка *f*

S640 **steam turbine** || Dampfturbine *f* || turbine *f* à vapeur || stoomturbine *f* (*m*) || паровая турбина *f*

S641 **steel** || Stahl *m* || acier *m* || staal *n* || сталь *f*

S642 **steel casting** || Stahlgußstück *n* || moulage *m* d'acier || staalgietwerk *n*, staalgietblok *n* || стальная отливка *f*

S643 **steepest descent method** || Methode *f* des steilsten Abstieges [des schnellsten Abstieges] || méthode *f* de la plus rapide descente || methode *f* van de snelste afdaling || метод *m* наискорейшего спуска

S644 **steepness** || Steilheit *f* *(einer Kennlinie)* || pente *f* || steilheid *f* || крутизна *f* *(характеристики)*

S645 **steering** || Lenkung *f* || direction *f*, mécanisme *m* de direction || besturing *f* || рулевое управление *n*

S646 **steering arm** || Lenkhebel *m* || levier *m* de manœuvre [de direction] || stuurarm *m*; stuurhefboom *m* || рычаг *m* управления, рулевой рычаг *m*

S647 **steering column** || Lenksäule *f* || colonne *f* de direction || stuurkolom *f* (*m*) || рулевая колонка *f*

S648 **steering gear** || **1.** Lenkgetrieberad *n* **2.** Lenkgetriebe *n* || engrenage *m* de direction; mécanisme *m* de direction || **1.** tandwiel *n* van het stuurmechanisme **2.** stuurinrichting *f* || **1.** шестерня *f* рулевого механизма **2.** рулевой механизм *m*

S649 **step-down transformer** || Abspanntransformator *m* || transformateur *m* abaisseur || dalingstransformator *m* || понижающий трансформатор *m*

S650 **step function** || Treppenfunktion *f* || fonction *f* en escalier || trapfunctie *f* || ступенчатая функция *f*

S651 **stepless transmission** || stufenloses Wechselgetriebe *n*, stufenloses Schaltgetriebe *f* || transmission *f* continue; variateur *m* || traploos regelbare overbrenging *f* || бесступенчатая передача *f*; бесступенчатая коробка *f* передач

S652 **stepped pulley** || Stufenscheibe *f* || poulie *f* étagée; poulie *f* à gradins || trapschijf *f* (*m*) || ступенчатый шкив *m*

S653 **step response** || Einheitsübergangsfunktion *f*, Sprungantwort *f* || caractéristique *f* transitoire || eenheidsovergangsfunctie *f* || переходная характеристика *f*

S654 **step-up transformer** || Aufspanntransformator *m* || transformateur *m* élévateur || verhogingstransformator *m* || повышающий трансформатор *m*

S655 **stereoisomerism** || Stereoisomerie *f*, Raumisomerie *f* || isomérie *f* stéréochimique, stéréo-isomérie *f* || stereo-isomerie *f* || стереоизомерия *f*, пространственная изомерия *f*

S656 **stereophony** || Stereophonie *f* || stéréophonie *f* || stereofonie || стереофония *f*, стереофонический эффект *m*

S657 **stereoscopy** || Stereoskopie *f* || stéréoscopie *f* || stereoscopie *f* || стереоскопия *f*, стереоскопический эффект *m*

S658 **steric hindrance** || sterische Behinderung *f*, sterische Hinderung *f* || difficulté *f* spatiale || ruimtemoeilijkheid *f* || стерическое препятствие *n*

S659 **Stieltjes integral** || Stieltjessches Integral *n* || intégrale *f* de Stieltjes || integraal *f* (*m*) van Stieltjes || интеграл *m* Стилтьеса

S660 **stiffness** || Steifigkeit *f*, Steifheit *f* || rigidité *f* || stevigheid, stijfheid *f* || жёсткость *f*

S661 **stiffness coefficient** || Steifigkeitszahl *f*, Steifigkeitsbeiwert *m*; Steifigkeitsmaß *n* || coefficient *m* d'élasticité *(p. ex. d'un appui élastique)* || stijfheidscoëfficiënt *m* || коэффициент *m* жёсткости [упругости]

S662 **stigmatic** || stigmatisch || stigmatique || stigmatisch || стигматический

S663 **stilb** || Stilb *n* || stilb *m* || stilb *n* || стильб *m* *(10⁴ кд/м²)*

S664 **stimulated emission** || induzierte Emission *f* || émission *f* provoquée [stimulée] || geïnduceerde straling *f* || стимулированное [индуцированное, вынужденное] излучение *n*

S665 **stimulated transition** || induzierter [erzwungener] Übergang *m* || transition *f* forcée || geïnduceerde overgang *m* || стимулированный [вынужденный] переход *m*

S666 **stochastic** || stochastisch || stochastique || stochastisch || стохастический; случайный

S667 **stochasticity** || Zufälligkeit *f*, Zufallscharakter *m* || stochasticité *f* || toevalligheid *f* || стохастичность *f*, стохастический характер *m*

S668 **stochastic process** || stochastischer [zufälliger] Prozeß *m* || processus *m* stochastique || stochastisch proces *n* || стохастический [случайный] процесс *m*

S669 **stochastic variable** || Zufallsvariable *f* || variable *f* stochastique || stochastische variabele *f* || случайная переменная *f*

S670 **stokes** || Stokes *n* || stokes *m* || stokes *n* || стокс *m* *(10^{-4} м²/с)*

S671 **stopping power** || Bremsvermögen *n* || pouvoir *m* d'arrêt || remvermogen *n* || тормозная способность *f*

S672 **stop valve** || Absperrventil *n*; Absperrhahn *m* || soupape *f* de fermeture || afsluiter *m*; afsluitkraan *f* (*m*) || запорный клапан *m*

S673 **storage** || Speicher *m* || mémoire *f* || geheugen *n* || запоминающее устройство *n*, память *f*

S674 **storage capacity** || Speicherkapazität *f* || volume *m* de mémoire || geheugencapaciteit *f* || ёмкость *f* памяти

S675 **storage cell** || Speicherzelle *f* || accumulateur *m* || accumulator *m* || аккумулятор *m*

S676 **store** *see* **storage**

S677 **stored energy** || aufgespeicherte Energie *f* || énergie *f* emmagasinée || achtergebleven energie *f*, opgezamelde energie *f* || запас *m* энергии, запасённая энергия *f*

S678 **straight bevel gear** || Geradzahnkegelrad *n*, geradverzahntes Kegelrad *n* || pignon *m* conique à denture droite; roue *f* conique à dents droites || conisch tandwiel *n* met rechtstandige tanden || прямозубое коническое колесо *n*; прямозубая коническая шестерня *f*

S679 **straightener** *see* **straightening machine**

S680 **straight-flute drill** || geradnutiger Bohrer *m* || foret *m* à rainures droites || (metaal)boor *f* (*m*) met rechte groeven || сверло *n* с прямыми канавками

S681 **straight-flute reamer** || geradgenutete [geradzahnige] Reibahle *f* || alésoir *m* à dents droites || ruimer *m* met rechte groeven || развёртка *f* с прямыми зубьями

S682 **straight line** || Gerade *f* || ligne *f* droite || rechte *f* || прямая линия *f*

S683 **straight-line motion** || geradlinige Bewegung *f* || mouvement *m* rectiligne || rechtlijnige beweging *f* || прямолинейное движение *n*

S684 **straight path** || geradlinige Bahn *f* || trajectoire *f* rectiligne || rechtlijnige baan *f* (*m*) || прямолинейная траектория *f*

S685 **straight spur gear** *see* **spur gear**

S686 **straightening machine** || Richtmaschine *f* || dresseuse *f*; machine *f* à dresser || richtmachine *f* || правильная машина *f*

S687 **straight tool** || gerader Meißel *m* || outil *m* droit || rechte beitel *m* || прямой резец *m*

S688 **strain** || Verformung *f*, Formänderung *f* || déformation *f* || vervorming *f*, deformatie *f* || деформация *f*

S689 **strain gauge** || Dehnungsmesser *m*, Tensometer *n* || capteur *m* de déformation; extensomètre *m* || rekstrookopnemer *m* || тензодатчик *m*; тензометр *m*

S690 **strangeness** || Strangeness *f*, Fremdheitsquantenzahl *f* étrangeté *f* || strangeness *f* || странность *f* (*квантовое число*)

S691 **strange particle** || «seltsames» Teilchen *n* || particule *f* étrange || zonderling deeltje *n* || странная частица *f*

S692 **strategy** || Strategie *f* || stratégie *f* || strategie *f* || стратегия *f*

S693 **stratification** || Schichtung *f* || stratification *f* || gelaagdheid *f* || стратификация *f*, расслоение *n*

S694 **stray** || **1.** unerwünscht, parasitär, Stör ... **2.** vagabundierend (*Strom*) **3.** Streu ... (*Feld*) || **1.** parasite, parasitaire **2.** errant **3.** diffusé || **1.**ongewenst **2.**vagebonderend (*stroom*) **3.** verspreid || **1.** паразитный **2.** блуждающий (*о токах*) **3.** рассеянный (*о поле*)

S695 **stray coupling** || Streukopplung *f* || couplage *m* parasitaire || parasitaire koppeling *f* || паразитная связь *f*

S696 **stray currents** || Irrströme *m pl* || courants *m pl* vagabonds || vagebonderende stromingen *f pl* || блуждающие токи *pl*

S697 **stream** || Strom *m*, Strömung *f* || courant *m*; jet *m* || stroming *f* || поток *m*; струя *f*

S698 **streamer** || Streamer *m* || streamer *m* || streamer *m* || стример *m* (*разряда*)

S699 **streamline** || Stromlinie *f*, Strömungslinie *f* (*inkompressible Strömungen*) || ligne *f* de courant || stroomlijn *f* (*m*) || линия *f* тока

S700 **streamline flow** || laminare [wirbelfreie] Strömung *f*, Laminarströmung *f* || écoulement *m* laminaire || laminaire stroming *f* || ламинарное течение *n*

S701 **strength** || **1.** Festigkeit *f* **2.** Festigkeitsgrenze *f*, Bruchfestigkeit *f* **3.** Stärke *f*, Feldstärke *f* || **1.** résistance *f* **2.** résistance *f* limite **3.** intensité *f* (de champ) || **1.** stevigheid *f* **2.** stevigheidsgrens *f* (*m*) **3.** intensiteit *f* || **1.** прочность *f* **2.** предел *m* прочности **3.** напряжённость *f* (*поля*)

S702 **strengthening** || Verstärkung *f*; Verfestigung *f*, Erhöhung *f* der mechanischen Eigenschaften || renforcement *m*; raidissement *m* || verstijving *f*; versterking *f* усиление *n*; укрепление *n*; повышение *n* механических свойств, упрочнение *n*

S703 **strengthening rib** || Verstärkungsrippe *f*, Versteifungsrippe *f* || raidisseur *m*; nervure *f* de raidissement [de renforcement] || verstijvingsrib *f* (*m*) || ребро *n* жёсткости

S704 **strength test** || Festigkeitsprüfung *f* || essai *m* de résistance || vastheidsproef *f* (*m*) || испытание *n* на прочность

S705 **stress** || Spannung *f*; Beanspruchung *f* || contrainte *f*, tension *f* || spanning *f* || (механическое) напряжение *n*

S706 **stress concentration** || Spannungskonzentration *f*; Spannungsspitze *f* || concentration *f* de contraintes || spanningsconcentratie *f* || концентрация *f* напряжений

S707 **stress cycle** || Spannungszyklus *m*, Schwingbeanspruchungsperiode *f* || cycle *m* de contraintes || spanningscyclus *m* || цикл *m* напряжений

S708 **stress range** || Spannungsamplitude *f*; Schwingbreite *f* der Spannung || domaine *m* [portée *f*] d'action des contraintes alternées || spanningsamplitude *f* || размах *m* напряжений цикла

S709 **stress relief** || Spannungsbeseitigung *f*, Beseitigung *f* der Eigenspannungen; Spannungsentlastung *f* || élimination *f* des contraintes internes || ontspanning *f* || снятие *n* внутренних напряжений

S710 **stress state** *see* **state of stress**

S711 **stretching force** *see* **tensile force**

S712 **strict** || strikt || strict, rigoureux || strikt || строгий

S713 **striking gear** || **1.** Riemenausrücker *m*, Riemenschalter *m*, Riemenumleger *m* || fourche *f* à courroie; guide-courroi *m* || riemaflegger *m* || ремённая отводка *f*

S714 **string** || **1.** Saite *f* **2.** Stromfaden *m* **3.** Kette *f* || **1.** corde *f* **2.** fil *m* **3.** chaîne *f* || **1.** snaar *f* **2.** draad *m* **3.** ketting *f* || **1.** струна *f* **2.** (электропроводная) нить *f* **3.** цепочка *f*

S715 **stripped atom** || hochionisiertes Atom *n* || atome *m* dépouillé d'électrons, atome *m* nucléaire || sterk geïoniseerd atoom *n* || полностью иониз(ир)ованный атом *m*

S716 **strip rolling mill** || Bandwalzwerk *n* || laminoir *m* à feuillards || stripwals *f* (*m*), stripwalserij *f* || полосовой (прокатный) стан *m*

S717 **strip steel** || Bandstahl *m* || feuillard *m* (en acier) || stripstaal *n*; bandstaal *n* || полосовая [ленточная] сталь *f*

S718 **stroboscope** || Stroboskop *n* || stroboscope *m* || stroboscoop *m* || стробоскоп *m*

S719 **stroboscopic effect** || stroboskopischer Effekt *m* || effet *m* stroboscopique || stroboscopisch effect *n* || стробоскопический эффект *m*

S720 **stroke** || Hub *m*; Hubweg *m*, Hublänge *f* (*z. B. eines Kolbens*) || course *f* (*p. ex. de piston*) || slag *m*; slaglengte *f* (*bijv. zuiger*) || ход *m*; длина *f* хода (*напр. поршня*)

S721 **strong interaction** || starke Wechselwirkung *f* || interaction *f* forte || sterke wisselwerking *f* || сильное [ядерное] взаимодействие *n*

S722 **strongly monotonic function** || streng monotone Funktion *f* || fonction *f* strictement monotone || streng monotone functie *f* || строго монотонная функция *f*

S723 **strontium, Sr** || Strontium *n* || strontium *m* || strontium *n* || стронций *m*

S724 **strophoid** || Strophoide *f* || strophoïde *f* || strophoïde *f* || строфоида *f*

S725 **structural analysis** || Strukturanalyse *f* || analyse *f* de structure [structurale] || structuuranalyse *f* || структурный анализ *m*

S726 **structure** || Struktur *f*, Gefüge *n*; Aufbau *m*; Konstruktion *f* || structure *f*; construction *f* || structuur *f*, bouw *m*; gebouw *n*; bouwwerk *n*, constructie *f* || структура *f*; строение *n*; конструкция *f*

S727 **stud bolt** || **1.** Stiftschraube *f* **2.** (kurze) Achse *f*; Zapfen *m* || boulon *m* prisonnier, goujon *m*, pivot *m* || **1.** stift *f* (*m*) **2.** tapeind *n*; tap *m* || **1.** шпилька *f* **2.** (короткая) ось *f*; цапфа *f*

S728 **Student's t, Students's t-distribution** || Student-Verteilung *f*, t-Verteilung *f* || distribution *f* de Student, distribution *f* t || verdeling *f* van Student || распределение *n* Стьюдента, t-распределение *n*

S729 **stuffing** || Packung *f*; Dichtung *f*, Liderung *f* || garniture *f*; presse-étoupe *m* || dichting *f*; dichtstopping *f* || набивка *f*; прокладка *f*

S730 **stuffing box** || Stopfbuchse *f* || boîte *f* à garniture [à étoupe], garniture *f* d'étanchéité || pakkingbus *f* (*m*) || сальник *m*, сальниковое уплотнение *n*

S731 **subalgebra** || Unteralgebra *f* || sous-algèbre *f* || onderalgebra *f* || подалгебра *f*

S732 **subcritical** || unterkritisch || sous-critique, subcritique || onderkritisch || докритический, подкритический

S733 **subfield** || Unterkörper *m* || sous-corps *m* || onderlichaam *n* || подполе *n*

S734 **subgroup** || Untergruppe *f* || sous-groupe *m* || ondergroep *f* (*m*) || подгруппа *f*

S735 **subintervall** || Teilintervall *n* || sous-intervalle *m* || deelinterval *n* || подинтервал *m*

S736 **sublevel** || Unterniveau *n* || sous-niveau *m* || onderniveau *n* || подуровень *m* (*энергии*)

S737 **sublimation** || Sublimation *f* || sublimation *f* || sublimatie *f* || сублимация *f*, возгонка *f*

S738 **submatrix** || Untermatrix *f*, Teilmatrix *f* || sous-matrice *f* || ondermatrix *f*, deelmatrix *f* || подматрица *f*

S739 **submerged weight** || Gewicht *n* des untergetauchten Körpers || poids *m* du corps submergé || gewicht *n* van een ondergedompeld lichaam || вес *m* тела в жидкости

S740 **subprogram** || Unterprogramm *n* || sous-programme *m* || subprogramma *n* || подпрограмма *f*

S741 **subprogram library** || Unterprogrammbibliothek *f* || bibliothèque *f* des sous-programmes || bibliotheek van subprogrammas *f* || библиотека *f* подпрограмм

S742 **subring** || Unterring *m*, Teilring *m* || sous-anneau *m* || ondering *m* || подкольцо *n*

S743 **subroutine** *see* **subprogramm**

S744 **subscriber** || Teilnehmer *m* || utilisateur *m*, abonné *m* || abonné *m*, *f* || абонент *m*

S745 **subset** || Untermenge *f* || sous-ensemble *m* || ondermenigte *f* || подмножество *n*

S746 **subsonic aerodynamics** || Unterschallaerodynamik *f* || aérodynamique *f* subsonique || subsonische aërodynamica *f* || аэродинамика *f* дозвуковых скоростей

S747 **subsonic velocity** || Unterschallgeschwindigkeit *f* || vitesse *f* subsonique || subsonische snelheid *f* || дозвуковая скорость *f*

S748 **subspace** || Unterraum *m*, Teilraum *m* || sous-espace *m* || onderruimte *f* || подпространство *n*

S749 **substance** || Stoff *m*; Substanz *f* || substance *f* || stof *f* (*m*) || вещество *n*

S750 **substitution** || Substitution *f*; Einsetzung *f* || substitution *f* || substitutie *f* || **1.** *мат.* подстановка *f* **2.** *хим.* замещение *n*

S751 **substitution group** || Substitutionsgruppe *f* || groupe *m* des substitutions || substitutiegroep *f* (*m*) || группа *f* подстановок

S752 **substitution reaction** || Substitutionsreaktion *f*, Austauschreaktion *f* || réaction *f* de substitution [de remplacement] || substitutiereactie *f* || реакция *f* замещения

S753 **substructure** || Unterstruktur *f* || sous-structure *f* || onderstructuur *f* || подструктура *f*

S754 **subtracter** || Subtraktor *m*, Subtraktionseinheit *f* || soustracteur *m* || aftrekker *m* || вычитатель *m*

S755 **subtraction** || Subtraktion *f* || soustraction *f* || aftrekken *n* || вычитание *n*

S756 **subtrahend** || Subtrahend *m* || nombre *m* à soustraire || aftrekker *m* || вычитаемое *n*

S757 **successive approximations method** || Methode *f* der sukzessiven Approximation [der schrittweisen Näherung] || méthode *f* des approximations successives || methode *f* van successieve benaderingen || метод *m* последовательных приближений

S758 **suction** || Einsaugen *n*; Ansaugung *f*; Absaugen *n* || succion *f*; aspiration *f* || zuigen *n*, zuiging *f*; opzuigen *n*; afzuiging *f* || всасывание *n*; подсос *n*; отсасывание *n*
S759 **suction pressure** || Saugdruck *m* || pression *f* de succion; pression à l'aspiration || zuigdruk *m* || давление *n* всасывания
S760 **suction pump** || Saugpumpe *f* || pompe *f* aspirante [à aspiration] || zuigpomp *f* (*m*) || всасывающий насос *m*
S761 **suction valve** || Saugventil *n* || clapet *m* [soupape *f*] d'aspiration || zuigklep *f* (*m*) || всасывающий клапан *m*
S762 **sufficient** || hinreichend || suffisant || voldoende || достаточный
S763 **sufficient estimate** || effektive [leistungsfähige] Schätzung *f* || estimation *f* suffisante || effectieve schatting *f* || эффективная оценка *f*
S764 **sulphate** || Sulfat *n* || sulfate *m* || sulfaat *n* || сульфат *m*
S765 **sulphide** || Sulfid *n* || sulfure *m* || sulfide *n* || сульфид *m*
S766 **sulphite** || Sulfit *n* || sulphite *m* || sulfiet *n* || сульфит *m*
S767 **sulphonic acid** || Sulfiet *f* || acide *m* sulfonique, sulfacide *m*, sulfo-acide *m* || sulfonzuur *n* || сульфокислота *f*
S768 **sulphur**, S || Schwefel *m* || soufre *m* || zwavel *m* || сера *f*
S769 **sulphur dioxide**, SO_2 || Schwefeldioxid *n* || bioxyde *m* de soufre, anhydride *m* sulfureux || zwaveldioxyde *m* || двуокись *f* серы, сернистый ангидрид *m*
S770 **sulphuric acid**, H_2SO_4 || Schwefelsäure *f* || acide *m* sulfurique || zwavelzuur *n* || серная кислота *f*
S771 **sulphuric anhydride** *see* **sulphur trioxide**
S772 **sulphurous acid**, H_2SO_3 || schweflige Säure *f* || acide *m* sulfureux || zwaveligzuur *n* || сернистая кислота *f*
S773 **sulphur trioxide**, SO_3 || Schwefeltrioxid *n*, Schwefelsaüreanhydrid *n* || trioxyde *m* de soufre || zwaveltrioxyde *n* || трёхокись *f* серы, серный ангидрид *m*
S774 **sum** || Summe *f*, Betrag *m* || somme *f* || som *f* (*m*), bedrag *n* || сумма *f*
S775 **summable** || summierbar || sommable || sommeerbaar || суммируемый *(о функции)*
S776 **summation** || Summation *f*, Summierung *f* || sommation *f* || sommering *f* || суммирование *n*
S777 **sunk key** || Nutenkeil *m*, Einlegekeil *m* || clavette *f* encastrée; clavette *f* à rainure || sleufspie *f* (*m*) || врезная шпонка *f*; закладная шпонка *f*
S778 **sunk screw** || versenkte [eingelassene] Schraube *f* || vis *f* à tête noyée || verzonken schroef *f* (*m*) || винт *m* с потайной головкой
S779 **sun wheel** || Sonnenrad *n* *(Umlaufgetriebe)* || engrenage *m* fixe *(d'un planétaire)* || centraal tandwiel *n* *(planeetdrijfwerk)* || солнечное [центральное] зубчатое колесо *n* *(планетарной передачи)*
S780 **supercharged engine** || Kompressormotor *m*, aufgeladener Motor *m* || moteur *m* suralimenté || compressormotor *m* || двигатель *m* (внутреннего сгорания) с наддувом
S781 **supercharge load** || Überbelastung *f*, Überlast *f* || (sur)charge *f* excessive || overbelasting *f* || чрезмерная нагрузка *f*, перегрузка *f*
S782 **supercharger** || Gebläse *n*; Verdichter *m*, Kompressor *m* || compresseur *m* de suralimentation || aanjager *m*; compressor *m* || нагнетатель *m*; компрессор *m* наддува
S783 **supercharging** || Aufladung *f* || suralimentation *f* || aanjaging *f* || наддув *m* *(ДВС)*
S784 **superconducting** || supraleitend || superconductif || supraconductief || сверхпроводящий
S785 **superconducting** state || supraleitender Zustand *m*, Supraleitungszustand *m* || état *m* superconductif || supraconductieve toestand *m* || сверхпроводящее состояние *n*
S786 **superconductivity** || Supraleitung *f* || superconductivité *f*

supraconductiviteit *f* || сверхпроводимость *f*

S787 **superconductor** || Supraleiter *m* || superconducteur *m* || supergeleider *m* || сверхпроводник *m*

S788 **supercooled liquid** unterkühlte Flüssigkeit *f* || liquide *m* surfondu || onderkoelde vloeistof *f* (*m*) || переохлаждённая жидкость *f*

S789 **supercooled state** || Unterkühlung *f* || état *m* de sous-refroidissement || onderkoeling *f* || переохлаждённое состояние *n*; переохлаждение *n*

S790 **supercooling** || Unterkühlung *f* || sous-refroidissement *m* || onderkoeling *f* || переохлаждение *n*

S791 **supercritical** || überkritisch || su(pe)rcritique || overkritisch || надкритический, сверхкритический

S792 **superfinish** || Fein(st)ziehschleifen *n*, Superfinish *n* || superfinition *f* || superfinishen *n* || суперфиниширование *n*, суперфиниш *m*, чистовая обработка *f*

S793 **superfinish machine** || Fein(st)ziehschleifmaschine *f* || machine *f* à superfinir || superfinishmachine *f* || станок *m* для суперфиниширования

S794 I **superfluid** || Supraflüssigkeit *f* || superfluide *m* || supravloeistof *f* (*m*) || сверхтекучая жидкость *f*

S795 II **superfluid** || suprafluid || superfluide || supravloeibaar || сверхтекучий

S796 **superfluidity** || Suprafluidität *f* || superfluidité *f* || supravloeibaarheid *f* || сверхтекучесть *f*

S797 **superheated steam** || überhitzter Dampf *m*, Heißdampf *m* || vapeur *f* surchauffée || oververhitte stoom *m* || перегретый пар *m*

S798 **superheating** || Überhitzung *f* || surchauffage *m* || oververhitting *f* || перегрев *m*, перегревание *n*

S799 **superhigh frequency** || Superhochfrequenz *f* || hyperfréquence *f* || superfrequentie *f* || сверхвысокая частота *f*, СВЧ

S800 **superionic conductivity** || Supraionenleitfähigkeit *f* || conductivité *f* superionique || supra-ionengeleidbaarheid *f* || суперионная проводимость *f*

S801 **superlattice** || Überstruktur *f* (*eines Kristalls*); Übergitter *n* || surstructure *f* || overstructuur *f* || сверхструктура *f* (*кристалла*); сверхрешётка *f*

S802 **superparamagnetism** || Superparamagnetismus *m* || superparamagnétisme *m* || superparamagnetisme *n* || суперпарамагнетизм *m*, магнетизм *m* дисперсных частиц

S803 **superposition** || Superposition *f*; Überlagerung *f* || superposition *f* || superpositie *f* || суперпозиция *f*, наложение *n*

S804 **superposition method** || Superpositionsmethode *f* || méthode *f* de superposition || superpositiemethode *f* || метод *m* суперпозиции

S805 **superposition of waves** || Wellenüberlagerung *f* || superposition *f* des ondes || golvensuperpositie *f* || суперпозиция *f* волн

S806 **superposition principle** || Superpositionsprinzip *n* || principe *m* de superposition || principe *n* van de superpositie || принцип *m* суперпозиции

S807 **superposition theorem** || Superpositionssatz *m* || théorème *m* de superposition || superpositietheorema *n* || теорема *f* о суперпозиции

S808 **superpressure** || Überdruck *m* || surpression *f* || overdruk *m* || избыточное (*над атмосферным*) давление *n*

S809 **supersaturated vapo(u)r** || übersättigter Dampf *m* || vapeur *f* sursaturée || oververzadigde stoom *m* || пересыщенный пар *m*

S810 **supersaturation** || Übersättigung *f* || sursaturation *f* || oververzadiging *f* || пересыщение *n*

S811 **supersonic aerodynamics** || Überschallaerodynamik *f* || aérodynamique *f* supersonique || supersonische aërodynamica *f* || аэродинамика *f* сверхзвуковых скоростей

S812 **supersonics** || **1.** Überschallaerodynamik *f* **2.** Ultraschallakustik *f* || **1.** aérodynamique *f* supersonique **2.** acoustique *f* d'ultra-son || **1.** supersonische aërodynamica *f* **2.** ultrasoonakoestik *f* || **1.** сверхзвуковая аэродинамика *f* 2. акустика *f* ультразвука, ультраакустика *f*

S813 **supersonic velocity** || Überschallgeschwindigkeit *f* || vitesse *f* supersonique || supersonische snelheid *f* || сверхзвуковая скорость *f*

S814 **supervisor** || Supervisor *m*, Überwacher *m*, Überwachungsprogramm *n* || superviseur *m* || supervisor *m* || супервизор *m*

S815 **supervoltage** || Höchstspannung *f* || surtension *f* || hoogspanning *f* || сверхвысокое напряжение *n*

S816 **supply** || Speisung *f*, Versorgung *f* || alimentation *f* || voeding *f* || питание *n*, подача *f*

S817 **supply source** || Stromquelle *f*, Speisequelle *f* || source *f* d'alimentation || voedingsbron *f* (*m*) || источник *m* питания

S818 **supply unit** || Netzteil *m* || bloc *m* d'alimentation || voedingseenheid *f* || блок *m* питания

S819 **support** || **1.** Stütze *f*, Abstützung *f*; Untersatz *m* **2.** Support *m* **3.** Kragarm *m*; Träger *m* || support *m*; appui *m* || steun *m*; ondersteuning *f* **2.** support *n* **3.** stut *m* || **1.** опора *f*; подставка *f* **2.** суппорт *m* **3.** кронштейн *m*

S820 **support force** || Stützkraft *f* || réaction *f* d'appui || steunreactie *f* || реакция *f* опоры

S821 **supporting bearing** || Stützlager *n*, Traglager *n* || palier *m* d'appui; palier *m* de charge; palier *m* radial || steunlager *n* || опорный подшипник *m*

S822 **supporting surface** || Stützfläche *f* || Auflagerfläche *f* || surface *f* d'appui || draagvlak *n* || опорная поверхность *f*

S823 **suppression** || Unterdrückung *f* || suppression *f* || onderdrukking *f* || подавление *n*

S824 I **surd** || irrationale Zahl *f*; irrationale Größe *f* || nombre *m* irrationnel || irrationeel getal *n* || иррациональное число *n*; иррациональная величина *f*

S825 II **surd** || irrational || irrationnel || irrationeel || иррациональный

S826 **surface** || Oberfläche *f*; Fläche *f* || surface *f* || vlak *n*, oppervlakte *f* || поверхность *f*

S827 **surface active agent** || oberflächenaktiver Stoff *m* || substance *f* tensio-active || oppervlak-actieve stof *f* (*m*) || поверхностно-активное вещество *n*

S828 **surface activity** || Oberflächenaktivität *f* || tensio-activité *f*, activité *f* superficielle || oppervlakteactiviteit *f* || поверхностная активность *f*

S829 **surface density** || Oberflächendichte *f* || densité *f* superficielle || oppervlaktedichtheid *f* || поверхностная плотность *f*

S830 **surface energy** || Oberflächenenergie *f* || énergie *f* superficielle || oppervlakte-energie *f* || поверхностная энергия *f*; энергия *f* границы раздела

S831 **surface finish** || Oberflächengüte *f*, Bearbeitungsgüte *f* || degré *m* de finition de la surface; propreté *f* de la surface || oppervlaktekwaliteit *f*; afwerkingskwaliteit *f* || чистота *f* поверхности, чистота *f* обработки

S832 **surface grinding machine** || Flachschleifmaschine *f* || rectifieuse *f* plane; machine *f* à rectifier les surfaces planes || vlakslijpmachine *f* || плоскошлифовальный станок *m*

S833 **surface integral** || Oberflächenintegral *n* || intégrale *f* de surface || oppervlakte-integraal *f* (*m*) || интеграл *m* по поверхности

S834 **surface milling machine** || Langfräsmaschine *f*, Planfräsmaschine *f* || raboteuse *f*; machine *f* à raboter || vlakfreesmachine *f* || продольно-фрезерный станок *m*

S835 **surface normal** || Flächennormale *f* || normale *f* à une surface || vlaktenormaal *f* (*m*) || нормаль *f* к поверхности

S836 **surface of revolution** || Rotationsfläche *f* || surface *f* de révolution || rotatievlak *n* || поверхность *f* вращения

S837 **surface stress** || Oberflächenspannung *f* || contrainte *f* superficielle || oppervlaktespanning *f* || поверхностное напряжение *n*

S838 **surface tension** || Oberflächenspannung *f* || tension *f* superficielle || oppervlaktespanning *f* || поверхностное натяжение *n*

S839 **surface viscosity** || Oberflächenzähigkeit *f* || viscosité *f* superficielle || oppervlakteviscositeit *f* || вязкость *f* поверхностной плёнки

S840 **surface wave** || Oberflächenwelle *f* || onde *f* superficielle || oppervlakgolf *f* *(m)* || поверхностная волна *f*

S841 **surfacing** || **1.** Oberflächenbearbeitung *f* **2** Plandrehen *m*, Planen *n* || surfaçage *m* || **1.** oppervlaktebehandeling *f*; oppervlakte-afwerking *f* **2.** plandraaien *n*, vlakdraaien *n*, voorhoofddraaien *n* || **1.** обработка *f* поверхности **2.** лобовое точение *n*

S842 **surfacing lathe** *see* **facing lathe**

S843 **surge-crest ammeter** || Maximumstrommesser *m* || ampèremètre *m* maximal || maximumstroommeter *m* || максимальный амперметр *m*

S844 **surge current** || Stromstoß *m* || accroissement *m* subit de courant || stroomstoot *m* || бросок [толчок] *m* тока

S845 **surroundings** || Umgebung *f* || environnement *m*, milieu *m* ambiant || omgeving *f* || окружающая [внешняя] среда *f*

S846 **susceptance** || Blindleitwert *m*, Suszeptanz *f* || admittance *f* imaginaire, susceptance *f* || susceptantie *f* || реактивная проводимость *f*

S847 **susceptibility** || Suszeptibilität *f* || susceptibilité *f* || susceptibiliteit *f* || восприимчивость *f*

S848 **suspension** || **1.** Suspension *f* **2.** Aufhängung *f* || suspension *f* || **1.** suspensie *f* **2.** ophanging *f* || **1.** суспензия *f*, взвесь *f* **2.** подвес *m*; подвеска *f*

S849 **sustained oscillations** || ungedämpfte Schwingungen *f pl* || oscillations *f pl* entretenues || ongedempte trillingen *f pl* || незатухающие колебания *pl*

S850 **swage** || **1.** Gesenk *n* **2.** Stempel *m* **3.** Matrize *f* || matrice *f* (de forge); étampe *f*; emboutisseuse *f* || **1.** smeedhamer *m* **2.** pons *m* **3.** matrijs *f* *(m)* || **1.** (ковочный) штамп *m* **2.** пуансон *m* **3.** матрица *f*

S851 **swaging** || Gesenkschmieden *n*; Formschmieden *n* || estampage *m* [matriçage] à chaud || matrijssmeden *n* || ковка *f* в штампах; фасонная ковка *f*

S852 **sweat cooling** || Schwitzkühlung *f*, Transpirationskühlung *f* || refroidissement *m* par paroi poreuse || verdampingskoeling *f* *(door poreuse materialen)* || охлаждение *n* испарением *(через пористые материалы)*

S853 **sweep** || Abtastung *f* || balayage *m* || zwaai *m* || развёртка *f*

S854 **switch** || Schalter *m* || interrupteur *m*; commutateur *m* || schakelaar *m* || выключатель *m*; переключатель *m*

S855 **switchboard** || Schalttafel *f* || pupitre *m* de commande || schakelbord *n* || пульт *m* [щит *m*] управления; распределительный [коммутационный] щит *m*; коммутационная панель *f*

S856 **switchgear** || Schaltgerät *n* || appareillage *m* de distribution || schakelapparaten *n pl* || коммутационная аппаратура *f*; распределительное устройство *n*

S857 **switching** || Schalten *n*; Umschalten *n* || interruption *f*; commutation *f* || schakeling *f* || включение *n*; выключение *n*; переключение *n*

S858 **switching algebra** || Schaltalgebra *f* || algèbre *f* de commutation || schakelalgebra *f* *(m)* || алгебра *f* релейных схем

S859 **switching circuit** || Schaltkreis *m* || circuit *m* de commutation || schakelcircuit *n* || переключающая схема *f*

S860 **switch key** || Schalter *m* || commutateur *m* || schakelaar *m* || переключатель *m*, коммутационный ключ *m*

S861 **swivel** || Wirbel *m*, Drehwirbel *m*; Wirbel *m* *(einer Kette)* || tête *f* pivotante; joint *m* articulé || wartel *m*, wervel *m*; kettingwartel *m* || вертлюг *m*; вертлюжное звено *n* *(цепи)*

S862 **swivelling tool holder** || Schwenkmeißelhalter *m* || porte-outil *m* pivotant || draaibare beitelhouder *m* || поворотный резцедержатель *m*

S864 **symbol** || Symbol *n* || symbole *m* || symbool *n* || символ *m*

S865 **symbolic address** || symbolische Adresse *f* || adresse *f* symbolique || symbolisch adres *n* || символьный адрес *m*

S866 **symbolic language** || Symbolsprache *f*, symbolische Programmiersprache *f* || langage *m* symbolique || symbolische taal || символьный язык *m* *(программирования)*

S867 **symbolic logic** || symbolische Logik *f* || logique *f* symbolique || symbolische logica *f* || символьная логика *f*

S868 **symmetrical** || symmetrisch || symétrique || symmetrisch || симметричный; симметрический

S869 **symmetric group** || symmetrische Gruppe *f* || groupe *m* symétrique || symmetrische groep *f* *(m)* || симметрическая группа *f*

S870 **symmetric matrix** || symmetrische Matrix *f* || matrice *f* symétrique || symmetrische matrix *f* || симметрическая матрица *f*

S871 **symmetric relation** || symmetrisches Verhältnis *n* || relation *f* symétrique || symmetrische verhouding *f* || симметричное отношение *n*

S872 **symmetrization** || Symmetrierung *f*, Ausgleich *m* || symétrisation *f* || symmetrisatie *f* || симметрирование *n*

S873 **symmetry** || Symmetrie *f* || symétrie *f* || symmetrie *f* || симметрия *f*

S874 **symmetry operation** || Deckoperation *f*, Decktransformation *f*, Deckbewegung *f* || opération *f* de symétrie || symmetrieoperatie *f* || операция *f* симметрии

S875 I **symplex** || Symplex *m* || simplex *m* || simplex *m* || симплекс *m*

S876 II **symplex** || Symplex... || simplex || simplex... || симплексный

S877 **synchrocyclotron** || Synchrozyklotron *n* || synchrocyclotron *m* || synchrocyclotron *n* || синхроциклотрон *m*

S878 **synchronism** || Synchronismus *m* || synchronisme *m* || synchronisme *m* || синхронизм *m*, синхронность *f*

S879 **synchronization** || Synchronisierung *f*, Synchronisation *f* || synchronisation *f* || synchronisatie *f* || синхронизация *f*

S880 **synchronizer** || Synchronisator *m* || synchronisateur *m* || synchronisator *m* || синхронизатор *m*

S881 **synchronizing voltmeter** || Synchronisierungsvoltmeter *n*, Nullvoltmeter *n* || voltmètre *m* de synchronisation || synchroniseervoltmeter *m* || вольтметр *m* для синхронизации, нулевой вольтметр *m*

S882 **synchronous** || synchron, gleichlaufend || synchrone || synchroon || синхронный, одновременный

S883 **synchronous generator** || Synchrongenerator *m* || alternateur *m* synchrone || synchroongenerator *m* || синхронный генератор *m*

S884 **synchronous motor** || Synchronmotor *m* || moteur *m* synchrone || synchroonmotor *m* || синхронный электродвигатель *m*

S885 **synchrophasotron** || Synchrophasotron *n* || synchrophasotron *m* || synchrofasotron *n* || синхрофазотрон *m*

S886 **synchrotron** || Synchrotron *n* || synchrotron *m* || synchrotron *n* || синхротрон *m*

S887 **synchrotron radiation** || Synchrotronstrahlung *f* || rayonnement *m* synchrotron || synchrotronstraling *f* || синхротронное излучение *n*

S888 **syneresis** || Synärese *f*, Synäresis *f* || synérèse *f* || syneresis *n* || синерезис *m*

S889 **syntax** || Syntax *f* || syntaxe *f* || syntaxis *f* || синтаксис *m*

S891 **synthesis** || Synthese *f* || synthèse *f* || synthese *f* || синтез *m*

S892 **synthetic fuel** || Synthesetreibstoff *m* || combustible *m* synthétique || synthetische brandstof *f* *(m)* || синтетическое топливо *n*

S893 **syphon** || Saugheber *m*, Heber *m* || siphon *m* || sifon *m* || сифон *m*

S894 **system** || System *n* || système *m* || systeem *n*, stelsel *n* || система *f*

S895 **systematic error** || systematischer Fehler *m* || erreur *f* systématique || systematische fout *f (m)* || систематическая погрешность *f*

S896 **system of differential equations** || differentiales Gleichungssystem *n* || système *m* d'équations différentielles || differentiaalvergelijkingssysteem *n* || система *f* дифференциальных уравнений

S897 **system of linear equations** || lineares Gleichungssystem *n* || système *m* d'équations linéaires || lineair vergelijkingssysteem *n* || система *f* линейных уравнений

S898 **system of mass points** || Massenpunktsystem *n* || système *m* de points matériels || massapuntsysteem *n* || система *f* материальных точек

S899 **system of reference** || Bezugssystem *n* || système *m* de référence || aftelsysteem *n* || система *f* отсчёта

S900 **system of simultaneous equations** || kompatibles Gleichungssystem *n* || système *m* d'équations compatibles || compatibel vergelijkingssysteem *n* || совместная система *f* уравнений

S901 **system of units** || Einheitensystem *n* || système *m* d'unités || eenhedenstelsel *n* || система *f* единиц

T

T1 **table** || 1. Tisch *m* 2. Tabelle *f*, Tafel *f* || table *f* || 1. tafel *f (m)* 2. tabel *f (m)* || 1. стол *m (напр. станка)* 2. таблица *f*

T2 **tabulator** || Tabelliermaschine *f* || tabulateur *m*, tabulatrice *f* || tabelleermachine *f* || табулятор *m*

T3 **tab washer** || Sicherungsblech *n* mit abbiegbaren Lappen || rondelle *f* à ailettes (rabattues) || schijf *f (m)* met uitbuigbare einden || шайба *f* с отгибаемыми концами [лапками]

T4 **tacho-generator** || Tachogenerator *m* || tachygénérateur *m*, génératrice *f* tachymétrique || tachogenerator *m* || тахогенератор *m*

T5 **tachometer** || Tachometer *n*, Drehzahlmesser *m* || tachymètre *m* || tachometer *m* || тахометр *m*

T6 **tackle** || Flaschenzug *m*, Rollenzug *m*; Hebezeug *n* || palan *m*; appareil *m* de levage || talie *f*; takel *m*, *n*; hijswerktuig *n* || таль *f*; полиспаст *m*; грузоподъёмное устройство *n*

T7 **tailstock** || Reitstock *m (Werkzeugmaschine)* || contre-poupée *f* || achterkop *m*, losse draaibankkop *m* || задняя бабка *f (станка)*

T8 **tangent** || 1. Tangens *m* 2. Tangente *f* || tangente *f* || 1. tangens *f (m)* 2. tangent *f (m)* || 1. тангенс *m* 2. касательная *f*

T9 **tangential** || tangential, Tangential... || tangentiel || tangentiaal || касательный, тангенциальный

T10 **tangential acceleration** || Tangentialbeschleunigung *f* || accélération *f* tangentielle || tangentiale versnelling *f* || тангенциальное [касательное] ускорение *n*

T11 **tangential force** || Tangentialkraft *f* || force *f* tangentielle || tangentiale kracht *f (m)* || тангенциальная [касательная] сила *f*, тангенциальная составляющая *f* силы

T12 **tangential key** || Tangentialkeil *m* || clavette *f* tangente || tangentiale spie *f (m)* || тангенциальная шпонка *f*

T13 **tangential stress** || Tangentialspannung *f*; Tangentialbeanspruchung *f* || contrainte *f* tangentielle; cission *f* || tangentiale spanning *f* || тангенциальное [касательное, скалывающее] напряжение *n*

T14 **tangential velocity** || Tangentialgeschwindigkeit *f* || vitesse *f* tangentielle || tangentiale snelheid *f* || тангенциальная [касательная] скорость *f*, тангенциальная составляющая *f* скорости

T15 **tangent plane** || Tangentialebene *f* || plan *m* tangent || tangentenvlak *n* || касательная плоскость *f*

T16 **tannic acid** || Gerbsäure *f* || acide *m* ta(n)nique || looizuur *n* || дубильная кислота *f*, танин *m*

T17 **tantalum,** Ta || Tantal *n* || tantale *m* || tantaal *n*, tantalium *n* || тантал *m*

T18 **tap** || 1. Gewindebohrer *m* 2. Hahn *m* 3. Auslauföffnung *f*; Auslaßöffnung *f* || 1. taraud *m* 2. robinet *m* 3. orifice *m* de vidange [d'écoulement, d'évacuation] || 1. tap *m* 2. tap *m*; kraan *f* (*m*) 3. uitlaatopening *f* || 1. метчик *m* 2. кран *m* 3. выпускное отверстие *n*

T19 **tape punch** || Lochstreifenlocher *m* || perforateur *m* de ruban || bandperforator *m* || ленточный перфоратор *m*

T20 **tape reader** || Lochstreifenabtaster *m* || lecteur *m* de bande perforée || ponsbandlezer *m* || устройство *n* для считывания с перфоленты; устройство *n* ввода с перфоленты

T21 **taper key** || Keilfeder *f* || clavette *f* en coin || tapse spie *f* (*m*) || клиновая шпонка *f*

T22 **taper screw thread** || Kegelgewinde *n* || filet *m* [filetage *m*] conique || tapse [conische] schroefdraad *m* || коническая резьба *f*

T23 **taper tap** || Vorschneider *m* || taraud *m* ébaucheur || voorsnijtap *m* || черновой метчик *m*

T24 **tapper** *see* **tapping machine**

T25 **tappet** || Schaft *m*; Stößel *m* (*eines Ventils*) || poussoir *m* (*de soupape*); came *f* || staaf *f* (*m*); klepstoter *m*, kleplichter *m* || стержень *m*; толкатель *m* клапана; кулачок *m*; эксцентрик *m*

T26 **tappet valve** || Nockenventil *n* || soupape commandée par came || nok(ken)ventiel *n* || клапан *m* с кулачковым приводом

T27 **tapping** || Gewindeschneiden *n* || taraudage *m* || tappen *n* || нарезание *n* резьбы метчиком

T28 **tapping head** || Gewindeschneideinheit *f*, Gewindeschneidkopf *m* || tête *f* à tarauder || schroefdraadsnijkop *m* || резьбонарезная головка *f*

T29 **tapping machine** || Gewindebohrmaschine *f* || taraudeuse *f* || tapmachine *f* || станок *m* для нарезания резьбы метчиком; гайконарезной станок *m*

T30 **tap wrench** || Windeisen *n* || tourne-à-gauche *m* || draadsnijder *m* || вороток *m*

T31 **target** || Target *n* || cible *f* || trefplaat *f*, trefplaats *f* || мишень *f* (*для частиц*)

T32 **target language** || Objektsprache *f* || langage *m* d'objet || objecttaal *f* || объектный язык *m*

T33 **target nucleus** || Targetkern *m* || noyau-cible *m* || trefkern *m* || ядро-мишень *n*, бомбардируемое ядро *n*

T34 **tartaric acid,** HOOC·CHOH · · COOH || Weinsäure *f*, Dihydroxybutandisäure *f* || acide *m* tart(a)rique || wijn(steen)zuur *n* || винная кислота *f*

T35 **task** || Task *n*, Aufgabe *f* || tâche *f* || opgave *f* || задача *f*

T36 **task management** || Task-Management *n*, Aufgabenverwaltung *f* || gestion *f* des tâches || besturen *n* van opgaven || управление *n* задачами

T37 **tautology** || Tautologie *f* || tautologie *f* || tautologie *f* || тавтология *f*

T38 **t-average** || zeitliches Mittel *n* || moyenne temporelle || tijdelijke middelwaarde *f* || средняя *f* по времени

T39 **Taylor's series expansion** || Taylorsche Entwicklung *f* || développement *m* (en série) de Taylor || ontwikkeling *f* van Taylor || разложение *n* в ряд Тейлора

T40 **T-bar** || T-Träger *m* || profilé *m* en T || T-balk *m* || тавровая балка *f*; тавровый профиль *m*

T41 **t-distribution** || t-Verteilung *f*, Student-Verteilung *f* || distribution *f* t || verdeling *f* van Student || распределение *n* Стьюдента, t-распределение *n*

T42 **technetium,** Tc || Technetium *n* || technétium *m* || technetium *n* || технеций *m*

T43 **technics** || Technik *f* || technique *f* || techniek *f* || технические науки *pl*, техника *f*

T44 **technique** || Technik *f*; Methode *f*, Verfahren *n* || procédé *m*; mode *m*; méthode *f*; technique *f* || techniek *f* || техника *f*, способ *m*, метод *m*, технические приёмы *m pl*

T45 **tee** || 1. T-Stück *n* 2. T-Träger *m* || 1. raccord *m* en T 2. poutre *f* en T || T-stuk *n* || 1. тройник *m* 2. тавровый профиль *m*, тавровая балка *f*

T46 **teeth** || Zähne *m pl*; Hieb *m* *(Feile)* || dents *f pl*; denture *f* || tanden *m pl (tandwiel)*; kerving *f (vijl)* || зубья *pl (напр. шестерни)*; насечка *f (напильника)*

T47 **tee-valve** || Dreiwegeventil *n*; Dreiwegehahn *m* || soupape *f* à trois voies || driewegafsluiter *m*; driewegkraan *f* *(m)* || трёхходовой клапан *m*; трёхходовой кран *m*

T48 **telecommunication** || Fernverkehr *m* || télécommunication *f* || telecommunicatie *f* || дистанционная [дальняя] связь *f*, телесвязь *f*

T49 **telecontrol** || Fernsteuerung *f* || télécommande *f* || telebesturing *f* || дистанционное управление *n*, телеуправление *n*

T50 **telegraph** || Telegraf *m* || télégraphe *m* || telegraaf *m* || телеграф *m*

T51 **telegrapher's equation** || Telegraphengleichung *f* || équation *f* des télégraphistes || telegraafvergelijking *f* || телеграфное уравнение *n*

T52 **telemechanics** || Fernwirktechnik *f* || télémécanique *f* || telemechanica *f* || телемеханика *f*

T53 **telemetering, telemetry** || Fernmessung *f*, Telemetrie *f* || télémesure *f*, télémétrie *f* || telemetrie *f* || дистанционное измерение *n*, телеметрия *f*

T54 **telephone** || Telephon *n* || téléphone *m* || telefoon *m* || телефон *m*

T55 **telescope** || 1. Teleskop *n*, Fernrohr *n* 2. Sehrohr *n* || 1. télescope *m* 2. lunette *f* d'observation || 1. telescoop *m* 2. (verre)kijker *m* || 1. телескоп *m* 2. зрительная труба *f*

T56 **television** || Fernsehen *n* || télévision *f* || televisie *f* || телевидение *n*

T57 **tellurium,** Te || Tellur *n* || tellure *m* || telluur *n*, tellurium *n* || теллур *m*

T58 **temper** || Anlassen *n*; Vergüten *n (beim Anlassen)* || revenu *m* || tempering *f*, ontlating *f* || отпуск *m* *(стали)*; улучшение *n (режим отпуска)*

T59 **temperature** || Temperatur *f* || température *f* || temperatuur *f* || температура *f*

T60 **temperature conductivity** || Temperaturleitfähigkeit *f* || conductibilité *f* thermique || temperatuurgeleidingsvermogen *n* || температуропроводность *f*

T61 **temperature expansion** || Wärmedehnung *f* || dilatation *f* thermique || warmte-uitzetting *f* || тепловое расширение *n*

T62 **temperature gradient** || Temperaturgefälle *n* || gradient *m* de température || temperatuurgradiënt *m* || градиент *m* температур, температурный градиент *m*

T63 **temperature-resistant** || temperaturbeständig || résistant à la température || vuurvast || температуростойкий

T64 **temperature scale** || Temperaturskale *f* || échelle *f* des températures || temperatuurschaal *f* *(m)* || температурная шкала *f*

T65 **temperature stress** || Wärmespannung *f* || contrainte *f* thermique || warmtespanning *f* || температурное напряжение *n*

T66 **temper brittleness** || Anlaßsprödigkeit *f* || fragilité *f* de revenu || temperbroosheid *f* || отпускная хрупкость *f*

T67 **tempering** *see* **temper**

T68 **tenacity** || 1. Zähigkeit *f*, Duktilität *f* 2. Zugfähigkeit *f*, Reißfestigkeit *f* || ténacité *f* || 1. taaiheid *f* 2. breukvastheid *f* || 1. вязкость *f*, тягучесть *f* 2. сопротивление *n* разрыву

T69 **tensile fatigue test** || Dauerzugprüfung *f* || essai *m* de fatigue par tractions répétées || duurproef *f* *(m)* *(bij oprekken)* || испытание *n* на усталость *(при растяжении)*

T70 **tensile force** || Dehnkraft *f* || force *f* [effort *m*] de traction || trekkracht *f* *(m)* || растягивающее усилие *n*, растягивающая сила *f*

T71 **tensile strain** || Zugverformung *f*, Dehnverformung *f* || déformation *f* de traction || relatieve rek *m* || (относительная) деформация *f* растяжения

T72 **tensile strength** || Zugfestigkeit(sgrenze) *f* || **résistance** *f* à la traction || treksterkte *f*, trekvastheid *f*; trekvastheidsgrens *f* (*m*) || предел *m* прочности при растяжении, временное сопротивление *n* растяжению

T73 **tensile stress** || Zugspannung *f*, Dehnungsspannung *f* || contrainte *f* de traction || trekspanning *f* || растягивающее напряжение *n*

T74 **tensile test** || Zugversuch *m*, Zugprobe *f* || essai *m* de traction || trekproef *f* (*m*) || испытание *n* на растяжение; испытание *n* на разрыв

T75 **tensile-testing machine** || Zugprüfmaschine *f* || machine *f* d'essai de traction || trekproefmachine *f* || машина *f* для испытания на разрыв

T76 **tension** || **1.** Spannung *f*; Zugdehnung *f* **2.** (elektrische) Spannung *f* || tension *f* || **1.** uitrekking *f* **2.** spanning *f* || **1.** натяжение *n*; растягивающее напряжение *n* **2.** (электрическое) напряжение *n*

T77 **tension rod** || auf Zug beanspruchter Stab *m*, Zugstab *m* || barre *f* tendue || trekstaaf *f* (*m*) || растянутый стержень *m*, стержень *m* (*конструкции*), работающий на растяжение

T78 **tensor analysis** || Tensoranalysis *f* || analyse *f* tensorielle || tensoranalyse *f* || тензорный анализ *m*

T79 **tensor calculus** || Tensorrechnung *f*, Tensorkalkül *m* || calcul *m* tensoriel || tensorrekening *f* || тензорное исчисление *n*

T80 **tera...** || Tera... || téra... || tera... || тера... (*приставка в наименовании кратной единицы, соответствующая* 10^{12})

T81 **term** || **1.** Glied *n* **2.** Term *m* || terme *m* || term *m* || **1.** член *m* **2.** терм *m*, энергетический уровень *m*

T82 **terminal** || **1.** Terminal *n*, Datenendgerät *n* **2.** Klemme *f* || **1.** terminal *m* **2.** borne *f* || **1.** terminal *m* **2.** klem *f* (*m*) || **1.** терминал *m*, абонентский пункт *m* **2.** зажим *m*, клемма *f*, (электрический) вывод *m*

T83 **terminal state** || Finalzustand *m* || état *m* final || eindtoestand *m* || конечное [терминальное] состояние *n*

T84 **terminal velocity** || Endgeschwindigkeit *f* || vitesse *f* uniforme; vitesse stabilisée [finale] || eindsnelheid *f* || установившаяся [конечная] скорость *f*

T85 **terminal voltage** || Klemmenspannung *f* || tension *f* aux bornes || klemspanning *f* || напряжение *n* на зажимах (*источника тока*)

T86 **termination** || Abschluß *m* || terminaison *f* || afsluiting *f* || замыкающая (электрическая) цепь *f*

T87 **ternary logic** || ternäre [dreiwertige] Logik *f* || logique *f* ternaire || ternaire logica *f* || трёхзначная логика *f*

T88 **terrestrial magnetism** || Erdmagnetismus *m* || magnétisme *m* terrestre || aardmagnetisme *n* || земной магнетизм *m*, геомагнетизм *m*

T89 **terrestrial weight** || Erdgewicht *n*; Gewicht *n* an der Erdoberfläche || poids *m* terrestre || aardgewicht *n* || земной вес *m*; вес *m* на поверхности Земли

T90 **tertium non datur** || tertium non datur, Satz *m* vom ausgeschlossenen Dritten || loi *f* du tiers exclu || tertium non datur || wet *f* (*m*) van een uitgesloten derde || закон *m* исключённого третьего

T91 **tesla**, T || Tesla *n* || tesla *m* || tesla *m* || тесла *f*, Тл

T92 **test** || **1.** Test *m*; Versuch *m*; Probe *f* **2.** *Math.* Test *m*, Kriterium *n* || **1.** essai *m*; vérification *f* **2.** *math.* test *m*, critère *m* || **1.** proef *f* (*m*), test *m* **2.** test *m* || **1.** испытание *n*; проверка *f*; **2.** *мат.* критерий *m*

T93 **test certificate** || Prüfschein *m*; Prüfprotokoll *n* || certificat *m* d'épreuve [d'essai] || proefcertificaat *n* || свидетельство *n* об испытании; паспорт *m* (*прибора*)

T94 **tester** || **1.** Testgerät *n* **2.** Tester *m*, Prüfer *m* || appareil *m* d'essai || testinrichting *f* || **1.** испытательный прибор *m* **2.** тестер *m*

T95 **testing machine** || Prüfstand *m*, Versuchsstand *m*; Prüfmaschine *f*, Materialprüfungsmaschine *f* || machine *f* d'essai || beproevingsmachine *f*

|| испытательная машина *f*, машина *f* для испытаний

T96 **test load** || Prüflast *f*; Probebelastung *f* || surcharge *f* d'essai; surcharge *f* d'épreuve || proefbelasting *f* || испытательная нагрузка *f*; пробная нагрузка *f*

T97 **test of significance** || Signifikanztest *m* || test *m* de signification || testv *m* van de betekenis || критерий *m* значимости

T98 **test routine** || Prüfprogramm *n*, Testprogramm *n* || programme *m* [routine *f*] de test || testprogramma *n* || испытательная программа *f*, тест-программа *f*

T99 **test specimen** || Probestück *n*, Probekörper *m*, Prüfstück *n*, Prüfling *m* || éprouvette *f*, échantillon *m* d'essai || proefmonster *n* || образец *m* для испытаний

T100 **test unit** || Versuchsanlage *f* || installation *f* d'essai || testinstallatie *f* || испытательная установка *f*

T101 **tetrahedron** || Tetraeder *n* || tétraèdre *m* || tetraëder *m* || тетраэдр *m*

T102 **texture** || Textur *f* || texture *f* || textuur *f* || текстура *f*

T103 **thallium**, Tl || Thallium *n* || thallium *m* || thallium *n* || таллий *m*

T104 **theorem** || Theorem *n*, Satz *m* || théorème *m* || theorema *n* || теорема *f*

T105 **theoretical efficiency** || theoretischer Wirkungsgrad *m* || efficacité *f* théorique || theoretisch rendement *n* || теоретический коэффициент *m* полезного действия

T106 **theoretical physics** || theoretische Physik *f* || physique *f* théorique || theoretische fysica *f* || теоретическая физика *f*

T107 **theory** || Theorie *f* || théorie *f* || theorie *f* || теория *f*

T108 **theory of distributions** || Distributionstheorie *f* || théorie *f* des distributions || theorie *f* van distributies || теория *f* распределений

T109 **theory of games** || Spieltheorie *f* || théorie *f* des jeux || speltheorie *f* || теория *f* игр

T110 **theory of numbers** || Zahlentheorie *f* || théorie *f* des nombres || getaltheorie *f* || теория *f* чисел

T111 **thermal** || Wärme..., thermisch || thermique || thermisch || тепловой, термический

T112 **thermal barrier** || Hitzebarriere *f*, Hitzemauer *f*, Wärmemauer *f* || barrière *f* [mur *m*, seuil *m*] thermique || warmtebarrière *f* (*m*) || тепловой барьер *m*

T113 **thermal conduction** || Wärmeleitung *f*, Wärmeleitfähigkeit *f* || conductibilité *f* calorifique [de la chaleur, thermique] || warmtegeleiding *f* || теплопроводность *f*

T114 **thermal conductivity** || (spezifische) Wärmeleitfähigkeit *f* || conductivité *f* thermique || soortelijke warmtegeleiding *f* || удельная теплопроводность *f*

T115 **thermal current** || Wärmestrom *m*, Wärmefluß *m* || flux *m* de chaleur || warmtestroom *m* || тепловой поток *m*

T116 **thermal diffusion** || Thermodiffusion *f* || thermodiffusion *f* || thermodiffusie *f* || термодиффузия *f*

T117 **thermal efficiency** || thermischer Wirkungsgrad *m*, Wärmewirkungsgrad *m* || rendement *m* thermique || thermische werkingsgraad *m* || термический [тепловой] кпд *m*

T118 **thermal energy** || Wärmeenergie *f* || énergie *f* thermique || warmte-energie *f* || тепловая энергия *f*

T119 **thermal equilibrium** || Wärmegleichgewicht *n* || équilibre *m* thermique || warmte-evenwicht *n* || тепловое равновесие *n*

T120 **thermal expansion** || Wärmedehnung *f* || dilatation *f* (thermique) || warmte-uitzetting *f* || тепловое расширение *n*

T121 **thermal fluctuations** || thermische Schwankungen *f pl* || fluctuations *f pl* thermiques || thermische fluctuaties *f pl* || тепловые флуктуации *pl*

T122 **thermal junction** || Lötstelle *f* (*Thermoelement*) || jonction *f* du couple thermo-électrique [du thermocouple] || soldering *f* || спай *m* термопары

T123 **thermal lag** || thermische Trägheit *f*, Wärmeträgheit *f* || inertie *f* thermique || warmtetraagheid *f* || тепловая инерция *f*

T124 **thermal motion** || Wärmebewegung *f* || agitation *f* thermocinétique [thermique] || warmtebeweging *f* || тепловое движение *n*

T125 **thermal neutron** || thermisches Neutron *n* || neutron *m* (d'énergie) thermique || thermisch neutron *n* || тепловой нейтрон *m*

T126 **thermal neutron reactor** || thermischer Reaktor *m* || réacteur *m* à neutrons thermiques || thermische reactor *m* || реактор *m* на тепловых нейтронах, тепловой реактор *m*

T127 **thermal noise** || Wärmerausch *m* || bruit *m* (d'agitation) thermique || warmteruis *m* || тепловой шум *m*

T128 **thermal radiation** || Wärmestrahlung *f*, thermische Strahlung *f* || rayonnement *m* thermique, émission *f* calorifique || warmtestraling *f*, thermische straling *f* || тепловое излучение *n*

T128a **thermal resistance** || Wärmewiderstand *m* || résistance *f* thermique || warmteweerstand *m* || термическое сопротивление *n*

T129 **thermal stress** *see* **heat stress**

T130 **thermal treatment** || Wärmebehandlung *f* || traitement *m* thermique || warmtebehandeling *f* || термическая обработка *f*, термообработка *f*

T131 **thermal velocity** || thermische Geschwindigkeit *f* || vitesse *f* thermique || thermische snelheid *f* || скорость *f* теплового движения

T132 **thermal vibration** || Wärmeschwingungen *f pl* || oscillations *f pl* thermiques || thermische trillingen *f pl* || тепловые колебания *n pl* (*в твёрдых телах*)

T133 **thermionic emission** || Glühemission *f* || émission *f* thermo-électronique || thermo-elektronische emissie *f* || термоэлектронная эмиссия *f*

T134 **thermistor** || Thermistor *m* || thermistance *f*, thermistor *m* || thermistor *m* || термистор *m*

T135 **thermobattery** || Thermobatterie *f*; Thermosäule *f*, thermoelektrische Säule *f* || thermobatterie *f*, pile *f* thermo-électrique, thermopile *f* || thermobatterij *f* || термоэлектрическая батарея *f*, термобатарея *f*

T136 **thermochemistry** || Thermochemie *f*, chemische Thermodynamik *f* || thermochimie *f*, thermodynamique *f* chimique || thermochemie *f*, chemische thermodynamica *f* || термохимия *f*, химическая термодинамика *f*

T137 **thermocouple** || Thermoelement *n* || thermocouple *m*, couple *m* thermo-électrique || thermokoppel *n* || термопара *f*; термоэлемент *m*

T138 **thermocouple thermometer** || thermo-elektrisches Thermometer *n* || thermomètre *m* à thermocouple || thermoelektrische thermometer *m* || термоэлектрический термометр *m*

T139 **thermocurrent** || thermoelektrischer Strom *m*, Thermostrom *m* || courant *m* thermo-électrique || thermo-elektrische stroom *m* || термоэлектрический ток *m*, термоток *m*

T140 **thermodiffusion** *see* **thermal diffusion**

T141 **thermodynamic** || thermodynamisch || thermodynamique || thermodynamisch || термодинамический

T142 **thermodynamic cycle** || thermodynamischer Kreisprozeß *m* || cycle *m* thermodynamique || thermodynamische cyclus *m* || термодинамический цикл *m*

T143 **thermodynamic equilibrium** || thermodynamisches Gleichgewicht || || équilibre *m* thermodynamique || thermodynamisch evenwicht *n* || термодинамическое равновесие *n*

T144 **thermodynamic potential** || thermodynamisches Potential *n* || potentiel *m* thermodynamique || thermodynamische potentiaal *m* || термодинамический потенциал *m*

T145 **thermodynamic process** || thermodynamischer Prozeß *m* || processus *m* thermodynamique || thermodynamisch proces *n* || термодинамический процесс *m*

T146 **thermodynamics** || Thermodynamik *f* || thermodynamique *f* || thermodynamica *f* || термодинамика *f*

T147 **thermodynamic state** || thermodynamischer Zustand *m* || état *m* thermodynamique || thermodynamische toestand *m* || термодинамическое состояние *n*

T148 **thermodynamic system** || thermodynamisches System *n* || système *m* thermodynamique || thermodynamisch systeem *n* || термодинамическая система *f*

T149 **thermodynamic temperature** || thermodynamische Temperatur *f* || température *f* thermodynamique [absolue] || thermodynamische temperatuur *f* || термодинамическая [абсолютная] температура *f*

T150 **thermodynamic temperature scale** || thermodynamische Temperaturskala *f* || échelle *f* absolue [thermodynamique] des températures || thermodynamische temperatuurschaal *f* (*m*) || термодинамическая шкала *f* температур

T151 **thermoelectric** || thermoelektrisch || thermo-électrique || thermoelektrisch || термоэлектрический

T152 **thermoelectricity** || Thermoelektrizität *f*, Wärmeelektrizität *f* || thermo-électricité *f* || thermo-elektriciteit *f* || термоэлектричество *n*

T153 **thermoelectromotive force** || Thermo-EMK *f*, thermoelektromotorische Kraft *f*, Thermospannung *f* || force *f* thermoélectromotrice || thermo-elektromotorische kracht *f* (*m*) || термоэлектродвижущая сила *f*, термо-эдс *f*

T154 **thermoelectron** || Glühelektron *n*, Thermoelektron *n* || thermoélectron *m* || thermo-elektron *n* || термоэлектрон *m*

T155 **thermoelement** || Thermoelement *n* || thermo-élément *m*, thermocouple *m* || thermokoppel *n* || термоэлемент *m*

T156 **thermogenerator** || thermoelektrischer Generator *m* || thermogénérateur *m*, générateur *m* thermoélectrique || thermo-elektrische generator *m* || термоэлектрический генератор *m*

T157 **thermogram** || Thermogramm *n* || thermogramme *m* || thermogram *n* || термограмма *f*

T158 **thermograph** || Thermograph *m*, Temperaturschreiber *m* || thermographe *m*, thermomètre *m* enregistreur || thermograaf *m* || термограф *m*, самопишущий термометр *m*

T159 **thermography** || Thermographie *f* || thermographie *f* || thermografie *f* || термография *f*

T160 **thermojunction** *see* **thermal junction**

T161 **thermoluminescence** || Thermolumineszenz *f* || thermoluminescence *f* || thermoluminescentie *f* || термолюминесценция *f*

T162 **thermomagnetism** || Thermomagnetismus *m* || thermomagnétisme *m* || thermomagnetisme *m* || термомагнетизм *m*

T163 **thermometer** || Thermometer *n* || thermomètre *m* || thermometer *m* || термометр *m*

T164 **thermometry** || Thermometrie *f* || thermométrie *f* || thermometrie *f* || термометрия *f*

T165 **thermonuclear fusion** || thermonukleare Kernfusion *f* || fusion *f* thermonucléaire || thermonucleaire kernfusie *f* || термоядерный синтез *m*

T166 **thermonuclea reaction** || thermonukleare Reaktion *f* || réaction *f* thermonucléaire || thermonucleaire reactie *f* || термоядерная реакция *f*

T167 **thermonuclear reactor** || thermonuklearer Reaktor *m* || réacteur *m* thermonucléaire || thermonucleaire kernreactor *m* || термоядерный реактор *m*

T168 **thermophysics** || Thermophysik *f* || thermophysique *f* || thermofysica *f* || теплофизика *f*

T169 **thermopile** || **1.** Thermosäule *f* **2.** Thermobatterie *f* || thermopile *f*, pile *f* thermo-électrique; thermobatterie *f* || **1.** thermozuil *f* (*m*) **2.** thermobatterij *f* || **1.** термостолбик *m* **2.** термобатарея *f*

T170 **thermoplastics materials** || Thermoplaste *m pl* || thermoplastiques *m pl* || thermoplasten *m pl* || термопласты *pl*, термопластичные пластмассы *pl*

T171 **thermostability** || Wärmebeständigkeit *f* || thermostabilité *f* || thermische stabiliteit *f* || теплостойкость *f*, термостойкость *f*

T172 **thermostat** || Thermostat *m* || thermostat *m* || thermostaat *m* || термостат *m*

T173 **thesaurus** || Thesaurus *m* || thésaurus *m* || thesaurus *m* || тезаурус *m*

T174 **theta function** || Theta-Funktion *f* || fonction *f* thêta || theta-functie *f* || тета-функция *f*

T175 **thickness** || **1.** Dicke *f*, Stärke *f* **2.** Dicke *f* || épaisseur *f* || **1.** dikte *f* **2.** dichtheid *f* || **1.** толщина *f* **2.** густота *f*

T176 **thickness gauge** || Dickenmesser *m* || jauge *f* d'épaisseur || diktemeter *m* || толщиномер *m*

T177 **thorium,** Th || Thorium *n* || thorium *m* || thorium *n* || торий *m*

T178 **thread** *see* **screw thread**

T179 **thread chasing** || Gewindestrehlen *n* || filetage *m* par [au] peigne || schroefdraadsnijden *n* met meerdraadsbeitel || нарезание *n* резьбы гребёнкой

T180 **thread cutting lathe** *see* **screw-cutting lathe**

T181 **thread cutting machine** *see* **screw-thread cutting machine**

T182 **threaded coupling** || Gewindeverbindung *f* || raccord *m* fileté; raccordement *m* par vissage || schroefdraadtverbinding *f* || резьбовое соединение *n*

T184 **thread grinding machine** *see* **screw thread grinding machine**

T185 **threading** || Gewindeschneiden *n* || filetage *m* || schroefdraadsnijden *n* || нарезание *n* резьбы

T186 **threading die** || Schneideisen *n* || filière *f* || (draad)snijplaat *f* (*m*) || (резьбо)нарезная плашка *f*

T187 **threading lathe** *see* **screw-cutting lathe**

T188 **threading machine** *see* **screw-tread cutting machine**

T189 **thread milling cutter** || Gewindefräser *m* || fraise *f* à fileter || schroefdraadfrees *f* || резьбовая [резьбонарезная] фреза *f*

T190 **thread milling machine** *see* **screw thread milling machine**

T191 **thread rolling head** || Gewindewalzkopf *m*, Gewinderollkopf *m* || tête *f* pour roulage des filets || schroefdraadwalskop *m* || (резьбо)накатная головка *f*

T192 **thread rolling machine** *see* **screw thread rolling machine**

T193 **thread rolling tool** || Gewinderollwerkzeug *n* || outil *m* pour roulage des filets || werktuig *n* voor het draadrollen *n* || резьбонакатный инструмент *m*

T194 **thread template** Gewindeschablone *f* || jauge *f* de filetage || schroefdraadsjabloon *f* (*m*) || резьбовой шаблон *m*

T195 **thread whirling** || Gewindewirbeln *n* || fraisage *m* tourbillonnaire || schroefdraadwervelen *n* || вихревой метод *m* нарезания резьбы, вихревое резьбофрезерование *n*

T196 **three-body problem** || Dreikörperproblem *n* || problème *m* des trois corps || drielichamenprobleem *n* || задача *f* трёх тел

T197 **three-colour** || Dreifarben... || de trois couleurs || driekleurig || трёхцветный

T198 **three-dimensional** || dreidimensional || triedimensioneel || driedimensionaal || трёхмерный, пространственный, объёмный

T199 **three-high mill** || Triowalzwerk *n* || laminoir *m* [train *m*] trio || wals *m* met drie rollen || стан-трио *m*, трёхвалковый прокатный стан *m*

T200 **three-lobe rotor** || dreieckiger Läufer *m (z. B. bei Wankelmotor)* || rotor *m* trilobé *(p. ex. du moteur Wankel)* || driekante rotor *m* || трёхгранный ротор *(напр. двигателя Ванкеля)*

T201 **three-phase current** || Dreiphasenstrom *m*, Drehstrom *m* || courant *m* triphasé || driefasenstroom *m* || трёхфазный ток *m*

T202 **threshold** || Schwelle *f*, Schwellenwert *m* || seuil *m* || drempel *m* || порог *m*

T203 **threshold energy** || Energieschwelle *f* || énergie de seuil, seuil *m* énergétique || energiedrempel *m* || пороговая энергия *f*, энергетический порог *m*

T204 **threshold intensity** || Intensitätsschwelle *f* || intensité *f* de seuil ||

intensiteitsdrempel *m* || пороговая интенсивность *f*

T205 **threshold of reaction** || Reaktionsschwelle *f* || seuil *m* de réaction || reactiedrempel *m* || (энергетический) порог *m* реакции

T206 **threshold sensitivity** || Empfindlichkeitsschwelle *f* || sensibilité *f* de seuil || gevoeligheidsdrempel *m* || пороговая чувствительность *f*; порог *m* чувствительности

T207 **threshold value** || Schwellenwert *m* || valeur *f* de seuil || drempelwaarde *f* || пороговое [предельное] значение *n*

T208 **throttle** || **1.** Regler *m* **2.** Drosselklappe *f* || **1.** régulateur *m* d'admission *(p. ex. d'une turbine à vapeur)* **2.** papillon *f* d'étranglement, volet *m* des gaz || **1.** regelaar *m* **2.** gas-klep *f* (*m*), smoorklep *f* (*m*) || **1.** регулятор *m* **2.** дроссельная заслонка *f*, дроссель *m*

T209 **throttle valve** || **1.** Drosselventil *n*; Steuerschieber *m* **2.** Drosselklappe *f* *(eines Vergasers)* || **1.** soupape *f* d'admission **2.** papillon *m* de gaz *(d'un carburateur)* || **1.** smoorklep *f* (*m*) **2.** regelklep *f* (*m*) || **1.** дроссельный клапан *m*; золотник *m* регулятора **2.** дроссельная заслонка *f* *(карбюратора)*

T210 **throttling** Drosselung *f* || étranglement *m* || smoren *n* || дросселирование *n*

T211 **throughput** || Durchsatz *m*; Durchfluß *m*, Durchflußleistung *f* || débit *m*, rendement *m* || doorlaatvermogen *n* || пропускная способность *f*

T212 **thrust** || **1.** axiale Kraft *f* **2.** Schub *m* || **1.** effort *m* axial **2.** poussée *f* || **1.** axiale kracht *f* (*m*) **2.** duwkracht *f* (*m*) || **1.** осевое усилие *n* **2.** реактивная тяга *f*

T213 **thrust-and-torque combination** || Winder *m*, Dyname *f* || visseur *m* || duw-en-draaiingcombinatie *f* || динамический винт *m*, динама *f*

T214 **thrust bearing** || Drucklager *n*, Längslager *n*, Spurlager *n* || palier *m* de butée || druklager *n*; drukkogellager *n* || упорный подшипник *m*

T215 **thrust collar** || Stoßring *m*; Druckring *m* || anneau *m* de butée; bague *f* de butée || drukring *m* || упорное нажимное кольцо *n*

T216 **thrust load** *see* **axial load**

T217 **thrust screw** || Anschlagschraube *f*; Druckschraube *f* || vis *f* de butée || drukschroef *f* (*m*) || упорный [нажимной] винт *m*

T218 **thrust shaft** || *(achsrechte Belastung aufnehmende)* Druck(lager)-welle *f* || arbre *m* de butée || drukas *f* (*m*) || упорный вал *m* *(воспринимающий осевую нагрузку)*, вал *m* на упорных подшипниках (скольжения)

T219 **thrust washer** || Druckscheibe *f*; Begrenzungsscheibe *f* || rondelle *f* d'appui [de butée] || druksluitring *m* || упорная шайба *f*

T220 **thyratron** || Thyratron *n* || thyratron *m* || thyratron *n* || тиратрон *m*

T221 **thyristor** || Thyristor *m* || thyristor *m* || thyristor *m* || тиристор *m*

T222 **tidal power plant** || Gezeitenkraftwerk *n* || usine *f* marémotrice || getijde-krachtcentrale *f* (*m*) || приливная электростанция *f*

T223 **tie bolt** *see* **coupling bolt**

T224 **tight binding** || starke Kopplung *f* *(Teilchen)* || couplage *m* fort || sterke koppeling *f* || сильная связь *f* *(между частицами)*

T225 **tightening pulley** || Spannscheibe *f* || poulie *f* de tension, poulie-tendeur *f* || spanschijf *f* (*m*) || натяжной шкив *m*

T226 **tightness** || Undurchlässigkeit *f*, Dichtheit *f* || étanchéité *f* || dichtheid *f* || герметичность *f*, непроницаемость *f*, плотность *f*

T227 **tilt** || Neigung *f*; Schräge *f* || inclinaison *f* || helling *f* || наклон *m*

T228 **tilting** || **1.** Neigung *f* **2.** Kippen *n* || **1.** inclinaison *f* **2.** basculement *m*, renversement *m* || **1.** helling *f* **2.** kanteling *f* || **1.** наклон *m* **2.** опрокидывание *n*

T229 **time** || Zeit *f*; Periode *f* || temps *m*; période *f*; instant *m* || tijd *m*; periode *f*; moment *n* || время *n*; период *m*; момент *m*

T230 **time average** || zeitliches Mittel *n* || moyenne *f* temporelle || tijdelijke middeleenheid *f* || средняя *f* по времени
T231 **time-average** || Zeitmittel... || moyen en temps || tijdmiddel... || средний по времени
T232 **time constant** || Zeitkonstante *f* || constante *f* de temps || tijdconstante *f* (*m*) || постоянная *f* времени
T233 **time delay** || Zeitverzögerung *f*, zeitlche Verzögerund *f* || délai *m*, retard *m* || tijdvertraging *f* || запаздывание *n*, задержка *f* *(по времени)*
T234 **time dilation** || Zeitdilatation *f*, Zeitdehnung *f* *(Relativitätstheorie)* || étalement *m* en temps || tijdspreiding *f* || растяжение *n* времени *(в теории относительности)*
T235 **time interval** || Zeitintervall *n*, Zeitabstand *m* || intervalle *m* de temps || tijdspan *f* || временной интервал *m*, промежуток *m* времени
T236 **time inversion** || Zeitumkehr *f* || inversion *f* de temps || tijdomkering *f* || обращение *n* *(течения, хода)* времени
T237 **time lag** *see* **time delay**
T238 **time-mean value** || Zeitmittelwert *m* || valeur *f* de temps moyenne || tijdgemiddelde *n* || среднее по времени (значение *n*)
T239 **time of vibration** || **1.** Schwingungsperiode *f* 2. Schwingungsdauer *f* || **1.** période *f* de vibrations 2. durée *f* de vibrations || **1.** trillingstijd *m* 2. trillingsduur *m* || **1.** период *m* колебаний 2. длительность *f* [продолжительность *f*] колебаний
T240 **timer** || Zeitrelais *n*, Timer *m* || relais *m* temporisé, timer *m* || tijdrelais *n*, timer *m* || реле *n* времени, таймер *m*
T241 **time resolution** || Zeitauflösung *f* || résolution *f* en temps || tijdoplossing *f* || временно́е разрешение *n*
T242 **time response** || Zeitcharakteristik *f* || réponse *f* temporelle || tijddiagram *n* || временна́я характеристика *f*
T243 **time sharing** || time sharing, Zeitteilung *f* || partage *m* de temps || tijdverdeling *f* || разделение *n* времени
T244 **time symmetry** || Zeitsymmetrie *f* || symétrie *f* par rapport au temps || tijdsymmetrie *f* || симметрия *f* относительно (обращения знака) времени
T245 **time variation** || Zeitveränderung *f*, zeitliche Veränderung *f* || variation *f* en temps || tijdverandering *f* || изменение *n* во времени
T246 **timing** || Zeitsteuerung *f* || synchronisation *f* || synchronisatie *f* || **1.** синхронизация *f* **2.** хронометраж *m*
T247 **timing belt** || Zahntreibriemen *m* || courroie *f* cranée || tandriem *m* || зубчатый (приводной) ремень *m*
T248 **timing chain** || Steuerkette *f* || chaîne *f* du mécanisme de distribution || stuurketting *m*, *f* || цепь *f* привода распределительного механизма
T249 **timing gear** || Steuerung *f*, Steuermechanismus *m*; Steuerrad *n* || pignon *m* distributeur || distributietandwiel *n* || распределительный механизм *m*; шестерня *f* распределительного вала
T250 **tin** || Zinn *n* || étain *m* || tin *n* || олово *n*
T251 **tinning** || Verzinnen *n* || étamage *m*; métallisation *f* à l'étain || vertinning *f* || **1.** лужение *n* **2.** полуда *f*
T252 **tin plate** || Weißblech *n* || tôle *f* étamée; fer-blanc *m* || blik *n* || лужёная [белая] жесть *f*
T253 **tip circle** *see* **addendum circle**
T254 **tip cone** || Kopfkegel *m* *(Kegelradverzahnung)* || cône *m* au sommet *(de l'engrenage conique)* || voorsprongkegel *m*, kopkegel *m* *(van kegelradvertanding)* *f* || конус *m* выступов *(конического зубчатого зацепления)*
T255 **tip diameter** || Spitzendurchmesser *m* *(Zähne)* || diamètre *m* de tête *(d'engrenage)* || tandpuntdiameter *m* || диаметр *m* вершин *(зубьев)*
T256 **tip vortex** || Randwirbel *m*, Wirbelzopf *m* || tourbillon *m* marginal [à l'extrémité] || randwervel *m* || концевой [сбегающий] вихрь *m*
T257 **tire rolling mill** || Radreifenwalzwerk *n* || laminoir *m* [train *m*]

à bandages || bandagewals *f* (*m*) || бандажепрокатный стан *m*

T258 **titanium,** T || Titan *n* || titane *m* || titanium *n*, titaan *n* || титан *m*

T259 **titration** || Titration *f*, Titrieren *n* || titrage *m* || titratie *f* || титрование *n*

T260 **tolerance** || Toleranz *f* || tolérance *f* || tolerantie *f* || допуск *m* *(на размер)*

T261 **tolerance (dose)** || zulässige Strahlungsdosis *f* || dose *f* tolérée || tolerantiedosis *f* || допустимая доза *f* облучения

T262 **tolerance interval** || Toleranzbereich *m*, Toleranzgebiet *n* || intervalle *m* de tolérance || tolerantiebereik *n* || интервал *m* допуска, пределы *pl* допусков

T263 **tone** || Ton *m* || ton *m* || toon *m* || тон *m*

T264 **tone generator** || Tongenerator *m* || générateur *m* de fréquence sonore || toongenerator *m* || генератор *m* звуковой частоты

T265 **tool** || **1.** Werkzeug *n*; Schneidwerkzeug *n*; Meißel *m* **2.** Werkzeugmaschine *f* || **1.** outil *m* **2.** machine-outil *f* || **1.** snijgereedschap *n*; beitel *m* **2.** werktuigmachine *f* || **1.** (режущий) инструмент *m*; резец *m* **2.** станок *m*

T266 **tool bit** *see* **tool holder bit**

T267 **tool block** || Vielmeißelblock *m*, Werkzeugblock *m*; Meißelhalter *m* || bloc *m* porte-outil || beitelhouder *m* || резцовый блок *m*; резцедержатель *m*

T268 **tool clamping** *see* **tool post**

T269 **tool coolant** || Kühlflüssigkeit *f*, Kühlemulsion *f* || liquide *m* refroidissant [de refroidissement] || koelvloeistof *f* (*m*); koelemulsie *f* || охлаждающая (инструмент) жидкость *f*

T271 **tool head** || Meißelkopf *m*; Meißelhalter *m* || tête *f* à tailler; tête *f* d'usinage || beitelhouder *m*, beitelklem *f* (*m*) || резцовая головка *f*; резцедержатель *m*

T272 **tool holder** *see* **tool post**

T273 **tool holder bit** || Einsteckmeißel *m*, Einsatzschneide *f* || lame *f* amovible [rapportée] || snijstaaf *f*, beitelstaaf *f* || вставной резец *m*

T274 **tooling quality** || Zerspanbarkeit *f* || usinabilité *f* des métaux; aptitude *f* à l'usinage *(par enlèvement de copeaux)* || bewerkbaarheid *f* door snijden || обрабатываемость *f* резанием

T275 **tool life** || Standzeit *f* *(Schneidwerkzeug)* || vie *f* d'un outil || standtijd *m* || стойкость *f* (режущего) инструмента

T276 **tool post** || Meißelhalter *m* || porte-outil *m*, porte-couteau *m* || beitelhouder *m* || резцедержатель *m*

T277 **tool side clearance** || Freiwinkel *m* || dépouille *f* latérale de l'outil || vrijloophoek *m* || задний угол *m* резца

T278 **tool slide** || Meißelschlitten *m* || chariot *m* porte-outil || beitelslede *f* (*m*), beitelsupport *n* || резцовые салазки *pl*

T279 **tool steel** || Werkzeugstahl *m* || acier *m* à outils || gereedschapstaal *n*, kroezenstaal *n* || инструментальная сталь *f*

T280 **tool turret** || Revolverkopf *m* || tourelle *f* (porte-outil) || revolverkop *m* || револьверная головка *f*

T281 **tooth crest** || Zahnkopfscheitel *m* || sommet *m* de la dent || tandpunt *n* || вершина *f* зуба

T282 **toothed gearing** || Zahn(rad)-getriebe *n* || engrenage *m* || tandwieloverbrenging *f* || зубчатая передача *f*

T283 **toothed wheel** || Zahnrad *n* || roue *f* (dentée), pignon *m* (denté) || tandwiel *n*, tandrad *n* || зубчатое колесо *n*; шестерня *f*

T284 **toothing** || Verzahnung *f*; Eingriff *m* || engrenage *m* || vertanding *f* || зубчатое зацепление *n*

T285 **tooth profile** || Zahnprofil *n* || profil *m* de dent || tandprofiel *n* || профиль *m* зуба

T286 **tooth space** || Zahnlücke *f* || espace *m* interdentaire; entredent *f* || tandkuil *m* || впадина *f* между зубьями; ширина *f* впадины по дуге

T287 **tooth tip** *see* **tooth crest**

T288 **top** || Kreisel *m* || toupie *f* ||

tol *m* || волчок *m*; молекулярный волчок *m*

T289 **top of band** || obere Grenze *f* eines Energiebandes || limite *f* supérieure d'une zone énergétique || bovengrens *f* (*m*) van een energiezone *f* || верхняя граница *f* энергетической зоны

T290 **topological algebra** || topologische Algebra *f* || algèbre *f* topologique || topologische algebra *f* || топологическая алгебра *f*

T291 **topological isomorphism** || Homeomorphie *f* || isomorphisme *m* topologique || homeomorfie *f*, topologische isomorfie *f* || гомеоморфизм *m*

T292 **topological space** || topologischer Raum *m* || espace *m* topologique || topologische ruimte *f* || топологическое пространство *n*

T293 **topological transformation** || topologische Transformation *f*, Homeomorphie *f* || transformation *f* topologique || topologische transformatie *f*, homeomorfie *f* || топологическое отображение *n*, гомеоморфизм *m*

T294 **topology** || Topologie *f* || topologie *f* || topologie *f* || топология *f*

T295 **top rake** || Spanwinkel *m* (*Meißel*) || pente *f* effective de coupe; (angle *m* de) pente *f* d'affûtage || spaanhoek *m* (*snijwerktuig*) || передний угол *m* (*режущего инструмента*)

T296 **top surface** *see* **face**

T297 **toroid** || Toroid *n* || toroïde *m* || toroïde *f* (*m*) || тороид *m*

T298 **toroidal coordinates** || Ringkoordinaten *f pl*, Toruskoordinaten *f pl* || coordonnées *f pl* toroïdales || toroïdale coördinaten *m pl* || тороидальные координаты *pl*

T299 **torque** || Kräftemoment *n*; Drehmoment *n* || couple *m* des forces; moment *m* de rotation || krachtmoment *n*, koppel *n*; draaimoment *n*; torsiekoppel *n* || момент *m* сил; вращающий момент *m*

T300 **torque balance** || Drehmomentengleichgewicht *n* || équilibre *m* des moments de rotation || draaimomentenevenwicht *n* || равновесие *n* вращающих моментов

T301 **torque converter** || Drehmomentwandler *m* || convertisseur *m* de couple || koppelomvormer *m* || гидротрансформатор *m*

T302 **torque force** || Torsionskraft *f* || effort *m* de torsion || torsiekracht *f* (*m*) || крутящее [закручивающее] усилие *n*

T303 **torque meter** || Torsions(moment)messer *m* || torsiomètre *m*, dynamomètre *m* de torsion || torsiemeter *m* || торсиометр *m*, крутильный динамометр *m*

T304 **torque wrench** || Drehmomentenschlüssel *m* || clef *f* [clé *f*] dynamométrique || draaimomentsleutel *m* || динамометрический гаечный ключ *m*; гайковерт *m* предельного момента

T305 **torr** || Torr *n* || torr *m* || torr *m* || торр *m*, миллиметр *m* ртутного столба (*133,3 Па*)

T306 **torsion** || Torsion *f*, Verdrehung *f* || torsion *f* || torsie *f* || кручение *n*

T307 **torsional** || Torsions..., Verdreh(ungs)... || de torsion || torsie... || крутильный; торсионный

T308 **torsional angle** || Torsionswinkel *m*, Verdrehungswinkel *m* || angle *m* de torsion (total) || torsiehoek *m* || угол *m* закручивания

T309 **torsional moment** || Torsionsmoment *n* || moment *m* de torsion || wringmoment *n*, torsiemoment *n* || крутящий момент *m*

T310 **torsional oscillation** || Torsionsschwingungen *f pl* || oscillations *f pl* de torsion || torsietrillingen *f pl* || крутильные колебания *pl*

T311 **torsional pendulum** || Torsionspendel *n* || pendule *m* de torsion || torsiependel *n* || крутильный маятник *m*

T312 **torsional spring** || **1.** verdrehbeanspruchte Feder *f* **2.** Drehstabfeder *f*, Torsionsfeder *f* || ressort *m* de torsion || torsieveer *f* (*m*), wringveer *f* (*m*) || **1.** пружина *f*, работающая на кручение **2.** торсионная рессора *f*

T313 **torsional strain** || Verdrehverformung *f* || déformation *f* de torsion || torsievervorming *f*, wringdeformatie *f*

|| деформация *f* кручения [закручивания]

T314 **torsional strength** || Verdreh(ungs)festigkeit *f*, Torsionsfestigkeit *f* || résistance *f* à la torsion || torsievastheid *f*, wringweerstand *m* || предел *m* прочности при кручении; сопротивление *n* закручиванию

T315 **torsional stress** || Verdrehspannung *f*, Torsionsspannung *f* || contrainte *f* de torsion || torsiespanning *f*, wringspanning *f* || напряжение *n* при кручении

T316 **torsion balance** || Torsionswaage *f* || balance *f* de torsion || torsiebalans *f* (*m*) || крутильные весы *pl*

T317 **torsion dynamometer** || Torsionsdynamometer *n* || torsiomètre *m*, dynamomètre *m* de torsion || torsiedynamometer *m*, wringingsdinamometer *m* || крутильный динамометр *m*

T318 **torus** || Torus *m* || tore *m* || torr *m* || тор *m*

T319 **total curvature** || Gesamtkrümmung *f*, Totalkrümmung *f* || courbure *f* totale || totaalkromming *f* || полная кривизна *f*

T320 **total energy** || Gesamtenergie *f* || énergie *f* totale || totaalenergie *f* || полная [суммарная] энергия *f*

T321 **total function** || überall definierte Funktion *f* || fonction *f* totale || totale functie *f* || всюду определённая функция *f*

T322 **total head** || Gesamtdruck *m* || chute *f* totale || totaaldruk *m* || полный напор *m*

T323 **total heat** || Enthalpie *f* || enthalpie *f* || enthalpie *f* || энтальпия *f*

T324 **total intensity** || Gesamtintensität *f* || intensité *f* totale || totaalintensiteit *f* || полная [суммарная] интенсивность *f*

T325 **total internal reflection** || Totalreflexion *f* || réflexion *f* interne totale || totale inwendige reflectie *f* || полное внутреннее отражение *n*

T326 **totally reflecting prism** || Totalreflexionsprisma *n*, Reflexionsprisma *n* || prisme *m* à réflexion interne totale || totaalreflectieprisma *n* || призма *f* полного внутреннего отражения

T327 **total output** || Gesamtausgangsleistung *f* || puissance *f* totale de sortie || totaal uitgangsvermogen *n* || полная выходная мощность *f*

T328 **total quantum number** || innere Quantenzahl *f* || nombre *m* quantique total || totaalquantumgetal *n* || полное [внутреннее] квантовое число *n*

T329 **total recursive function** || allgemein-rekursive Funktion *f* || fonction *f* récursive totale || algemeenrecursieve functie *f* || общерекурсивная функция *f*

T330 **toughness** || Viskosität *f*; Zähigkeit *f*; Kerbschlagzähigkeit *f* || ductilité *f*; ténacité *f* || taaiheid *f*; slagsterkte *f* || вязкость *f*; ударная вязкость *f*

T331 **toughness test** *see* **notch-bar test**

T332 **touth flank** || Zahnfußflanke *f* || flanc *m* de la dent || tandvoetprofiel *n* || профиль *m* ножки зуба

T333 **trace** || Spur *f* || trace *f* || spoor *n* || след *m*

T334 **tracer** || Tracer *m*, radioaktiver Indikator *m* || indicateur *m*, traceur *m* || indicator *m*, speurder *m* || радиоактивный [радиоизотопный] индикатор *m*, радиоизотоп *m*

T335 **track** || **1.** Spur *f* **2.** Aufzeichnung *f* || **1.** trace *f* **2.** enregistrement *m* || **1.** baan *f*, spoor *n* **2.** opname *f* (*m*), registratie || **1.** след *m*, трек *m* (*напр. частицы*) **2.** запись *f* (*самописца*)

T336 **traction** || **1.** Zug *m*; Zugkraft *f* **2.** Haftkraft *f* (*z. B. zwischen Reifen und Boden*) || **1.** traction *f* **2.** adhérence *f* (*p. ex. du pneu*) || **1.** tractie *f*, trek *m*, trekbeweging *f*; trekkracht *f* (*m*) **2.** hechtkracht *f* (*m*) || **1.** тяга *f*, сила *f* тяги **2.** сила *f* сцепления (*напр. шины с дорогой*)

T337 **tractive force** || Zugkraft *f*, Zug *m* || force *f* de traction, traction *f* || tractie *f*, trekkracht *f* (*m*) || тяга *f*, тяговое усилие *n*

T338 **tractrix** || Traktrix *f* || tractrice *f* || tractrix *f* || трактриса *f*

T340 **trailing wave** || nachfolgende Welle *f* || onde *f* de suite || opvolgende golf *f* (*m*) || хвостовая волна *f*; спутная волна *f*

T341 **train of gearing** || Wechselrädersatz *m (in der Schere einer Werkzeugmaschine)* || train *m* d'engrenages || wisselwielenstel *n*, wisselwielengarnituur *n* || набор *m* сменных зубчатых колёс *(установленный в гитаре станка)*
T342 **train of waves** || Wellenzug *m* || train *m* [groupe *m*] d'ondes || golventrein *m* || цуг *m* [группа *f*] волн
T343 **trajectory** || Bahn *f*; Flugbahn *f*; Bewegungsbahn *f* || trajectoire *f* || baan *f* || траектория *f*
T344 **transceiver** || Sende-Empfangsgerät *n* || émetteur-récepteur *m*, poste *m* émetteur-récepteur || zenden ontvangapparaat *n* || приёмопередатчик *m*
T345 **transcendental numbers** || transzendente Zahlen *f pl* || nombres *m pl* transcendants || transcendentale getallen *n pl* || трансцендентные числа *pl*
T346 **transconductance** || Steilheit *f* der Voltamperecharakteristik || pente *f* de la caractéristique voltampère || steilheid *f* van voltampèrekarakteristiek || крутизна *f* вольт-амперной характеристики
T347 **transducer** || Meßwandler *m*; Meßgeber *m* || capteur *m*; convertisseur *m*; tra(ns)ducteur *m* || transductor *m*; omvormer *m*; gever *m* || преобразователь *m*; датчик *m*
T348 **transduction** || Umformung *f* || transduction *f* || transductie *f*, omvorming *f* || преобразование *n*
T349 **transductor** || magnetischer Verstärker *m*, Transduktor *m* || tra(ns)ducteur *m*; amplificateur *m* magnétique || magnetische versterker *m*, transductor *m* || магнитный усилитель *m*, трансдуктор *m*
T350 **transfer** || Übertragung *f*, Transfer *m* || transfert *m*, transport *m* || transport *n*, overdragen *n* || перенос *m*; передача *f*
T351 **transfer equation** || Transportgleichung *f* || équation *f* de transfert || overdrachtsvergelijking *f* || уравнение *n* переноса, транспортное уравнение *n*
T352 **transfer function** || Übertragungsfunktion *f* || fonction *f* de transfert || overdrachtsfunctie *f* || передаточная функция *f*
T353 **transfer line** || Transfer(maschinen)straße *f*, Maschinenfließstraße *f* || chaîne *f* transfert || transferstraat *f* || автоматическая (станочная) линия *f*
T354 **transfer matrix** || Übertragungsmatrix *f* || matrice *f* de transfert || overdrachtsmatrix *f* (*m*) || передаточная матрица *f*
T355 **transfer ratio** || Übertragungsfaktor *m* || facteur *m* de transmission || overdrachtsfactor *m* || коэффициент *m* передачи
T356 **transfinite number** || transfinite Zahl *f* || nombre *m* transfini || transfiniet getal *n* || трансфинитное число *n*
T357 **transformation** || Umwandlung *f*, Wandlung *f* || transformation *f* || transformatie *f* || **1.** преобразование *n* **2.** превращение *n*
T358 **transformation of coordinates** || Koordinatentransformation *f* || transformation *f* des coordonnées || coördinatentransformatie *f* || преобразование *n* координат
T359 **transformation of similitude** || Ähnlichkeitstransformation *f*, Ähnlichkeitsabbildung *f* || transformation *f* de similitude || gelijkheidstransformatie *f* || преобразование *n* подобия
T360 **transformation ratio** || Umwandlungsfaktor *m* || rapport *m* de transformation || overbrengings-verhouding *f*, transformatieverhouding *f* || коэффициент *m* трансформации; коэффициент *m* преобразования
T361 **transformation temperature** || Umwandlungstemperatur *f*, Umwandlungspunkt *m* || température *f* de transformation || transformatiepunt *n* || температура *f* [точка *f*] фазового превращения
T362 **transformer** || **1.** Transformator *m* **2.** Wandler *m* || **1.** transformateur *m* **2.** convertisseur *m* || **1.** transformator *m* **2.** stroomrichter *m* || **1.** трансформатор *m* **2.** преобразователь *m*
T363 **transformer coupling** || Transformatorkopplung *f*, induktive Kopplung *f* || couplage *m* par transformateur, couplage inductif || transfor-

matorkoppeling *f* || трансформаторная [индуктивная] связь *f*

T364 I **transient** || Übergangsprozeß *m* || phénomène *m* [processus *m*] transitoire || overgangsproces *n* || переходный [неустановившийся] процесс *m*

T365 II **transient** || Übergangs..., nichtstationär || transitoire; non stationnaire || overgangs...; instabiel || переходный; неустановившийся

T366 **transient conditions** || Übergangszustand *m* || régime *m* transitoire || overgangstoestand *m* || неустановившийся [переходный] режим *m*

T367 **transient motion** || nichtstationäre [inkonstante] Bewegung *f*; Einschwingbewegung *f* || mouvement *m* non stationnaire || instabiele beweging *f* || неустановившееся движение *n*

T368 **transient response** || Übergangscharakteristik *f* || caractéristique *f* transitoire || overgangskarakteristiek *f* || переходная характеристика *f*

T369 **transient stability** || dynamische Stabilität *f* || stabilité *f* dynamique || dynamische stabiliteit *f* || динамическая устойчивость *f*

T370 **transistor** || Transistor *m* || transistor *m* || transistor *m* || полупроводниковый триод *m*, транзистор *m*

T371 **transition** || **1.** Übergang *m* **2.** Phasenumwandlung *f*, Phasenübergang *m* **1.** transition *f* **2.** transformation *f* de phase || **1.** overgang *m* **2.** fasetransformatie *f* || **1.** переход **2.** фазовое превращение *n*, фазовый переход *m*

T372 **transition probability** || Übergangswahrscheinlichkeit *f* || probabilité *f* de passage [de transition] || overgangswaarschijnlijkheid *f* || вероятность *f* перехода

T373 **transition process** || Übergangsprozeß *m* || phénomène *m* [processus *m*] transitoire || overgangsproces *n* || переходный [неустановившийся] процесс *m*

T374 **transition time** || Übergangszeit *f* || durée *f* de processus transitoire || overgangstijd *m* || длительность *f* переходного процесса

T375 **transitive relation** || transitives Verhältnis *n* || relation *f* transitive || transitieve verhouding *f* || транзитивное отношение *n*

T376 **transitivity** || Transitivität *f* || transitivité *f* || transitiviteit *f* || транзитивность *f*

T377 **translation** || **1.** translatorische Bewegung *f* **2.** Parallelverschiebung *f* **3.** Übertragung *f* **4.** Übersetzung *f* || translation *f* || **1.** voorwaartsbeweging *f* **2.** paralleloverdracht *f* (*m*) **3.** translatie *f* || **1.** поступательное движение *n* **2.** параллельный перенос *m* **3.** трансляция *f* **4.** перевод *m*; трансляция *f*

T378 **translational motion** || translatorische Bewegung *f*, Fortschreitung *f*; Translationsbewegung *f* || mouvement *m* de translation; mouvement *m* d'entraînement || voorwaartsbeweging *f*; translatiebeweging *f* || поступательное движение *n*; переносное движение *n*

T379 **translational symmetry** || Translationssymmetrie *f* || symétrie *f* de translation || translatieve symmetrie *f* || трансляционная симметрия *f*

T380 **translator** || Übersetzer *m* || translator *m*, translateur *m* || translator *m* || транслятор *m*; программа-транслятор *m*

T381 **translucent screen** || durchscheinender Schirm *m* || écran *m* translucide || halfdoorzichtig scherm *n* || светопрозрачный экран *m*; просвечивающий экран *m*

T382 **transmission** || **1.** Übertragung *f* *(z. B. Energie)* **2.** Durchlassung *f* *(der Strahlung)*; Durchlaufen *n* *(von Teilchen)* **3.** Schaltgetriebe *n* **4.** Transmission *f*; Kraftübertragung *f* **5.** Übersetzung *f* || transmission *f* || **1.** overdracht *f* (*m*) **2.** doorlaat *m*; doordringing *f* **3.** tandwielkast *f* (*m*) **4.** transmissie *n* **5.** (tandwiel)overbrenging *f* || **1.** передача *f* *(напр. энергии)* **2.** пропускание *n* *(излучения)*; прохождение *n* *(частиц)* **3.** коробка *f* передач **4.** трансмиссия *f*; силовая передача *f* **5.** (зубчатая) передача *f*

T383 **transmission line** || Übertragungsleitung *f* || ligne *f* de transmission [de transport] || overdrachtslijn *f* (*m*) || линия *f* (электро)передачи

T384 **transmission of energy** || Energieübertragung *f* || transmission *f* [transport *m*] d'énergie || energieoverdracht *f* || передача *f* энергии

T385 **transmissivity** || Durchlässigkeitsgrad *m* || facteur *m* de transmission lumineuse; transparence *f* || **1.** doorlaatbaarheid *f* **2.** doorzichtigheidsgraad *m*; doorlaatbaarheidsgraad *m* || **1.** проницаемость *f*, пропускаемость *f* **2.** коэффициент *m* прозрачности; коэффициент *m* пропускания

T386 **transmittance** || Durchlässigkeitsfaktor *m* || facteur *m* de transmission lumineuse, transparence *f* || doorlaatbaarheidsgraad *m*, doorzichtigheidsgraad *m* || коэффициент *m* пропускания, коэффициент *m* прозрачности

T387 **transmitted light** || Durchlicht *n* || lumière *f* transmise || doorlicht *n* || проходящий свет *m*; пропущенный свет *m*

T388 **transmitter** || **1.** Sender *m* **2.** Mikrophon *n* **3.** Geber *m* || **1.** émetteur *m* **2.** microphone *m* **3.** transmetteur *m* || **1.** zender *m* **2.** microfoon *m* **3.** gever *m* || **1.** передатчик *m* **2.** микрофон *m* **3.** датчик *m*

T389 **transmutation** || Umwandlung *f*, Kernumwandlung *f* || transmutation *f* || transmutatie *f* || превращение *n*, распад *m*

T390 **transonic aerodynamics** || Aerodynamik *f* im Schallgrenzbereich || aérodynamique *f* transsonique || transsonische aërodynamica *f* || аэродинамика *f* околозвуковых скоростей, околозвуковая аэродинамика *f*

T391 **transparency** || **1.** Durchsichtigkeit *f*, Transparenz *f*; Durchlässigkeit *f* **2.** Schablone *f* *(Holografie)*; Diapositiv *n* || **1.** transparence *f* **2.** panneau *m* lumineux; diapositive *f* || **1.** doorzichtigheid *f* **2.** sjabloon *f* (*m*); diapositief *n* || **1.** прозрачность *f* **2.** трафарет *m*, транспарант *m* *(в голографии)*; диапозитив *m*

T392 **transparent** || durchsichtig, transparent; durchlässig || transparent || doorzichtig || прозрачный

T393 **transport number** || Überführungszahl *f* || nombre *m* de transport || transportgetal *n* || число *n* переноса

T394 **transpose** *see* **transposition**

T395 **transposed matrix** || transponierte [gestürzte] Matrix *f* || matrice *f* transposée || getransponeerde matrix *f* (*m*) || транспонированная матрица *f*

T396 **transposition** || Transponierung *f* || transposition *f* || transponeren *n* || транспонирование *n*

T397 **transversal** || transversal, Quer... || transversal || transversaal..., dwars... || поперечный, трансверсальный

T398 **transversality condition** || Transversalitätsbedingung *f* || condition *f* de transversalité || transversaliteitsvoorwaarde *f* || условие *n* трансверсальности

T399 **transverse** || Quer...; normal || transversal; normal || dwars..., normaal || поперечный; нормальный, перпендикулярный

T400 **transverse acceleration** || Querbeschleunigung *f*, Transversalbeschleunigung *f* || accélération *f* transversale || dwarsversnelling *f* || поперечное ускорение *n*

T401 **transverse displacement** || Querverschiebung *f*, Transversalverschiebung *f* || déplacement *m* transversal || dwarsverschuiving *f* || поперечное смещение *n*

T402 **transverse pitch** || Stirnteilung *f* *(Zahnrad)* || pas *m* apparent; pas *m* circonférentiel || transversale tandsteek *m* || торцевой шаг *m* *(зубчатого колеса)*

T403 **transverse section** *see* **cross section**

T404 **transverse strain** *see* **lateral strain**

T405 **transverse vibration** || Transversalschwingungen *f pl* || vibrations *f pl* [oscillations *f pl*] transversales || dwarstrillingen *f pl* || поперечные колебания *pl*

T406 **transverse wave** || Transversalwelle *f* || onde *f* transversale || dwarsgolf *f* (*m*) || поперечная волна *f*

T407 **trap** || Haftstelle *f*, Falle *f* || piège *m*; trappe *f* || val *m*, valstrik *m* || ловушка *f*

T408 **trapezoid** || Trapez *n* || trapèze *m* || trapezium *n* || трапеция *f*

T409 **trapezoidal rule** || Trapezregel *f*, Trapezformel *f* || formule *f* des trapèzes || trapeziumformule *f* (*m*) || формула *f* трапеций

T410 **trapping** || Einfang *m* || piégeage *m* || vang *m* || захват *m* (в ловушку)

T411 **travel** || Wanderung *f*; Lauf *m*, Hub *m* || avance *f*, avancement *m* || overplaatsing *f* || перемещение *n*, передвижение *n*; ход *m*; подача *f*

T412 **travelling steady** *see* **movable rest**

T413 **travelling wave** || Wanderwelle *f* || onde *f* mobile || progressieve [lopende] golf *f* (*m*) || бегущая волна *f*

T414 **travelling-wave ration** || Wanderwellenkoeffizient *m*, Wanderwellenverhältnis *n* || taux *m* d'ondes progressives || lopende-golvencoëfficiënt *m*, lopende-golvenverhouding *f* || коэффициент *m* бегущей волны

T415 **tread** || **1.** Wälzfläche *f* **2.** Lauffläche *f* (*Reifen*) || surface *f* de roulement; bande *f* de roulement, roulement *m* || **1.** walsvlak *n* **2.** loopvlak *n* (*band*) || **1.** поверхность *f* качения **2.** протектор *m* (*шины*)

T416 **treatment** || Behandlung *f* (*z. B. thermisch*) || traitement *m* (*p. ex. thermique*) || behandeling *f* (*bijv. thermische*) || обработка *f* (*напр. термическая*)

T417 **tree** || Baum *m* || arbre *m* || boom *m* || **1.** дерево *n* **2.** вал *m*, ось *f* **3.** стойка *f*, подпорка *f*

T418 **triad** || Triade *f* || triade *f* || triade *f* || триада *f*

T419 **trial** || Versuch *m*; Prüfung *f*; Probe *f* || épreuve *f*, essai *m* || proef *m*, test *m* || испытание *n*; проба *f*

T420 **triangle** || Dreieck *n* || triangle *m* || driehoek *m* || треугольник *m*

T421 **triangle axiom** || Dreiecksungleichung *f*, Dreiecksaxiom *n* || axiome *m* du triangle || driehoeksaxioma *n* || аксиома *f* треугольника

T422 **triangle commutations** || Dreiecksberechnung *f* || **1.** trigonométrie *f* **2.** résolution *f* des triangles || **1.** trigonometrie *f* **2.** driehoeksberekening *f* || **1.** исчисление *n* треугольников; тригонометрия *f* **2.** решение *n* треугольников

T42 **triangular matrix** || Dreiecksmatrix *f* || matrice *f* trigonale [triangulaire] || driehoeksmatrix *f* || треугольная матрица *f*

T424 **triangular thread** *see* **angular thread**

T425 **triboelectricity** || Reibungselektrizität *f*, Triboelektrizität *f* || tribo-électricité *f* || tribo-elektriciteit *f* || трибоэлектричество *n*

T426 **triboelectrification** || Elektrisieren *n* durch Reiben || tribo-électrisation *f* || tribo-elektriseren *n* || электризация *f* трением

T427 **triboluminescence** || Triboluminesczenz *f* || triboluminescence *f* || triboluminescentie *f* || триболюминесценция *f*

T428 **tribometry** || Tribometrie *f* || tribométrie *f* || tribometrie *f* || трибометрия *f*

T429 **tribonics** || Tribonik *f* || tribonique *f* || tribonica *f* || трибоника *f*, физика *f* и техника *f* трения

T430 **tribophysics** || Tribophysik *f* || tribophysique *f* || tribofysica *f* || трибофизика *f*, физика *f* трения

T431 **trichromatic** || Dreifarben... || à trois couleurs || driekleurig || трёхцветный

T432 **trigger** || Trigger *m* || trigger *m* || trigger *m* || триггер *m*

T433 **trigger circuit** || Trigger *m*, Triggerschaltung *f* || circuit *m* trigger, trigger *m* || trigger *m* || спусковая схема *f*, триггер *m*

T434 **trigonometric functions** || trigonometrische Funktionen *f pl* || fonctions *f pl* trigonométriques || trigonometrische functies *f pl* || тригонометрические функции *pl*

T435 **triode** || Triode *f* || triode *f* || triode *f* || триод *m*

T436 **trip gear** || Auslösevorrichtung *f*, Auslöser *m*, Ausklinkvorrichtung *f* || mécanisme *m* disjoncteur || uitklinkmechanisme *n* || расцепляющий механизм *m*; расцепляющее устройство *n*

T437 I **triple** || Tripel *n* || triplet *m* || triplet *m* || тройка *f*

T438 II **triple** || Tripel... || triple || tripel... || тройной

T439 **triple bond** || Dreifachbindung *f*, dreifache Bindung *f* || triple liaison *f* || drievoudige binding *f* || тройная связь *f*

T440 **triple point** || Tripelpunkt *m* || point *m* triple || tripelpunt *n* || тройная точка *f*

T441 **triple product** || Tripelprodukt *n* || produit *m* triple || tripelprodukt *n* || тройное произведение *n* *(векторов)*

T442 **triplet** || Triplett *n* || triplet *m* || triplet *m* || триплет *m*

T443 **triple thread** || dreigängiges Gewinde *n* || filet *m* triple || drievoudige [driegangige] schroefdraad *m* || трёхзаходная резьба *f*

T444 **trip lever** || Ausrückhebel *m*, Entkupplungshebel *m*; Schalthebel *m* || levier *m* disjoncteur [de libération] || uitkoppelhefboom *m*, ontkoppelhefboom *m* || расцепляющий рычаг *m*; выключающий рычаг *m*

T445 **trisectrix** || Trisektrix *f* || trisectrice *f* || trisectrix *f* *(m)* || трисектрис(с)а *f*

T446 **trisonic** || Unterschall..., Schallgrenz... und Überschall... || sub-, trans et supersonique || sub-, trans- en supersoon || до-, около- и сверхзвуковой

T447 **true address** || absolute Adresse *f* || adresse *f* vraie || absoluut adres *n* || абсолютный [истинный] адрес *m*

T448 **true representation** || wahrheitsgetreue Darstellung *f* || représentation *f* vraie || ware voorstelling *f* || точное [истинное] представление *n*

T449 **truncated normal distribution** || gestufte Normalverteilung *f* || distribution *f* normale tronquée || afgeknotte normaalverdeling *f* || усечённое нормальное распределение *n*

T450 **truncation** || 1. Abbrechen *n* *(Reihe)* 2. Abschneiden *n*; Abstumpfen *n* || troncature *f* || 1. afbreken *n* *(reeks)* 2. afknotting *f* || 1. отбрасывание *n* *(напр. членов ряда)*; обрыв *m* 2. усечение *n*

T451 **truncation error** || Abbrechfehler *m* || erreur *f* de troncature || afbraakfout *f* *(m)* || ошибка *f* отбрасывания; погрешность *f* обрыва

T452 **trunnion** || Zapfen *m*; Drehachse *f*, Schwenkachse *f* || tourillon *m* || tap *m*; taats *f* *(m)* || цапфа *f*; ось *f* поворота

T453 **truth** || Wahrheit *f* || vérité *f*; valeur *f* de vérité || waarheid *f* || истинностное значение *n*, истина *f*; истинность *f*

T454 **truth table** || Wahrheitstafel *f*, Wahrheitstabelle *f* || table *f* de vérité || waarheidstabel *f* *(m)* || истинностная таблица *f*, таблица *f* значений истинности

T455 **truth value** || Wahrheitswert *m* || valeur *f* de vérité || waarheidswaarde *f* || истинностное значение *n*, значение *n* истинности

T456 **t-test** || t-Test *m*, Student-Test *m* || test *m* de Student || Studenttest *m* || критерий *m* Стьюдента, t-критерий *m*

T457 **tube** || 1. Rohr *n* 2. Röhre *f*, Elektronenröhre *f* || 1. tube *m* 2. tube *m* électronique || buis *f* *(m)* || 1. труба *f*; трубка *f* 2. электронная лампа *f*; электроннолучевая трубка *f*

T459 **tube drawing** || Rohrziehen *n* || étirage *m* des tubes || buistrekken *n* || волочение *n* труб

T460 **tube expander** || Rohraufweitevorrichtung *f* || dudgeon *m* || pijpruitzetter *m* || вальцовка *f*, развальцовочный станок *m* *(для труб)*

T461 **tube-rolling mill** *see* **pipe mill**

T462 **tube voltmeter** || Röhrenvoltmeter *n* || voltmètre *m* à lampes || buizenvoltmeter *m* || ламповый вольтметр *m*

T463 **tumbling** || Rommeln *n*, Trommeln *n* || polissage *m* au tambour [au tonneau] || trommelen *n* || галтовка *f* *(обработка деталей во вращающемся барабане)*

T464 **tunable laser** || abstimmbarer Laser *m* || laser *m* réglable [adaptable] || afstembare laser *m* || перестраиваемый лазер *m*

T465 **tuned load** || angepaßter Abschluß *m* || charge *f* adaptée ||

aangepaste lading *f* || согласованная нагрузка *f*

T466 **tuner** || Tuner *m* || tuner *m* || tuner *m* || тюнер *m*

T467 **tungsten,** W || Wolfram *n* || tungstène *m* || wolfra(a)m *n* || вольфрам *m*

T468 **tungsten carbide** || Wolframkarbid *n* || carbure *m* de tungstène || wolframcarbide *n* || карбид *m* вольфрама

T469 **tungsten steel** || Wolframstahl *m* || acier *m* au tungstène || wolframstaal *m* || вольфрамовая сталь *f*

T470 **tuning** || Abstimmung *f* || accord *m* || afstemming *f* || настройка *f*

T471 **tunnel current** || Tunnelstrom *m* || courant *m* (d'effet) tunnel || tunnelstroom *m* || туннельный ток *m*

T472 **tunneling** || Tunneleffekt *m*, Tunneln *n* || effet *m* (de) tunnel; passage *m* tunnel || tunneleffect *n* || туннельный эффект *m*, туннелирование *n*

T473 **turbidimetry** || Streulichtmessung *f* || turbidimétrie *f* || turbidimetrie *f* || турбидиметрия *f*

T474 **turbidity** || Trübung *f* || turbidité *f* || troebelheid *f* || мутность *f*; помутнение *n*; тусклость *f*

T475 **turbid medium** || trübes Medium *n* || milieu *m* trouble || troebel medium *n* || мутная среда *f*

T476 **turbine** || Turbine *f* || turbine *f* || turbine *f* || турбина *f*

T477 **turbine pump** || Turbopumpe *f* || turbo-pompe *f* || turbopomp *f* (*m*), turbinepomp *f* (*m*) || турбонасос *m*

T478 **turbine wheel** || Turbinenlaufrad *n* || roue *f* de turbine || turbinerad *n*, turbinewiel *n* || рабочее колесо *n* турбины

T479 **turbogenerator** || Turbogenerator *m* || turbogénérateur *m* || turbogenerator *m* || турбогенератор *m*

T480 **turbojekt engine** || Turbinenluftstrahltriebwerk *n* || turboréacteur *m* || turbostraalmotor *m* || турбореактивный двигатель *m*

T481 **turboprop engine** || Propellerturbinentriebwerk *n* || turbopropulseur *m* || schroefturbinemotor *m* || турбовинтовой двигатель *m*

T482 **turbosupercharged engine** || Turboaufladungsmotor *m* || moteur *m* à turbocompresseur || motor *m* met turboopllading || двигатель *m* с турбонаддувом

T483 **turbosupercharger** || Turboverdichter *m*; Turbolader *m* || turbocompresseur *m* || turbocompressor *m* || турбокомпрессор *m*; турбонагнетатель *m*

T484 **turbulence** || Turbulenz *f* || turbulence *f* || turbulentie *f* || турбулентность *f*

T485 **turbulent** || turbulent, Turbulenz..., Wirbel... || turbulent || turbulent, wervel... || турбулентный, вихревой

T486 **turbulent flow** || turbulente Strömung *f* || écoulement *m* turbulent || turbulente stroming *f* || турбулентное течение *n*, турбулентный поток *m*

T487 **turbulization** || Verwirbelung *f* || tourbillonnement *m* || werveling *f* || завихрение *n*, турбулизация *f*

T488 **Turing machine** || Turingmaschine *f* || machine *f* de Turing || Turingmachine *f* || машина *f* Тьюринга

T489 **turn** || Windung *f*; Umdrehung *f* || spire *f*; tour *m*; révolution *f* || winding *f*, omwenteling *f* || виток *m*; оборот *m*

T490 **turnbuckle** || Spannschloß *n*, Spannschraube *f*; Gewindekupplung *f* || tendeur *m* à vis || spanschroef *f* (*m*); schroefkoppeling *f* || винтовая стяжка *f*, винтовая муфта *f*

T491 **turning** || **1.** Drehen *n*, Drehbearbeitung *f* **2.** Wendung *f*; Schwenkung *f* **3.** Drehung *f*, Umlauf *m*, Rotation *f* || **1.** tournage *m* **2.** rotation *f* || **1.** draaien *n* **2.** bocht *f* (*m*) **3.** draaïng *f*; omwenteling *f* || **1.** точение *n*, токарная обработка *f* **2.** поворот *m* **3.** вращение *n*

T492 **turning machine** *see* **lathe**

T493 **turning mill** || Karusselldrehmaschine *f* || tour *m* vertical [à plateau horizontal] || carrousel(draai)-bank *f* (*m*) || (токарно)-карусельный станок *m*

T494 **turning tool** *see* **lathe tool**

T495 **turn-table** || Drehtisch *m* ||

plate-forme *f* tournante || draaitafel *f* (*m*) || поворотный стол *m*

T496 **turret head** *see* **capstan tool head**

T497 **turret head drilling machine** || Revolverbohrmaschine *f* || perceuse *f* (à tourelle) revolver || revolverboormachine *f* || сверлильно-револьверный станок *m*

T498 **turret lathe** *see* **capstan lathe**

T499 **twin paradox** || Zwillingsparadoxon *n* *(Relativitätstheorie)* || paradoxe *m* des horloges || tweelingsparadox *m* || парадокс *m* близнецов, парадокс *m* часов *(в теории относительности)*

T500 **twist** *see* **twisting**

T501 **twist drill** || Spiralbohrer *m* || foret *m* hélicoïdal || spiraalboor *f* || спиральное сверло *n*

T502 **twisting** || Torsion *f*, Verdrehung *f* || torsion *f* || torsie *f* || кручение *n*, закручивание *n*

T503 **two-dimensional** || zweidimensional || bidimensionnel || tweedimensionaal || двухмерный; плоский; поверхностный

T504 **two-high mill** || Duowalzwerk *n*, Duo(walz)straße *f*, Duo *n* || laminoir *m* [train *m*] duo || duowalsstraat *f* (*m*)|| стан-дуо, двухвалковый прокатный стан *m*

T505 **two-person game** || Zweipersonenspiel *n* || jeu *m* à deux personnes || twee-personenspel *n* || игра *f* с двумя партнёрами

T506 **two-start thread** *see* **double thread**

T507 **two-stroke cycle** || Zweitaktarbeitsspiel *n*, Zweitaktprozeß *m* || cycle *m* à deux temps || tweeslagscyclus *m* || двухтактный цикл *m*

T508 **two-terminal network** || Zweipol *m* || réseau *m* dipôle, dipôle *m* || tweepool *f* (*m*) || двухполюсник *m*

T509 **two-throw pump** || Zweistrompumpe *f*, Zweikreispumpe *f*, Zwillingspumpe *f* || pompe *f* double [duplex] || dubbelstroom-pomp *f* (*m*) || сдвоенный насос *m*

T510 **two-valued logic** || zweiwertige Logik *f* || logique *f* à deux valeurs || tweewaardige logica *f* || двузначная логика *f*

T511 **two-wire** || Zweidraht..., Zweileiter... || bifilaire || tweedraads || двухпроводный, двухжильный

T512 **type** || Typ *m* || type *m* || type *n* || тип *m*

U

U1 **ultimate analysis** || Elementaranalyse *f* || analyse *f* ultime, analyse *f* ultième || elementaire analyse *f* || элементарный анализ *m*

U2 **ultimate factor of safety** || Grenzsicherheitsfaktor *m* || coefficient *m* de sécurité par rapport à la charge de rupture || grensveiligheidsfactor *m*, eindige veiligheidsfactor *m* || коэффициент *m* запаса прочности *(по отношению к разрушающей нагрузке)*

U3 **ultimate load** || kritische Belastung *f*, Grenzlact *f* || (sur)charge *f* de rupture || breukbelasting *f* || разрушающая нагрузка *f*

U4 **ultimate strength** || Festigkeitsgrenze *f*; Reißfestigkeit *f*, Zugfestigkeit *f* || résistance *f* à la rupture || breuksterkte *f*; breukvastheid *f* || предел *m* прочности; сопротивление *n* разрыву

U5 **ultimate stress** *see* **breaking stress**

U6 **ultimate tensile stress** || Grenzzugspannung *f*; Zugfestigkeit *f* || résistance *f* à la traction || grenstrekspanning *f* || предельное напряжение *n* при растяжении; сопротивление *n* разрыву

U7 **ultimate velocity** || Grenzgeschwindigkeit *f* || vitesse *f* limite || grenssnelheid *f* || предельная [максимальная] скорость *f*

U8 **ultrahigh** || ultrahoch || ultrahaut || ultrahoog || сверхвысокий, ультравысокий

U9 **ultrahigh frequency** || Höchstfrequenz *f*, UHF || ultra-haute fréquence *f*, UHF || ultrahoogfrequentie *f*, UHF || ультравысокая частота *f*, УВЧ

U10 **ultrahigh temperature** || Höchsttemperatur *f* || ultra-haute température *f* || ultrahoogtemperatuur *f* || сверхвысокая температура *f*

U11 **ultralow frequency** || Ultratieffrequenz *f* || ultra-basse fréquence *f* || ultralaagfrequentie *f* || ультранизкая частота *f*, УНЧ

U12 **ultramicroscope** || Ultramikroskop *n*, Elektronenübermikroskop *n* || ultramicroscope *m* || ultramicroscoop *m* || ультрамикроскоп *m*

U13 **ultrasonic** || Ultraschall... || ultra-sonore, superaudible, supersonique || ultrasonoor || ультразвуковой

U14 **ultrasonic waves** || Ultraschallwellen *f pl* || ondes *f pl* ultra-sonores [ultra-soniques] || ultrageluidgolven *f (m) pl*, ultrageluid *n* || ультразвуковые волны *f pl*, ультразвук *m*

U15 **ultrasound** || Ultraschall *m* || ultra-son *m* || ultrageluid *n* || ультразвук *m*

U16 I **ultraviolet** || Ultraviolettbereich *m* || ultraviolet *m* || ultraviolet *n* || ультрафиолетовая область *f* *(спектра)*

U17 II **ultraviolet** || Ultraviolet..., UV- || ultraviolet || ultraviolet || ультрафиолетовый

U18 **ultraviolet radiation** || Ultraviolettstrahlung *f*, UV-Strahlung *f* || radiation *f* ultraviolette || ultravioletstraling *f* || ультрафиолетовое излучение *n*

U19 **ultraviolet region** *see* **ultraviolet**

U20 **umbilical point** || Nabelpunkt *m* || point *m* ombilical || navelpunt *n* || омбилическая точка *f*, омбилик *m*

U21 **unalloyed steel** || unlegierter Stahl *m* || acier *m* non allié || ongelegeerd staal *n* || нелегированная сталь *f*

U22 **unambiguous language** || eindeutige Sprache *f* || langage *m* non ambigu || eenduidige taal *f* || однозначный язык *m*

U23 **unbalance** || Unbalance *f*; Unwucht *f* || balourd *m*, déséquilibre *m* || onbalans *f (m)*, onevenwichtigheid *f* || дисбаланс *m*, неуравновешенность *f*

U24 **unbalanced load** || unsymmetrische Belastung *f* || charge *f* déséquilibrée || onevenwichtige lading *f* || || неуравновешенная нагрузка *f*

U25 **unbiased estimate** || biasfreie [unverzerrte, erwartungsfreie] Schätzung *f* || estimation *f* sans biais || schuinvrije schatting *f* || несмещённая оценка *f*

U26 **uncertainty** || Unsicherheit *f*; Unschärfe *f* || indétermination *f* || onbepaaldhei *f* || неопределённость *f*

U27 **uncertainty principle** || (Heisenbergsche) Unschärferelation *f* || principe *m* d'incertitude [d'indétermination] || onbepaaldheidsprincipe *n* || принцип *m* неопределённости (Гейзенберга)

U28 **unconditional branch** *see* **unconditional jump**

U29 **unconditional convergence** || absolute Konvergenz *f* || convergence *f* inconditionnelle || absolute convergentie *f* || абсолютная сходимость *f*

U30 **unconditional jump** || unbedingter Sprung *m* || saut *m* inconditionnel, branchement *m* inconditionnel || onvoorwaardelijke sprong *m* || безусловный переход *m*

U31 **uncorrelated functions** || nichtkorrelierte Funktionen *f pl* || fonctions *f pl* incorrélées || ongecorreleerde functies *f pl* || некоррелированные функции *f pl*

U32 **undecidability** || Unentscheidbarkeit *f* || indécidabilité *f* || onoplosbaarheid *f* || неразрешимость *f*

U33 **undecidable** || unentscheidbar || indécidable || onoplosbaar || неразрешимый

U34 **undercritical** || unterkritisch || sous-critique, subcritique || onderkritisch || докритический; подкритический

U35 **undercutting tool** || Plandrehmeißel *m* || outil *m* à saigner || voorsnijbeitel *m* || подрезной резец *m*

U36 **underload** || Unterbelastung *f* || charge *f* incomplète [réduite] || onderbelasting *f* || неполная нагрузка *f*

U37 **underpressure** || Unterdruck *m* || sous-pression *f*, dépression *f* ||

onderdruk *m* || давление *n* ниже атмосферного; разрежение *n*

U38 **undervoltage** || Unterspannung *f* || sous-tension *f* || onderspanning *f* || напряжение *n* ниже номинального

U39 **undisturbed orbit** || ungestörte Bahn *f* || orbite *f* imperturbée || ongestoorde baan *f* (*m*) || невозмущённая орбита *f*

U40 **undular** || wellenförmig, Wellen... || ondulatoire || golf... || волнообразный, волновой

U41 **undulation** || Wellenbewegung *f* || mouvement *m* ondulatoire || golfbeweging *f* || волнообразное [волновое] движение *n*

U42 **ungrounded** || nicht geerdet || non mis à la terre || ongeaard || незаземлённый

U43 **uniaxial** || einachsig || monoaxial || eenassig || одноосный

U44 **unidirectional** einseitig gerichtet || monodirectionnel, unidirectionnel || eenzijdig gericht || однонаправленный, остронаправленный

U45 **unified** || Einheits... || unifié || uniforme || унифицированный

U46 **unifilar** || unifilar, Einfaden... || unifilaire || unifilair || унифилярный

U47 **uniform** || **1.** gleichartig **2.** gleichförmig || uniforme || **1.** gelijksoortig **2.** uniform || **1.** однородный **2.** равномерный

U48 **uniform acceleration** || gleichförmige Beschleunigung *f* || accélération *f* uniforme || uniforme versnelling *f* || постоянное ускорение *n*, ускорение *n* равнопеременного движения

U49 **uniform compression** || allseitiger Druck *m* || compression *f* triaxiale [volumétrique] || alzijdige druk *m* || всестороннее сжатие *n*

U50 **uniform convergence** || gleichmäßige Konvergenz *f* || convergence *f* uniforme || gelijkmatige convergentie *f* || равномерная сходимость *f*

U51 **uniform distribution** || Gleichverteilung *f* || distribution *f* uniforme || gelijkmatige [gelijkvormige] verdeling *f*, gelijkmatige [gelijkvormige] distributie *f* || равномерное распределение *n*

U52 **uniform extension** || allseitige Zugspannung *f* || traction *f* triaxiale || alzijdig trekken *n* || всестороннее растяжение *n*

U53 **uniform load** || gleichmäßige Belastung *f* || charge *f* uniforme || gelijkmatige belasting *f* || равномерная нагрузка *f*

U54 **uniformly accelerated motion** || gleichförmig beschleunigte Bewegung *f* || mouvement *m* uniformément accéléré || eenparig versnelde beweging *f* || равномерно-ускоренное [равноускоренное] движение *n*

U55 **uniformly retarded motion** || gleichförmig verzögerte Bewegung *f* || mouvement *m* uniformément retardé || eenparig vertraagde beweging *f* || равномерно-замедленное [равнозамедленное] движение *n*

U56 **uniform motion** || gleichförmige Bewegung *f* || mouvement *m* uniforme || eenparige beweging *f* || равномерное движение *n*

U57 **uniform pressure** || allseitiger Druck *m* || pression *f* uniforme || alzijdige druk *m* || всестороннее давление *n*

U58 **uniform velocity** || gleichförmige Geschwindigkeit *f* || vitesse *f* uniforme || gelijkvormige snelheid *f* || постоянная скорость *f*

U59 **unilateral limit** || einseitiger Grenzwert *m* || limite *f* unilatérale || eenzijdige grenswaarde *f* || односторонний предел *m*

U60 **unimodal distribution** || unimodale [eingipflige] Verteilung *f* || distribution *f* unimodale || unimodale verdeling *f* || одномодальное распределение *n*

U61 **unimodular matrix** || unimodulare Matrix *f* || matrice *f* unimodulaire || unimodulaire matrix *f* || унимодулярная матрица *f*

U62 **unimolecular film** monomolekulare Schicht *f*, Monomolekularfilm *m* || couche *f* monomoléculaire || monomoleculaire laag *f* (*m*) || мономолекулярный слой *m*

U63 **union** || Vereinigung (smenge) *f* || union *f* || vereniging *f* || объединение *n*, логическая сумма *f*

U64 **unipolar induction** || Unipolarinduktion *f* || induction *f* unipolaire || unipolaire inductie *f* || униполярная индукция *f*

U65 **unipolar inductor** || Unipolargenerator *m* || inducteur *m* unipolaire || unipolaire generator *m* || униполярный генератор *m*

U66 **uniqueness theorem** || Unitätssatz *m* || théorème *m* d'unicité || uniteitstheorema *n* || теорема *f* единственности

U67 **unit** || **1.** Einheit *f*; Maßeinheit *f* **2.** Anlage *f*; Einrichtung *f*; Baueinheit *f*, Einheit *f*; Aggregat *n*; Block *m*, Einheit *f* **3.** Bauteil *n* || **1.** unité *f* **2.** installation *f* **3.** unité *f* || **1.** eenheid *f*; maateenheid *f* **2.** inrichting *f*; knooppunt *n*; aggregaat *n*; blok *m* **3.** deel *n*, detail *n*, stuk *n* || **1.** единица *f* (измерения) **2.** установка *f*; устройство *n*; узел *m*; агрегат *m*; блок *m* **3.** деталь *f*

U68 **unitary** *math.* || unitär || unitaire || unitair || унитарный *(напр. о матрице, операторе, преобразовании)*

U69 **unit cell** || Elementarzelle *f* || cellule *f* cristallographique [élémentaire] || elementaire cel *f* (*m*) || элементарная ячейка *f* *(кристалла)*

U70 **unit length** || **1.** Einheitslänge *f* **2.** Längeneinheit *f* || **1.** longueur *f* unitaire **2.** unité *f* de longueur || **1.** eenheidslengte *f* **2.** lengteëenheid *f* || **1.** единичная длина *f* **2.** единица *f* длины

U71 **unit mass** || Masseneinheit *f* || unité *f* de masse || massaeenheid *f* || единица *f* массы, единичная масса *f*

U72 **unit measure** || Maßeinheit *f* || unité *f* de mesure || maateenheid *f* || единица *f* измерения

U73 **unit-step function** || Einheitssprungfunktion *f* || fonction *f* unitaire || eenheidssprongfunctie *f* || функция *f* единичного скачка

U74 **unit-step response** || Sprungantwort *f* || réponse *f* unitaire || eenheidsstap-antwoord *n* || реакция *f* на единичное (ступенчатое) воздействие

U75 **unit vector** || Einheitsvektor *m* || vecteur *m* unité || eenheidsvector *m* || единичный вектор *m*, орт *m*

U76 **unit weight** || Gewichtseinheit *f* || unité *f* de poids || gewichtseenheid *f* || единица *f* веса

U77 **unity** || Einheit *f* || unité *f* || eenheid *f* || единица *f*

U78 **universal algorithm** || Universalalgorithmus *m* || algorithme *m* universel || universeel algorithme *n* || универсальный алгоритм *m*

U79 **universal chuck** *see* **scroll chuck**

U80 **universal gas constant** || universelle Gaskonstante *f* || constante *f* universelle des gaz || universele gasconstante *f* (*m*) || универсальная газовая постоянная *f*

U81 **universal gravitation** || Newtonsche Gravitation *f* || gravitation *f* [attraction *f*] universelle || gravitatie *f* || всемирное тяготение *n*

U82 **universal joint** || Kardangelenk *n*, Kreuzgelenkkupplung *f* || joint *m* universel [de cardan]; charnière *f* universelle || kruis(scharnier)koppeling *f* || карданный шарнир *m*

U83 **universal-joint drive** || Gelenkwellenübertragung *f*; Kardanantrieb *m* || transmission *f* à articulation [par cardan] || cardanoverbrenging *f*, cardanaandrijving *f* || карданная передача *f*, карданный привод *m*

U84 **universal quantifier** || Allloperator *m*, Allquantor *m*, Generalisator *m* || quantificateur *m* universel || generalisator *m* || квантор *m* всеобщности

U85 **universal set** || Universalmenge *f* || ensemble *m* universel || universele menigte *f* || универсальное множество *n*

U86 **universal time** || Weltzeit *f* || temps *m* universel || wereldtijd *m* || всемирное время *n*

U87 **universe** || **1.** Universum *n*, All *n* **2.** *Math.* Grundgesamtheit *f* || **1.** univers *m* **2.** *math.* ensemble *m* universel || heelal *n* || **1.** вселенная *f* **2.** *мат.* генеральная совокупность *f*

U88 **unknown** || Unbekannte *f* || inconnue *f* || onbekende *f* || неизвестная (величина) *f*

U89 **unlike charges** || ungleichnamige Ladungen *f pl* || charges *f pl* de noms [de signes] contraires || ongelijknamige ladingen *f pl* || разноимённые заряды *m pl*

U90 **unlike poles** || ungleichnamige Pole *m pl* || pôles *m pl* opposables [contraires, de noms contraires] || ongelijknamige polen *f (m) pl* || разноимённые полюса *m pl*

U91 **unmagnetized** || nichtmagnetisiert || non aimanté || niet-gemagnetiseerd || ненамагниченный

U92 **unoccupied** || unbesetzt || disponible || onbezet || незанятый, вакантный *(напр. уровень энергии)*

U93 **unpolarized light** || unpolarisiertes Licht *n* || lumière *f* non polarisée || niet-gepolariseerd licht *n* || неполяризованный свет *m*

U94 **unsaturated** || ungesättigt || non saturé || onverzadigde || ненасыщенный *(напр. о паре, химическом соединении)*

U95 **unstable** || instabil, unstabil || instable || onstabiel, labiel || неустойчивый, нестабильный; неуравновешенный

U96 **unstable equilibrium** || instabiles [labiles] Gleichgewicht *n* || équilibre *m* instable || labiel evenwicht *n* || неустойчивое равновесие *n*

U97 **unstable nucleus** || instabiler [radioaktiver] Kern *m* || noyau *m* instable [labile] || labiele [onstabiele] kern *f* || неустойчивое [радиоактивное] ядро *n*

U98 **unsteady** || nicht stationär || instable || niet-stationair || неустановившийся, нестационарный

U99 **unsteady oscillations** || nichtstationäre Schwingungen *f pl* || oscillations *f pl* non stationnaires [transitoires] || niet-stationaire trillingen *f pl* || неустановившиеся колебания *n pl*

U100 **unsteady state** || nicht stationärer Zustand *m* || état *m* non stationnaire || niet-stationaire toestand *m* || неустановившееся состояние *n*; нестационарное состояние *n*

U101 **upper bound, upper limit** || obere Schranke *f*, oberer Limes *m* || borne *f* supérieure; limite *f* supérieure || bovengrens *f (m)* || верхняя граница *f*; верхний предел *m*

U102 **upright drilling machine** *see* **pillar drill**

U103 **upset forging** || Stauchen *n* || refoulement *m* || stuiking *f*, opstuiking *f* || высадка *f*, осадка *f (в металлообработке)*

U104 **upset pressure** || Stauchdruck *m* || pression *f* de refoulement || stuikdruk *m*, opstuikdruk *m* || давление *n* высадки [осадки]

U105 **upward transition** || Übergang *m* nach oben || transition *f* au niveau plus haut || overgang *m* naar boven || переход *m* на более высокий (энергетический) уровень

U106 **uranim, U** || Uran *n* || uranium *m* || uranium *n*, uraan *n* || уран *m*

U107 **useful load** || Nutzlast *f* || charge *f* utile || nuttige belasting *f* || полезная нагрузка *f*

U108 **useful power** || **1.** Nutzleistung *f* **2.** Nutzenergie *f* || **1.** puissance *f* utile **2.** énergie *f* utile || **1.** nuttig vermogen *n* **2.** nuttige energie *f* || **1.** полезная мощность *f* **2.** полезная энергия *f*

V

V1 **vacancy, vacant (lattice) site** || Leerstelle *f*, Gitterleerstelle *f* || vacance *f*, lacune *f* du réseau (cristallin) || (rooster) vacantie *f* || вакансия *f (дефект кристаллической решётки)*

V2 **vacuum** || Vakuum *n* || vide *m*, vacuum *m* || vacuüm *n* || вакуум *m*

V3 **vacuum brake** || Saugluftbremse *f*, Vakuumbremse *f* || frein *m* à vide || vacuümrem *f (m)* || вакуумный тормоз *m*

V4 **vacuum gauge, vacuum manometer** || Vakuummeter *n*, Vakuummeßgerät *n*, Unterdruckmesser *m* || vacuomètre *m*, manomètre *m* à vide ||

vacuümmeter *m* || вакуумметр *m*, вакуумный манометр *m*

V5 **vacuum pump** || Vakuumpumpe *f* || pompe *f* à vide || vacuümpomp *f* (*m*) || вакуумный насос *m*

V6 **vacuum technique** || Vakuumtechnik *f* || technique *f* de vide || vacuümtechniek *f* || вакуумная техника *f*

V7 **vagabond(ing) currents** || vagabundierende Ströme *m pl*, Irrströme *m pl* || courants *m pl* vagabonds || vagebonderende stromen *m pl* || блуждающие токи *m pl*

V8 **valence band** || Valenzband *n* || bande *f* de valence || valentieband *m* || валентная (энергетическая) зона *f*

V9 **valence electron** || Valenzelektron *n* || électron *m* de valence || valentie-elektron *n* || валентный электрон *m*

V10 **valency** || Valenz *f*, Wertigkeit *f* || valence *f* || valentie *f* || валентность *f*

V11 **value** || Wert *m* || valeur *f* || waarde *f* || значение *n*, величина *f*

V12 **valve** || 1. Ventil *n* 2. Röhre *f*, Katodenstrahlröhre *f* || 1. valve *f* 2. tube *m* électronique || 1. ventiel *n*; klep *f* (*m*) 2. buis *f* (*m*) || 1. клапан *m*; вентиль *m* 2. электронная лампа *f*

V13 **valve lifter** *see* **valve tappet**

V14 **valve rod** || 1. Ventilstoßstange *f* 2. Schieber(schub)stange *f* || tige *f* de soupape; tige *f* de vanne; tige *f* de tiroir || 1. klepstoterstang *f* (*m*) 2. schuifstang *f* (*m*) || 1. штанга *f* толкателя клапана 2. золотниковый шток *m*

V15 **valve seat** || Ventilsitz *m* || siège *m* de soupape || klepzitting *f* || седло *n* [гнездо *n*] клапана

V16 **valve stem** || Ventilschaft *m* || tige *f* de soupape || klepsteel *m* || стержень *m* клапана

V17 **valve tappet** || Ventilstößel *m* || poussoir *m* de soupape || klepstoter *m*, kleplichter *m* || толкатель клапана

V18 **vanadium,** V || Vanadin *n* || vanadium *m* || vanadium *n* || ванадий *m*

V19 **Van de Graaf generator** || Van-de-Graaf-Generator *m* || générateur *m* [accélérateur *m*] Van de Graaf || vandegraaffgenerator *m* || генератор *m* Ван-де-Граафа

V20 **Van der Waals forces** || Van-der-Waals-Kräfte *f pl* || forces *f pl* de Van der Waals || vanderwaalskrachten *f* (*m*) *pl* || вандерваальсовы силы *f pl*

V21 **vane** || Schaufel *f*, Schaufelblatt *n*; Flügel *m* || aile *f*, ailette *f*; aube *f*; palette *f* || schoep *f* (*m*), vleugel *m* || лопатка *f*; лопасть *f* (*напр. насоса*)

V22 **vane compressor** || Kreiselverdichter *m* || compresseur *m* centrifuge radial || centrifugaalcompressor *m*, radiaalcompressor *m* || центробежный компрессор *m*; осевой лопаточный компрессор *m*

V23 **vane pump** || Flügelzellenprumpe *f* || pompe *f* à palettes; pompe *f* à ailettes || vleugelpomp *f* (*m*) || 1. лопастный насос *m*; крыльчатый насос *m* 2. (роторный) пластинчатый насос *m*

V24 **vane supercharger** || Kreiselgebläse *n* || compresseur *m* rotatif à palettes || vleugelaanjager *m* || лопаточный нагреватель *m*

V25 **vapo(u)r** || Dampf *m* || vapeur *f* || damp *m* || пар *m*

V26 **vapo(u)rizability** || Verdampfbarkeit *f*, Verdampfungsfähigkeit *f* || pouvoir *m* vaporifère, capacité *f* d'évaporation || verdampingsvermogen *n* || 1. испаряемость *f* 2. паропроизводительность *f*

V27 **vapo(u)rization** || Verdampfung *f* || vaporisation *f*, évaporation *f* || verdamping *f* || испарение *n*; парообразование *n*

V28 **vapo(u)rizer** || Verdampfer *m* || évaporateur *m*, vaporisateur *m* || verdamper *m* || испаритель *m*

V29 **vapo(u)r-liquid equilibrium** || Dampf-Flüssigkeit-Gleichgewicht *n* || équilibre *m* de vapeur-liquide || damp-vloeistof-evenwicht || равновесие *n* между жидкостью и паром, парожидкостное равновесие *n*

V30 **vapo(u)rous state** || dampfförmiger Zustand *m* || état *m* de vapeur || dampvormige toestand *m* || парообразное состояние *n*

V31 **vapo(u)r phase** || Dampfphase *f* || phase *f* de vapeur || dampfase *f* || паровая фаза *f*

V32 **vapo(u)r pressure** || Dampfdruck *m* || pression *f* de vapeur || dampdruk *m* || давление *n* [упругость *f*] пара

V33 **varactor** || Varaktor *m*, Varikap *f* || varactor *m*, diode *f* paramétrique || varactor *m* || варактор *m*, параметрический диод

V34 **variable** || Variable *f* || variable *f* || veranderlijke *f* (*m*), variabele *f* (*m*) || переменная (величина) *f*

V35 **variable capacitor** || Drehkondensator *m* || condensateur *m* variable || draaicondensator *m*, variabele condensator *m* || переменный конденсатор *m*, конденсатор *m* переменной ёмкости

V36 **variable-focal-length lens** *see* **zoom lens**

V37 **variable force** || veränderliche Kraft *f* || force *f* variable || veranderlijke kracht *f* (*m*) || переменная сила *f*

V38 **variable load** || veränderliche Belastung *f*, Wechselbelastung *f*, wechselnde Last *f* || (sur)charge *f* variable || wisselende [veranderliijke] belasting *f* || переменная нагрузка *f*

V39 **variable mass** || veränderliche Masse *f* || masse *f* variable || veranderlijke massa *f* (*m*) || переменная масса *f*

V40 **variable motion** || veränderliche Bewegung *f* || mouvement *m* variable || veranderlijke beweging *f* || переменное движение *n*

V41 **variable resistor** || Stellwiderstand *m*, Regelwiderstand *m* || résistance *f* variable || regelweerstand *m* || переменный резистор *m*, реостат *m*

V42 **variable-speed drive** || Antrieb *m* mit stufenloser Drehzahlregelung || entraînement *m* à vitesse variable; commande *f* par variateur; transmission *f* continue || regelbare aandrijving *f* || привод *m* с плавно регулируемой частотой вращения

V43 **variable-speed gear** || stufenloses [stufenlos regelbares] Getrieb *n*, Variator *m* || transmission *f* continue; variateur *m* continu || traploz variator *m*; **traploos** drijfwerk *n* бесступенчатый редуктор *m*; бесступенчатый вариатор *m* скоросте

V44 **variac** || Stelltransformatc *m* || autotransformateur *m* variable variac *m* || steltransformator *m* variator *m* || вариак *m*

V45 **variance** || Streuung *f*; Varian *f* || variance *f* || variantie *f* || дисперсия *f*

V46 **variance law** || Varianzenadditionsgesetz *n* || loi *f* de variance wet *f* (*m*) van de variantie-additie закон *m* сложения дисперсий

V47 **variance of an estimate** Schätzungsvarianz *f* || variance d'une estimation || schattingsvariantie *f* || дисперсия *f* оценки

V48 **variate** || Zufallsvariable *f* variable *f* aléatoire || toevalsvariabel *f* (*m*), toevalsveranderlijke *f* (*m*) случайная переменная *f*

V49 **variation** || Variation *f* || variation *f* || variatie *f* || вариация

V50 **variational calculus** || Variationsrechnung *f* || calcul *m* des variations || variatierekening *f* || вариационное исчисление *n*

V51 **varicap** || **1.** Varikap *f* **2.** Varikond *m* || **1.** varicap *m* **2.** varicond *m* || **1.** varicap *m* **2.** variconde *m* **1.** варикап *m* **2.** варикoнд *m*

V52 **varifocal lens** *see* **zoom len**

V53 **varistor** || Varistor *m*, VDR Widerstand *m* || varistance *f* || varistor *m* || варистор *m*

V54 **varmeter** || Blindleistungsmesser *m* || varmètre *m* || varmete *m* || варметр *m*, реактивный ваттметр *m*

V55 **varying acceleration** || veränderliche Beschleunigung *f*, Wechselbeschleunigung *f* || accélération variable || veranderlijke versnelling *f* || переменное ускорение *n*

V56 **varying component** || veränderliche Komponente *f* || composante alternative || veranderlijke component *m* || переменная составляющая *f*

V57 **vector** || Vektor *m* || vecteur *m* || vector *m* || вектор *m*
V58 **vector algebra** || Vektoralgebra *f* || algèbre *f* vectorielle || vectoralgebra *f* || векторная алгебра *f*
V59 **vector calculus** || Vektorrechnung *f* || calcul *m* vectoriel || vectorrekening *f* || векторное исчисление *n*
V60 **vector field** || Vektorfeld *n* || champ *m* vectoriel || vectorveld *n* || векторное поле *n*
V61 **vector flux** || Vektorfluß *m* || flux *m* de vecteur || vectorstroom *m* || поток *m* вектора
V62 **vector of force** || Kraftvektor *m* || vecteur *m* (de la) force || krachtvector *m* || вектор *m* силы
V63 **vector potential** || Vektorpotential *n* || vecteur-potentiel *m* || vectorpotentiaal *m* || вектор-потенциал *m*
V64 **vector product** || vektorielles Produkt *n*, Vektorprodukt *n* || produit *m* vectoriel || vectorprodukt *n* || векторное произведение *n*
V65 **vector space** || Vektorraum *m* || espace *m* vectoriel || vectorruimte *f* || векторное пространство *n*
V66 **vee-belt** || Keilriemen *m* || courroie *f* en coin || V-snaar *f* (*m*) || клиновой ремень *m*
V67 **vee-thread** *see* **angular thread**
V68 **velocimeter** || Geschwindigkeitsmesser *m* || compteur *m* [détecteur *m*] de vitesse || snelheidsmeter *m* || скоростемер *m*; спидометр *m*; тахометр *m*
V69 **velocity** || Geschwindigkeit *f* || vitesse *f* || snelheid *f* || скорость *f*
V70 **velocity jump** || Geschwindigkeitssprung *m* || saut *m* de vitesse || snelheidssprong *m* || скачок *m* [разрыв *m*] скорости
V71 **velocity of light** || Lichtgeschwindigkeit *f* || vitesse *f* de la lumière || lichtsnelheid *f* || скорость *f* света
V72 **velocity potential** || Geschwindigkeitspotential *n* || potentiel *m* de vitesses || snelheidspotentiaal *m* || потенциал *m* скоростей
V73 **velocity profile** || Geschwindigkeitsprofil *n* || profil *m* de vitesses || snelheidsprofiel *n* || профиль *m* [распределение *n*] скоростей
V74 **velocity ratio** || Übersetzungsverhältnis *n* || rapport *m* de transmission [de réduction] || overbrengings-verhouding *f* || передаточное отношение *n*, передаточное число *n*
V75 **velocity summation** || Geschwindigkeitsaddition *f* || addition *f* des vitesses || snelheidsadditie *f* || сложение *n* скоростей
V76 **ventilator** || Lüfter *m*, Ventilator *m*; Gebläse *n* || ventilateur *m* || ventilateur *m*, ventilator *m* || вентилятор *m*
V77 **venturi tube** || Venturi-Rohr *n*, Venturi-Düse *f* || tube *f* Venturi || venturi(buis) *f* (*m*), venturimeter *m* || труба *f* [расходомер *m*] Вентури
V78 **verification** || Verifizierung *f*, Vergleichsprüfung *f* || vérification *f* || verificatie *f* || верификация *f*
V79 **verifier** || Prüfer *m*, Kontrollgerät *n* || vérificateur *m*; vérificatrice *f* || verificatie-inrichting *f* || верификатор *m*
V80 **vernier** || Nonius *m* || vernier *m*, nonius *m* || nonius *m* || нониус *m*, верньер *m*
V81 **vernier calliper** || Schieblehre *f*, Schublehre *f*; Meßschieber *m* || compas *m* à coulisse avec vernier || schuifpasser *m* met nonius; schuifmaat *f* (*m*) met nonius || штангенциркуль *m* с нониусом
V82 **vertex** || **1.** Eckpunkt *m*; Spitze *f*; Scheitel *m*, Ecke *f* (*Graph*) **2.** Zenit *m* || **1.** sommet *m* **2.** zénith *m* || **1.** top *m*, toppunt *n* **2.** zenit *n* || **1.** вершина *f* **2.** зенит *m*
V83 **vertical** || vertikal, Vertikal..., senkrecht || vertical || verticaal, loodrecht || вертикальный
V84 **vertical boring mill** || Karusselldrehmaschine *f* || tour *m* vertical || carrouselbank *f* || карусельный станок *m*
V85 **vertical drilling machine** || Senkrechtbohrmaschine *f*, Vertikalbohrmaschine *f* || perceuse *f* verticale, perceuse *f* à montant [à colonne] || verticale boormachine *f* || вертикально-сверлильный станок *m*

V86 **vertical feed** || Senkrechtvorschub *m* || avance *f* verticale || loodrechte [verticale] aanzet *f*, verticale voeding *f* || вертикальная подача *f*

V87 **vertical milling machine** || Senkrechtfräsmaschine *f*, Vertikalfräsmaschine *f* || fraiseuse *f* verticale || verticale freesmachine *f* || вертикально-фрезерный станок *m*

V88 **vibrating system** || Schwingungssystem *n* || système *m* oscillatoire || trillingssysteem *n* || колебательная система *f*

V89 **vibration** || Vibration *f*, (mechanische) Schwingung *f* || vibration *f* || trilling *f*, vibratie *f*; slingering *f* || вибрация *f*, (механическое) колебание *n*

V90 **vibrational** || Schwing(ungs)...|| vibratoire, oscillatoire || trillings... || колебательный

V91 **vibrational energy** || Schwingungsenergie *f* || énergie *f* oscillatore [d'oscillations, de vibrations] || trillingsenergie *f* || энергия *f* колебательного движения, энергия *f* колебания

V92 **vibrational spectrum** || Schwingungsspektrum *n* || spectre *m* de vibration || trillingsspectrum *n* || колебательный спектр *m*

V93 **vibration damper** || Schwingungsdämpfer *m* || antivibreur *m*; amortisseur *m* des vibrations, damper *m* || trillingsdemper *m* || гаситель *m* колебаний, антивибратор *m*; амортизатор *m*

V94 **vibration isolation** || Schwingungsdämpfung *f* || isolation *f* contre les vibrations || trillingsisolatie *f* || виброизоляция *f*

V95 **vibration isolator** || Schwingungsdämpfter *m* || antivibreur *m*; étouffeur (des vibrations) || trillingsdemper *m* || гаситель *m* колебаний, антивибратор *m*

V96 **vibration node** || Schwingungsknoten *m* || nœud *m* de vibration || trillingsknoop *m* || узел *m* колебаний

V97 **vibration pickup** || Schwingungsgeber *m* || transmetteur *m* de vibrations || trillingsgever *m* || вибродатчик *m*

V98 **vibration-rotational spectrum** || Rotationsschwingungsspektrum *n* || spectre *m* de rotation-vibration || rotatie-trillingsspectrum *n* || вращательно-колебательный спектр *m*

V99 **vibration test** || Schwingungsversuch *m* || essai *m* de vibration [vibratoire] || tril(lings)proef *f* (*m*) || вибрационное испытание *n*, испытание *n* на вибропрочность

V100 **vibration-testing machine** || Schwingungsprüfmaschine *f* || machine *f* d'essai vibratoire || trilproefmachine *f* || машина *f* для вибрационных испытаний

V101 **vibration theory** || Schwingungstheorie *f* || théorie *f* des ondulations || trillingstheorie *f* || теория *f* колебаний

V102 **vibrator** || **1.** Vibrator *m*, Rüttler *m*; Rüttelmaschine *f* **2.** Schwingungswandler *m* || **1.** vibrateur *m*, oscillateur *m* **2.** redresseur *m* à lame vibrante [à vibreur] || vibrator *m* || **1.** вибратор *m*; вибрационная машина *f* **2.** вибропреобразователь *m*

V103 **vibratory motion** || Schwingbewegung *f*, oszillatorische [schwingende] Bewegung *f* || mouvement *m* oscillatoire || slingerbeweging *f* || колебательное движение *n*, колебание *n*

V104 **vibrometer** || Vibrometer *n*, Schwingungsmesser *m* || vibromètre *m* || trillingsmeter *m* || **1.** виброметр *m* **2.** сейсмограф *m*

V105 **vice** || **1.** Schraubstock *m*, Spannstock *m* **2.** Zange *f* || **1.** étau *m* **2.** tenaille *f*, pince *f* || **1.** bankschroef *f* (*m*) **2.** tang *f* (*m*) || **1.** тиски *pl* **2.** клещи *pl*

V106 **vice chuck** *see* **machine vice**

V107 **vicinity** || Nähe *f*; Umgebung *f* || voisinage *m* || nabijheid *f* || близость *f*; окрестность *f*

V108 **Vickers hardness** || Vickershärte *f*, HV || dureté *f* Vickers || Vickershardheid *f* || твёрдость *f* по Виккерсу

V109 **Vickers hardness number** || Vickers-Härtezahl *f* || nombre *m* de (dureté) Vickers || Vickershardheidsgraad *m* || число *n* твёрдости по Виккерсу

V110 **video frequency** || Videofrequenz *f* || vidéofréquence *f*, fréquence *f* d'image || videofrequentie *f* || видеочастота *f*

V111 **virial theorem** || Virialsatz *m* || théorème *m* de viriel || virialtheorema *n* || теорема *f* вириала

V112 **virtual** || **1.** virtuell **2.** scheinbar, Schein... || **1.** virtuel **2.** imaginaire || **1.** virtueel **2.** imaginair || **1.** виртуальный; возможный **2.** мнимый *(напр. об изображении)*

V113 **virtual address** || virtuelle Adresse *f* || adresse *f* virtuelle || virtueele adres *n* || виртуальный адрес *m*

V114 **virtual center** *see* **instantaneous center**

V115 **virtual displacement** || virtuelle Verschiebung *f* || déplacement *m* virtuel || virtuele verschuiving *f* || виртуальное [возможное] перемещение *n*

V116 **virtual focus** || virtueller Brennpunkt *m* || foyer *m* virtuel || virtueel brandpunt *n* || мнимый фокус *m*

V117 **virtual image** || virtuelles Bild *n* || image *f* virtuelle || virtueel beeld *n* || мнимое изображение *n*

V118 **virtual level** || virtuelles Niveau *n*, virtuelles Energieniveau *n* || niveau *m* virtuel || virtueel niveau *n* || виртуальный уровень *m* (энергии)

V119 **virtual mass** || virtuelle [scheinbare] Masse *f* || masse *f* virtuelle || virtuele massa *f* *(m)* || присоединённая масса *f*

V120 **virtual memory** || virtueller Speicher *m* || mémoire *f* virtuelle || virtueel geheugen *n* || виртуальная память *f*

V121 **virtual pair** || virtuelles Paar *n* || paire *f* virtuelle || virtueel paar *n* || виртуальная пара *f* *(частицы и античастицы)*

V122 **virtual particle** || virtuelles Teilchen *n* || particule *f* virtuelle || virtueel deeltje *n* || виртуальная частица *f*

V123 **virtual source** || virtuelle Quelle *f* || source *f* virtuelle || virtuele bron *f* || мнимый [виртуальный] источник *m*

V124 **virtual work principle** || Prinzip *n* der virtuellen Verschiebungen || principe *m* des déplacements virtuels || principe *n* van virtuele verschuivingen || принцип *m* возможных [виртуальных] перемещений

V125 **viscometer** || Viskosimeter *n* || viscosimètre *m* || viscosimeter *m* || вискозиметр *m*

V126 **viscosity** || Viskosität *f*; innere Reibung || viscosité *f* || viscositeit *f* || (динамическая) вязкость *f*; внутреннее трение *n*

V127 **viscous** || viskos || visqueux || visceus || вязкий; вязкостный

V128 **viscous flow** || viskose Strömung *f* || écoulement *m* visqueux || viskeuse stroming *f* || вязкое течение *n*, течение *n* вязкой среды

V129 **viscous force** || Viskositätskraft *f* || force *f* de viscosité || viscositeitskracht *f* *(m)* || сила *f* внутреннего трения, вязкостная сила *f*

V130 **viscous friction** || innere Reibung *f* || frottement *m* visqueux || binnenwrijving *f* || внутреннее трение *n*, вязкость *f*

V131 **viscous losses** || viskoser Verlust *m* || pertes *f pl* par viscosité || viscositeitsverliezen *n pl* || вязкие потери *pl*, потери *pl* на внутреннее трение

V132 **visibility** || **1.** spektrale Hellempfindlichkeit *f* **2.** Sichtbarkeit *f*; Sicht *f* || **1.** efficacité *f* lumineuse **2.** visibilité *f* || **1.** zicht *n* **2.** zichtbaarheid *f* || **1.** видность *f* **2.** видимость *f*; расстояние *n* видимости

V133 **visibility threshold** || Sichtbarkeitsschwelle *f* || seuil *m* de visibilité || zichtbaarheidsdrempel *m* || порог *m* видимости, порог *m* зрительного ощущения

V134 **visible** || sichtbar; im sichtbaren Spektralbereich liegend || visible || zichtbaar || видимый; в видимой части спектра

V135 **visible light** || sichtbares Licht *n* || lumière *f* visible || zichtbaar licht *n* || видимый свет *m*

V136 **visible radiation** || sichtbare Strahlung *f* || radiation *f* visible || zichtbare straling *f* || видимое [оптическое] излучение *n*

V137 **visible spectrum** || sichtbares Spektrum *n*, Lichtspektrum *n* || spectre *m* visible [lumineux] || zichtbaar spectrum *n* || видимый [оптический] спектр *m*

V138 **vision** || Sehen *n*; Sehvorgang *m*; visueller Eindruck *m* || vision *f* || gezicht *n* || зрение *n*; зрительное восприятие *n*

V139 **visualization** 1. Sichtbarmachung *f* 2. Modelldarstellung *f* || 1. visualisation *f* 2. représentation *f* en modèle || 1. zichtbaarmaken *n* 2. modelvoorstelling *f* || 1. визуализация *f* 2. модельное представление *n* *(явления)*

V140 **visual perception** || Gesichtswahrnehmung *f*, optische Wahrnehmung *f* || perception *f* visuelle || gezichtswaarneming *f* || зрительное восприятие *n*

V141 **volatility** || Flüchtigkeit *f* || volatilité *f* || vluchtigheid *f* || летучесть *f*

V142 **volt**, V || Volt *n* || volt *m* || volt *m* || вольт *m*, В

V143 **voltage** || Spannung *f*; Potentialunterschied *m* || tension *f*, voltage *m* || spanning *f*, voltage *f* || (электрическое) напряжение *n*; разность *f* потенциалов

V144 **voltage amplification** || Spannungsverstärkung *f* || amplification *f* en voltage || spanningsversterking *f* || усиление *n* по напряжению

V145 **voltage-dependent resistance** || nichtlinearer [spannungsabhängiger] Widerstand *m* || résistance *f* (à variation) non linéaire || niet-lineaire weerstand *m* || нелинейное сопротивление *n*

V146 **voltage divider** || Spannungsteiler *m* || diviseur *m* de tension || spanningsdeler *m* || делитель *m* напряжения

V147 **voltage drop** || Spannungsabfall *m* || chute *f* de tension || spanningsval *m* || падение *n* напряжения

V148 **voltage gain** || Spannungsverstärkungsfaktor *m*, Spannungsverstärkung *f* || amplification *f* en voltage [en tension] || spanningsversterking *f* || усиление *n* по напряжению

V149 **voltage overload** || Spannungsüberbelastung *f* || surtension *f* || spanningsoverbelasting *f* || перегрузка *f* по напряжению

V150 **voltage pulse** || Spannungsimpuls *m*, Spannungsstoß *m* || impulsion *f* de tension || spanningsimpuls *m* || импульс *m* напряжения

V151 **voltage ratio** || Spannungsverhältnis *n* || rapport *m* de transformation en tension || spanningsverhouding *f* || коэффициент *m* трансформации по напряжению

V152 **voltage resonance** || Spannungsresonanz *f*, Serienresonanz *f*, Reihenresonanz *f* || résonance *f* de tension || spanningsresonantie *f* || резонанс *m* напряжений, последовательный резонанс *m*

V153 **voltage stabilization** || Spannungsstabilisierung *f* || stabilisation *f* de tension || spanningsstabilisatie *f* || стабилизация *f* напряжения

V154 **voltage supply** || 1. Spannungsquelle *f* 2. Spannungszuführung *f* || 1. source *f* de tension 2. application *f* de tension || 1. spanningsbron *f (m)* 2. spanningstoevoer *m* || 1. источник *m* напряжения 2. подача *f* напряжения

V155 **voltage transformer** || Spannungswandler *m*, Spannungsumsetzer *m* || transformateur *m* de tension [de potentiel] || spanningstranformator *m* || трансформатор *m* напряжения

V156 **voltammeter** || Strom- und Spannungsmesser *m*, Volt- und Amperemeter *n* || voltampèremètre *m* || voltampèremeter *m* || вольтамперметр *m*, ампервольтметр *m*

V157 **volt-ampere**, VA || Voltampere *n* || voltampère *m* || voltampère *m* || вольт-ампер *m*, В·А

V158 **voltmeter** || Voltmeter *n* || voltmètre *m* || voltmeter *m* || вольтметр *m*

V159 **volume** || Volum(en) *n*; Umfang *m* || volume *m* || volume *n*; || omvang *m* || объём *m*

V160 **volume density** || Raumdichte *f*, Volumendichte *f* || densité *f* cubique || ruimtedichtheid *f* || объёмная плотность *f*

V161 **volume expansion** || (thermische) Volumenausdehnung *f* || dilatation *m* cubique [de volume] || (thermische) volumevergroting *f* || объёмное (тепловое) расширение *n*

V162 **volume force** || Volum(en)-kraft *f* || force *f* volumique || volumekracht *f* (*m*) || массовая [объёмная] сила *f*

V163 **volume integral** || Volum(en)-integral *n* || intégrale *f* de volume || volume-integraal *f* (*m*) || интеграл *m* по объёму

V164 **volume percent** || Volum(en)-prozent *n* || pourcentage *m* en volume || volumepercentage *n* || объёмный процент *m*, процент *m* по объёму

V165 **volute spring** || Schraubenfeder *f* || ressort *m* en volute || schroefveer *f* (*m*) || винтовая [витая] пружина *f*

V166 **vortex** || Wirbel *m* || tourbillon *m* || wervel *m* || вихрь *m*, завихрение *n*

V167 **vortex field** || Wirbelfeld *n* || champ *m* rotationnel; champ tourbillonnaire || wervelveld *n* || вихревое поле *n*

V168 **vortex flow** || Wirbelströmung *f* || écoulement *m* tourbillonnaire || wervelstroming *f* || вихревое течение *f*

V169 **vortex-free flow** || wirbelfreie Strömung *f* || écoulement *m* non tourbillonnaire || wervelvrije stroming *f* || потенциальное [безвихревое] течение *n*

V170 **V-thread** *see* **angular thread**

W

W1 **waiting-line theory** || Warteschlangentheorie *f*, Bedienungstheorie *f* || théorie *f* de l'encombrement, théorie *f* des files [des queues] d'attente || bedieningstheorie *f* || теория *f* массового обслуживания, теория *f* очередей

W2 **waiting time** || Wartezeit *f* || temps *m* d'attente || wachttijd *m* || время *n* ожидания (*в теории массового обслуживания*)

W3 **wake** || Nachlauf *m*, Nachstrom *m* || sillage *m* || naloop *m* || спутная струя *f*

W4 **wall** || Wand *f*; Mauer || paroi *f*; mur *m* || Wand *m* || стенка *f*; стена *f*

W5 **wall-reflected** || wandreflektiert || réfléchi d'une paroi || wandgereflecteerd || отражённый от стенки

W6 **Wankel engine** || Wankelmotor *m*, Rotationskolbenmotor *m* || moteur *m* Wankel || Wankelmotor *m* || двигатель *m* Ванкеля, роторно-поршневой двигатель *m*

W7 **washer** || Unterlegscheibe *f*; Beilage *f*, Einlage *f* || rondelle *f* || onderlegplaatje *n* || шайба *f*; прокладка *f*

W8 **water-cooled enigne** || wassergekühlter Motor *m* || moteur *m* à refroidissement par eau || motor *m* met waterkoeling || двигатель *m* с водяным охлаждением

W9 **water drag** || hydrodynamischer Widerstand *m* || résistance *f* hydrodynamique || hydrodynamische weerstand *m* || гидродинамическое сопротивление *n*

W10 **water jacket** || Wassermantel *m*; Kühl(wasser)mantel *m* || chemise *f* d'eau || watermantel *m* || водяная рубашка *f*

W11 **water-pressure engine** *see* **hydraulic engine**

W12 **water turbine** *see* **hydraulic turbine**

W13 **watt,** W || Watt *n* || watt *m* || watt *m* || ватт *m*, Вт

W14 **wattage** || Wattzahl *f* || puissance *f* en watts || wattage *f* || мощность *f* в ваттах

W15 **wattful current** || Wirkstrom *m* || courant *m* actif [watté] || werkstroom *m* || активный ток *m*

W16 **wattless current** || Blindstrom *m* || courant réactif [déwatté] || blinde [reactieve, wattloze] stroom *m* || реактивный ток *m*

W17 **wattmeter** || Wattmeter *n* || wattmètre *m* || wattmeter *m* || ваттметр *m*

W18 **wave** || Welle *f* || onde *f* || golf *f* (*m*) || волна *f*
W19 **wave amplitude** || Wellenamplitude *f* || amplitude *f* d'onde || golfamplitude *f* || амплитуда *f* волны
W20 **wave-corpuscle duality** *see* **wave-particle duality**
W21 **wave crest** || Wellenkamm *m* || crête *f* [sommet *m*] d'une onde || golfberg *m* || гребень *m* волны
W22 **wave drag** || Wellenwiderstand *m* || traînée *f* d'onde || golfweerstand *m* || волновое сопротивление *n*
W23 **wave equation** || Wellengleichung *f* || équation *f* des ondes || golfvergelijking *f* || волновое уравнение *n*
W24 **wave front** || Wellenfront *f* || front *m* d'onde || golffront *n* || волновой фронт *m*, фронт *m* волны
W25 **wave function** | Wellenfunktion *f* || fonction *f* d'onde || golffunctie *f* || волновая функция *f*; ψ-функция *f*
W26 **waveguide** || Wellenleiter *m*; Hohlleiter *m* || guide *m* [conduit *m*] d'ondes || golfgeleider *m* || волновод *m*
W27 **wave impedance** || (charakteristischer) Wellenwiderstand *m* || impédance *f* caractéristique || (karakteristieke) golfweerstand *m* || волновое (характеристическое) сопротивление *n*
W28 **wave interference** || Welleninterferenz *f* || interférence *f* d'ondes || golfinterferentie *f* || интерференция *f* волн
W29 **wavelength** || Wellenlänge *f* || longueur *f* d'onde || golflengte *f* || длина *f* волны
W30 **wavelength standard** || Wellenlängennormal *n* || étalon *m* de longueur photométrique || golflengte-etalon *n* || световой [волновой] эталон *m* длины
W31 **wavelet** || Elementarwelle *f* || onde *f* élémentaire || elementaire golf *f* (*m*) || элементарная волна *f*
W32 **wave mechanics** || Wellenmechanik *f*; Quantenmechanik *f* || mécanique *f* ondulatoire; mécanique quantique || golfmechanica *f*; quantummechanica *f* || волновая механика *f*; квантовая механика *f*
W33 **wavemeter** || Wellenmesser *m* || ondemètre *m* || golflengtemeter *m*, golfmeter *m* || волномер *m*
W34 **wave motion** || Wellenbewegung *f* || mouvement *m* ondulatoire || golfbeweging *f* || волновое движение *n*
W35 **wave node** || Wellenknoten *m* || nœud *m* d'onde || golfknoop *m* || узел *m* (стоячей) волны
W36 **wave normal** || Wellen(front)-normale *f* || normale *f* au front d'onde || golfnormaal *f* (*m*) || нормаль *f* к фронту волны
W37 **wave number** || Wellenzahl *f* || nombre *f* d'onde || golfgetal *n* || волновое число *n*
W38 **wave optics** || Wellenoptik *f* || optique *f* d'ondes || golfoptiek *f* || волновая оптика *f*
W39 **wave packet** || Wellenpaket *n* || paquet *m* [train *m*] d'ondes || golftrein *m* || волновой пакет *m*, группа *f* волн
W40 **wave-particle duality** || Dualismus *m* Welle-Teilchen *n* || dualisme *m* onde-corpuscule || golf-corpusculair dualisme *n* || корпускулярно-волновой дуализм *m*
W41 **wave phase** || Wellenphase *f* || phase *f* d'onde || golffase *f* || фаза *f* волны
W42 **wave resistance** || Wellenwiderstand *m* || résistance *f* d'onde || golfweerstand *m* || волновое сопротивление *n*
W43 **wave surface** || Wellenfront *f*, Wellenfläche *f* || surface *f* d'onde || golfoppervlak *n* || фронт *m* (объёмной) волны, волновая поверхность *f*
W44 **wave train** || Wellenzug *m* || train *m* d'ondes || golftrein *m* || цуг *m* [группа *f*] волн
W45 **wave vector** || Wellenvektor *m* || vecteur *m* d'onde || golfvector *m* || волновой вектор *m*
W46 **waviness** || Welligkeit *f* || ondulation *f* || golving *f* || волнистость *f* (*поверхности*)
W47 **weak binding** || schwach

Kopplung *f* || couplage *m* faible || zwakke koppeling *f* || слабая связь *f* *(между частицами)*

W48 **weak interaction** || schwache Wechselwirkung *f* || interaction *f* faible || zwakke wisselwerking *f* || слабое взаимодействие *n*

W49 **weak lens** || schwache Linse *f* || lentille *f* faible || zwakke lens *f* (*m*) || линза *f* с малой оптической силой, слабая линза *f*

W50 **weakly monotonic function** || nicht streng monotone Funktion *f* || fonction *f* faiblement monotone || niet-streng monotone functie *f* || нестрого монотонная функция *f*

W51 **wear** || Verschleiß *m*, Abnutzung *f*; Abrieb *m* || usure *f* || slijtage *f*, afslijting *f* || износ *m*; истирание *n*

W52 **wear hardness** || Abnutz(ungs)härte *f*, Verschleißhärte *f* || dureté *f* à l'usure [à l'abrasion] || slijtvastheid *f* || износостойкость *f*, износоустойчивость *f*

W53 **wear-in** || Einlaufen *n* || rodage *m* || inloop *m* || приработка *f* *(двигателя)*

W54 **wear resistance** || Verschleißfestigkeit *f*, Abnutzungswiderstand *m* || résistance *f* à l'usure || slijtbestendigheid *f* || сопротивление *n* износу, износостойкость *f*, износоустойчивость *f*

W55 **weber,** Wb || Weber *n* || weber *m* || weber *m* || вебер *m*, Вб

W56 **weighted average** || gewogenes Mittel *n*, gewogener Mittelwert *m* || valeur *m* pesée moyenne || gewogen gemiddelde *n* || взвешенное среднее (значение) *n*

W57 **weighting function** || Gewichtsfunktion *f*, Gewicht *n* || fonction *f* de poids || gewichtsfunctie *f* || весовая функция *f*

W58 **weightlessness** || Schwerelosigkeit *f* || apesanteur *f*, non-pesanteur *f*, impondérabilité *f* || gewichtsloosheid *f* || невесомость *f*; состояние *n* невесомости

W59 **weight percent** || Masseprozent *n* || pourcentage *m* en masse || massapercent *n* || массовый процент *m*, процент *m* по массе

W60 **weld** || Schweißnaht *f*; Schweißverbindung *f* || soudure *f* || lasnaad *m* || сварной шов *m*

W61 **welding** || Schweßen *n*, Schweißung *f* || soudage *m* || lassen *n* || сварка *f*

W62 **welding machine** || Schweißmaschine *f* || soudeuse *f*, machine *f* à souder || lasmachine *f*, lasapparaat *n* || сварочный аппарат *m*

W63 **well-ordered set** || wohlgeordnete Menge *f* || ensemble *m* bien ordonné || volledig geregelde menigte *f* || вполне упорядоченное множество *n*

W64 **wet grinding** || Naßschleifen *n* || meulage *m* à l'eau; [en mouillant, en arrosant] || natslijpen *n* || мокрое шлифование *n*

W65 **wettability** || Netzfähigkeit *f*, Netzbarkeit *f* || mouillabilité *f* || bevochtigbaarheid *f* || смачиваемость *f*

W66 **wetting** || Benetzung *f* || mouillage *m*, humectage *m*, humidification *f* || bevochtiging *f* || смачивание *n*

W67 **wetting angle** || Randwinkel *m*, Benetzungswinkel *m* || angle *m* de mouillage || kanthoek *m* || краевой угол *m*

W68 **Wheatstone bridge** || Wheatstone-Brücke *f* || pont *m* de Wheatstone || brug *f* (*m*) van Wheatstone || мост *m* Уитстона

W69 **wheel** || Rad *n*; Zahnrad *n* || roue *f* || rad *n*; wiel *n*; tandwiel *n* || колесо *n*; зубчатое колесо *n*

W70 **wheel cutter** *see* **gear cutter**

W71 **whirl** || Wirbel *m* || tourbillon *m* || wervel *m* || вихрь *m*, завихрение *n*

W72 **whirling** || Wirbelbewegung *f* || mouvement *m* tourbillonnaire || wervelbeweging *f*, werveling *f* || вихревое движение *n*

W73 **white iron** || weißes Roheisen *n* || fonte *f* blanche || wit ruwijzer *n* || белый чугун *m*

W74 **white metal** *see* **antifriction metal**

W75 **white-metal bearing** *see* **babbit-metal bearing**

W76 **white noise** || weißes Rauschen *n* || bruit *m* blanc || witte ruis *m* || белый [сплошной] шум *m*

W77 **white radiation** || weiße [kontinuierliche] Strahlung *f* || rayonnement *m* blanc, radiation *f* blanche || witte straling *f* || белое [сплошное] излучение *n*

W78 **Whitworth (screw) thread** || Zollgewinde *n*, Whitworth-Gewinde *n* || filet *m* [filetage *m*] (au pas) Whitworth || Whitworth-draad *m* || дюймовая резьба *f*, резьба *f* Витворта

W79 **wide-band** || Breitband... || à bande large || breedband... || 1. широкополосный *(об усилителе, фильтре)* 2. широкозонный *(о полупроводнике)*

W80 **winch** || Winde *f*; Haspel *f* || treuil *m* || windas *n*, lier *f* (*m*); haspel *m* || лебёдка *f*; ворот *m*

W82 **winding** || Wicklung *f* || enroulement *m* || wikkeling *f* || обмотка *f*

W83 **wing profile** || Tragflügelprofil *n* || profil *m* de l'aile || vleugelprofiel *n* || профиль *m* (обтекаемого) крыла

W84 **wire** || **1.** Draht *m* **2.** Leitung *f*, Leiter *m*, Draht *m* || fil *m*, conducteur *m* || draad *m* || **1.** проволока *f* **2.** провод *m*

W85 **wire drawing** || Drahtziehen *n* || filage *m*; tréfilage *m* du fil || draadtrekkerij *f* || волочение *n* проволоки

W86 **wiring** || **1.** Leitung *f* **2.** (elektrische) Schaltung *f* || **1.** câblage *m* 2. circuit *m* || **1.** leiding *f* **2.** schakeling *f* || **1.** (электро)проводка *f* 2. (электрическая) схема *f*

W87 **wiring scheme** || Verdrahtungsplan *m* || schéma *m* de câblage || schakelschema *n* || схема *f* (электрических) соединений, монтажная схема *f*

W88 **wolfram steel** *see* **tungsten steel**

W89 **Woodruff key** || Scheibenfeder *f* || clavette *f* Woodruff || schijfspie *f* (*m*) || сегментная шпонка *f*

W90 **work function** || Austrittsarbeit *f* || travail *m* de sortie [d'extraction] || uittreedarbeid *m* || работа *f* выхода *(электрона)*

W91 **working aperture** || Objektivöffnung *f* || ouverture *f* de l'objectif || objectiefopening *f* || действующее отверстие *n* *(объектива)*

W92 **working fluid** || Betriebsflüssigkeit *f* || fluide *m* hydraulique || bedrijfswerkvloeistof *f* (*m*) || рабочая жидкость *f*

W93 **working gas** || Arbeitsgas *n* || gaz *m* moteur [de propulsion] || werkgas *n* || рабочий газ *m*

W94 **working life** || Betriebsdauer *f*, Lebensdauer *f* || vie *f*, durée *f* d'emploi || bedrijfsduur *m*, levensduur *m* || срок *m* службы; долговечность *f*

W95 **working pressure** || Betriebsdruck *m* || pression *f* de travail || werkdruk *m* || рабочее давление *n*

W96 **working range** || Arbeitsbereich *m* || bande *f* de fonctionnement [de travail] || arbeidsbereik *n* || рабочий диапазон

W97 **working volume** || Arbeitsvolumen *n* || volume *m* de travail [utile] || arbeidsvolume *n* || рабочий объём *m*

W98 **workpiece** || Werkstück *n* || pièce *f* à usiner || werkstuk *n* || (обрабатываемая) деталь *f*, заготовка *f*

W99 **work speed** || Arbeitsgeschwindigkeit *f* || vitesse *f* de travail || arbeidssnelheid *f* || рабочая скорость *f*

W100 **world curve** || Weltlinie *f* || ligne *f* [trajectoire *f*] d'Univers || wereldlijn *f* (*m*) || мировая линия *f*

W101 **world point** || Raum-Zeit-Punkt *m* || point *m* d'Univers || wereldpunt *n* || мировая точка *f*

W102 **worm** || Schnecke *f* || vis *f* sans fin [tangente] || worm *m* || червяк *m*, червячный винт *m*

W103 **worm gear, worm gearing** || Schnecken(rad)getriebe *n* ,Schneckentrieb *m* || engrenage *m* à vis sans fin || wormoverbrenging *f* || червячная передача *f*

W104 **worm screw** *see* **worm**

W105 **worm wheel** || Schneckenrad *n* || roue *f* à vis [de vis tangente] || wormwiel *n* || червячное колесо *n*

W106 **wrench** *see* **spanner**

W107 **Wronskian** || Wronskische Determinante *f* || déterminant *m* wronskien || wronskiaanse determinant *m* || вронскиан *m*, определитель *m* Вронского

W108 **wye** ' Sternschaltung *f* || couplage *m* en étoile [en Y] || sterschakeling *f* || соединение *n* звездой

W109 **wye-delta** || Stern-Dreieck-Schaltung *f* || couplage *m* étoile-triangle || sterdriehoekschakeling *f* || соединение *n* «звезда — треугольник»

X

X1 **X-axis** Abszißenachse *f*, X-Achse *f* || axe *m* X, axe *m* d'abscisses || X-as *f* (*m*), abscis-as *f* (*m*) || ось *f* абсцисс

X2 **X-coordinate** || Abszisse *f* || abscisse *f*, coordonnée *f* X || abscis *f* (*m*) || абсцисса *f*

X3 **xenon**, Xe || Xenon *n* || xénon *m* || xenon *n* || ксенон *m*

X4 **X-ray diffraction analysis** || Röntgenstrukturanalyse *f* || analyse *f* (structurale) aux rayons X, radiocristallographie *f* (de rayons X) || röntgen-structuuranalyse *f* || рентгеноструктурный анализ *m*

X5 **X-rays** || Röntgenstrahlung *f* || radiation *f* [rayonnement *m*, rayons *m pl*] X || röntgenstraling *f* || рентгеновское излучение *n*

X6 **X-ray spectroscopy** || Röntgenspektroskopie *f* || spectroscopie *f* par rayons X || röntgenspectroscopie *f* || рентгеновская спектроскопия *f*

X7 **X-ray spectrum** || Röntgenspektrum *n* || spectre *m* des rayons X || röntgenspectrum *n* || рентгеновский спектр *m*

Y

Y1 **Y-axis** || Ordinatenachse *f*, Y-Achse *f* || axe *m* Y, axe *m* d'ordonnées || ij-as *f* (*m*), ordinaatas *f* (*m*), ordinatenas *f* (*m*) || ось *f* ординат

Y2 **Y-coordinate** || Ordinate *f* || coordonnée *f* Y, ordonnée *f* || ij-coördinaat *m*, ordinaat *f* || ордината *f*

Y3 **yield** || **1.** Fließen *n* *(im festen Zustand)* **2.** Ausbeute *f* *(bei einem Prozeß)* || **1.** fluidité *f* *(en état solide)* **2.** rendement *m* *(d'un processus)* || **1.** vloeien *n* *(in vaste toestand)* **2.** rendement *n* *(van een proces)* || **1.** текучесть *f* *(в твёрдом состоянии)* **2.** выход *m* *(процесса)*

Y4 **yield point, yield strength** || Fließgrenze *f* || limite *f* de fluage; limite *f* apparente d'élasticité || vloeigrens *f* (*m*) || предел *m* текучести

Y5 **Young's modulus** || Youngscher Elastizitätsmodul *m*, E-Modul *m*, Youngscher Modul *m* || module *m* d'elasticité en traction, module *m* d'allongement || elasticiteitsmodulus *m* || модуль *m* (нормальной) упругости, модуль *m* Юнга

Z

Z1 **Z-axis** || Applikatenachse *f*, Z-Achse *f* || axe *m* des Z, axe *m* des appliques || z-as *f* (*m*), applicaatas *f* (*m*) || ось *f* аппликат

Z2 **Z-coordinate** || Applikate *f* || appliquée *f*, coordonnée *f* Z || z-coördinaat *m*, applicaat *m* || аппликата *f*

Z3 **Z-distribution** || z-Verteilung *f* || distribution *f* z || z-verdeling *f* || z-распределение *n*

Z4 **Zeno's paradox** || Zenonsches Paradoxon *n* || paradoxe *m* de Zenon || zenonse paradox *m* || парадокс *m* Зенона

Z5 **zero** || Null *f*; Nullpunkt *m* || zéro *m* || nul *f* (*m*); nulpunt *n* || нуль *m*

Z6 **zero divisor** || Nullteiler *m* || diviseur *m* de zéro || nuldeler *m* || делитель *m* нуля

Z7 **zero gravity** || Schwerelosigkeit *f* || apesanteur *f*, non pesanteur *f*, impondérabilité *f* || gewichtsloosheid *f* || невесомость *f*

Z8 **zero-gravity state** || Schwerelosigkeitszustand *m* || état *m* d'apesanteur || gewichtsloosheidstoestand *m* || состояние *n* невесомости

Z9 **zero-loss** || verlustfrei || sans pertes || verliesvrij || без потерь, лишённый потерь

Z10 **zero-point energy** || Nullpunktsenergie *f* || énergie *f* au point zéro || nulpuntenergie *f* || энергия *f* нулевых колебаний, нулевая энергия *f*

Z11 **zero spin** || Nullspin *m* || spin *m* nul; spin zéro || nulspin *m* || нулевой спин *m*

Z12 **zero-sum game** || Nullsummenspiel *n* || jeu *m* à somme nulle || nulsomspel *n* || игра *f* с нулевой суммой

Z13 **zero time** || Nullzeitpunkt *m* || moment *m* nul [zéro] || nultijdstip *f* (*m*) || нулевой момент *m* *(времени)*

Z14 **zeta function** || Zeta-Funktion *f* || fonction *f* zêta || zeta-functie *f* || дзета-функция *f*

Z15 **Zinc,** Zn || Zink *n* || zinc *m* || zink *n* || цинк *m*

Z16 **zinc(-base) alloy** || Zinklegierung *f* || alliage *m* à base de zinc || zinklegering *f* || цинковый сплав *m*, сплав *m* на основе цинка

Z17 **zinc-plating** || Verzinken *n*, Verzinkung *f* || zincage *m*, zingage *m*; galvanisation *f* || verzinking *f* || цинкование *n*

Z18 **zone** || Zone *f*, Gebiet *n*, Bereich *m* || zone *f* || zone *f* || зона *f*

Z19 **zone melting** || Zonenschmelzen *n* || fusion *f* à zone, zone *f* fondue || zonesmelting *f* || зонная плавка *f*

Z20 **zoom lens** || Varioobjektiv *n*, Gummilinse *f* || objectif *m* à (distance) focale variable, zoom *m* || zoomlens (*m*) || объектив *m* с переменным фокусным расстоянием; вариообъектив *m*

DEUTSCH

FRANÇAIS

NEDERLANDS

РУССКИЙ

FSC
www.fsc.org
MIX
Papier aus verantwortungsvollen Quellen
Paper from responsible sources
FSC® C105338

If you have any concerns about our products,
you can contact us on
ProductSafety@springernature.com

In case Publisher is established outside the EU,
the EU authorized representative is:
Springer Nature Customer Service Center GmbH
Europaplatz 3, 69115 Heidelberg, Germany

Printed by Libri Plureos GmbH
in Hamburg, Germany